ELECTRONIC COMMUNICATION SYSTEMS

ELECTRONIC COMMUNICATION SYSTEMS

THIRD EDITION

GEORGE KENNEDY

Supervising Engineer, Overseas Telecommunications Commission, Australia

Tata McGraw-Hill Publishing Company Limited
NEW DELHI

McGraw-Hill Offices

New Delhi New York St Louis San Franciso Auckland Bogotá
Guatemala Hamburg Lisbon London Madrid Mexico Milan
Montreal Panama Paris San Juan São Paulo
Singapore Sydney Tokyo Toronto

ELECTRONIC COMMUNICATION SYSTEMS, THIRD EDITION
TATA McGRAW-HILL EDITION 1991
Third Reprint 1992

Rs. 150.00

Sponsoring Editor: George Z. Kuredjian
Editing Supervisor: James Fields
Design and Art Supervisors: Meri Shardin and Nancy Axelrod
Production Supervisor: Priscilla Taguer
Interior Design: Edward F. Butler

Library of Congress Cataloging in Publication Data

Kennedy, George, (date)
Electronic communication systems.

Includes index.
1. Telecommunication. I. Title.
TK5101.K39 1984 621.38'0413 84-7920
ISBN 0-07-034054-4

When ordering this title use ISBN 0-07-462182-3

Published by Tata McGraw-Hill Publishing Company Limited
4/12 Asaf Ali Road, New Delhi 110 002 and printed by
Mohan Makhijani at Rekha Printers Pvt. Ltd, New Delhi 110 020

RZDCRRCERZZZB

CONTENTS

PREFACE

This book has its origins in printed notes produced by the author at a time when he taught communications at a technical college in Sydney and textbooks written at this level apparently did not exist. The production of these notes got out of hand when the courses spread to other technical colleges across the state, and so they were converted to a textbook published in Australia. The next step was the publication of the first edition of this book, which was aimed primarily but not exclusively at North American students and covered a much wider field. The second edition soon followed, and now, after what seems a very short time—but one that has seen a large number of developments—this third edition has been written.

This edition is designed for communications students at the advanced level, and it presents information about the basic philosophies, processes, circuits and other building blocks of communications systems. It is intended for use as text material, but for greatest effect it should be backed up by demonstrations and practical work in which students participate directly. The author tries to sit on the fence dividing rigorous mathematical proof and a general acceptance on faith. At times this is a very uncomfortable position. Nonetheless, a sustained effort has been made to verify all concepts as they are introduced. The methods used for this are any of the following, in combination if required: complete mathematical proof, partial but adequate mathematical proof, graphical analysis or synthesis, analogy and logical deduction.

Following from the above, the mathematical prerequisites are an understanding of the *j* operator, trigonometric formulas of the product-of-two-sines form, very basic differentiation and integration, and binary arithmetic. While not exacting, these are essential.

The basic electrical/electronic prerequisite is a knowledge of some circuit theory and common active circuits. On the electrical side, this involves familiarity with dc and ac circuit theory, including resonance, filters, mutually coupled circuits and transformers. On the electronics side, it is necessary to understand thoroughly the operation of common solid-state devices. Some knowledge of thermionic devices and electron ballistics is helpful in the understanding of microwave tubes. Finally, communications prerequisites are restricted to a working knowledge of tuned voltage and power amplifiers, oscillators, flip-flops and gates.

An examination of the contents shows that the text is logically subdivided into a number of parts. The first two chapters introduce the fundamental subjects of communications systems and noise. The next three chapters deal with the theory and generation of analog modulation systems: amplitude modulation, the various forms of single-sideband, frequency and phase modulation. The next section consists of one large chapter devoted to radio receivers used for the modulation systems covered in the preceding three chapters.

Transmission systems, together with radiation and propagation of waves, form the next major portion of the text. This is a part in which frequencies become higher as the section progresses, culminating in work on microwave antennas and waveguides. These topics form a bridge to the next main section, which deals with microwaves almost exclusively, covering microwave tubes and solid-state devices, lasers and other optical components, including fibers.

Digital systems are introduced in the next two chapters, which are in turn devoted to pulse and data communications. The last three chapters discuss systems which use a mixture of analog and digital techniques. These are broadband communications systems (including multiplexing), radar and television.

Where there are different means of achieving the same ends, the book tries to compare and contrast them, so that the treatment ensures that their interrelations are clearly demonstrated. For example, the various modulation systems, analog and digital, are compared with each other, as are different types of antennas, microwave tubes and semiconductor devices, microwave links and coaxial cables, and so on. Where applicable, system summaries are provided. The aim is to ensure that the student understands not only what a particular device or system does, and how it does this, but also where it fits into "the scheme of things" and what its future prospects are. On the other hand, coverage is sufficiently thorough to make it possible to "plot a course" through the book. One such course could be "Modulation and Demodulation," another might be "Propagation and Transmission," while a third could be "Microwave and Optical Techniques."

Users of the second edition will find that the third edition contains a substantial number of changes. Updating has been thorough, although of course it is more evident in the later chapters. These deal with state-of-the-art topics, none less so than Semiconductor Microwave Devices and Broadband Systems. A new chapter (Digital Communications) has been added, while one (the somewhat specialized Special Communications Circuits) was removed, though some parts of it were incorporated in other chapters. Additions have been made to several chapters, notably Radio Receivers, Antennas and Semiconductor Microwave Devices. End-of-chapter and multiple-choice questions have been revised and updated, as have the recommended references and many illustrations. Finally, the chapter order has been changed, to achieve a more logical and cohesive presentation.

The author is indebted to Mr. Noel T. Smith of Central Texas College, who provided the material for the chapter on Digital Communications and assisted in the revision of the chapters on radar and television.

Like its predecessor, this book is dedicated to some of the author's Abyssinian cats. They are three American imports: Negus, Nulli and Bambi—all of whom really have much fancier names and titles!

George Kennedy

LIST OF SYMBOLS AND ABBREVIATIONS

PHYSICAL CONSTANTS

e = charge of an electron = 1.602×10^{-19} C
= base of natural logarithm system = 2.718
h = Planck's constant = 6.626×10^{-34} J · s
k = Boltzmann's constant = 1.380×10^{-23} J/K
T_0 = standard (noise) temperature = 17°C = 290 K
v_c = velocity of light in vacuo = 299,792,500 ± 300 m/s
$\mathfrak{Z}$ = characteristic impedance of free space = 120π = 377 Ω
ϵ = electric permittivity of space = 8.854×10^{-12} F/m
μ = permeability of space = 1.257×10^{-6} H/m

PREFIXES

T = tera = 10^{12}
G = giga = 10^{9}
M = mega = 10^{6}
k = kilo = 10^{3}
c = centi = 10^{-2}
m = milli = 10^{-3}
μ = micro = 10^{-6}
n = nano = 10^{-9}
p = pico = 10^{-12}

FREQUENCY BANDS

For the names of allocated frequency and radar bands see Fig. 8-11 and Table 16-1.

ACRONYMS

balun = balance-to-unbalance transformer
bit = binary digit

codan = carrier-operated device, antinoise
coho = coherent oscillator (in radar)
compandor = compressor-expander
dit = decimal digit
domsat = domestic satellite (system)
laser = light amplification by stimulated emission of radiation
lincompex = linked compressor and expander
maser = microwave amplification by stimulated emission of radiation
modem = modulator-demodulator
paramp = parametric amplifier
radar = radio detection and ranging
stalo = stable oscillator (in radar)

ABBREVIATIONS

CHEMICAL SUBSTANCES

GaAlAs = gallium aluminum arsenide
GaAs = gallium arsenide
GaSb = gallium antimonide
Ge = germanium
InGaAsP = indium gallium arsenide phosphide
InP = indium phosphide
Si = silicon
YIG = yttrium-iron garnet

DEVICES

APD = avalanche photodiode
BWO = backward-wave oscillator
CFA = crossed-field amplifier
ESBAR = epitaxial Schottky barrier (diode)
FET = field-effect transistor
IMPATT = impact avalanche and transit time (diode)
LED = light-emitting diode
MESFET = mesa field-effect transistor
MIC = microwave integrated circuit
PIN = *P*-intrinsic-*N* (diode)
RIMPATT = Read-IMPATT (diode)
SAW = surface acoustic wave
TED = transferred electron device
TEO = transferred electron oscillator
TRAPATT = trapped plasma avalanche triggered transit (diode)
TWT = traveling-wave tube
VTM = voltage-tunable magnetron

MODULATION AND MULTIPLEXING

A3E = double-sideband, full-carrier AM
AM = amplitude modulation
B8E = independent-sideband AM
C3F = vestigial-sideband AM
FDM = frequency-division multiplex
FM = frequency modulation
FMVFT = frequency-modulated voice-frequency telegraph
FSK = frequency-shift keying
H3E = single-sideband, full-carrier AM
ISB = independent sideband (AM)
J3E = single-sideband, suppressed-carrier AM
LSB = lower sideband
PAM = pulse-amplitude modulation
PCM = pulse-code modulation
PM = phase modulation
PPM = pulse-position modulation
PSK = phase-shift keying
PTM = pulse-time modulation
PWM = pulse-width modulation
R3E = single-sideband, reduced-carrier AM
SSB = single sideband (AM)
TDM = time-division multiplex
USB = upper sideband

ORGANIZATIONS

CCIR = Comité Consultatif International de Radio
CCITT = Comité Consultatif International de Télégraphie et Téléphonie
COMSAT = Communications Satellite Corporation
FCC = Federal Communications Commission
INMARSAT = International Maritime Satellite Organization
INTELSAT = International Telecommunications Satellite Organization
ITU = International Telecommunication Union
NASA = National Aeronautics and Space Administration
NTSC = National Television Standards Committee

OTHERS

AF = audio frequency
AFC = automatic frequency correction (*or* control)
AGC = automatic gain control
AOS = Atlantic Ocean satellite
APC = automatic phase correction
ARQ = automatic request for repetition
ASCII = American Standard Code for Information Interchange

BFO = beat-frequency oscillator
BMEWS = Ballistic Missile Early Warning System
CTE = channel translating equipment
CW = continuous wave
DF = direction finding
EBCDIC = Extended Binary Coded Decimal Interchange Code
ECM = electronic countermeasures
GTE = group translating equipment
IF = intermediate frequency
IFF = identification, friend or foe
IOS = Indian Ocean satellite
MEWS = Missile Early Warning System
MFC = multifrequency coding
MTBF = mean time between failures
MTI = moving-target indication
MUF = maximum usable frequency
PAL = phase alternation by line (color TV system)
PFN = pulse-forming network
POS = Pacific Ocean satellite
PPI = plan-position indicator
PPM = periodic permanent magnet
PRF = pulse repetition frequency
RF = radio frequency
SAGE = semiautomatic ground environment
SMO = stabilized master oscillator
S/N = signal-to-noise ratio
SOLAS = safety of life at sea
SWR = standing-wave ratio
TE = transverse electric
TEM = transverse electromagnetic
TM = transverse magnetic
TR = transmit-receive
TRF = tuned radio frequency
TT & C = telemetry, tracking and command
TWS = track-while-scan
VCO = voltage-controlled oscillator
VFO = variable-frequency oscillator

SYMBOLS

A = cross-sectional area of antenna
A_0 = capture area of antenna
A_p = power gain
C = channel capacity
C_{eq} = equivalent capacitance

D = directivity (*or* mouth diameter) of antenna
$\mathscr{E}$ = field strength
F = noise figure
f_c = carrier frequency
= resistive cutoff frequency (of diode)
f_d = Doppler frequency
f_i = intermediate frequency
= idler frequency
f_m = modulation frequency
f_{max} = (transistor) maximum oscillating frequency
f_p = parallel-resonant frequency
= pump frequency
f_s = signal frequency
= series-resonant frequency
f_{si} = image frequency
f_T = (transistor) gain-bandwidth frequency
$f_{\alpha b}$ = alpha cutoff frequency
$f_{\alpha e}$ = beta cutoff frequency
f_∞ = frequency of infinite attenuation
g_c = conversion transconductance
g_m = transconductance
J_n = Bessel function
k = dielectric constant
= any constant quantity
L_{eq} = equivalent inductance
m = modulation index in AM
= number of half-wavelengths of field intensity between the side walls of a rectangular waveguide
m_f = modulation index in FM
m_p = modulation index in PM
n = any integer
= number of half-wavelengths of field intensity between the top and bottom walls of a waveguide
$\mathscr{P}$ = power density
P_{min} = minimum receivable power
P_n = noise power
P_r = received power
P_t = transmitted power
R_{eq} = equivalent noise resistance
r_{max} = (radar) maximum range
S = (crystal) pole-zero separation
= radar target cross section
S/N = signal-to-noise ratio
T = (specific) time
= temperature (absolute)
T_{eq} = equivalent noise temperature

V_c = carrier voltage
v_g = group velocity
V_m = modulation voltage
V_n = noise voltage
v_n = normal velocity
v_p = phase velocity
$\mathcal{Z}_0$ = characteristic impedance
α = rejection or attenuation
= transistor current gain in common base
β = transistor current gain in common emitter
δ = frequency deviation in FM
= small change in some quantity
δf = bandwidth
η = efficiency
λ = wavelength
λ_n = normal wavelength
λ_o = cutoff wavelength
λ_p = waveguide wavelength
ρ = reflection coefficient
= symbol used in simplifying formula
ϕ = beamwidth
ω_c = carrier angular frequency
ω_m = modulation angular frequency

ELECTRONIC COMMUNICATION SYSTEMS

1 INTRODUCTION TO COMMUNICATIONS SYSTEMS

This chapter serves to introduce the subject of communications systems, and also this book as a whole. In studying it, you will be introduced to information, a basic communications system, transmitters, receivers and noise. Importantly, modulation is introduced, and the absolute need to use it in conveying information will be made clear. The final section briefly discusses bandwidth requirements and shows that the bandwidth needed to transmit some signals and waveforms is a great deal more than might be expected.

1-1 COMMUNICATIONS

In its basic electrical sense, the term *communications* refers to the sending, receiving and processing of information by electric means. As such, it started with wire telegraphy in the eighteen-forties, developing with telephony some decades later and radio at the beginning of this century. *Radio* communication, made possible by the invention of the triode tube, was greatly stimulated by the work done during World War II. It subsequently became even more widely used and refined through the invention and use of the transistor, integrated circuits and other semiconductor devices [1]. More recently, the use of satellites and *fiber optics* has made communications even more widespread, with an increasing emphasis on computer and other data communications.

A modern *communications system* is first concerned with the sorting, processing and storing of information before its transmission. The actual transmission then follows, with further processing and the filtering of noise. Finally we have reception, which may include processing steps such as decoding, storage and interpretation. In this context, forms of communications include radio telephony and telegraphy, broadcasting, point-to-point and mobile communications (commercial or military), computer communications, radar, radiotelemetry and radio aids to navigation. All these are treated in turn, in following chapters.

In order to become familiar with these systems, it is necessary first to know about amplifiers and oscillators, the building blocks of all electronic processes and equipment. With these as a background, the everyday communications concepts of *noise, modulation* and *information theory,* as well as the various systems themselves, may be approached. Any logical order may be used, but the one adopted here, that is, basic systems, communications processes and circuits, and more complex systems, is considered most suitable. It is also important to consider the human factors influencing a particular system, since they must always affect its design, planning and use.

1-2 COMMUNICATIONS SYSTEMS

Before investigating individual systems, we have to define and discuss important terms such as *information, message* and *signal, channel, noise* and *distortion, modulation* and *demodulation,* and finally *encoding* and *decoding*. To correlate these concepts, a block diagram of a general communications system is shown in Fig. 1-1.

1-2.1 Information

The communications system exists to communicate a message. This message comes from the information source, which originates it, in the sense of selecting one message from a group of messages. Although this applies more to telegraphy than to entertainment broadcasting, for example, it may nevertheless be shown to apply to all forms of communications. The *set,* or total number of messages, consists of individual messages which may be distinguished from one another. These may be words, groups of words, code symbols or any other prearranged units.

Information itself is *that which is conveyed.* The amount of information contained in any given message is measured in *bits* or in *dits,* which are dealt with in Chap. 13, and depends on the number of choices that must be made. The greater the total number of possible selections, the larger the amount of information conveyed. For example, to indicate the position of a word on this page, it may be sufficient to say that it is on the top or bottom, left or right side, i.e., two consecutive choices of one out of two possibilities. If this word may appear in any one of two pages, it is now necessary to say which one, and hence more information must be given. The meaning (or lack of

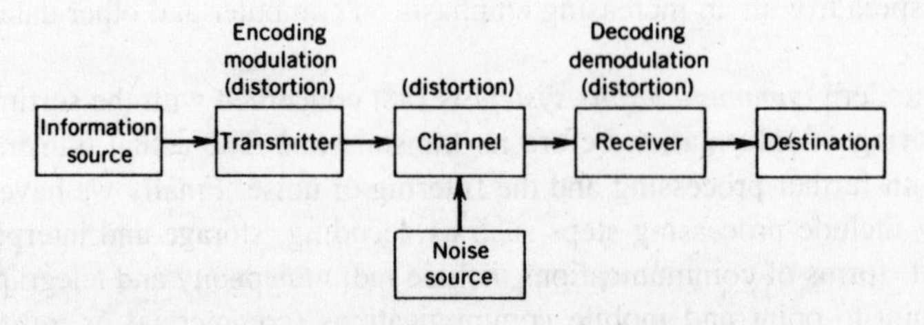

FIGURE 1-1 Block diagram of communications system.

meaning) of the information does not matter, from this point of view; only the quantity is important. However, it must be realized that no real information is conveyed by a redundant (i.e., totally predictable) message. Redundancy is not wasteful under all conditions, however. Apart from its obvious use in entertainment, teaching and any appeal to the emotions, it also helps a message to remain intelligible under difficult or noisy conditions.

1-2.2 Transmitter

Unless the message that comes from the information source is electrical in nature, it will be unsuitable for immediate sending. Even then, a lot of work must be done to make such a message suitable. This may be demonstrated in *single-sideband modulation* (treated in Chap. 4), where it is necessary to convert the incoming sound signals into electrical variations, to restrict the range of the audio frequencies and then to *compress* their amplitude range. All this is done before any *modulation*. In wire telephony no processing may be required, but in long-distance communications, a transmitter is required to process, and possibly encode, the incoming information so as to make it suitable for transmission and subsequent reception.

Eventually, in a transmitter, the information modulates the *carrier*, i.e., is impressed on a high-frequency sine wave. The actual method of modulation varies from one system to another. Thus modulation may be *high level* or *low level*, and the system itself may be *amplitude modulation, frequency modulation, pulse modulation* or any variation or combination of these, depending on the requirements. Figure 1-2 shows a high-level amplitude-modulated broadcast transmitter of a type that will be discussed in detail in Chap. 6.

1-2.3 Channel—Noise

The acoustic channel (i.e., shouting!) is not used for long-distance communications, and neither was the visual channel until the advent of the *laser*. "Communications," in this context, will be restricted to radio, wire and fiber optic channels. Quite separately, it should also be noted that the term *channel* is often used to refer to the frequency range allocated to a particular service or transmission, such as a *television channel*.

It is inevitable that the signal will deteriorate during the process of transmission and reception as a result of some distortion in the system, or because of the introduc-

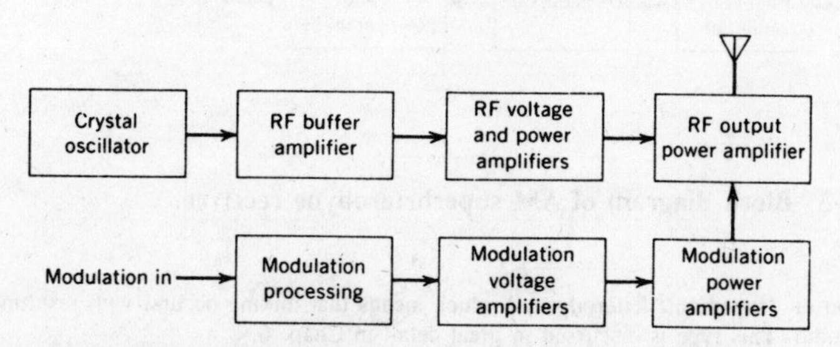

FIGURE 1-2 Block diagram of typical radio transmitter.

tion of noise, which is *unwanted energy, usually of random character, present in a transmission system, due to any cause*. Since noise will be received together with the signal, it obviously places a limitation on the transmission system as a whole. When noise is severe, it may mask a given signal so much that the signal becomes unintelligible and therefore useless. In Fig. 1-1, only one source of noise is shown, not because only one exists, but to simplify the block diagram. Noise may interfere with signal at any point in a communications system, but *it will have its greatest effect when the signal is weakest;* this means that noise in the channel or at the input to the receiver is the most noticeable. It is treated in detail in Chap. 2.

1-2.4 Receiver

There are a great variety of receivers in communications systems, since the exact form of a particular receiver is influenced by a great many requirements. Among the more important requirements are the modulation system used, the operating frequency and its range and the type of display required, which in turn depends on the destination of the intelligence received. However, most receivers do conform broadly to the *superheterodyne*[1] type, as does the simple broadcast receiver whose block diagram is shown in Fig. 1-3.

Receivers run the whole range of complexity from a very simple crystal receiver, with headphones, to a far more complex radar receiver, with its involved antenna arrangements and visual display system. Whatever the receiver, its most important function is demodulation (and sometimes also decoding). Both these processes are the reverse of the corresponding transmitter processes and are discussed in the ensuing chapters.

As stated initially, the purpose of a receiver and the form of its output influence its construction as much as the type of modulation system used. Thus the output of a receiver may be variously fed to a loudspeaker, video display unit, teletypewriter, various radar displays, television picture tube, pen recorder or computer; in

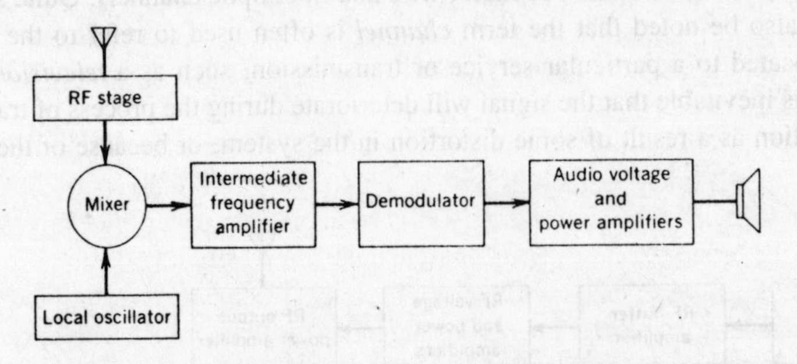

FIGURE 1-3 Block diagram of AM superheterodyne receiver.

[1] A contraction of "supersonic heterodyne," which means that mixing occurs, with resulting frequencies higher than audio. This type is discussed in great detail in Chap. 6.

each instance different arrangements must be made, each affecting the receiver design. Note also that the transmitter and receiver must be in agreement with the modulation and coding methods used (and also timing or synchronization in some systems).

1-3 MODULATION

1-3.1 Description

In the process of modulation, some characteristic of a high-frequency sine wave (the carrier) is varied in accordance with the instantaneous value of the (modulating) signal. Such a sine wave may be represented by the equation $e = E \sin(\omega t + \phi)$, where e is the instantaneous value of the sine wave, called the *carrier*; E is its maximum amplitude, ω is the angular velocity or angular frequency, while ϕ is the phase relation with respect to some reference. Any of these last three characteristics, or *parameters*, of the carrier may be varied by the modulating signal, giving rise to amplitude, frequency or phase modulation, respectively.

Such a process is fairly complicated and would obviously not be used without some very sound and compelling reasons. These will now be discussed, because they are very important but not immediately apparent.

1-3.2 Need for Modulation

There are two alternatives to the use of a modulated carrier for the transmission of messages over long distances in the radio channel: one could try to send the (modulating) signal itself, or else use an unmodulated carrier. The impossibility of transmitting the signal itself will be demonstrated first.

Although the topic has not yet been discussed, several difficulties are involved in the propagation of electromagnetic waves at frequencies corresponding to the audio spectrum, i.e., below 20 kilohertz (20 kHz) (see also Chaps. 8 and 9). The greatest of these is that for efficient radiation and reception the transmitting and receiving antennas would have to have heights comparable to a quarter-wavelength of the frequency used. This is 75 meters (75 m) at 1 megahertz (1 MHz), in the broadcast band, but at 15 kHz it has increased to 5000 m (or just over 16,000 feet)! A vertical antenna of this size is unthinkable.

There is an even more important argument against transmitting signal frequencies directly: all sound is concentrated within the range from 20 Hz to 20 kHz, so that all signals from the different sources would be hopelessly and inseparably mixed up. In any city, the broadcasting stations alone would completely blanket the "air," and yet they represent a very small proportion of the total number of transmitters in use.

In order to separate the various signals, it is necessary to translate them all to different portions of the electromagnetic spectrum; each must be given its own "pigeonhole." This also overcomes the difficulties of poor radiation at low frequencies. Once signals have been translated, a tuned circuit is employed in the front end of the receiver to make sure that the desired section of the spectrum is admitted and all the unwanted ones are rejected. The tuning of such a circuit is normally made variable and connected to the tuning control, so that the receiver can select any desired transmission within a predetermined range, such as the very high frequency (VHF) broadcast band used for frequency modulation (FM).

Although this separation of signals has removed a number of the difficulties encountered in the absence of modulation, the fact still remains that unmodulated carriers of various frequencies cannot, by themselves, be used to transmit intelligence. An unmodulated carrier has a constant maximum amplitude, a constant frequency and a constant phase relationship with respect to some reference; in fact, all its parameters are constant. A message, however, consists of ever-varying quantities: speech, for instance, is made up of rapid and unpredictable variations in *amplitude* (volume) and *frequency* (pitch). Since it is impossible to represent these two variables by a set of three constant parameters, an unmodulated carrier cannot be used to convey information. In a continuous-wave-modulation system (amplitude or frequency modulation, but *not* pulse modulation) one of the parameters of the carrier is varied by the message. Thus at any instant its deviation from the unmodulated value is proportional to the instantaneous value of the modulating voltage, and the rate at which this deviation takes place is equal to the modulating frequency. In this fashion, enough information about the instantaneous amplitude and frequency is transmitted to enable the receiver to recreate the original message.

1-4 BANDWIDTH REQUIREMENTS

It is reasonable to expect that the frequency range (i.e., bandwidth) required for a given transmission should depend on the bandwidth occupied by the modulation signals themselves. For example, noting that a high-fidelity audio signal occupies the range 50 to 15,000 Hz, and a bandwidth of 300 to 3400 Hz is adequate for a telephone conversation, when a carrier has been similarly modulated with each, a greater bandwidth will be required for the high-fidelity (hi-fi) transmission. At this point, it is worth injecting the thought that the transmitted bandwidth need not be exactly the same as the bandwidth of the original signal, for reasons connected with the properties of the different modulation systems. This will be made abundantly clear in Chaps. 3 to 5.

Before trying to estimate the bandwidth of a modulated transmission, it is essential to know the bandwidth occupied by the modulating signal itself. If this consists of sinusoidal signals, there is no problem, and the occupied bandwidth will simply be the frequency range between the lowest and the highest sine-wave signal. However, if the modulating signals are nonsinusoidal, a much more complex situation results. Since such nonsinusoidal waves occur very frequently as modulating signals in communications, their frequency requirements will now be investigated.

1-4.1 Frequency Spectra of Nonsinusoidal Waves

If any nonsinusoidal waves, such as square waves, are to be transmitted by a communications system, then it is important to realize that each such wave may be broken down into its constituent sine waves. The bandwidth required will therefore be considerably greater than might have been expected if only the repetition rate of such a wave had been taken into account. A more formal statement now follows.

It may be shown that any nonsinusoidal, single-valued repetitive waveform consists of sine waves and/or cosine waves. The frequency of the lowest-frequency, or fundamental, *sine wave is equal to the repetition rate of the nonsinusoidal waveform, and all others are harmonics of the fundamental. There are an infinite number of such harmonics.* Thus, some non-sine wave recurring at a rate of 200 times per second will consist of a 200-Hz fundamental sine wave, and harmonics at 400, 600 and 800 Hz, and so on. No other frequencies will be present, but for some waveforms only the even (or perhaps only the odd) harmonics will be present. As a general rule, it may be added that the higher the harmonic, the lower its relative amplitude, so that in bandwidth calculations the highest harmonics are often ignored.

The preceding statement may be verified in any one of three different ways. It may be proved mathematically by a process called *Fourier analysis* [2]. Graphical synthesis may be used, in which case adding the appropriate sine-wave components, taken from a formula derived by Fourier analysis, demonstrates the truth of the statement. An added advantage of this method is that it makes it possible for us to see the effect on the overall waveform of the absence of some of the constituents (for instance, the higher harmonics). (Question 1-5 illustrates the use of this method.)

Finally, the presence of the component sine waves in the correct proportions may be demonstrated with a wave analyzer, which is basically a high-gain tunable amplifier with a narrow passband, enabling it to tune to each component sine wave and measure its amplitude. Some formulas for frequently encountered nonsinusoidal waves are now given, and more may be found in handbooks. If the amplitude of the nonsinusoidal wave is A and its repetition rate is $\omega/2\pi$ per second, then it may be represented as follows:

Square wave:

$$e = \frac{4A}{\pi}\left(\cos \omega t - \tfrac{1}{3} \cos 3\omega t + \tfrac{1}{5} \cos 5\omega t - \tfrac{1}{7} \cos 7\omega t + \ldots\right) \tag{1-1}$$

Triangular wave:

$$e = \frac{4A}{\pi^2}\left(\cos \omega t + \tfrac{1}{9} \cos 3\omega t + \tfrac{1}{25} \cos 5\omega t + \ldots\right) \tag{1-2}$$

Sawtooth wave:

$$e = \frac{2A}{\pi}\left(\sin \omega t - \tfrac{1}{2} \sin 2\omega t + \tfrac{1}{3} \sin 3\omega t - \tfrac{1}{4} \sin 4\omega t + \ldots\right) \tag{1-3}$$

In each case several of the harmonics will be required, in addition to the fundamental frequency, if the wave is to be represented adequately (i.e., with acceptably low distortion). This, of course, will greatly increase the required bandwidth.

QUESTIONS

For self-testing questions on this chapter, see p. 677.

1-1. The carrier performs certain functions in radio communications. What are they?

1-2. Define noise. Where is it most likely to affect the signal?

1-3. What does modulation actually do to a message and carrier?
1-4. List the basic functions of a radio transmitter and the corresponding functions of the receiver.
1-5. Ignoring the constant relative amplitude component, plot and add the appropriate sine waves graphically, in each case using the first four components, so as to synthesize *(a)* a square wave, *(b)* a sawtooth wave.

REFERENCES

1. A longer historical introduction is given in Gregg, W. D., *Analog and Digital Communication,* John Wiley & Sons, Inc., New York, 1977, pp. 1–5.
2. For example, International Telephone and Telegraph Corporation, *Reference Data for Radio Engineers,* 6th ed., Howard W. Sams and Co. Inc., Indianapolis, 1975, pp. 44-12–44-14.

2
NOISE

Noise is probably the only topic in electronics and telecommunications with which everyone must be familiar, no matter what his or her specialization. It is ever present and limits the performance of virtually every system. Also, measuring it is very contentious—almost everybody has a different pet method of quantifying noise and its effects.

Despite this, after studying this chapter you should be familiar with the types and sources of noise. The methods of calculating the noise produced by various sources will be learned, and so will be the ways of adding such noise. The very important noise quantities—*signal-to-noise ratio, noise figure*, and *noise temperature*—will have been covered in detail, as will methods of measuring noise.

Noise may be defined, in an electrical sense, as any unwanted form of energy tending to interfere with the proper and easy reception and reproduction of wanted signals. Many disturbances of an electrical nature produce noise in receivers, modifying the signal in an unwanted manner. In radio receivers, for example, noise may produce hiss in the loudspeaker output, whereas in television receivers ''snow'' or ''confetti'' (colored snow) becomes superimposed on the picture. In pulse communications systems, noise may produce unwanted pulses or perhaps cancel out the wanted ones; it may cause serious errors in this fashion. Noise is thus seen as limiting the range of systems, for a given transmitted power. It affects the sensitivity of receivers, by placing a limit on the weakest signals that can be amplified. It may sometimes even force a reduction in the bandwidth of a system, as will be seen in *radar*.

There are numerous ways of classifying noise. It may be subdivided according to type, source, effect, or relation to the receiver, depending on circumstances. It is most convenient here to divide noise into two broad groups: noise whose sources are external to the receiver, and noise created within the receiver itself. On the one hand, external noise is difficult to treat quantitatively, and furthermore there is often little that can be done about it, short of moving the system to another location. Note how radiotelescopes are always located away from industry, whose processes create so much electrical noise. International satellite earth stations are also located in noise-free valleys, where possible. On the other hand, internal noise is both more quantifiable and capable of being reduced by appropriate receiver design.

Because noise has such a limiting effect, and also because it is often possible to reduce its effects through intelligent circuit use and design, it is most important for all those connected with communications to be well informed about noise.

2-1 EXTERNAL NOISE

The various forms of noise created outside the receiver come under the heading of external noise and include atmospheric and extraterrestrial noise and industrial noise.

2-1.1 Atmospheric Noise

Perhaps the best way to become acquainted with atmospheric noise is to listen to shortwaves on a receiver which is not well equipped to receive them. An astonishing variety of strange sounds will be heard, all tending to interfere with the program. Most of these sounds are the result of spurious radio waves which induce voltages in the antenna. The majority of these radio waves come from natural sources of disturbance; they represent atmospheric noise, generally called *static*.

Static is caused by lightning discharges in thunderstorms and other natural electric disturbances occurring in the atmosphere. Owing to its origin it is in the form of impulses, and because such processes are random in nature, it is spread over all the radio spectrum normally used for broadcasting. Atmospheric noise thus consists of spurious radio signals with components distributed over a wide range of frequencies. It is propagated over the earth in the same way as ordinary radio waves of the same frequencies, so that at any point on the ground, static will be received from all thunderstorms, local and distant. The static from the former is likely to be more severe but, obviously, less frequent. Field strength is approximately inversely proportional to frequency, so that this noise will interfere more with the reception of radio than that of television. Such noise consists of impulses, and (as shown in Chap. 1) these nonsinusoidal waves have harmonics whose amplitude falls off with increase in the harmonic. Static from distant sources will vary in intensity according to the variations in propagating conditions. Hence we have the usual increase in its level at night, at both broadcast and shortwave frequencies.

Atmospheric noise becomes less severe at frequencies above about 30 MHz because of two separate factors. First, the higher frequencies are limited to line-of-sight propagation (as will be seen in Chap. 8), i.e., less than 80 kilometers or so. Second, the nature of the mechanism generating this noise is such that very little of it is created in the VHF range and above.

2-1.2 Extraterrestrial Noise

It is only a slight exaggeration to say that there are almost as many types of space noise as there are sources. For convenience, however, a division into two subgroups will suffice.

Solar noise The sun throws so many things our way that we should not be too surprised to find that noise is noticeable among them; again there are two types. Under normal "quiet" conditions, there is a constant noise radiation from the sun, simply because it is a large body at a very high temperature (over 6000°C on the surface). It therefore radiates over a very broad frequency spectrum which includes the frequencies we use for communications. However, the sun is a variable star and undergoes cycles at the peak of which electrical disturbances erupt, such as corona flares and sunspots. Even though the additional noise thus produced comes from limited portions of the sun's disk, it may still be orders of magnitude greater than that received during periods of quiet sun.

The solar cycle repeats these periods of great electrical disturbance approximately every 11 years. In addition, if a line is drawn to join these 11-year peaks, it is seen that a supercycle is in operation, with the peaks reaching an even higher maximum every 100 years or so. Finally, these 100-year peaks appear to be increasing in intensity. Since there is a correlation between peaks in solar disturbance and growth rings in trees, it has been possible to trace them back to the beginning of the eighteenth century. Evidence shows that the year 1957 was not only a peak but the highest such peak on record.

Cosmic noise Since distant stars are also suns and have high temperatures, they radiate noise in the same manner as our sun, and what they lack in nearness they nearly make up in numbers. The noise thus received is called *thermal*[1] (or *black-body*) noise and is distributed fairly uniformly over the entire sky. We also receive noise from the center of our own galaxy (the Milky Way), from other galaxies, and from other virtual point sources such as "quasars" and "pulsars." This *galactic* noise is very intense, but it comes from sources which are only points in the sky. Two of the strongest sources, which were also two of the earliest discovered, are Cassiopeia A and Cygnus A. Note that it is inadvisable to refer to the foregoing as "noise" sources when talking with radio astronomers!

Summary Space noise is observable at frequencies in the range from about 8 MHz to somewhat above 1.43 gigahertz (1.43 GHz), the latter frequency corresponding to the 21-cm hydrogen "line." Apart from man-made noise it is the strongest component over the range of about 20 to 120 MHz. Not very much of it below 20 MHz penetrates down through the *ionosphere*,[2] while its eventual disappearance at frequencies in excess of 1.5 GHz is probably governed by the mechanisms generating it, and its absorption by hydrogen in interstellar space.

2-1.3 Industrial Noise

Between the frequencies of approximately 1 to 600 MHz, in urban, suburban and other industrial areas, the intensity of noise made by humans easily outstrips that created by any other source, internal or external to the receiver. Under this heading, sources such as automobile and aircraft ignition, electric motors and switching gear, leakage from high-voltage lines and a multitude of other heavy electric machines are all included. Fluorescent lights are another powerful source of such noise and therefore should not be used where sensitive receiver reception or testing is being conducted.

[1] Meaning it is uniformly distributed with frequency—see Sec. 2-2.1.
[2] See Sec. 8-2.2.

The noise is produced by the arc discharge present in all these operations, and under these circumstances it is not surprising that this noise should be most intense in industrial and densely populated areas. (Given some encouragement, industrial noise due to spark discharge may even span oceans, as demonstrated by Marconi in 1901 at St. John's, Newfoundland.)

The nature of industrial noise is so variable that it is difficult to analyze it on any basis other than the statistical. It does, however, obey the general principle that received noise increases as the receiver bandwidth is increased (Sec. 2-2.1).

2-2 INTERNAL NOISE

Under the heading of internal noise, we discuss noise created by any of the active or passive devices found in receivers. Such noise is generally random, that is, impossible to treat on an individual voltage basis, but easy to describe statistically. Because the noise is randomly distributed over the entire radio spectrum there is, on the average, as much of it at one frequency as at any other. *Thus random noise power is proportional to the bandwidth over which it is measured.*

2-2.1 Thermal Agitation Noise

The noise generated in a resistance or the resistive component of any impedance is random and is referred to variously as *thermal, agitation, white* or *Johnson* noise. It is due to the rapid and random motion of the molecules. atoms and electrons of which any such resistor is made up.

In thermodynamics, kinetic theory shows that the temperature of a particle is a way of expressing its internal kinetic energy. Thus the "temperature" of a body is the statistical root mean square (rms) value of the velocity of motion of the particles in the body. As the theory states, the kinetic energy of these particles becomes approximately zero (i.e., their motion ceases) at the temperature of absolute zero, which is 0 K (kelvins, formerly called degrees Kelvin) and very nearly equals −273° C. It is thus apparent that the noise power generated by a resistor is proportional to its absolute temperature, in addition to being proportional to the bandwidth over which the noise is to be measured. Thus

$$P_n \propto T\,\delta f = \mathrm{k}T\,\delta f \qquad \text{(2-1)}$$

where k = Boltzmann's constant = 1.38×10^{-23} J(joules) /K the appropriate proportionality constant in this case

T = absolute temperature, K = 273 + °C

δf = bandwidth of interest

P_n = maximum noise power output of a resistor

If an ordinary resistor at the standard temperature of 17°C (290 K) is not connected to any voltage source, it might at first be thought that there is no voltage to be measured across it. That is correct if the measuring instrument is a direct current (dc) voltmeter, but it is incorrect if a very sensitive electronic voltmeter is used. The resistor is a noise generator, and there may even be quite a large voltage across it; however, since it is random and therefore has a definite rms value but no dc component, only the alternating current (ac) meter will register a reading. This noise voltage is caused by the random movement of electrons within the resistor, which constitutes a current. It is true that as many electrons arrive at one end of the resistor as at the other over any long period of time. However, at any instant of time, there are bound to be more electrons arriving at one particular end than at the other because their movement is random. The rate of arrival of electrons at either end of the resistor thus varies randomly, and so does the potential difference between the two ends. *A random voltage across the resistor definitely exists* and may be both measured and calculated.

It must be realized that all formulas referring to random noise are applicable only to the rms value of such noise, not to its instantaneous value, which is quite unpredictable. So far as peak noise voltages are concerned, all that may be stated is that they are unlikely to have values in excess of 10 times the rms value.

Using Eq. (2-1), the equivalent circuit of a resistor as a noise generator may be drawn as in Fig. 2-1, and from this the resistor's equivalent noise voltage V_n may be calculated. Assume that R_L is noiseless and is receiving the maximum noise power generated by R; under these conditions of maximum power transfer, R_L must be equal to R. Then

$$P_n = \frac{V^2}{R_L} = \frac{V^2}{R} = \frac{(V_n/2)^2}{R} = \frac{V_n^2}{4R}$$

$$V_n^2 = 4RP_n = 4RkT\ \delta f$$

and

$$V_n = \sqrt{4kT\ \delta f\ R} \tag{2-2}$$

It is seen from Eq. (2-2) that the square of the rms noise voltage associated with a resistor is proportional to the absolute temperature of the resistor, the value of its resistance, and the bandwidth over which the noise is measured. Note especially that the generated noise voltage is quite independent of the frequency at which it is measured; this stems from the fact that it is random and therefore evenly distributed over the frequency spectrum.

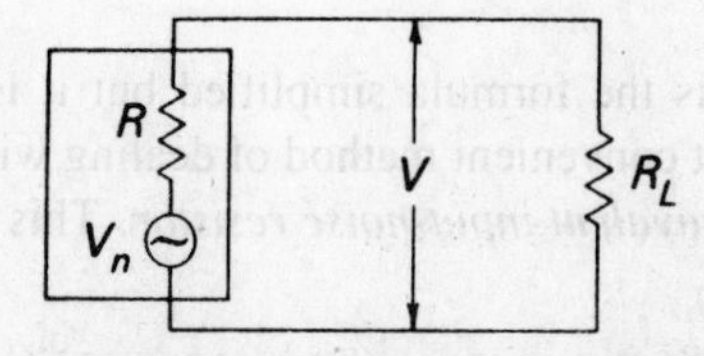

FIGURE 2-1 Resistance noise generator.

Example 2-1 An amplifier operating over the frequency range from 18 to 20 MHz has a 10-kilohm (10-kΩ) input resistor. What is the rms noise voltage at the input to this amplifier if the ambient temperature is 27°C?

$$V_n = \sqrt{4kT\,\delta f\,R}$$
$$= \sqrt{4 \times 1.38 \times 10^{-23} \times (27 + 273) \times (20 - 18) \times 10^6 \times 10^4}$$
$$= \sqrt{4 \times 1.38 \times 3 \times 2 \times 10^{-11}} = 1.82 \times 10^{-5}$$
$$= 18.2 \text{ microvolts } (18.2\ \mu\text{V})$$

As can be seen from this example, it would be futile to expect this amplifier to handle signals unless they were considerably larger than 18.2 μV; a low voltage fed to this amplifier would be masked by the noise and lost.

2-2.2 Shot Noise

Thermal agitation is by no means the only source of noise in receivers. The most important of all the other sources is the *shot* effect, which leads to shot noise in all amplifying devices and virtually all active devices. It is caused by random variations in the arrival of electrons (or holes) at the output electrode of an amplifying device and thus appears as a randomly varying noise current superimposed on the output. When amplified, it is supposed to sound as though a shower of lead shot were falling on a metal sheet; hence the name *shot* noise.

Although the average output current of a device is governed by the various bias voltages, at any instant of time there may be more or fewer electrons arriving at the output electrode. In bipolar transistors, for example, this is mainly a result of the random drift of the discrete current carriers across the junctions. The paths taken are random and therefore unequal, so that although the average collector current is constant, minute variations nevertheless occur. Shot noise behaves in the same manner as thermal agitation noise, apart from the fact that it has a different source.

Many variables are involved in the generation of this noise in the various amplifying devices, and so it is customary to use approximate equations for it. In addition, shot-noise *current* is a little difficult to add to thermal-noise *voltage* in calculations, so that for all devices with the exception of the diode, shot-noise formulas used are generally simplified. For a diode, the formula is exactly[3]

$$i_n = \sqrt{2ei_p\,\delta f} \tag{2-3}$$

where i_n = rms shot-noise current
e = charge of an electron = 1.6×10^{-19}C
i_p = direct diode current
δf = bandwidth of system

In all other instances not only is the formula simplified but it is not even a formula for shot-noise current. The most convenient method of dealing with shot noise is to find the value or formula for an *equivalent input-noise resistor*. This precedes the

[3] It may be shown that, for a vacuum tube diode, Eq. (2-3) holds only under so-called temperature-limited conditions, under which the "virtual cathode" has not been formed.

device, which is now assumed to be noiseless, and has a value such that the same amount of noise is present at the output of the equivalent system as in the practical amplifier. Thus the noise current has been replaced by a resistance so that it is now easier to add shot noise to thermal noise. Furthermore, it has also been referred to the input of the amplifier, which is a much more convenient place, as will be seen.

The value of the equivalent shot-noise resistance R_{eq} of a device is generally quoted in the manufacturer's specifications. Approximate formulas for equivalent shot-noise resistances are also available [1]; they all show that such noise is inversely proportional to transconductance and also directly proportional to output current. So far as the use of R_{eq} is concerned, the important thing to realize is that it is a completely fictitious resistance, whose sole function is to simplify calculations involving shot noise. Accordingly, for noise only, this resistance is treated as though it were an ordinary noise-creating resistor, at the same temperature as all the other resistors, and located in series with the input electrode of the device.

2-2.3 Transit-Time Noise

If the time taken by an electron to travel from (say) the emitter to the collector of a transistor becomes comparable to the period of the signal being amplified, i.e., at frequencies in the upper VHF range and beyond, the so-called *transit-time effect* takes place, and the noise input admittance of the transistor increases. This is treated fully in Sec. 12-1.1. The minute currents induced in the input of the device by random fluctuations in the output current become of great importance at such frequencies and create random noise.

Once this high-frequency noise makes its presence felt, it goes on increasing with frequency at a rate that soon approaches 6 decibels (6 dB) per octave, and this random noise then quickly predominates over the other forms. The result of all this is that it is preferable to measure noise at such high frequencies, instead of trying to calculate an input equivalent noise resistance for it. All in all, though, radio frequency (RF) transistors are remarkably low-noise. A noise *figure* (see Sec. 2-4) as low as 1 dB is possible with transistor amplifiers well into the UHF range.

2-2.4 Miscellaneous Noise

Flicker At low audio frequencies, a poorly understood form of noise called *flicker* or *modulation noise* is found in transistors. It is proportional to emitter current and junction temperature, but since it is also inversely proportional to frequency, it may be completely ignored above about 500 Hz. It is no longer very serious.

Resistance Thermal noise, sometimes called *resistance* noise, is also present in transistors. It is due to the base, emitter, and collector internal resistances, and in most circumstances the base resistance makes the largest contribution.

From about 500 Hz up to about $f_{\alpha b}/5$, transistor noise remains relatively constant, so that an equivalent input resistance for shot and thermal noise may be freely used.

Noise in mixers Mixers are much noisier than amplifiers using identical devices, except at microwave frequencies, where the situation is rather complex. This high value of noise in mixers is caused by two separate effects. First, *conversion transconductance*[4] of mixers is much lower than the transconductance of amplifiers. Second, if *image frequency rejection*[4] is inadequate, as often happens at shortwave frequencies, noise associated with the image frequency will also be accepted.

2-3 NOISE CALCULATIONS

2-3.1 Addition of Noise due to Several Sources

Let there be two sources of thermal agitation noise in series: thus $V_{n1} = \sqrt{4kT\,\delta f\,R_1}$ and $V_{n2} = \sqrt{4kT\,\delta f\,R_2}$. The sum of two such rms voltages in series is given by the square root of the sum of their squares, so that we have

$$V_{n,\text{tot}} = \sqrt{V_{n1}^2 + V_{n2}^2} = \sqrt{4kT\,\delta f\,R_1 + 4kT\,\delta f\,R_2}$$
$$= \sqrt{4kT\,\delta f\,(R_1 + R_2)} = \sqrt{4kT\,\delta f\,R_{\text{tot}}} \tag{2-4}$$

where

$$R_{\text{tot}} = R_1 + R_2 + \cdots \tag{2-5}$$

It is seen from the foregoing equations that in order to find the total noise voltage caused by several sources of thermal noise in series, the resistances are added and the noise voltage is calculated using this total resistance. The same procedure applies if one of those resistances is, in fact, an equivalent input-noise resistance.

Example 2-2 Calculate the noise voltage at the input of a television RF amplifier, using a device that has a 200-ohm (200-Ω) equivalent noise resistance and a 300-Ω input resistor. The bandwidth of the amplifier is 6 MHz, and the temperature is 17°C.

$$V_{n,\text{tot}} = \sqrt{4kT\,\delta f\,R_{\text{tot}}}$$
$$= \sqrt{4 \times 1.38 \times 10^{-23} \times (17 + 273) \times 6 \times 10^6 \times (300 + 200)}$$
$$= \sqrt{4 \times 1.38 \times 2.9 \times 6 \times 5 \times 10^{-13}} = \sqrt{48 \times 10^{-12}}$$
$$= 6.93 \times 10^{-6} = 6.93\ \mu\text{V}$$

To calculate the noise voltage due to several resistors in parallel, find the total resistance by standard methods, and then substitute this resistance into Eq. (2-4) as before. This means that the total noise voltage is less than that due to any of the individual resistors, but, as shown in Eq. (2-1), the noise power remains constant.

2-3.2 Addition of Noise due to Several Amplifiers in Cascade

The situation that occurs in receivers is illustrated in Fig. 2-2. It shows a number of amplifying stages in cascade, each having a resistance at its input and output. The first such stage is very often an RF amplifier, while the second is a mixer. The problem is to find their combined effect on the receiver noise.

[4] See Sec. 6-2 for the explanation of these terms.

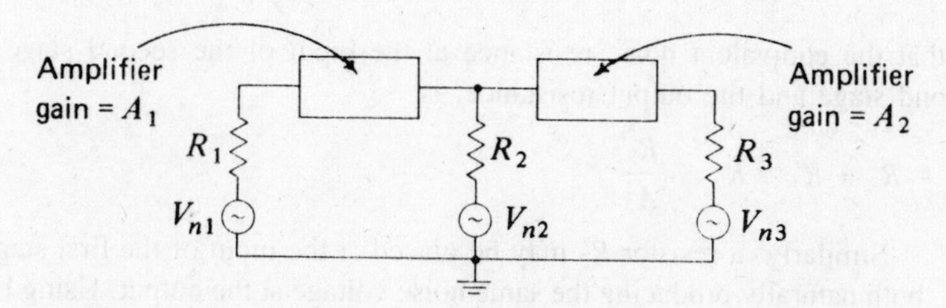

FIGURE 2-2 Noise of several amplifying stages in cascade.

It might appear logical that the procedure should be to combine all the noise resistances at the input, calculate their noise voltage, multiply it by the gain of the first stage and add this voltage to the one generated at the input of the second stage. The process might then be continued, and the noise voltage at the output, due to all the intervening noise sources, would thus be found. Admittedly, there is nothing wrong with such a procedure. However, the result is useless because the argument assumed that it is important to find the total output noise voltage, whereas the important thing is to find the equivalent *input* noise voltage. Actually, it is even better to go one step further and find an equivalent resistance for such an input voltage, i.e., the equivalent noise resistance for the whole receiver. This is the resistance that will produce the same random noise at the output of the receiver as does the actual receiver, so that we have succeeded in replacing an actual receiver by an ideal noiseless one with an equivalent noise resistance R_{eq} located across its input. This greatly simplifies subsequent calculations, gives a good figure for comparison with other receivers, and permits a quick calculation of the lowest input signal which this receiver may amplify without drowning it with noise.

Consider the two-stage amplifier of Fig. 2-2. The gain of the first stage is A_1, and that of the second is A_2. The first stage has a total input-noise resistance R_1, the second R_2 and the output resistance is R_3. The rms noise voltage at the output due to R_3 is

$$V_{n3} = \sqrt{4kT\,\delta f\,R_3}$$

The same noise voltage would be present at the output if there were no R_3 there; instead R_3' was present at the input of stage 2, such that

$$V_{n3}' = \frac{V_{n3}}{A_2} = \frac{\sqrt{4kT\,\delta f\,R_3}}{A_2} = \sqrt{4kT\,\delta f\,R_3'}$$

where R_3' is the resistance which if placed at the input of the second stage would produce the same noise voltage at the output as does R_3. Thus

$$R_3' = \frac{R_3}{A_2^2} \tag{2-6}$$

Equation (2-6) shows that when a noise resistance is "transferred" from the output of a stage to its input, it must be divided by the square of the voltage of the stage. Now the noise resistance actually present at the input of the second stage is R_2, so that the equivalent noise resistance at the input of the second stage, due to the second stage and the output resistance, is

$$R'_{eq} = R_2 + R'_3 = R_2 + \frac{R_3}{A_2^2}$$

Similarly, a resistor R'_2 may be placed at the input of the first stage to replace R'_{eq}, both naturally producing the same noise voltage at the output. Using Eq. (2-6) and its conclusion, we have

$$R'_2 = \frac{R'_{eq}}{A_1^2} = \frac{R_2 + R_3/A_2^2}{A_1^2} = \frac{R_2}{A_1^2} + \frac{R_3}{A_1^2 A_2^2}$$

The noise resistance actually present at the input of the first stage is R_1, so that the equivalent noise resistance of the whole cascaded amplifier, at the input of the first stage, will be

$$R_{eq} = R_1 + R'_2$$

$$= R_1 + \frac{R_2}{A_1^2} + \frac{R_3}{A_1^2 A_2^2} \tag{2-7}$$

It is possible to extend Eq. (2-7) by induction to apply to an n-stage cascaded amplifier, but this is not normally necessary. As Example 2-3 will show, the noise resistance located at the input of the first stage is by far the greatest contributor to the total noise, and only in broadband (i.e., low-gain) amplifiers is it necessary to consider a resistor past the output of the second stage.

Example 2-3 The first stage of a two-stage amplifier has a voltage gain of 10, a 600-Ω input resistor, a 1600-Ω equivalent noise resistance and a 27-kΩ output resistor. For the second stage, these values are 25, 81 kΩ, 10 kΩ and 1 megaohm (1 MΩ), respectively. Calculate the equivalent input-noise resistance of this two-stage amplifier.

$$R_1 = 600 + 1600 = 2200\ \Omega$$

$$R_2 = \frac{27 \times 81}{27 + 81} + 10 = 20.2 + 10 = 30.2\ \text{k}\Omega$$

$$R_3 = 1\ \text{M}\Omega \qquad \text{(as given)}$$

$$R_{eq} = 2200 + \frac{30{,}200}{10^2} + \frac{1{,}000{,}000}{10^2 \times 25^2} = 2200 + 302 + 16$$

$$= 2518\ \Omega$$

Note that the 1-MΩ output resistor has the same noise effect as a 16-Ω resistor at the input.

2-3.3 Noise in Reactive Circuits

If a resistance is followed by a tuned circuit which is theoretically noiseless, then the presence of the tuned circuit does not affect the noise generated by the resistance at the resonant frequency. To either side of resonance, however, the presence of the tuned circuit affects noise in just the same way as any other voltage, so that the tuned circuit limits the bandwidth of the noise source by not passing noise outside its own passband. The more interesting case, however, is a tuned circuit which is not ideal, i.e., one in which the inductance has a resistive component, which naturally generates noise.

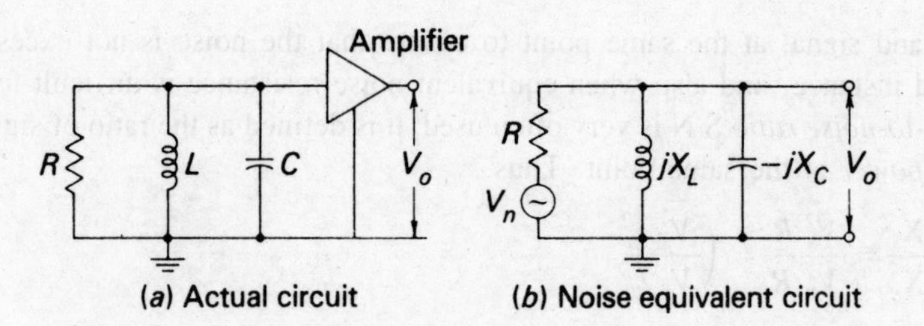

FIGURE 2-3 Noise in a tuned circuit.

In the preceding sections dealing with noise calculations, an input (noise) resistance has been used. It must be stressed here that this need not necessarily be an actual resistor. If all the resistors shown in Fig. 2-2 had been tuned circuits with equivalent parallel resistances equal to R_1, R_2, and R_3, respectively, the results obtained would have been identical. This may be shown as follows. Consider Fig. 2-3, which shows a parallel-tuned circuit. The series resistance of the coil, which is the noise source here, is shown as a resistor in series with a noise generator and with the coil. It is required to determine the noise voltage across the capacitor, i.e., at the input to the amplifier. This will allow us to calculate the resistance which may be said to be generating the noise.

The noise current in the circuit will be

$$i_n = \frac{v_n}{Z}$$

where $Z = R_s + j(X_L - X_C)$. Thus $i_n = v_n/R_s$ at resonance.

The magnitude of the voltage appearing across the capacitor, due to v_n, will be

$$v = i_n X_C = \frac{v_n X_C}{R_s} = \frac{v_n Q R_s}{R_s} = Q v_n \tag{2-8}$$

since $X_C = QR_s$ at resonance.

Equation (2-8) should serve as a further reminder that Q is called the *magnification factor!* Continuing, we have

$$v^2 = Q^2 v_n^2 = Q^2 4kT\,\delta f\,R_s = 4kT\,\delta f\,(Q^2 R_s) = 4kT\,\delta f\,R_p$$

$$v = \sqrt{4kT\,\delta f\,R_p} \tag{2-9}$$

where v is the noise voltage across a tuned circuit due to its internal resistance, and R_p is the equivalent parallel impedance of the tuned circuit at resonance.

Equation (2-9) shows that the equivalent parallel impedance of a tuned circuit is its equivalent resistance for noise (as well as for other purposes).

2-4 NOISE FIGURE

2-4.1 Signal-to-Noise Ratio

The calculation of the equivalent noise resistance of an amplifier, receiver or device may have either of two purposes (sometimes both). The first purpose is comparison of two kinds of equipment in evaluating their performance. The second is comparison of noise and signal at the same point to ensure that the noise is not excessive. In the second instance, and also when equivalent noise resistance is difficult to obtain, the *signal-to-noise ratio* S/N is very often used; it is defined as the ratio of signal *power* to noise *power* at the same point. Thus

$$\frac{\mathrm{S}}{\mathrm{N}} = \frac{X_s}{X_n} = \frac{V_s^2/R}{V_n^2/R} = \left(\frac{V_s}{V_n}\right)^2 \tag{2-10}$$

Equation (2-10) is a simplification that applies whenever the resistance across which the noise is developed is the same as the resistance across which signal is developed, and this is almost invariable. An effort is naturally made to keep the signal-to-noise ratio as high as practicable under a given set of conditions.

2-4.2 Definition of Noise Figure

For comparison of receivers or amplifiers working at different impedance levels the use of the equivalent noise resistance is misleading. For example, it is hard to determine at a glance whether a receiver with an input impedance of 50 Ω and R_{eq} = 90 Ω is better, from the point of view of noise, than another receiver whose input impedance is 300 Ω and R_{eq} = 400 Ω. As a matter of fact, the second receiver is the better one, as will be seen. Instead of equivalent noise resistance, a quantity known as *noise figure*, sometimes called *noise factor*, is defined and used. The noise figure F is defined as the ratio of the signal-to-noise power supplied to the input terminals of a receiver or amplifier to the signal-to-noise power supplied to the output or load resistor. Thus

$$F = \frac{\text{input S/N}}{\text{output S/N}} \tag{2-11}$$

It can be seen immediately that a practical receiver will generate some noise, and the S/N will deteriorate as one moves toward the output. Consequently, in a practical receiver, the output S/N will be lower than the input value, and so the noise figure will exceed 1. However, the noise figure will be 1 for an ideal receiver, which introduces no noise of its own. Hence we have the alternative definition of noise figure, which states that F is equal to the S/N of an ideal system divided by the S/N at the output of the receiver or amplifier under test, both working at the same temperature over the same bandwidth and fed from the same source. In addition, both must be linear. The noise figure may be expressed as an actual ratio or in decibels. The noise figure of practical receivers can be kept to below a couple of decibels up to frequencies in the lower gigahertz range by a suitable choice of the first transistor, combined with proper circuit design and low-noise resistors. At frequencies higher than that, equally low-noise figures may be achieved (lower, in fact) by devices which themselves use the transit-time effect or are relatively independent of it. This will be shown in Chap. 12.

2-4.3 Calculation of Noise Figure

Noise figure may be calculated for an amplifier or receiver in the same way by treating either as a whole; that is, each is treated as a four-terminal network having an input impedance R_t, an output impedance R_L, and an overall voltage gain A. It is fed from a

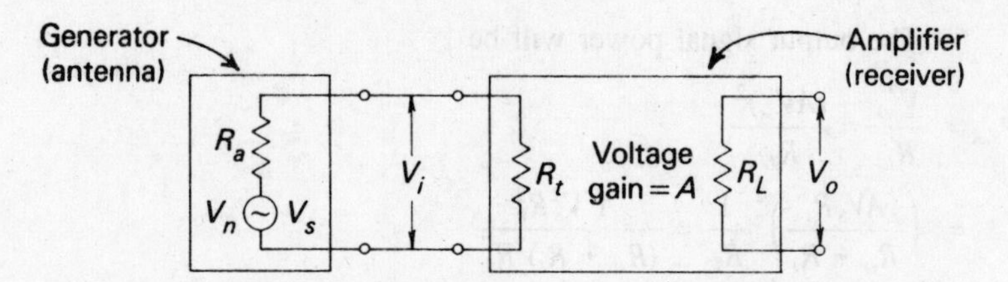

FIGURE 2-4 Block diagram for noise figure calculation.

source (antenna) of internal impedance R_a, which may or may not be equal to R_t as the circumstances warrant. A block diagram of such a four-terminal network (with the source feeding it) is shown in Fig. 2-4.

The calculation procedure may be broken down into a number of general steps. Each is now shown, followed by the number of the corresponding equations(s) to follow:

1. Determine the signal input power P_{si} (2-12, 2-13).
2. Determine the noise input power P_{ni} (2-14, 2-15).
3. Calculate the input signal-to-noise ratio S/N$_i$ from the ratio of P_{si} and P_{ni} (2-16).
4. Determine the signal output power P_{so} (2-17).
5. Write P_{no} for the noise output power to be determined later (2-18).
6. Calculate the output signal-to-noise ratio S/N$_o$ from the ratio of P_{so} and P_{no} (2-19).
7. Calculate the generalized form of noise figure from steps 3 and 6 (2-20).
8. Calculate P_{no} from R_{eq} if possible (2-21, 2-22), and substitute into the general equation for F to obtain the actual formula (2-23, 2-24); *or* determine P_{no} from measurement (2-3, 2-25, 2-26), and substitute to obtain the formula for F (2-27, 2-28, 2-29).

It is seen from Fig. 2-4 that the signal input voltage and power will be

$$V_{si} = \frac{V_s R_t}{R_a + R_t} \tag{2-12}$$

$$V_{si} = \frac{V_{si}^2}{R_t} = \left(\frac{V_s R_t}{R_a + R_t}\right)^2 \frac{1}{R_t} = \frac{V_s^2 R_t}{(R_a + R_t)^2} \tag{2-13}$$

Similarly, the noise input voltage and power will be

$$V_{ni}^2 = 4kT\,\delta f \frac{R_a R_t}{R_a + R_t} \tag{2-14}$$

$$P_{ni} = \frac{V_{ni}^2}{T_t} = 4kT\,\delta f \frac{R_a R_t}{R_a + R_t}\frac{1}{R_t} = \frac{4kT\,\delta f\,R_a}{R_a + R_t} \tag{2-15}$$

The input signal-to-noise ratio will be

$$\frac{\mathrm{S}}{\mathrm{N}_i} = \frac{P_{si}}{P_{ni}} = \frac{V_s^2 R_t}{(R_a + R_t)^2} \div \frac{4kT\,\delta f\,R_a}{R_a + R_t} = \frac{V_s^2 R_t}{4kT\,\delta f\,R_a(R_a + R_t)} \tag{2-16}$$

The output signal power will be

$$P_{so} = \frac{V_{so}^2}{R_L} = \frac{(AV_{si})^2}{R_L}$$

$$= \left(\frac{AV_sR_t}{R_a + R_t}\right)^2 \frac{1}{R_L} = \frac{A^2V_s^2R_t^2}{(R_a + R_t)^2R_L} \tag{2-17}$$

The noise output power may be difficult to calculate; for the time being, it may simply be written as

$$P_{no} = \text{noise output power} \tag{2-18}$$

Thus the output signal-to-noise ratio will be

$$\frac{\text{S}}{\text{N}_o} = \frac{P_{so}}{P_{no}} = \frac{A^2V_2^2R_t^2}{(R_a + R_t)^2R_LP_{no}} \tag{2-19}$$

Finally, the general expression for the noise figure is

$$F = \frac{\text{S/N}_i}{\text{S/N}_o} = \frac{V_s^2R_t}{4kT\ \delta f\ R_a(R_a + R_t)} \div \frac{A^2V_s^2R_t^2}{(R_a + R_t)^2R_LP_{no}}$$

$$= \frac{R_LP_{no}(R_a + R_t)}{4kT\ \delta f\ A^2R_aR_t} \tag{2-20}$$

Note that Eq. (2-20) is an intermediate result only. An actual formula for F may now be obtained by substitution for the output noise power, or from a knowledge of the equivalent noise resistance, or from measurement.

2-4.4 Noise Figure from Equivalent Noise Resistance

As derived in Eq. (2-7), the equivalent noise resistance of an amplifier or receiver is the sum of the input terminating resistance and the equivalent noise resistance of the first stage, together with the noise resistances of the previous stages referred to the input. Putting it another way, we see that all these resistances are added to R_t, giving a lumped resistance which is then said to concentrate all the "noise making" of the receiver. The rest of it is now assumed to be noiseless. All this applies here, with the minor exception that these noise resistances must now be added to the parallel combination of R_a and R_t. In order to correlate noise figure and equivalent noise resistance, it is thus convenient to define R'_{eq}, which is a noise resistance that does not incorporate R_t and which is given by

$$R'_{\text{eq}} = R_{\text{eq}} - R_t \tag{2-7'}$$

The total equivalent noise resistance for this receiver will now be

$$R = R'_{\text{eq}} + \frac{R_aR_t}{R_a + R_t} \tag{2-21}$$

Thus the equivalent noise voltage (said to be) generated at the input of the receiver will be

$$V_{ni} = \sqrt{4kT\ \delta f\ R}$$

Since the amplifier has an overall voltage gain A and may now be treated as though it were noiseless, the noise output will be

$$P_{no} = \frac{V_{no}^2}{R_L} = \frac{(AV_{ni})^2}{R_L} = \frac{A^2 4kT\ \delta f\ R}{R_L} \tag{2-22}$$

When Eq. (2-22) is substituted into the general Eq. (2-20), the result is an expression for the noise figure in terms of the equivalent noise resistance, namely,

$$F = \frac{R_L(R_a + R_t)}{4kT\ \delta f\ A^2 R_a R_t} P_{no} = \frac{R_L(R_a + R_t)}{4kT\ \delta f\ A^2 R_a R_t} \frac{A^2 4kT\ \delta f\ R}{R_L}$$

$$= R\frac{R_a + R_t}{R_a R_t} = \left(R'_{eq} + \frac{R_a R_t}{R_a + R_t}\right)\frac{R_a + R_t}{R_a R_t}$$

$$= 1 + \frac{R'_{eq}(R_a + R_t)}{R_a R_t} \tag{2-23}$$

It can be seen from Eq. (2-23) that if the noise is to be a minimum for any given value of the antenna resistance R_a, the ratio $(R_a + R_t)/R_t$ must also be a minimum, so that R_t must be much larger than R_a. This is, in fact, a situation exploited very often in practice, and it may now be applied to Eq. (2-23). Under these *mismatched* conditions, $(R_a + R_t)/R_t$ approaches unity, and the formula for the noise figure reduces to

$$F = 1 + \frac{R'_{eq}}{R_a} \tag{2-24}$$

This is a most important relationship, but it must be remembered that it applies under mismatched conditions only. Under matched conditions ($R_t = R_a$) or when the mismatch is not severe, Eq. (2-23) must be used instead.

Example 2-4 Calculate the noise figure of the amplifier of Example 2-3 if it is driven by a generator whose output impedance is 50 Ω. (Note that this constitutes a large enough mismatch.)

$$R'_{eq} = R_{eq} - R_t = 2518 - 600 = 1918\Omega$$

$$F = 1 + \frac{R'_{eq}}{R_a} = 1 + 38.4$$

$$= 39.4 \qquad (=15.84\ \text{dB})$$

Note that if an "equivalent noise resistance" is given without any other comment in connection with noise figure calculations, it may be assumed to be R'_{eq}.

2-4.5 Noise Figure from Measurement

The preceding section showed how the noise figure may be found if the equivalent noise resistance is easy to calculate. When this is not practicable, as under transit-time conditions, it is possible to make measurements that lead to the determination of the noise figure. A simple method, using the *diode noise generator*, is often employed. It is shown in Fig. 2-5 in circuit-block form.

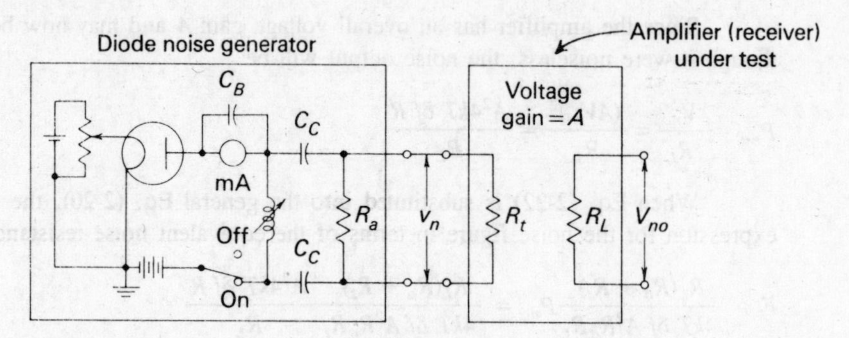

FIGURE 2-5 Noise figure measurement.

Equation (2-3) gave the formula for the exact plate noise current of a vacuum-tube diode, and this can now be used. As shown, the anode current is controlled by means of the potentiometer which varies filament voltage, and that is how shot-noise current is adjusted.

The output capacitance of the diode and its associated circuit is resonated at the operating frequency of the receiver by means of the variable inductance, so that it may be ignored. The output impedance of the noise generator will now simply be R_a. Thus the noise voltage supplied to the input of the receiver by the diode will be given by

$$v_n = i_n Z_n = i_n \frac{R_a R_t}{R_a + R_t} = \frac{R_a R_t \sqrt{2 v i_p \, \delta f}}{R_a + R_t} \tag{2-25}$$

The noise generator is connected to the receiver (or amplifier) under test, and the noise output power of the receiver is measured with zero diode plate current, i.e., with the diode plate voltage supply switched off. The diode plate voltage supply is now switched on, and the filament potentiometer is adjusted so that diode plate current begins to flow. It is further adjusted until the noise power developed in R_L is twice as large as the noise power in the absence of diode plate current. The plate current at which this happens, i_p, is measured with the milliammeter and noted. The additional noise power output is now equal to the normal noise power output, so that the latter can be expressed in terms of the diode plate current. We now have

$$P_{no} = \frac{V_{no}^2}{R_L} = \frac{(A v_n)^2}{R_L} = \frac{A^2 R_a^2 R_t^2 2 v i_p \, \delta f}{R_L (R_a + R_t)^2} \tag{2-26}$$

As already outlined, Eq. (2-26) may be substituted into Eq. (2-20). This yields

$$F = \frac{R_L (R_a + R_t) P_{no}}{A^2 4kT \, \delta f \, R_a R_t} = \frac{R_L (R_a + R_t)}{A^2 4kT \, \delta f \, R_a R_t} \, \frac{A^2 R_a^2 R_t^2 2 v i_p \, \delta f}{R_L (R_a + R_t)^2}$$

$$= \frac{v i_p R_a R_t}{2kT (R_a + R_t)} \tag{2-27}$$

If it is assumed once again that the system is mismatched and $R_t \gg R_a$, Eq. (2-27) is simplified to

$$F = \frac{R_a v i_p}{2kT} \tag{2-28}$$

If the above procedure is repeated right from the beginning for a system under matched conditions, it may then be proved that Eq. (2-28) applies exactly to such a system instead of being merely a good approximation, as it is here. Such a result emphasizes the value of the noise diode measurement.

As a final simplification, we substitute into Eq. (2-28) the values of the various constants it contains. These include the standard temperature at which such measurements are made, which is 17°C or 290 K.[5] This gives a formula which is very often quoted:

$$F = \frac{R_a v i_p}{2kT} = \frac{(R_a i_p)\ (1.6 \times 10^{-19})}{2 \times 290 \times 1.38 \times 10^{-23}}$$
$$= (R_a i_p)\ (2 \times 10)$$
$$= 20\ R_a i_p \tag{2-29}$$

where R_a is measured in ohms and i_p in amperes.

2-5 NOISE TEMPERATURE

The concept of noise figure, although frequently used, is not always the most convenient measure of noise, particularly in dealing with UHF and microwave low-noise antennas, receivers or devices. Controversy exists regarding which is the better all-around measurement, but noise temperature, derived from early work in radio astronomy, is employed extensively for antennas and low-noise microwave amplifiers. Not the least reason for its use is convenience, in that it is additive like noise power. This may be seen from reexamining Eq. (2-1), as follows:

$$Pt = kT\ \delta f$$
$$= P_1 + P_2 = kT_1\ \delta f + kT_2\ \delta f$$
$$kT_t\ \delta f = kT_1\ \delta f + kT_2\ \delta f$$
$$T_t = T_1 + T_2 \tag{2-30}$$

where P_1 and P_2 = two individual noise powers (e.g., received by the antenna and generated by the antenna, respectively) and P_t is their sum

T_1 and T_2 = the individual noise temperatures

T_t = the "total" noise temperature

Yet another advantage of the use of noise temperature for low noise levels is that it shows a greater variation for any given noise-level change than does the noise figure; therefore, changes are easier to grasp in their true perspective.

It will be recalled that the equivalent noise resistance introduced in Sec. 2-3 is quite fictitious, but it is often employed because of its convenience. Similarly, T_{eq}, the equivalent noise temperature, may also be utilized if it proves convenient. In defining the equivalent noise temperature of a receiver or amplifier, it is assumed that $R'_{eq} = R_a$.

[5] Standard temperature has also at times been quoted as 20 or 27°C, but seems now to have been standardized at the value given here.

If this is to lead to the correct value of noise output power, then obviously R'_{eq} must be at a temperature other than the standard one at which all the components (including R_a) are assumed to be. It is then possible to use Eq. (2-24) to equate noise figure and equivalent noise temperature, as follows:

$$F = 1 + \frac{R'_{eq}}{R_a} = 1 + \frac{kT_{eq}\ \delta f\ R'_{eq}}{kT_0\ \delta f\ R_a}$$

$$= 1 + \frac{T_{eq}}{T_0} \tag{2-31}$$

where $R'_{eq} = R_a$, as postulated in the definition of T_{eq}
$T_0 = 17°C = 290$ K
T_{eq} = equivalent noise temperature of the amplifier or receiver whose noise figure is F

Note that F here is a ratio and is not expressed in decibels. Also, T_{eq} may be influenced by (but is certainly not equal to) the actual ambient temperature of the receiver or amplifier. It must be repeated that the equivalent noise temperature is just a convenient fiction. If all the noise of the receiver were generated by R_a, its temperature would have to be T_{eq}. Finally we have, from Eq. (2-31),

$$T_0F = T_0 + T_{eq}$$
$$T_{eq} = T_0\,(F - 1) \tag{2-32}$$

Once noise figure is known, equivalent noise temperature may be calculated from Eq. (2-32), or a nomograph may be constructed if use is frequent enough to justify it. Graphs of noise temperature of various sources versus frequency [2] and sky temperature versus frequency [3] are also available.

Example 2-5 A receiver connected to an antenna whose resistance is 50 Ω has an equivalent noise resistance of 30 Ω. Calculate the receiver's noise figure in decibels and its equivalent noise temperature.

$$F = 1 + \frac{R_{eq}}{R_a} = 1 + \frac{30}{50} = 1 + 0.6 = 1.6$$
$$= 10 \log 1.6 = 10 \times 0.204 = 2.04 \text{ dB}$$
$$T_{eq} = T_0\,(F - 1) = 290\,(1.6 - 1) = 290 \times 0.6$$
$$= 174\text{K}$$

Finally, note Schwartz [4] as an excellent general reference on noise.

PROBLEMS

For self-testing questions on this chapter, see p. 677.

2-1. An amplifier operating over the frequency range of 455 to 460 kHz has a 200-kΩ input resistor. What is the rms noise voltage at the input to this amplifier if the ambient temperature is 17°C?
[Ans.: 4.00 μV]

2-2. The noise output of a resistor is amplified by a noiseless amplifier having a gain of 60 and a bandwidth of 20 kHz. A meter connected to the output of the amplifier reads 1 mV rms. *(a)*

The bandwidth of the amplifier is reduced to 5 kHz, its gain remaining constant. What does the meter read now? *(b)* If the resistor is operated at 80°C, what is its resistance?
[Ans.: (a) 0.5 mV, (b) 715 kΩ]

2-3. A parallel-tuned circuit, having a Q of 20, is resonated to 200 MHz with a 10-picafarad (10-pF) capacitor, If this circuit is maintained at 17°C, what noise voltage will a wideband voltmeter measure when placed across it?
[Ans.: 16.0 μV]

2-4. The front end of a television receiver, having a bandwidth of 7 MHz and operating at a temperature of 27°C, consists of an amplifier having a gain of 15 followed by a mixer whose gain is 20. The amplifier has a 300-Ω input resistor and a shot-noise equivalent resistance of 500 Ω; for the converter, these values are 2.2 and 13.5 kΩ, respectively, and the mixer load resistance is 470 kΩ. Calculate R_{eq} for this television receiver.
[Ans.: 876 Ω]

2-5. Calculate the minimum signal voltage that the receiver of Prob. 2-4 can handle for good reception, given that the input signal-to-noise ratio must be not less than 300/1.
[Ans.: 174 μV]

2-6. The RF amplifier of a receiver has an input resistance of 1000 Ω, and equivalent shot-noise resistance of 2000 Ω, a gain of 25, and a load resistance of 125 kΩ. Given that the bandwidth is 1.0 MHz and the temperature is 20°C, calculate the equivalent noise voltage at the input to this RF amplifier. If this receiver is connected to an antenna with ar. impedance of 75 Ω, calculate the noise figure.
[Ans.: 7.19 μV, 30.3]

QUESTIONS

2-1. List, separately, the various sources of random noise and impulse noise external to a receiver. How can some of them be avoided or minimized? What is the strongest source of extraterrestrial noise?

2-2. Discuss the types, causes and effects of the various forms of noise which may be created within a receiver or an amplifier.

2-3. Describe briefly the forms of noise to which a transistor is prone.

2-4. Define signal-to-noise ratio and noise figure of a receiver. When might the latter be a more suitable piece of information than the equivalent noise resistance?

2-5. A receiver has an overall gain A, an output resistance R_L, a bandwidth δf, and an absolute operating temperature T. If the receiver's input resistance is equal to the antenna resistance R_a, derive a formula for the noise figure of this receiver. One of the terms of this formula will be the noise output power. Describe briefly how this can be measured using the diode generator.

2-6. Define the equivalent noise temperature of a receiver or amplifier. Under what conditions could this be a more useful quantity than the noise figure? Why?

REFERENCES

1. This subject is well treated in Sturley, K.R., *Radio Receiver Design*, p I, 3d ed., Chapman & Hall, Ltd., London, 1965, pp. 193–198.
2. Karbowiak, A. E., "Optical Waveguides," in L. Young (ed.), *Advances in Microwaves*, vol. 1, Academic Press, Inc., New York, 1966, p. 110.
3. Stremler, G., *Introduction to Communication Systems*, Addison-Wesley Publishing Company, Reading, Mass., 1977, p. 175.
4. Schwartz, M., *Information Transmission, Modulation, and Noise*, 2d ed., McGraw-Hill Book Company, New York, 1970.

3 AMPLITUDE MODULATION

The definition and meaning of modulation in general, as well as the need for modulation, were introduced in Chap. 1. This chapter deals with amplitude modulation in detail and to this end is subdivided into two sections. Having studied amplitude modulation (AM) theory, students will be able to appreciate that an amplitude-modulated wave is made up of a number of sinusoidal components having a specific relation to one another. They will be able to visualize the AM wave and calculate the frequencies present in it, as well as their power or current relations to each other. The second part of the chapter will show several practical methods of generating AM, treating them from a circuit-waveform, rather than mathematical, point of view. Both vacuum-tube amplitude modulators, which are the common ones where high powers are involved, and transistor AM generators will be discussed.

3-1 AMPLITUDE MODULATION THEORY

In amplitude modulation, the amplitude of a *carrier* signal is varied by the *modulating voltage,* whose frequency is invariably lower than that of the carrier. In practice, the carrier may be high-frequency (HF) while the modulation is audio. Formally, *AM* is defined as a system of modulation in which the *amplitude of the carrier is made proportional to the instantaneous amplitude of the modulating voltage*.

Let the carrier voltage and the modulating voltage, v_c and v_m, respectively, be represented by

$$v_c = V_c \sin \omega_c t \tag{3-1}$$

$$v_m = V_m \sin \omega_m t \tag{3-2}$$

Note that phase angle has been ignored in both expressions since it is unchanged by the *amplitude* modulation process. Its inclusion here would merely complicate the proceedings, without affecting the result. However, it will certainly not be possible to ignore phase angle when we deal with frequency and phase modulation in Chap. 4.

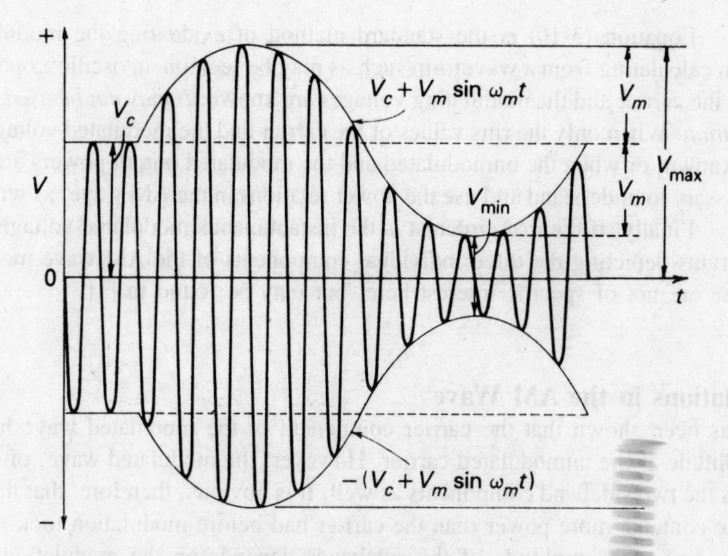

FIGURE 3-3 Amplitude-modulated wave.

The appearance of the amplitude-modulated wave is of great interest, and it is shown in Fig. 3-3 for one cycle of the modulating sine wave. It is derived from Fig. 3-1, which showed the amplitude, or what may now be called the top *envelope* of the AM wave, given by the relation $A = V_c + V_m \sin \omega_m t$. Similarly, the maximum negative amplitude, or bottom envelope, is given by $-A = -(V_c + V_m \sin \omega_m t)$. The modulated wave extends between these two limiting envelopes and has a repetition rate equal to the unmodulated carrier frequency.

It will be recalled that $V_m = mV_c$, and it is now possible to use this relation to calculate the index (or depth) of modulation from the waveform of Fig. 3-3 as follows:

$$V_m = \frac{V_{max} - V_{min}}{2} \tag{3-8}$$

and

$$V_c = V_{max} - V_m = V_{max} - \frac{V_{max} - V_{min}}{2}$$

$$= \frac{V_{max} + V_{min}}{2} \tag{3-9}$$

Dividing Eq. (3-8) by (3-9), we have

$$m = \frac{V_m}{V_c} = \frac{(V_{max} - V_{min})/2}{(V_{max} + V_{min})/2}$$

$$= \frac{V_{max} - V_{min}}{V_{max} + V_{min}} \tag{3-10}$$

Equation (3-10) is the standard method of evaluating the modulation index when calculating from a waveform such as may be seen on an oscilloscope, i.e., when both the carrier and the modulating voltages are known. *It may not be used in any other situation*. When only the rms values of the carrier and the modulated voltage or current are known, or when the unmodulated and the modulated output powers are given, it is necessary to understand and use the power relations in the AM wave, as will be shown.

Finally, if the main interest is the instantaneous modulated voltage, the phasor diagrams depicting the three individual components of the AM wave may be drawn. These are not of specific interest here, but may be found in [1].

3-1.3 Power Relations in the AM Wave

It has been shown that the carrier component of the modulated wave has the same amplitude as the unmodulated carrier. However, the modulated wave, of course, contains the two sideband components as well. It is obvious, therefore, that the modulated wave contains more power than the carrier had before modulation took place. Moreover, since the amplitude of the sidebands depends on the modulation index, it is anticipated that the total power in the modulated wave will depend on the modulation index also. This relation may now be derived.

The total power in the modulated wave will be

$$P_t = \frac{V_{\text{carr}}^2}{R} + \frac{V_{\text{LSB}}^2}{R} + \frac{V_{\text{USB}}^2}{R} \tag{3-11}$$

where all three voltages are rms values, and R is the resistance (e.g., antenna resistance) in which the power is dissipated. The first term of Eq. (3-11) is the unmodulated carrier power and is given by

$$P_c = \frac{V_{\text{carr}}^2}{R} = \frac{(V_c/\sqrt{2})^2}{R}$$

$$= \frac{V_c^2}{2R} \tag{3-12}$$

Similarly,

$$P_{\text{LSB}} = P_{\text{USB}} = \frac{V_{\text{SB}}^2}{\text{R}} = \left(\frac{mV_c/2}{\sqrt{2}}\right)^2 \div R = \frac{m^2V_c{}^2}{8R}$$

$$= \frac{m^2}{4}\frac{V_c^2}{2R} \tag{3-13}$$

Substituting Eqs. (3-12) and (3-13) into (3-11), we have

$$P_t = \frac{V_c^2}{2R} + \frac{m^2}{4}\frac{V_c^2}{2R} + \frac{m^2}{4}\frac{V_c^2}{2R} = P_c + \frac{m^2}{4}P_c + \frac{m^2}{4}P_c$$

$$\frac{P_t}{P_c} = 1 + \frac{m^2}{2} \tag{3-14}$$

Equation (3-14) relates the total power in the amplitude-modulated wave to the unmodulated carrier power. This is the equation which must be used to determine, among other quantities, the modulation index in instances not covered by Eq. (3-10) of the preceding section. The methods of doing this, as well as solutions to other problems, will be shown in exercises to follow.

It is interesting to note from Eq. (3-14) that the maximum power in the AM wave is $P_t = 1.5P_c$ when $m = 1$. This is important, because it is the maximum power that relevant amplifiers must be capable of handling without distortion.

Example 3-2 A 400-watt (400-W) carrier is modulated to a depth of 75 percent. Calculate the total power in the modulated wave.

$$P_t = P_c\left(1 + \frac{m^2}{2}\right) = 400\left(1 + \frac{0.75^2}{2}\right) = 400 \times 1.281$$

$$= 512.5 \text{ W}$$

Example 3-3 A broadcast radio transmitter radiates 10 kilowatts (10 kW) when the modulation percentage is 60. How much of this is carrier power?

$$P_c = \frac{P_t}{1 + m^2/2} = \frac{10}{1 + 0.6^2/2} = \frac{10}{1.18} = 8.47 \text{ kW}$$

Current calculations The situation which very often arises in AM is that the modulated and unmodulated currents are easily measurable, and it is then necessary to calculate the modulation index from them. This occurs when the antenna current of the transmitter is metered, and the problem may be resolved as follows. Let I_c be the unmodulated current and I_t the total, or modulated, current of an AM transmitter, both being rms values. If R is the resistance in which these currents flow, then

$$\frac{P_t}{P_c} = \frac{I_t^2 R}{I_c^2 R} = \left(\frac{I_t}{I_c}\right)^2 = 1 + \frac{m^2}{2}$$

$$\frac{I_t}{I_c} = \sqrt{1 + \frac{m^2}{2}} \quad \text{or} \quad I_t = I_c\sqrt{1 + \frac{m^2}{2}} \tag{3-15}$$

Example 3-4 The antenna current of an AM transmitter is 8 amperes (8 A) when only the carrier is sent, but it increases to 8.93 A when the carrier is modulated by a single sine wave. Find the percentage modulation. Determine the antenna current when the depth of modulation changes to 0.8.

$$\left(\frac{I_t}{I_c}\right)^2 = 1 + \frac{m^2}{2}$$

$$\frac{m^2}{2} = \left(\frac{I_t}{I_c}\right)^2 - 1$$

$$m = \sqrt{2\left[\left(\frac{I_t}{I_c}\right)^2 - 1\right]} \tag{3-16}$$

Here

$$m = \sqrt{2\left[\left(\frac{8.93}{8}\right)^2 - 1\right]} = \sqrt{2[(1.116)^2 - 1]}$$

$$= \sqrt{2(1.246 - 1)} = \sqrt{0.492} = 0.701 = 70.1\%$$

For the second part we have

$$I_t = I_c\sqrt{1 + \frac{m^2}{2}} = 8\sqrt{1 + \frac{0.8^2}{2}} = 8\sqrt{1 + \frac{0.64}{2}}$$

$$= 8\sqrt{1.32} = 8 \times 1.149 = 9.19\text{A}$$

Although Eq. (3-16) is merely (3-15) in reverse, it will be found useful in other problems.

Modulation by several sine waves In practice, modulation of a carrier by several sine waves simultaneously is the rule rather than the exception. Accordingly, a way has to be found to calculate the resulting power conditions. The procedure consists of calculating the total modulation index and then substituting it into Eq. (3-14), from which the total power may be calculated as before. There are two methods of calculating the total modulation index.

1. Let V_1, V_2, V_3, etc., be the simultaneous modulation voltages. Then the total modulating voltage V_t will be equal to the square root of the sum of the squares of the individual voltages; that is,

$$V_t = \sqrt{V_1^2 + V_2^2 + V_3^2 + \cdots}$$

Dividing both sides by V_c, we get

$$\frac{V_t}{V_c} = \frac{\sqrt{V_1^2 + V_2^2 + V_3^2 + \cdots}}{V_c}$$

$$= \sqrt{\frac{V_1^2}{V_c^2} + \frac{V_2^2}{V_c^2} + \frac{V_3^2}{V_c^2} + \cdots}$$

that is,

$$m_t = \sqrt{m_1^2 + m_2^2 + m_3^2 + \cdots} \tag{3-17}$$

2. Equation (3-14) may be rewritten to emphasize that the total power in an AM wave consists of carrier power and sideband power. This yields

$$P_t = P_c\left(1 + \frac{m^2}{2}\right) = P_c + \frac{P_c m^2}{2} = P_c + P_{\text{SB}}$$

where P_{SB} is the total sideband power and is given by

$$P_{\text{SB}} = \frac{P_c m^2}{2} \tag{3-18}$$

If several sine waves simultaneously modulate the carrier, the carrier power will be unaffected, but the total sideband power will now be the sum of the individual sideband powers. Hence we have

$$P_{SB_T} = P_{SB_1} + P_{SB_2} + P_{SB_3} + \cdots$$

Substitution gives

$$\frac{P_c m_t^2}{2} = \frac{P_c m_1^2}{2} + \frac{P_c m_2^2}{2} + \frac{P_c m_3^2}{2} + \cdots$$

$$m_t^2 = m_1^2 + m_2^2 + m_3^2 + \cdots$$

If the square root of both sides is now taken, Eq. (3-17) will once again be the result.

It is seen that the two approaches both yield the same result: To calculate the total modulation index, take the square root of the sum of the squares of the individual modulation indices. Note also that this total modulation index must still not exceed unity, or distortion will result as with overmodulation by a single sine wave. Whether modulation is by one or many sine waves, the output of the modulated amplifier will be zero during part of the negative modulating voltage peak if overmodulation is taking place. This point is discussed further in Chap. 6, in conjunction with distortion in AM demodulators.

Example 3-5 A certain transmitter radiates 9 kW with the carrier unmodulated, and 10.125 kW when the carrier is sinusoidally modulated. Calculate the modulation index. If another sine wave, corresponding to 40 percent modulation, is transmitted simultaneously, determine the total radiated power.

$$\frac{m^2}{2} = \frac{P_t}{P_c} - 1 = \frac{10.125}{9} - 1 = 1.125 - 1 = 0.125$$

$$m^2 = 0.125 \times 2 = 0.250$$

$$m = \sqrt{0.25} = 0.50$$

For the second part, the total modulation index will be

$$m_t = \sqrt{m_1^2 + m_2^2} = \sqrt{0.5^2 + 0.4^2} = \sqrt{0.25 + 0.16} = \sqrt{0.41} = 0.64$$

$$P_t = P_c\left(1 + \frac{m_t^2}{2}\right) = 9\left(1 + \frac{0.64^2}{2}\right) = 9(1 + 0.205) = 10.84 \text{ kW}$$

Example 3-6 The antenna current of an AM broadcast transmitter, modulated to a depth of 40 percent by an audio sine wave, is 11 A. It increases to 12 A as a result of simultaneous modulation by another audio sine wave. What is the modulation index due to this second wave?

From Eq. (3-15) we have

$$I_c = \frac{I_t}{\sqrt{1 + m^2/2}} = \frac{11}{\sqrt{1 + 0.4^2/2}} = \frac{11}{\sqrt{1 + 0.08}} = 10.58 \text{ A}$$

Using Eq. (3-16) and bearing in mind that here the modulation index is the total modulation index m_t, we obtain

$$m_t = \sqrt{2\left[\left(\frac{I_T}{I_c}\right)^2 - 1\right]} = \sqrt{2\left[\left(\frac{12}{10.58}\right)^2 - 1\right]} = \sqrt{2(1.286 - 1)}$$

$$= \sqrt{2 \times 0.286} = 0.757$$

From Eq. (3-17), we obtain

$$m^2 = \sqrt{m_t^2 - m_1^2} = \sqrt{0.757^2 - 0.4^2} = \sqrt{0.573 - 0.16} = \sqrt{0.413}$$
$$= 0.643$$

3-2 GENERATION OF AM

There are, basically speaking, two types of devices in which it may be necessary to generate amplitude modulation. The first of these, the AM transmitter, generates such high powers that its prime requirement is efficiency; hence quite complex means of AM generation may be employed. The other device is the (laboratory) AM generator. Here AM is produced at such a low power level that simplicity is a more important requirement than efficiency. Although the methods of generating AM described here relate to both applications, emphasis will be put on methods of generating high powers.

3-2.1 Basic Requirements—Comparison of Levels

In order to generate the AM wave of Fig. 3-4*b*, it is necessary merely to apply the series of current pulses of Fig. 3-4*a* to a tuned circuit. Each pulse, if it were the only one, would initiate a damped oscillation in the tuned circuit. The oscillation would have an initial amplitude proportional to the size of the current pulse and a decay rate dependent on the time constant of the circuit. Since a train of pulses is fed to the tank circuit here, each pulse will cause a complete sine wave proportional in amplitude to the size of this pulse. This will be followed by the next sine wave, proportional to the size of the next applied pulse, and so on. Bearing in mind that at least 10 times as many pulses per audio cycle are fed to a practical circuit as are shown in Fig. 3-4, we see that an extremely good approximation of an AM wave will result if the original current pulses are made proportional to the modulating voltage. The process is known as the *flywheel effect* of the tuned circuit, and it works best with a tuned circuit whose Q is not too low.

It is possible to make the output current of a class C amplifier proportional to the modulating voltage by applying this voltage in series with any of the dc supply voltages for this amplifier. Accordingly, *cathode* (or emitter), *grid* (or base) and *anode* (or collector) modulation of a class C amplifier are all possible. So is any combination of the methods just listed. Each has its own applications, advantages and drawbacks.

In an AM transmitter, amplitude modulation can be generated at any point after the radio frequency source. As a matter of fact, even a crystal oscillator could be amplitude-modulated, except that this would be an unnecessary interference with its frequency stability. If the output stage in a transmitter is plate-modulated (or collector-modulated in a lower-power transmitter), the system is called *high-level* modulation. If modulation is applied at any other point, including some other electrode of the output amplifier, then so-called *low-level* modulation is produced. Naturally the end product of both systems is the same, but the transmitter circuit arrangements are different.

It is not practicable to use anode modulation of the output stage in a television transmitter, because of the difficulty in generating high video (modulating) powers at

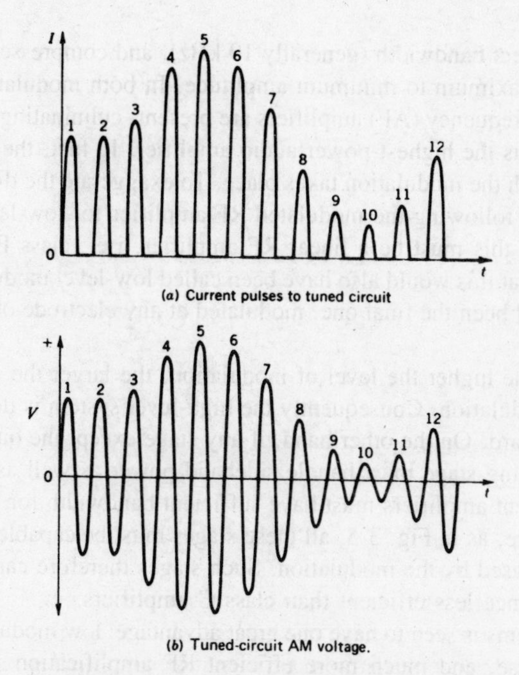

(a) Current pulses to tuned circuit

(b) Tuned-circuit AM voltage.

FIGURE 3-4 Current requirements for AM.

the large bandwidths required. Accordingly, grid modulation of the output stage is the highest level of modulation employed in TV transmitters. It is called "high-level" modulation in TV broadcasting, and anything else is then called "low-level" modulation.

Figure 3-5 shows a typical block diagram of an AM transmitter, which may be either low-level- or high-level-modulated. It is seen that there are a lot of common features. Both have a stable RF source, buffer amplifiers and subsequent RF power amplifiers. In both types of transmitters the audio voltage is processed, that is, filtered

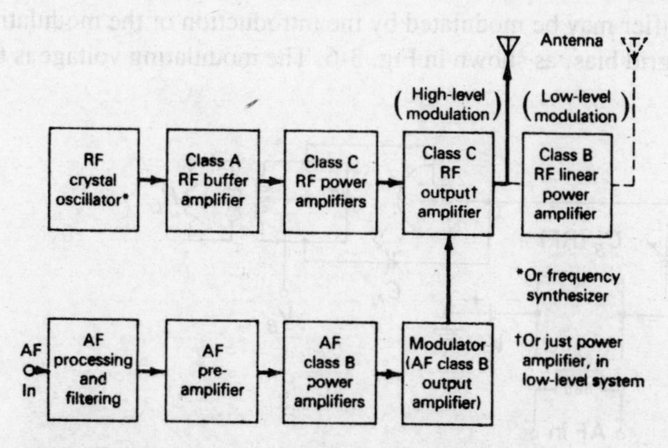

FIGURE 3-5 AM transmitter block diagram.

so as to occupy the correct bandwidth (generally 10 kHz), and compressed somewhat to reduce the ratio of maximum to minimum amplitude. In both modulation systems audio and power audio frequency (AF) amplifiers are present, culminating in the modulator amplifier, which is the highest-power audio amplifier. In fact, the only difference is the point at which the modulation takes place. To exaggerate the difference, an amplifier is shown here following the modulated RF amplifier for low-level modulation, and it is seen that this must be a linear RF amplifier, i.e., class B. However, students are reminded that this would also have been called low-level modulation if the modulated amplifier had been the final one, modulated at any electrode other than the plate (or collector).

It follows that the higher the level of modulation, the larger the audio power required to produce modulation. Consequently the high-level system is definitely at a disadvantage in this regard. On the other hand, if any stage except the output stage is modulated, each following stage must handle sideband power as well as the carrier. Also, all these subsequent amplifiers must have sufficient bandwidth for the sideband frequencies. Furthermore, as in Fig. 3-5, all these stages must be capable of handling amplitude variations caused by the modulation. Such stages therefore cannot be class C, and are in consequence less efficient than class C amplifiers.

Each of the systems is seen to have one great advantage: low modulating power requirements in one case, and much more efficient RF amplification with simpler circuit design in the other. Finally, it has been found in practice that an anode-modulated class C amplifier tends to have better efficiency, lower distortion and much better power-handling capabilities than a grid-modulated amplifier. Because of these considerations, broadcast AM transmitters today almost invariably use high-level modulation, and TV transmitters use grid modulation of the final stage. Other methods may be used in low-power and miscellaneous applications, AM generators and test instruments.

Broadcasting is the major application of AM, with typical output powers ranging from dozens to hundreds of kilowatts. Consequently, vacuum-tube systems are emphasized in the remainder of this chapter.

3-2.2 Grid-Modulated Class C Amplifier

A class C amplifier may be modulated by the introduction of the modulating voltage in series with the grid bias, as shown in Fig. 3-6. The modulating voltage is thus superim-

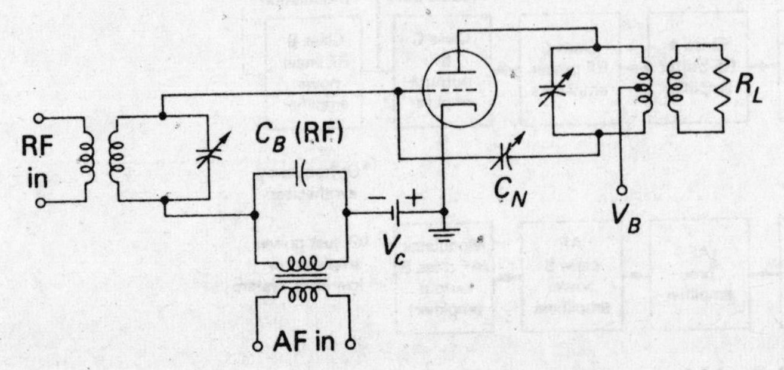

FIGURE 3-6 Grid-modulated class C amplifier.

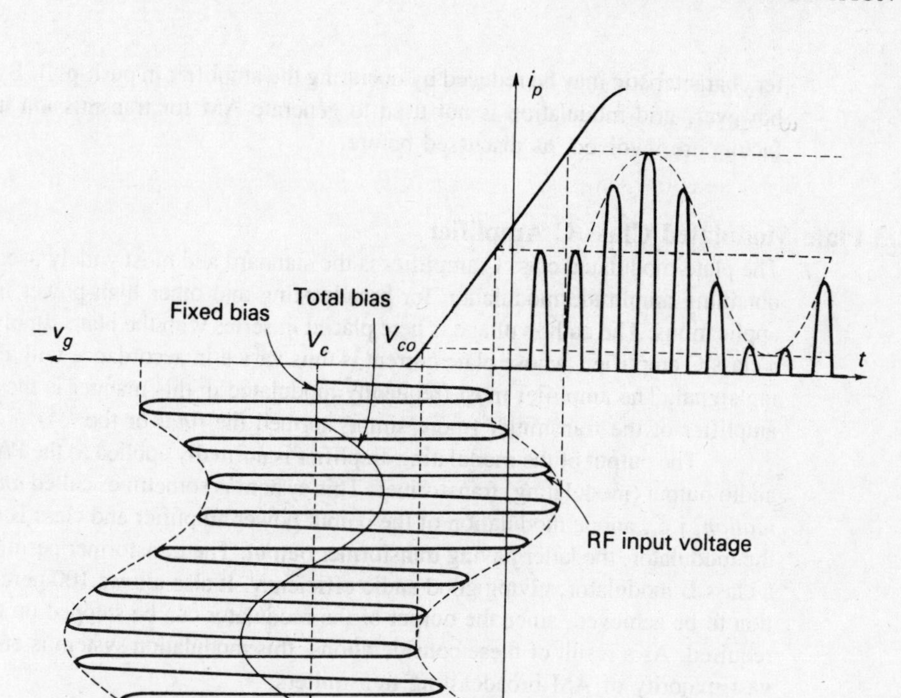

FIGURE 3-7 Grid-voltage–plate current waveforms for grid-modulated class C amplifier.

posed on the fixed negative grid bias. Hence the amplitude of the total bias is proportional to the amplitude of the modulating signal and varies at a rate equal to the modulating frequency. This is shown in Fig. 3-7, as is the RF input voltage superimposed on the total bias. The resulting plate current flows in pulses, as shown, the amplitude of each pulse being proportional to the instantaneous bias and therefore to the instantaneous modulating voltage. As in Fig. 3-4, the application of these pulses to the tuned tank circuit will yield amplitude modulation.

As the waveform shows, this system will operate without distortion only if the transfer characteristic of the triode is perfectly linear. Because this can never be so, the output must be somewhat distorted (more so, in fact, than from a plate-modulated amplifier). It turns out also that the plate efficiency is lower as a result of the need to arrange the input conditions, as shown in Fig. 3-7, so that grid current does not flow in the absence of modulation (this is so that full modulation may be obtained when desired, and it is caused by the geometry of the figure). Reference to the operation of the class C amplifier shows that it operates at maximum efficiency only when the grid is driven to the limit, and as this is not the case here, efficiency suffers.

Because of these bias conditions, the maximum output power from a grid-modulated amplifier is much less than that obtainable from the same tube if it is unmodulated (or plate-modulated), as has been shown [2]. The disadvantages of grid modulation are counterbalanced by the lower modulating power needed in comparison with plate modulation. The harmonics generated as a result of the nonlinearity of the transfer characteristic may be reduced by operating the amplifier in push-pull. By and large, however, grid modulation is not used to generate AM for transmission unless other factors are involved, as discussed before.

3-2.3 Plate-Modulated Class C Amplifier

The plate-modulated class C amplifier is the standard and most widely used method of obtaining amplitude modulation for broadcasting and other high-power transmission applications. The audio voltage is here placed in series with the plate-supply voltage of a class C amplifier, whose plate current is thus varied in accordance with the modulating signal. The amplifier most frequently modulated in this manner is the final power amplifier of the transmitter (more simply termed the *final* or the *PA*).

The output of the modulating amplifier is normally applied to the PA through an audio output (modulating) transformer. This system is sometimes called *anode-B modulation,* i.e., anode modulation of the output power amplifier and class B operation of the modulator, the latter having transformer output. The transformer permits the use of a class B modulator, giving good audio efficiency. It also allows 100 percent modulation to be achieved, since the output of the modulator can be stepped up to any value required. As a result of these considerations, this modulation system is employed in a vast majority of AM broadcasting transmitters.

Transformer modulation using triode The equivalent circuit of Fig. 3-8 has been transformed into the practical circuit of Fig. 3-9 by the inclusion of the modulator with its transformer output. Neutralization is also shown, as it was for the grid-modulated amplifier, since it will certainly be necessary to avoid the Miller effect in a triode at high frequencies. Shunt feed of the class C amplifier is shown mainly to simplify the explanation; it may or may not be used in practice. Note also that the secondary winding of the modulating transformer is bypassed for RF and prevented from going anywhere else than the output tank by the RF choke.

It is seen that the same supply voltage V_{bb} is used for both the PA and the modulator. This means that the peak modulator output voltage (per tube) is made less than V_{bb} to avoid distortion; a value of $V_{AF} = 0.7V_{bb}$ is optimum. Since the peak-to-peak modulator primary voltage is now $1.4V_{bb}$, and an audio voltage equal to V_{bb} is required to produce full modulation, the transformer turns ratio in this case will be 1.4:1. Finally, although a fixed bias supply V_{cc} is shown in Fig. 3-9, self-bias may be used in the form of leak-type bias. This fact results in better operation, as will be shown.

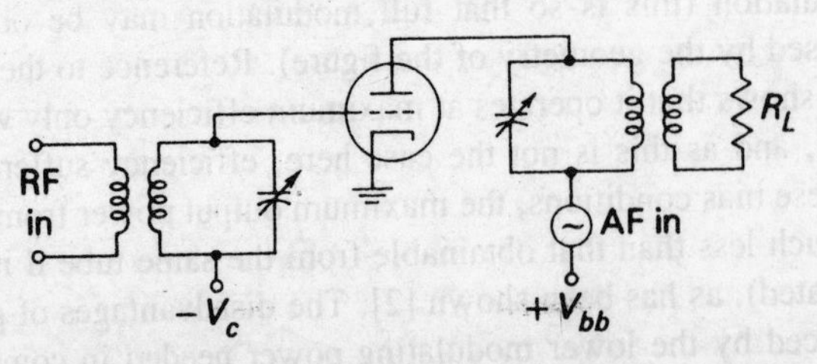

FIGURE 3-8 Plate-modulation equivalent circuit.

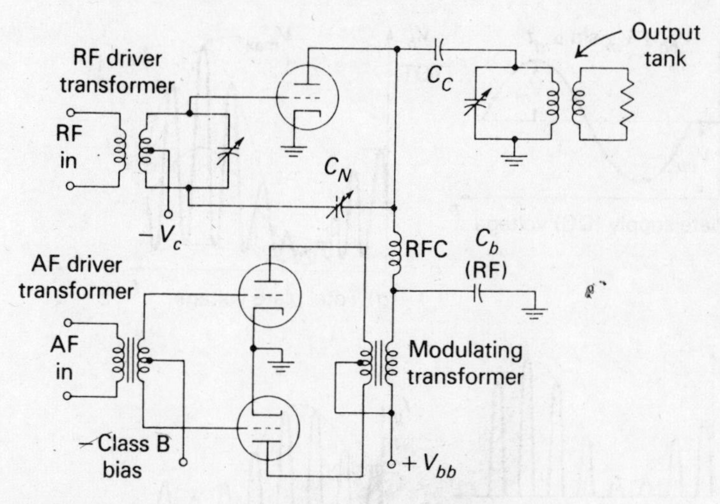

FIGURE 3-9 Plate-modulated triode class C amplifier.

The first waveform of Fig. 3-10 shows the total plate voltage applied to the class C amplifier, and the second shows the resulting plate current. This is seen to be a series of pulses governed by the size of the modulating voltage, and when these are applied to the tank circuit, the modulated wave of the waveform of Fig. 3-10*c* results. What happens here is obviously very similar to what happened with the grid-modulated amplifier, except that the appropriate train of current pulses is obtained by a different method, which also has the virtue of being more linear.

The last three waveforms of Fig. 3-10 are used not so much to explain how the circuit generates AM, but to describe the behavior of the circuit itself. The waveform of Fig. 3-10*d* shows the total voltage appearing between plate and cathode and is obtained by superimposing the RF-modulated voltage of Fig. 3-10*c* on the combined supply voltage of Fig. 3-10*a*. On the circuit of Fig. 3-9, the RF voltage appears across the primary coil of the output tank, while the supply voltage is dropped across the coupling capacitor C_c. The purpose of Fig. 3-10*d* is to show how much higher than V_{bb} the peak plate voltage may rise. At 100 percent modulation, in fact, the peak of the unmodulated RF voltage is nearly $2V_{bb}$, and the positive peak of the modulated cycle rises to twice that. Thus the maximum plate voltage may rise to nearly $4V_{bb}$ in the plate-modulated amplifier. Care must be taken to ensure that the tube used is capable of withstanding such a high voltage.

The waveform of Fig. 3-10*e* demonstrates that the grid current of the modulated amplifier varies during the modulating cycle, although the RF driving voltage does not. When the plate supply voltage falls at the negative modulation peak, the plate is then only moderately positive, and since the grid is also driven positive, the current in the grid now increases rather significantly. At the positive peak of modulation, the reverse applies, and grid current falls. Note that the change is not sinusoidal since the current rise greatly exceeds the fall. If this situation is left unchecked, two unpleasant effects result. First, the driver may become overloaded when grid current rises, resulting in the distortion of the output wave. Second, the grid of the PA will almost certainly melt!

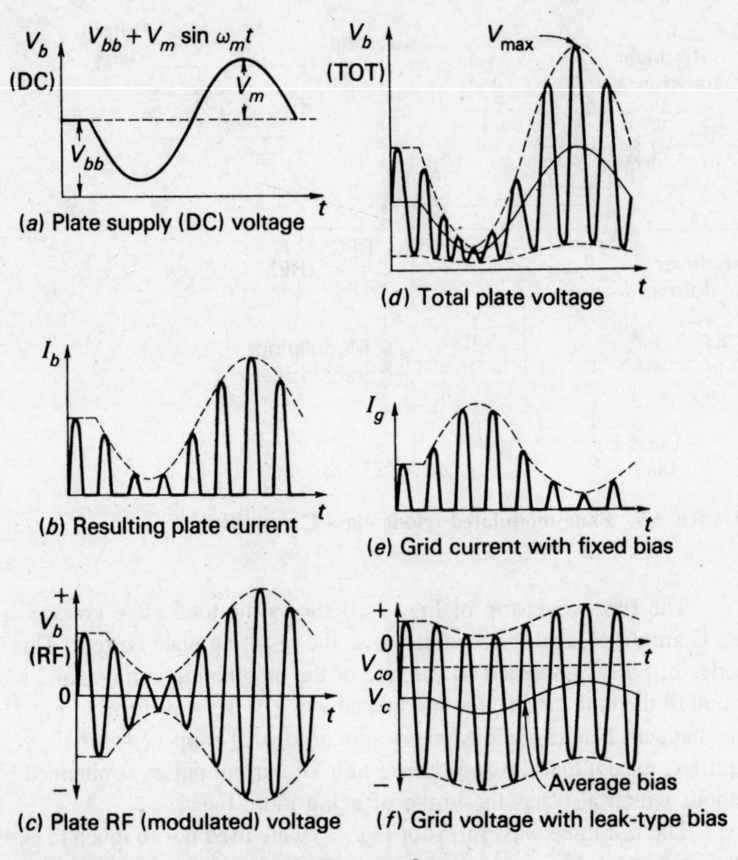

FIGURE 3-10 Plate-modulation waveforms.

Two cures are available, and Fig. 3-10*f* shows the result of the better of the two. The first is to have a driver with poor regulation so that grid current is incapable of rising, whereas the better cure is to use leak-type bias to achieve limiting similar to that used by the amplitude limiter in FM receivers (see Sec. 6-4.2 for a detailed description). As grid current now tends to rise, grid bias also rises, becoming more negative and thus tending to reduce grid current. For any original value of grid current a balance is now struck so that although grid current still rises, this rise is much less marked and can be well within acceptable limits. At the positive modulation peak, grid current would tend to fall, but this fall is now reduced by a simultaneous reduction in negative bias, so that the grid-current waveform now looks like a much-flattened version of Fig. 3-10*e*.

Instead of being constant, grid voltage now varies as in Fig. 3-10*f*. This aids the modulation process because the grid is now made more positive at the time of peak plate current. It now becomes easier to obtain such a high value of plate current, and so distortion at the positive modulation peak is prevented. Considered from another point of view, the waveform of Fig. 3-10*f* is seen to be very similar to the input wave for a grid-modulated amplifier, as in Fig. 3-7. We thus have a grid- and plate-modulated amplifier which is thereby more efficient and less distorted.

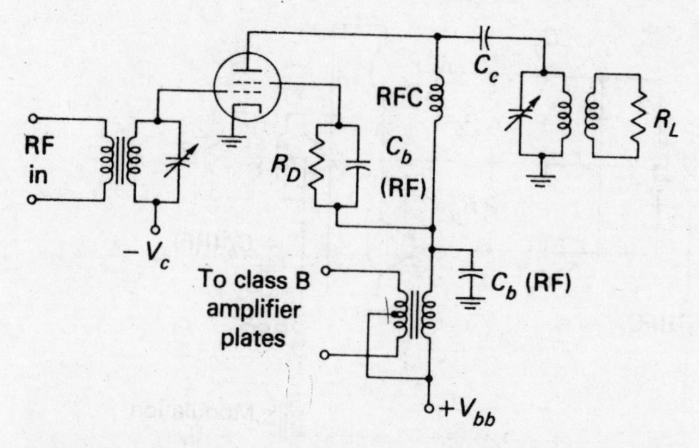

FIGURE 3-11 Plate-modulated tetrode class C amplifier.

Plate modulation of tetrode If a screen-grid tube is used as a plate-modulated class C amplifier, the same good results are obtainable, provided that the extra factors are considered. The fact that screen current increases enormously, if screen voltage exceeds plate voltage, must certainly be taken into account. If it were not, this situation could occur here once every cycle of modulating voltage (if the screen were merely connected to V_{bb} through a dropping resistor). This is illustrated in Fig. 3-12. The system would become very inefficient and the tube short-lived, but the situation is easily avoided by modulation of the screen simultaneously with the plate, as shown on the circuit of Fig. 3-11.

The same stratagem avoids the difficulty of trying to vary the plate current of a screen-grid tube by adjusting the plate voltage only (while keeping the screen voltage constant). We thus have a high-level system, which is simultaneously plate-, screen- and even grid-modulated when leak-type bias is used and has good characteristics, properties and a plate efficiency that can exceed 90 percent in practice.

3-2.4 Modulated Transistor Amplifiers

Modern high-power AM transmitters tend to use transistors at the lower power levels, so that transistor RF and AF exciters are common. The output stages, and generally the drivers, of such transmitters use tubes; hence we had the prior treatment of modulated tube amplifiers. However, all-transistor transmitters are used for lower-power applica-

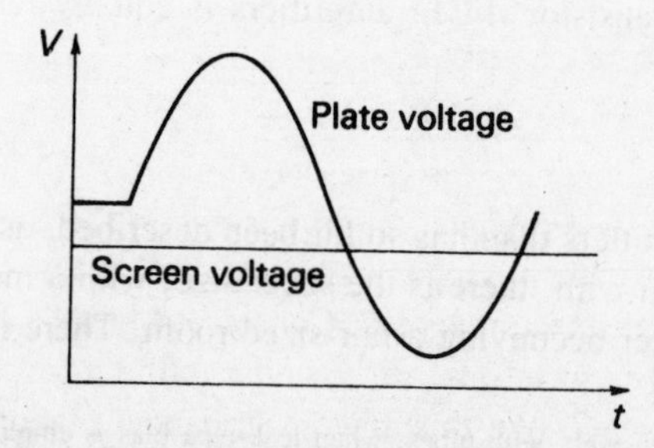

FIGURE 3-12 Plate-modulated tetrode voltages (see text).

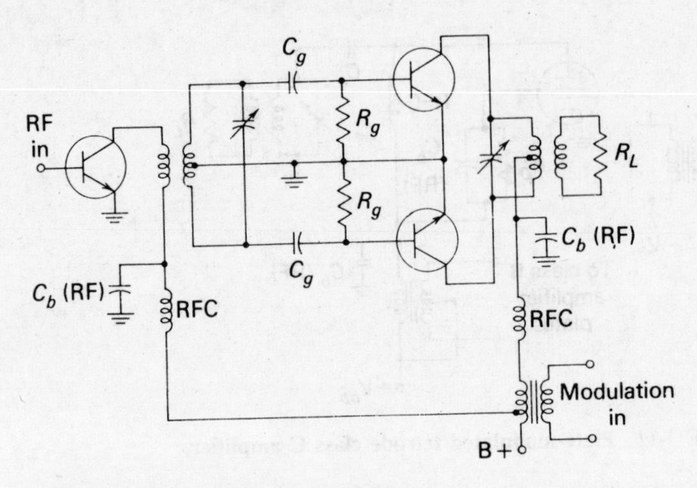

FIGURE 3-13 Collector modulation.

tions with a few kilowatts output obtainable if transistors in parallel are employed. As a result, modulated transistor amplifiers almost all have a push-pull final amplifier for maximum power output.

The modulation methods for transistor amplifiers are, understandably, counterparts of those used with tubes. Thus collector and base modulation of class C transistor amplifiers are both popular, having the same properties and advantages as the corresponding tube circuits. The result is that once again collector modulation is generally preferred. A typical circuit is shown in Fig. 3-13.

Collector modulation has the advantages over base modulation of better linearity, higher collector efficiency and higher power output per transistor; however, as expected, it requires more modulating power. In addition, *collector saturation prevents 100 percent modulation from being achieved with just the collector being modulated,* so that a compound form of modulation is used in a number of cases. Figure 3-13 shows one of the alternatives; here the driver as well as the output amplifier is collector-modulated, but apart from that the circuit is analogous to the tube circuit. The other alternative is to employ collector and base modulation of the same amplifier.[1] Although leak-type bias might again be employed for this purpose, there is the danger that if sufficient bias action is used to permit base modulation, the bias will become excessive, and power output will drop. As a result, simultaneous base and collector modulation, akin to that shown in Fig. 3-13, is preferable. Note finally that drain- and gate-modulation of field-effect-transistor (FET) amplifiers is equally feasible and is used in some systems.

3-2.5 System Summary

There is a lot more to AM transmitters than has so far been described, as anyone who has seen one will agree. To begin with, there is the sheer size, with a medium-power (say, 50-kW) broadcast transmitter occupying a fair-sized room. There is a large and

[1] Note that this method is also used extensively with tubes, when leak-type bias is employed.

complex power supply, generally three-phase on the ac side and regulated on the dc side to prevent unwanted output variations, with many outputs for the low- and high-power sections of the transmitter and carefully implemented isolation and de-coupling to prevent unwanted feedback and interference. A standby diesel generator is normally provided in order to minimize the effects of main supply variations or failures. Some transmitters even have a battery room and a dc-ac inverter, with complex automatic switchover circuitry, thus ensuring that there is no interruption in output immediately after a power supply failure before the standby generator takes over. Similarly, a standby transmitter is also often provided, with a power output of the order of one-fifth of the main transmitter, to take over during transmitter failures or maintenance. Transmitters with power outputs in the kilowatt range also employ forced cooling of the power stages, ensuring a reduced, constant operating temperature and maximizing operating life and power output.

With air cooling in operation, a large AM transmitter sounds as though it is operating; with the metering that is invariably provided, it also *looks* as though it is operating. Metering is provided for virtually all aspects of operation and is used as an operational, maintenance and diagnostic tool, so that trouble may be anticipated as much as possible. In older transmitters (broadcast transmitters are designed to have and operating life of at least 20 years), most of the metering will be in the form of dials for each stage, with switches to display different quantities. More recent transmitters are likely to have switchable digital displays, whereas the most modern ones may well have microprocessor-controlled metering functions, displayed on a screen by the operator, as required. The most important functions are the various voltages and currents, the transmit frequency, audio and RF output and the modulation percentage. It is important to prevent overmodulation; otherwise spurious frequencies will be transmitted, interfering with other transmissions.

Amplitude modulation itself is a well-established, mature art, used for broadcasting almost exclusively. It is easy to generate and, even more importantly, to receive and demodulate (as will be seen in Chap. 6). This last requirement is very important, because it leads to simple and inexpensive receivers; it must not be forgotten that, in broadcasting, the receiver/transmitter ratio is enormous, running into thousands if not millions of receivers for each transmitter. This is most significant, because it implies not only that the usual transmitter-receiver complexity tradeoffs are fairly limited, but also that major changes are difficult to implement if they affect receivers. So, although AM is by no means the best or most efficient modulation system, as will be seen in following chapters, its worldwide use for broadcasting is so entrenched that changes are not contemplated. However, whereas until comparatively recently AM was used for mobile communications—ship-to-shore and land-mobile—it is no longer used for those purposes, and so its major application is in sound broadcasting. As will be seen in Chap. 5, AM is more likely to be affected by noise than is FM. Similarly, an analysis of the components of an AM wave—the carrier and the two sidebands—is made in Chap. 4, showing that significant bandwidth and power savings can be made if the carrier and one of the sidebands are suppressed before transmission. The penalties, as might be expected, are increased receiver cost and complexity; such systems are therefore more likely to be used in applications, such as point-to-point links, where the number of receivers approximates that of the transmitters.

PROBLEMS

For self-testing questions on this chapter, see p. 678.

3-1. A 1000-kHz carrier is simultaneously modulated with 300-Hz, 800-Hz and 2-kHz audio sine waves. What will be the frequencies present in the output?
[Ans.: 998.0, 999.2, 999.7, 1000.3, 1000.8 and 1002.0 kHz]

3-2. A broadcast AM transmitter radiates 50 kW of carrier power. What will be the radiated power at 85 percent modulation?
[Ans.: 68.1 kW]

3-3. When the modulation percentage is 75, an AM transmitter produces 10 kW. How much of this is carrier power? What would be the percentage power saving if the carrier and one of the sidebands were suppressed before transmission took place?
[Ans.: 7.81 kW, 89 percent]

3-4. A 360-W carrier is simultaneously modulated by two audio waves with modulation percentages of 55 and 65, respectively. What is the total sideband power radiated?
[Ans.: 130 W]

3-5. A transistor class C amplifier has maximum permissible collector dissipation of 20 W and a collector efficiency of 75 percent. It is to be collector-modulated to a depth of 90 percent. *(a)* Calculate (i) the maximum unmodulated carrier power and (ii) the sideband power generated. *(b)* If the ***maximum*** depth of modulation is now restricted to 70 percent, calculate the new maximum sideband power generated.
[Ans.: (a) (i) 42.7 W, (ii) 17.3 W; (b) 11.9 W]

3-6. When a broadcast AM transmitter is 50 percent modulated, its antenna current is 12 A. What will the current be when the modulation depth is increased to 0.9?
[Ans.: 13.4 A]

3-7. The output current of a 60 percent modulated AM generator is 1.5 A. To what value will this current rise if the generator is modulated additionally by another audio wave, whose modulation index is 0.7? What will be the percentage power saving if the carrier and one of the sidebands are now suppressed?
[Ans.: 1.65 A, 85 percent]

QUESTIONS

3-1. Define amplitude modulation and modulation index. Use a sketch of a sinusoidally modulated AM waveform to help explain the definition.

3-2. Derive the relation between the output power of an AM transmitter and the depth of modulation, and plot it as a graph for values of the modulation index from zero to maximum. *Note:* a suppressed-zero graph is misleading in this instance and must not be used.

3-3. From the expression for the instantaneous voltage of an AM wave, derive a formula for the rms value of this wave.

3-4. Explain, with the aid of waveforms, how a grid-modulated class C amplifier generates AM.

3-5. Use a circuit diagram and appropriate waveforms to explain how an anode-modulated triode class C amplifier is able to generate AM.

3-6. Why is leak-type bias used with anode modulation? Explain, using waveforms to show what would happen if this self-bias were not used. What is a subsidiary advantage of using leak-type bias?

3-7. With the aid of the circuit of an anode-modulated class C tetrode amplifier, explain the precaution that must be taken to ensure the proper operation of a screen-grid tube in such a circuit.

3-8. What is the basic limitation of modulated transistor amplifiers? When are they used? Are there any circuits which are similar to comparable tube circuits? What appears to be the preferred transistor modulation circuit?

3-9. The collector-modulated class C transistor amplifier may experience a certain difficulty. What is this difficulty? How can it be solved? Show, with a circuit diagram, one of the solutions to this problem.

3-10. With waveforms, explain how a suitable train of current pulses fed to a tuned circuit will result in an amplitude-modulated output wave.

REFERENCES

1. For example, see Herrick, C.N., *Introduction to Electronic Communications*, Charles E. Merrill Publishing Company, Columbus, Ohio, 1969, p. 198.

2. Ibid., pp. 236–242.

4 SINGLE-SIDEBAND TECHNIQUES

The theory of AM, discussed in Chap. 3, showed that a carrier and two sidebands are produced in AM generation. This chapter will show that it is not necessary to transmit *all* those signals to provide the receiver with enough information to reconstruct the original modulation. Thus, it will be seen, the carrier may be removed or attenuated, and so can one of the two sidebands. The resulting signals will require less transmitted power and will occupy less bandwidth, and yet perfectly acceptable communications will be possible.

This chapter will show the method by which a carrier is removed and the three standard methods of removing the unwanted sideband. *(Attenuated) carrier reinsertion* and *independent sideband (ISB)* transmissions will be discussed, as will transmitter and receiver block diagrams. Finally, *vestigial sideband* transmission, used for video transmissions in all TV systems, will be treated in some detail.

Single-sideband (SSB) modulation has been quite possibly the fastest-spreading form of analog modulation in the second half of this century. It offers so many advantages that a very large number of communications systems have been changed to it, have used it from their beginning or in some cases have been made possible only because of the existence of SSB. Among its great advantages is the ability to transmit good communications-quality signals by using a very narrow bandwidth, with relatively low power for the distances involved. In order to investigate this form of modulation and its properties, it is now necessary to review some of the work on amplitude modulation from the early parts of Chap. 3.

4-1 EVOLUTION AND DESCRIPTION OF SSB

Equation (3-7) showed that when a carrier is amplitude-modulated by a single sine wave, the resulting signal consists of three frequencies: the original carrier frequency, the upper sideband frequency $(f_c + f_m)$ and the lower sideband frequency $(f_c - f_m)$. This is an automatic consequence of the amplitude-modulation process, and will always happen unless steps are taken to prevent it. As it happens, steps may be taken either during or after the modulation process, to remove or attenuate any combination of the components of the AM wave. It is the purpose of this chapter to deal with the factors involved in, and the advantages and disadvantages of, suppressing or removing the carrier and/or either of the sidebands. Generation of various forms of single-sideband modulation will also be considered.

It is quite apparent that the carrier of "standard" or double sideband, full carrier (DSBFC) AM (officially known as A3E modulation [1]) conveys no information. This is obvious from the fact that the carrier component remains constant in amplitude and frequency, no matter what the modulating voltage does. Just as obvious is the fact that the two sidebands are images of each other, since each is affected by changes in the modulating voltage amplitude via the exponent $mV_c/2$. It is seen therefore that all the information can be conveyed by the use of one sideband only. The carrier is superfluous, and the other sideband is redundant. The main reason for the widespread use of A3E is the relative simplicity of the modulating and demodulating equipment. Furthermore, A3E is the form used for broadcasting. The fact that radical (i.e., expensive) changes in home receivers would be required if SSB were introduced on a large scale has so far prevented any such changes, although "compatible" SSB has been proposed from time to time [2]. This form of SSB would require no changes in domestic receivers.

The AM power equation states that the ratio of total power to carrier power is given by $(1 + m^2/2)$:1. If the carrier is suppressed, only the sideband power remains. As this is only $P_c(m^2/2)$, a two-thirds saving is effected at 100 percent modulation, and even more is saved as the depth of modulation is reduced. If one of the sidebands is now also suppressed, the remaining power is $P_c(m^2/4)$, a further saving of 50 percent over carrier-suppressed AM. The situation is best illustrated with an example, as follows:

Example 4-1 Calculate the percentage power saving when the carrier and one of the sidebands are suppressed in an AM wave modulated to a depth of (*a*) 100 percent and (*b*) 50 percent.

$$(a)\quad P_t = P_c\left(1 + \frac{m^2}{2}\right) = P_c\left(1 + \frac{1^2}{2}\right) = 1.5P_c$$

$$P_{SB} = P_c\frac{m^2}{4} = P_c\frac{1^2}{4} = 0.25P_c$$

$$\text{Saving} = \frac{1.5 - 0.25}{1.5} = \frac{1.25}{1.5} = 0.833 = 83.3\%$$

$$(b)\quad P_t = P_c\left(1 + \frac{0.5^2}{2}\right) = 1.125P_c$$

$$P_{SB} = P_c\frac{0.5^2}{4} = 0.0625P_c$$

$$\text{Saving} = \frac{1.125 - 0.0625}{1.125} = \frac{1.0625}{1.125} = 0.944 = 94.4\%$$

Example 4-1 indicates how wasteful of power it is to send the carrier and both sidebands in situations in which only one sideband would suffice. A further check shows that the use of SSB immediately halves the bandwidth required for transmission, as compared with A3E.

In practice, SSB is used to save power in applications where such a power saving is warranted, i.e., in mobile systems, in which weight and power consumption must naturally be kept low. Single-sideband modulation is also used in applications in which bandwidth is at a premium. Point-to-point communications; land, air and maritime mobile communications; television; telemetry; military communications; radio navigation and amateur radio are the greatest users of SSB in one form or another.

Waveforms showing SSB are naturally of interest, and these are presented in Fig. 4-1, together with the modulating voltage, the corresponding AM voltage, and a wave with only the carrier removed. Two different modulating amplitudes and frequencies are shown for comparison. This demonstrates clearly that here the SSB wave is but a single radio frequency; its amplitude is proportional to the amplitude of the modulating voltage, and its frequency varies with the frequency of the modulating signal. An

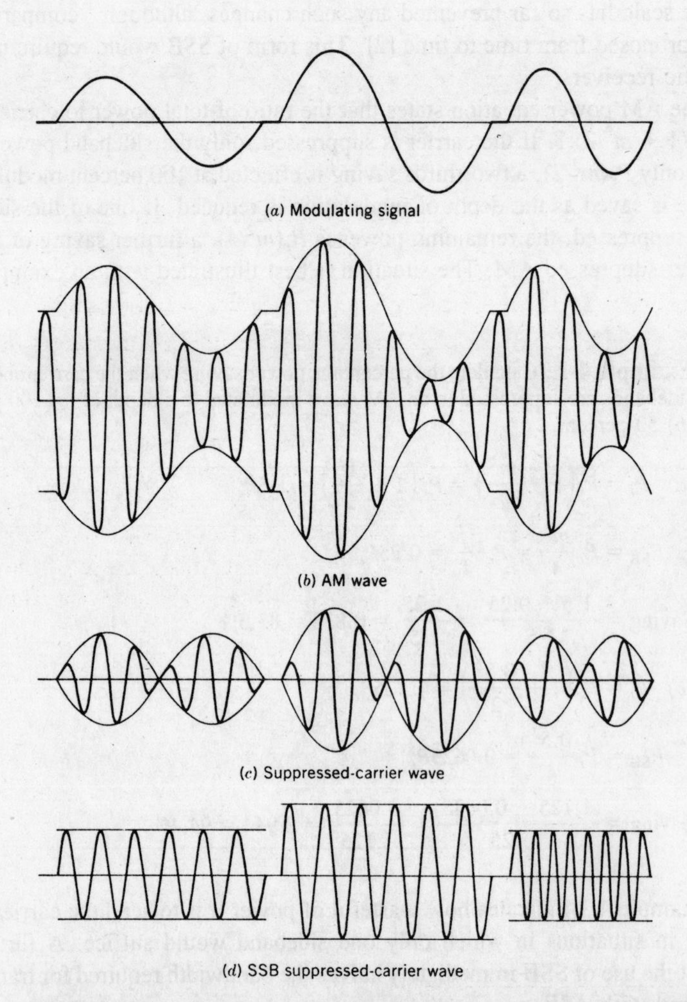

FIGURE 4-1 Waveforms for various types of amplitude modulation. (*a*) Modulating signal; (*b*) AM wave; (*c*) suppressed-carrier wave; (*d*) SSB suppressed-carrier wave.

upper sideband is shown so that its frequency increases with that of the modulation, but please note that this frequency increase is exaggerated here to indicate the effect clearly.

4-2 SUPPRESSION OF CARRIER

Three main systems are employed for the generation of SSB: the filter method, the phase cancellation method and the "third method." They differ from one another in the way of suppressing the unwanted sideband, but all use some form of balanced modulator to suppress the carrier. The balanced modulator is thus seen to be the key circuit in single-sideband generation; the next section leads up to the proof of its operation.

4-2.1 Effect of Nonlinear Resistance on Added Signals

The relationship between voltage and current in a linear resistance is given by

$$i = bv \tag{4-1}$$

where b is some constant of proportionality. If Eq. (4-1) refers to a resistor, then b is obviously its conductance. If, on the other hand, Eq. (4-1) is made to apply to the collector current and base voltage of a transistor, i will be the collector current, and v will be the voltage applied to the base. If the amplifier operates in class A, there will also be a dc component of collector current (a), which is not dependent on the signal voltage at the base. We may thus write

$$i = a + bv \tag{4-2}$$

where a is the dc component of the collector current, and b is the transconductance of the transistor.

In a nonlinear resistance, the current is still to a certain extent proportional to the applied voltage, but no longer directly as before. If the curve of current versus voltage is plotted, as in Fig. 4-2, it is found that there is now some curvature in it. The

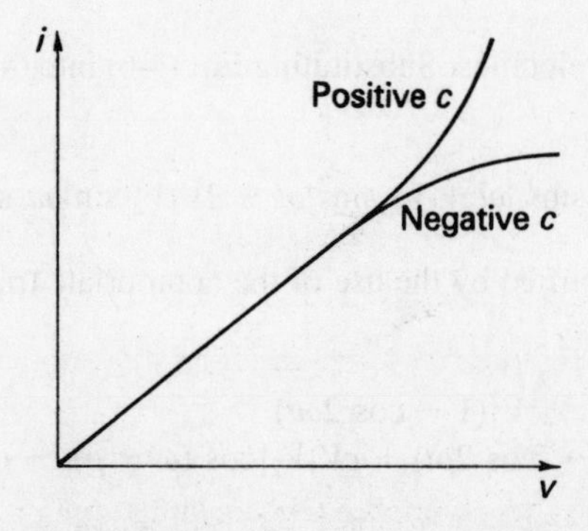

FIGURE 4-2 Nonlinear resistance characteristics.

previous linear relation seems to apply up to a certain point, after which current increases more (or less) rapidly with voltage, as shown. Whether the increase is more or less rapid depends on whether the device begins to saturate, or else some sort of avalanche current multiplication takes place. Current now obviously becomes proportional not only to voltage but also to the square, cube and higher powers of voltage. This nonlinear relation is most conveniently expressed as

$$i = a + bv + cv^2 + dv^3 + \text{higher powers} \tag{4-3}$$

The reason that the initial portion of the graph is linear is simply that the coefficient c is much smaller than b. For example, a typical numerical equation might well be something like $i = 5 + 15v + 0.2v^2$, in which case curvature is insignificant until v equals at least 3. Moreover, c in practical nonlinear resistances is much greater than d, which is in turn larger than the constants preceding the higher-power terms. In fact, only the square term is large enough to be taken into consideration for most applications, so that we are left with

$$i = a + bv + cv^2 \tag{4-4}$$

where a and b have the meanings previously given, and c is the *coefficient of nonlinearity*.

Since Eq. (4-4) is generally adequate in relating the output current to the input voltage of a nonlinear resistance, it may now be applied to the gate-voltage–drain-current characteristic of a FET. If two voltages are applied simultaneously to the gate, then

$$\begin{aligned} i &= a + b(v_1 + v_2) + c(v_1 + v_2)^2 \\ &= a + b(v_1 + v_2) + c(v_1^2 + v_2^2 + 2v_1v_2) \end{aligned} \tag{4-5}$$

Let the two input voltages be sinusoidal. We may then write

$$v_1 = V_1 \sin \omega t \tag{4-6a}$$

and

$$v_2 = V_2 \sin pt \tag{4-6b}$$

where ω and p are the two angular velocities. Substituting Eq. (4-6) into (4-5), we have

$$\begin{aligned} i = a &+ b(V_1 \sin \omega t + V_2 \sin pt) \\ &+ c(V_1^2 \sin^2 \omega t + V_2^2 \sin^2 pt + 2V_1V_2 \sin \omega t \sin pt) \end{aligned} \tag{4-7}$$

Equation (4-7) may be simplified by the use of the appropriate trigonometrical expressions,[1] giving

$$\begin{aligned} i = a &+ bV_1 \sin \omega t + bV_2 \sin pt + \tfrac{1}{2}cV_1^2(1 - \cos 2\omega t) \\ &+ \tfrac{1}{2}cV_2^2(1 - \cos 2pt) + cV_1V_2[\cos (\omega - p)t - \cos (\omega + p)t] \end{aligned}$$

[1] These expressions are $\sin x \sin y = \tfrac{1}{2}[\cos (x - y) - \cos (x + y)]$ and $\sin^2 x = \tfrac{1}{2}(1 - \cos 2x)$.

$$= \underset{\text{(I)}}{(a + \tfrac{1}{2}cV_1^2 + \tfrac{1}{2}cV_2^2)} + \underset{\text{(II)}}{bV_1 \sin \omega t} + \underset{\text{(III)}}{bV_2 \sin pt}$$
$$- \underset{\text{(IV)}}{(\tfrac{1}{2}cV_1^2 \cos 2\omega t + \tfrac{1}{2}cV_2^2 \cos 2pt)}$$
$$+ \underset{\text{(V)}}{cV_1V_2 \cos (\omega - p)t} - \underset{\text{(VI)}}{cV_1V_2 \cos (\omega + p)t} \quad \textbf{(4-8)}$$

The foregoing derivation, with the resulting Eq. (4-8), is possibly the most important of all in communications. It is the proof (1) that harmonic and intermodulation distortion may occur in audio and RF amplifiers, (2) that sum and difference frequencies will be present in the output of a mixer, (3) that the diode detector has audio frequencies in its output, (4) of the operation of the *beat-frequency oscillator* (BFO) and the *product detector;* finally, (5) it is part of the proof that the balanced modulator produces AM with the carrier suppressed.

If in Eq. (4-8) ω is taken as the carrier angular frequency, and p as the modulating angular frequency, then term I is the dc component, term II is the carrier, term III is the modulating signal, term IV consists of harmonics of the carrier and the modulation, term V represents the lower sideband voltage and term VI is the upper sideband. The equation shows that when two frequencies are passed together through a nonlinear resistance, the process of amplitude modulation takes place. In a practical modulation circuit, the voltages in Eq. (4-8) would be developed across a circuit tuned to the carrier frequency, and with a bandwidth just large enough to pass the two sideband frequencies but no others.

4-2.2 The Balanced Modulator

Two circuits of the balanced modulator are shown in Fig. (4-3). Each utilizes the nonlinear principles just discussed. The modulation voltage v_2 is fed in push-pull, and the carrier voltage v_1 in parallel, to a pair of identical diodes or class A (transistor or FET) amplifiers. In the FET circuit, the carrier voltage is thus applied to the two gates in phase; whereas the modulating voltage appears 180° out of phase at the gates, since these are at the opposite ends of a center-tapped transformer. The modulated output currents of the two FETs are combined, as shown, in the center-tapped primary of the push-pull output transformer. They therefore subtract, as indicated by the direction of the arrows in Fig. 4-3*b*. If this system is made completely symmetrical, the carrier frequency will be completely canceled. No system can of course be perfectly symmetrical in practice, so that the carrier will be heavily suppressed rather than completely removed (a 45-dB suppression is normally regarded as acceptable). The output of the balanced modulator thus contains the two sidebands and some of the miscellaneous components which are taken care of by the tuning of the output transformer's secondary winding. Thus the final output consists only of sidebands.

Since it is not immediately obvious how and why only the carrier is suppressed, a mathematical analysis of the balanced modulator is now given.

As indicated, the input voltage will be $v_1 + v_2$ at the gate of T_1 and $v_1 - v_2$ at the gate of T_2. If perfect symmetry is assumed,[2] the proportionality constants will be

[2] It follows that the two devices used in the balanced modulator, whether diodes or transistors, must be well matched.

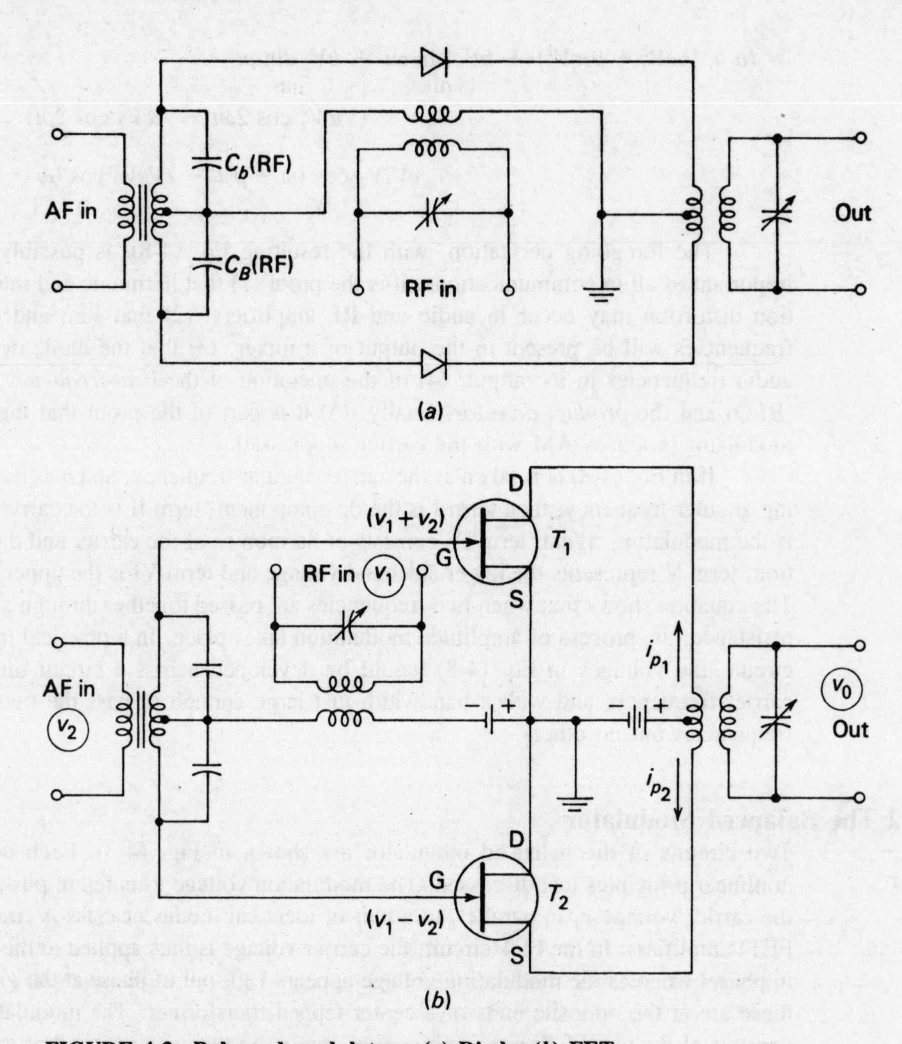

FIGURE 4-3 Balanced modulators. (*a*) Diode; (*b*) FET.

the same for both FETs and may be called a, b, and c as before. The two drain currents, calculated as in the preceding section, will be

$$
\begin{aligned}
i_{d_1} &= a + b(v_1 + v_2) + c(v_1 + v_2)^2 \\
&= a + bv_1 + bv_2 + cv_1^2 + cv_2^2 + 2cv_1v_2 \qquad (4\text{-}9)\\
i_{d_2} &= a + b(v_1 - v_2) + c(v_1 - v_2)^2 \\
&= a + bv_1 - bv_2 + cv_1^2 + cv_2^2 - 2cv_1c_2 \qquad (4\text{-}10)
\end{aligned}
$$

As previously indicated, the primary current is given by the difference between the individual drain currents. Thus

$$i_1 = i_{d_1} - i_{d_2} = 2bv_2 + 4cv_1v_2 \qquad (4\text{-}11)$$

when Eq. (4-10) is subtracted from (4-9).

We may now represent the carrier voltage v_1 by $v_c \sin \omega_c t$ and the modulating voltage v_2 by $V_m \sin \omega_m t$. Substituting these into Eq. (4-11) gives

$$\begin{aligned} i_1 &= 2bV_m \sin \omega_m t + 4cV_m V_c \sin \omega_c t \sin \omega_m t \\ &= 2bV_m \sin \omega_m t + 4cV_m V_c \tfrac{1}{2}[\cos (\omega_c - \omega_m)t - \cos (\omega_c + \omega_m)t] \end{aligned} \quad \textbf{(4-12)}$$

The output voltage v_O is proportional to this primary current. Let the constant of proportionality be α. Then

$$\begin{aligned} v_O &= \alpha i_1 \\ &= 2\alpha b V_m \sin \omega_m t + 2\alpha c V_m V_c[\cos (\omega_c - \omega_m)t - \cos (\omega_c + \omega_m)t] \end{aligned}$$

Simplifying, we let $P = 2\alpha b V_m$ and $Q = 2\alpha c V_m V_c$. Then

$$v_O = \underbrace{P \sin \omega_m t}_{\text{Modulation frequency}} + \underbrace{Q \cos(\omega_c - \omega_m)t}_{\text{Lower sideband}} - \underbrace{Q \cos (\omega_c + \omega_m)t}_{\text{Upper sideband}} \quad \textbf{(4-13)}$$

Equation (4-13) shows that (under ideally symmetrical conditions) the carrier has been canceled out, leaving only the two sidebands and the modulating frequencies; the proof applies equally to both diagrams of Fig. 4-3. The tuning of the output transformer will remove the modulating frequencies from the output, but it is also possible to suppress them by the action of the circuit. The addition of two more diodes to the diode balanced modulator is required for this result, and the circuit then becomes known as the *ring modulator*.

4-3 SUPPRESSION OF UNWANTED SIDEBAND

As stated earlier, the three practical methods of SSB generation all use the balanced modulator to suppress the carrier, but each uses a different method of removing the unwanted sideband. All three systems will remove either the upper or the lower sideband with equal ease, depending on the specific circuit arrangement. Each of the systems will now be studied in turn.

4-3.1 The Filter System

The filter system is the simplest system of the three—after the balanced modulator the unwanted sideband is removed (actually heavily attenuated) by a filter. The filter may be *LC*, crystal, ceramic or mechanical, depending on the carrier frequency and other requirements. See [3] for a description of the operation and applications of mechanical filters. A block diagram of an SSB transmitter employing this system is shown in Fig. 4-4.

The key circuits in this transmitter are the balanced modulator and the sideband-suppression filter; the special considerations involving the latter must now be examined.

Basically, such a filter must have a flat passband and extremely high attenuation outside the passband. There is no limit on this: the higher the attenuation, the better. In radio communications systems, the frequency range used for voice is 300 to

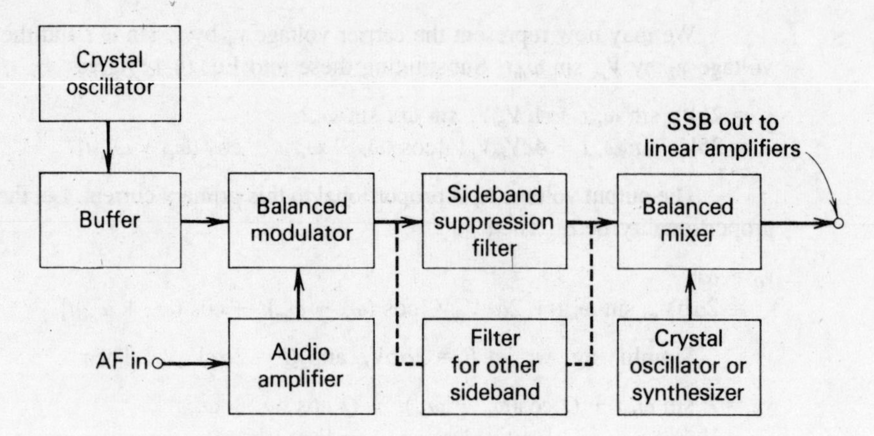

FIGURE 4-4 Filter method of sideband suppression.

about 2800 Hz in most cases. If it is required to suppress the lower sideband and if the transmitting frequency is f, then the lowest frequency that this filter must pass without attenuation is $f + 300$ Hz, whereas the highest frequency that must be fully attenuated is $f - 300$ Hz. In other words, the filter's response must change from zero attenuation to full attenuation over a range of only 600 Hz; if the transmitting frequency is much above 10 MHz, this is virtually impossible. Furthermore, it can be appreciated that the situation becomes even worse if lower modulating frequencies are employed, such as the 50-Hz minimum in AM broadcasting. In order to obtain a filter response curve with skirts as steep as those suggested above, the Q of the tuned circuits used must be very high. Again, as the transmitting frequency is raised, so must the Q be raised, until a situation is reached where the necessary Q is so high that there is no practicable method of achieving it.

Looking at the situation from the other end, we find that there must be an upper frequency limit for any type of filter circuit used. It has been found, for instance, that multistage *LC* filters cannot be used for RF values much greater than about 100 kHz; above this frequency the attenuation outside the passband is insufficient. *LC* filters may still be encountered in currently used HF equipment, but they have otherwise tended to be superseded by crystal, ceramic or mechanical filters, mainly because of the bulky size of indicators on the one hand, and great improvements in mechanical filters on the other. Mechanical filters have been used at frequencies up to 500 kHz, and crystal or ceramic filters up to about 20 MHz.

Of the three major types of SSB filters, the mechanical filter seems to be the one with the best all-around properties; small size, good bandpass, very good attenuation characteristics and an adequate upper frequency limit are its chief advantages. Crystal or ceramic filters may be cheaper, but are technically preferable only at frequencies in excess of 1 MHz.

All these filters (even the crystal) have the disadvantage that their maximum operating frequency is below the usual transmitting frequencies. This is a reason for the balanced mixer shown in Fig. 4-4. (It is very much like a balanced modulator, except that the sum frequency, as used here, is much farther from the crystal oscillator frequency than the USB was from the carrier, so that it can be selected with a tuned circuit.) In this mixer, the frequency of the crystal oscillator or synthesizer is added to the SSB signal from the filter, the frequency thus being raised to the value desired for transmission. Such an arrangement also allows the transmitter to be tunable. However, if the transmitting frequency is much higher than the operating frequency of the sideband filter, then two lots of mixing will be required; otherwise, it becomes too difficult to filter out the unwanted frequencies in the output of the mixer.

It will be noted that the mixer is followed by linear amplifiers. The reason, of course, is that the amplitude of the SSB signal is variable and must not be fed to a class C amplifier, which would distort it. A class B RF amplifier is used instead (being much more efficient than a class A amplifier) and bears the name *linear amplifier*. Linear amplifiers are of course not confined to SSB systems. They are used in any AM system in which an amplifier other than the final one is modulated.

4-3.2 The Phase-Shift Method

The phase-shift method avoids filters and some of their attendant disadvantages, and instead makes use of two balanced modulators and two phase-shifting networks, as shown in Fig. 4-5. As indicated, one of the modulators, M_1, receives the carrier voltage (shifted by 90°) and the modulating voltage; whereas the other, M_2, is fed the modulating voltage (shifted through 90°) and the carrier voltage. Sometimes the modulating voltage phase shift is arranged slightly differently. It is made +45° for one of the balanced modulators and −45° for the other, but the result is the same.

Both modulators produce an output consisting only of sidebands. It will be shown, however, that whereas both upper sidebands lead the input carrier voltage by 90°, one of the lower sidebands leads the reference voltage by 90° and the other lags it by 90°. The two lower sidebands are thus out of phase, and when combined in the adder, they cancel each other. The upper sidebands are in phase at the adder and thus add, giving SSB in which the lower sideband has been canceled. The foregoing may be proved as follows.

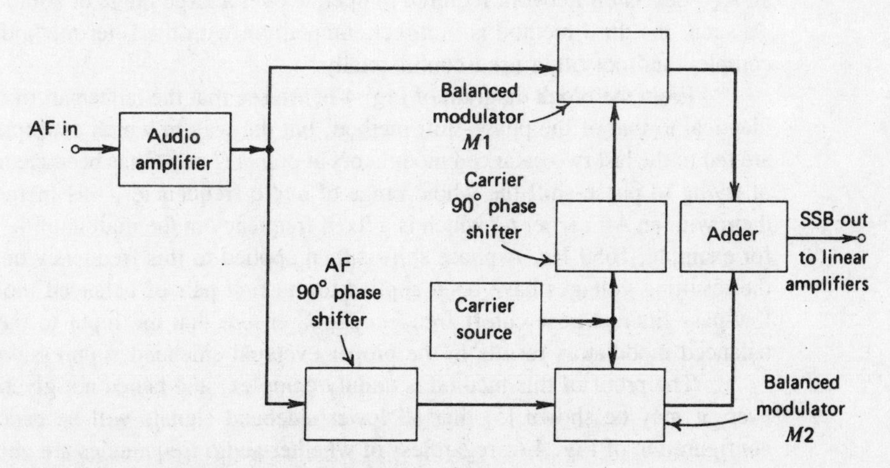

FIGURE 4-5 Single sideband by phase shift.

If it is taken for granted that the two balanced modulators are also balanced with respect to each other, then amplitudes may be ignored as they do not affect the result. Note also that both balanced modulators are fed from the same sources. As before, taking $\sin \omega_c t$ as the carrier and $\sin \omega_m t$ as the modulation, we see that the balanced modulator M_1 will receive $\sin \omega_m t$ and $\sin (\omega_c t + 90°)$, whereas M_2 takes $\sin (\omega_m t + 90°)$ and $\sin \omega_c t$. Following the reasoning in the proof of the balanced modulator, we know that the output of M_1 will contain sum and difference frequencies. Thus

$$v_1 = \cos [(\omega_c t + 90°) - \omega_m t] - \cos [(\omega_c t + 90°) + \omega_m t]$$
$$= \underset{\text{(LSB)}}{\cos (\omega_c t - \omega_m t + 90°)} - \underset{\text{(USB)}}{\cos (\omega_c t + \omega_m t + 90°)} \tag{4-14}$$

Similarly, the output of M_2 will contain

$$v_2 = \cos [\omega_c t - (\omega_m t + 90°)] - \cos [\omega_c t + (\omega_m t + 90°)]$$
$$= \cos (\omega_c t - \omega_m t - 90°) - \cos (\omega_c t + \omega_m t + 90°) \tag{4-15}$$

The output of the adder is

$$v_O = v_1 + v_2 = 2 \cos (\omega_c t + \omega_m t + 90°) \tag{4-16}$$

This output is obtained by adding Eqs. (4-14) and (4-15) and observing that the first term of the first equation is 180° out of phase with the first term of the second. We have thus proved that one of the sidebands in the adder is canceled, whereas the other is reinforced. The system as shown yields the upper sideband. A similar analysis shows that SSB with the lower sideband present will be obtained if both signals are fed (phase-shifted) to the one balanced modulator.

4-3.3 The "Third" Method

The third method of generating SSB was developed by Weaver [4] as a means of retaining the advantages of the phase-shift method, such as its ability to generate SSB at any frequency and use low audio frequencies, without the associated disadvantage of an AF phase-shift network required to operate over a large range of audio frequencies. As such, the third method is in direct competition with the filter method, but is very complex and not often used commercially.

From the block diagram of Fig. 4-6, we see that the latter part of this circuit is identical to that of the phase-shift method, but the way in which appropriate voltages are fed to the last two balanced modulators at points C and F has been changed. Instead of trying to phase-shift the whole range of audio frequencies, this method combines them with an AF carrier f_O, which is a fixed frequency in the middle of the audio band, for example, 1650 Hz. A phase shift is then applied to this frequency only, and after the resulting voltages have been applied to the first pair of balanced modulators, the low-pass filters whose cutoff frequency is f_O ensure that the input to the last pair of balanced modulators results in the proper eventual sideband suppression.

The proof of this method is unduly complex, and hence not given here. However, it may be shown [5] that all lower sideband signals will be canceled for the configuration of Fig. 4-6, regardless of whether audio frequencies are above or below f_O. If a lower sideband signal is required, the phase of the carrier voltage applied to M_1 may be changed by 180°.

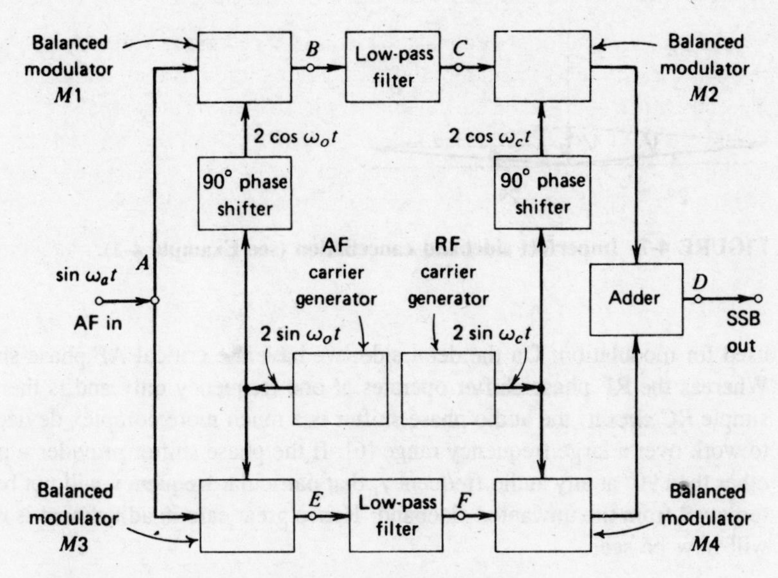

FIGURE 4-6 Third method of SSB generation.

4-3.4 System Evaluation and Comparison

The same SSB is, of course, generated regardless of which method of generation is used. An acceptable single-sideband suppressed-carrier signal is obtained, with either sideband removed, as required. It has also been found from subjective listening tests that the quality is much the same from all three systems. However, to put each system into the proper perspective, it is necessary to examine the technical differences of the three methods of generation.

The filter system gives more than adequate sideband suppression (50 dB possible), and the sideband filter also helps to attenuate the carrier, adding a safety feature which is absent from the two phasing systems. The bandwidth is sufficiently flat and wide, except possibly with crystal filters at the lowest frequencies, where it tends to be restricted and to give a "tinny" quality. The big disadvantage of this system had been its bulk, but this has been overcome with the advent of small mechanical filters of excellent quality and crystal filters of reduced size. The main disadvantage now is the inability of this system to generate SSB at high radio frequencies, so that repeated mixing is required in conjunction with extremely stable crystal oscillators. Also, there are the limitations that low audio frequencies cannot be used, and that two rather expensive filters are required with each transmitter to make it capable of suppressing either sideband at will. Nonetheless, this is an excellent means of generating SSB of communications quality. It is used in a vast majority of commercial systems, particularly with mechanical filters, except in multichannel equipment, where crystal or even *LC* filters are often used.

The phase-cancellation method was originally introduced to overcome the bulk of the filter system using *LC* filters. Since these have now been replaced by much smaller filters, this initial advantage no longer applies, but there are still two others: the ease of switching from one sideband to the other and the ability to generate SSB at any frequency, rendering mixing unnecessary. In addition, low audio frequencies may be

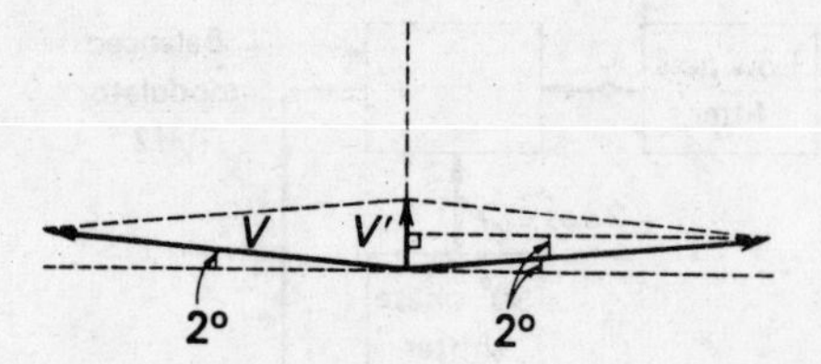

FIGURE 4-7 Imperfect sideband cancellation (see Example 4-2).

used for modulation. On the debit side, we have the critical AF phase-shift network. Whereas the RF phase shifter operates at one frequency only and is therefore a very simple *RC* circuit, the audio phase shifter is a much more complex device since it has to work over a large frequency range [6]. If the phase shifter provides a phase change other than 90° at any audio frequency, that particular frequency will not be completely removed from the unwanted sideband. Hence great care in adjustment is necessary, as will now be seen.

Example 4-2 In a phase-shift SSB system, the phase shift at the audio frequency of 500 Hz is only 88°. To what extent will this frequency be present in the unwanted lower sideband?

From Eq. (4-16) it follows that, ideally, the relative size of the output upper sideband voltage should be $2V$, and that of the lower sideband voltage should be 0. The cancellation of the LSB will not be complete here, and, as illustrated in Fig. 4-7, the amplitude of this unwanted component will be $V(2 \sin 2°)$. The relative attenuation of this signal will therefore be

$$\alpha = \frac{2V}{2V \sin 2°} = \frac{1}{\sin 2°} = \frac{1}{0.0349} = 28.6$$
$$= 20 \log 28.6 = 20 \times 1.457 = 29.14 \text{ dB}$$

The attenuation under the conditions of Example 4-2 is inadequate for commercial operation, in which such attenuation of the unwanted sideband should be at least 40 dB. The example illustrates the problem involved and also suggests that variation in phase shift in practice must be somewhat less than 1°.

In addition, the system has two balanced modulators, which must both give exactly the same output (this assumption was made in the proof), or the cancellation will again be incomplete. Finally, it has been found in practice that layout is quite critical with this system. The result of all these considerations is that the phase-shift system is not used commercially, but it is employed widely by amateurs since filters tend to be rather expensive.

The third method requires neither a sideband-suppression filter nor a wideband audio phase-shift network; correct output can be maintained simply without critical parts or adjustments. Low audio frequencies may be transmitted if desired, and since the majority of the circuitry is at AF, layout and component tolerances are not critical. Sidebands may also be switched quite easily, but an extra crystal may be required for this procedure. On the other hand, dc coupling may be needed to avoid the loss of signal components close to the audio carrier frequency; also, whistle will exist at that frequency if the balance of the low-frequency balanced modulators deteriorates. The system is the most complex of the three, but its biggest disadvantage is that the filter method works so well for present requirements. Although the third method has been used commercially, present indications are that it is unlikely to replace the filter method.

4-4 EXTENSIONS OF SSB

4-4.1 Forms of Amplitude Modulation

This section on forms of amplitude modulation defines, describes and lists the main applications of the various forms of AM used for telephony and TV, particularly the various forms of SSB. The International Telecommunications Union (I.T.U.) Radio Regulations [1] also define and describe all forms of emission, however modulated.

A3E (previously A3) Double sideband, full carrier. As already discussed, this is "standard" AM, used for broadcasting and by now for very little else.

R3E (previously A3A) Single-sideband, reduced carrier. This is a *pilot carrier system,* treated in Sec. 4-4.2. An attenuated carrier is reinserted into the SSB signal, to facilitate receiver tuning and demodulation. Except for so-called maritime mobile distress frequencies (also known as SOLAS—Safety of Life at Sea), especially 2182 kHz, it is being steadily replaced by J3E, on a worldwide basis.

H3E (previously A3H) Single-sideband, full carrier. This could be used as a compatible AM broadcasting system, with A3E receivers. Distortion not exceeding 5 percent is claimed for H3E transmissions received by an A3E receiver.

J3E (previously A3J) Single-sideband, suppressed-carrier. This is the system so far referred to as "SSB," in which the carrier is suppressed by at least 45 dB in the transmitter. It was at first slow to take off, because of the high receiver stabilities required. However, with the advent of acceptable synthesizer-driven receivers, it swiftly became the standard form of SSB for radio communications.

B8E (previously A3B) Two independent sidebands, with a carrier that is most commonly attenuated or suppressed. This form of modulation is also known as *independent-sideband* emission (ISB) and is treated fully in Sec. 4-4.3. It is used for HF point-to-point radiotelephony, in which more than one channel is required.

C3F (previously A5C) Vestigial sideband (used for television video transmissions). A system in which a vestige, i.e., a trace, of the unwanted sideband is transmitted, usually with a full carrier. This system is treated in Sec. 4-4.4. It is used for video transmissions in all the world's various TV systems to conserve bandwidth.

Lincompex This is an acronym which stands for "*lin*ked *comp*ressor and *ex*pander." It is basically a system in which all audio frequencies above 2.7 kHz are filtered out to allow for the presence of a control tone at 2.9 kHz with a bandwidth of 120 Hz. Lincompex may be used with any form of AM but is most common with B8E, R3E, and J3E. Before transmission, the signal is passed through an amplitude limiter and a compressor, so that the amplitude of the transmitter output remains virtually constant. This means that there can also be an amplitude limiter in the receiver, akin to the amplitude limiter in an FM receiver (see also Sec. 5-2), to minimize the effects of noise and *fading* (see Sec. 8-2.2). Thus amplitude variations due to spurious effects will be largely removed, and the received signal will be much better than is usual in HF. The control tone is frequency-modulated with a signal derived from the transmitter compressor. After reception and demodulation, this signal is applied to the receiver expander, which ensures that the signal out of the receiver is expanded just as it was compressed in the transmitter. The compressor and expander have thus been linked, and normal signal amplitude variations have been reproduced acceptably. This system is quite popular for commercial HF radiotelephony, as a cheaper alternative to satellite communications on "thin" routes, such as Australia–Antarctica. The quality can be almost as good as that of satellite or undersea cable communications. Interested students are referred to Sharp [6] for a more detailed description.

4-4.2 Carrier Reinsertion—Pilot-Carrier Systems

As can be appreciated, J3E requires excellent frequency stability on the part of both transmitter and receiver, because any frequency shift, anywhere along the chain of events through which the information must pass, will cause an equal frequency shift to the received signals. Imagine a 40-Hz frequency shift in a system through which three signals are being transmitted at 200, 400 and 800 Hz. Not only will they all be shifted in frequency to (say) 160, 360 and 760 Hz, respectively, but their relation to one another will also stop being harmonic. The result is that good-quality music is obviously difficult to transmit via J3E. Speech will also be impaired (although it suffers less than music) unless long-term stabilities of the order of 1 part in 10^7 (or better) are attained.

Such frequency stability has been available from good-quality temperature-stabilized crystal oscillators for many years. That is fine for fixed-frequency transmitters, but receivers are a different proposition altogether, since they must be tunable. Until the advent of frequency synthesizers of less-than-monstrous bulk it was, in fact, simply not possible to produce receiver variable-frequency oscillators stable enough for J3E. The technique that was used to solve this problem, and which is still widely used, is to transmit a *pilot carrier* with the wanted sideband. The block diagram of such a transmitter is very similar to those already shown, with the one difference that an attenuated carrier signal is added to the transmission after the unwanted sideband has been removed. The technique of carrier reinsertion for a filter system is illustrated in Fig. 4-8.

The carrier is normally reinserted at a level of 16 or 26 dB below the value it would have had if it had not been suppressed in the first place, and it provides a reference signal to help demodulation in the receiver. The receiver can then use automatic frequency control (AFC) similar to that described in Sec. 5-3.3; this topic is further described in Sec. 6-5.2.

Since the frequency stability obtainable over long-term periods with R3E is of the order of 1 part in 10^7, such systems are widely employed. They are particularly found in transmarine point-to-point radiotelephony and in maritime mobile communications, especially at the distress frequencies. For high-density traffic, short- or long-haul, different modulation techniques are employed. They are known as *frequency-* or *time-division multiplex* and are treated in Chaps. 13 through 15.

4-4.3 Independent Sideband (ISB) Systems

As mentioned in the preceding section, multiplex techniques are used for high-density point-to-point communications. For low- or medium-density traffic, ISB transmission is often employed. The growth of modern communications on many routes has been from a single HF channel, through a four-channel ISB system (with or without Lincompex) to satellite or submarine cable communication.

As shown in the block diagram of Fig. 4-8, ISB essentially consists of R3E with two SSB channels added to form two sidebands around the reduced carrier.

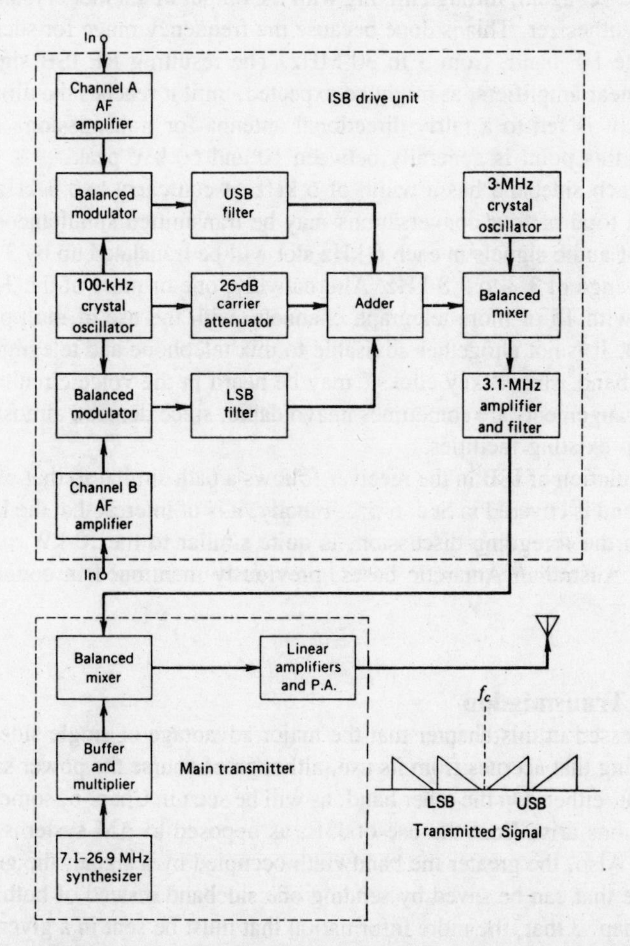

FIGURE 4-8 ISB Transmitter. ***(By permission, based on the Post Office Electrical Engineers' Journal.)***

However, each sideband is quite independent of the other. It can simultaneously convey a totally different transmission, to the extent that the upper sideband could, for example, be used for telephony while the lower sideband carries telegraphy.

Each 6-kHz channel is fed to its own balanced modulator, each balanced modulator also receiving the output of the 100-kHz crystal oscillator. The carrier is suppressed (by 45 dB or more) in the balanced modulator and the following filter, the main function of the filter still being the suppression of the unwanted sideband, as in all other SSB systems. The difference here is that while one filter suppresses the lower sideband, the other suppresses the upper sideband. Both outputs are then combined in the adder with the −26-dB carrier, so that a low-frequency ISB signal exists at this point, with a pilot carrier also present. Through mixing with the output of another crystal oscillator, the frequency is then raised to the standard value of 3.1 MHz. Note the use of balanced mixers, to permit easier removal of unwanted frequencies by the output filter.

The signal now leaves the *drive unit* and enters the main transmitter. Its frequency is raised yet again, through mixing with the output of another crystal oscillator, or frequency synthesizer. This is done because the frequency range for such transmissions lies in the HF band, from 3 to 30 MHz. The resulting RF ISB signal is then amplified by linear amplifiers, as might be expected, until it reaches the ultimate level, at which point it is fed to a fairly directional antenna for transmission. The typical power level at this point is generally between 10 and 60 kW peak.

Since each sideband has a width of 6 kHz, it can carry two 3-kHz voice circuits, so that a total of four conversations may be transmitted simultaneously. Naturally, one set of audio signals in each 6-kHz slot will be translated up by 3 kHz, so as to occupy the range of 3.3 to 5.8 kHz. Alternatively, one or more of the 3-kHz bands may be filled with 15 or more telegraph channels, with the use of multiplexing (see also Sec. 13-3). It is not altogether advisable to mix telephone and telegraph channels in the one sideband, since ''key clicks'' may be heard in the voice circuit. However, such hybrid arrangements are sometimes unavoidable, since demand almost invariably tries to outstrip existing facilities.

Demodulation of ISB in the receiver follows a path similar to that of the modulation process and is covered in Sec. 6-5.2. Finally, it is of interest that the ISB system, as described in the foregoing discussion, is quite similar to the 10-kW circuits from Sydney to the Australian Antarctic bases, previously mentioned in connection with Lincompex.

4-4.4 Vestigial Sideband Transmission

It has been stressed in this chapter that the major advantage of single sideband is the bandwidth saving that accrues from its use, although of course the power saving is not to be sneezed at, either. On the other hand, as will be seen in Chap. 6, some demodulation complications arise from the use of J3E, as opposed to AM systems in which a carrier is sent. Also, the greater the bandwidth occupied by a signal, the greater is the spectrum space that can be saved by sending one sideband instead of both. Finally, it was seen in Chap. 3 that, the more information that must be sent in a given time (i.e., per second), the larger the bandwidth required to send it.

After these preliminaries, we may now turn to the question of transmitting the video signals required for the proper reception of television, noting that the bandwidth occupied by such signals is at least 4 MHz. Bearing in mind filter characteristics, a transmitted bandwidth of 9 MHz would be the minimum required if video transmissions used A3E (which, for that precise reason, they do not). The use of some form of SSB is clearly indicated here to ensure spectrum conservation. So as to simplify video demodulation in the receiver, the carrier is, in practice, sent undiminished. Because the phase response of filters, near the edges of the flat passband, would have a harmful effect on the received video signals in a TV receiver, a portion of the unwanted (lower) sideband must also be transmitted. The result is vestigial sideband transmission, or C3F, as shown in Fig. 4-9*a*. Please note that the frequencies shown there, like the ones used in text, refer strictly only to the NTSC TV system in use in the United States, Canada and Japan. The principles are the same, but the frequencies are somewhat different in the PAL TV system as used in Europe, Australia and elsewhere, and are again different in the French SECAM system.

By sending the first 1.25 MHz of the lower sideband (the first 0.75 MHz of it undiminished), it is possible to make sure that the lowest frequencies in the wanted upper sideband are not distorted in phase by the vestigial-sideband filter. Because only

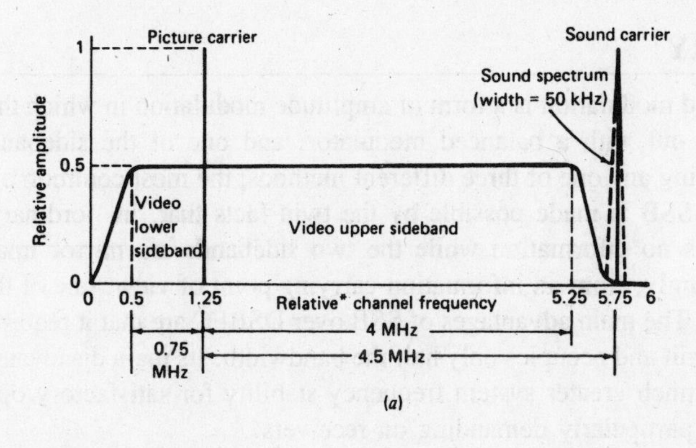

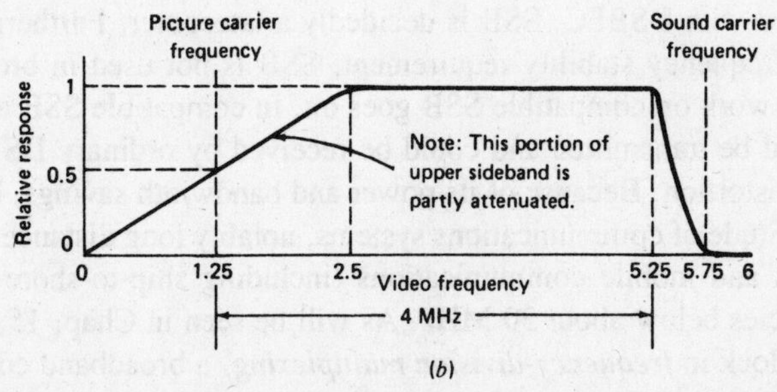

* That is, 0 corresponds to 82 MHz in Channel 6, 174 MHz in Channel 7, and so on.

FIGURE 4-9 Vestigial sideband as used for TV video transmission. (*a*) Spectrum of transmitted signals (NTSC); (*b*) corresponding receiver video amplifier frequency response.

the first 1.25 MHz of the lower sideband is transmitted, 3 MHz of spectrum is saved for every TV channel. Since the total bandwidth requirement of a television channel is now 6 MHz instead of 9 MHz, clearly a great saving has been made, and more channels consequently can be accommodated.

For completeness, Fig. 4-9*a* shows also the location, in frequency, of the frequency-modulated sound transmissions that accompany the video. It should be noted that these transmissions have nothing to do with the fact that the modulation system for video is C3F, and would have been there regardless of the video modulation system. All these signals occupy frequencies near the video transmissions simply because sound is required with the pictures, and it would not be very practical to have a completely separate receiver for the sound, operating at some frequency remote from the video transmitted frequencies.

Fig. 4-9*b* shows the video frequency response of the television receiver. Attenuation is, as can be seen, purposely provided for the video frequencies from 0 to 1.25 MHz. The reason is quite simple: Extra power is transmitted at these frequencies (since they are sent in both sidebands, whereas the remaining video frequencies are not only in the upper sideband). Accordingly, these frequencies would be unduly emphasized in the video output of the receiver if they were not attenuated appropriately.

4-5 SYSTEM SUMMARY

Single-sideband modulation is a form of amplitude modulation in which the carrier has been canceled out with a balanced modulator, and one of the sidebands has been removed by using any one of three different methods; the most common of these is the filter method. SSB is made possible by the twin facts that, in "ordinary" AM, the carrier contains no information while the two sidebands are mirror images of each other. Accordingly, from an information-carrying point of view, one of the sidebands is superfluous. The main advantages of SSB over DSBFC are that it requires much less power to transmit and occupies only half the bandwidth. Its main disadvantage is that it requires very much greater system frequency stability for satisfactory operation, and this feature is particularly demanding on receivers.

Compared with DSBFC, SSB is decidedly a latecomer. Furthermore, because of the receiver frequency stability requirement, SSB is not used in broadcasting, although sporadic work on compatible SSB goes on. In compatible SSB a proportion of the carrier would be transmitted and could be received by ordinary DSBFC receivers without undue distortion. Because of its power and bandwidth savings, however, SSB is used in a multitude of communications systems, notably long distance point-to-point communications and mobile communications (including ship-to-shore and shore-to-ship) at frequencies below about 30 MHz. As will be seen in Chap. 15, SSB is also a basic building block in *frequency-division multiplexing*, a broadband communications technique used for multichannel communications in microwave links, terrestrial and submarine coaxial cables and satellite up- and down-links.

A number of versions of SSB exist in addition to the form in which the carrier has been fully suppressed. One such is the reduced-carrier form, R3E, in which a pilot carrier is sent along with the wanted sideband. In another form, B8E, two sidebands are sent, but they are quite independent of each other, making it a favorite form of transmission for two- or four-channel HF point-to-point communication links. In the vestigial sideband system, C3F, the wanted sideband is accompanied by a full carrier and a portion of the unwanted sideband. Such a system is used in all of the world's TV systems for their video transmissions. A portion of the unwanted sideband is sent because it would be too difficult to suppress in a television system.

Physically, SSB transmitters are akin to the AM transmitters described in Chap. 3, except that they are generally smaller, for reasons described above. In addition, because SSB is used mostly for (two-way) communications, transceivers—combined transmitter-receivers—are the rule rather than the exception. As a general sort of rule, such transceivers are used for *simplex* rather than *duplex* communications, so that the transceiver is used alternately to transmit or receive, but not for both simultaneously. Externally, this is most apparent by the presence of a press-to-talk facility, generally on the microphone. Internally, this kind of operation has the beneficial effect of reducing the total number of circuits needed in the equipment, in that quite a number of them can double up as parts of either the transmitter or the receiver.

PROBLEMS

For self-testing questions on this chapter, see p. 680.

4-1. A 400-W carrier is modulated on a depth of 75 percent; calculate the total power in the modulated wave in the following forms of AM: (*a*) A3E (*b*) double-sideband, suppressed carrier (*c*) J3E.
[Ans.: 512 W, 112 W, 56 W]

4-2. An AM broadcast station has a modulation index which is 0.75 on the average. What would be its average power saving it if could go over to J3E transmissions, while having to maintain the same signal strength in its reception area?
[Ans.: 89 percent]

4-3. A signal voltage $v_s = 25 \sin 1000t$ and a carrier voltage $v_c = 5 \sin (4 \times 10^6 t)$ are applied to the input of a nonlinear resistance whose input-voltage–output-current characteristic can be represented by the expression $i_o = (10 + 2v_i - 0.2v_i^2)$ mA. By substitution into Eq. (4-8), determine the amplitudes and frequencies of all the components in the output current from this nonlinear resistance. Calculate the percentage modulation if the output current is passed through a tuned circuit, so that only the components appropriate to AM are produced.
[Ans.: 50 percent]

4-4. A J3E transmitter operating at 16 MHz has a frequency stability of 1 part per million. If its transmission is reproduced by a receiver whose frequency stability is 8 parts per million, what is the maximum frequency error that the output of this receiver could have in reproducing this transmission?
[Ans.: 144 Hz]

QUESTIONS

4-1. What is single-sideband suppressed-carrier (J3E) modulation? What are its advantages with respect to "ordinary" AM (A3E)?

4-2. What are the disadvantages of J3E with respect to A3E?

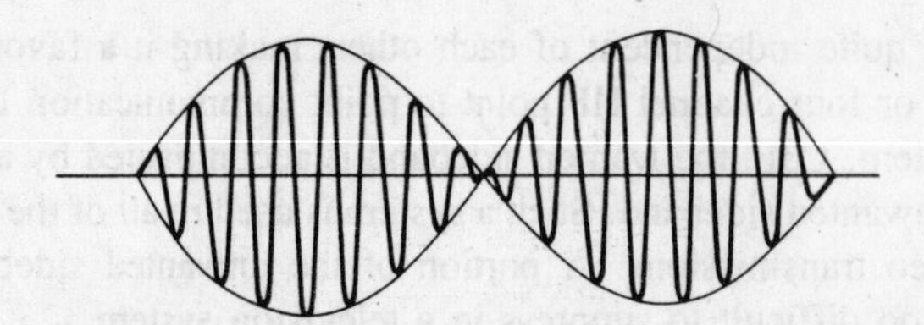

FIGURE 4-10 Modulated voltage of Question 4-3.

4-3. A modulated waveform has the appearance of Fig. 4-10. How would you determine whether it is the waveform of a suppressed-carrier signal, rather than that of a 100 percent modulated A3E wave?

4-4. Show how to derive the equation normally used to describe a nonlinear resistance.

4-5. Show mathematically what happens when two frequencies are added and then passed through a nonlinear resistance. List the various circuits and processes which make use of this state of affairs.

4-6. Prove that the balanced modulator produces an output consisting of sidebands only, with the carrier removed. Other than in SSB generation, what applications can this circuit have?

4-7. Draw the block diagram of an SSB transmitter using the filter system. Why must the filter have such sharp cutoff outside the passband? In a transmitter, must this cutoff be equally sharp on each side of the filter's passband?

4-8. What are linear amplifiers? Why are they used in SSB transmitters?

4-9. Draw the block diagram of a phase cancellation SSB generator and explain how the carrier and the unwanted sideband are suppressed. What change is necessary to suppress the other sideband?

4-10. Compare the three main systems of SSB generation by drawing up a table of the outstanding characteristics of each system.

4-11. Why could J3E not be used for "compatible" broadcasting? What form of SSB might be so used?

4-12. Explain briefly what *Lincompex* is. Could it be used with any form of SSB?

4-13. In what way does the use of a pilot-carrier system help to reduce the difficulties of receiving and demodulating J3E? What is the alternative to the use of R3E if high receiver frequency stability is needed?

4-14. Define and describe independent-sideband transmission.

4-15. Draw the block diagram of an ISB transmitter, operating at 22.275 MHz, without a pilot carrier, and with two 3-kHz sidebands.

4-16. Define and describe vestigial sideband transmission. What is its only application?

4-17. Show the frequency spectrum of an NTSC video transmission. Why is a portion of the lower sideband transmitted at all?

REFERENCES

1. See Sec. 4-4.1, and also "Radio Regulations," *Addition of 1982*, book 1, art. 4 (Designation of Emissions), General Secretariat of the International Telecommunication Union, Geneva, 1982, pp. 4-1–4-4.

2. Taub, H., and D. L. Schilling, *Principles of Communication Systems*, McGraw-Hill Book Company, New York, 1971, pp. 107–108 and 144–145.

3. Johnson, Robert A., *Mechanical Filters in Electronics,* John Wiley and Sons, Inc., New York, 1982.

4. (*Historical*) Weaver, D. K., ''A Third Method of Generation and Detection of SSB Signals,'' *Proc. IRE,* December 1956, pp. 1703–1705.

5. See the first edition of this book.

6. See A. R. R. L. Headquarters Staff, *The Radio Amateur's Handbook,* 58th ed., American Radio Relay League, Inc., Newington, Conn., 1981, pp. 12-8–12-9, for a circuit of an AF phase shifter and details of its operation.

7. Sharp, E. S., ''High-Quality Radiotelephone Communication,'' *Electron. Eng.*, April 1970, pp. 74–77.

5 FREQUENCY MODULATION

Following the pattern set in Chap. 3, this chapter treats first the theory of frequency modulation, and then the generation. However, both the theory and the generation of FM are a good deal more complex to think about and visualize than those of AM. This is mainly because FM involves minute frequency variations of the carrier, whereas AM results in large-scale amplitude variations of the carrier. FM is more difficult to treat mathematically and has sideband behavior that is decidedly complex, as well.

Having studied this chapter, students will understand that FM is a form of angle modulation, of which phase modulation is another, tantalizingly similar form. The theory of both is treated in detail, as are their similarity and important differences. It will be seen that frequency modulation is the preferred form for most applications. Frequency and amplitude modulation are then compared, on the basis that both are widely used practical systems; their relative merits are dealt with in detail.

Unlike amplitude modulation, FM is, or can be made, relatively immune to the effects of noise. This point is discussed at length. It will be seen that the effect of noise in FM depends on the noise sideband frequency, a point that is brought out under the heading of noise triangle. It will be shown that processing of the modulating signals, known as pre-emphasis and de-emphasis, plays an important part in making FM relatively immune to noise.

The final topic studied in this chapter is the generation of FM. It will be shown that two basic methods of generation exist. The first is direct generation, in which a voltage-dependent reactance varies the frequency of an oscillator. The second method is one in which basically phase modulation is generated, but circuitry is used to convert this to frequency modulation. Both methods are used in practice.

5-1 THEORY OF FREQUENCY AND PHASE MODULATION

Frequency modulation is a system in which the amplitude of the modulated carrier is kept constant, while its frequency is varied by the modulating signal. The first practical system was put forward in 1936 as an alternative to AM in an effort to make radio transmissions more resistant to noise. *Phase modulation* is a similar system in which the phase of the carrier is varied instead of its frequency; as in FM, the amplitude of the carrier remains constant.

5-1.1 Description of Systems

The general equation of an unmodulated wave, or carrier, may be written as

$$x = A \sin (\omega t + \phi) \qquad (5\text{-}1)$$

where x = instantaneous value (of voltage or current)
A = (maximum) amplitude
ω = angular velocity, radians per second (rad/s)
ϕ = phase angle, rad

Note that ωt represents an angle in radians.

If any one of these three parameters is varied in accordance with another signal, normally of a lower frequency, then the second signal is called the *modulation,* and the first is said to be *modulated* by the second. Amplitude modulation, already discussed, is achieved when the amplitude A is varied; alteration of the phase angle ϕ will yield phase modulation. Finally, if the frequency of the carrier is made to vary, frequency-modulated waves are obtained.

For simplicity, it is again assumed that the modulating signal is sinusoidal. This signal has two important parameters which must be represented by the modulation process without distortion: namely, its amplitude and frequency. It is assumed that the phase relations of a complex modulation signal will be preserved. By the definition of frequency modulation, the amount by which the carrier frequency is varied from its unmodulated value, called the *deviation,* is made *proportional to* the *instantaneous* value of the *modulating voltage*. The rate at which this frequency variation or oscillation takes place is naturally equal to the modulating frequency.

The situation is illustrated in Fig. 5-1, which shows the modulating voltage and the resulting frequency-modulated wave. Figure 5-1 also shows the frequency variation with time, which is seen to be identical to the variation with time of the modulating voltage. The result of using that modulating voltage to produce AM is also shown for comparison. As an example of FM, all signals having the same amplitude will deviate the carrier frequency by the same amount, say 45 kHz, no matter what their frequencies. Similarly, all signals of the same frequency, say 2 kHz, will deviate the carrier at the same rate of 2000 times per second, no matter what their individual amplitudes. *The amplitude of the frequency-modulated wave remains constant at all times;* this is, in fact, the greatest single advantage of FM.

5-1.2 Mathematical Representation of FM

From Fig. 5-1*d*, it is seen that the instantaneous frequency f of the frequency-modulated wave is given by

$$f = f_c (1 + kV_m \cos \omega_m t) \qquad (5\text{-}2)$$

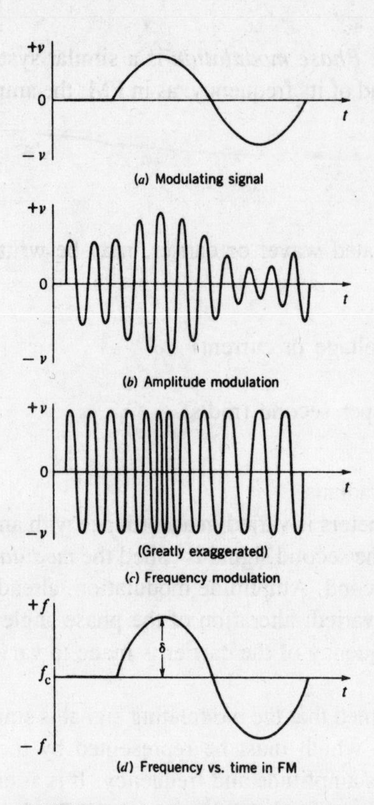

FIGURE 5-1 Basic modulation waveforms.

where f_c = unmodulated (or average) carrier frequency
k = proportionality constant
$V_m \cos \omega_m t$ = instantaneous modulating voltage (cosine being preferred for simplicity in calculations)

The maximum deviation for this particular signal will occur when the cosine term has its maximum value, that is, ±1. Under these conditions, the instantaneous frequency will be

$$f = f_c (1 \pm kV_m) \tag{5-3}$$

so that the maximum deviation δ will be given by

$$\delta = kV_m f_c \tag{5-4}$$

The instantaneous amplitude of the FM signal will be given by a formula of the form

$$v = A \sin [F(\omega_c, \omega_m)] = A \sin \theta \tag{5-5}$$

where $F(\omega_c, \omega_m)$ is some function, as yet undetermined, of the carrier and modulating frequencies. This function represents an angle and will be called θ for convenience. The problem now is to determine the instantaneous value (i.e., formula) for this angle.

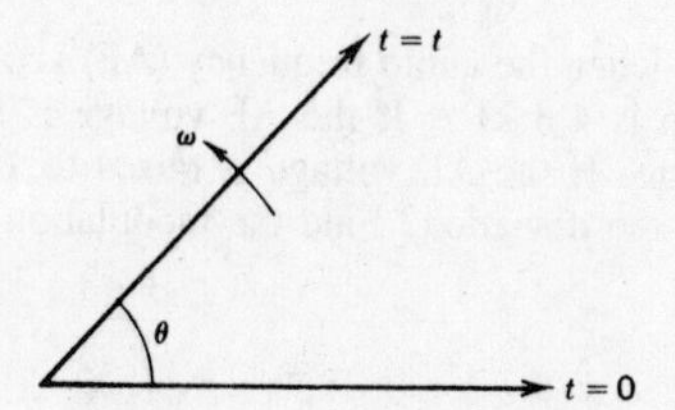

FIGURE 5-2 Frequency-modulated vectors.

As Fig. 5-2 shows, θ is the angle traced out by the vector A in time t. If A were rotating with a constant angular velocity, say p, this angle θ would be given by pt (in radians). In this instance, however, the angular velocity is anything but constant. It is, in fact, governed by the formula for ω obtained from Eq. (5-2), that is, $\omega = \omega_c\,(1 + kV_m \cos \omega_m t)$. In order to find θ, ω must be integrated with respect to time. Thus

$$\theta = \int \omega\, dt = \int \omega_c\,(1 + kV_m \cos \omega_m t)\, dt = \omega_c \int (1 + kV_m \cos \omega_m t)\, dt$$

$$= \omega_c\left(t + \frac{kV_m \sin \omega_m t}{\omega_m}\right) = \omega_c t + \frac{kV_m \omega_c \sin \omega_m t}{\omega_m}$$

$$= \omega_c t + \frac{kV_m f_c \sin \omega_m t}{f_m}$$

$$= \omega_c t + \frac{\delta}{f_m} \sin \omega_m t \qquad \textbf{(5-6)}$$

The derivation utilized, in turn, the fact that ω_c is constant, the formula $\int \cos nx\, dx = (\sin nx)/n$ and Eq. (5-4), which had shown that $kV_m f_c = \delta$. Equation (5-6) may now be substituted into (5-5) to give the instantaneous value of the FM voltage; thus

$$v = A \sin\left(\omega_c t + \frac{\delta}{f_m} \sin \omega_m t\right) \qquad \textbf{(5-7)}$$

The modulation index for FM, m_f, is defined as

$$m_f = \frac{\text{(maximum) frequency deviation}}{\text{modulating frequency}} = \frac{\delta}{f_m} \qquad \textbf{(5-8)}$$

Substituting Eq. (5-8) into (5-7), we obtain

$$v = A \sin\,(\omega_c t + m_f \sin \omega_m t) \qquad \textbf{(5-9)}$$

It is important to note that as the modulating frequency decreases and the modulating voltage amplitude (that is, δ) remains constant, the modulation index increases. This will be the basis for distinguishing frequency modulation from phase modulation. Note also that m_f, which is the ratio of two frequencies, is measured in radians.

Example 5-1 In an FM system, when the audio frequency (AF) is 500 Hz and the AF voltage is 2.4 V, the deviation is 4.8 kHz. If the AF voltage is now increased to 7.2 V, what is the new deviation? If the AF voltage is raised to 10 V while the AF is dropped to 200 Hz, what is the deviation? Find the modulation index in each case.

As $\delta \propto V_m$, we may write

$$\frac{\delta}{V} = \frac{4.8}{2.4} = 2 \text{ kHz/V} \qquad \text{(of modulating signal)}$$

Thus, when $V_m = 7.2$ V,

$$\delta = 2 \times 7.2 = 14.4 \text{ kHz}$$

Similarly, when $V_m = 10$V,

$$\delta = 2 \times 10 = 20 \text{ kHz}$$

Note that the change in modulating frequency made no difference to the deviation since it is independent of the modulating frequency. Calculation of the modulation indices gives

$$m_{f1} = \frac{\delta_1}{f_{m1}} = \frac{4.8}{0.5} = 9.6$$

$$m_{f2} = \frac{\delta_2}{f_{m1}} = \frac{14.4}{0.5} = 28.8$$

$$m_{f3} = \frac{\delta_3}{f_{m2}} = \frac{20}{0.2} = 100$$

The modulating frequency change did have to be taken into account in the modulation index calculation.

Example 5-2 Find the carrier and modulating frequencies, the modulation index, and the maximum deviation of the FM wave represented by the voltage equation $v = 12 \sin (6 \times 10^8 t + 5 \sin 1250t)$. What power will this FM wave dissipate in a 10-Ω resistor?

Comparing the numerical equation with Eq. (5-9), we have

$$f_c = \frac{6 \times 10^8}{2\pi} = 95.5 \text{ MHz} \qquad f_m = \frac{1250}{2\pi} = 199 \text{ Hz}$$

$$m_f = 5 \qquad \text{(as given)}$$

$$\delta = m_f f_m = 5 \times 199 = 995 \text{ Hz}$$

$$P = \frac{V_{\text{rms}}^2}{R} = \frac{(12/\sqrt{2})^2}{10} = \frac{72}{10} = 7.2 \ W$$

5-1.3 Frequency Spectrum of the FM Wave

When a comparable stage was reached with AM theory, i.e., when Eq. (3-7) had been derived, it was possible to tell at a glance what frequencies were present in the modulated wave. Unfortunately, the situation is far more complex, mathematically speaking, for FM. Since Eq. (5-9) is the sine of a sine, the only solution involves the use of *Bessel functions* [1]. Using these, it may then be shown [2] that Eq. (5-9) may be expanded to yield

$$
\begin{aligned}
v = A\{&J_0(m_f)\sin\omega_c t \\
&+ J_1(m_f)\,[\sin(\omega_c+\omega_m)t - \sin(\omega_c-\omega_m)t] \\
&+ J_2(m_f)\,[\sin(\omega_c+2\omega_m)t + \sin(\omega_c-2\omega_m)t] \\
&+ J_3(m_f)\,[\sin(\omega_c+3\omega_m)t - \sin(\omega_c-3\omega_m)t] \\
&+ J_4(m_f)\,[\sin(\omega_c+4\omega_m)t + \sin(\omega_c-4\omega_m)t]\cdots\}
\end{aligned}
\tag{5-10}
$$

It is seen that the output consists of a carrier and an apparently infinite number of pairs of sidebands, each preceded by J coefficients. These are Bessel functions. Here they happen to be of the first kind and of the order denoted by the subscript, with the argument m_f. $J_n(m_f)$ may be shown to be a solution of an equation of the form

$$
(m_f)^2\frac{d^2y}{dm_f^2} + m_f\frac{dy}{dm_f} + (m_f^2 - n^2)y = 0 \tag{5-11}
$$

This solution, i.e., the formula for the Bessel function, is

$$
J_n(m_f) = \left(\frac{m_f}{2}\right)^n\left[\frac{1}{n!} - \frac{(m_f/2)^2}{1!\,(n+1)!} + \frac{(m_f/2)^4}{2!(n+2)!} - \frac{(m_f/2)^6}{3!(n+1)!} + \cdots\right] \tag{5-12}
$$

In order to evaluate the value of a given pair of sidebands or, for that matter, the value of the carrier, it is necessary to know the value of the corresponding Bessel function. However, separate calculation from Eq. (5-12) for each case is not required since information of this type is freely available in table form, as in Table 5-1, or graphical form, as in Fig. 5-3.

Observations The mathematics of the foregoing discussion may be reviewed in a series of observations as follows:

TABLE 5-1 Bessel Functions of the First Kind*

x	*n* or Order																
(m_f)	J_0	J_1	J_2	J_3	J_4	J_5	J_6	J_7	J_8	J_9	J_{10}	J_{11}	J_{12}	J_{13}	J_{14}	J_{15}	J_{16}
0.00	1.00	—	—	—	—	—	—	—	—	—	—	—	—	—	—	—	—
0.25	0.98	0.12	—	—	—	—	—	—	—	—	—	—	—	—	—	—	—
0.5	0.94	0.24	0.03	—	—	—	—	—	—	—	—	—	—	—	—	—	—
1.0	0.77	0.44	0.11	0.02	—	—	—	—	—	—	—	—	—	—	—	—	—
1.5	0.51	0.56	0.23	0.06	0.01	—	—	—	—	—	—	—	—	—	—	—	—
2.0	0.22	0.58	0.35	0.13	0.03	—	—	—	—	—	—	—	—	—	—	—	—
2.5	−0.05	0.50	0.45	0.22	0.07	0.02	—	—	—	—	—	—	—	—	—	—	—
3.0	−0.26	0.34	0.49	0.31	0.13	0.04	0.01	—	—	—	—	—	—	—	—	—	—
4.0	−0.40	−0.07	0.36	0.43	0.28	0.13	0.05	0.02	—	—	—	—	—	—	—	—	—
5.0	−0.18	−0.33	0.05	0.36	0.39	0.26	0.13	0.05	0.02	—	—	—	—	—	—	—	—
6.0	0.15	−0.28	−0.24	0.11	0.36	0.36	0.25	0.13	0.06	0.02	—	—	—	—	—	—	—
7.0	0.30	0.00	−0.30	−0.17	0.16	0.35	0.34	0.23	0.13	0.06	0.02	—	—	—	—	—	—
8.0	0.17	0.23	−0.11	−0.29	−0.10	0.19	0.34	0.32	0.22	0.13	0.06	0.03	—	—	—	—	—
9.0	−0.09	0.24	0.14	−0.18	−0.27	−0.06	0.20	0.33	0.30	0.21	0.12	0.06	0.03	0.01	—	—	—
10.0	−0.25	0.04	0.25	0.06	−0.22	−0.23	−0.01	0.22	0.31	0.29	0.20	0.12	0.06	0.03	0.01	—	—
12.0	0.05	−0.22	−0.08	0.20	0.18	−0.07	−0.24	−0.17	0.05	0.23	0.30	0.27	0.20	0.12	0.07	0.03	0.01
15.0	−0.01	0.21	0.04	−0.19	−0.12	0.13	0.21	0.03	−0.17	−0.22	−0.09	0.10	0.24	0.28	0.25	0.18	0.12

*From E. Cambi, "Bessel Functions," Dover Publications, Inc., New York, 1948. Reprinted through permission of the publisher.

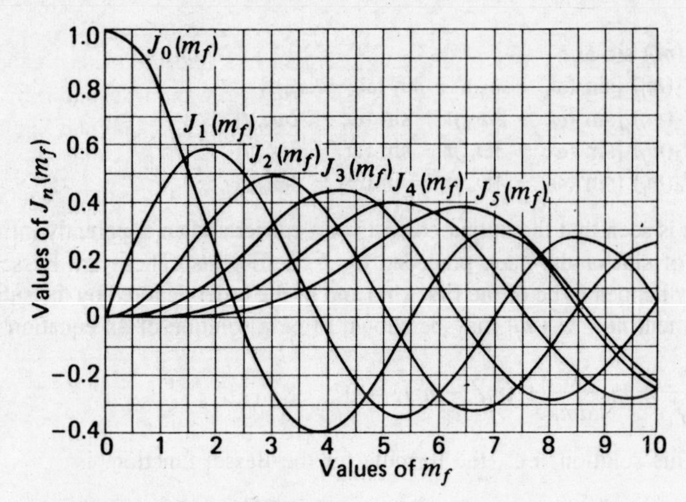

FIGURE 5-3 Bessel functions.

1. Unlike AM, where there are only three frequencies (the carrier and the first two sidebands), *FM has an infinite number of sidebands,* as well as the carrier. They are separated from the carrier by f_m, $2f_m$, $3f_m$, . . . , and thus have a recurrence frequency of f_m.
2. The J coefficients eventually decrease in value as n increases, but not in any simple manner. As seen in Fig. 5-3, the value fluctuates on either side of zero, gradually diminishing. Since each J coefficient represents the amplitude of a particular pair of sidebands, these also eventually decrease, but only past a certain value of n. *The modulation index thus determines how many sideband components have significant amplitudes.*
3. The sidebands at equal distances from f_c have equal amplitudes, so that the sideband distribution is symmetrical about the carrier frequency. The J coefficients occasionally have negative values, signifying a 180° phase change for that particular pair of sidebands.
4. Looking down Table 5-1, we see that, as m_f increases, so does the value of a particular J coefficient, such as (say) J_{12}. Bearing in mind that m_f is inversely proportional to the modulating frequency, we see that the relative amplitude of distant sidebands increases when the modulation frequency is lowered. The foregoing assumes that deviation (i.e., the modulating voltage) has remained constant.
5. In AM, increased depth of modulation increases the sideband power and therefore the total transmitted power. In FM, the total transmitted power always remains constant, but with increased depth of modulation the required bandwidth is increased. To be quite specific, what increases is the bandwidth required to transmit a relatively undistorted signal. This is true because increased depth of modulation means increased deviation, and therefore an increased modulation index, so that more distant sidebands acquire significant amplitudes.
6. As evidenced by Eq. (5-10), the theoretical bandwidth required in FM is infinite. In practice, the bandwidth used is one that has been calculated to allow for all

significant amplitudes of sideband components under the most exacting conditions. This really means ensuring that, with maximum deviation by the highest modulating frequency, no significant sideband components are lopped off.

7. In FM, unlike in AM, the amplitude of the carrier component does not remain constant. Its J coefficient is J_0, which is, of course, a function of m_f. In fact, it is quite logical and necessary that this should be so. Since the overall amplitude of the FM wave remains constant, it would be very odd indeed if the amplitude of the carrier were not reduced when the amplitude of the various sidebands is increased, and vice versa.
8. It is possible for the carrier component of the FM wave to disappear completely. This happens for certain values of the modulation index, called *eigenvalues*. Figure 5-3 shows that these are approximately 2.4, 5.5, 8.6, 11.8, and so on. These disappearances of the carrier for specific values of m_f form a handy basis for measuring deviation, as will be seen.

Bandwidth and required spectra Using Table 5-1, it is possible to evaluate the size of the carrier and each sideband for each specific or interesting value of the modulation index. When this is done, the frequency spectrum of the FM wave for that particular value of m_f may be plotted. This is done in Fig. 5-4, which shows these spectrograms first for increasing deviation (f_m constant), and then for decreasing modulating frequency (δ constant). Both the table and the spectrograms illustrate the observations, especially points 2, 3, 4, and 5. It is seen, for example, that as modulation depth increases, so does bandwidth (Fig. 5-4*a*), and also that reduction in modulation frequency increases the number of sidebands, though not necessarily the bandwidth (Fig. 5-4*b*). Another point shown very clearly is that although the number of sideband components is theoretically infinite, in practice a lot of the higher sidebands have insignificant relative amplitudes, and this is why they are not shown in the spectrograms. Thus their exclusion in a practical system will not distort the modulated wave unduly.

In order to calculate the required bandwidth accurately, one need merely glance at the table to see which is the last J coefficient shown for that value of modulation index.

Example 5-3 What is the bandwidth required for an FM signal in which the modulating frequency is 2 kHz and the maximum deviation is 10 kHz?

$$m_f = \frac{\delta}{f_m} = \frac{10}{2} = 5$$

From Table 5-1, it is seen that the highest J coefficient included for this value of m_f is J_8. This means that all higher values of Bessel functions for that modulation index have values less than 0.01 and may therefore be ignored. Thus the eighth pair of sidebands is the furthest from the carrier to be included in this instance This gives

$$\Delta = f_m \times \text{highest needed sideband} \times 2$$
$$= 2\text{ kHz} \times 8 \times 2 = 32\text{kHz}$$

A rule of thumb (Carson's rule) states that (as a good approximation) the bandwidth required to pass an FM wave is twice the sum of the deviation and the highest

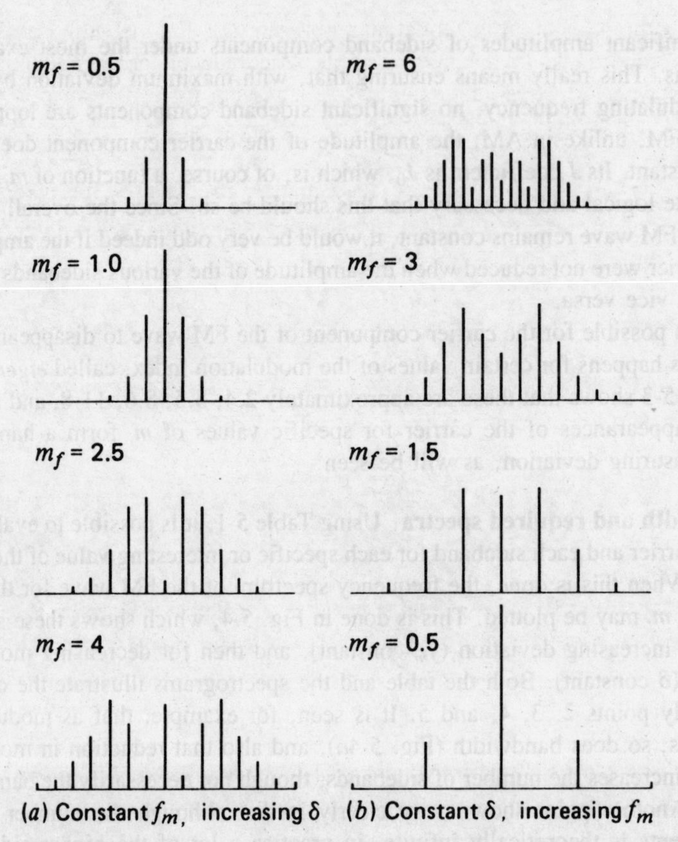

FIGURE 5-4 FM spectrograms. ***(After K. R. Sturley, Frequency-Modulated Radio, 2d ed., George Newnes Ltd., London, 1958, by permission of the publisher.)***

modulating frequency, but it must be remembered that this is only an approximation. Actually, it does give a fairly accurate result if the modulation index is in excess of about 6.

5-1.4 Phase Modulation

Strictly speaking, there are two types of continuous-wave[1] modulation: amplitude modulation and angle modulation. Angle modulation may be subdivided into two distinct types: frequency modulation and phase modulation (PM). Thus, PM and FM are closely allied, and this is the first reason for considering PM here. The second reason is somewhat more practical: it is possible to obtain frequency modulation from phase modulation by the so-called *Armstrong system*. It must be stressed, however, that phase modulation as such is not used in practical analog transmission systems.

[1] Also called "analog" (as opposed to "digital") modulation.

If the phase ϕ in the equation $v = A \sin(\omega_c t + \phi)$ is varied so that its magnitude is proportional to the instantaneous amplitude of the modulating voltage, the resulting wave is phase-modulated. The expression for a PM wave is

$$v = A \sin(\omega_c t + \phi_m \sin \omega_m t) \tag{5-13}$$

where ϕ_m is the maximum value of phase change introduced by this particular modulating signal and is proportional to the maximum amplitude of this modulation. For the sake of uniformity, this is rewritten as

$$v = A \sin(\omega_c t + m_p \sin \omega_m t) \tag{5-14}$$

where $m_p = \phi_m$ = modulation index for phase modulation.

To visualize phase modulation, consider a horizontal metronome or pendulum placed on a rotating record turntable. As well as rotating, the arm of this metronome is swinging sinusoidally back and forth about its mean point. If the maximum displacement of this swing can be made proportional to the size of the "push" applied to the metronome, and if the frequency of swing can be made equal to the number of "pushes" per second, then the motion of the arm is *exactly* the same as that of a phase-modulated vector. Actually, PM seems easier to visualize than FM.

Equation (5-13) was obtained directly, without recourse to the derivation required for the corresponding expression for FM, Eq. (5-9). This occurs because in FM an equation for the *angular velocity* was postulated, from which the phase angle for $v = A \sin(\theta)$ had to be derived, whereas in PM the phase relationship is defined and may be substituted directly. Comparison of Eqs. (5-14) and (5-9) shows them to be identical, except for the different definitions of the modulation index. It is obvious, therefore, that these two forms of angle modulation are indeed similar. They will now be compared and contrasted.

5-1.5 Intersystem Comparisons

Frequency and phase modulation From the purely theoretical point of view, the difference between FM and PM is quite simple—the modulation index is defined differently in each system. However, this is not nearly as obvious as the difference between AM and FM, and it must be developed further. First, however, the similarity will be stressed.

In phase modulation, the phase deviation is proportional to the amplitude of the modulating signal and therefore independent of its frequency. Also, since the phase-modulated vector sometimes leads and sometimes lags the reference carrier vector, its instantaneous angular velocity must be continually changing between the limits imposed by ϕ_m; thus some form of frequency change must be taking place. In frequency modulation, the frequency deviation is proportional to the amplitude of the modulating voltage. Also, if we take a reference vector, rotating with a constant angular velocity which corresponds to the carrier frequency, then the FM vector will have a phase lead or lag with respect to the reference, since its frequency oscillates between $f_c - \delta$ and $f_c + \delta$. Hence FM must be a form of PM. With this close similarity of the two forms of angle modulation established, it now remains to explain the difference.

If we consider FM as a form of phase modulation, we must determine what it is that causes the phase change in FM. Obviously, the larger the frequency deviation, the larger the phase deviation, so that the latter depends at least to a certain extent on the amplitude of the modulation, just as PM. The difference is shown by comparing the definition of PM, which states in part that the modulation index is proportional to the modulating voltage *only*, with that of the FM, which states that the modulation index *is also inversely proportional to the modulation frequency*. This means that under identical conditions *FM and PM are indistinguishable for a single modulating frequency*. When the modulating frequency is changed, however, the PM modulation index will remain constant, whereas the FM modulation index will increase as modulation frequency is reduced, and vice versa. This is best illustrated with an example.

Example 5-4 A 25-MHz carrier is modulated by a 400-Hz audio sine wave. If the carrier voltage is 4 V and the maximum deviation is 10 kHz, write the equation of this modulated wave for *(a)* FM and *(b)* PM. If the modulating frequency is now changed to 2 kHz, all else remaining constant, write a new equation for *(c)* FM and *(d)* PM.

Calculating the frequencies in radians, we have

$$\omega_c = 2\pi \times 25 \times 10^6 = 1.57 \times 10^8 \text{ rad/s}$$

$$\omega_m = 2\pi \times 400 = 2513 \text{ rad/s}$$

The modulation index will be

$$m = m_f = m_p = \frac{\delta}{f_m} = \frac{10{,}000}{400} = 25$$

This yields the equations

$$(a)\ v = 4 \sin (1.57 \times 10^8 t + 25 \sin 2513t) \text{ (FM)}$$

$$(b)\ v = 4 \sin (1.57 \times 10^8 t + 25 \sin 2513t) \text{ (PM)}$$

Note that the two expressions are identical, as should have been anticipated. Now, when the modulating frequency is multiplied by 5, the equation will show a fivefold increase in the (angular) modulating frequency. However, while the modulation index in FM is reduced fivefold, for PM the modulation index remains constant. Hence

$$(c)\ v = 4 \sin (1.57 \times 10^8 t + 5 \sin 12{,}565t) \text{ (FM)}$$

$$(d)\ v = 4 \sin (1.57 \times 10^8 t + 25 \sin 12{,}565t) \text{ (PM)}$$

Note, once again, that the difference between FM and PM is not apparent at a single modulating frequency. Rather, it reveals itself in the *differing behavior of the two systems when the modulating frequency is varied*.

The practical effect of all these considerations is that if an FM transmission were received on a PM receiver, the bass frequencies would have considerably more deviation (*of phase*) than a PM transmitter would have given them. Since the output of a PM receiver would be proportional to phase deviation (or modulation index, which is the same thing), the signal would appear unduly bass-boosted. Alternatively (and this is far more of a practical situation), PM received by an FM system would appear to be *lacking in bass*. This deficiency could, of course, be *corrected by bass boosting the modulating signal prior to phase modulation*. This is the practical difference between phase and frequency modulation, but it is seen quite clearly that *one can be obtained from the other very simply*.

Frequency and amplitude modulation Frequency and amplitude modulation are compared on a different basis from that for FM and PM. These are both practical systems, quite different from each other, and so the performance and characteristics of the two systems will be compared. To begin with, frequency modulation has the following advantages:

1. The amplitude of the frequency-modulated wave is constant. It is thus independent of the modulation depth, whereas in AM modulation depth governs the transmitted power. This means that, in FM transmitters, low-level modulation may be used but all the subsequent amplifiers can be class C and therefore more efficient. Since all these amplifiers will handle constant power, they need not be capable of managing up to four times the average power, as they must in AM. Finally, *all* the transmitted power in FM is useful, whereas in AM most of it is in the transmitted carrier, which does not indicate any modulation changes.
2. FM receivers can be fitted with amplitude limiters to remove the amplitude variations caused by noise, as shown in Sec. 5-2.2; this makes FM reception a good deal more immune to noise than AM reception.
3. It is possible to reduce noise still further by increasing the deviation (see Sec. 5-2.2). This is a feature which AM does not have, since it is not possible to exceed 100 percent modulation without causing severe distortion.
4. Commercial FM broadcasts began in 1940, decades after their AM counterparts. Consequently, they have a number of advantages due to better planning and other circumstances. The following are the most important ones:
   a. Standard[2] frequency allocations provide a guard band between commercial FM stations, so that there is less adjacent-channel interference than in AM;
   b. FM broadcasts operate in the upper VHF and UHF frequency ranges, at which there happens to be less noise than in the MF and HF ranges occupied by AM broadcasts;
   c. At the FM broadcast frequencies, the *space wave* is used for propagation, so that the radius of operation is limited to slightly more than line of sight, as shown in Sec. 8-2.3. It is thus possible to operate several independent transmitters on the same frequency with considerably less interference than would be possible with AM. Reference [3] quotes a cochannel interference reduction factor of 30:1.

The advantages are not all one-sided, or there would be no AM transmissions left. The following are some of the disadvantages of FM:

1. A much wider channel is required by FM, up to 10 times as large as that needed by AM. This is the most significant disadvantage of FM.
2. FM transmitting and receiving equipment tends to be more complex, particularly for modulation and demodulation.
3. Since reception is limited to line of sight, the area of reception for FM is much smaller than for AM. This may be an advantage for cochannel allocations, but it

[2] Allocated worldwide by the International Radio Consultative Committee (CCIR) of the I.T.U.

is a disadvantage for FM mobile communications over a wide area. Note that this is due not so much to the intrinsic properties of FM, but rather to the frequencies employed for its transmission.

5-2 NOISE AND FREQUENCY MODULATION

Frequency modulation is much more immune to noise than amplitude modulation and is significantly more immune than phase modulation. In order to establish the reason for this and to determine the extent of the improvement, it is necessary to examine the effect of noise on a carrier.

5-2.1 Effects of Noise on Carrier—Noise Triangle

A single-noise frequency will affect the output of a receiver only if it falls within its passband; the carrier and noise voltages will mix, and if the difference is audible, it will naturally interfere with the reception of wanted signals. If such a single-noise voltage is considered vectorially, it is seen that the noise vector is superimposed on the carrier, rotating about it with a relative angular velocity $\omega_n - \omega_c$. This is shown in Fig. 5-5. The maximum deviation in amplitude from the average value will be V_n, whereas the maximum phase deviation will be $\phi = \sin^{-1} (V_n / V_c)$.

Let the noise voltage amplitude be one-quarter of the carrier voltage amplitude. Then the modulation index for this amplitude modulation by noise will be $m = V_n / V_c = 0.25/1 = 0.25$, and the maximum phase deviation will be $\phi = \sin^{-1} 0.25/1 = 14.5°$. For voice communication, an AM receiver will not be affected by the phase change; whereas the FM receiver will not be bothered by the amplitude change, which can be removed with an amplitude limiter, as will be seen in Chap. 6. It now remains to discover whether or not the phase change affects the FM receiver more than the amplitude change affects the AM receiver.

The comparison will initially be made under conditions that will prove to be the worst case for FM. Consider that the modulating frequency (by a proper signal, this time) is 15 kHz, and, for convenience, the modulation index for both AM and FM is unity. Under such conditions the relative noise-to-signal ratio in the AM receiver will be $0.25/1 = 0.25$. For FM, we first convert the unity modulation index from radians to degrees (1 rad = 57.3°) and then calculate the noise-to-signal ratio. Here the ratio is $14.5°/57.3° = 0.253$, just slightly worse than in the AM case.

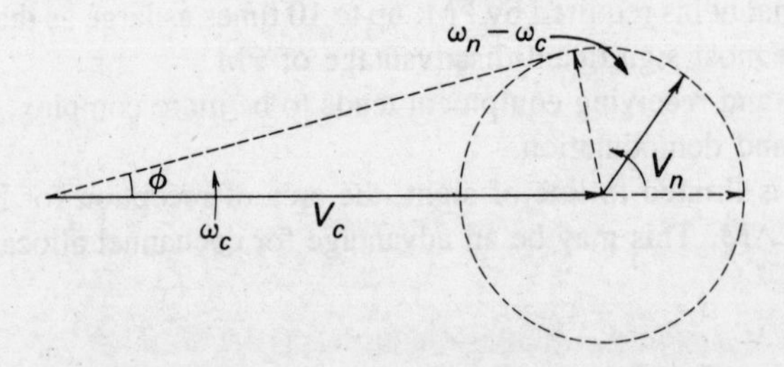

FIGURE 5-5 Vector effect of noise on carrier.

The effects of noise frequency change must now be considered. In AM, there is no difference in the relative noise, carrier, and modulating voltage amplitudes, when both the noise difference and modulating frequencies are reduced from 15 kHz to the normal minimum audio frequency of 30 Hz (in high-quality broadcast systems). That is to say, changes in the noise and modulating frequency do not affect the noise-to-signal ratio in AM. In FM, however, the picture is entirely different. As the ratio of noise to carrier voltage remains constant, so does the value of the modulation index (i.e., maximum phase deviation) due to noise. It is thus seen that *the noise voltage phase-modulates the carrier*. So, while the modulation index due to noise remains constant (as the noise sideband frequency is reduced), the modulation index caused by the signal will go on increasing in proportion to the reduction in frequency, as already explained. The noise-to-signal ratio in FM goes on reducing with frequency, therefore, until it reaches its lowest value when both signal and noise have an audio output frequency of 30 Hz. At this point the noise-to-signal ratio is $0.253 \times 30/15{,}000 = 0.000505$, a reduction from 25.3 percent at 15 kHz to 0.05 percent at 30 Hz.

Assuming noise frequencies to be evenly spread across the bandpass of the receiver, we can see that noise output from the receiver decreases uniformly with noise sideband frequency for FM. In AM it remains constant; the situation is illustrated in Fig. 5-6*a*. The triangular noise distribution for FM is called the *noise triangle;* the corresponding AM distribution, also shown, is of course a rectangle. It might be supposed from the figure that the average voltage improvement for FM under these conditions would be 2:1. Such a supposition might be made by considering the average audio frequency, at which FM noise appears to be relatively half the size of the AM noise. However, the picture is more complex, and in fact the FM improvement is only $\sqrt{3}$:1 as a voltage ratio. This is, nonetheless, a worthwhile improvement—it represents an increase of 3:1 in the (power) signal-to-noise ratio for FM compared with AM. Such a 4.75-dB improvement is certainly worth having.

It will be noted that this discussion began with noise voltage that was definitely lower than the signal voltage. This was done on purpose. The amplitude limiter previously mentioned is a device that is actuated by the stronger signal and tends to reject the weaker signal, if two simultaneous signals are received. That being the case, if peak noise voltages exceeded signal voltages, the signal would be excluded by the limiter. Under conditions of *very low* signal-to-noise ratio, therefore, AM is the supe-

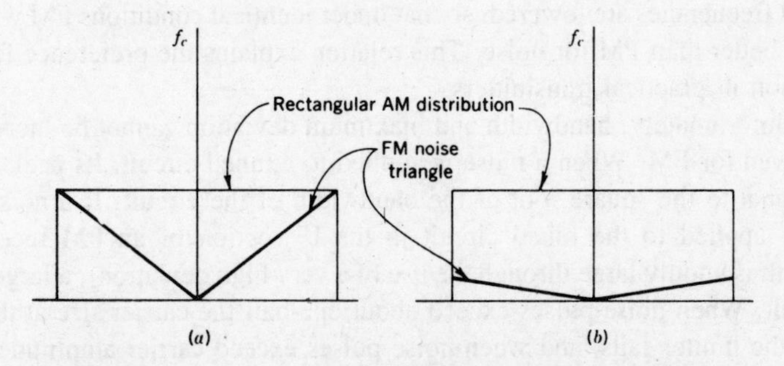

FIGURE 5-6 Noise sideband distribution (noise triangle). (*a*) $m_f = 1$ at the maximum frequency; (*b*) $m_f = 5$ at the maximum frequency.

rior system. The precise value of signal-to-noise ratio at which this becomes apparent depends on the value of the FM modulation index. As a general guide, it may be said that FM becomes superior to AM at the signal-to-noise ratio level used in the example (voltage ratio = 4, power ratio = 16 = 12 dB) at the amplitude limiter input.

A number of other considerations must now be taken into account. The first of these is that $m = 1$ is the maximum permissible modulation index for AM, whereas in FM there is no such limit. It is the maximum frequency deviation that is limited in FM, to 75 kHz in the wideband VHF broadcasting service. Thus, even at the highest audio frequency of 15 kHz, the modulation index in FM is permitted to be as high as 5. It may of course be much higher than that at lower audio frequencies: for example, 75 when the modulating frequency is 1 kHz. If a given ratio of signal voltage to noise voltage exists at the output of the FM amplitude limiter when $m = 1$, this ratio will be reduced in proportion to an increase in modulation index. Thus, when m is made equal to 2, the ratio of signal voltage to noise voltage at the limiter output in the receiver will be doubled; it will be tripled when $m = 3$, and so on. This ratio is thus proportional to the modulation index, and so the signal-to-noise (power) ratio in the output of an FM receiver is proportional to the *square* of the modulation index. Hence, when $m = 5$ (highest permitted when $f_m = 15$ kHz), there will be a 25:1 (14-dB) improvement for FM, whereas no such improvement for AM is possible. Assuming an adequate initial signal-to-noise ratio at the receiver input (as mentioned earlier), an overall improvement of 18.75 dB at the receiver output is shown at this point by wideband FM compared with AM. Figure 5-6*b* shows the relationship when $m = 5$ is used at the highest frequency.

This leads us to the second consideration, that FM has properties which permit the trading of bandwidth for signal-to-noise ratio, which cannot be done in AM. In connection with this, one fear should be allayed. Just because the deviation (and consequently the system bandwidth) is increased in an FM system, this does not necessarily mean that more *random* noise will be admitted. This extra random noise has no effect if the noise sideband frequencies lie outside the passband of the receiver. From this particular point of view, maximum deviation (and hence bandwidth) may be increased without fear.

Phase modulation also has this property and, in fact, all the noise-immunity properties of FM except the noise triangle. Since noise phase-modulates the carrier (like the signal), there will naturally be no improvement as modulating and noise sideband frequencies are lowered, so that under identical conditions FM will always be 4.75 dB better than PM for noise. This relation explains the preference for frequency modulation in practical transmitters.

Unfortunately, bandwidth and maximum deviation cannot be increased indefinitely, even for FM. When a pulse is applied to a tuned circuit, its peak amplitude is proportional to the square root of the bandwidth of the circuit. If a noise *impulse* is similarly applied to the tuned circuit in the IF section of an FM receiver (whose bandwidth is unduly large through the use of a very high deviation), a large noise pulse will result. When noise pulses exceed about one-half the carrier size at the amplitude limiter, the limiter fails; and when noise pulses exceed carrier amplitude, the limiter goes one better and limits the signal, having been "captured" by noise, as will be seen. The normal maximum deviation permitted, 75 kHz, is a compromise between the two effects described.

It may be shown that under ordinary circumstances (that is, $2V_n < V_c$) impulse noise is reduced in FM to the same extent as random noise. The amplitude limiter found in AM communications receivers does not limit random noise at all, and it limits impulse noise by only about 10 dB; thus FM is better off in this regard also.

5-2.2 Pre-emphasis and De-emphasis

The noise triangle showed that noise has a greater effect on the higher modulating frequencies than on the lower ones. Thus, if the higher frequencies were artificially boosted at the transmitter and correspondingly cut at the receiver, an improvement in noise immunity could be expected. This boosting of the higher modulating frequencies, in accordance with a prearranged curve, is termed *pre-emphasis,* and the compensation at the receiver is called *de-emphasis*. An example of a circuit used for each function is shown in Fig. 5-7.

Take two modulating signals having the same initial amplitude, with one of them pre-emphasized to (say) twice this amplitude, whereas the other is unaffected (being at a much lower frequency). The receiver will naturally have to de-emphasize the first signal by a factor of 2, to ensure that both signals have the same amplitude in the output of the receiver. Before demodulation, i.e., while susceptible to noise interference, the emphasized signal had twice the deviation it would have had without pre-emphasis and was thus more immune to noise. Alternatively, it is seen that when this signal is de-emphasized, any noise sideband voltages are de-emphasized with it and therefore have a correspondingly lower amplitude than they would have had without emphasis; again, their effect on the output is reduced.

The amount of pre-emphasis in U.S. FM broadcasting, and in the sound transmissions accompanying television, has been standardized at 75 μs, whereas a number of other services, notably European and Australian broadcasting and TV sound transmission, use 50 μs. The usage of microseconds for defining emphasis is standard. A 75-μs de-emphasis corresponds to a frequency response curve that is 3 dB down at the frequency whose time constant RC is 75 μs. This frequency is given by $f = 1/2\pi RC$ and is therefore 2120 Hz; with 50-μs de-emphasis it would be 3180 Hz. Figure 5-8 shows pre-emphasis and de-emphasis curves for a 75-μs emphasis, as used in the United States.

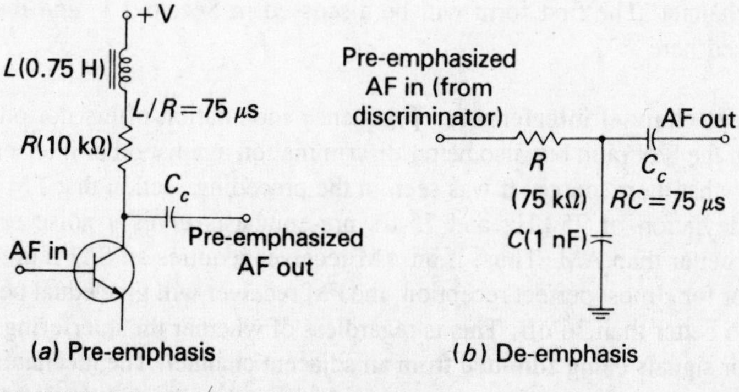

+V
L(0.75 H)
L/R = 75 μs
R(10 kΩ)
C_c
AF in
Pre-emphasized AF out
(a) Pre-emphasis

Pre-emphasized AF in (from discriminator)
AF out
R
C_c
(75 kΩ) RC = 75 μs
C(1 nF)
(b) De-emphasis

FIGURE 5-7 75-μs emphasis circuits.

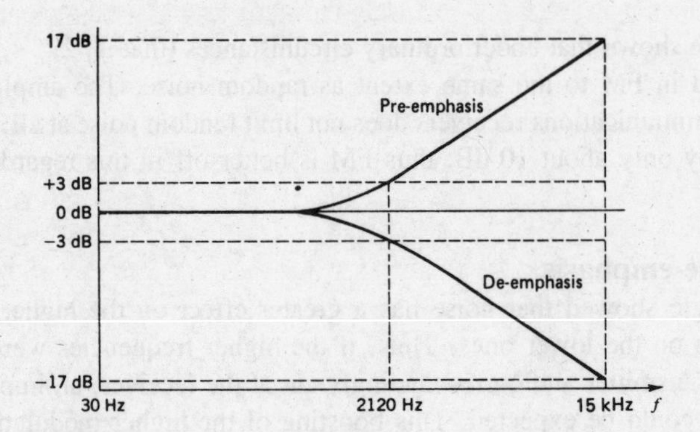

FIGURE 5-8 75-μs emphasis curves.

It is a little more difficult to estimate the benefits of emphasis than it is to evaluate the other FM advantages, but subjective BBC tests with 50 μs give a figure of about 4.5 dB; American tests have shown an even higher figure with 75 μs. However, there is a danger that must be considered; the higher modulating frequencies must not be overemphasized. The curves of Fig. 5-8 show that a 15-kHz signal is pre-emphasized by about 17 dB; with 50 μs this figure would have been 12.6 dB. It must be made certain that when such boosting is applied, the resulting signal cannot overmodulate the carrier by exceeding the 75-kHz deviation, since distortion will be introduced. It is thus seen that a limit for pre-emphasis exists, and any practical value used is always a compromise between protection for high modulating frequencies on the one hand and the risk of overmodulation on the other.

If emphasis were applied to amplitude modulation, some improvement would also result, but it is not as great as in FM because the highest modulating frequencies in AM are no more affected by noise than any others. Apart from that, it would be difficult to introduce pre-emphasis and de-emphasis in existing AM services since extensive modifications would be needed, particularly in view of the huge numbers of receivers in use.

5-2.3 Other Forms of Interference

In addition to noise, other forms of interference found in radio receivers include the image frequency, transmitters operating on an adjacent channel and those using the same channel. The first form will be discussed in Sec. 6-2.1, and the other two are discussed here.

Adjacent-channel interference Frequency modulation offers not only an improvement in the S/N ratio but also better discrimination against other interfering signals, no matter what their source. It was seen in the preceding section that FM having a maximum deviation of 75 kHz and 75-μs pre-emphasis gives a noise rejection at least 24 dB better than AM. Thus, if an AM receiver requires an S/N ratio of 60 dB at the detector for almost perfect reception, the FM receiver will give equal performance for a ratio no better than 36 dB. This is regardless of whether the interfering signal is due to noise or signals being admitted from an adjacent channel. The mechanism whereby the

FM limiter reduces interference is precisely the same as that used to deal with random noise; its discussion here would be mere repetition.

One more factor should be included in this discussion of adjacent-channel interference. When FM broadcasting systems began, AM systems had been in operation for nearly 30 years, a lot of experience with broadcasting systems had been obtained, and planners could profit from earlier mistakes. Thus, as already mentioned, each wideband FM broadcasting channel occupies 200 kHz (of which only 180 kHz is used), and the remaining 20-kHz guard band goes a long way toward reducing adjacent-channel interference even further.

Cochannel interference—capture effect The amplitude limiter works on the principle of passing the stronger signal and eliminating the weaker. This was the reason for mentioning earlier that noise reduction is obtained only when the signal is at least twice the noise peak amplitude. A relatively weak interfering signal from another transmitter will also be attenuated in this manner, as much as any other form of interference. This applies even if the other transmitter operates on the same frequency as the desired transmitter.

In mobile receivers, traveling from one transmitter toward another (cochannel) one, the interesting phenomenon of *capture* occurs. However, it must first be mentioned that the effect would be very straightforward with AM transmitters. The nearer transmitter would always predominate, but the other one would be heard as quite significant interference although it might be very distant.

The situation is far more interesting with FM. Until the signal from the second transmitter is less than about half of that from the first, the second transmitter is virtually inaudible, causing practically no interference. After this point, the transmitter toward which the receiver is moving becomes quite audible as a background and eventually predominates, finally *excluding* the first transmitter. The moving receiver has been *captured* by the second transmitter. If a receiver is between the two transmitters (roughly in the center zone) and fading conditions prevail, first one signal, and then the other, will be the stronger. As a result, the receiver will be captured alternatively by either transmitter. This switching from one program to the other is most distracting, of course (once the initial novelty has worn off!), and would not happen in an AM system.

5-2.4 Comparison of Wideband and Narrowband FM

By convention, *wideband FM* has been defined as that in which the modulation index normally exceeds unity. This is the type so far discussed. Since the maximum permissible deviation is 75 kHz[3] and modulating frequencies range from 30 Hz to 15 kHz, the maximum modulation index ranges from 5 to 2500. The modulation index in narrowband FM is near unity, since the maximum modulating frequency there is usually 3 kHz, and the maximum deviation is typically 5 kHz.

[3] The maximum permissible deviation for the sound accompanying TV transmissions is 25 kHz in the United States. NTSC system and 50 kHz in the PAL system used in Europe and Australia. Both are thus wideband systems.

The proper bandwidth to use in an FM system depends on the application. With a large deviation, noise will be better suppressed (as will other interference), but care must be taken to ensure that impulse noise peaks do not become excessive. On the other hand, the wideband system will occupy up to 15 times the bandwidth of the narrowband system. These considerations have resulted in wideband systems being used in entertainment broadcasting, while narrowband systems are employed for communications.

Thus narrowband FM is used by the so-called *FM mobile* communications services. These include police, ambulances, taxicabs, radio-controlled appliance repair services, short-range VHF ship-to-shore services and the Australian ''Flying Doctor'' service. The higher audio frequencies are attenuated, as indeed they are in most carrier (long-distance) telephone systems, but the resulting speech quality is still perfectly adequate. Maximum deviations of 5 to 10 kHz are permitted, and the channel space is not much greater than for AM broadcasting, i.e., of the order of 15 to 30 kHz. Narrowband systems with even lower maximum deviations are envisaged. Pre-emphasis and de-emphasis are used, as indeed they are with all FM transmissions.

5-2.5 Stereophonic FM Multiplex System

Stereo FM transmission is a modulation system in which sufficient information is sent to the receiver to enable it to reproduce original stereo material. It became commercially available in 1961, several years after commercial monaural transmissions. Like color TV (which of course came after monochrome TV), it suffers from the disadvantage of having been made more complicated than it needed to be, to ensure that it would be compatible with the existing system. Thus, in stereo FM, it is not possible to have a two-channel system with a *left* channel and a *right* channel transmitted simultaneously and independently, because a monaural system would not receive all the information in an acceptable form.

As shown in the block diagram of Fig. 5-9, the two channels in the FM stereo multiplex system are passed through a matrix which produces two outputs. The sum $(L + R)$ modulates the carrier in the same manner as the signal in a monaural transmission, and this is the signal which is demodulated and reproduced by a mono receiver

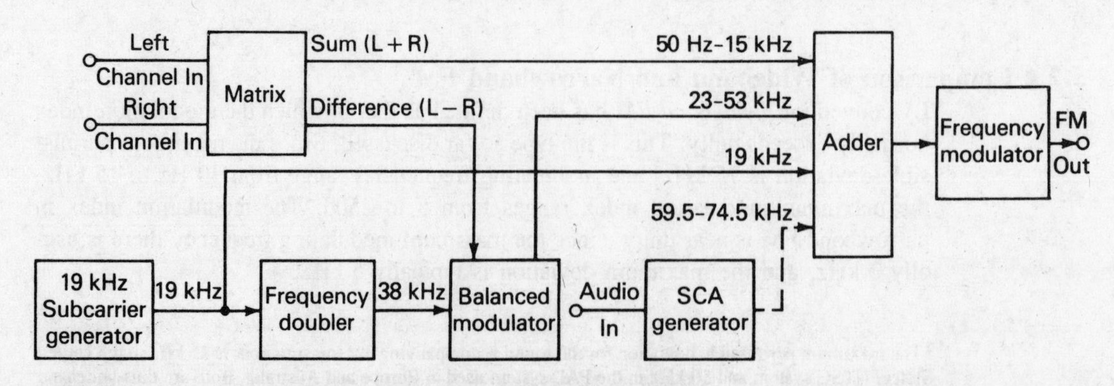

FIGURE 5-9 Stereo FM multiplex generator with optional SCA.

tuned to a stereo transmission. The other output of the matrix is the difference signal $(L - R)$. After demodulation in a stereo receiver, $(L - R)$ will be added to $(L + R)$ to produce the left channel, while the difference between the two signals will produce the right channel. This will be explained further in Chap. 6. In the meantime, however, it is necessary to understand how the difference signal is impressed on the carrier.

What happens, in essence, is that the difference signal is shifted in frequency from the 50- to 15,000-Hz range (which it would otherwise co-occupy with the sum signal) to a higher frequency. As will be seen in Chap. 15, such signal "stacking" is known as *multiplexing*, hence the name of the system. In this case, as in other multiplexing, a form of single sideband suppressed carrier (SSBSC) is used, with the signals to be multiplexed up being modulated onto a subcarrier at a high audio or supersonic frequency. However, there is a snag here, which makes this form of multiplexing different from the more common ones. The problem is that the lowest audio frequency is 50 Hz, much lower than the normal minimum of 300 Hz encountered in communications voice channels. This makes it difficult to suppress the unwanted sideband without affecting the wanted one; pilot carrier extraction in the receiver is equally difficult. And yet some form of carrier must be transmitted, to ensure that the receiver has a stable reference frequency for demodulation; otherwise, distortion of the difference signal will result.

The two problems are solved in associated but separate ways. In the first place, the difference signal is applied to a balanced modulator (as it would be in any multiplexing system) which, in orthodox fashion, suppresses the carrier. However, both sidebands are then used as modulating signals and duly transmitted, whereas normally one might expect one of them to be removed prior to transmission. Since the subcarrier frequency is 38 kHz, the sidebands produced by the difference signal occupy the frequency range from 23 to 53 kHz. It is thus seen that they do not interfere with the sum signal, which occupies the range of 50 Hz to 15 kHz.

The reason that the 38-kHz subcarrier is generated by a 19-kHz oscillator whose frequency is then doubled may now be explained. Indeed, this is the trick used to avoid the difficulty of having to extract the pilot carrier from among the close sideband frequencies in the receiver. As shown in the block diagram, the output of the 19-kHz subcarrier generator is added to the sum and difference signals in the output adder preceding the modulator. In the receiver, as will be seen in Chap. 6, the frequency of the 19-kHz signal is doubled, and it can then be reinserted as the carrier for the difference signal. It should be noted that the subcarrier is inserted at a level of 10 percent, which is both adequate and not so large as to take undue power from the sum and difference signals (or to cause overmodulation). Also, the frequency of 19 kHz fits neatly into the space between the top of the sum signal and the bottom of the difference signal—it is far enough from each of them so that no difficulty in extracting it is experienced in the receiver.

The FM stereo multiplex system described here is the one used in the United States, and is in accordance with the standards promulgated by the Federal Communications Commission (FCC) in 1961. Stereo FM has by now spread to broadcasting in most other parts of the world, where the systems in use are either identical or quite similar to the above. A Subsidiary Communications Authorization (SCA) signal may also be transmitted in the U.S. stereo multiplex system; it is the remaining signal

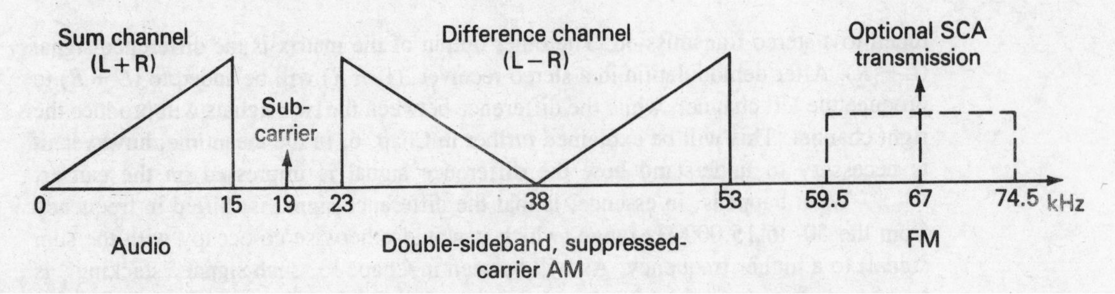

FIGURE 5-10 Spectrum of stereo FM multiplex modulating signal (with optional SCA).

feeding in to the output adder. It is shown dashed in the diagram because it is not always present. Some stations provide SCA as a second, medium-quality transmission, used as background music in stores, restaurants and the like.

SCA uses a subcarrier at 67 kHz, modulated to a depth of ±7.5 kHz by the audio signal. Frequency modulation is used, and any of the methods described in Sec. 5-3 can be employed. The frequency band thus occupied ranges from 59.5 to 74.5 kHz and fits sufficiently above the difference signal as not to interfere with it. The overall frequency allocation within the modulating signal of an FM stereo multiplex transmission with SCA is shown in Fig. 5-10. As can be appreciated, the amplitude of the sum and difference signals must be reduced (generally by 10 percent) in the presence of SCA; otherwise, overmodulation of the main carrier could result.

5-3 GENERATION OF FREQUENCY MODULATION

The prime requirement of a frequency modulation generator is a variable output frequency, with the variation proportional to the instantaneous amplitude of the modulating voltage. The subsidiary requirements are that the unmodulated frequency should be constant, and the deviation independent of the modulating frequency. However, if the system proper does not produce these characteristics, corrections can be introduced during the modulation process.

5-3.1 FM Methods

One method of FM generation suggests itself immediately; if either the capacitance or inductance of an *LC* oscillator tank is varied, frequency modulation of some form will result. If this variation can be made directly proportional to the voltage supplied by the modulation circuits, true FM will be obtained.

There are several controllable electrical and electronic phenomena which provide a variation of capacitance as a result of a voltage change; there are also some in which an inductance may be similarly varied. Generally, if such a system is used, a voltage-variable reactance is placed across the tank, and the tank is tuned so that (in the absence of modulation) the oscillating frequency is equal to the desired carrier frequency. The capacitance (or inductance) of the variable element is changed with the modulating voltage, increasing (or decreasing) as the modulating voltage increases positively, and going the other way when the modulation becomes negative. The larger the departure of the modulating voltage from zero, the larger the reactance variation and therefore the frequency variation. When the modulating voltage is zero, the variable reactance will have its average value. Thus, at the carrier frequency, the oscillator inductance is tuned by its own (fixed) capacitance in parallel with the average reactance of the variable element.

There are a number of devices whose reactance can be varied by the application of voltage. The three-terminal ones include the reactance field-effect transistor (FET), the bipolar transistor and the tube. Each of them is a normal device which has been biased so as to exhibit the desired property. By far the most common of the two-terminal devices is the varactor diode. Methods of generating FM that do not depend on varying the frequency of an oscillator will be discussed under the heading "Indirect Method." A prior generation of phase modulation is involved, as will be shown.

5-3.2 Direct Methods

Of the various methods of providing a voltage-variable reactance which can be connected across the tank circuit of an oscillator, the most common are the reactance modulator and the varactor diode. These will now be discussed in turn.

Basic reactance modulator Provided that certain simple conditions are met, the impedance z, as seen at the input terminals A–A of Fig. 5-11, is almost entirely reactive. The circuit shown is the basic circuit of a FET reactance modulator, which behaves as a three-terminal reactance[4] that may be connected across the tank circuit of the oscillator to be frequency-modulated. It can be made inductive or capacitive by a simple component change. More importantly, the value of this reactance is proportional to the transconductance of the device, which in turn can be made to depend on the gate bias and its variations. Note that an FET is used in the explanation here for simplicity only. Identical reasoning would apply to a bipolar transistor or a vacuum tube, or indeed to any other amplifying device.

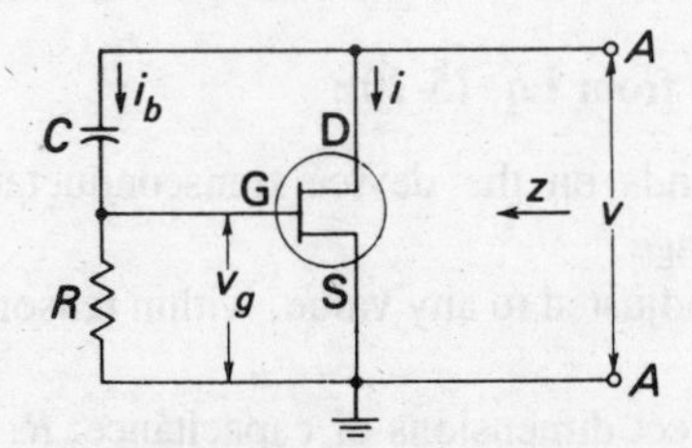

FIGURE 5-11 Basic reactance modulator.

[4] **The reactance appears between the terminals A–A, that is, between the drain and the source; its value may be controlled by a signal at the third terminal, i.e., the gate.**

Theory of reactance modulators In order to determine z, a voltage v is applied to the terminals A–A between which the impedance is to be measured, and the resulting current i is calculated. The applied voltage is then divided by this current, giving the impedance seen when looking into the terminals. In order for this impedance to be a pure reactance—it is capacitive here, as will be proved—two requirements must be fulfilled. The first is that the bias network current i_b must be negligible compared to the drain current. That is to say, the impedance of the bias network must be large enough to be ignored. The second requirement is that the drain-to-gate impedance (X_C here) must be greater than the gate-to-source impedance (R in this case), preferably by more than 5:1. The following analysis may then be applied:

$$v_g = i_b R = \frac{Rv}{R - jX_C} \tag{5-15}$$

The FET drain current is

$$i = g_m v_g = \frac{g_m Rv}{R - jX_C} \tag{5-16}$$

Therefore, the impedance seen at the terminals A–A is

$$z = \frac{v}{i} = v \div \frac{g_m Rv}{R - jX_C} = \frac{R - jX_C}{g_m R} = \frac{1}{g_m}\left(1 - \frac{jX_C}{R}\right) \tag{5-17}$$

If $X_C \gg R$ in Eq. (5-17), the equation will reduce to

$$z = -j\frac{X_C}{g_m R} \tag{5-18}$$

This impedance is quite clearly a capacitive reactance, which may, therefore, be written as

$$X_{\text{eq}} = \frac{X_C}{g_m R} = \frac{1}{2\pi f g_m RC} = \frac{1}{2\pi f C_{\text{eq}}} \tag{5-19}$$

From Eq. (5-19) it is seen that under such conditions the input impedance of the device at A–A is a pure reactance and is given by

$$C_{\text{eq}} = g_m RC \tag{5-20}$$

The following should be noted from Eq. (5-20):

1. This equivalent capacitance depends on the device transconductance and can therefore be varied with bias voltage.
2. The capacitance can be originally adjusted to any value, within reason, by varying the components R and C.
3. The expression $g_m RC$ has the correct dimensions of capacitance; R, measured in ohms, and g_m, measured in siemens (s), cancel each other's dimensions, leaving C as required.
4. It was stated earlier that the gate-to-drain impedance must be much larger than the gate-to-source impedance. This is illustrated by Eq. (5-17); if X_C/R had not been much greater than unity, z would have had a resistive component as well.

If R is not much less than X_C (in the particular reactance modulator treated), the gate voltage will no longer be exactly 90° out of phase with the applied voltage v, nor will the drain current i. Thus the input impedance will no longer be purely reactive. As shown in Eq. (5-17), the resistive component for this particular FET reactance modulator will be $1/g_m$. Since this component contains g_m, it will vary with the applied modulating voltage. This variable resistance (like the variable reactance) will appear directly across the tank circuit of the master oscillator, varying its Q and therefore its output voltage. Hence a certain amount of amplitude modulation will be created; this applies to all the forms of reactance modulator. If the situation is unavoidable, the oscillator being modulated must be followed by an amplitude limiter.

The gate-to-drain impedance is, in practice, made five to ten times the gate-to-source impedance. Let $X_C = nR$ (at the carrier frequency) in the capacitive RC reactance FET so far discussed. Then

$$X_C = \frac{1}{\omega C} = nR$$

$$C = \frac{1}{\omega nR} = \frac{1}{2\pi fnR} \tag{5-21}$$

Substituting Eq. (5-21) into (5-20) gives

$$C_{eq} = g_m RC = \frac{g_m R}{2\pi fnR}$$

$$C_{eq} = \frac{g_m}{2\pi fn} \tag{5-22}$$

Equation (5-22) is a very useful formula; in practical situations the frequency of operation and the ratio of X_C to R are the usual starting data from which other calculations are made.

Example 5-5 Determine the value of the capacity reactance obtainable from a reactance FET whose g_m is 12 millisiemens (12 mS). Assume that the gate-to-source resistance is one-ninth of the reactance of the gate-to-drain capacitor and that the frequency is 5 MHz. Thus

$$C_{eq} = \frac{g_m}{2\pi fn} \qquad \therefore \qquad 2\pi f C_{eq} = \frac{g_m}{n} = \frac{1}{X_{C_{eq}}}$$

$$X_{C_{eq}} = \frac{n}{g_m} = \frac{9}{12 \times 10^{-3}} = 750\ \Omega$$

Example 5-6 The mutual conductance of an FET varies linearly with gate voltage between the limits of 0 and 9 mS.[5] The FET is used as a capacitive reactance modulator, with $X_{C_{gd}} = 8R_{gs}$, and is placed across an oscillator circuit which is tuned to 50 MHz by a 50-pF fixed capacitor. What will be the total frequency variation when the

[5] This variation is made unrealistically large here so as to simplify the arithmetic somewhat. For the same reason, in this example and the next, the two extreme frequencies will be used rather than the center frequency and one of the extreme frequencies.

transconductance of the FET is varied from zero to maximum by the modulating voltage?

For this example and the next, let

C_n = minimum equivalent capacitance of reactance FET
C_x = maximum equivalent capacitance of reactance FET
f_n = minimum frequency
f_x = maximum frequency
f = average frequency
δ = maximum deviation

Then

$$C_n = 0$$

$$C_x = \frac{g_m}{2\pi f n} = \frac{9 \times 10^{-3}}{2\pi \times 5 \times 10^7 \times 8} = \frac{9 \times 10^{-11}}{8\pi}$$

$$= 3.58 \times 10^{-12} = 3.58 \text{ pF}$$

$$\frac{f_x}{f_n} = \frac{1/2\pi\sqrt{LC}}{1/2\pi\sqrt{L(C + C_x)}} = \sqrt{\frac{C + C_x}{C}} = \sqrt{1 + \frac{C_x}{C}}$$

$$= \sqrt{1 + \frac{3.58}{50}} = \sqrt{1.0716} = 1.0352$$

Now

$$\frac{f_x}{f_n} = \frac{f + \delta}{f - \delta}$$

$$f + \delta = (f - \delta) \times 1.0352$$
$$= 1.0352f - 1.0352\delta$$
$$2.0352\delta = 0.0352f$$
$$\delta = 0.0352f/2.0352 = 0.0352 \times 50 \times 10^6/2.0352$$
$$= 0.865 \times 10^6 = 0.865 \text{ MHz}$$

Total frequency variation is $2\delta = 2 \times 0.865$
$= 1.73$ MHz

Example 5-7 It is required to provide a maximum deviation of 75 kHz for the 88-MHz carrier frequency of a VHF FM transmitter. A FET is used as a capacitive reactance modulator, and the linear portion of its $g_m - v_{gs}$ curve lies from 320 μS (at which $V_{gs} = -2$V) to 830 μS (at which $V_{gs} = -0.5$V). Assuming that R_{gs} is one-tenth of X_{Cgd}, calculate

(a) The rms value of the required modulating voltage.

(b) The value of the fixed capacitance and inductance of the oscillator tuned circuit across which the reactance modulator is connected.

(a) $V_m$peak to peak $= 2 - 0.5 = 1.5V$

$$V_m \text{ rms} = 1.5/2\sqrt{2} = 0.53V$$

(b) $$C_n = \frac{g_{m,min}}{2\pi f n} = \frac{3.2 \times 10^{-4}}{2\pi \times 8.8 \times 10^7 \times 10}$$

$$= \frac{3.2 \times 10^{12}}{2\pi \times 8.8} = 5.8 \times 10^{-14}$$

$$= 0.058 \text{ pF}$$

$$C_x = \frac{C_n g_{m,max}}{g_{m,min}} = 0.058 \times \frac{830}{320}$$

$$= 0.15 \text{ pF}$$

Now

$$\frac{f_x}{f_n} = \frac{1}{2\pi\sqrt{L(C + C_n)}} \div \frac{1}{2\pi\sqrt{L(C + C_x)}}$$

$$= \sqrt{\frac{C + C_x}{C + C_n}}$$

$$\left(\frac{f_x}{f_n}\right)^2 = \frac{C + C_x}{C + C_n}$$

$$\frac{f_x^2}{f_n^2} - 1 = \frac{C + C_x}{C + C_n} - 1$$

$$\frac{f_x^2 - f_n^2}{f_n^2} = \frac{C + C_x - C - C_n}{C + C_n}$$

$$\frac{(f_x + f_n)(f_x - f_n)}{f_n^2} = \frac{4f\delta}{f_n^2} \approx \frac{4f\delta}{f^2} = \frac{C_x - C_n}{C + C_n}$$

$$C + C_n = \frac{(C_x - C_n)f^2}{4f\delta}$$

$$C = \frac{(C_x - C_n)f}{4\delta} - C_n \qquad \text{(5-23)}$$

$$= \frac{(0.150 - 0.058) \times 88}{4 \times 0.075} - 0.058$$

$$\approx \frac{0.092 \times 88}{0.3} = 27 \text{ pF}$$

$$f = \frac{1}{2\pi\sqrt{L(C + C_{av})}} \approx \frac{1}{2\pi\sqrt{LC}}$$

$$L = \frac{1}{4\pi^2 f^2 C} = \frac{1}{4\pi^2 \times 8.8^2 \times 10^{14} \times 2.7 \times 10^{-11}}$$

$$= \frac{10^{-3}}{39.5 \times 77.4 \times 2.7} = \frac{10^{-5}}{82.5} = 1.21 \times 10^{-7}$$

$$= 0.121\ \mu\text{H}$$

Example 5-7 is typical of reactance modulator calculations. Note, therefore, how approximations were used where they were warranted, i.e., when a small quantity was to be subtracted from or added to a large quantity. On the other hand, a ratio of two almost identical quantities, f_x/f_n, was expanded for maximum accuracy. It will also be noted that the easiest possible units were employed for each calculation. Thus, to evaluate C, picofarads and megahertz were used, but this was not done in the inductance calculation since it would have led to confusion. Note finally that Eq. (5-23) is universally applicable to this type of situation, whether the reactance modulator is an FET, a tube, a junction transistor or a varactor diode.

Types of reactance modulators There are four different arrangements of the reactance modulator (including the one initially treated) which will yield useful results. Their data are shown in Table 5-2, together with their respective prerequisites and output reactance formulas. The general prerequisite for all of them is that drain current must be much greater than bias network current. It is seen that two of the arrangements give a capacitive reactance, and the other two give an inductive reactance.

TABLE 5-2

NAME	Z_{gd}	Z_{gs}	CONDITION	REACTANCE FORMULA
RC capacitive	C	R	$X_C \gg R$	$C_{eq} = g_m RC$
RC inductive	R	C	$R \gg X_C$	$L_{eq} = \dfrac{RC}{g_m}$
RL inductive	L	R	$X_L \gg R$	$L_{eq} = \dfrac{L}{g_m R}$
RL capacitive	R	L	$R \gg X_L$	$C_{eq} = \dfrac{g_m L}{R}$

In the reactance modulator shown in Fig. 5-12, an *RC* capacitive transistor reactance modulator, quite a common one in use, operates on the tank circuit of a Clapp-Gouriet oscillator. Provided that the correct component values are employed, any reactance modulator may be connected across the tank circuit of any *LC* oscillator (not crystal, of course) with one proviso: the oscillator used must not be one that requires two tuned circuits for its operation, such as the tuned-base-tuned-collector oscillator. The Hartley and Colpitts (or Clapp-Gouriet) oscillators are most commonly used, and each should be isolated with a buffer. Note the RF chokes in the circuit shown; they are used to isolate various points of the circuit for alternating current while still providing a dc path.

Varactor diode modulator Varactor diodes[6] may also be used to produce frequency modulation; they are certainly employed frequently, together with a reactance modulator, to provide automatic frequency correction for an FM transmitter. The circuit of Fig. 5-13 shows such a modulator. It is seen that the diode has been back-biased to

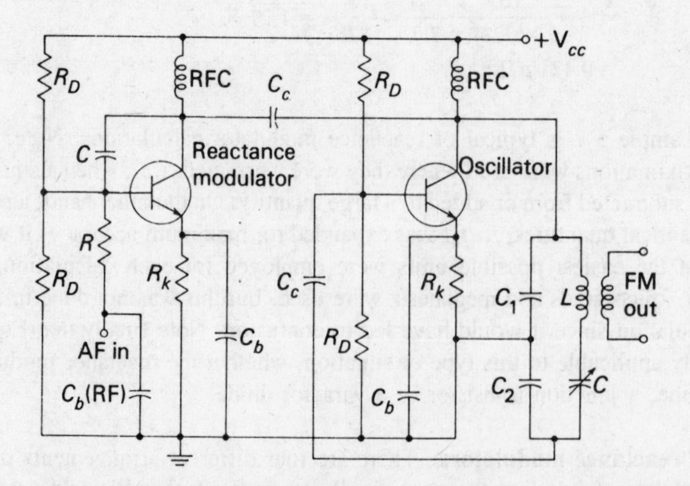

FIGURE 5-12 Transistor reactance modulator.

[6] A varactor diode is a semiconductor diode whose junction capacitance varies linearly with applied bias when the diode is back-biased. The diode and its properties are described fully in Sec. 12-2.1.

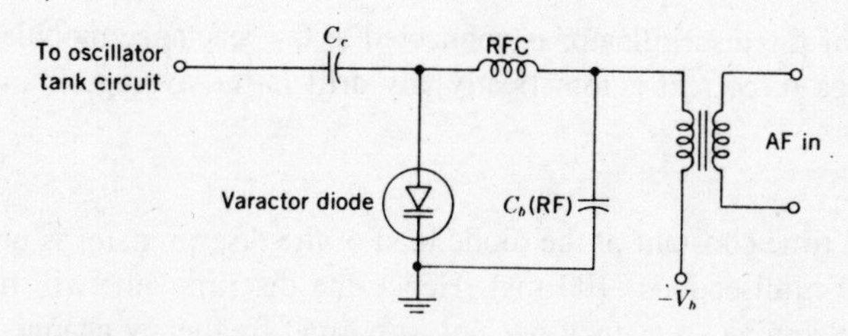

FIGURE 5-13 Varactor diode modulator.

provide the junction capacitance effect, and since this bias is varied by the modulating voltage which is in series with it, the junction capacitance will also vary, causing the oscillator frequency to change accordingly. Although this is the simplest reactance modulator circuit, it does have the disadvantage of using a two-terminal device; its applications are thus somewhat limited. However, it is often used for automatic frequency control and remote tuning.

5-3.3 Stabilized Reactance Modulator—AFC

Although the oscillator on which a reactance modulator operates cannot be crystal-controlled, it must nevertheless have the stability of a crystal oscillator if it is to be part of a commercial transmitter. This suggests that frequency stabilization of the reactance modulator is required, and since this is very similar to an automatic frequency control system, AFC will also be considered. The block diagram of a typical system is shown in Fig. 5-14. [However, to acquire a full understanding of the operation of this circuit, it is necessary to be familiar with the phase (Foster-Seeley) discriminator, which is treated in Sec. 6-4.3]

As can be seen, the reactance modulator operates on the tank circuit of an *LC* oscillator. It is isolated by a buffer, whose output goes through an amplitude limiter to power amplification by class *C* amplifiers (not shown). A fraction of the output is taken from the limiter and fed to a mixer, which also receives the signal from a crystal oscillator. The resulting difference signal, which has a frequency usually about one-twentieth of the master oscillator frequency, is amplified and fed to a phase discrimina-

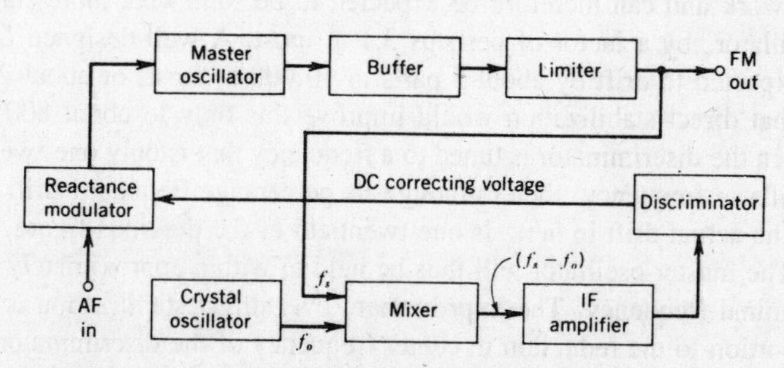

FIGURE 5-14 A typical transmitter AFC system.

tor. The output of the discriminator is connected to the reactance modulator and provides a dc voltage to correct automatically any drift in the average frequency of the master oscillator.

Operation The time constant of the diode load of the discriminator is quite large, in the order of 100 milliseconds (100 ms). Hence the discriminator will react to slow changes in the incoming frequency but not to normal frequency changes due to frequency modulation (since they are too fast). Note also that the discriminator must be connected to give a positive output if the input frequency is higher than the discriminator tuned frequency, and a negative output if it is lower.

Consider what happens when the frequency of the master oscillator drifts high. A higher frequency will eventually be fed to the mixer, and since the output of the crystal oscillator may be considered as stable, a somewhat higher frequency will also be fed to the phase discriminator. Since the discriminator is tuned to the correct frequency difference which should exist between the two oscillators, and its input frequency is now somewhat higher, the output of the discriminator will be a positive dc voltage. This voltage is fed in series with the input of the reactance modulator and therefore increases its transconductance. The output capacitance of the reactance modulator is given by $C_{eq} = g_m RC$, and it is of course, increased, therefore lowering the oscillator's center frequency. The frequency rise which caused all this activity has thus been corrected. When the master oscillator drifts low, a negative correcting voltage is obtained from this circuit, and the frequency of the oscillator is increased correspondingly.

This correcting dc voltage may instead be fed to a varactor diode connected across the oscillator tank and be used for AFC only. Alternatively, a system of amplifying the dc voltage and feeding it to a servomotor which is connected to a trimmer capacitor in the oscillator circuit may be used. The setting of the capacitor plates is then altered by the motor and in turn corrects the frequency.

Reasons for mixing If it were possible to stabilize the oscillator frequency directly instead of first mixing it with the output of a crystal oscillator, the circuit would be much simpler but the performance would suffer. It must be realized that the stability of the whole circuit depends on the stability of the discriminator; if its frequency drifts, the output frequency of the whole system must drift equally. The discriminator is a passive network and can therefore be expected to be somewhat more stable than the master oscillator, by a factor of perhaps 3:1 at most. A well-designed *LC* oscillator could be expected to drift by about 5 parts in 10,000 at most, or about 2.5 kHz at 5 MHz, so that direct stabilization would improve this only to about 800 Hz at best.

When the discriminator is tuned to a frequency that is only one-twentieth of the master oscillator frequency, then (although its percentage frequency drift may still be the same) the actual drift in hertz is one-twentieth of the previous figure, or 40 Hz in this case. The master oscillator will thus be held to within approximately 40 Hz of its 5-MHz nominal frequency. The improvement over direct stabilization is therefore in direct proportion to the reduction in center frequency of the discriminator, or twentyfold here.

Unfortunately, it is not possible to make the frequency reduction much greater than 20:1, although the frequency stability would undoubtedly be improved even further. The reason for this is a practical one; the bandwidth of the discriminator's S curve could then become insufficient (see Sec. 6-4.3) to encompass the maximum possible frequency drift of the master oscillator, so that stabilization could be lost. There is a cure for this also; if the frequency of the output of the mixer is divided, the frequency drift will be divided with it. Thus the discrimination can now be tuned to this divided frequency, and stability can be improved without theoretical limit.

Although the foregoing discussion is concerned directly with the stabilization of the center frequency of an FM transmitter, it applies equally to the frequency stabilization of any oscillator which, for some reason, cannot be crystal-controlled. The only difference in such an AFC system is that now no modulation is fed to the reactance modulator, and the discriminator load time constant may now be faster. It is also most likely that a varactor diode would then be used for AFC.

5-3.4 Indirect Method

Because a crystal oscillator cannot be successfully frequency-modulated, the direct modulators have the disadvantage of being based on an *LC* oscillator which is not stable enough for communications or broadcast purposes. In turn, this requires stabilization of the reactance modulator with attendant circuit complexity. It is possible, however, to generate FM through phase modulation, where a crystal oscillator can be used. Since this method is often used in practice, it will now be described. It is called the *Armstrong system* after its inventor, and it historically precedes the reactance modulator, as shown by the original references given [4,5].

The block diagram of an Armstrong system is shown in Fig. 5-15. Actually, the system proper terminates at the output of the combining network; the remaining blocks are included to show how wideband FM might be obtained. Also, as will be explained

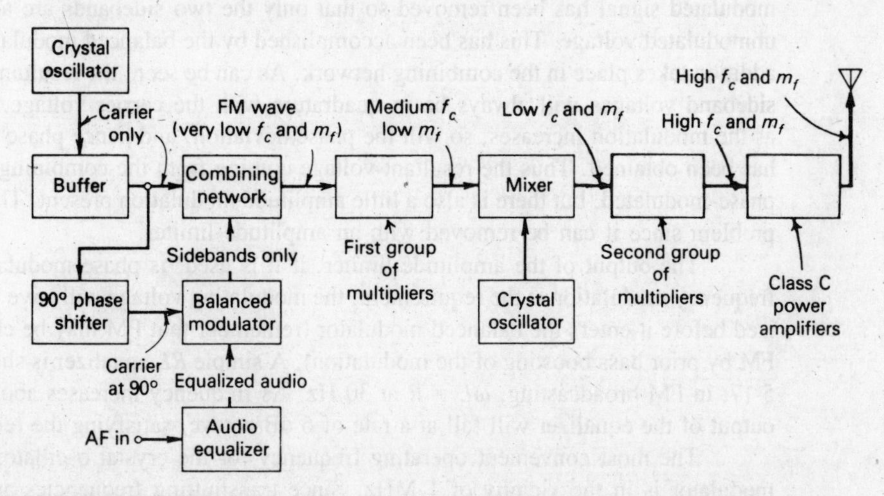

FIGURE 5-15 Block diagram of the Armstrong frequency-modulation system.

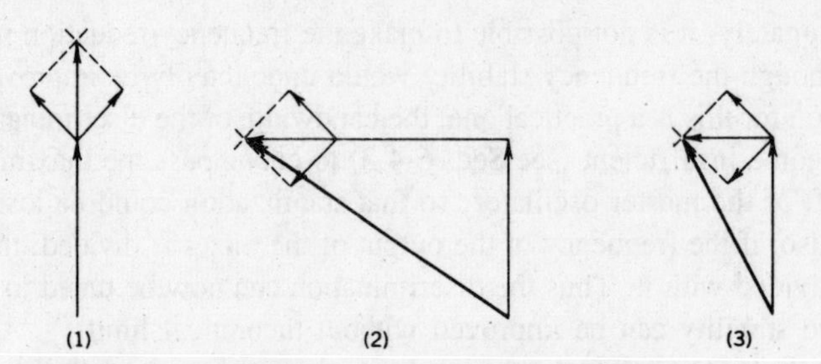

FIGURE 5-16 Phase-modulation vector diagrams.

further, the effect of mixing on an FM signal is to change the center frequency only, whereas the effect of frequency multiplication is to multiply center frequency and deviation equally.

The vector diagrams of Fig. 5-16 illustrate the principles of operation of this modulation system. Diagram 1 shows an amplitude-modulated signal. It will be noted that the resultant of the two sideband frequency vectors is always in phase with the unmodulated carrier vector, so that there is amplitude variation but no phase (or frequency) variation. Since it is phase change that is needed here, some arrangement must be found which ensures that this resultant of the sideband voltages is always out of phase (preferably by 90°) with the carrier vector. If an amplitude-modulated voltage is added to an unmodulated voltage of the same frequency and the two are kept 90° apart in phase, as shown by diagram 2, some form of phase modulation will be achieved. Unfortunately, it will be a very complex and nonlinear form having no practical use; however, it does seem like a step in the right direction. Note that the two frequencies must be identical (suggesting the one source for both) with a phase-shifting network in one of the channels.

Diagram 3 shows the solution to the problem. The carrier of the amplitude-modulated signal has been removed so that only the two sidebands are added to the unmodulated voltage. This has been accomplished by the balanced modulator, and the addition takes place in the combining network. As can be seen, the resultant of the two sideband voltages will always be in quadrature with the carrier voltage. Moreover, as the modulation increases, so will the phase deviation, and hence phase modulation has been obtained. Thus the resultant voltage coming from the combining network is phase-modulated, but there is also a little amplitude modulation present. The AM is no problem since it can be removed with an amplitude limiter.

The output of the amplitude limiter, if it is used, is phase modulation. Since frequency modulation is the requirement, the modulating voltage will have to be equalized before it enters the balanced modulator (remember that PM may be changed into FM by prior bass boosting of the modulation). A simple *RL* equalizer is shown in Fig. 5-17; in FM broadcasting, $\omega L = R$ at 30 Hz. As frequency increases above that, the output of the equalizer will fall at a rate of 6 dB/octave, satisfying the requirements.

The most convenient operating frequency for the crystal oscillator and phase modulator is in the vicinity of 1 MHz. Since transmitting frequencies are normally much higher than this, frequency multiplication must be used, and so multipliers are shown in the block diagram of Fig. 5-15.

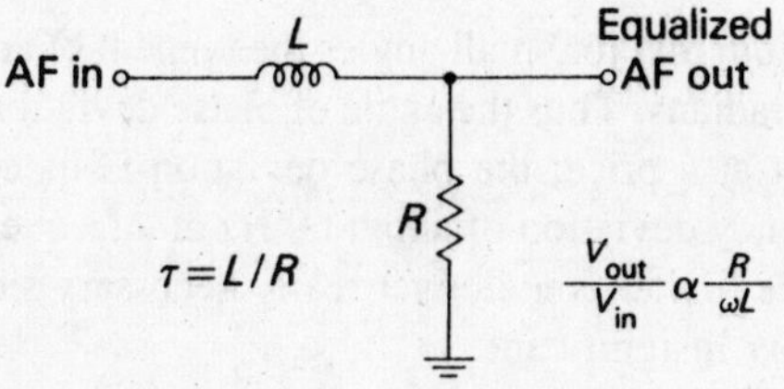

AF in — L — Equalized AF out; R to ground; $\tau = L/R$; $\frac{V_{out}}{V_{in}} \propto \frac{R}{\omega L}$

FIGURE 5-17 RL equalizer.

Effects of frequency changing on an FM Signal The foregoing has shown that frequency changing of an FM signal is essential in the Armstrong system. Indeed, for convenience it is very often used with the reactance modulator also. Investigation will show that the modulation index is multiplied by the same factor as the center frequency, whereas frequency translation (changing) does not affect the modulation index.

If a frequency-modulated signal $f_c \pm \delta$ is fed to a frequency doubler, the output signal will contain twice each input frequency; for the extreme frequencies here, this will be $2f_c - 2\delta$ and $2f_c + 2\delta$. Thus the frequency deviation has quite clearly doubled to $\pm 2\delta$, with the result that the modulation index has also doubled.[7] In this fashion, both center frequency and deviation may be increased by the same factor or, if frequency division should be used, reduced by the same factor.

When a frequency-modulated wave is mixed, the resulting output contains difference frequencies (among others). The original signal might again be $f_c \pm \delta$; when mixed with a frequency f_0, it will yield $f_c - f_0 - \delta$ and $f_c - f_0 + \delta$ as the two extreme frequencies in its output. It is seen that the FM signal has been translated to a lower center frequency $f_c - f_0$, but the maximum deviation has remained at $\pm\delta$. It is thus possible to reduce (or increase, if desired) the center frequency of an FM signal without affecting the maximum deviation.

Since the modulating frequency has obviously remained constant in the two cases treated, the modulation index will be affected in the same manner as the deviation. It will thus be multiplied together with the center frequency or unaffected by mixing. Also, it is possible to raise the modulation index without affecting the center frequency by multiplying both by (say) 9 and mixing the result with a frequency eight times the original frequency. The difference will be equal to the initial frequency, but the modulation index will have been multiplied ninefold.

Further consideration in the Armstrong system One of the characteristics phase modulation is that the angle of phase deviation must be proportional to the modulating voltage. A careful look at diagram 3 of Fig. 5-16 shows that this is not so in this case, although this fact was carefully glossed over in the initial description. In fact, it is the *tangent* of the angle of phase deviation that is *proportional* to the amplitude of the *modulating voltage*, not the angle itself. However, the difficulty is not impossible to

[7] Although both this explanation and its conclusion are correct and valid, the situation in practice is a little more complicated. For further explanation, the interested student is referred to pp. 20–22 in [3].

resolve: it is a trigonometric axiom that for small angles the tangent of an angle is equal to the angle itself, measured in radians. Thus the angle of phase deviation is kept small, and the problem is solved, but at a price; the phase deviation is indeed tiny, corresponding to a maximum frequency deviation of about 60 Hz at a frequency of 1 MHz. On the other hand, an amplitude limiter is no longer really necessary since the amount of amplitude modulation is now insignificant.

To achieve sufficient deviation for broadcast purposes, both mixing and multiplication are necessary, whereas for narrowband FM, multiplication may be sufficient by itself. In the latter case, operating frequencies are in the vicinity of 180 MHz. Therefore, starting with an initial f_c = 1 MHz and δ = 60 Hz, it is possible to achieve a deviation of 10.8 kHz at 180 MHz, which is more than adequate for FM mobile work.

The FM broadcasting station uses a higher maximum deviation with a lower center frequency, so that both mixing and multiplication must be used. For instance, if the starting conditions are as above and 75 kHz deviation is required at 100 MHz, f_0 must be multiplied by 100/1 = 100 times, whereas δ must be increased 75,000/60 = 1250 times. The mixer and crystal oscillator in the middle of the multiplier chain are used to reconcile the two multiplying factors. After being raised to about 6 MHz, the frequency-modulated carrier is mixed with the output of a crystal oscillator, whose frequency is such as to produce a difference of 6 MHz/12.5. The center frequency has been reduced, but the deviation is left unaffected. Thus both can now be multiplied by the same factor to give the desired center frequency and maximum deviation.

5-4 SYSTEM SUMMARY

FM and PM are the two forms of angle modulation, which is a form of continuous-wave or analog modulation whose chief characteristics are as follows:

1. The amplitude of the modulated carrier is kept constant.
2. The frequency of the modulated carrier is varied by the modulating voltage.

In frequency modulation, the carrier's frequency deviation is proportional to the instantaneous amplitude of the modulating voltage. In phase modulation, the carrier's phase deviation is proportional to the instantaneous amplitude of the modulating voltage. This is equivalent to saying that, in PM, the *frequency* deviation is proportional to the instantaneous amplitude of the modulating voltage, but it is also proportional to the modulating frequency. It follows, therefore, that PM played through an FM receiver would be intelligible but would sound as though a uniform bass cut (or treble boost) had been applied to all the audio frequencies. It also follows that FM could be generated from an essentially PM process, provided that the modulating frequencies were first passed through a suitable bass-boosting network.

The major advantages of angle modulation over amplitude modulation are:

1. The transmitted amplitude is constant, and thus the receiver can be fitted with an efficient amplitude limiter (since, by definition, all amplitude variations are spurious); this characteristic has the advantage of significantly improving immunity to noise and interference.

2. Since there is no natural limit to the modulation index, as in AM, the modulation index can be increased to provide additional noise immunity, but there is a tradeoff involved—system bandwidth must be increased.

Frequency modulation additionally has the advantage, over both AM and PM, of providing greater protection from noise for the lowest modulating frequencies; the resulting noise-signal distribution is here seen as a triangle, whereas it is rectangular in both AM and PM. A consequence of this is that FM is used for analog transmissions, whereas PM is not. Because FM broadcasting is a latecomer compared with AM broadcasting, the system design has benefited from the experience gained with AM. Two of the most notable benefits are the provision of guard bands between adjacent transmissions and the use of pre-emphasis and de-emphasis. With emphasis, the highest modulating frequencies are artificially boosted before transmission and correspondingly attenuated after reception, to reduce the effects of noise.

Wideband FM is used for broadcast transmissions, with or without stereo multiplex, and for the sound accompanying TV transmissions. Narrowband FM is used for communications, in competition with SSB, having its main applications in various forms of mobile communications, generally at frequencies above 30 MHz. It is also used in conjunction with SSB in *frequency division multiplexing* (FDM), as will be seen in Chap. 15. FDM is a technique for combining large numbers of channels in broadband links used for terrestrial or satellite communications.

Two basic methods of generating FM are in general use. The reactance modulator is a direct method of generating FM, in which the tank circuit reactance, and thus the frequency of an *LC* oscillator, is varied electronically by the modulating signal. To ensure adequate frequency stability, the output frequency is then compared with that of a crystal oscillator and corrected automatically as required. The alternative means of generating FM, the Armstrong system, is one in which PM is initially generated, but the modulating frequencies are correctly bass-boosted; thus FM results in the output. Because only small frequency deviations are possible in the basic Armstrong system, extensive frequency multiplication and mixing are used to increase deviation to the wanted value. The power and auxiliary stages of FM transmitters are akin to those in AM transmitters, except that FM has an advantage here: since it is a constant-amplitude modulation system, all the power amplifiers can be operated in class C, i.e., very efficiently.

PROBLEMS

For self-testing questions on this chapter, see p. 680.

5-1. A 500-Hz modulating voltage fed into a PM generator produces a frequency deviation of 2.25 kHz. What is the modulation index? If the amplitude of the modulating voltage is kept constant, but its frequency is raised to 6 kHz, what is the new deviation?
[Ans.: 4.5, 54 kHz]

5-2. When the modulating frequency in an FM system is 400 Hz and the modulating voltage is 2.4 V, the modulation index is 60. Calculate the maximum deviation. What is the modulation index when the modulating frequency is reduced to 250 Hz and the modulating voltage is simultaneously raised to 3.2 V?
[Ans.: 24 kHz, 128]

5-3. The equation of an angle-modulated voltage is $v = 10 \sin (10^8 t + 3 \sin 10^4 t)$. What form of angle modulation is this? Calculate the carrier and modulating frequencies, the modulation index and deviation, and the power dissipated in a 100-Ω resistor.
[Ans.: FM or PM, 15.9 MHz, 1.59 kHz, 3, 4.77 kHz, 0.5 W]

5-4. The center frequency of an *LC* oscillator, to which a capacitive reactance FET modulator is connected, is 70 MHz. The FET has a g_m which varies linearly from 1 to 2 mS, and a bias capacitor whose reactance is 10 times the resistance of the bias resistor. If the fixed tuning capacitance across the oscillator coil is 25 pF, calculate the maximum available frequency deviation.
[Ans.: ±157 kHz]

5-5. An *RC* capacitive reactance modulator is used to vary the frequency of a 10-MHz oscillator by ±100 kHz. An FET whose transconductance varies linearly with gate voltage from 0 to 0.628 mS, is used in conjunction with a resistance whose value is one-tenth of the capacitive reactance used. Calculate the inductance and capacitance of the oscillator tank circuit.
[Ans.: 25 pF, 10.1 μH]

QUESTIONS

5-1. Describe frequency and phase modulation, giving mechanical analogies for each.

5-2. Derive the formula for the instantaneous value of an FM voltage and define the modulation index.

5-3. In an FM system, if m_f is doubled by halving the modulating frequency, what will be the effect on the maximum deviation?

5-4. Describe an experiment designed to calculate by measurement the maximum deviation in an FM system, which makes use of the disappearance of the carrier component for certain values of the modulation index. Draw the block diagram of such a setup.

5-5. With the aid of Table 5-1, estimate the total bandwidth required by an FM system whose maximum deviation is 3 kHz, and in which the modulating frequency may range from 300 to 2000 Hz. Note that any sideband with a relative amplitude of 0.01 or less may be ignored.

5-6. On graph paper, draw to scale the frequency spectrum of the FM wave of Question 5-5 for (a) $f_m = 300$ Hz; (b) $f_m = 2000$ Hz. The deviation is to be 3 kHz in each case.

5-7. Explain fully the difference between frequency and phase modulation, beginning with the definition of each type and the meaning of the modulation index in each case.

5-8. Of the various advantages of FM over AM, identify and discuss those due to the intrinsic qualities of frequency modulation.

5-9. With the aid of vector diagrams, explain what happens when a carrier is modulated by a single noise frequency.

5-10. Explain the effect of random noise on the output of an FM receiver fitted with an amplitude limiter. Develop the concept of the noise triangle.

5-11. What is pre-emphasis? Why is it used? Sketch a typical pre-emphasis circuit and explain why de-emphasis must be used also.

5-12. What determines the bandwidth used by any given FM communications system? Why are two different types of bandwidth used in frequency-modulated transmissions?

5-13. Using a block diagram and a frequency spectrum diagram, explain the operation of the stereo multiplex FM transmission system. Why is the difference subcarrier originally generated at 19 kHz?

5-14. Explain, with the aid of a block diagram, how you would design an FM stereo transmission system which does not need to be compatible with monaural FM systems.

5-15. Showing the basic circuit sketch and stating the essential assumptions, derive the formula for the capacitance of the *RL* reactance FET.

5-16. Why is it not practicable to use a reactance modulator in conjunction with a crystal oscillator? Draw the equivalent circuit of a crystal in your explanation and discuss the effect of changing the external parallel capacitance across the crystal.

5-17. With the aid of a block diagram, show how an AFC system will counteract a downward drift in the frequency of the oscillator being stabilized.

5-18. Why should the discriminator tuned frequency in the AFC system be as low as possible? What lower limit is there on its value? What part can frequency division play here?

5-19. What is the function of the balanced modulator in the Armstrong modulation system?

5-20. Draw the complete block diagram of the Armstrong frequency modulation system and explain the functions of the mixer and multipliers shown. In what circumstances can we dispense with the mixer?

5-21. Starting with an oscillator working near 500 kHz and using a maximum frequency deviation not exceeding ± 30 Hz at that frequency, calculate the following for an Armstrong system which is to yield a center frequency precisely 97 MHz with a deviation of exactly 75 kHz: *(a)* starting frequency; *(b)* exact initial deviation; *(c)* frequency of the crystal oscillator; *(d)* amount of frequency multiplication in each group. Note that there are several possible solutions to this problem.

REFERENCES

1. See Wylie, C. R., *Advanced Engineering Mathematics,* 3d ed., McGraw-Hill Book Company, New York, 1966.

2. Stremler, F. G., *Introduction to Communications Systems,* Addison-Wesley Publishing Co., Reading, Mass., 1977, pp. 261–275.

3. Sturley, K. R., *Frequency-modulated Radio,* George Newnes Ltd., London, 1958, pp. 34–35.

4. *(Historical)* Armstrong, E. H., ''A Method of Reducing Disturbances in Radio Signalling by a System of Frequency Modulation,'' *Proc. IRE* **24**(5):680 (1936).

5. *(Historical)* Crosby, M. G., Reactance Tube Frequency Modulation, *RCA Rev.*, vol. 5, p. 89, July 1940.

6. Mandl, M., *Principles of Electronic Communications,* Prentice-Hall, Inc., Englewood Cliffs, N.J., 1973.

6 RADIO RECEIVERS

As shown in Chaps. 3 to 5, a signal to be transmitted is impressed onto a carrier wave in any of the modulation methods thus far described, and it is then suitably treated, amplified and applied to a transmitting antenna. As will be shown in Chaps. 8 and 9, the modulated signal is radiated, propagated and (a little of it) collected by a receiving antenna. What must a receiver do? Bearing in mind that the signal at this point is generally quite weak, powers of the order of picowatts being common, the receiver must first amplify the received signal. Since the signal is quite likely to be accompanied by lots of other (unwanted) signals, probably at neighboring frequencies, it must be selected and the others rejected. Finally, since modulation took place in the transmitter, the reverse process of *demodulation* must be performed in the receiver, to recover the original modulating voltages.

This chapter will deal with radio receivers in general, showing why their format has been to a certain extent standardized. Each block of the receiver will be discussed in detail, as will its functions and design limitations. This will be done for receivers corresponding to all the modulation systems so far studied, be they for domestic or professional purposes. For ease of understanding, each block will be treated as though consisting of discrete circuits. In practice, of course, many circuits or indeed whole receivers are likely to be integrated in current equipment.

It is apparent that a receiver has the function of selecting the desired signal from all the other unwanted signals, amplifying and demodulating it, and displaying it in the desired manner. This outline of functions that must be performed shows that the major difference between receivers of various types is likely to be in the way in which they demodulate the received signal. In turn, this will depend on the type of modulation employed, be it AM, FM, SSB, or any of the forms treated in later chapters. However, it does appear that the same *type* of receiver should be capable of dealing with the basic requirements, and this will indeed be seen.

6-1 RECEIVER TYPES

Of the various forms of receivers proposed at one time or another, only two have any real practical or commercial significance: the tuned radio-frequency (TRF) receiver and the superheterodyne receiver. Only the second of these is used to a large extent today, but it is convenient to explain the operation of the TRF receiver first since it is the simpler of the two. Also, perhaps the best way of justifying the existence and overwhelming popularity of the superheterodyne receiver is by showing the shortcomings of the TRF type.

6-1.1 Tuned Radio-Frequency (TRF) Receiver

Until shortly before World War II, most radio receivers were of the TRF type, whose block diagram is shown in Fig. 6-1.

The TRF receiver is a simple ''logical'' receiver; a person with just a little knowledge of communications would probably expect all radio receivers to have this form. The virtues of this type, which is now not used except as a fixed-frequency receiver in special applications, are its simplicity and high sensitivity. It must also be mentioned that when the TRF receiver was first introduced, it was a great improvement on the types used previously: mainly crystal, regenerative and superregenerative receivers.

Two or perhaps three RF amplifiers, all tuning together, were employed to select and amplify the incoming frequency and simultaneously to reject all others. After the signal was amplified to a suitable level, it was demodulated (detected) and fed to the loudspeaker after being passed through the appropriate audio amplifying stages. Such receivers were simple to design and align at broadcast frequencies (535 to 1640 kHz), but they presented difficulties at higher frequencies. This was mainly because of the risks of instability associated with high gain being achieved at one frequency by a multistage amplifier. If such an amplifier has a gain of 40,000, all that is needed is 1/40,000 of the output of the last stage to find itself back at the input to the first stage with correct polarity (having got there through some stray path), and oscillations will occur, at the frequency at which the polarity of this spurious feedback is

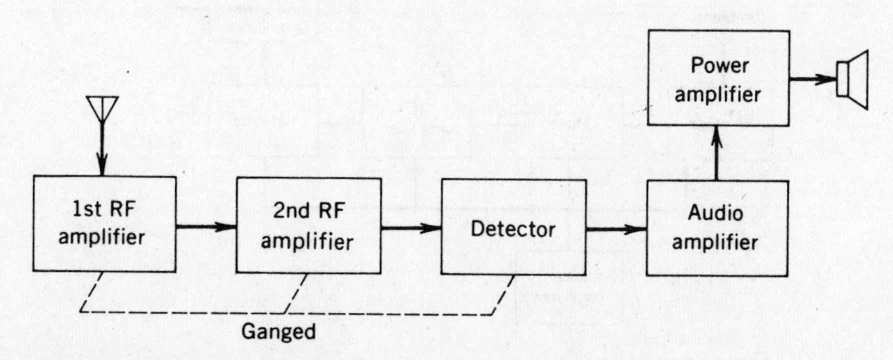

FIGURE 6-1 The TRF receiver.

positive. Such a state of affairs is almost unavoidable at high frequencies and is certainly not conducive to good receiver operation. In addition the TRF receiver suffered from a variation in bandwidth over the tuning range. Also, it was unable to achieve sufficient selectivity at high frequencies, partly as a result of the enforced use of single-tuned circuits. It was not possible to use double-tuned RF amplifiers in this receiver, although it was realized that they would naturally yield better selectivity. This was due to the fact that all such amplifiers had to be tunable, and the difficulties of making several double-tuned amplifiers tune in unison were too great (see also Sec. 6-2.2).

Consider a tuned circuit required to have a bandwidth of 10 kHz at a frequency of 535 kHz. The Q of this circuit must be $Q = f/\Delta f = 535/10 = 53.5$. At the other end of the broadcast band, i.e., at 1640 kHz, the inductive reactance (and therefore the Q) of the coil should in theory have increased by a factor of 1640 /535 to 164. In practice, however, various losses dependent on frequency will prevent so large an increase. Thus the Q at 1640 kHz is unlikely to be in excess of 120, giving a bandwidth of $\Delta f = 1640/120 = 13.7$ kHz and ensuring that the receiver will pick up adjacent stations as well as the one to which it is tuned. Consider again a TRF receiver required to tune to 36.5 MHz, the upper end of the shortwave band. If the Q required of the RF circuits is again calculated, still on the basis of a 10-kHz bandwidth, we have $Q = 36{,}500/10 = 3650$! It is obvious that such a Q is impossible to obtain with ordinary tuned circuits.

The problems of instability, insufficient adjacent-frequency rejection, and bandwidth variation can all be solved by the use of a superheterodyne receiver, which introduces relatively few problems of its own.

6-1.2 Superheterodyne Receiver

The block diagram of Fig. 6-2 shows a basic superheterodyne receiver and is a more practical version of Fig 1-3. There are slightly different versions, but they are logical modifications of Fig. 6-2, and most are treated in this chapter. In the superheterodyne

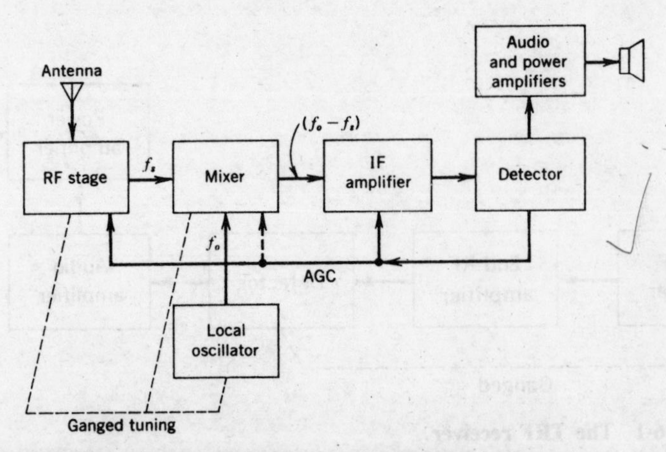

FIGURE 6-2 The superheterodyne receiver.

receiver, the signal voltage is combined with the local oscillator voltage and normally converted into a signal of a lower fixed frequency. The signal at this *intermediate frequency* contains the same modulation as the original carrier, and it is now amplified and detected to reproduce the original information. The *superhet* thus has the same essential components as the TRF receiver, in addition to the mixer, local oscillator and intermediate-frequency (IF) amplifier.

A constant frequency difference is maintained between the local oscillator and the RF circuits, normally through capacitance tuning, in which all the capacitors are *ganged* together and operated in unison by one control knob. The IF amplifier generally uses two or three transformers, each consisting of a pair of mutually coupled tuned circuits. With this large number of double-tuned circuits operating at a constant, specially chosen frequency, the IF amplifier provides most of the gain (and therefore sensitivity) and bandwidth requirements of the receiver. Since the characteristics of the IF amplifier are independent of the frequency to which the receiver is tuned, the selectivity and sensitivity of the superhet are usually fairly uniform throughout its tuning range and not subject to the variations that beset the TRF receiver. The RF circuits are now used mainly to select the wanted frequency, to reject interference such as the *image frequency* and (especially at high frequencies) to reduce the noise figure of the receiver.

A simplified form of the superheterodyne receiver is also in existence, in which the mixer output is in fact audio. Such a *direct conversion receiver* has been used by amateurs, with good results (1).

The advantages of the superheterodyne receiver make it the most suitable type for the great majority of radio receiver applications; AM, FM, communications, single-sideband, television and even radar receivers all use it, with only slight modifications in principle. It may be considered as today's standard form of radio receiver, and as such it will now be examined in some detail, section by section.

6-2 AM RECEIVERS

Since the type of receiver is much the same for the various forms of modulation, it has been found most convenient to explain the principles of a superheterodyne receiver in general while dealing with AM receivers in particular. In this way, a basis is formed with the aid of a simple example of the use of the superheterodyne principle, so that more complex versions can be compared and contrasted with it afterwards; at the same time the overall system is being treated from a practical point of view.

6-2.1 RF Section and Characteristics

A radio receiver always has an RF section, which is a tunable circuit connected to the antenna terminals. It is there to select the wanted frequency and reject some of the unwanted frequencies. However, such a receiver need not have an RF amplifier following this tuned circuit. If there is an amplifier, its output is fed to the mixer, at whose input another tunable circuit is present. In many instances, however, the tuned circuit connected to the antenna is the actual input circuit of the mixer; the receiver is then said to have no RF amplifier or, more simply, no RF stage.

Reasons for use and functions of RF amplifier The receiver having an RF stage is undoubtedly superior in performance to the receiver without one, all else being equal. On the other hand, there are some instances in which an RF amplifier is uneconomical, i.e., where its inclusion would increase the cost of the receiver significantly while improving performance only marginally. The best example of this kind of receiver is a domestic one used in a high-signal-strength area, such as the metropolitan area of any large city.

The advantages of having an RF amplifier are as follows (reasons 4 to 7 are either more specialized or less important):

1. Greater gain, i.e., better sensitivity
2. Improved image-frequency rejection
3. Improved signal-to-noise ratio
4. Improved rejection of adjacent unwanted signals, i.e., better selectivity
5. Better coupling of the receiver to the antenna (important at VHF and above)
6. Prevention of spurious frequencies from entering the mixer and heterodyning there to produce an interfering frequency equal to the IF from the desired signal
7. Prevention of reradiation of the local oscillator through the antenna of the receiver (relatively rare)

The single-tuned, transformer-coupled type is the amplifier most commonly employed for RF amplification, as illustrated in Fig. 6-3. Both diagrams in the figure are seen to have an RF gain control, which is very rare with domestic receivers but quite common in communications receivers. Whereas the medium-frequency amplifier of Fig. 6-3*a* is quite straightforward, the VHF amplifier of Fig. 6-3*b* contains a number of refinements. Feedthrough capacitors are used as bypass capacitors and, in conjunction with the RF choke, to decouple the output from the V_{CC}. As indicated in Fig. 6-3*b*, one of the electrodes of a feedthrough capacitor is the wire running through it. This is surrounded by the dielectric, and around that is the grounded outer electrode; this arrangement minimizes stray inductance in series with the bypass capacitor. Feedthrough capacitors are almost invariably provided for bypassing at VHF and often have a value of 1000 pF. In addition, a single-tuned circuit is used at the input and is coupled to the antenna by means of a trimmer (the latter being manually adjustable for matching to different antennas). Such coupling is used here because of the high frequencies involved. Finally, RF amplifiers in practice have the input and output tuning capacitors ganged to each other and to the one tuning the local oscillator.

Sensitivity The sensitivity of a radio receiver is its ability to amplify weak signals. It is often defined in terms of the voltage that must be applied to the receiver input terminals to give a standard output power, measured at the output terminals. For AM broadcast receivers, several of the relevant quantities have been standardized. Thus 30 percent modulation by a 400-Hz sine wave is used, and the signal is applied to the receiver through a standard coupling network known as a *dummy antenna*. The standard output is 50 milliwatts (50 mW), and for all types of receivers the loudspeaker is replaced by a load resistance of equal value.

Sensitivity is often expressed in microvolts or in decibels below 1 V and measured at three points along the tuning range when a production receiver is lined up. It

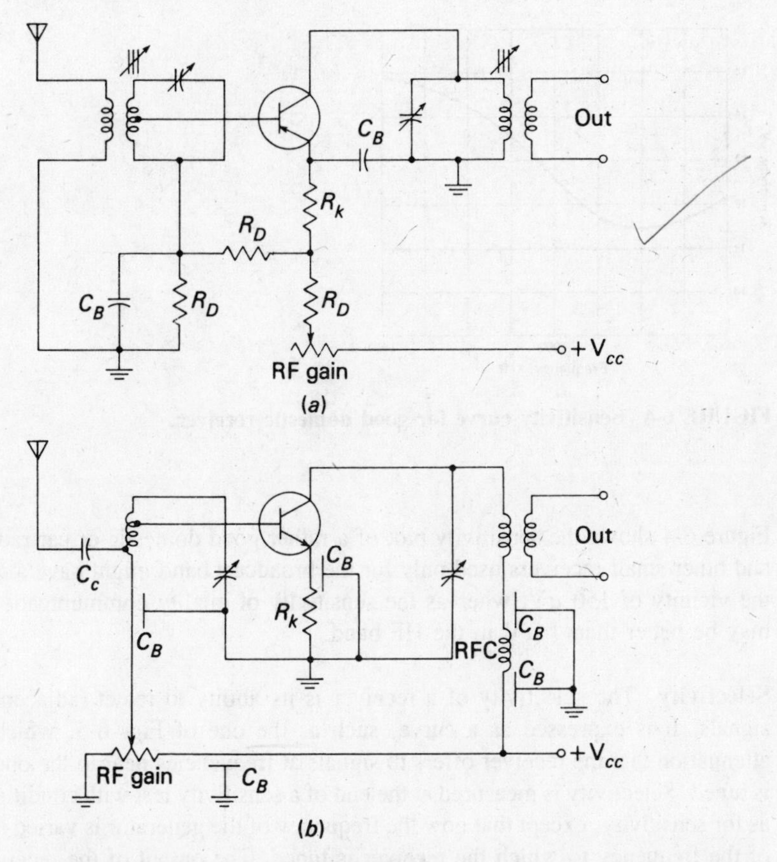

FIGURE 6-3 Transistor RF amplifiers. *(a)* Medium-frequency; *(b)* VHF.

is seen from the sensitivity curve in Fig. 6-4 that sensitivity varies over the tuning band. At 1000 kHz, this particular receiver has a sensitivity of 12.7 μV, or −98 dBV (dB below 1 V). Sometimes the sensitivity definition is extended, and the manufacturer of this receiver may quote it to be, not merely 12.7 μV, but "12.7 μV for a signal-to-noise ratio of 20 dB in the output of the receiver."

For professional receivers, there is a tendency to quote the sensitivity in terms of signal power required to produce a minimum acceptable output signal with a minimum acceptable signal-to-noise ratio. The measurements are made under the conditions described, and the minimum input power is quoted in dB below 1 mW or dBm. Thus, under the heading of "sensitivity" in the specifications of a receiver, a manufacturer might quote, "a −85-dBm 1-MHz signal, 30 percent modulated with a 400-Hz sine wave will, when applied to the input terminals of this receiver through a dummy antenna, produce an output of at least 50 mW with a signal-to-noise ratio not less than 20 dB in the output."

The most important factors determining the sensitivity of a superheterodyne receiver are the gain of the IF amplifier(s) and that of the RF amplifier, if there is one. It is also obvious from the foregoing that the noise figure plays an important part.

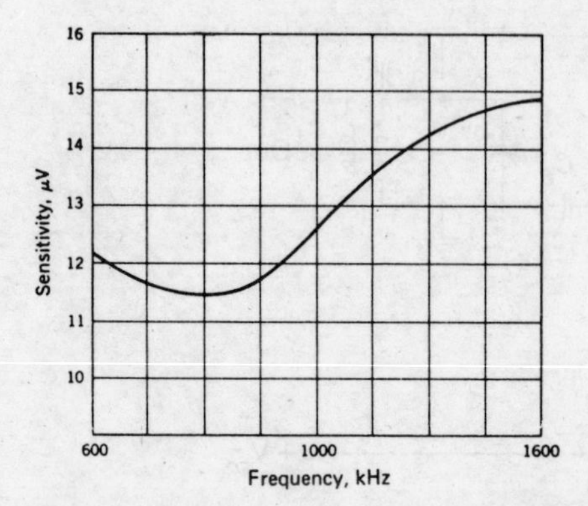

FIGURE 6-4 Sensitivity curve for good domestic receiver.

Figure 6-4 shows the sensitivity plot of a rather good domestic or car radio. Portable and other small receivers used only for the broadcast band might have a sensitivity in the vicinity of 150 μV, whereas the sensitivity of quality communications receivers may be better than 1 μV in the HF band.

Selectivity The selectivity of a receiver is its ability to reject (adjacent) unwanted signals. It is expressed as a curve, such as the one of Fig. 6-5, which shows the attenuation that the receiver offers to signals at frequencies near to the one to which it is tuned. Selectivity is measured at the end of a sensitivity test with conditions the same as for sensitivity, except that now the frequency of the generator is varied to either side of the frequency to which the receiver is tuned. The output of the receiver naturally falls, since the input frequency is now incorrect. Thus the input voltage must be increased until the output is the same as it was originally. The ratio of the voltage required of resonance to the voltage required when the generator is tuned to the receiver's frequency is calculated at a number of points and then plotted in decibels to give a

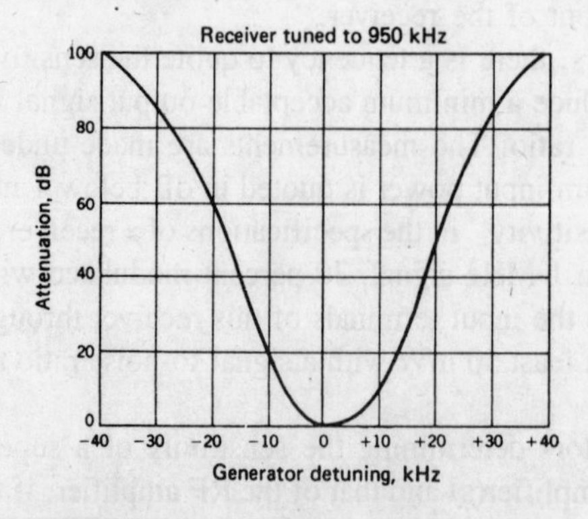

FIGURE 6-5 Typical selectivity curve.

curve, of which the one in Fig. 6-5 is representative. Looking at the curve, we see that, for example, at 20 kHz below the receiver tuned frequency, an interfering signal would have to be 60 dB greater than the wanted signal to come out with the same amplitude.

Selectivity varies with receiving frequency if ordinary tuned circuits are used in the IF section, and becomes somewhat worse when the receiving frequency is raised. In general, it is determined by the response of the IF section, with the mixer and RF amplifier input circuits playing a small but significant part. It should be noted that it is selectivity that determines the adjacent-channel rejection of a receiver.

Image frequency and its rejection In a standard broadcast receiver (and, in fact, in the vast majority of all receivers made) the local oscillator frequency is made higher than the incoming signal frequency for reasons that will become apparent. It is made equal at all times to the signal frequency plus the intermediate frequency. Thus $f_o = f_s + f_i$, or $f_s = f_o - f_i$, no matter what the signal frequency may be. When f_s and f_o are mixed, the difference frequency, which is one of the by-products, is equal to f_i. As such, it is the only one passed and amplified by the IF stage.

If a frequency f_{si} manages to reach the mixer, such that $f_{si} = f_o + f_i$, that is, $f_{si} = f_s + 2f_i$, then this frequency will also produce f_i when mixed with f_o. The relationship of these frequencies is shown in Fig. 3-2, though in a different context. Unfortunately, this spurious intermediate-frequency signal will also be amplified by the IF stage and will therefore provide interference. This has the effect of two stations being received simultaneously and is naturally undesirable. The term f_{si} is called the *image frequency* and is defined as the signal frequency plus twice the intermediate frequency. Reiterating, we have

$$f_{si} = f_s + 2f_i \tag{6-1}$$

The rejection of an image frequency by a single-tuned circuit, i.e., the ratio of the gain at the signal frequency to the gain at the image frequency, is given by

$$\alpha = \sqrt{1 + Q^2\rho^2} \tag{6-2}$$

where

$$\rho = \frac{f_{si}}{f_s} - \frac{f_s}{f_{si}} \tag{6-3}$$

Q = loaded Q of tuned circuit

If the receiver has an RF stage, then there are two tuned circuits, both tuned to f_s; the rejection of each will be calculated by the same formula, and the total rejection will be the product of the two. Whatever applies to gain calculations applies also to those involving rejection.

Image rejection depends on the front-end selectivity of the receiver and *must be achieved before the IF stage*. Once the spurious frequency enters the first IF amplifier, it becomes impossible to remove it from the wanted signal. It can be seen that if f_{si}/f_s is large, as it is in the AM broadcast band, the use of an RF stage is not essential for good image-frequency rejection, but it does become necessary above about 3 MHz.

Example 6-1 In a broadcast superheterodyne receiver having no RF amplifier, the loaded Q of the antenna coupling circuit (at the input to the mixer) is 100. If the intermediate frequency is 455 kHz, calculate *(a)* the image frequency and its rejection ratio at 1000 kHz and *(b)* the image frequency and its rejection ratio at 25 MHz.

(a) $f_{si} = 1000 + 2 \times 455 = 1910$ kHz

$$\rho = \frac{1910}{1000} - \frac{1000}{1910} = 1.910 - 0.524 = 1.386$$

$$\alpha = \sqrt{1 + 100^2 \times 1.386^2} = \sqrt{1 + 138.6^2} = 138.6$$

This is 42 dB and is considered adequate for domestic receivers in the MF band.

(b) $f_{si} = 25 + 2 \times 0.455 = 25.91$ MHz

$$\rho = \frac{25.91}{25} - \frac{25}{25.91} = 1.0364 - 0.9649 = 0.0715$$

$$\alpha = \sqrt{1 + 100^2 \times 0.0715^2} = \sqrt{1 + 7.15^2} = 7.22$$

It is obvious that this rejection will be insufficient for a practical receiver in the HF band.

Example 6-1 shows, as it was meant to, that although image rejection need not be a problem for an AM broadcast receiver without an RF stage, special precautions must be taken at HF. This will be seen in Sec. 6-3, but two possibilities can be explored now, in Example 6-2.

Example 6-2 In order to make the image-frequency rejection of the receiver of Example 6-1 as good at 25 MHz as it was at 1000 kHz, calculate *(a)* the loaded Q which an RF amplifier for this receiver would have to have and *(b)* the new intermediate frequency that would be needed (if there is to be no RF amplifier).

(a) Since the mixer already has a rejection of 7.22, the image rejection of the RF stage will have to be

$$\alpha' = \frac{138.6}{7.22} = 19.2 = \sqrt{1 + Q'^2 \times 0.0715^2}$$

$$Q'^2 = \frac{19.2^2 - 1}{0.0715}$$

$$Q' = \frac{\sqrt{367.6}}{0.0715} = 268$$

Understandably, of course, a well-designed receiver would have the same Q for both tuned circuits. Here this works out to 164 each, that being the geometric mean of 100 and 268.

(b) If the rejection is to be the same as initially, through a change in the intermediate frequency, it is apparent that ρ will have to be the same as in Example 6-1 *a*, since the Q is also the same. Thus

$$\frac{f'_{si}}{f'_s} - \frac{f'_s}{f'_{si}} = 138.6 = \frac{1910}{1000} - \frac{1000}{1910}$$

$$\frac{f'_s}{f'_{si}} = \frac{1910}{1000} = 1.91$$

$$\frac{25 + 2f'_i}{25} = 1.91$$

$$25 + 2f'_i = 1.91 \times 25$$

$$f'_i = \frac{1.91 \times 25 - 25}{2} = \frac{0.91 \times 25}{2} = 11.4 \text{ MHz}$$

Double spotting This is a well-known phenomenon, which manifests itself by the picking up of the same shortwave station at two nearby points on the receiver dial. It is caused by poor front-end selectivity, i.e., inadequate image-frequency rejection. That is to say, the front end of the receiver does not select different adjacent signals very well, but, of course, the IF stage takes care of eliminating almost all of them. This being the case, it is obvious that the precise tuning of the local oscillator is what determines which signal will be amplified by the IF stage. Within broad limits, the setting of the tuned circuit at the input of the mixer is far less important (it being assumed that there is no RF amplifier in a receiver which badly suffers from double spotting). Consider such a receiver at HF, having an IF of 455 kHz. If there is a strong station at (say) 14.7 MHz, the receiver will naturally pick it up—note that, when it does, the local oscillator frequency will be 15.155 MHz. However, the receiver will also pick up this strong station when it (the receiver) is tuned to 13.790 MHz. When the receiver is tuned to the second frequency, its local oscillator will be adjusted to 14.245 MHz. Since this is exactly 455 kHz below the frequency of the strong station, the two signals will produce 455 kHz when they are mixed, and of course the IF amplifier will not reject this signal. If there had been an RF amplifier, the 14.7-MHz signal might have been rejected before reaching the mixer, but without an RF amplifier this receiver cannot adequately reject 14.7 MHz when it is tuned to 13.79 MHz.

Double spotting is harmful because a weak station may be masked by the reception of a nearby strong station at the spurious point on the dial. As a matter of interest, double spotting may be used to calculate the intermediate frequency of an unknown receiver, since the spurious point on the dial is precisely $2f_i$ *below* the correct frequency.

As expected, an improvement in image-frequency rejection will produce a corresponding reduction in double spotting.

6-2.2 Frequency Changing and Tracking

Generally speaking, a frequency changer[1] is a nonlinear resistance having two sets of input terminals and one set of output terminals. The signal from the antenna or from the preceding RF amplifier is fed to one set of input terminals, and the output of the local oscillator is fed to the other set. As was shown in Eq. (4-8), such a nonlinear resistance will have several frequencies present in its output, including the difference between the two input frequencies—in AM this was called the lower sideband. The difference frequency here is the intermediate frequency and is the one to which the output circuit of the mixer is tuned.

The most common types of mixers are the bipolar transistor, FET, dual-gate metal-oxide-semiconductor field-effect transistor (MOSFET) and integrated circuit. All are generally self-excited in domestic receivers, so that the device acts as both oscillator and mixer. When tubes were common, the pentagrid and triode-hexode were made specially for self-excited mixer duty. At UHF and above, crystal (i.e., silicon) diodes have been used as mixers since before World War II, because of their low noise figures. These and other diodes, with even lower noise figures, are still so used, as will be seen in the microwave chapters. Naturally, diode mixers are separately excited.

[1] More commonly called a *mixer*, sometimes a *converter*, and, in the early days of radio, *the first detector*.

Conversion transconductance It will be recalled that the coefficient of nonlinearity of most nonlinear resistances is rather low, so that the IF output of the mixer will be very low indeed unless some preventive steps are taken. The usual step is to make the local oscillator voltage quite large, 1 V rms or more to a mixer whose signal input voltage might be 100 μV or less. That this has the desired effect is shown by term (V) of Eq. (4-8). It is then said that the local oscillator *varies the bias* on the mixer from zero to cutoff, thus varying the transconductance in a nonlinear manner. The mixer amplifies the signal with this varying g_m, and an IF output results.

Like any other amplifying device, a mixer has a transconductance. However, the situation here is a little more complicated, since the output frequency is different from the input frequency. *Conversion transconductance* is defined as

$$g_c = \frac{\Delta i_p \text{ (at the intermediate frequency)}}{\Delta v_g \text{ (at the signal frequency)}} \tag{6-4}$$

The conversion transconductance of a transistor mixer is of the order of 6 mS, which is decidedly lower than the g_m of the same transistor used as an amplifier. Since g_c depends on the size of the local oscillator voltage, the above value refers to optimum conditions.

Separately excited mixer In this circuit, which is shown in Fig. 6-6, one device acts as a mixer while the other supplies the necessary oscillations. In this case, T_1, the FET, is the mixer, to whose gate is fed the output of T_2, the bipolar transistor Hartley oscillator. An FET is well suited for mixer duty, because of the square-law characteristic of its drain current. If T_1 were a dual-gate MOSFET, the RF input would be applied to one of the gates, rather than to the source as shown here, with the local oscillator

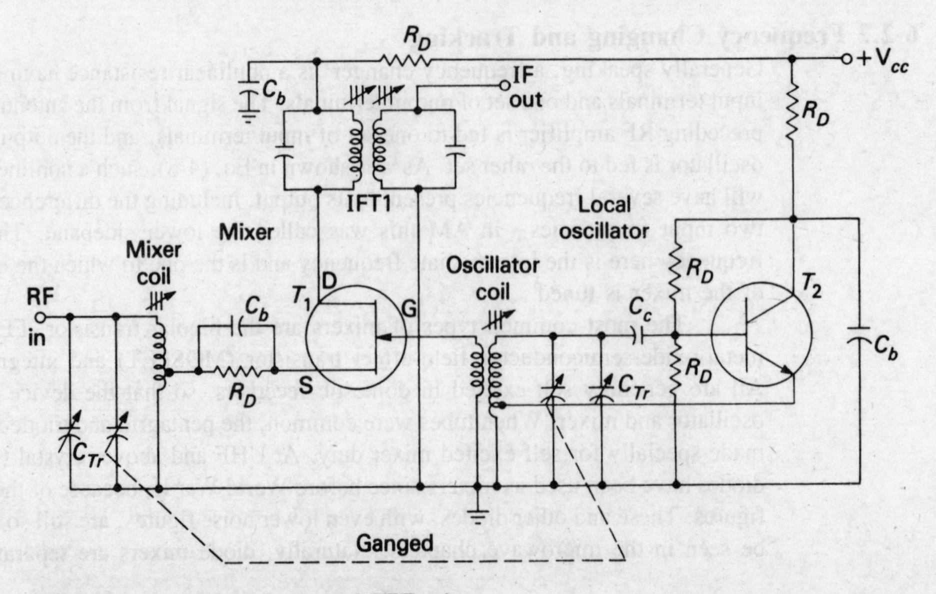

FIGURE 6-6 Separately excited FET mixer.

output going to the other gate, just as it goes to the single gate here. Note the ganging together of the tuning capacitors across the mixer and oscillator coils, and that each in practice has a trimmer (C_{Tr}) across it for fine adjustment by the manufacturer. Note further that the output is taken through a double-tuned transformer (the first IF transformer) in the drain of the mixer and fed to the IF amplifier. The arrangement as shown is most common at higher frequencies, whereas in domestic receivers a self-excited mixer is more likely to be encountered.

Self-excited mixer[2] The circuit of Fig. 6-7 is best considered at each frequency in turn, but the significance of the $L_5 - L_3$ arrangement must first be explained. To begin, it is necessary that the tuned circuit $L_3 - C_G$ be placed between collector and ground, but only for ac purposes. Also, the construction of a ganged capacitor (C_G is one of its sections) is such that in all the various sections the rotating plates are connected to one another by the rotor shaft. Naturally, therefore, the rotor of the gang is grounded. Thus one end of C_G must go to ground, and yet there has to be a continuous path for direct current from HT to collector. One of the solutions to this problem would be the use of an RF choke instead of L_4, and the connection of a coupling capacitor from the bottom of L_6 to the top of L_3; however, the arrangement as shown is equally effective and happens to be simpler and cheaper. It is merely inductive coupling instead of a coupling capacitor, and an extra transformer winding instead of an RF choke.

Now, at the signal frequency, the collector and emitter tuned circuits may be considered as being effectively short-circuited so that (at the RF) we have an amplifier

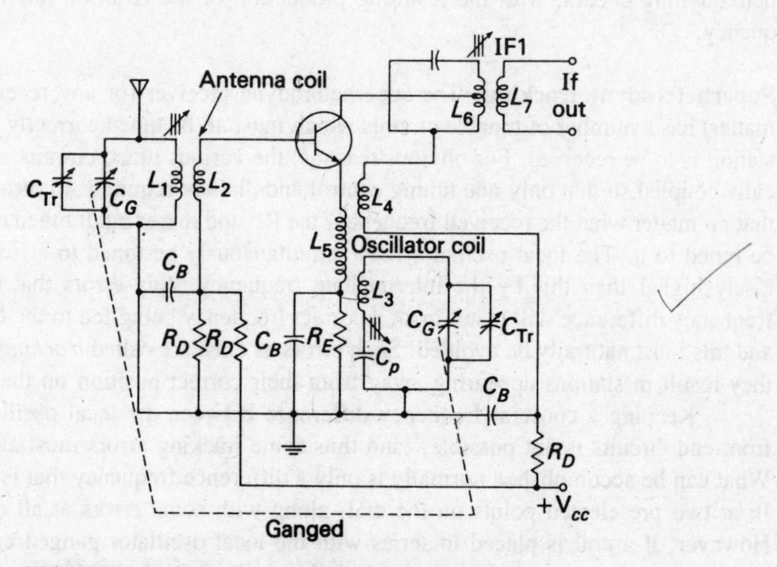

FIGURE 6-7 Self-excited bipolar transistor mixer.

[2] After material in the publication "Germanium and Silicon Transistors and Diodes," by permission of Philips Industries Pty. Ltd.

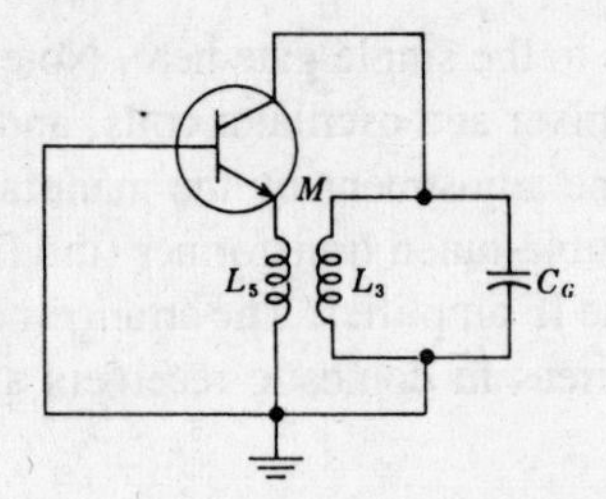

FIGURE 6-8 Mixer equivalent at f_o.

with an input tuned circuit and an output that is indeterminate. At the IF, on the other hand, the base and emitter circuits are the ones which may be considered short-circuited. Thus, at the IF, we have an amplifier whose input comes from an indeterminate source, and whose output is tuned to the IF. Both these ''amplifiers'' are common-emitter amplifiers.

At the local oscillator frequency, the RF and IF tuned circuits may both be considered as though they were short-circuited, so that the equivalent circuit of Fig. 6-8 results (at f_o only). This is seen to be a tuned-collector Armstrong oscillator of the common-base variety.

We have considered each function of the mixer individually, but the circuit performs them all simultaneously, of course. Thus, the circuit oscillates, the transconductance of the transistor is varied in a nonlinear manner at the local oscillator rate and this variable g_m is used by the transistor to amplify the incoming RF signal. Hence heterodyning occurs, with the resulting production of the required intermediate frequency.

Superheterodyne tracking The superheterodyne receiver (or any receiver for that matter) has a number of tunable circuits which must all be tuned correctly if any given station is to be received. For obvious reasons, the various tuned circuits are mechanically coupled so that only one tuning control and dial are required. In turn, this means that no matter what the received frequency, the RF and mixer input tuned circuits must be tuned to it. The local oscillator must simultaneously be tuned to a frequency precisely higher than this by the intermediate frequency. Any errors that exist in this frequency difference will result in an incorrect frequency being fed to the IF amplifier, and this must naturally be avoided. Such errors as exist are called *tracking errors,* and they result in stations appearing away from their correct position on the dial.

Keeping a constant frequency difference between the local oscillator and the front-end circuits is not possible,[3] and thus some tracking errors must always occur. What can be accomplished normally is only a difference frequency that is equal to the IF at two preselected points on the dial, along with some errors at all other points. However, if a coil is placed in series with the local oscillator ganged capacitor, or, more commonly, a capacitor in series with the oscillator coil, then *three-point tracking*

[3] Except with circuitry that is too elaborate and expensive for domestic receivers; this circuitry will be treated under the topic communications receivers (Sec. 6–3).

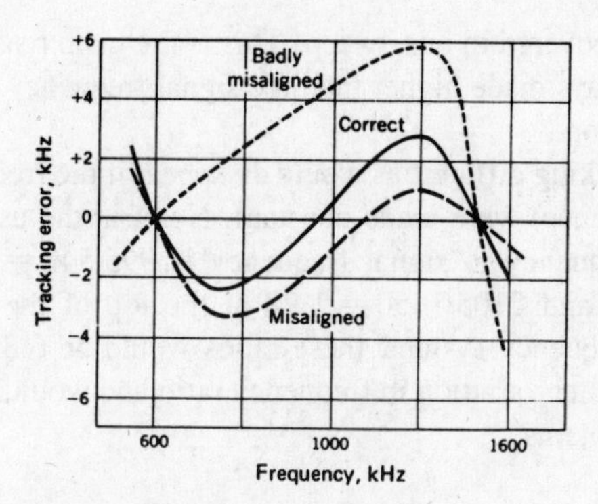

FIGURE 6-9 Tracking curves.

results and has the appearance of the solid curve of Fig. 6-9. The capacitor in question is called a *padding capacitor* or a *padder* and is shown (labeled C_p) in Figs. 6-6 and 6-7. The wanted result has been obtained because the variation of the local oscillator coil reactance with frequency has been altered. The three frequencies of correct tracking may be chosen in the design of the receiver and are often as shown in Fig. 6-9, that is, just above the bottom end of the band (600 kHz), somewhat below the top end (1500 kHz), and at the geometric mean of the two (950 kHz).

It is entirely possible to keep maximum tracking error below 3 kHz, as shown; a value as low as that is generally considered quite acceptable. However, since the padder has a fixed value, it provides correct three-point tracking only if the adjustable local oscillator coil has been preadjusted, i.e., *aligned,* to the correct value. If this has not been done, then incorrect three-point tracking will result, or the center point may disappear completely, as shown in Fig. 6-9.

Local oscillator In receivers operating up to the limit of shortwave broadcasting, that is, 36 MHz, the most common types of local oscillators are the Armstrong and the Hartley. The Colpitts, Clapp, or ultra-audion oscillators are used at the top of this range and above, with the Hartley also having some use if frequencies do not exceed about 120 MHz. Note that all these oscillators are *LC* and that each employs only one tuned circuit to determine its frequency. Where, for some reason, the frequency stability of the local oscillator must be particularly high, AFC (see Secs. 5-3.3 and 6-3.2) or a frequency synthesizer (see Secs. 6-3.2 and 6-5.2) may be used. Ordinary local oscillator circuits are shown in Figs. 6-6 and 6-7.

The frequency range of a broadcast receiver local oscillator is calculated on the basis of a signal frequency range from 540 to 1650 kHz, and an intermediate frequency which is generally 455 kHz. For the usual case of local oscillator frequency above signal frequency, this range is 995 to 2105 kHz, giving a ratio of maximum to minimum frequencies of 2.2:1. If the local oscillator had been designed to be below signal frequency, the range would have been 85 to 1195 kHz, and the ratio would have been 14:1. The normal tunable capacitor has a capacitance ratio of approximately 10:1, giving a frequency ratio of 3.2:1. Hence the 2.2:1 ratio required of the local oscillator operating above signal frequency is well within range, whereas the other system has a frequency range that cannot be covered in one sweep. This is the main reason why the local oscillator frequency is always made higher than the signal frequency in receivers with variable-frequency oscillators.

It may be shown that tracking difficulties would disappear if the frequency ratio (instead of the frequency difference) were made constant. Now, in the usual system, the ratio of local oscillator frequency to signal frequency is 995/540 = 1.84 at the bottom of the broadcast band, and 2105/1650 = 1.28 at the top of the band. In a local-oscillator-below-signal-frequency system, these ratios would be 6.35 and 1.38, respectively. This is a much greater variation in frequency ratio and would result in far more troublesome tracking problems.

6-2.3 Intermediate Frequencies and IF Amplifiers

Choice of frequency The intermediate frequency (IF) of a receiving system is usually a compromise, since there are reasons why it should be neither low nor high, nor in a certain range between the two. The following are the major factors influencing the choice of the intermediate frequency in any particular system:

1. If the intermediate frequency is too high, poor selectivity and poor adjacent-channel rejection result unless sharp cutoff (e.g., crystal or mechanical) filters are used in the IF stages.
2. A high value of intermediate frequency increases tracking difficulties.
3. As the intermediate frequency is lowered, image-frequency rejection becomes poorer. Equations (6-1), (6-2) and (6-3) showed that rejection is improved as the ratio of image frequency to signal frequency is increased, and this, naturally, requires a high IF. Extrapolating, it is seen that image-frequency rejection becomes worse as signal frequency is raised, as was shown by Example 6-1*a* and *b*.
4. A very low intermediate frequency can make the selectivity too sharp, cutting off the sidebands. This problem arises because the Q must be low when the IF is low, unless crystal or mechanical filters are used, and hence the gain per stage is low. Thus a designer is more likely to raise the Q than to increase the number of IF amplifiers.
5. If the IF is very low, the frequency stability of the local oscillator must be made correspondingly higher because any frequency drift is now a larger proportion of the low IF than of a high IF.
6. The intermediate frequency must not fall within the tuning range of the receiver, or else instability will occur and heterodyne whistles will be heard, making it impossible to tune to the frequency band immediately adjacent to the intermediate frequency.

Frequencies used As a result of many years' experience, the foregoing requirements have been translated into specific frequencies, whose use is fairly well standardized throughout the world (but by no means compulsory). These are as follows:

1. Standard broadcast AM receivers [tuning to 540 to 1650 kHz, perhaps 6 to 18 MHz, and possibly even the European long-wave band (150 to 350 kHz)] use

an IF within the 438- to 465-kHz range, with 455 kHz by far the most popular frequency.

2. AM, SSB and other receivers employed for shortwave or VHF reception have a first IF often in the range from about 1.6 to 2.3 MHz, or else above 30 MHz. (Such receivers have two or more different intermediate frequencies. See Sec. 6-3.1.)
3. FM receivers using the standard 88- to 108-MHz band have an IF which is almost always 10.7 MHz.
4. Television receivers in the VHF band (54 to 223 MHz) and in the UHF band (470 to 940 MHz) use an IF between 26 and 46 MHz, with approximately 36 and 46 MHz the two most popular values.
5. Microwave and radar receivers, operating on frequencies in the 1- to 10-GHz range, use intermediate frequencies depending on the application, with 30, 60 and 70 MHz among the most popular.

By and large, services covering a wide frequency range have IFs somewhat below the lowest receiving frequency, whereas other services, especially fixed-frequency microwave ones, may use intermediate frequencies as much as 40 times lower than the receiving frequency.

Intermediate-frequency amplifiers The IF amplifier is a fixed-frequency amplifier, with the very important function of rejecting adjacent unwanted frequencies. It should thus have a frequency response with steep skirts. When the desire for a flat-topped response is added, the resulting recipe is for a double-tuned or stagger-tuned amplifier. Whereas FET and integrated circuit IF amplifiers generally are (and vacuum-tube ones always were) double-tuned at the input and at the output, bipolar transistor amplifiers often are single-tuned. A typical bipolar IF amplifier for a domestic receiver is shown in Fig. 6-10. It is seen to be a two-stage amplifier, with all IF transformers single

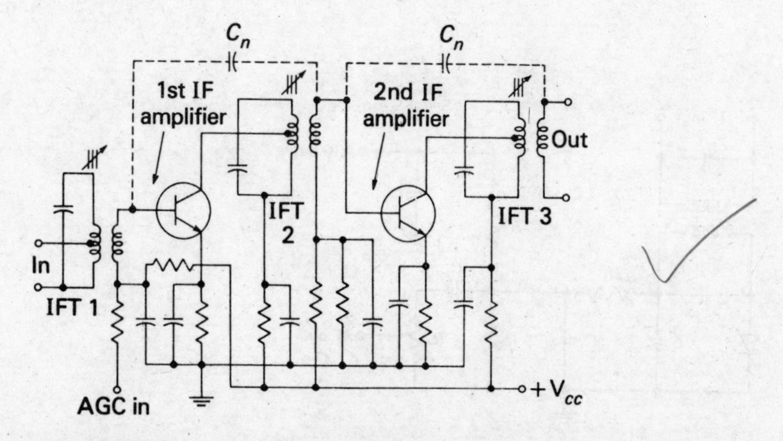

FIGURE 6-10 Two-stage IF amplifier.

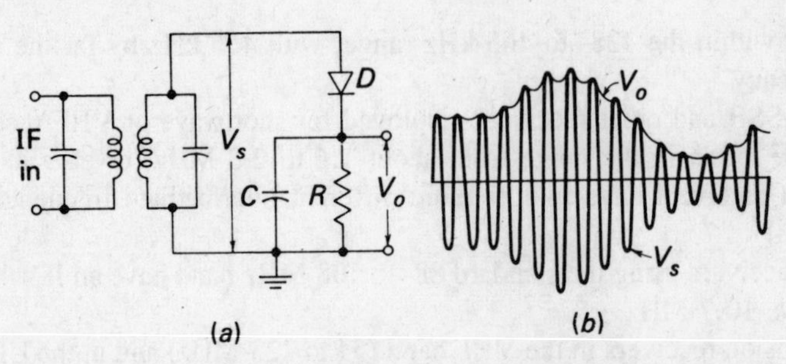

FIGURE 6-11 Simple diode detector. (*a*) Circuit diagram; (*b*) input and output voltages.

tuned. This departure from a single-stage, double-tuned amplifier is for the sake of extra gain, and hence receiver sensitivity.

Although a double-tuned circuit, such as those shown in Figs. 6-11 and 6-12, rejects adjacent frequencies far better than a single-tuned circuit, bipolar transistor amplifiers, on the whole, use single-tuned circuits for interstage coupling. The reason is that greater gain is achieved in this way because of the need for tapping coils in tuned circuits. This tapping may be required to obtain maximum power transfer and a reduction of tuned circuit loading by the transistor. Since transistor impedances may be low, tapping is employed, together with somewhat lower inductances than would have been used with tube circuits. If a double-tuned transformer were used, both sides of it might have to be tapped, rather than just one side as with a single-tuned transformer. Thus a reduction in gain would result. Note also that neutralization may have to be used (capacitors C_n in Fig. 6-10) in the transistor IF amplifier, depending on the frequency and the type of transistor employed.

When double tuning is used, the coefficient of coupling varies from 0.8 times critical to critical; overcoupling is not normally used without a special reason. Finally, the IF transformers are often all made identical so as to be interchangeable.

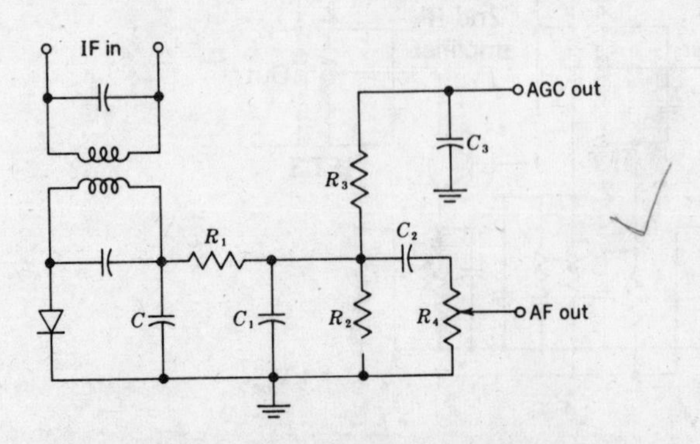

FIGURE 6-12 Practical diode detector.

6-2.4 Detection and Automatic Gain Control (AGC)

Operation of diode detector The diode is by far the most common device used for AM demodulation (or detection), and its operation will now be considered in detail. On the circuit of Fig. 6-11*a*, *C* is a small capacitance and *R* is a large resistance; the parallel combination of *R* and *C* is the load resistance across which the rectified output voltage V_o is developed. At each positive peak of the RF cycle, *C* charges up to a potential almost equal to the peak signal voltage V_s. The difference is due to the diode drop, since the forward resistance of the diode is small (but not zero). Between peaks a little of the charge in *C* decays through *R*, to be replenished at the next positive peak. The result is the voltage V_o, which reproduces the modulating voltage accurately, except for the small amount of RF ripple. Note that the time constant of *RC* combination must be slow enough to keep the RF ripple as small as possible, but sufficiently fast for the detector circuit to follow the fastest modulation variations.

This simple diode detector has the disadvantages that V_o, in addition to being proportional to the modulating voltage, also has a dc component, which represents the average envelope amplitude (i.e., carrier strength), and a small RF ripple. However, the unwanted components are removed in a practical detector, leaving only the intelligence and some second harmonic of the modulating signal.

Practical diode detector A number of additions have been made to the simple detector, and its practical version is shown in Fig. 6-12. The circuit operates in the following manner. The diode has been reversed, so that now the negative envelope is demodulated. This has no effect on detection, but it does ensure that a negative AGC voltage will be available, as will be shown. The resistor *R* of the basic circuit has been split into two parts (R_1 and R_2) to ensure that there is a series dc path to ground for the diode, but at the same time a low-pass filter has been added, in the form of $R_1 - C_1$. This has the function of removing any RF ripple that might still be present. Capacitor C_2 is a coupling capacitor, whose main function is to prevent the diode dc output from reaching the volume control R_4. Whereas it is by no means compulsory to have the volume control immediately after the detector, nevertheless, that is a favorite and convenient place for it. The combination $R_3 - C_3$ is a low-pass filter designed to remove AF components, thus providing a dc voltage whose amplitude is proportional to the carrier strength, and which may be used for automatic gain control.

It can be seen from Fig. 6-12 that the dc diode load is equal to $R_1 + R_2$, whereas the audio load impedance Z_m is equal to R_1 in series with the parallel combination of R_2, R_3 and R_4, assuming that the capacitors have reactances which may be ignored. This will be true at medium frequencies, but at high and low audio frequencies Z_m may have a reactive component, causing a phase shift and distortion as well as an uneven frequency response.

Principles of simple automatic gain control Simple AGC is a system by means of which the overall gain of a radio receiver is varied automatically with the changing strength of the received signal, to keep the output substantially constant. A dc bias voltage, derived from the detector as shown and explained in connection with Fig.

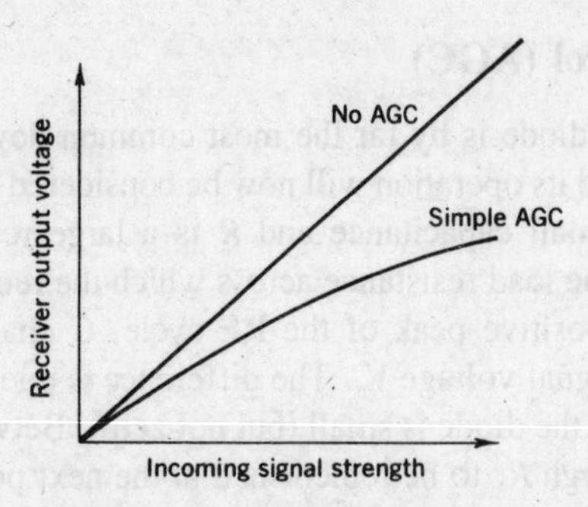

FIGURE 6-13 Simple AGC characteristics.

6-12, is applied to a selected number of the RF, IF and mixer stages. The devices used in those stages are ones whose transconductance and hence gain depends on the applied bias voltage or current. It may be noted in passing that, for correct AGC operation, this relationship between applied bias and transconductance need not be strictly linear, as long as transconductance drops significantly with increased bias. The overall result on the receiver output is seen in Fig. 6-13.

All modern receivers are furnished with AGC, which enables tuning to stations of varying signal strengths without appreciable change in the size of the output signal. Thus AGC "irons out" input signal amplitude variations; and the gain control does not have to be readjusted every time the receiver is tuned from one station to another, except when the change in signal strengths is enormous. In addition, AGC helps to smooth out the rapid fading which may occur with long-distance shortwave reception and prevents the overloading of the last IF amplifier which might otherwise have occurred.

Simple AGC in bipolar transistor receivers The significant difference between FET and bipolar transistor receivers, from the point of view of AGC application, is that in the bipolar case bias current is fed back, so that some power is required. Various methods are used for the application of AGC in transistor receivers. A common one is analogous to that once employed in tube circuits, in that the gain of the relevant amplifiers is controlled by the adjustment (with AGC bias current) of emitter current.

The emitter current is most easily controlled by variation of the base current, provided that sufficient AGC power is available. Since this power must be larger if the stage to be controlled has been stabilized against slow collector current variation, it is preferable to make this stabilization less effective in a stage controlled by AGC. The method of applying automatic gain control of this type is shown at the base of the "first IF amplifier" in Fig. 6-10.

It is possible to increase the available control power by using dc amplification after the detector. Whereas a separate amplifier would be employed for this purpose in an elaborate receiver, the first audio amplifier is much more likely to be used in a broadcast receiver. In such an arrangement, the first AF amplifier must be dc-coupled, in which case care must be taken to ensure that its bias is not upset unduly; otherwise, the amplifier will distort.

Distortion in diode detectors Two types of distortion may arise in diode detectors. One is caused by the ac and dc diode load impedances being unequal, and the other by the fact that the ac load impedance acquires a reactive component at the highest audio frequencies.

Just as modulation index of the modulated wave was defined as the ratio V_m/V_c in Eq. (3-4) and Fig. 3-1, so the modulation index in the *demodulated* wave is defined as

$$m_d = \frac{I_m}{I_c} \tag{6-5}$$

The two currents are shown in Fig. 6-14, and it is to be noted that the definition is in terms of currents because the diode is a current-operated device. Bearing in mind that all these are peak (rather than rms) values, we see that

$$I_m = \frac{V_m}{Z_m} \quad \text{and} \quad I_c = \frac{V_c}{R_c} \tag{6-6}$$

where Z_m = audio diode load impedance, as described previously, and is assumed to be resistive

R_c = dc diode load resistance

The audio load resistance is smaller than the dc resistance. Hence it follows that the AF current I_m will be larger, in proportion to the dc current, than it would have been if both load resistances had been exactly the same. This is another way of saying that *the modulation index in the demodulated wave is higher than it was in the modulated wave applied to the detector*. This, in turn, suggests that it is possible for overmodulation to exist in the output of the detector, despite a modulation index of the applied voltage of less than 100 percent. The resulting diode output current, when the input modulation index is too high for a given detector, is shown in Fig. 6-14*b*. It exhibits *negative peak clipping*. The maximum value of applied modulation index which a diode detector will handle without negative peak clipping is calculated as follows.

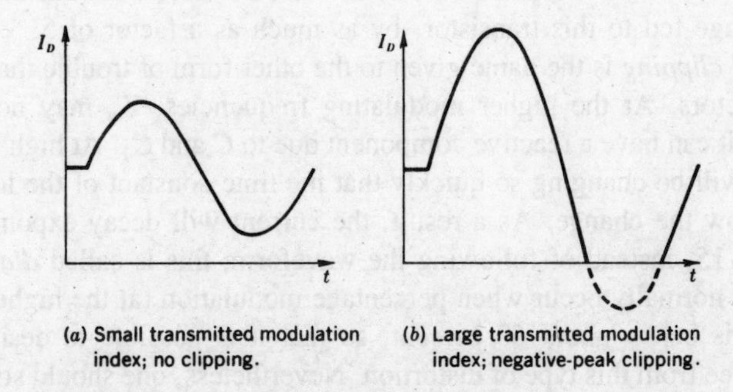

(*a*) Small transmitted modulation index; no clipping.

(*b*) Large transmitted modulation index; negative-peak clipping.

FIGURE 6-14 Detector diode currents.

The modulation index in the demodulated wave will be

$$m_d = \frac{I_m}{I_c} = \frac{V_m/Z_m}{V_c/R_c} = m\frac{R_c}{Z_m} \tag{6-7}$$

Since the maximum tolerable modulation index in the diode output is unity, the maximum permissible transmitted modulation index will be

$$m_{\max} = m_{d,\max}\frac{Z_m}{R_c} = 1\frac{Z_m}{R_c}$$

$$= \frac{Z_m}{R_c} \tag{6-8}$$

Example 6-3 Let the various resistances in Fig. 6-12 be $R_1 = 110\ \text{k}\Omega$, $R_2 = 220\ \text{k}\Omega$, $R_3 = 470\ \text{k}\Omega$ and $R_4 = 1\ \text{m}\Omega$. What is the maximum modulation index which may be applied to this diode detector without causing negative peak clipping?

We have

$$R_c = R_1 + R_2 = 110 + 220 = 330\ \text{k}\Omega$$

$$Z_m = \frac{R_2R_3R_4}{R_2R_3 + R_3R_4 + R_4R_2} + R_1$$

$$= \frac{220 \times 470 \times 1000}{220 \times 470 + 470 \times 1000 + 1000 \times 220} + 110 = 130 + 110$$

$$= 240\ \text{k}\Omega$$

Then

$$m_{\max} = \frac{Z_m}{R_c} = \frac{240}{330} = 0.73 = 73\%$$

Because the modulation percentage in practice (in a broadcasting system at any rate) is very unlikely to exceed 70 percent, this can be considered a well-designed detector. Since bipolar transistors may have a rather low input impedance, which would be connected to the wiper of the volume control and would therefore load it and reduce the diode audio load impedance, the first audio amplifier could well be made a field-effect transistor. Alternatively, a resistor may be placed between the moving contact of the volume control and the base of the first transistor, but this unfortunately reduces the voltage fed to this transistor, by as much as a factor of 5.

Diagonal clipping is the name given to the other form of trouble that may arise with diode detectors. At the higher modulating frequencies, Z_m may no longer be purely resistive; it can have a reactive component due to C and C_1. At high modulation depths, current will be changing so quickly that the time constant of the load may be too slow to follow the change. As a result, the current will decay exponentially, as shown in Fig. 6-15, instead of following the waveform; this is called *diagonal clipping*. It does not normally occur when percentage modulation (at the highest modulation frequency) is below about 60 percent, so that it is possible to design a diode detector that is free from this type of distortion. Nevertheless, one should still be aware of its existence as a limiting factor on the size of the RF filter capacitors.

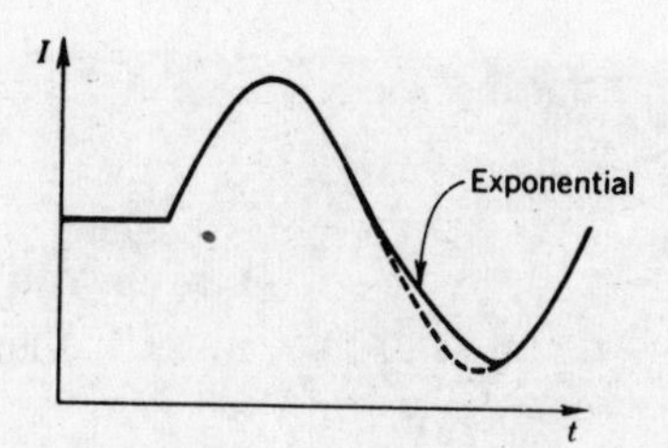

FIGURE 6-15 Diagonal clipping.

6-3 COMMUNICATIONS RECEIVERS

A communications receiver is one whose main function is the reception of signals used for communications rather than for entertainment. It is a radio receiver designed to perform the tasks of low- and high-frequency reception much better than the type of set found in the average household. In turn, this makes the communications receiver useful in other applications, such as the detection of signals from high-frequency impedance bridges (where it is used virtually as a high-sensitivity tuned voltmeter), signal-strength measurement, frequency measurement and even the detection and display of individual components of a high-frequency wave (such as an FM wave with its many sidebands). It is often operated by electronically qualified people, so that any added complications in its tuning and operation are not necessarily detrimental, as they would be in a receiver to be used by the general public.

The communications receiver has evolved from the ordinary home receiver, as the block diagram of Fig. 6-16 and the photograph of Fig. 6-17 demonstrate. Both are, for example, superheterodyne receivers, but in order to perform its tasks the communi-

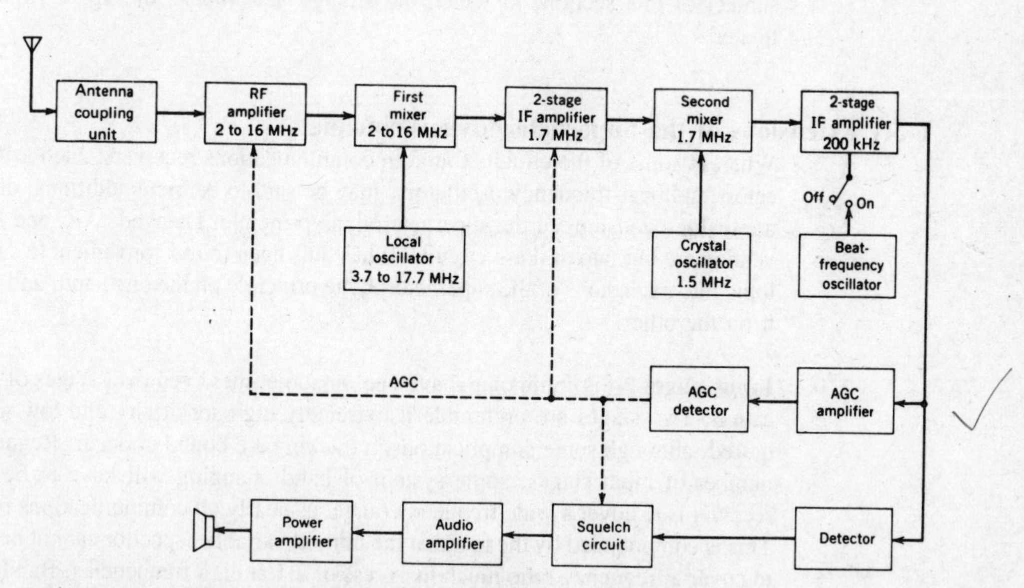

FIGURE 6-16 Basic block diagram of communication receiver.

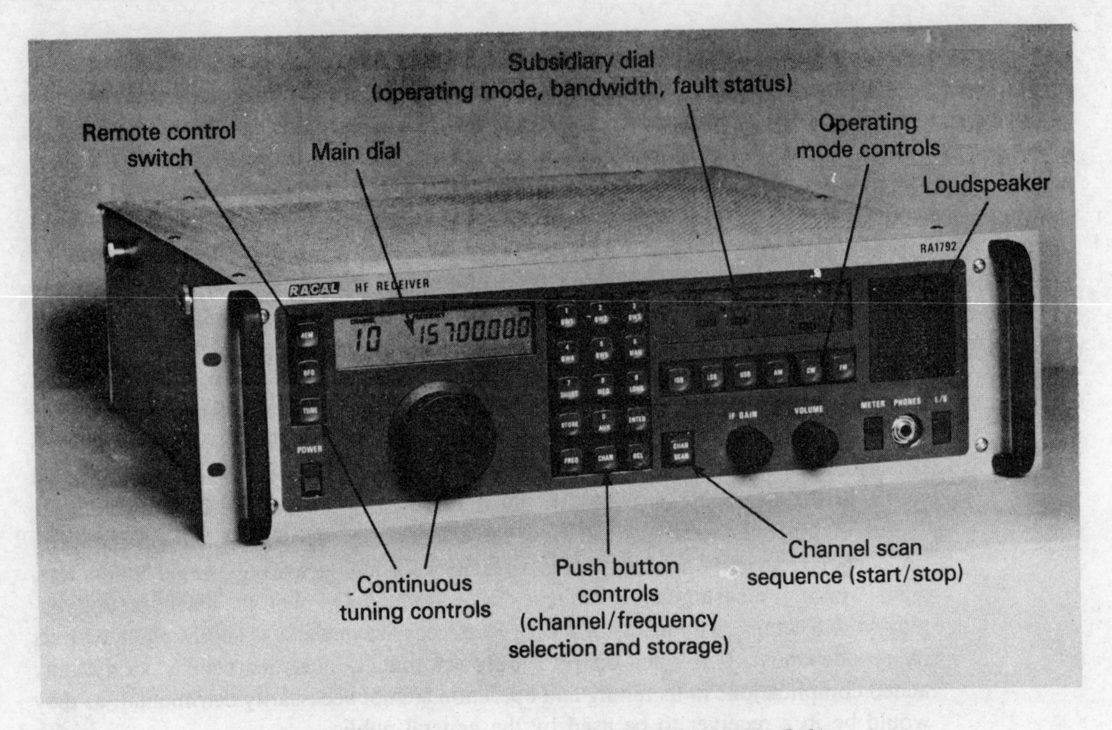

FIGURE 6-17 **Communications receiver.** *(Courtesy of Racal Electronics Pty. Ltd.)*

cations receiver has a number of modifications and added features. These are the subject of this section, in which the strange new blocks of Fig. 6-16 will also be treated.

6-3.1 Extensions of the Superheterodyne Principle

Whereas some of the circuits found in communications receivers, such as tuning indicators and beat-frequency oscillators, may be said to be mere additions, other circuits are really extensions of the superheterodyne principle. Delayed AGC and double conversion are but two of these circuits. It has thus been found convenient to subdivide the topic into extensions of the superheterodyne principle on the one hand, and additions to it on the other.

Input stages It is common to have one, or sometimes even two, stages of RF amplification. Two stages are preferable if extremely high sensitivity and low noise are required, although some complications in tracking are bound to occur. Regardless of the number of input stages, some system of band changing will have to be used if the receiver is to cover a wide frequency range, as nearly all communications receivers do. This is compounded by the fact that the normal variable capacitor cannot be relied upon to cover a frequency ratio much in excess of 2:1 at high frequencies. Band changing is accomplished in either of two ways: by switching in the required RF, mixer and local oscillator coils, or by frequency synthesis. The use of synthesis is a clear favorite in modern receivers.

Fine tuning As the name implies, fine tuning permits stations transmitting on frequencies very close to each other to be separated, or resolved, by the receiver. This ability is essential in a communications receiver. Some good-quality domestic receivers provide an electrical fine-tuning control, using a separate dial. In this method, the local oscillator frequency is varied by means of a trimmer on either side of the main tuning control setting, and in this fashion close stations are separated. Similar results can be achieved with a mechanical vernier control, and indeed this is how fine tuning, or *bandspread,* used to be provided in communications receivers. These days, the use of frequency synthesis, as in the receiver of Fig. 6-17, is almost invariable in commercial communications receivers. See Sec. 6-3.3 for details of this method.

Double conversion Communications receivers and some high-quality domestic AM receivers have more than one intermediate frequency—generally two, but sometimes even more. When a receiver has two different IFs, as does the one shown in block diagram form in Fig. 6-16, it is then said to be a *double-conversion* receiver. The first IF is high, several megahertz or even higher, and sometimes not even a fixed frequency. The second IF is quite low, of the order of 1 MHz or even less. After leaving the RF amplifier, the signal in such a receiver is still mixed with the output of a local oscillator. This is similar to the local oscillator of a domestic receiver, except that now the resulting frequency difference is a good deal higher than 455 kHz. The high intermediate frequency is then amplified by the high-frequency IF amplifier, and the output is fed to a second mixer and mixed with that of a second local oscillator. Since the second local oscillator frequency is normally fixed, this could be a crystal oscillator, and in fact very often is, in nonsynthesized receivers. The low second intermediate frequency is amplified by an LF IF amplifier and then detected in the usual manner.

Double conversion is desirable in communications receivers. As will be recalled from Sec. 6-2.3, the intermediate frequency selected for any receiver is bound to be a compromise since there are equally compelling reasons why it should be both high and low. Double conversion avoids this compromise. The high first intermediate frequency pushes the image frequency farther away from the signal frequency and therefore permits much better attenuation of it. The low second IF, on the other hand, has all the virtues of a low fixed operating frequency, particularly sharp selectivity and hence good adjacent-channel rejection.

Please note that the *high intermediate frequency must come first.* If this does not happen, the image frequency will be insufficiently rejected at the input and will become mixed up with the proper signal, so that no amount of high IF stages will make any difference afterward.

Having two such intermediate frequencies provides a combination of higher image and adjacent-frequency rejection than could be achieved with the simple superheterodyne system. It should be noted, on the other hand, that double conversion offers no great advantages for broadcast or other medium-frequency receivers. However, it is

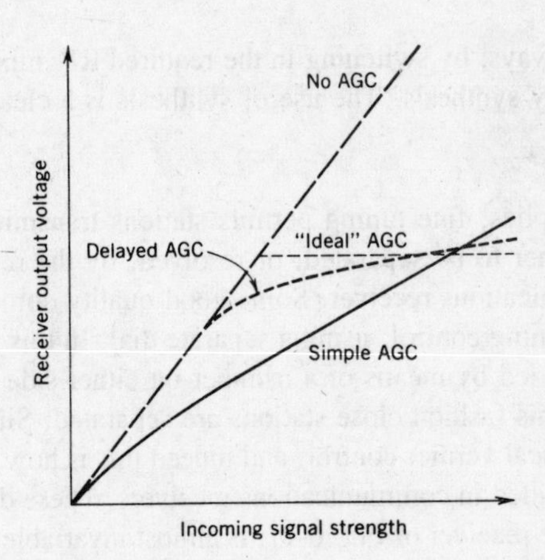

FIGURE 6-18 Various AGC characteristics.

essential for receivers operating in crowded HF and other bands. See also Sec. 6-5.1 and [2] for a further discussion of double conversion and its achievement with novel techniques.

Delayed AGC Simple AGC, as treated in Sec. 6-2.4, is clearly an improvement on no AGC at all, in that the gain of the receiver is reduced for strong signals. Unfortunately, as Figs. 6-13 and 6-18 both show, even weak signals do not escape this reduction. Figure 6-18 also shows two other AGC curves. The first is an "ideal" AGC curve. Here, no AGC would be applied until signal strength was considered adequate, and after this point a constant average output would be obtained no matter how much more the signal strength rose. The second is the *delayed* AGC curve. This shows that AGC bias is not applied until the signal strength has reached a predetermined level, after which bias is applied as with normal AGC, but more strongly. As the signal strength then rises, receiver output also rises, but relatively slightly. The problem of reducing the gain of the receiver for weak signals has thus been avoided, as with "ideal" AGC.

A very common method of obtaining delayed AGC is shown in Fig. 6-19. It uses two separate diodes: the detector and the AGC detector. These can be connected either to separate transformer windings, as shown, or both may be connected to the secondary without too much interference. As indicated, a positive bias is applied to the cathode of the AGC diode, to prevent conduction until a predetermined signal level has been reached. A control is often provided, as shown, to allow manual adjustment of the bias on the AGC diode, and hence of the signal level at which AGC is applied. If weak stations are mostly likely to be received, the delay control setting may be quite high (i.e., no AGC until signal level is fairly high). Nevertheless, it should be made as low as possible, to prevent overloading of the last IF amplifier by unexpected stronger signals.

The method just described works well with FETs, and also with bipolar transistors if the number of stages controlled is large enough. If in the latter case fewer than three stages are being controlled, it may not be possible to reduce the gain of the

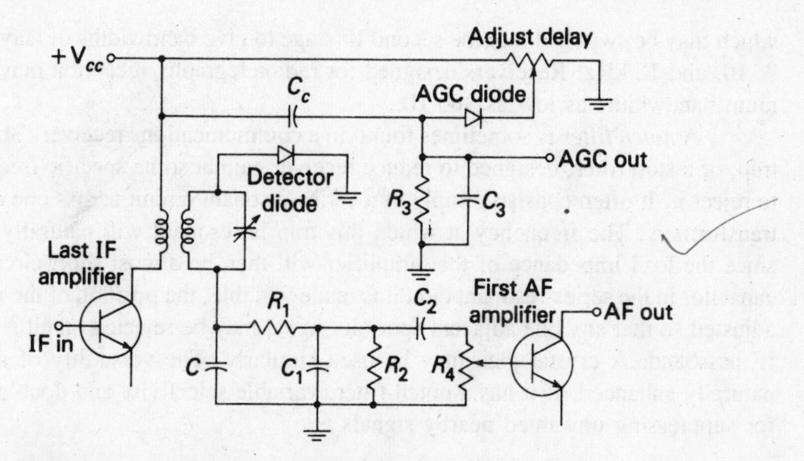

FIGURE 6-19 Delayed AGC circuit.

receiver sufficiently for very strong signals, because of collector leakage current. If that is so, a secondary method of AGC is sometimes used together with simple AGC, the overall result being not unlike delayed AGC. A diode is here employed for variable damping, in a manner akin to that used in the ratio detector, as described in Sec. 6-4.4.

Variable sensitivity and selectivity The ratio of the highest to the lowest signal strengths which a communications receiver may have to cope with could be as high as 10^5:1. This means that the receiver must have sufficient sensitivity to amplify fully very weak signals, while also being capable of having its gain reduced by ACG action by a ratio of 10^5:1, or 100 dB, so as not to overload on the strongest signal. Even the best ACG system is not capable of this performance. Apart from the alarming variations in output that could occur, there is also the risk of overloading several of the IF amplifiers, especially the last one, and also the demodulator diode. To prevent the distortion which would follow, as well as possible permanent damage, the most sensitive communications receivers incorporate a sensitivity control.[4] This control generally consists of a potentiometer which varies the bias on the RF amplifier and is, in fact, an RF gain control. The AGC is still present, but it now acts to keep the sensitivity of the receiver to the level determined by the setting of this control. The receiver is now considerably more versatile in handling varying input signal levels.

The selectivity, or, to be more precise, the bandwidth, of the low-frequency IF amplifier may be made variable over a range that is commonly 1 to 12 kHz. The largest bandwidth permits reception of high-quality broadcasts; whereas the smallest (although it greatly impairs this quality) reduces noise and therefore increases intelligibility and will also reduce adjacent-channel interference. Variable selectivity is achieved in practice by switching in crystal, ceramic, or mechanical filters. A set is provided, any of

[4]Unfortunately, this statement does not work in reverse. Mere possession of a sensitivity control by a receiver does not guarantee that it is, in fact, a *sensitive* receiver.

which may be switched into the second IF stage to give bandwidths of (say) 1, 2, 4, 6, 8, 10, and 12 kHz. Receivers designed for radiotelegraphy reception may have minimum bandwidths as low as 300 Hz.

A *notch filter* is sometimes found in a communications receiver. This is a wave trap, or a stop filter, designed to reduce receiver gain at some specific frequency so as to reject it. It often çonsists simply of a series-resonant circuit across one of the LF IF transformers. The frequency at which this trap is resonant will naturally be rejected since the load impedance of that amplifier will then be almost short-circuited. If the capacitor in the series-resonant circuit is made variable, the position of the notch can be adjusted so that any one adjacent spurious signal may be rejected on either side of the IF passband. A crystal gate may be used similarly. The versatility of a receiver is naturally enhanced, if it has a notch filter, variable selectivity and double conversion for suppressing unwanted nearby signals.

Blocking If a radio receiver is tuned to a weak signal, naturally the developed AGC will be low and the front-end gain high. If a strong signal not too distant in frequency is now received, then unless it is properly rejected, it could develop substantial AGC voltage. Such a high AGC, caused by a spurious signal, could reduce the gain of the receiver, perhaps to the point of making the wanted signal inaudible. This situation is unwelcome; if the interfering signal is intermittent, it is intolerable. A receiver whose AGC system has very little reaction to the nearby spurious signals is said to have good *blocking*. A good way of showing how blocking is defined and measured is to state how it is quoted in receiver specifications. The Redifon R 551 is a receiver with very good blocking performance, quoted by the manufacturers as follows: ''With a 1 mV EMF A0 (SSB, 1000 Hz tone) wanted signal, a simultaneous 6 V EMF A0 unwanted signal (at least 20 kHz from wanted signal) will not reduce the wanted AF output by more than 3 dB.''

Needless to say, very high IF rejection of adjacent signals is needed to produce such excellent blocking performance. Yet this performance is required in SSB receivers, and in all other instances of working in crowded frequency bands.

6-3.2 Additional Circuits

Whereas the foregoing circuits and characteristics were most easily classified as extensions of the superheterodyne system, the following are best thought of as additions. It must however be admitted that the subdivision, although convenient, is at times a little artificial.

Tuning calibration Tuning calibration consists of having a built-in crystal oscillator, usually nonsinusoidal, operating at 500 to 1000 kHz, whose output may be fed to the input of the receiver by throwing the appropriate switch. With the *beat-frequency oscillator* in operation (to follow), whistles will now be heard at 500- or 1000-kHz intervals, especially since the crystal oscillator works into a resistive load, so as not to attenuate harmonics of the fundamental frequency. (Refer also to Sec. 1-4.1.) The calibration of the receiver may now be corrected by adjustment of the pointer or cursor, which must, of course, be movable independently of the gang. An elaborate receiver, which is also tunable to frequencies above 30 MHz, may have a built-in crystal amplifier, whose function is to amplify the higher harmonics of the crystal oscillator to make frequency calibration easier at those frequencies. Synthesized receivers do not require this facility.

Beat-frequency oscillator A communications receiver should be capable of receiving transmissions of Morse code, i.e., pulse-modulated RF carrier. In the diode detector of a normal receiver, there is no provision for registering the difference between the presence and the absence of a carrier.[5] Accordingly, such pulse-modulated dots, dashes and spaces would produce no output whatever from the detector.

In order to make Morse code audible, the receiver has a built-in BFO, normally at the detector, as shown on the block diagram of Fig. 6-16. The BFO is not really a beat-frequency oscillator at all; it is merely a simple *LC* oscillator. The Hartley BFO is one of the favorites, operating at a frequency of 1 kHz or 400 Hz above or below the last intermediate frequency. When the IF is present, a whistle is heard in the loudspeaker, so that it is the combination of the receiver, detector, input signal and this extra oscillator which has now become a beat-frequency oscillator. Since signal is present only during a dot or a dash in Morse code, only these are heard; thus the code can be received satisfactorily, as can radiotelegraphy. To prevent interference, the BFO is switched off when nontelegraph reception is resumed.

Noise limiter A fair proportion of communications receivers are provided with noise limiters. The name is a little misleading since it is patently not possible to do anything about random noise in an AM receiving system (it is possible to reduce random noise in FM, as will be seen). An AM noise limiter is really an *impulse-noise limiter,* a circuit for reducing the interfering noise pulses created by ignition systems, electrical storms or electrical machinery of various types. This is often done by automatically silencing the receiver for the duration of a noise pulse, this being preferable to a loud, sharp noise in the loudspeaker or headphones. In a common type of noise limiter, a diode is used in conjunction with a differentiating circuit. The limiter circuit provides a negative voltage as a result of the noise impulse or any very sharp voltage rise, and this negative voltage is applied to the detector, which is thus cut off. The detector then remains cut off for the duration of the noise pulse, a period that generally does not exceed a few hundred milliseconds. It is essential to provide a facility for switching off the noise limiter, or else it will interfere with Morse code or radiotelegraphy reception.

There are many different types of noise limiters for impulse noise; cf. [3].

Squelch (muting) When no carrier is present at the input, i.e., in the absence of transmissions on a given channel or between stations, a sensitive receiver will produce a disagreeable amount of loud noise. This is because AGC disappears in the absence of any carrier; the receiver acquires its maximum sensitivity and amplifies the noise

[5] This is not strictly true since there are two ways of registering the difference, but neither is satisfactory for Morse code and dial calibration. First, there is the fact that noise comes up strongly when the carrier disappears; second, a signal-strength meter or tuning indicator would show the presence of a carrier, but much too slowly.

present at its input. In some circumstances this is not particularly important, but in many others it can be annoying and tiring. Systems such as those used by the police, ambulances and coast radio stations, in which a receiver must be monitored at all times but transmission is sporadic, are the principal beneficiaries of squelch. It enables the receiver's output to remain cut off unless the carrier is present. Apart from eliminating inconvenience, such a system must naturally increase the efficiency of the operator. Squelch is also called *muting* or *quieting*. Quiescent (or quiet) AGC and Codan (carrier-operated device, antinoise) are similar systems.

The squelch circuit, as shown in Fig. 6-20, consists of a dc amplifier to which AGC is applied and which operates upon the first audio amplifier of the receiver. When the AGC voltage is low or zero, the dc amplifier, T_2, draws current so that the voltage drop across its load resistor R_1 cuts off the audio amplifier, T_1; thus no signal or noise is passed. When the AGC voltage becomes sufficiently negative to cut off T_2, this dc amplifier no longer draws collector current, so that the only bias now on T_1 is its self-bias, furnished by the bypassed emitter resistor R_2 and also by the base potentiometer resistors. The audio amplifier now functions as though the squelch circuit were not there.

Resistor R_3 is a dropping resistor, whose function it is to ensure that the dc voltage supplied to the collector and base potentiometer of T_1 is higher than the dc voltage supplied (indirectly) to its emitter. Manual adjustment of R_3 will allow the cut-in bias of T_2 to be varied so that quieting may be applied for a selected range of AGC values. This facility must be provided; otherwise weak stations, not generating sufficient AGC, might be cut off. The squelch circuit is normally inserted immediately after the detector, as shown in Figs. 6-16 and 6-20.

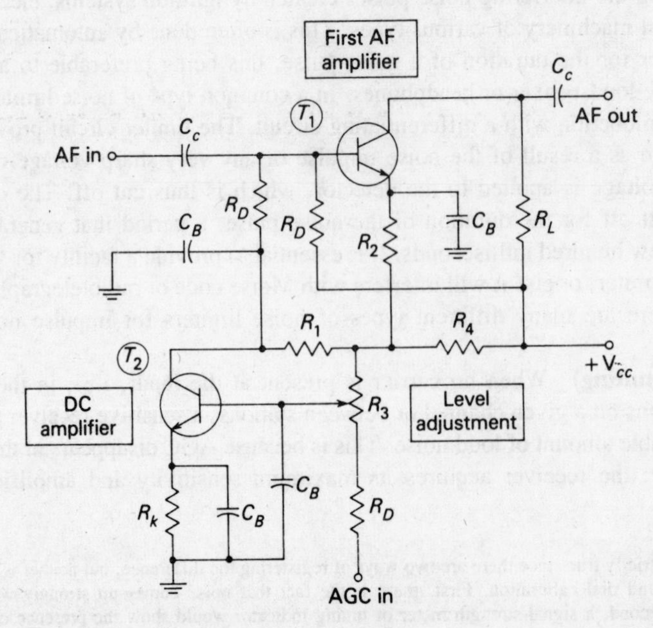

FIGURE 6-20 Typical squelch circuit.

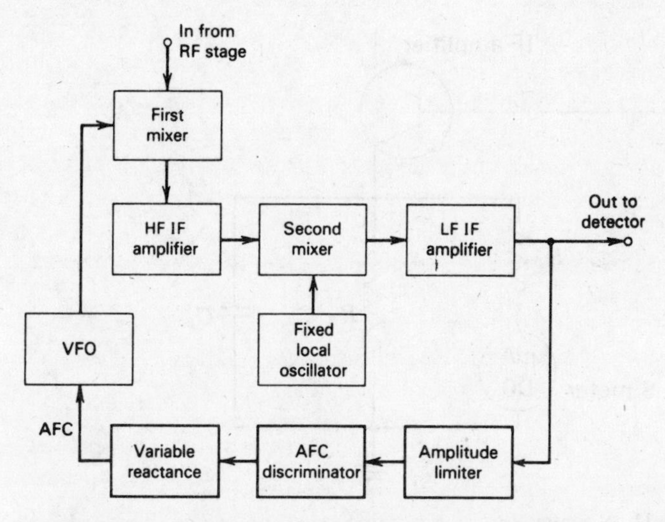

FIGURE 6-21 Block diagram of receiver with AFC.

Automatic frequency control As will be recalled from Sec. 6-3.3, the heart of an AFC circuit is a frequency-sensitive device, such as the phase discriminator, which produces a dc voltage whose amplitude and polarity are proportional to the amount and direction of the local oscillator frequency error. This dc control voltage is then used to vary, automatically, the bias on a variable-reactance device, whose output capacitance is thus changed. This variable capacitance appears across the (first) local oscillator coil, and (in the manner described in Sec. 5-3.3) the frequency of this variable-frequency oscillator (VFO)[6] is automatically kept from drifting with temperature, line voltage changes or component aging. A block diagram of a receiver AFC system is shown in Fig. 6-21.

It is worth noting that the number of extra stages required to provide AFC is much smaller in a double-conversion receiver than in the stabilized reactance modulator, since most of the functions required are already present. On the other hand, not all receivers require AFC, especially not synthesized ones. Those that benefit most from its inclusion are undoubtedly nonsynthesized SSB receivers, whose local oscillator stability must be exceptionally good to prevent drastic frequency variations in the demodulated signal.

Metering A built-in meter with a function switch is very often provided. It is very helpful diagnosing any faults that may occur, since it measures voltages at key points in the receiver. One of the functions (sometimes the sole function) of this meter is to measure the incoming signal strength. It is then called an *S meter,* and very often reads the collector current of an IF amplifier to which AGC is applied, as shown in Fig. 6-22. Since this collector current decreases as the AGC goes up, the meter has its zero on the

[6]This term is commonly used in such a situation.

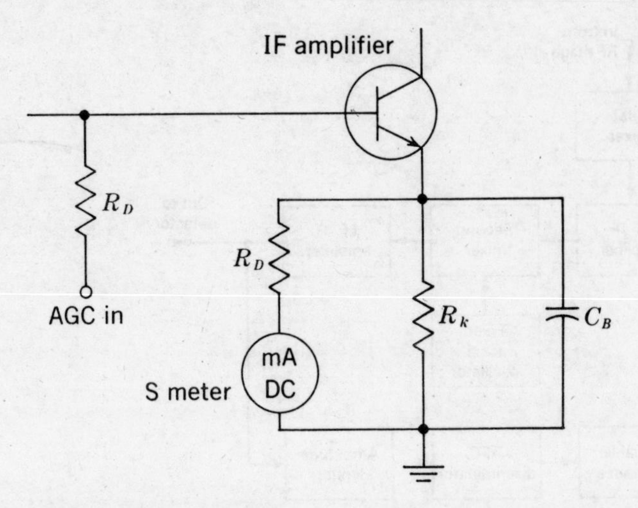

FIGURE 6-22 S meter.

right-hand side. The S meter may sometimes be in an unbalanced bridge and hence forward-reading. In either case, the calibration of the meter is likely to be quite arbitrary because of the great variation of the sensitivity of the receiver through the bands, especially if there is a sensitivity control or adjustable delayed AGC.

A receiver with an S meter is definitely more versatile than one without. Not only can tuning to a wanted signal now be more accurate, but the receiver can also be used as a relative signal-strength meter and even as the detector for an RF impedance bridge. Moreover, it can also be used for applications such as tuning individually to the various sideband frequencies of an FM signal. This can determine the presence of those components and demonstrate the disappearance of the carrier for certain values of modulation index, from which readings deviation and linearity of the FM source may be determined (see Sec. 5-1.3). As will be seen in Sec. 6-3.3, microprocessor-controlled receivers can have very elaborate metering, with digital display of many functions, including the relative signal strength.

FM and SSB reception Many receivers have provision for the reception of FM, either the narrowband FM used by mobile networks or the high-quality broadcast transmissions in the 88- to 108-MHz band. To allow FM reproduction, a receiver requires broadband IF stages, an FM demodulator and an amplitude limiter; these are described later in this chapter.

Most present-day communications receivers have facilities for single-sideband reception. Basically this means that a product detector (see Sec. 6-5) must be provided, but it is also very helpful if there is an AFC system present, as well as variable selectivity (preferably with crystal or mechanical filters), since the bandwidth used for SSB is a good deal narrower than for ordinary AM.

Diversity reception Diversity reception is not so much an additional circuit in a communications receiver as a specialized method of using such receivers. There are two forms: *space diversity* and *frequency diversity*.

Whereas AGC helps greatly to minimize some of the effects of fading, it cannot help when the signal fades into the noise level. Diversity-reception systems make use of the fact that although fading may be severe at some instant of time, some frequency, and some point on earth, it is extremely unlikely that signals at different points or different frequencies will fade simultaneously. (See also Sec. 8-2.2 for a detailed description of fading, its various causes and its effects upon reception.)

Both systems are in constant use, by communications authorities, commercial point-to-point links and the military. In space diversity, two or more receiving antennas are employed, separated by about nine or more wavelengths. There are as many receivers as antennas, and arrangements are made to ensure that the AGC from the receiver with the strongest signal at the moment cuts off the other receivers. Thus only the signal from the strongest receiver is passed to the common output stages.

Frequency diversity works in much the same way, but now the same antenna is used for the receivers, which work with simultaneous transmissions at two or more frequencies. Since frequency diversity is more wasteful of the frequency spectrum, it is used only where space diversity cannot be employed, such as in restricted spaces where receiving antennas could not have been separated sufficiently. Ship-to-shore and ship-to-ship communications are the greatest users of frequency diversity at HF.

As described, both systems are known as *double-diversity* systems, in that there are two receivers, employed in a diversity pattern. Where conditions are known to be critical, as in *tropospheric scatter* communications, *quadruple diversity* is used. This is a space-diversity system which has receiver arrangements as just described, with two transmitters at each end of the link arranged in the same way as the receivers. This ensures that signals of adequate quality will be received under even the worst possible conditions. (See Sec. 8-2.4, where tropospheric scatter is described fully and the use of diversity with it is discussed, and also [4].)

There is one snag, unfortunately, that applies to diversity systems and limits their use in voice communications. Since, in general, each signal travels over a slightly different path, the audio output will have a phase difference when compared with that of the other receiver(s). As a result, diversity reception is used very often for telegraph or data transmission (i.e., pulses); however, present diversity systems for voice communications leave much to be desired, unless some form of pulse modulation is employed, as described in Sec. 13-2.

6-3.3 Additional Systems

The preceding section dealt with additional circuits that may be included in communications receivers in order to improve their performance. This section will cover additional, large-scale systems similarly included in professional communications receivers. They are frequency synthesis and microprocessor control, and will now be discussed in turn. It should be added that these systems can also be found elsewhere in communications equipment, for example in transmitters or signal generators. Thus their treatment here is mostly for convenience of presentation.

Frequency synthesizers Synthesis is the making up of a whole by combining the elements, and this is just how a frequency synthesizer produces its output frequency. It is a piece of equipment capable of generating a very large number of extremely stable frequencies within some design range, while generally employing only one crystal. After the first synthesizers of the early 1940s left the laboratory, their first common application was in military multichannel single-sideband generating equipment. Their use then spread to commercial SSB tunable transmitters, and then to high-quality SSB receivers and test-bench signal generators. With advances in integrated circuits and digital techniques, the use of frequency synthesizers has spread to most applications where a range of stable frequencies may be required, including most HF commercial receivers and signal generators. Also, the application of synthesis has outgrown the HF range to very much higher frequencies. Indeed, the frequency range of the synthesizer shown in Fig. 6-24 extends to 500 MHz, and still higher frequencies are common.

Original synthesizers were of the multiple-crystal variety. They had as many oscillators as decades, and each oscillator was furnished with 10 crystals. The wanted frequency was obtained by switching the appropriate crystal into each oscillator, and the synthesizer then mixed the outputs to produce the desired output. This was done to avoid the bulk of filters and vacuum-tube multipliers, dividers and *spectrum generators* (see next section). The biggest problem of such a synthesizer was trying to keep all the crystals to the accuracy and stability demanded by modern standards; the spare parts situation was a major headache, as well.

The advent of transistors, integrated circuits, and then miniature filters ensured that the multicrystal synthesizer was superseded. Modern types use generally only one crystal oscillator (and one crystal) on which all due care may be lavished to ensure the required stability. Current synthesizers fall into a number of categories, as now treated.

Direct synthesizers A direct synthesizer is a piece of apparatus in which multiples and submultiples of a single crystal oscillator frequency are combined by addition and/or subtraction to provide a very wide selection of stable output frequencies. The great advantage of such a system is that the accuracy and frequency stability of the output signal are equal to those of the crystal oscillator. The problems involved in the design and maintenance of a single-frequency oscillator of extreme precision and stability are much simpler than those associated with multifrequency oscillators. Furthermore, as crystals and techniques improve, the stability of this synthesizer is improved (if desired) simply by replacing the master oscillator. This is the reason why many synthesizers have master oscillators as separate modules.

As shown by the block diagram of Fig. 6-23 and the illustration of Fig. 6-24, the direct synthesizer lends itself ideally to modular construction, with extra self-contained decades added as required. The circuit has also lent itself to the use of large-scale integration (LSI).

A convenient method of explaining the operation and reasons for certain circuit arrangements is to show how a typical frequency is selected. Let this frequency be 34,970.265 kHz, as shown in Fig. 6-23. It will be seen immediately, for instance, that

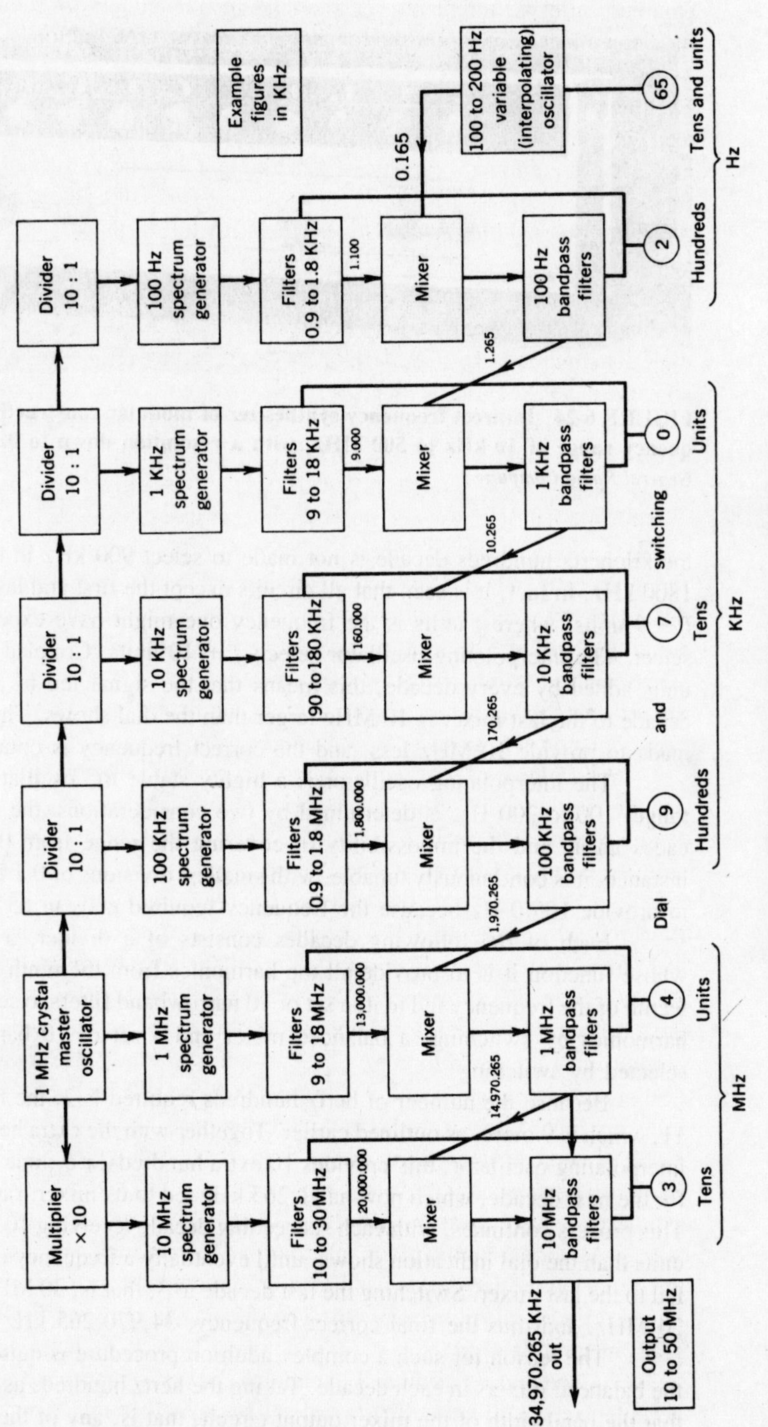

FIGURE 6-23 Direct synthesizer block diagram.

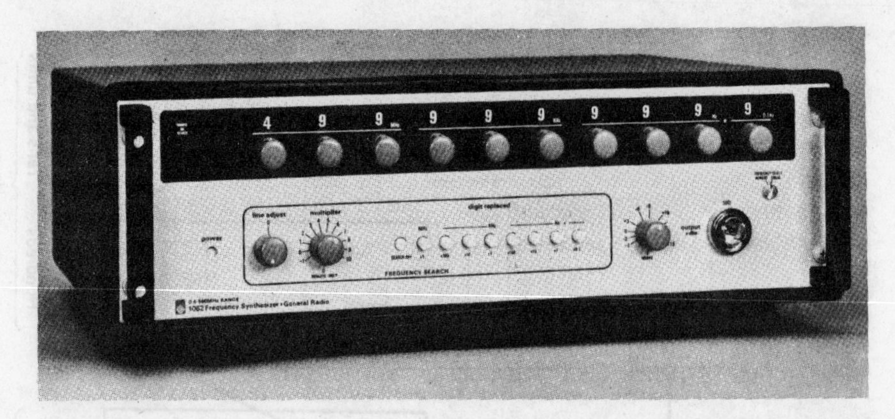

FIGURE 6-24 Indirect frequency synthesizer of modular construction, covering the frequency range of 10 kHz to 500 MHz, with a resolution down to 0.1 Hz. *(Courtesy of the General Radio Company.)*

the kilohertz hundreds decade is not made to select 900 kHz in this case, but rather 1800 kHz. In fact, it is seen that all circuits except the first and last are made to select $f + 9$ units, where f units is the frequency one might have expected that decade to select. The interpolating oscillator selects $f + 10$ units. Coupled with the extra nine units added by every decade, this means that the signal fed by the megahertz units decade to the last mixer is 10 MHz larger than the dial shows. Thus the last decade is made to provide 10 MHz less, and the correct frequency is obtained.

The interpolating oscillator is a highly stable *RC* oscillator whose frequency range, 100 to 200 Hz, is determined by two considerations: the operation of the decades ahead and the impossibility of covering the range from 0 to 100 Hz. In this instance, it is continuously tunable, with smallest divisions of 0.1 Hz, and is here made to provide 165.0 Hz because the frequency required ends in 65 Hz.

Each of the following decades consists of a divider, a spectrum generator whose function it is to provide all the harmonics from the ninth through to the eighteenth of the frequency fed to it, a set of 10 narrowband filters to select any one of these harmonics by switching, a balanced mixer and a set of 10 bandpass filters, again selected by switching.

Because the number of hertz hundreds required is 2, the first decade provides 11, which is 9 more, as outlined earlier. Together with the extra hertz hundred from the interpolating oscillator, this provides 10 extra hundreds, i.e., one extra thousand hertz for the next decade, which now has 1.265 kHz fed to its mixer, rather than 0.265 kHz. This process continues, with each succeeding decade receiving 10 more of the previous units than the dial indication shows, until eventually a frequency of 14,970.265 kHz is fed to the last mixer. Switching the last decade to 3, that is, 30 MHz, really brings only 20 MHz, and thus the final correct frequency, 34,970.265 kHz, is obtained.

The reason for such a complex addition procedure is quite simple: it concerns the balanced mixers in each decade. Taking the hertz hundreds as an example, we find that the bandwidth of the mixer output circuit, that is, any of the 10 bandpass filters, must be 100 Hz. This is so because the incoming frequency from the interpolating oscillator can be anywhere within the range from 100 to 200 Hz, depending on the frequency required. If the extra nine units were not added, the frequencies going into this mixer would be 65 Hz from the interpolating oscillator and 200 Hz from the filter above, yielding outputs of 265 and 135 Hz. These are fed to a filter whose bandpass stretches from 200 to 300 Hz, so that it might be pardoned for not suppressing the spurious frequency of 135 Hz at all well. In the system as used, however, the mixer receives 165 and 1100 Hz and puts out 1265 and 935 Hz. The filter now passes the 1200- to 1300-Hz range and is much better able to suppress the spurious frequency of 935 Hz.

The direct synthesizer is a good piece of equipment, but it does have some drawbacks. The first disadvantage is the possible presence of spurious frequencies generated by the many mixers used. Extensive filtering and extremely careful selection of operating frequencies are required, and it should be pointed out that the method shown here is but one of many. Spurious frequency problems increase as the output frequency range increases and channel spacing is reduced.

The second drawback is serious only in certain applications, such as military transceivers, in which an unstable output is better than none. It will be noted that, should the crystal oscillator in a direct synthesizer fail, no output will be available at all. In indirect synthesizers, on the other hand, an output of reduced stability will be available under these conditions, as will be shown. Finally, the direct synthesizer may suffer unduly from so-called *wideband phase noise*, as will be discussed in the next section.

The result of these drawbacks is that direct synthesizers have just about disappeared, as such. However, the circuitry described above is still in widespread use, but as an integral part of an indirect synthesizer rather than an instrument in its own right.

Indirect synthesizers The required frequency range in most synthesizers nowadays is obtained from a variable voltage-controlled oscillator (VCO), whose output is corrected by comparison with that of a reference source. This inbuilt source is virtually a direct synthesizer. As shown in Fig. 6-25, the phase comparator obtains an output from the VCO, compares it with the output of the stable reference source and produces a dc controlling output voltage whenever the VCO output is incorrect. The dc correcting voltage forms the basis of the *automatic phase-correction* (APC) loop, whose output is applied to a voltage-variable capacitance, which in turn pulls the VCO into line. The APC loop, or *phase-locked loop*, is explained in detail [5], but it can be quite similar to

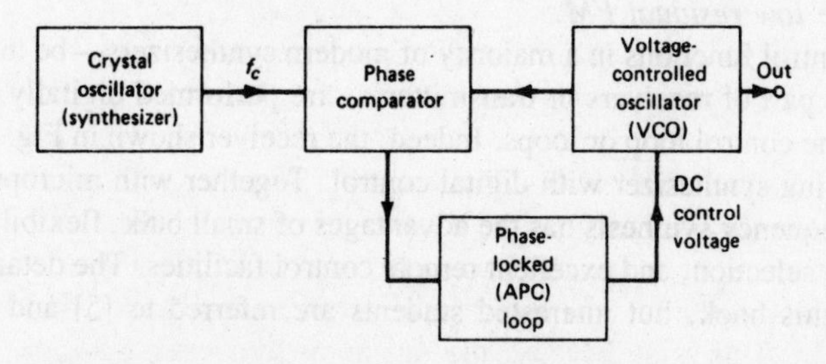

FIGURE 6-25 Basic block diagram of indirect synthesizer.

an *automatic frequency correction* (AFC) loop, treated here in Sec. 5-3.3. The overall arrangement is known variously as a stabilized master oscillator (SMO), a phase-lock synthesizer or even as a VCO synthesizer with a phase-locked loop.

There are various possible arrangements for comparing the output of the VCO with that of the reference, via one or more phase-locked loops. The simplest one to think about is that in which all the circuits of the reference source (i.e., direct synthesizer) are duplicated in the VCO. Whenever a frequency is selected, suitable capacitors/inductors are switched in to provide the VCO with an approximately correct frequency, which is finally corrected by the APC loop. Another possibility is to have each decade of the direct synthesizer drive its own SMO. Still another is to have a VCO that is switched in decade steps only, and whose frequency adjustment is sufficiently wideband to allow correction to any frequency within that decade frequency range.

It is pretty obvious that an indirect synthesizer is a good deal more complex than a direct one. Accordingly, there must be good reasons why it has ousted the direct synthesizer, and these reasons will now be discussed.

The first reason has already been mentioned. It deals with the ability of the indirect synthesizer to provide an *unlocked* output, should the crystal oscillator fail. In a number of applications, such as commercial transmitters, this facility is more a liability, in fact. However, in portable transceivers used by the military, police, and other similar users, this emergency facility is most desirable.

The second advantage of the indirect synthesizer is a major one and relates to reduced trouble with spurious frequencies, as compared with the direct synthesizer. As discussed in the preceding section, the various frequencies generated in the production of the wanted frequency in the direct synthesizer cannot be present in the output here, since that portion of the device is now used only for reference.

Finally, it so happens that the short-term (milliseconds) stability of a crystal oscillator is not as good as that of a free-running oscillator designed with comparable care. Of course, the long-term stability of the crystal oscillator is better by far. Thus the VCO represents a marriage, in which an oscillator that has an excellent short-term stability is controlled by an oscillator whose long-term stability is similarly excellent, and the result is thus the best of both worlds. The direct synthesizer by itself would produce rapid jittering in phase, which is interpreted as noise. Because the jitter is rapid, the noise can occupy a large bandwidth; hence the name *wideband phase noise*. Using the output of the VCO effectively overcomes the source of this noise, and long-term stability is guaranteed by comparison with the output of the crystal-derived frequency via the APC loop. With phase jitter thus minimized, a frequency synthesizer is said to have *low residual FM*.

The control functions in a majority of modern synthesizers—be they individual instruments or part of receivers or transmitters—are performed digitally. This applies especially to the control loop or loops. Indeed, the receiver shown in Fig. 6-17 features a very interesting synthesizer with digital control. Together with microprocessor control, digital frequency synthesis has the advantages of small bulk, flexibility, programmable channel selection, and excellent remote control facilities. The details are outside the scope of this book, but interested students are referred to [5] and [6] for more details.

Microprocessor control Whereas frequency synthesizers are found mostly in communications equipment, and thus there was need to explain their operation in the preceding section, microprocessors belong more in the realm of computers. In addition, they form a large subject of study on their own, and so in this section only their applications to receiver control will be discussed. It will be assumed that students understand the basics of microprocessor operation.

The application of microprocessor control to receivers (and also transmitters) arose out of the need for remote control, as well as the desire to increase operational flexibility and convenience. It was made possible by the advent of large-scale integration, which enabled microprocessors of small size and great power to be fitted conveniently into receivers, and the availability of frequency synthesizers, which avoided the need for manual tuning and calibration of receivers.

Fig. 6-26 shows, in simplified form, the main microprocessor control functions built into the communications receiver of Fig. 6-17. It is seen that, after receiving its instructions (by remote control or along the data bus from the manual controls), the microprocessor initiates an address sequence. The word or words in this sequence are decoded and applied to a number of key points in the receiver, which then generate the desired actions. One of the addresses is applied to a set of data latches (glorified flip-flops). These, in turn, select the appropriate AGC function, the wanted IF band-

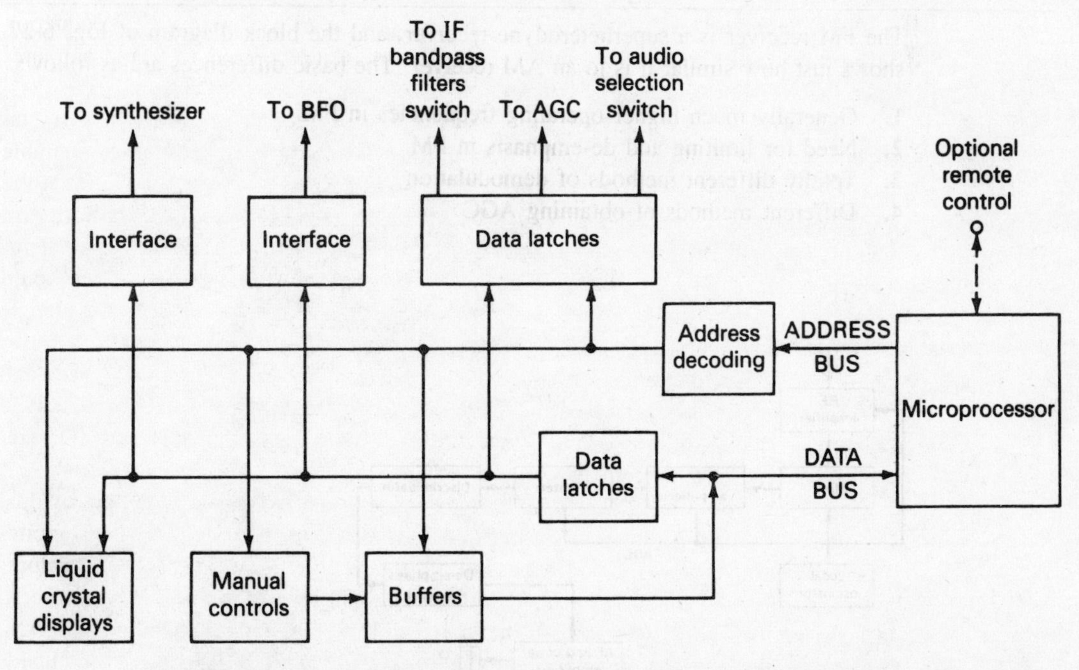

FIGURE 6-26 Functional diagram of communications receiver microprocessor control.
(Adapted by permission of Racal Electronics Pty. Ltd.)

pass filter from among the six provided, and the appropriate audio function (e.g., USB or LSB). The address is also applied to the manual controls, where appropriate voltages select the wanted channel, IF gain, form of frequency scanning, and so on. It is also applied to the liquid crystal displays, to ensure that the selected quantities are correctly indicated for the operator. Similarly, either on request by the microprocessor or because of the manual settings, other data latches operate the synthesizer interface and the BFO interface, adjusting their outputs to the desired values.

Microprocessor control very significantly increases the versatility of a communications receiver. To begin with, full remote control becomes available. This has many applications, such as, for example, switching between two distant receivers for best output under fading conditions, or even complete remote control of a coast radio station from a central point. Further, complex search and channel selection patterns can be stored in the microprocessor memory and used as required with very simple initiation procedures. Finally, test routines can be stored and simply used as required.

This has been a necessarily very brief treatment of microprocessor control. For interested students, [10] provides extensive additional information on microprocessor linkages with the equipment being controlled.

6-4 FM RECEIVERS

The FM receiver is a superheterodyne receiver, and the block diagram of Fig. 6-27 shows just how similar it is to an AM receiver. The basic differences are as follows:

1. Generally much higher operating frequencies in FM
2. Need for limiting and de-emphasis in FM
3. Totally different methods of demodulation
4. Different methods of obtaining AGC

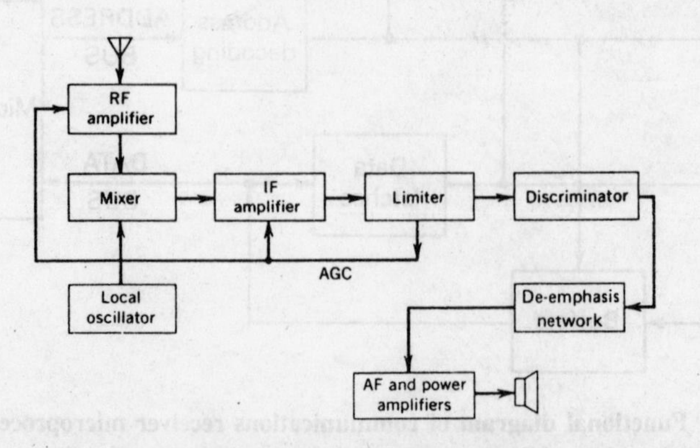

FIGURE 6-27 FM receiver block diagram.

6-4.1 Common Circuits—Comparison with AM Receivers

A number of sections of the FM receiver correspond exactly to those of other receivers already treated. For example, the same criteria apply in the selection of the intermediate frequency, and IF amplifiers are basically similar. Again, a number of concepts have very similar meanings so that only the differences and special applications need be pointed out.

RF amplifiers An RF amplifier is always used in an FM receiver. Its main purpose is to reduce the noise figure, which could otherwise be a problem because of the large bandwidths needed for FM. It is also required to match the input impedance of the receiver to that of the antenna. To meet the second requirement, grounded gate (or base) or cascode amplifiers are employed. Both types have the property of low input impedance and matching the antenna, while neither requires neutralization. This is because the input electrode is grounded in either type of amplifier, effectively isolating input from output. A typical FET grounded-gate RF amplifier is shown in Fig. 6-28. It has all the good points mentioned and the added features of low distortion and simple operation.

Frequency changers The oscillator circuit takes any of the usual forms, with the Colpitts and Clapp predominant, being suited to VHF operation. Tracking is not normally much of a problem in FM broadcast receivers. This is because the tuning frequency range is only 1.25:1, much less than in AM broadcasting.

A very satisfactory arrangement for the front end of an FM receiver consists of FETs for the RF amplifier and mixer, and a bipolar transistor oscillator. As implied by this statement, separately excited oscillators are normally used, with an arrangement as shown in Fig. 6-6.

Intermediate frequency and IF amplifiers Again, the types and operation do not differ much from their AM counterparts. It is worth noting, however, that the intermediate frequency and the bandwidth required are far higher than in AM broadcast receivers. Typical figures for receivers operating in the 88- to 108-MHz band are an IF of 10.7 MHz and a bandwidth of 200 kHz. As a consequence of the large bandwidth, gain per stage may be low. Hence two IF amplifier stages are often provided, in which case the shrinkage of bandwidth as stages are cascaded must be taken into account.

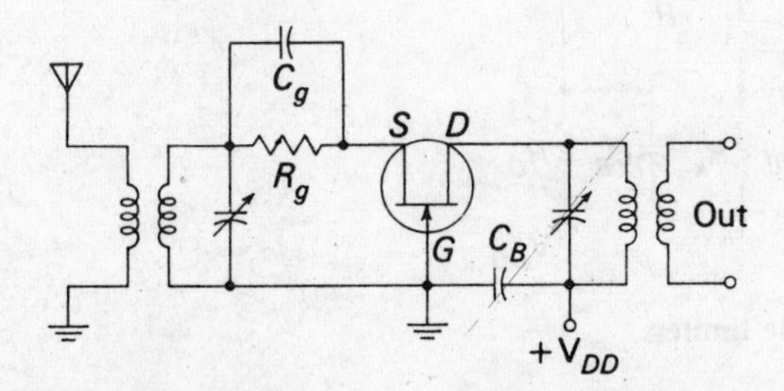

FIGURE 6-28 Grounded-gate FET RF amplifier.

6-4.2 Amplitude Limiting

In order to make full use of the advantages offered by FM, a demodulator[7] must be preceded by an amplitude limiter, as discussed in Chap. 5, on the grounds that any amplitude changes in the signal fed to the FM demodulator are spurious. They must therefore be removed if distortion is to be avoided. The point is significant, since most FM demodulators react to amplitude changes as well as frequency changes. As can be gathered, the limiter is a form of clipping device, a circuit whose output tends to remain constant despite changes in the input signal. Most limiters behave in this fashion, provided that the input voltage remains within a certain range. The common type of limiter uses two separate electrical effects to provide a relatively constant output. There are leak-type bias and early (collector) saturation.

Operation of the amplitude limiter Figure 6-29 shows a typical FET amplitude limiter. Examination of the dc conditions shows that the drain supply voltage has been dropped through resistor R_D. Also, the bias on the gate is leak-type bias supplied by the parallel R_g–C_g combination. Finally, the FET is shown neutralized by means of capacitor C_N, in consideration of the high frequency of operation.

Leak-type bias provides limiting, as shown in Fig. 6-30. When input signal voltage rises, current flows in the R_g–C_g bias circuit, and a negative voltage is developed across the capacitor. It is seen that the bias on the FET is increased in proportion to the size of the input voltage. As a result, the gain of the amplifier is lowered, and the output voltage tends to remain constant.

Although some limiting is achieved by this process, it is insufficient by itself—the action just described would occur only with rather large input voltages. To overcome this, early saturation of the output current is used, achieved by means of a low drain supply voltage. This is the reason for the drain dropping resistor of Fig. 6-29. The supply voltage for a limiter is typically one-half of the normal dc drain voltage. The result of early saturation is to ensure limiting for conveniently low input voltages. However, it is possible for the gate-drain section to become forward-biased under saturation conditions, causing a short circuit between input and output. To avert this, a

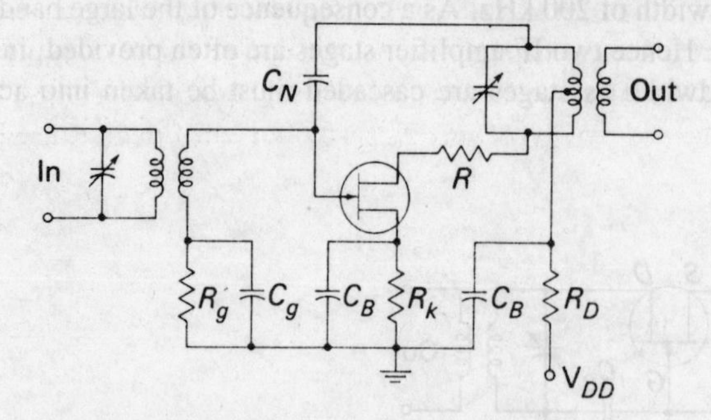

FIGURE 6-29 Amplitude limiter.

[7] This does not apply to a receiver with a ratio detector which (as is shown in Sec. 6-4.4) provides a fair amount of limiting.

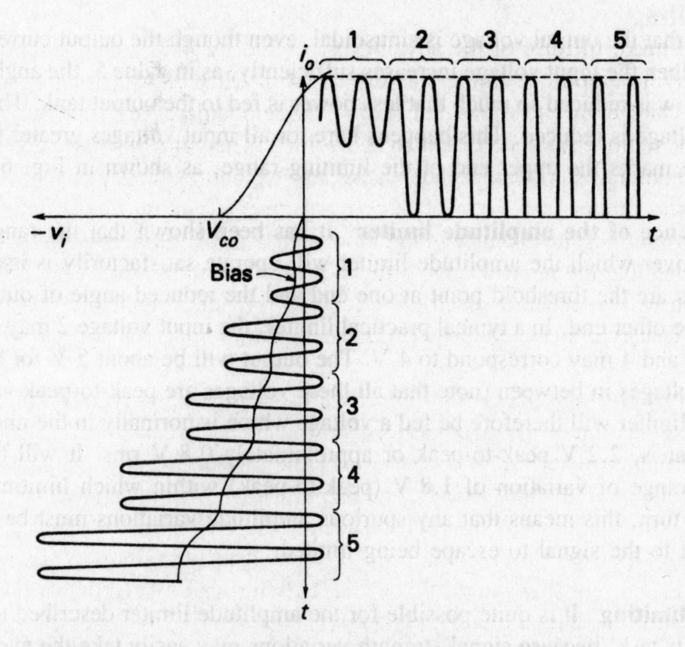

FIGURE 6-30 Amplitude limiter transfer characteristic.

resistance of a few hundred ohms is placed between the drain and its tank. This is *R* of Fig. 6-29.

Figure 6-31 shows the response characteristic of the amplitude limiter. It indicates clearly that limiting takes place only for a certain range of input voltages, outside which output varies with input. Referring simultaneously to Fig. 6-30, we see that as input increases from value 1 to value 2, output current also rises. Thus no limiting has yet taken place. However, comparison of 2 and 3 shows that they both yield the same output current and voltage. Thus limiting has now begun. Value 2 is the point at which limiting starts and is called the *threshold of limiting*. As input increases from 3 to 4, there is no rise in output; all that happens is that the output current flows for a somewhat shorter portion of the input cycle. This, of course, suggests operation like that of a class C amplifier. Thus the *flywheel effect* of the output tank circuit is used here also,

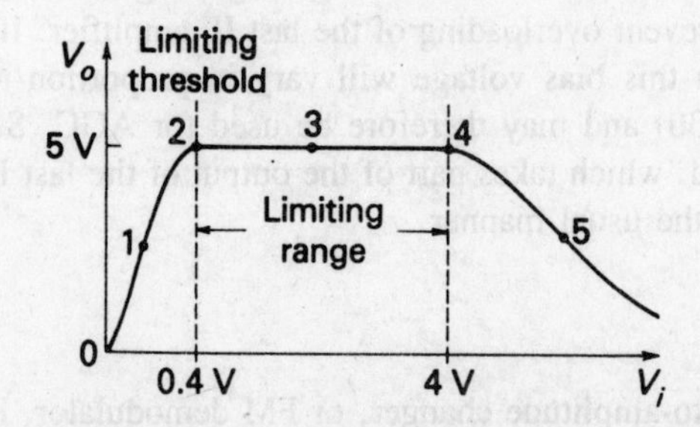

FIGURE 6-31 Typical limiter response characteristic.

to ensure that the output voltage is sinusoidal, even though the output current flows in pulses. When the input voltage increases sufficiently, as in value 5, the angle of output current flow is reduced so much that less power is fed to the output tank. Therefore the output voltage is reduced. This happens here for all input voltages greater than 4, and this value marks the upper end of the limiting range, as shown in Fig. 6-31.

Performance of the amplitude limiter It has been shown that the range of input voltages over which the amplitude limiter will operate satisfactorily is itself limited. The limits are the threshold point at one end and the reduced angle of output current flow at the other end. In a typical practical limiter, the input voltage 2 may correspond to 0.4 V, and 4 may correspond to 4 V. The output will be about 5 V for both values and all voltages in between (note that all these voltages are peak-to-peak values). The practical limiter will therefore be fed a voltage which is normally in the middle of this range, that is, 2.2 V peak-to-peak or approximately 0.8 V rms. It will thus have a possible range of variation of 1.8 V (peak-to-peak) within which limiting will take place. In turn, this means that any spurious amplitude variations must be quite large compared to the signal to escape being limited.

Further limiting It is quite possible for the amplitude limiter described to be inadequate to its task, because signal-strength variations may easily take the average signal amplitude outside the limiting range. As a result, further limiting is required in a practical FM receiver.

Double limiter

A double limiter consists of two amplitude limiters in cascade, an arrangement that increases the limiting range very satisfactorily. Numerical values given to illustrate limiter performance showed an output voltage (all values peak-to-peak, as before) of 5 V for any input within the 0.4- to 4-V range, above which output gradually decreases. It is quite possible that an output of 0.6 V is not reached until the input to the first limiter is about 20 V. If the range of the second limiter is 0.6 to 6 V, it follows that all voltages between 0.4 and 20 V fed to the double limiter will be limited. The use of the double limiter is thus seen to have increased the limiting range quite considerably.

Automatic gain control (AGC)

A suitable alternative to the second limiter is automatic gain control. This is to ensure that the signal fed to the limiter is within its limiting range, regardless of the input signal strength, and also to prevent overloading of the last IF amplifier. If the limiter used has leak-type bias, then this bias voltage will vary in proportion to the input voltage (as shown in Fig. 6-30) and may therefore be used for AGC. Sometimes a separate AGC detector is used, which takes part of the output of the last IF amplifier and rectifies and filters it in the usual manner.

6-4.3 Basic FM Demodulators

The function of a frequency-to-amplitude changer, or FM demodulator, is to change the frequency deviation of the incoming carrier into an AF amplitude variation (identical to the one that originally caused the frequency variation). This conversion should be done efficiently and linearly. In addition, the detection circuit should (if at all possible) be insensitive to amplitude changes and should not be too critical in its adjustment and operation. Generally speaking, this type of circuit converts the frequency-modulated IF voltage of constant amplitude into a voltage that is both frequency- and amplitude-modulated. This latter voltage is then applied to a detector which reacts to the amplitude change but ignores the frequency variations. It is now necessary to devise a circuit which has an output whose amplitude depends on the frequency deviation of the input voltage.

Slope detection Consider a frequency-modulated signal fed to a tuned circuit whose resonant frequency is to one side of the center frequency of the FM signal. The output of this tuned circuit will have an amplitude that depends on the frequency deviation of the input signal; that is illustrated in Fig. 6-32. As shown, the circuit is detuned by an amount δf, to bring the carrier center frequency to point *A* on the selectivity curve (note that *A*′ would have done just as well). Frequency variation produces an output voltage proportional to the frequency deviation of the carrier, as shown.

This output voltage is applied to a diode detector with an *RC* load of suitable time constant. The circuit is, in fact, identical to that of an AM detector, except that the secondary winding of the IF transformer is off-tuned. (In a desperate emergency, it is possible, after a fashion, to receive FM with an AM receiver, with the simple expedient of giving the slug of the coil to which the detector is connected two turns clockwise. Remember to reverse the procedure after the emergency is over!)

The slope detector does not really satisfy any of the conditions laid down in the introduction: it is inefficient, and it is linear only along a very limited frequency range. It quite obviously reacts to all amplitude changes. Moreover, it is relatively difficult to adjust, since the primary and secondary windings of the transformer must be tuned to slightly differing frequencies. Its only virtue is that it simplifies the explanation of the operation of the balanced slope detector.

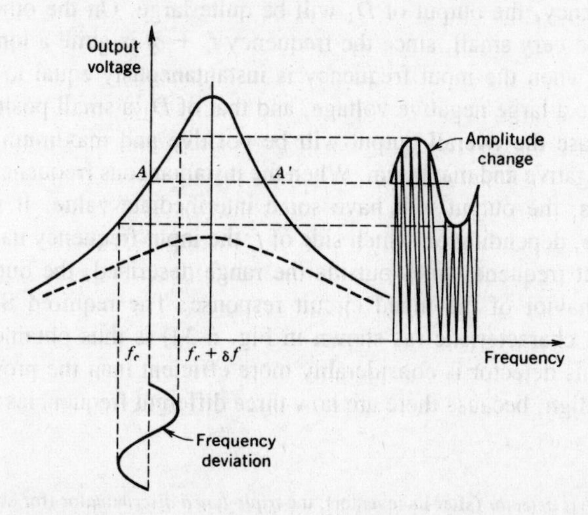

FIGURE 6-32 Slope detector characteristic curve. ***(From K. R. Sturley, Frequency-Modulated Radio, 2d ed., George Newnes Ltd., London, 1958, by permission of the publisher.)***

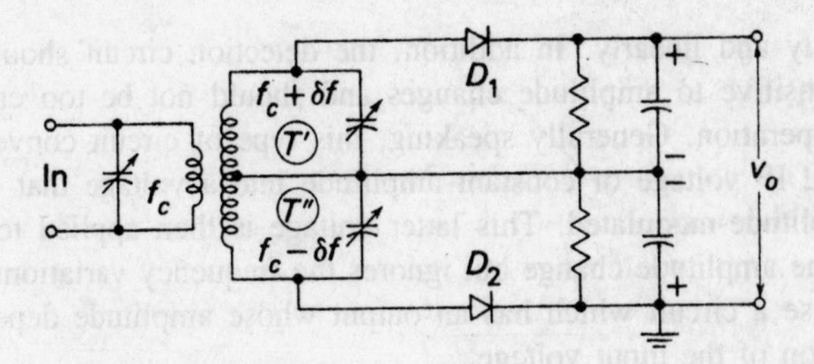

FIGURE 6-33 Balanced slope detector.

Balanced slope detector[8] As can be seen from Fig. 6-33, the circuit uses two slope detectors. They are connected back to back, to the opposite ends of a center-tapped transformer, and hence fed 180° out of phase. The top secondary circuit is tuned above the IF by an amount which, in FM receivers with a deviation of 75 kHz, is 100 kHz. The bottom circuit is similarly tuned below the IF by the same amount. Each tuned circuit is connected to a diode detector with an *RC* load. The output is taken from across the series combination of the two loads, so that it is the sum of the individual outputs.

Let f_c be the IF to which the primary circuit is tuned, and let $f_c + \delta f$ and $f_c - \delta f$ be the resonant frequencies of the upper secondary and lower secondary circuits T' and T'', respectively. When the input frequency is instantaneously equal to f_c, the voltage across T', that is, the input to diode D_1, will have a value somewhat less than the maximum available, since f_c is somewhat below the resonant frequency of T'. A similar condition exists across T''. In fact, since f_c is just as far from $f_c + \delta f$ as it is from $f_c - \delta f$, the voltages applied to the two diodes will be identical. The dc output voltages will also be identical, and thus the detector output will be zero, since the output of D_1 is positive and that of D_2 is negative.

Now consider the instantaneous frequency to be equal to $f_c + \delta f$. Since T' is tuned to this frequency, the output of D_1 will be quite large. On the other hand, the output of D_2 will be very small, since the frequency $f_c + \delta f$ is quite a long way from $f_c - \delta f$. Similarly, when the input frequency is instantaneously equal to $f_c - \delta f$, the output of D_2 will be a large negative voltage, and that of D_1 a small positive voltage. Thus in the first case the overall output will be positive and maximum, and in the second it will be negative and maximum. When the instantaneous frequency is between these two extremes, the output will have some intermediate value. It will then be positive or negative, depending on which side of f_c the input frequency happens to lie. Finally, if the input frequency goes outside the range described, the output will fall because of the behavior of the tuned circuit response. The required S-shaped frequency-modulation characteristic (as shown in Fig. 6-34) is thus obtained.

Although this detector is considerably more efficient than the previous one, it is even trickier to align, because there are now three different frequencies to which the

[8] Also known as the *Travis detector* (after its inventor), the *triple-tuned discriminator* (for obvious reasons), and as the *amplitude discriminator* (erroneously).

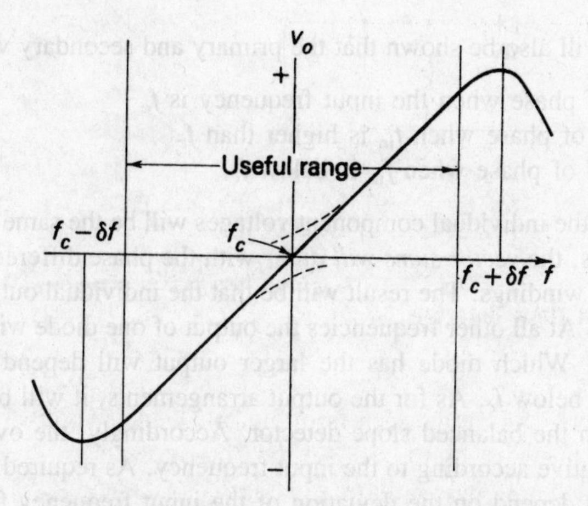

FIGURE 6-34 Balanced slope detector characteristic.

various tuned circuits of the transformer must be adjusted. Amplitude limiting is still not provided, and the linearity, although better than that of the single slope detector, is still not good enough.

Phase discriminator[9] It is possible to obtain the same S-shaped response curve from a circuit in which the primary and the secondary windings are both tuned to the center frequency of the incoming signal. This is desirable because it greatly simplifies alignment, and also because the process yields far better linearity than slope detection. In this new circuit, as shown in Fig. 6-35, the same diode and load arrangement is used as in the balanced slope detector because such an arrangement is eminently satisfactory. However, the method of ensuring that the voltages fed to the diodes vary linearly with the deviation of the input signal has been changed completely. It is true to say, nevertheless, that the Foster-Seeley discriminator is derived from the Travis detector.

A limited mathematical analysis [7] will now be given, to show that the voltage applied to each diode is the sum of the primary voltage and the corresponding half-

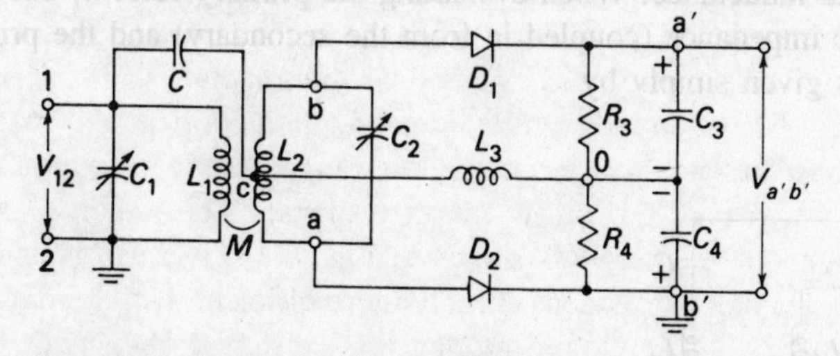

FIGURE 6-35 Phase discriminator.

[9]Among the various other names given to this circuit are the *discriminator*, the *center-tuned discriminator*, and the most popular name, the *Foster-Seeley discriminator*, after its inventors.

secondary voltage. It will also be shown that the primary and secondary voltages are:

1. Exactly 90° out of phase when the input frequency is f_c
2. Less than 90° out of phase when f_{in} is higher than f_c
3. More than 90° out of phase when f_{in} is below f_c

Thus, although the individual component voltages will be the same at the diode inputs at all frequencies, the *vector sums will differ* with the phase difference between primary and secondary windings. The result will be that the individual output voltages will be equal only at f_c. At all other frequencies the output of one diode will be greater than that of the other. Which diode has the larger output will depend entirely on whether f_{in} is above or below f_c. As for the output arrangements, it will be noted that they are the same as in the balanced slope detector. Accordingly, the overall output will be positive or negative according to the input frequency. As required, the magnitude of the output will depend on the deviation of the input frequency from f_c.

The resistances forming the load are made much larger than the capacitive reactances. It can thus be seen that the circuit composed of C, L_3 and C_4 is effectively placed across the primary winding. This is shown in Fig. 6-36. The voltage across L_3, V_L, will then be

$$V_L = \frac{V_{12}Z_{L_3}}{Z_C + Z_{C_4} + Z_{L_3}}$$

$$= V_{12}\frac{j\omega L_3}{j\omega L_3 - j(1/\omega C + 1/\omega C_4)} \qquad \textbf{(6-9)}$$

L_3 is an RF choke and is purposely given a large reactance. Hence its reactance will greatly exceed those of C and C_4, especially since the first of these is a coupling capacitor and the second is an RF bypass capacitor. Accordingly, Eq. (6-9) will reduce to

$$V_L \approx V_{12} \qquad \textbf{(6-10)}$$

The first part of the analysis has thus been achieved, that is, proof that the voltage across the RF choke is equal to the applied primary voltage.

The mutually coupled, double-tuned circuit has high primary and secondary Q and a low mutual inductance. When evaluating the primary current, one may, therefore, neglect the impedance (coupled in from the secondary) and the primary resistance. Then I_p is given simply by

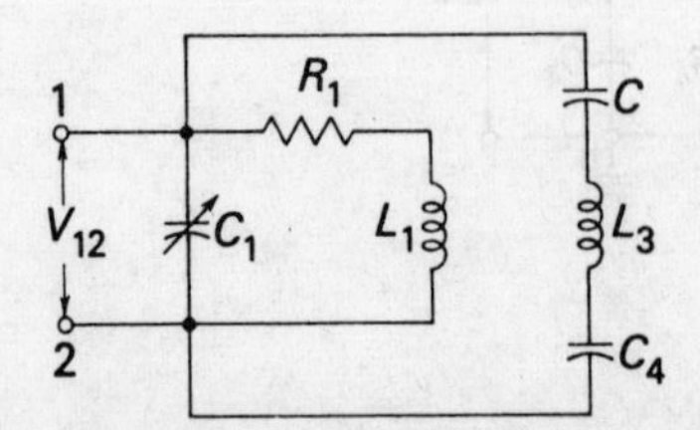

FIGURE 6-36 Discriminator primary voltage.

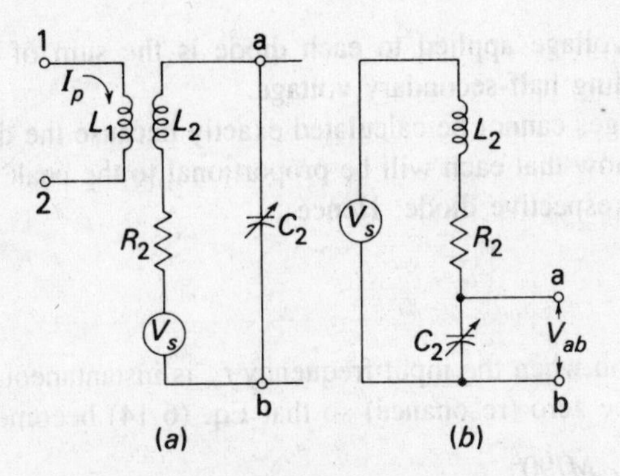

FIGURE 6-37 Discriminator secondary circuit and voltages. *(a)* Primary-secondary relations; *(b)* secondary redrawn.

$$I_p = \frac{V_{12}}{j\omega L_1} \tag{6-11}$$

It will be remembered from transformer circuit theory that a voltage is induced in series with the secondary as a result of the primary current and is given by

$$V_s = \pm j\omega M I_p \tag{6-12}$$

where the sign depends on the direction of winding.

As will be seen, it is simpler here to take the connection giving negative mutual inductance. The secondary circuit is shown in Fig. 6-37*a*, and we have

$$V_s = -j\omega M I_p = -j\omega M \frac{V_{12}}{j\omega L_1} = -\frac{M}{L_1} V_{12} \tag{6-13}$$

The voltage across the secondary winding, V_{ab}, can now be calculated with the aid of Fig. 6-37*b*, which shows the secondary redrawn for this purpose. Then

$$V_{ab} = V_s \frac{Z_{C_2}}{Z_{C_2} + Z_{L_2} + R_2} = \frac{-jX_{C_2}(-V_{12}M/L_1)}{R_2 + j(X_{L_2} - X_{C_2})}$$

$$= \frac{jM}{L_1} \frac{V_{12}X_{C_2}}{R_2 + jX_2} \tag{6-14}$$

where

$$X_2 = X_{L_2} - X_{C_2} \tag{6-15}$$

and may be positive, negative or even zero, depending on the frequency.

The total voltages applied to D_1 and D_2, V_{ao} and V_{bo}, respectively, may now be calculated. Thus

$$V_{ao} = V_{ac} + V_L = ½V_{ab} + V_{12} \tag{6-16}$$

$$V_{bo} = V_{bc} + V_L = -V_{ac} + V_L = -½V_{ab} + V_{12} \tag{6-17}$$

As predicted, the voltage applied to each diode is the sum of the primary voltage and the corresponding half-secondary voltage.

The dc output voltages cannot be calculated exactly because the diode drop is unknown. However, we know that each will be proportional to the peak value of the RF voltage applied to the respective diode. Hence

$$\begin{aligned} V_{a'b'} &= V_{a'o} - V_{b'o} \\ &\propto V_{ao} - V_{bo} \end{aligned} \tag{6-18}$$

Consider the situation when the input frequency f_{in} is instantaneously equal to f_c. In Eq. (6-15), X_2 will be zero (resonance) so that Eq. (6-14) becomes

$$V_{ab} = \frac{jM}{L_1} \frac{V_{12}X_{C_2}}{R_2} = \frac{V_{12}X_{C_2}M\angle 90^\circ}{R_2L_1} \tag{6-19}$$

From Eq. (6-19), it follows that the secondary voltage V_{ab} leads the applied primary voltage by 90°. Thus $\frac{1}{2}V_{ab}$ will lead V_{12} by 90°, and $-\frac{1}{2}V_{ab}$ will lag V_{12} by 90°. It is now possible to add the diode input voltages vectorially, as in Fig. 6-38*a*. It

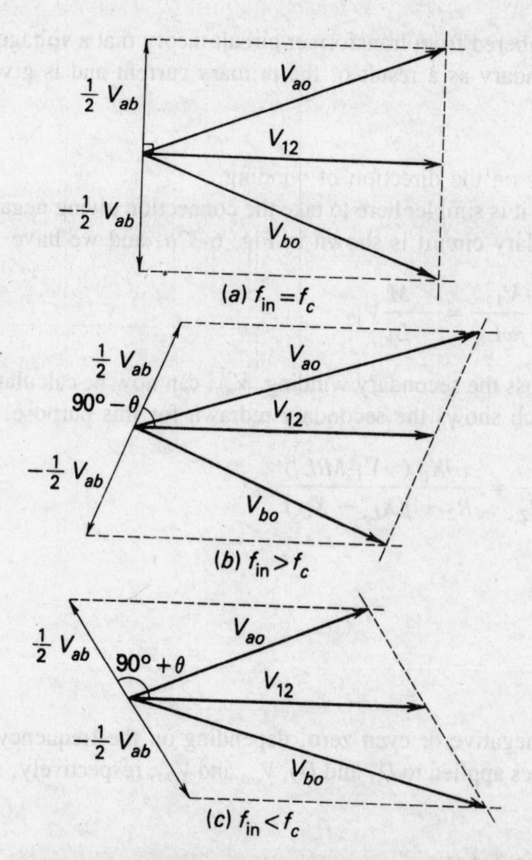

FIGURE 6-38 Phase discriminator phasor diagrams. ***(a)*** **f_{in} equal to f_c;** ***(b)*** **f_{in} greater than f_c;** ***(c)*** **f_{in} less than f_c.** ***(After Samuel Seely, Radio Electronics, McGraw-Hill Book Company, New York, 1956. Used with permission of McGraw-Hill Book Company.)***

is seen that since $V_{ao} = V_{bo}$, the discriminator output is zero. Thus there is no output from this discriminator when the input frequency is equal to the unmodulated carrier frequency, i.e., no output for no modulation. (Actually, this is not a particularly surprising result. The clever part, as will be seen, is that at any other frequency there *is* an output.)

Now consider the case when f_{in} is greater than f_c. In Eq. (6-15), X_{L_2} is now greater than X_{C_2} so that X_2 is positive. Equation (6-14) thus becomes

$$V_{ab} = \frac{jM}{L_1}\frac{V_{12}X_{C_2}}{R_2 + jX_2} = \frac{V_{12}X_{C_2}M\angle 90^\circ}{L_1|Z_2|\angle\theta^\circ} = \frac{V_{12}X_{C_2}M}{L_1|Z_2|}\angle(90-\theta)^\circ \tag{6-20}$$

From Eq. (6-20), it is seen that V_{ab} leads V_{12} by less than 90° so that $-\frac{1}{2}V_{ab}$ must lag V_{12} by more than 90°. It is apparent from the vector diagram of Fig. 6-38 that V_{ao} is now greater than V_{bo}. Thus the discriminator output will be positive when f_{in} is greater than f_c.

Similarly, when the input frequency is smaller than f_c, X_2 in Eq. (6-15) will be negative, and the angle of the impedance Z_2 will also be negative. Thus V_{ab} will lead V_{12} by more than 90°. This time V_{ao} will be smaller than V_{bo}, and the output voltage $V_{a'b'}$ will be negative. The appropriate vector diagram is shown in Fig. 6-38*c*.

If the frequency response is plotted for the phase discriminator, it will follow the required S shape, as in Fig. 6-39. As the input frequency moves farther and farther away from the center frequency, the disparity between the two diode input voltages becomes greater and greater. The output of the discriminator will increase up to the limits of the useful range, as indicated. The limits correspond roughly to the half-power points of the discriminator tuned transformer. Beyond these points, the diode input voltages are reduced because of the frequency response of the transformer, so that the overall output falls.

The phase discriminator is much easier to align than the balanced slope detector; there are now only two tuned circuits, and both are tuned to the same frequency.

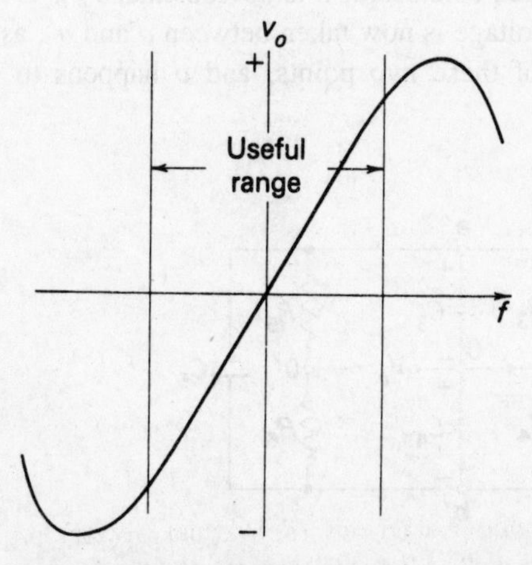

FIGURE 6-39 Discriminator response.

Linearity is also better, because the circuit relies less on frequency response and more on the primary-secondary phase relation, which is quite linear. The only defect of this circuit, if it may be called a defect, is that it does not provide any amplitude limiting.

6-4.4 Ratio Detector

In the Foster-Seeley discriminator, changes in the magnitude of the input signal will give rise to amplitude changes in the resulting output voltage. This makes prior limiting necessary. It is possible to modify the discriminator circuit to provide limiting, so that the amplitude limiter may be dispensed with. A circuit so modified is called a *ratio detector*.

If Fig. 6-38 is reexamined from a new point of view, it will be seen that, by and large, the sum $V_{ao} + V_{bo}$ remains constant, although the difference varies because of changes in input frequency. As a matter of fact, this assumption is not completely true. However, deviation from this ideal does not result in undue distortion in the ratio detector, although some distortion is undoubtedly introduced. It follows, therefore, that any variations in the magnitude of this sum voltage can be considered spurious here. Accordingly, their suppression will lead to a discriminator which is unaffected by the amplitude of the incoming signal; it will therefore not react to noise amplitude or spurious amplitude modulation.

It now remains to ensure that the sum voltage is kept constant. Unfortunately, this cannot be accomplished in the phase discriminator, and the circuit must be modified. This has been done in Fig. 6-40, which presents the ratio detector in its basic form; this is used to show how the circuit is derived from the discriminator and to explain its operation. It is seen that three important changes have been made: one of the diodes has been reversed, a large capacitor (C_5) has been placed across what used to be the output, and the output now is taken from elsewhere.

Operation With diode D_2 reversed, o is now positive with respect to b', so that $V_{a'b'}$ is now a sum voltage, rather than the difference it was in the discriminator. Accordingly, it is now possible to connect a large capacitor between a' and b' to keep this sum voltage constant. Once C_5 has been connected, it is obvious that $V_{a'b'}$ is no longer the output voltage; thus the output voltage is now taken between o and o', as shown. It is now necessary to ground one of these two points, and o happens to be the more

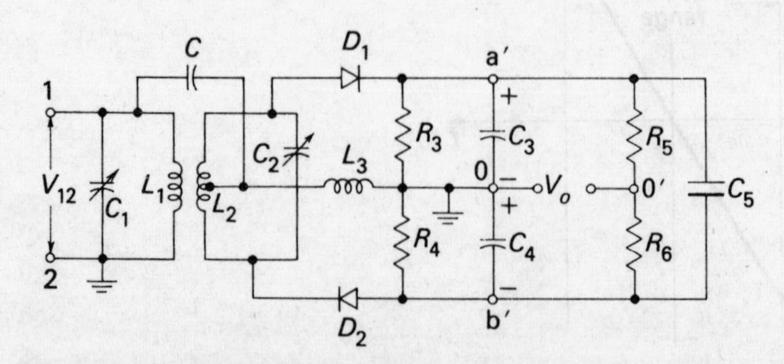

FIGURE 6-40 Basic ratio detector circuit.

convenient, as will be seen when dealing with practical ratio detectors. Bearing in mind that in practice $R_5 = R_6$, V_o is calculated as follows:

$$V_o = V_{b'o'} - V_{b'o} = \frac{V_{a'b'}}{2} - V_{b'o} = \frac{V_{a'o} + V_{b'o}}{2} - V_{b'o}$$

$$= \frac{V_{a'o} - V_{b'o}}{2} \qquad \textbf{(6-21)}$$

Equation (6-21) shows that the ratio detector output voltage is equal to half the difference between the output voltages from the individual diodes. Thus (as in the phase discriminator) the output voltage is proportional to the difference between the individual output voltages. The ratio detector therefore behaves identically to the discriminator for input frequency changes.The S curve of Fig. 6-39 applies equally to both circuits and need not be derived afresh.

Amplitude limiting by the ratio detector It is thus established that the ratio detector behaves in the same way as the phase discriminator when input frequency varies (but input voltage remains constant). The next step is to explain how the ratio detector reacts to amplitude changes. If the input voltage V_{12} is constant and has been so for some time, C_5 has been able to charge up to the potential existing between a' and b'. Since this is a dc voltage if V_{12} is constant, there will be no current either flowing in to charge the capacitor or flowing out to discharge it. In other words, the input impedance of C_5 is infinite. The total load impedance for the two diodes is therefore the sum of R_3 and R_4, since these are in practice much smaller than R_5 and R_6.

If V_{12} tries to increase, C_5 will tend to oppose any rise in V_o. The way in which it does this is not, however, merely to have a fairly long time constant, although this is certainly *part* of the operation. As soon as the input voltage tries to rise, extra diode current flows, but this excess current flows into the capacitor C_5, charging it. The voltage $V_{a'b'}$ remains constant at first because it is not possible for the voltage across a capacitor to change instantaneously. The situation now is that the current in the diodes' load has risen, but the voltage across the load has not changed; the conclusion is that the load impedance has decreased. This being so, the secondary of the ratio detector transformer is more heavily damped, the Q falls, and so does the gain of the amplifier driving the ratio detector. This neatly counteracts the initial rise in input voltage.

Should the input voltage fall, the diode current will fall, but the load voltage will not, at first, because of the presence of the capacitor. The effect is that of an increased diode load impedance; the diode current has fallen, but the load voltage has remained constant. Accordingly, damping is reduced, and the gain of the driving amplifier rises, this time counteracting an initial fall in the input voltage. The ratio detector provides what is known as *diode variable damping*. We have here a system of varying the gain of an amplifier by changing the damping of its tuned circuit. This maintains a constant output voltage despite changes in the amplitude of the input.

Practical circuits Many practical variations of the ratio detector are in use. Figure 6-40 is perhaps the best adapted to explain the principles involved and to show the similarity to the phase discriminator. However, it is by no means the most practical

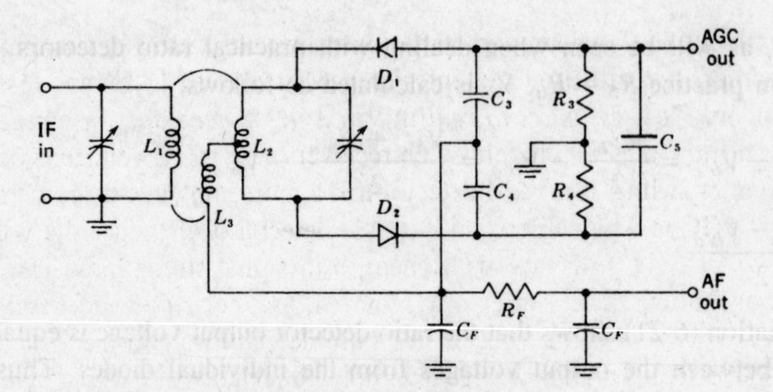

FIGURE 6-41 Balanced ratio detector.

circuit. Broadly, there are two types of ratio detector in use: balanced and unbalanced. The former is probably the better and more frequently employed version, and a form of it is shown in Fig. 6-41.

The tertiary winding L_3 serves the same purpose as did the coil L_3 in the basic circuit together with the capacitor C. The primary voltage is again connected to the center tap of L_2, and there is an impedance across which it is developed. This is actually an improvement on the original connection, because L_3 is also used to match the low-impedance secondary to the primary, whose operation is improved if its dynamic impedance is made high. In other words, L_3 gives a voltage step-down to prevent too-great damping of the primary by the ratio detector action. Such an arrangement may also be used with the phase discriminator, although the need is not so great there. However, this does mean that it is very easy to change a practical ratio detector circuit into a discriminator, and vice versa.

Resistors R_5 and R_6 of Fig. 6-40 have been dispensed with. They are replaced by an arrangement in which point o of the original circuit is still a single point for RF (C_F is an RF bypass capacitor connecting the C_3-C_4 junction to ground for RF), but for dc it has been split into two points. The output voltage is the same as before, and it is calculated in an identical manner. The two voltage dividers are now C_3-C_4 and R_3-R_4, instead of R_3-R_4 and R_5-R_6 as previously. Two resistors have been saved.

The circuit consisting of the two capacitors C_F and the resistor R_F is a low-pass filter designed to remove RF ripple from the audio signal, in just the same way as the corresponding filter in the AM detector. Both diodes have been reversed in the drawing so that the top of C_5 is now negative for dc. Automatic gain control may be taken for the rest of the receiver from this point.

Need for further limiting The time constant of the load resistors in parallel with the large capacitor is quite long. Hence the circuit will respond neither to fast amplitude changes due to noise impulses nor to the slower changes in amplitude due to spurious amplitude modulation. Typical component values are $R_3 + R_4 = 15\ \text{k}\Omega$ and $C_5 = 8\ \mu\text{F}$, giving a time constant of 120 ms. A time constant much slower than this would result in the circuit being disconcerting to align.

It is obvious, therefore, that the ratio detector will follow very slow amplitude changes of the input signal. The circuit will therefore not limit against changes in carrier strength due to signal strength variations caused by fading or changing from one station to another. Aircraft-produced interference, at rates of 15 Hz and less, also falls into this category. *It is essential to realize that AGC is necessary in a receiver which incorporates a ratio detector*. In television receivers, this AGC voltage is derived from the video detector, which is an AM detector and a more convenient source of AGC. In FM receivers AGC is obtainable from the ratio detector itself, since the voltage at the top of C_5 in Fig. 6-41 will vary with changes in signal strength, as just explained.

Further limiting is very often also required, particularly in wideband FM broadcast receivers. This is because the Q of the ratio-detector transformer tuned circuits is rather low. The effect is that the variable damping does not make as much difference to the gain of the driving amplifier as it would have in a narrowband system. This is especially true when the input signal increases and damping tries to reduce Q even further. A partial solution, often quite adequate, is to employ leak-type bias for the driving amplifier, in addition to a good AGC system. Alternatively, a complete stage of limiting may be used prior to the ratio detector.

6-4.5 FM Demodulator Comparison

The slope detectors—single or balanced—are not used in practice. They were described so that their disadvantages could be explained, and also as an introduction to practical discriminators. The Foster-Seeley discriminator is very widely used in practice, especially in FM radio receivers, wideband or narrowband. It is also used in satellite station receivers, especially for the reception of TV carriers (as will be explained in Chap. 15, most satellite transmissions are frequency-modulated).

The ratio detector is a good FM demodulator, also widely used in practice, especially in TV receivers, for the sound section, and sometimes also in narrowband FM radio receivers. Its advantage over the discriminator is that it provides both limiting and a voltage suitable for AGC, while the main advantage of the discriminator is that it is very linear. Thus, the discriminator is preferred in situations in which linearity is an important characteristic (e.g., high-quality FM receivers), whereas the ratio detector is preferred in applications in which linearity is not critical, but component and price savings are (e.g., in TV receivers).

It may be shown that, under critical noise conditions, even the discriminator is not the best FM demodulator. Such conditions are encountered in satellite station receivers, where noise reduction may be achieved by increasing signal strength, receiver sensitivity, or receiver antenna size. Since each of these can be an expensive solution, demodulator noise performance does become very significant. In these circumstances, so-called threshold extension demodulators are preferred, such as the *FM feedback demodulator* or the *phase-locked loop demodulator*. Both are outside the scope of this book, but they are well described in [8].

6-4.6 Stereo FM Multiplex Reception

Assuming there have been no losses or distortion in transmission, the demodulator output in a stereo FM multiplex receiver, tuned to a stereo transmission, will be exactly as shown in Fig. 5-10. Going up in frequency, the signal components will therefore be

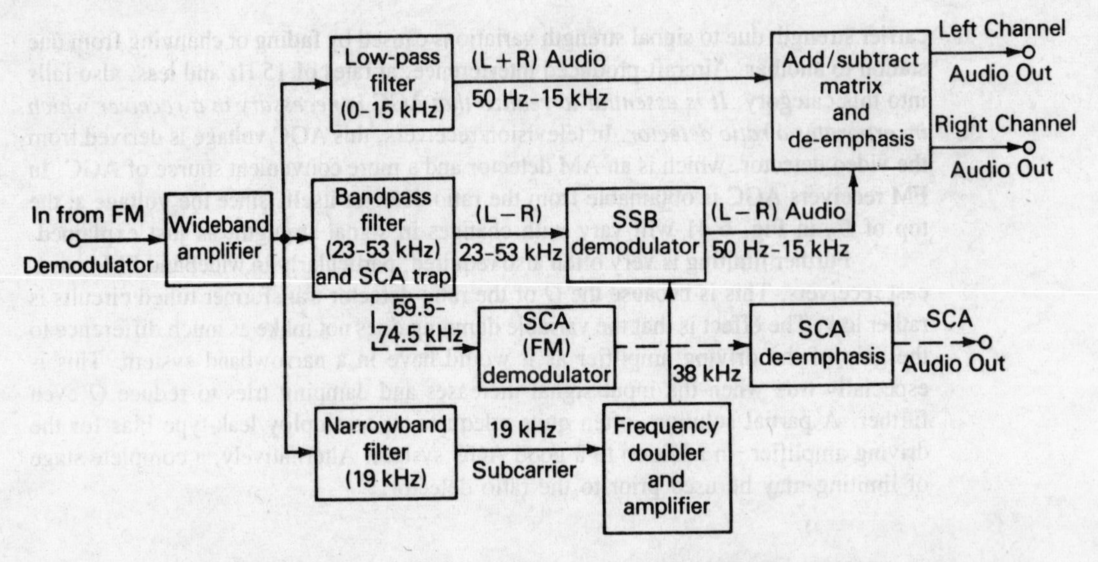

FIGURE 6-42 Stereo FM multiplex demodulation with optional SCA output.

sum channel ($L+R$), 19-kHz subcarrier, the lower and upper sidebands of the difference channel ($L-R$), and finally the optional SCA signal, frequency-modulating a 67-kHz subcarrier. Figure 6-42 shows how these signals are separated and reproduced.

As shown in this block diagram, the process of extracting the wanted information is quite straightforward. A low-pass filter removes all frequencies in excess of 15 kHz and thus has the sum signal ($L+R$) at its output. In a monaural receiver, this would be the only output treated further, through a de-emphasis network to audio amplification. The center row of Fig. 6-42 shows a bandpass filter selecting the sidebands which correspond to the difference signals ($L-R$) and also rejecting the (optional) SCA frequencies above 59.5 kHz. The sidebands are fed to a product detector or to a balanced modulator (see Sec. 6-5 for a discussion of their operation), which also receives the output of the frequency doubler. The doubler converts the transmitted 19-kHz subcarrier, which was selected with a narrowband filter, to the wanted 38-kHz carrier signal, which is then amplified. It will be recalled that the subcarrier had been transmitted at a much reduced amplitude. The two inputs to the SSB demodulator result in this circuit's producing the wanted difference signal ($L-R$), which, when fed to the matrix along with ($L+R$), produces the left channel from an adder and the right channel from a subtractor. After de-emphasis, these are ready for further audio amplification.

Finally, if present, the SCA signal is selected, demodulated, also de-emphasized, and produced as a separate audio output.

6-5 SINGLE- AND INDEPENDENT-SIDEBAND RECEIVERS

Single- and independent-sideband receivers are normally used for professional or commercial communications. There are of course also a lot of amateur SSB receivers, but this section will concentrate on the professional applications. Such receivers are almost invariably required to detect signals in difficult conditions and crowded frequency bands. Consequently, they are always multiple-conversion receivers, of a pattern similar to, or developed from, the block shown in Fig. 6-16. The special requirements of SSB and ISB receivers are:

1. High reliability (and simple maintenance), since such receivers may be operated continuously
2. Excellent suppression of adjacent signals
3. Ability to demodulate SSB
4. Good blocking performance
5. High signal-to-noise ratio
6. Ability to separate the independent channels (in the case of ISB receivers)

The specialized aspects of SSB and ISB receivers will now be investigated.

6-5.1 Demodulation of SSB

Demodulation of SSB must obviously be different from ordinary AM detection. The basic SSB demodulation device is the *product detector*, which is rather similar to an ordinary mixer. The balanced modulator may also be used. It almost always is so used in transceivers, in which it is naturally important to utilize as many circuits as possible for dual purposes. It is also possible to demodulate SSB with the complete phase-shift network; the complete third-method system can similarly be used for demodulation.

Product demodulator The product demodulator (or detector), as shown in Fig. 6-43, is virtually a mixer with audio output. It is popular for SSB, but is equally capable of demodulating all other forms of AM. In the circuit shown, the input SSB signal is fed to the base via a fixed-frequency IF transformer, and the signal from a crystal oscillator is applied to the unbypassed emitter. The frequency of this oscillator is either equal to the nominal carrier frequency or derived from the pilot frequency, as applicable.

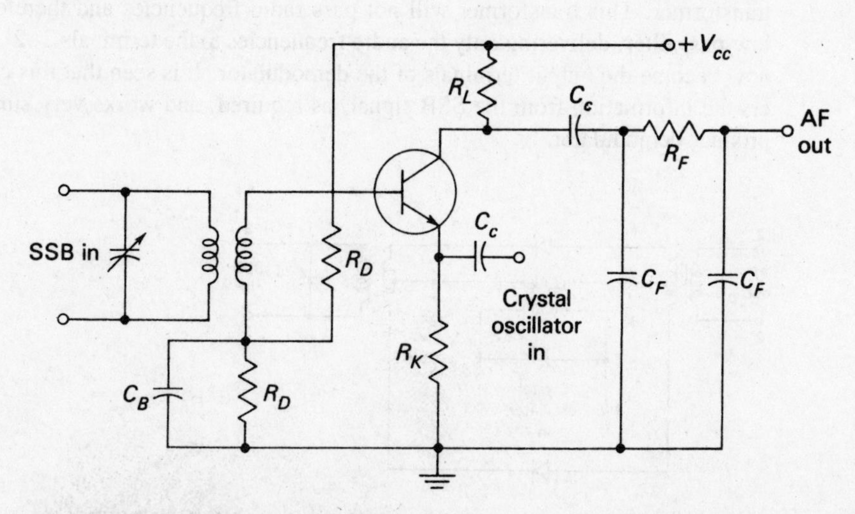

FIGURE 6-43 Product demodulator.

If this is a fairly standard double-conversion receiver, like the one shown in Fig. 6-46, the IF fed to the product detector will be 455 kHz. If the USB is being received, the signal will cover the frequency band from 455.3 to 458.0 kHz for J3E (in R3E 455 kHz would also be present at this point). This signal is mixed with the output of the crystal oscillator, at 455 kHz. Several frequencies will result in the output, including the difference frequencies (as derived in Sec. 4-21). These range from 300 to 3000 Hz and are the wanted audio frequencies. All other signals present at this point will be blocked by the low-pass filter consisting of capacitors C_F and resistor R_F in Fig. 6-43. It is seen that the circuit has recovered the wanted intelligence from the input signal and is therefore a suitable SSB demodulator.

If the lower sideband is being received, the missing carrier frequency is at 458 kHz, and the sideband stretches from 457.7 to 455 kHz. A new crystal must be switched in for the oscillator, but apart from that, the operation is identical.

Detection with the diode balanced modulator In a portable SSB transmitter-receiver, it is naturally desirable to employ as small a number of circuits as possible to save weight and power consumption. As already mentioned, if a particular circuit is capable of performing either function, it is always so used, with the aid of appropriate switching when changing from transmission to reception. (See [9], which details circuits that may be employed for both transmission and reception.) Since the diode balanced modulator can demodulate SSB, it is used for that purpose in transceivers, in preference to the product demodulator. A circuit of the balanced modulator is shown in Fig. 6-44, it is identical to the one in Fig. 4-3*a*, but the emphasis here is on demodulation.

As in carrier suppression, the output of the local crystal oscillator, having the same frequency as in the product detector (200 or 203 kHz, depending on the sideband being demodulated), is fed to the terminals 1–1′. Where the carrier-suppressed signal was taken from the modulator at terminals 3–3′, the SSB signal is now fed in. The balanced modulator now operates as a nonlinear resistance and, as in the product detector, sum and difference frequencies appear at the primary winding of the AF transformer. This transformer will not pass radio frequencies and therefore acts as a low-pass filter, delivering only the audio frequencies to the terminals 2–2′, which have now become the output terminals of the demodulator. It is seen that this circuit recovers the information from the SSB signal, as required, and works very similarly to the product demodulator.

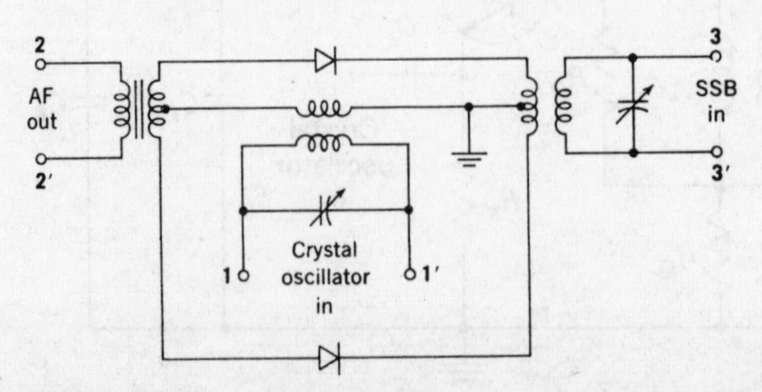

FIGURE 6-44 Balanced modulator used for demodulation of SSB.

6-5.2 Receiver Types

It is proposed to describe a pilot-carrier receiver and a suppressed-carrier receiver, the latter incorporating a frequency synthesizer for extra stability. The latter will also be used to show how ISB may be demodulated.

Pilot-carrier receiver As shown in Fig. 6-45, in block form, a pilot-carrier receiver is a fairly straightforward communications receiver with trimmings. It uses double conversion, as expected, and AFC based on the pilot carrier. AFC is needed to ensure good frequency stability, which must be at least 1 part of 10^7 (long-term) for long-distance telephone and telegraph communications. Note also the use of one local crystal oscillator, with multiplication by 9, rather than two separate oscillators; this also improves stability.

The output of the second mixer contains two components: the wanted sideband and the weak carrier. As shown, they are separated by means of filters, the sideband going to the product detector, and the carrier to AGC and AFC circuits via an extremely narrowband filter and amplifier. The output of the carrier amplifier is fed, together with the buffered output of the crystal oscillator, to a *phase comparator*. This is almost identical to the phase discriminator and works in a similar fashion. Here, however, the output depends on the *phase difference* between the two applied signals, which is zero or a positive or negative dc voltage, just as in the discriminator. Understandably, the phase difference between the two inputs to the phase-sensitive circuit can be zero only if the frequency difference is zero. Thus excellent frequency stability is obtainable. The output of the phase comparator actuates a varactor diode connected across the tank circuit of the VFO and pulls it into frequency as required.

Because a pilot carrier is transmitted, automatic gain control is not much of a problem, although that part of the circuit may look complicated. The output of the

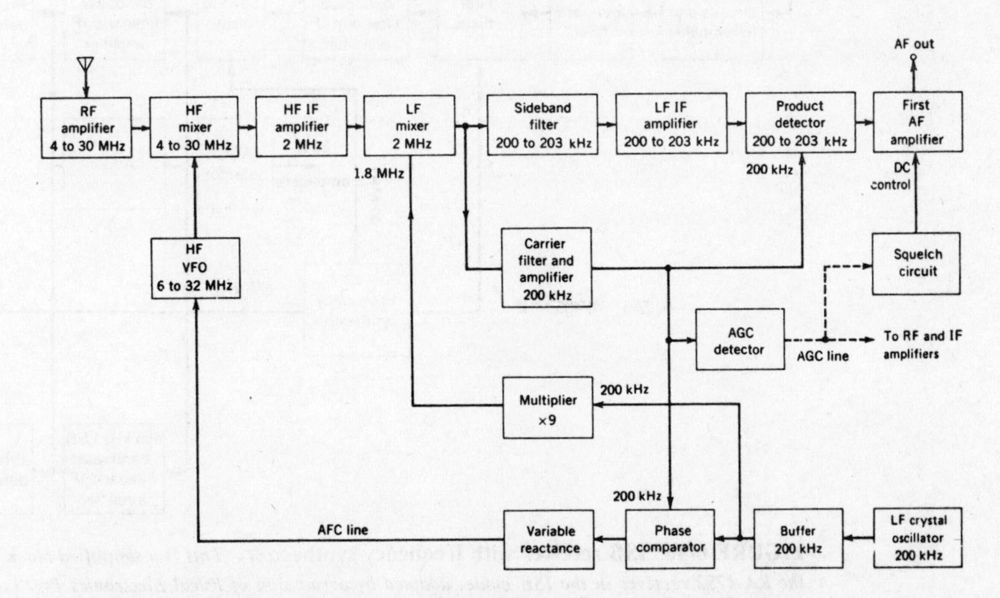

FIGURE 6-45 Block diagram of pilot-carrier single-sideband receiver.

carrier filter and amplifier is a carrier whose amplitude varies with the strength of the input signal, so that it may be used for AGC after rectification. Automatic gain control is also applied to the squelch circuit, as explained in Sec. 6-3.2. It should also be mentioned that receivers of this type often have AGC with two different time constants. This is helpful in telegraphy reception, and in coping to a certain extent with signal-strength variations caused by fading.

Suppressed-carrier receiver A typical block diagram is shown in Fig. 6-46. This is actually a much simplified version of the receiver of Fig. 6-17, which is capable of receiving all forms of AM but has here been shown in the ISB mode. The receiver has a number of very interesting features, of which the first is the fixed-frequency RF amplifier. This may be wideband, covering the entire 100-kHz to 30-MHz receiving range; or, optionally, a set of filters may be used, each covering a portion of this range. The second very interesting feature is the very high first intermediate frequency, 40.455 MHz. Such high frequencies have been made possible by the advent of VHF crystal bandpass filters. They are increasingly used by SSB receivers, for a number of reasons. One, clearly, is to provide image frequency rejections much higher than previously available. Another reason is to facilitate receiver tuning. In the RA 1792, which is typical of high-quality professional receivers, a variety of tuning methods are available, such as push-button selection, or even automatic selection of a series of wanted preset channels stored in the microprocessor memory. However, an important method is the orthodox continuous tuning method, which utilizes a tuning knob. Since receivers of this type are capable of remote tuning, the knob actually adjusts the

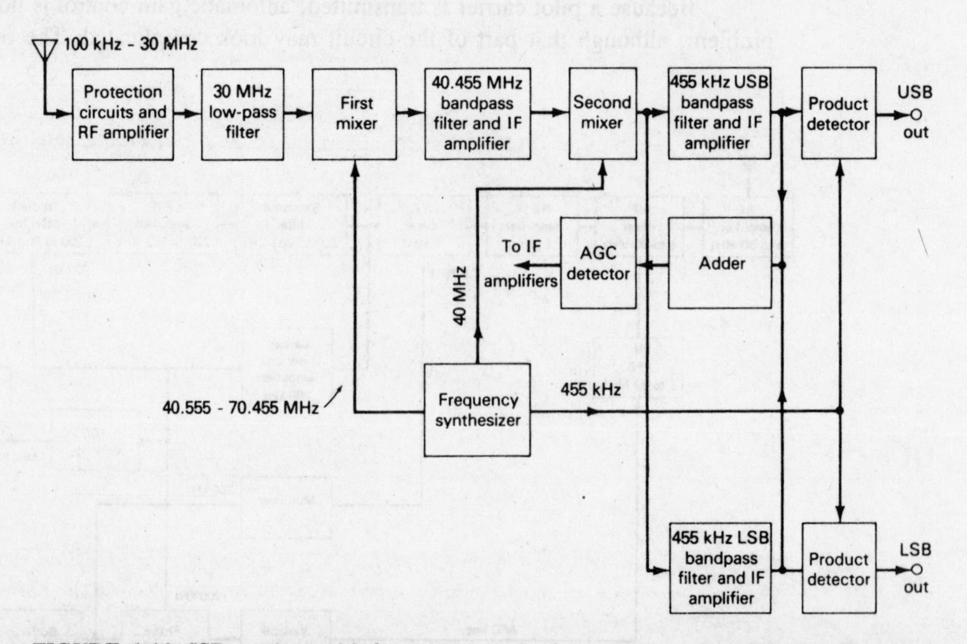

FIGURE 6-46 ISB receiver with frequency synthesizer. ***(This is a simplified block diagram of the RA 1792 receiver in the ISB mode, adapted by permission of Racal Electronics Pty. Ltd.)***

voltage applied to a varactor diode across the VFO in an indirect frequency synthesizer. There is a limit to the tuning range. If the first IF is high, the resulting range (70.455 MHz ÷ 40.555 MHz = 1.74:1) can be covered in a single sweep; with a much lower first IF it cannot be covered so readily.

It will be seen that this is, nonetheless, a double-conversion superheterodyne receiver, up to the low-frequency IF stages. After this the main differences are due to the presence of the two independent sidebands, which are separated at this point with mechanical filters. If just a single upper and a single lower sideband are transmitted, the USB filter will have a bandpass of 455.25 to 458 kHz, and the LSB filter 452 to 454.75 kHz. Since the carrier is not transmitted, it is necessary to obtain AGC by rectifying part of the combined audio signal. From this a dc voltage proportional to the average audio level is obtained. This requires an AGC circuit time constant of sufficient length to ensure that AGC is not proportional to the instantaneous audio voltage. Because of the presence of the frequency synthesizer, the frequency stability of such a receiver can be very high. For example, one of the frequency standard options of the RA 1792 will give a long-term frequency stability of 3 parts in 10^9 per day.

PROBLEMS

For self-testing questions on this chapter, see p. 683.

6-1. When a superheterodyne receiver is tuned to 555 kHz, its local oscillator provides the mixer with an input at 1010 kHz. What is the image frequency? The antenna of this receiver is connected to the mixer via a tuned circuit whose loaded Q is 40. What will be the rejection ratio for the calculated image frequency?

[Ans.: 1465 kHz, 116]

6-2. Calculate the image rejection of a receiver having an RF amplifier and an IF of 450 kHz, if the Qs of the relevant coils are 65, at an incoming frequency of *(a)* 1200 kHz; *(b)* 20 MHz.

[Ans.: (a) 5870, (b) 33.8]

6-3. A superheterodyne receiver having an RF amplifier and an IF of 450 kHz is tuned to 15 MHz. Calculate the Qs of the RF and mixer input tuned circuits, both being the same, if the receiver's image rejection is to be 120.

[Ans.: 93.6]

6-4. Calculate the image-frequency rejection of a double-conversion receiver which has a first IF of 2 MHz and a second IF of 200 kHz, an RF amplifier whose tuned circuit has a Q of 75 (the same as that of the mixer) and which is tuned to a 30-MHz signal. The answer is to be given in decibels.

[Ans.: 51 dB]

QUESTIONS

6-1. With the aid of the block diagram of a simple receiver, explain the basic superheterodyne principle.

6-2. Briefly explain the function of each of the blocks in the superheterodyne receiver.

6-3. What are the advantages that the superheterodyne receiver has over the TRF receiver? Are there any disadvantages?

6-4. How is the constant intermediate frequency achieved in the superheterodyne receiver?

6-5. Explain how the use of an RF amplifier improves the signal-to-noise ratio of a superheterodyne receiver.

6-6. Define the terms *sensitivity, selectivity* and *image frequency*.

6-7. Of all the frequencies that must be rejected by a superheterodyne receiver, why is the *image frequency* so important? What is the image frequency, and how does it arise? If the image-frequency rejection of a receiver is insufficient, what steps could be taken to improve it?

6-8. Explain what *double spotting* is and how it arises. What is its nuisance value?

6-9. Describe the general process of frequency changing in a superheterodyne receiver. What are some of the devices that can be used as frequency changers? Why must some of them be separately excited?

6-10. Using circuit diagrams, explain the operation of the self-excited transistor mixer by the three-frequency approach.

6-11. What is *three-point* tracking? How do tracking errors arise in the first place? What is the name given to the element that helps to achieve three-point tracking? Where is it placed?

6-12. What are the functions fulfilled by the intermediate-frequency amplifier in a radio receiver?

6-13. List and discuss the factors influencing the choice of the intermediate frequency for a radio receiver.

6-14. With the aid of a circuit diagram, explain the operation of a practical diode detector circuit, indicating what changes have been made from the basic circuit. How is AGC obtained from this detector?

6-15. What is *simple* automatic gain control? What are its functions?

6-16. Sketch a practical diode detector with typical component values and calculate the maximum modulation index it will tolerate without causing negative peak clipping.

6-17. Write a survey of the desirable features of communications receivers, briefly explaining the operation of each where necessary, and showing some of the more interesting circuit or block diagrams.

6-18. Explain in detail, with numerical examples if necessary, why the high intermediate frequency must come first in a double-conversion system.

6-19. Discuss the merits of delayed AGC as compared with simple AGC. Show AGC curves to illustrate the comparison and explain how delayed AGC may be obtained and applied. What does the "delayed AGC control" adjust?

6-20. What are the functions of variable selectivity and sensitivity? How is each achieved in practice?

6-21. What is *blocking* in a receiver? How is good blocking achieved? What will be the effect on a communications receiver if its blocking performance is poor?

6-22. Explain how tuning calibration may be provided in a communications receiver and also explain the functions and circuit arrangement of the beat-frequency oscillator.

6-23. What, exactly, does a noise limiter do in an AM receiver? How does it do this?

6-24. Discuss automatic frequency control and metering in communications receivers.

6-25. In addition to the usual superheterodyne circuits, a high-grade communications receiver has some further refinements and other features. Discuss the more important of these, illustrating your answer with circuit or block diagrams, as necessary.

6-26. What is frequency synthesis? What are the situations in which it is advantageous to use a frequency synthesizer? What advantages does it have over the alternatives?

6-27. Draw the block diagram of a direct frequency synthesizer covering the same frequency range as the one in Fig. 6-23, but having one decade less. What will now be the frequency coverage of the interpolating oscillator?

6-28. Explain how the indirect synthesizer is related to the direct synthesizer. What are the disadvantages of the latter which have resulted in its virtual replacement by indirect synthesizers in current equipment?

6-29. Discuss the functions and advantages of microprocessor control for communications receivers. With the aid of a block diagram, indicate what aspects of the receiver could profitably be controlled by a microprocessor.

6-30. Describe the differences between FM and AM receivers, bearing in mind the different frequency ranges and bandwidths over which they operate.

6-31. Draw the circuit of an FET amplitude limiter, and with the aid of the transfer characteristic explain the operation of this circuit.

6-32. What can be done to improve the overall limiting performance of an FM receiver? Explain, describing the need for, and operation of, the double limiter and also AGC in addition to a limiter.

6-33. Explain the operation of the balanced slope detector, using a circuit diagram and a response characteristic. Discuss, in particular, the method of combining the outputs of the individual diodes. In what ways is this circuit an improvement on the slope detector, and, in turn, what are its disadvantages?

6-34. Prove that the phase discriminator is an FM demodulator.

6-35. With circuits, explain how, and for what reason, the ratio detector is derived from the phase discriminator, listing the properties and advantages of each circuit.

6-36. Explain how the ratio detector demodulates an FM signal, proving that the output voltage is proportional to the difference between the individual input voltages to the diodes.

6-37. Draw the practical circuit of a balanced ratio detector, and show how it is derived from the basic circuit. Explain the improvement effected by each of the changes.

6-38. Using circuit diagrams, show how the Foster-Seeley discriminator is derived from the balanced slope detector, and how, in turn, the ratio detector is derived from the discriminator. In each step stress the common characteristics, and show what it is that makes each circuit different from the previous one.

6-39. Compare and contrast the performance and applications of the various types of frequency demodulators.

6-40. Draw the block diagram of that portion of a stereo FM multiplex receiver which lies between the main FM demodulator and the audio amplifiers. Explain the operation of the system, showing how each signal is extracted and treated.

6-41. List the various methods and circuits that can be used to demodulate J3E transmissions. Can demodulation also be performed with an AM receiver that has a BFO? If so, how?

6-42. Use a circuit diagram to help in an explanation of how a balanced modulator is able to demodulate SSB signals.

6-43. Explain the operation of an R3E receiver with the aid of a suitable block diagram. Stress, in particular, the various uses to which the weak transmitted carrier is put.

6-44. Compare the method of obtaining AGC in a pilot-carrier receiver with that employed in a J3E receiver.

6-45. Redraw the block diagram of Fig. 6-46, if this receiver is now required for USB J3E reception.

REFERENCES

1. Hawker, P., "Keep It Simple—Direct-Conversion HF Receivers," *Proc. Conf. Radio Receivers Assoc. Systs.*, The Institution of Electronic and Radio Engineers, London, July 1978, pp. 135–148.

2. Rohde, U.L., "Eight Ways to Better Radio Receiver Design," *Electronics,* Feb. 20, 1975, pp. 87–91.

3. A.R.R.L. Headquarters Staff, *The Radio Amateur's Handbook,* 59th ed., American Radio Relay League, Inc., Newington, Conn., 1981, chap. 8.

4. Parsons, J. D., M. Henze, P. A. Ratliff, and M. J. Withers, "Diversity Techniques for Mobile Radio Reception," *Radio Electron. Eng.*, July 1975, pp. 357–367 (48 further references given).
5. Gill, P. C., I. Hart, and J. E. Phillips, "The Real Cost of Synthesizers in Mobile Radio," *Proc. Conf. Radio Receivers Assoc. Syst.*, The Institution of Electronic and Radio Engineers, London, July 1978, pp. 195–206.
6. Ayre, D. C., and K. G. Woodard, "Digital Frequency Synthesis—A New Approach," *Wireless World*, May 1975.
7. Adapted from Seely, Samuel, "Radio Electronics," McGraw-Hill Book Company, New York, 1956.
8. Miya, K. (ed.), "Satellite Communications Engineering," Lattice Company, Ltd., Tokyo, 1975, pp. 307–311.
9. A.R.R.L. Headquarters Staff, *The Radio Amateur's Handbook*, 59th ed., American Radio Relay League, Inc., Newington, Conn., 1981, chap. 12.
10. Lin, W.C. (ed.), *Microprocessors: Fundamentals and Applications*, IEEE Press, New York, 1976, pp. 105–135.

7 TRANSMISSION LINES

In many communications systems, it is often necessary to interconnect points that are some distance apart from each other. The connection between a transmitter and its antenna is a typical example of this. If the frequency is high enough, such a distance may well become an appreciable fraction of the wavelength being propagated. It then becomes necessary to consider the properties of the interconnecting wires, since these no longer behave as short circuits. It is thus evident that the size, separation and general layout of the system of wires becomes significant under these conditions.

Wire systems which have properties that cannot be dismissed are generally called transmission lines. It is intended in this chapter to treat them with the attention they deserve. The treatment will begin with fundamentals and go on to such properties as the *characteristic impedance* of transmission lines. The *Smith chart* and its applications will be considered next and applied to the very many problems that can be solved with its aid. Finally, the chapter looks at the various transmission-line components in common use, notably *stubs, directional couplers* and *balance-to-unbalance transformers (baluns)*.

The treatment of transmission lines will be from a purely practical point of view. It will be assumed that a fundamental circuit theory approach is not required. This may be either because such a presentation has already been made to the students in a circuit theory subject (an approach which the author found very rewarding in his teaching), or because this had been decided against in a particular course. In either case, the coverage given here will be found self-sufficient.

7-1 BASIC PRINCIPLES

Transmission lines are a means of conveying signals or power from one point to another. From such a broad definition, any system of wires can be considered as forming one or more transmission lines. However, if the properties of these lines must be taken into account, the lines might as well be arranged in some simple, constant pattern. This will make the properties much easier to calculate, and it will also make them constant for any type of transmission line. Thus all practical transmission lines are arranged in some uniform pattern; this simplifies calculations, reduces costs and increases convenience.

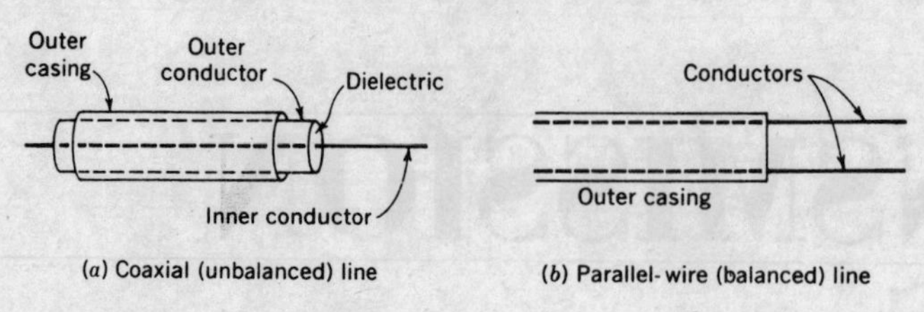

FIGURE 7-1 Transmission lines

7-1.1 Fundamentals of Transmission Lines

There are two types of commonly used transmission lines. The parallel-wire (balanced) line is shown in Fig. 7-1*b*, and the coaxial (unbalanced) line in Fig. 7-1*a*.

The parallel-wire line is employed where balanced properties are required: for instance, in connecting a *folded-dipole* antenna to a TV receiver or a *rhombic* antenna to an HF transmitter. On the other hand, the coaxial line is used when unbalanced properties are needed, as in the interconnection of a broadcast transmitter to its grounded antenna. It is also employed, as will be seen, at UHF and microwave frequencies, to avoid the risk of radiation from the transmission line itself.

Any system of conductors is likely to radiate if the conductor separation approaches a half-wavelength at the operating frequency. This is far more likely to occur in a parallel-wire line than in a coaxial line, whose outer conductor surrounds the inner one and is invariably grounded. Accordingly, parallel-wire lines are never used for microwaves, whereas coaxial lines may be employed for frequencies up to at least 18 GHz. It will be seen in Chap. 10 that *waveguides* also have frequency limitations. However, from the general point of view the limit is on the *lowest* usable frequency; below about 1 GHz, waveguide cross-sectional dimensions become inconveniently large. Thus, between 1 and 18 GHz, either waveguides or coaxial lines are used, depending on the requirements and application, whereas waveguides are not normally used below 1 GHz, and coaxial lines are not normally used above 18 GHz.

Description Within each broad grouping or type of transmission line there is an astonishing variety of different kinds, dictated by various applications. A browse through a manufacturer's illustrated catalog is most rewarding. It invariably shows, as does Table 7-1 for a number of lines, the different forms of lines and connectors, together with their characteristics, ratings and applications [1]. Lines may be rigid or flexible, air-spaced or filled with different dielectrics, with smooth or corrugated conductors as the circumstances warrant. Different diameters and properties are also available. Flexible lines are naturally more convenient than rigid ones, since they may be bent to follow any physical layout and are much easier to stow and transport. On the other hand, rigid cables can generally carry much higher powers, and it is easier to make them air-dielectric rather than filled with a solid dielectric. This consideration is important, especially for high powers, since all solid dielectrics have significantly higher losses than air, particularly as frequencies are increased.

Rigid coaxial air-dielectric lines consist of an inner and outer conductor with spacers of low-loss dielectric separating the two every few centimeters. There may be

TABLE 7-1 Selected Transmission Lines (Coaxial Unless Otherwise Stated)

JAN TYPE, NO.	OUTSIDE DIAMETER, mm	Z_o, ohms	DIELECTRIC MATERIAL	VELOCITY FACTOR	ATTENUATION PER METER, dB @ 100 MHz	ATTENUATION PER METER, dB @ 1 GHz	AVERAGE POWER RATING, kW @ 100 MHz	AVERAGE POWER RATING, kW @ 1 GHz	NOTES
RG-58C/U	5.0	53.5	PE[a]	0.66	0.18	0.72	0.18	0.05	Small flexible
RG-213/U	10.3	52.0	PE[a]	0.66	0.08	0.28	0.68	0.19	Medium flexible (previously RG-8/U)
RG-218/U	22.1	52.0	PE[a]	0.66	0.03	0.14	2.0	0.50	Large LA[b], HP[c] (previously RG-17A/U)
RG-11A/U	10.5	75.0	PE[a]	0.66	0.08	0.28	0.68	0.19	Medium flexible, video
RG-85A/U	39.8	75.0	PE[a]	0.66	0.03	0.14	2.0	0.50	Large LA[b], HP[c], armored
RG-55B/U	8.4	53.5	PE[a]	0.66	0.15	0.56	0.18	0.05	Small, microwave (previously RG-5B/U)
RG-211A/U	18.5	50.0	PTFE[d]	0.70	0.10	0.39	22.0	5.5	High temperature, semiflexible
(⅞ in)	22.2	50.0	PTFE[d] pins	0.81	0.01	0.04	4.8	1.6	Rigid } Essentially air dielectric,
(3⅛ in)	79.4	50.0	PTFE[d] pins	0.81	0.003	0.01	52	17	Rigid } with PTFE[d] separating pins
(6⅛ in)	155.6	75.0	PTFE[d] pins	0.81	0.001	0.005[e]	150	46[e]	Rigid } at regular intervals
(SLA12-50J)	12.7	50.0	PTFE[d] spiral	0.81	0.03	0.09	24	7	Air dielectric, flexible
RG-57A/U	15.9	95.0	PE[a]	0.66					Twin conductor, flexible
		300.0	Foam PE[a]	0.82	0.04[f]	—[g]			Twin lead, flexible

[a] Polyethylene.
[b] Low attenuation.
[c] High power.
[d] Polytetrafluoroethylene (Teflon).
[e] At 900 MHz.
[f] Approximately.
[g] Maximum frequency is 150 MHz.

a sheath around the outer conductor to prevent corrosion, but this is not always the case. A flexible air-dielectric cable generally has corrugations in both the inner and the outer conductor, running at right angles to its length, and a spiral of dielectric material between the two.

The power-handling ability of a transmission line is limited by flashover between the conductors due to a high-voltage gradient breaking down the dielectric. It depends on the type of dielectric material used, as well as the distance between the conductors. Thus, for the high-power cables employed in transmitters, nitrogen under pressure may be used to fill the cable and reduce flashover. Since nitrogen is less reactive than the oxygen component of air, corrosion is reduced as well. Dry air under pressure is also used as a means of keeping out moisture. Clearly, as the power transmitted is increased, so must be the cross-sectional dimensions of the cable. Thus, for example, rigid air-dielectric coaxial copper cable with an outer diameter of 22.5 mm has a peak power rating of 43 kW. This increases to 400 kW for an outer diameter of 80 mm and to 3 MW for a 230-mm outer diameter. As will be seen, the inner diameter must be changed along with the outer diameter, to ensure a constant value of all line properties.

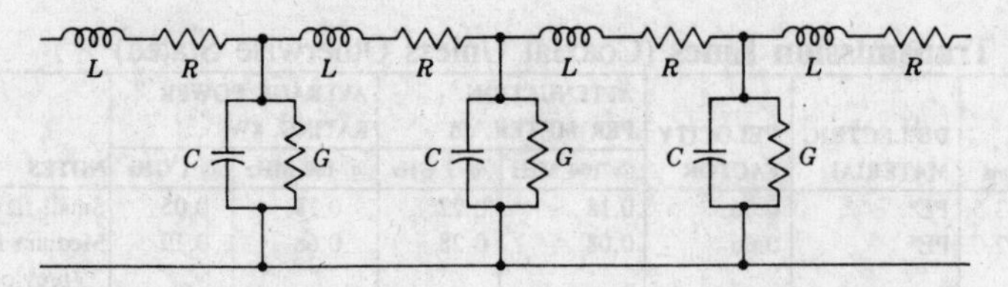

FIGURE 7-2 General equivalent circuit of transmission line.

Equivalent circuit representation Since each conductor has a certain length and diameter, it must have resistance and inductance; since there are two wires close to each other, there must be capacitance between them. Finally, the wires are separated by a medium called the *dielectric,* which cannot be perfect in its insulation; the current leakage through it can be represented by a shunt conductance. The resulting equivalent circuit is as shown in Fig. 7-2. Note that all the quantities shown are proportional to the length of the line, *and unless measured and quoted per unit length, they are meaningless.*

At radio frequencies, the inductive reactance is much larger than the resistance. The capacitive susceptance is also much larger than the shunt conductance. Thus both R and G may be ignored, resulting in a line that is considered lossless (as a very good approximation for RF calculations). The equivalent circuit is simplified as shown in Fig. 7-3.

It is to be noted that the quantities L, R, C, and G, shown in Figs. 7-2 and 7-3, are all measured per unit length, e.g., per meter, because they occur continuously along the line. They are thus distributed throughout the length of the line. Under no circumstances can they be assumed to be lumped at any one point.

7-1.2 Characteristic Impedance

Any circuit that consists of series and shunt impedances must have an input impedance. For the transmission line this input impedance will depend on the type of line, its length and the termination at the far end. To simplify description and calculation, the input impedance under certain standard, simple and easily reproducible conditions is taken as the reference and is called the *characteristic impedance* of that line. *By definition, the characteristic impedance of a transmission line, Z_0, is the impedance measured at the input of this line when its length is infinite*. Under these conditions the type of termination at the far end has no effect, and consequently is not mentioned in the definition.

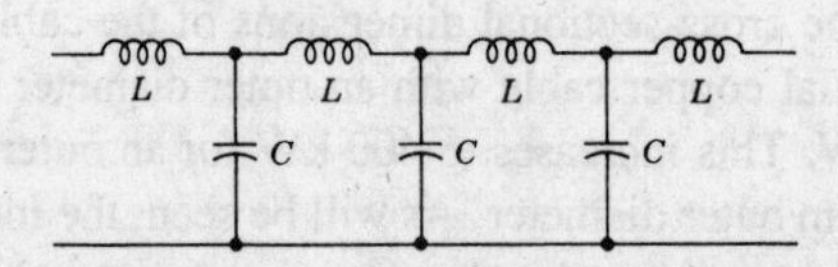

FIGURE 7-3 Transmission-line RF equivalent circuit.

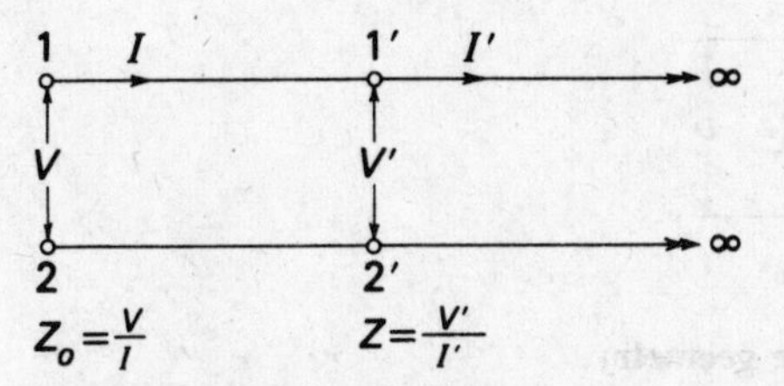

FIGURE 7-4 Infinite line.

Methods of calculation It can now be shown that the characteristic impedance of a line will be measured at its input when the line is terminated at the far end in an impedance equal to Z_0, no matter what length the line has. This is important, because such a situation is far easier to reproduce for measurement purposes than a line of infinite length.

If a line has infinite length, all the power fed into it will be absorbed. It should be fairly obvious that as one moves way from the input, voltage and current will decrease along the line, as a result of the voltage drops across the inductance and current leakage through the capacitance. From the meaning of infinity, the points 1′–2′ of Fig. 7-4 are just as far from the far end of this line as the points 1–2. Thus the impedance seen at 1′–2′ (looking to the right) is also Z_0, although the current and voltage are lower than at 1–2. We can thus say that the input terminals see a piece of line up to 1′–2′, followed by a circuit which has the input impedance equal to Z_0. It quite obviously does not matter what the circuit to the right of 1′–2′ consists of, provided that it has an input impedance equal to the characteristic impedance of the line. *Thus Z_0 will be measured at the input of a transmission line if the output is terminated in Z_0.*

It follows from filter theory that the characteristic impedance of an iterative circuit consisting of series and shunt elements is given by

$$Z_0 = \sqrt{\frac{Z}{Y}} \tag{7-1}$$

where Z = series impedance per section
$= R + j\omega L$ (Ω/m here) and is the series impedance per unit length
Y = shunt admittance per section
$= G + j\omega C$ (S/m here) and is the shunt admittance per unit length

Thus

$$Z_0 = \sqrt{\frac{R + j\omega L}{G + j\omega C}} \tag{7-2}$$

From Eq. (7-2) it follows that the characteristic impedance of a transmission line may be complex, and indeed it very often is, especially in line communications, i.e., telephony at voice frequencies. At radio frequencies, however, as already mentioned, the resistive components of the equivalent circuit become insignificant, and the

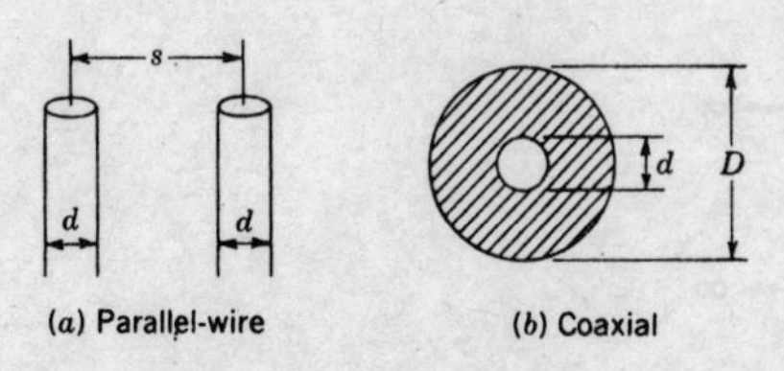

FIGURE 7-5 Transmission-line geometry.

expression for Z_0 reduces to

$$Z_0 = \sqrt{\frac{j\omega L}{j\omega C}} = \sqrt{\frac{L}{C}} \qquad \textbf{(7-3)}$$

L is measured in henrys per meter and C in farads per meter, whence it follows that Eq. (7-3) shows the characteristic impedance of a line in ohms and is dimensionally correct. It also shows that this *characteristic impedance is resistive at radio frequencies*.

Physically, characteristic impedance is determined by the geometry, size and spacing of the conductors, and by the dielectric constant of the insulator separating them. It may be calculated from the following formulas, the various terms having meanings as shown in Fig. 7-5:

For the parallel-wire line, we have

$$Z_0 = 276 \log \frac{2s}{d} \quad \Omega \qquad \textbf{(7-4)}$$

For the coaxial line, this is

$$Z_0 = \frac{138}{\sqrt{k}} \log \frac{D}{d} \quad \Omega^1 \qquad \textbf{(7-5)}$$

where k = dielectric constant of the insulation.

Equation (7-4) appears to take no account of the dielectric constant of the insulating material; this is so because this material is very often air for parallel-wire lines, and its dielectric constant is unity. The formula for the Z_0 of a balanced line with solid dielectric is almost identical, except that the first term becomes $276/\sqrt{k}$.

The usual range of characteristic impedances for balanced lines is 150 to 600 Ω, and 40 to 150 Ω for coaxial lines, both being limited by their geometry. This, as well as the method of using the characteristic impedance formulas, will be shown in the next three examples.

Example 7-1 A piece of RG-59B/U coaxial cable has a 75-Ω characteristic impedance and a nominal capacitance of 69 pF/m. What is its inductance per meter? If the

[1] The figure 138 is not as odd as it seems, being equal to $120\pi/e$, where $120\pi = 377\ \Omega$ happens to be the impedance of free space, and e is the base of the natural logarithm system; 276 is, of course, 2×138.

diameter of the inner conductor is 0.584 mm, and the dielectric constant of the insulation is 2.23, what is the outer conductor diameter?

$$Z_0 = \sqrt{\frac{L}{C}}$$

$$L = Z_0^2 C = 75^2 \times 69 \times 10^{-12} = 3.88 \times 10^{-7} = 0.388\ \mu\text{H/m}$$

$$Z_0 = \frac{138}{\sqrt{k}} \log \frac{D}{d}$$

$$\log \frac{D}{d} = \frac{Z_0}{138/\sqrt{k}} = \frac{75}{138/\sqrt{2.23}} = 0.81$$

$$D = D \times \text{antilog } 0.81 = 0.584 \times 6.457 = 3.77 \text{ mm}$$

Example 7-2 What is the minimum value that the characteristic impedance of an air-dielectric parallel-wire line could have?

Minimum impedance will occur when $2s/d$ is also minimum, and this is reached when the two wires of Fig. 7-5*a* just touch. It is then seen that $s = d$, so that we have

$$Z_{0,\min} = 276 \log 2 \times 1 = 276 \times 0.3010 = 83\ \Omega$$

Example 7-3 A coaxial cable, having an inner diameter of 0.025 mm and using an insulator with a dielectric constant of 2.56, is to have a characteristic impedance of 2000 Ω. What must be the outer conductor diameter?

$$\log \frac{D}{d} = \frac{Z_0}{138/\sqrt{k}} = \frac{2{,}000}{138/\sqrt{2.56}} = 2{,}000 \times \frac{1.6}{138}$$

$$= 23.1884$$

$$D = d \times \text{antilog } 23.1884 = 0.025 \times 10^{23} \times 1.543 = 3.86 \times 10^{21} \text{ mm}$$

$$= 3.86 \times 10^{15} \text{ km}$$

$$= \frac{3.86 \times 10^{15}}{9.44 \times 10^{12}} = 409 \text{ light-years}$$

A light-year, as the name suggests, is the distance covered by light in 1 year at a velocity of 300,000 km per second. The figure of 409 light-years is almost exactly 100 times the distance of the nearest star (Proxima Centauri) from the solar system, and this example tries to show conclusively that such a high value of characteristic impedance is just not possible!

If a high value of characteristic impedance is needed, it is seen that the conductors must be very small to give a large inductance per unit length. As well, the distance between them must be very large to yield as small a shunt capacitance per unit length as possible. One eventually runs out of distance. At the other end of the scale, the exact reverse applies. Distances between conductors become inconveniently small for coaxial lines. They become impossible for parallel-wire lines, since overlapping of conductors would occur if a Z_0 less than 83 Ω were attempted.

7-1.3 Losses in Transmission Lines

Types of losses There are three ways in which energy, applied to a transmission line, may become dissipated before reaching the load: radiation, conductor heating and dielectric heating.

Radiation losses arise because a transmission line may act as an antenna if the separation of the conductors is an appreciable fraction of a wavelength. This is treated in Sec. 9-1.1 and applies more to parallel-wire lines than to coaxial lines. Radiation losses are difficult to estimate, being normally measured rather than calculated. Understandably, they increase with frequency for any given transmission line, eventually ending that line's usefulness at some high frequency.

Conductor heating, or I^2R loss, is proportional to current and therefore inversely proportional to characteristic impedance. It also increases with frequency, this time because of the *skin effect*. Dielectric heating is proportional to the voltage across the dielectric and hence inversely proportional to the characteristic impedance for any power transmitted. It again increases with frequency (for solid dielectric lines) because of gradually worsening properties with increasing frequency for any given dielectric medium. For air, however, dielectric heating remains negligible. Since the last two losses are proportional to length, they are usually lumped together and given by manufacturers in charts, expressed in decibels per 100 meters. For practical coaxial lines at 1 GHz, these losses vary from as much as 200 dB/100 m for a solid-dielectric, flexible 6-mm line, to as little as 0.94 dB/100 m for a rigid, air-dielectric 127-mm line.

Velocity factor The velocity of light and all other electromagnetic waves depends on the medium through which they travel. It is very nearly 3×10^8 m/s in a vacuum and slower in all other media. The velocity of light in a medium is given by

$$v = \frac{v_c}{\sqrt{k}} \tag{7-6}$$

where v = velocity in the medium
v_c = velocity of light in a vacuum
k = dielectric constant of the medium (1 for a vacuum and very nearly 1 for air)

The *velocity factor* of a dielectric substance, and thus of a cable, is the velocity reduction ratio and is therefore given by

$$vf = \frac{1}{\sqrt{k}} \tag{7-7}$$

The dielectric constants of materials commonly used in transmission lines range from about 1.2 to 2.8, giving corresponding velocity factors from 0.9 to 0.6. Note also that since $v = f\lambda$ and f is constant, the wavelength λ is also reduced by a ratio equal to the velocity factor. This is of particular importance in *stub* calculations, as will be seen.

7-1.4 Standing Waves

If a lossless transmission line has infinite length or is terminated in its characteristic impedance, all the power applied to the line by the generator at one end is absorbed by the load at the other end. Conversely, if a finite piece of line is terminated in an impedance not equal to the characteristic impedance, it can be appreciated that some (but not all) of the applied power will be absorbed by the termination. The remaining power will be *reflected*.

Reflections from an imperfect termination When a transmission line is incorrectly terminated, the power not absorbed by the load is sent back toward the generator, so that an obvious inefficiency exists. The greater the difference between the load impedance and the characteristic impedance of the line, the larger is this inefficiency.

A line terminated in its characteristic impedance is called a *nonresonant,* or *flat,* line. The voltage and current in such a line are constant throughout its length if the line is lossless, or are reduced exponentially (as the load is approached) if the line has losses. When a line is terminated in a short circuit or an open circuit, none of the power will be dissipated in such a termination, and all of it will be reflected back to the generator. If the line is lossless, it should be possible to send a wave out and then quickly replace the generator by a short circuit. The power in the line would shunt back and forth, never diminishing because the line is lossless. The line is then called *resonant* because of its similarity to a resonant *LC* circuit, in which the power shunts back and forth between the electric and magnetic fields. If the load impedance has a value between 0 and Z_0 or between Z_0 and ∞, oscillations still take place. This time, however, the amplitude decreases with time, more sharply as the value of the load impedance approaches Z_0.

Standing waves When power is applied to a transmission line by a generator, a voltage and a current appear whose values depend on the characteristic impedance and the applied power. The voltage and current waves travel to the load at a speed slightly less than v_c, depending on the velocity factor. If $Z_L = Z_0$, the load absorbs all the power, and none is reflected. The only waves then present are the voltage and current *traveling waves* from generator to load.

If Z_L is not equal to Z_0, some power is absorbed, and the rest is reflected. We thus have one set of waves, *V* and *I*, traveling toward the load, and the reflected set traveling back to the generator. These two sets of traveling waves, going in opposite directions, set up an interference pattern known as *standing waves,* i.e., beats, along the line. This is shown in Fig. 7-6 for a short-circuited line. It is seen that *stationary* voltage and current minima (nodes) and maxima (antinodes) have appeared. They are separated by half the wavelength of the signal, as will be explained. Note that voltage nodes and current antinodes coincide on the line, as do current nodes and voltage antinodes.

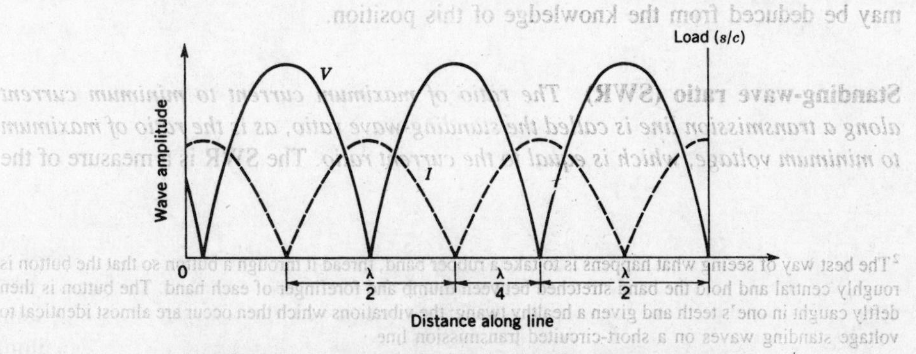

FIGURE 7-6 Lossless line terminated in a short circuit.

Consider only the forward traveling voltage and current waves for the moment. At the load, the voltage will be zero and the current a maximum because the load is a short circuit. Note that the current has a finite value since the line has an impedance. At that instant of time, the same conditions also apply at a point exactly one wavelength on the generator side of the load, and so on. The current at the load is always a maximum, although the size of this maximum varies cyclically with time, since the applied wave is sinusoidal.

The reflection that takes place at the short circuit affects both voltage and current. The current now starts traveling back to the generator, unchanged in phase, but *the voltage is reflected with a* 180° *phase reversal* [2]. Thus, at a point exactly a quarter-wavelength from the load, the current is *permanently* zero (as shown in Fig. 7-6). This is because the forward and reflected current waves are exactly 180° out of phase, as the reflected wave has had to travel a distance of $\lambda/4 + \lambda/4 = \lambda/2$ farther than the forward wave. The two thus cancel, and a current node is established. The voltage wave has also had to travel an extra distance of $\lambda/2$, but since it underwent a 180° phase reversal on reflection, its total phase change is 360°. Hence reinforcement will take place, resulting in a voltage antinode at precisely the same point as the current node.

A half-wavelength from the load is a point at which there will be a voltage zero and a current maximum. This arises because the forward and reverse current waves are now in phase (current has had to travel a total distance of one wavelength to return to this point). Simultaneously the voltage waves will cancel, because the 180° phase reversal on reflection must be added to the extra distance the reflected wave has to travel. In fact, all these conditions will repeat at half-wavelength distances, as shown in Fig. 7-6. Every time a point is considered that is $\lambda/2$ farther from the load than some previously considered point, the reflected wave has had to travel one whole wavelength farther. Therefore it has the same relation to the forward wave as it had at the first point.

It must be emphasized that this situation is permanent for any given load and is determined by it; such waves are truly *standing* waves. All the nodes are permanently fixed minima, and all the antinodes are maxima whose positions are constant (but they do have amplitudes which vary sinusoidally, just like the applied signal).[2] Much the same conditions apply if the load is an open circuit, except that the first current minimum (and voltage maximum) is now at the load, instead of a quarter-wavelength away from it. Since the load determines the position of the first current node, the type of load may be deduced from the knowledge of this position.

Standing-wave ratio (SWR) *The ratio of maximum current to minimum current along a transmission line is called the standing-wave ratio, as is the ratio of maximum to minimum voltage, which is equal to the current ratio*. The SWR is a measure of the

[2] The best way of seeing what happens is to take a rubber band, thread it through a button so that the button is roughly central and hold the band stretched between thumb and forefinger of each hand. The button is then deftly caught in one's teeth and given a healthy twang; the vibrations which then occur are almost identical to voltage standing waves on a short-circuited transmission line.

mismatch between the load and the line, and is the first and most important quantity calculated for a particular load. The SWR is equal to unity (a highly desirable state of affairs) when the load is perfectly matched. When the line is terminated in a purely resistive load, the standing-wave ratio is given by

$$\text{SWR} = Z_0/R_L \quad \text{or} \quad \text{SWR} = R_L/Z_0 \quad \text{(whichever is larger)} \tag{7-8}$$

where R_L is the load resistance.

It is customary to put the larger quantity in the numerator of the fraction, so that the ratio will be greater than 1. This does not lead to any confusion. Regardless of whether the load resistance is half as large or twice as large as the line characteristic impedance, the ratio of a voltage maximum to a voltage minimum is 2:1, and the degree of mismatch is the same in both cases.

If the load is purely reactive, SWR will be infinity; the same can be seen to apply for a short-circuit or an open-circuit termination. Since in all three cases no power is absorbed, the reflected wave has the same size as the forward wave. Somewhere along the line complete cancellation will occur, giving a voltage zero, and hence SWR must be infinite. When the load is complex, SWR can still be computed, but it is much easier to determine it from a transmission-line calculator, or to measure it.

The higher the SWR, the greater the mismatch between line and load or, for that matter, between generator and line. In practical lines, power loss increases with SWR, and so a low value of standing-wave ratio is always sought, except when the transmission line is being used as a pure reactance or as a tuned circuit. This will be shown in Sec. 7-1.5.

Normalization of impedance It is customary to *normalize* an impedance with respect to the line to which it is connected, i.e., to divide this impedance by the characteristic impedance of the line, as

$$z_L = \frac{Z_L}{Z_0} \tag{7-9}$$

thus obtaining the normalized impedance.[3] This is very useful because the behavior of the line depends not on the absolute magnitude of the load impedance, but on its value relative to Z_0. This fact can be seen from Eq. (7-8); the SWR on a line will be 2 regardless of whether $Z_0 = 75\ \Omega$ and $R_L = 150\ \Omega$ or $Z_0 = 300\ \Omega$ and $R_L = 600\ \Omega$. In addition, the normalizing of impedance opens up possibilities for transmission-line charts. It is in that respect similar to the process used to obtain the universal response curves for tuned circuits and *RC*-coupled amplifiers.

Consider a pure resistance connected to a transmission line, such that $R_L \neq Z_0$. Since the voltage and current vary along the line, as shown in Fig. 7-7, so will the resistance or impedance. However, conditions do repeat every half-wavelength, as already outlined. Hence the impedance at P will be equal to that of the load, if P is a half-wavelength away from the load and the line is lossless.

[3] Note that the normalized impedance is a dimensionless quantity, not to be measured or given in ohms.

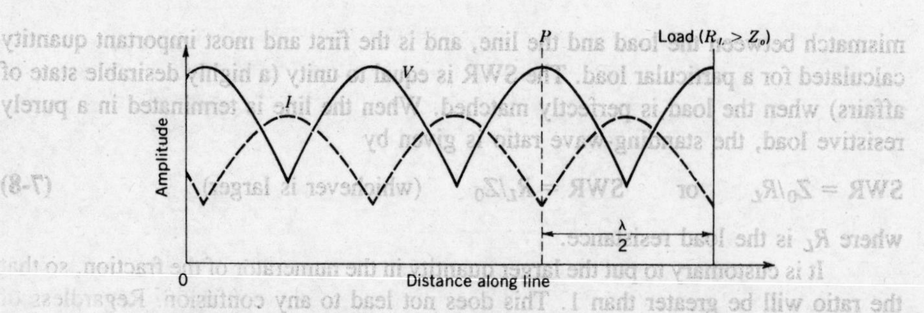

FIGURE 7-7 Lossless line terminated in a pure resistance greater than Z_0 (note that voltage SWR equals current SWR).

7-1.5 Quarter- and Half-Wavelength Lines

Sections of transmission lines that are exactly a quarter-wavelength or a half-wavelength long have important impedance-transforming properties, and are often used for this purpose at radio frequencies. Such lines will now be treated in turn.

Impedance inversion by quarter-wavelength lines Consider Fig. 7-8, which shows a load of impedance Z_L connected to a piece of transmission line of length s and having Z_0 as its characteristic impedance. When the length s is exactly a quarter-wavelength (or an odd number of quarter-wavelengths) and the line is lossless, then the impedance Z_s, seen when looking toward the load, is given by

$$Z_s = \frac{Z_0^2}{Z_L} \tag{7-10}$$

Equation (7-10) represents a very important and fundamental relation, which is somewhat too complex to derive here, but whose truth may be indicated as follows. Unless a load is resistive and equal to the characteristic impedance of the line to which it is connected, standing waves of voltage and current are set up along the line, with a node (and antinode) repetition rate of $\lambda/2$. This has already been shown and is indicated again in Fig. 7-9. Note that here the voltage and current minima are not zero; the load is not a short circuit, and therefore the standing-wave ratio is not infinite. Note also that the current nodes are separated from the voltage nodes by a distance of $\lambda/4$, as before.

It is obvious that at the point A (voltage node, current antinode) the line impedance is low, and at the point B (voltage antinode, current node) it is the reverse, i.e., high. In order to change the impedance at A, it would be necessary to change the SWR on the line. If the SWR were increased, the voltage minimum at A would be lower, and

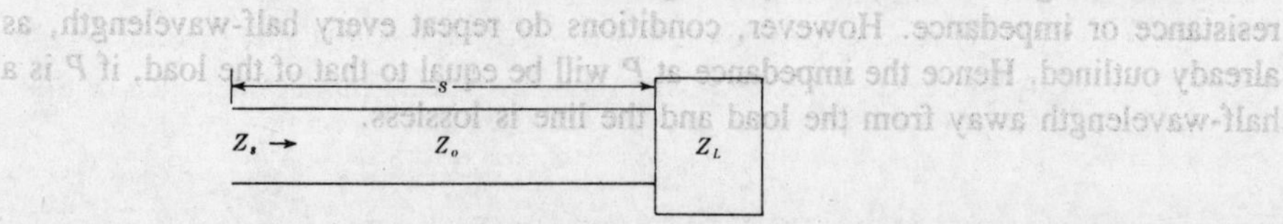

FIGURE 7-8 Loaded line.

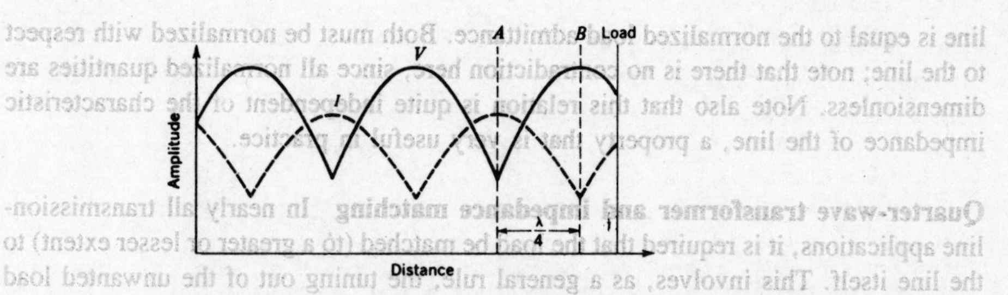

FIGURE 7-9 Standing waves along a mismatched transmission line; impedance inversion.

so would be the impedance at A. By the same token, however, the size of the voltage maximum at B would be increased, and so would the impedance at B. Thus an increase in Z_B is accompanied by a decrease in Z_A (if A and B are $\lambda/4$ apart). This amounts to saying that *the impedance at A is inversely proportional to the impedance at B*. Equation (7-10) states this relation mathematically and also supplies the proportionality constant; this happens to be the square of the characteristic impedance of the transmission line.[4] The relation holds just as well when the two points are not voltage nodes and antinodes, and a glance at Fig. 7-9 shows that it also applies when the distance separating the points is three, five, seven and so on, quarter-wavelengths.

Another interesting property of the quarter-wave line is seen if, in Eq. (7-10), the impedances are normalized with respect to Z_0. Dividing both sides by Z_0, we have

$$\frac{Z_s}{Z_0} = \frac{Z_0}{Z_L} \tag{7-11}$$

but

$$\frac{Z_s}{Z_0} = z_s$$

and

$$\frac{Z_L}{Z_0} = z_L$$

whence $Z_0/Z_L = 1/z_L$.

Substituting these results into Eq. (8-11) gives

$$z_s = \frac{1}{z_L}$$

$$= y_L \tag{7-12}$$

where y_L is the *normalized admittance* of the load.

Equation (7-12) is a very important relation. It states that if a quarter-wave-length line is connected to an impedance, then the normalized input impedance of this

[4]Note that this constant had to be the square of some impedance; otherwise the equation would have been dimensionally incorrect.

line is equal to the normalized load admittance. Both must be normalized with respect to the line; note that there is no contradiction here, since all normalized quantities are dimensionless. Note also that this relation is quite independent of the characteristic impedance of the line, a property that is very useful in practice.

Quarter-wave transformer and impedance matching In nearly all transmission-line applications, it is required that the load be matched (to a greater or lesser extent) to the line itself. This involves, as a general rule, the tuning out of the unwanted load reactance (if any) and the transformation of the resulting impedance to the value required. Ordinary RF transformers may be used up to the middle of the VHF range. However, their performance is not good enough at frequencies much higher than this, owing to excessive leakage inductance and stray capacitances. The quarter-wave line provides unique opportunities for impedance transformation up to the highest frequencies and is, furthermore, compatible with transmission lines.

Equation (7-10) shows that the impedance at the input of a quarter-wave line depends on two quantities: these are the load impedance (which is fixed for any load at a constant frequency) and the characteristic impedance of the interconnecting transmission line. If this Z_0 can be varied, the impedance seen at the input to the $\lambda/4$ transformer will be varied accordingly, and the load may thus be matched to the characteristic impedance of the main line. This is similar to varying the turns ratio of a transformer to obtain a required value of input impedance for any given value of load impedance. An example is now given to illustrate the procedure.

Example 7-4 It is required to match a 200-Ω load to a 300-Ω transmission line, to reduce the SWR along the line to 1. What must be the characteristic impedance of the quarter-wave transformer used for this purpose, if it is connected directly to the load?

Since the condition SWR = 1 is wanted along the main line, the impedance Z_s at the input to the $\lambda/4$ transformer must equal the characteristic impedance Z_0 of the main line. Let the transformer characteristic impedance be Z_0'; then, from Eq. (7-10),

$$Z_s = \frac{Z_0'^2}{Z_L} = Z_0 \qquad \text{(of main line)}$$

$$Z_0' = \sqrt{Z_0 Z_L} \tag{7-13}$$

$$= \sqrt{200 \times 300} = 245\Omega$$

Equation (7-13) was derived for this exercise, but it is universal in application and quite important.

It must be understood that a quarter-wave transformer has a length of $\lambda/4$ at only one frequency. It is thus highly frequency-dependent, and is in this respect akin to a high-Q tuned circuit. As a matter of fact, the difference between the transmission-line transformer and an ordinary tuned transformer is purely one of construction; the practical behavior is identical. This property of the quarter-wave transformer makes it useful as a filter, to prevent undesirable frequencies from reaching the load, often an antenna. On the other hand, if broadband impedance matching is required, the transformer must be constructed of high-resistance wire to lower its Q, thereby increasing bandwidth.

It should be mentioned that the procedure becomes somewhat more involved if the load is complex, rather than purely resistive as so far considered. The quarter-wave transformer can still be used, but it must now be connected at some precalculated distance from the load. It is generally connected at the nearest resistive point to the load, whose position may be found with the aid of a transmission-line calculator, such as a *Smith chart*.

Half-wavelength line It has already been shown that voltage and current nodes and antinodes recur every half-wavelength along a mismatched transmission line. Thus a half-wave transformer has the property that the input impedance to it is equal to the impedance of the load placed at the far end of the half-wave line. This property is independent of the characteristic impedance of this line, but once again it is frequency-dependent.

The advantages of this property are many. For instance, it is very often not practicable to measure the impedance of a load directly. This being the case, the impedance may be measured along a transmission line connected to the load, at a distance which is a half-wavelength (or a whole number of half-wavelengths) from the load. Again, it is sometimes necessary to short-circuit a transmission line at a point that is not physically accessible. The same results will be obtained if the short circuit is placed a half-wavelength (etc.) away from the load. Yet again, if a short-circuited half-wave transmission line is connected across the main line, the latter will be short-circuited at that point, but only at the frequency at which the shunt line is a half-wavelength. That frequency will not pass this point, but others will, especially if they are farther and farther away from the initial frequency. The short-circuited shunt half-wave line has thus become a band-stop filter. Finally, if the frequency of a signal is known, a short-circuited transmission line may be connected to the generator of this frequency, and a half-wavelength along this line may be measured very accurately. From the knowledge of frequency and wavelength, the velocity of the wave along the line can be calculated. Hence, we may determine the velocity factor, and therefore the dielectric constant of the insulation, as discussed in Sec. 7-1.3.

7-1.6 Reactance Properties of Transmission Lines

Just as a suitable piece of transmission line may be used as a transformer, so other chosen transmission-line configurations may be used as series or shunt inductive or capacitive reactances. This is very advantageous indeed; not only can such circuits be employed at the highest frequencies, unlike *LC* circuits, but also they are compatible with transmission lines.

Open- and short-circuited lines as tuned circuits The input impedance of a quarter-wave piece of transmission line, short-circuited at the far end, is infinity, and the line has transformed a short circuit into an open circuit. As just discussed, however, this applies only at the frequency at which the piece of line is exactly $\lambda/4$ in length. At some frequency near this, the line will be just a little longer or shorter than $\lambda/4$, so that at this frequency the impedance will not be infinity. The further we move, in frequency, away from the original, the lower will be the impedance of this piece of line. We therefore seem to have a parallel-tuned circuit, or at least something that behaves as one. Such a line is often used for this purpose at UHF, as an oscillator tank circuit (see Fig. 11-3) or in other applications.

If the quarter-wave line is open-circuited at the far end, then, by a similar process of reasoning, a series-tuned circuit is obtained. Similarly, a short-circuited half-wave line will behave as a series-tuned circuit, in the manner described in the preceding section. Such short- or open-circuited lines may be employed at high frequencies in place of LC circuits. In practice, however, short-circuited lines are used by preference, since open-circuited lines tend to radiate.

Properties of lines of various lengths Restating the position, we know that a piece of transmission line $\lambda/4$ long and short-circuited at the far end (or $\lambda/2$ long and open-circuited at the far end) looks like an open circuit and behaves *exactly* like a parallel-tuned circuit. If the frequency of operation is lowered, the shunt inductive reactance of this tuned circuit is lower and the shunt capacitive reactance is higher. Inductive current predominates, and therefore the impedance of the circuit is purely inductive. Now, this piece at the new frequency is less than $\lambda/4$ long, since the wavelength is now greater and the length of line is naturally unchanged. We thus have the important property that a short-circuited line less than $\lambda/4$ long behaves as a pure inductance. Similarly, an open-circuited line less than $\lambda/4$ long appears as a pure capacitance. The various possibilities are shown in Fig. 7-10, which is really a table of various line lengths and terminations and their equivalent LC circuits.

Stubs If a load is connected to a transmission line and matching is required, a quarter-wave transformer may be used if Z_L is purely resistive. If the load impedance is complex, one of the ways of matching it to the line is to tune out the reactance with an inductor or a capacitor, and then to match with a quarter-wave transformer. Short-cir-

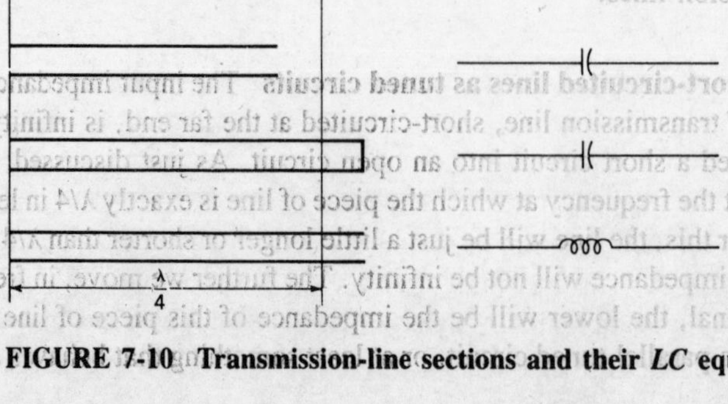

Line

Equivalent

$\frac{\lambda}{4}$

FIGURE 7-10 Transmission-line sections and their LC equivalents.

cuited transmission lines are more often used than lumped components at very high frequencies; a transmission line so used is called a *stub*. The procedure adopted is as follows:

1. Calculate load admittance.
2. Calculate stub susceptance.
3. Connect stub to load, the resulting admittance being the load conductance G.
4. Transform conductance to resistance, and calculate Z_0' of the quarter-wave transformer as before.

Example 7-5 A $(200 + j75)$-Ω load is to be matched to a 300-Ω line to give SWR = 1. Calculate the reactance of the stub and the characteristic impedance of the quarter-wave transformer, both connected directly to the load.

1. $$Y_L = \frac{1}{Z_L} = \frac{1}{200 + j75} = \frac{200 - j75}{40{,}000 + 5625}$$
$$= 4.38 \times 10^{-3} - j1.64 \times 10^{-3}$$

2. $$B_{\text{stub}} = +1.64 \times 10^{-3}\ \text{S} \qquad X_{\text{stub}} = \frac{-1}{1.64 \times 10^{-3}} = -610\ \Omega$$

3. With stub connected,
$$Y_L = G_L = 4.38 \times 10^{-3}\ \text{S}$$

4. $$R_L = \frac{1}{G_L} = \frac{1}{4.38 \times 10^{-3}} = 228\ \Omega$$

Then
$$Z_0' = \sqrt{Z_0 Z_L} = \sqrt{300 \times 228} = 262\ \Omega$$

Impedance variation along a mismatched line When a complex load is connected to a transmission line, standing waves result even if the magnitude of the load impedance is equal to the characteristic impedance of the line. If z_L is the normalized load impedance, then as impedance is investigated along the line, z_L will be measured $\lambda/2$ away from the load, and then at successive $\lambda/2$ intervals when the line is lossless. As already shown, a normalized impedance equal to y_L will be measured $\lambda/4$ away from the load (and at successive $\lambda/2$ intervals from then on). If $z_L = r + jx$, the normalized impedance measured $\lambda/4$ farther on will be given by

$$z_s = y_L = \frac{1}{r + jx} = \frac{r - jx}{r^2 + x^2} \tag{7-14}$$

The normalized load impedance was inductive, and yet, from Eq. (7-14), the normalized impedance seen $\lambda/4$ away from the load is capacitive; it is obvious that, somewhere between these two points, it must have been purely resistive. This point is not necessarily $\lambda/8$ from the load, but the fact that it exists at all is of great importance, as will be seen. The position of the purely resistive point is very difficult to calculate without a chart such as the Smith chart previously mentioned. Many transmission-line calculations are made easier by the use of charts, and none more so than those involving lines with complex loads.

7-2 THE SMITH CHART AND ITS APPLICATIONS

The various properties of transmission lines may be represented graphically on any of a large number of charts. The most useful representations are those that give the impedance relations along a lossless line for different load conditions. The most widely used calculator of this type is the Smith chart [3,4,5].

7-2.1 Fundamentals of the Smith Chart

Description The polar impedance diagram, or Smith chart as it is more commonly known, is illustrated in Fig. 7-11. It consists of two sets of circles, or arcs of circles, which are so arranged that various important quantities connected with mismatched transmission lines may be plotted and evaluated fairly easily. The complete circles, whose centers lie on the only straight line on the chart, correspond to various values of normalized resistance ($r = R/Z_0$) along the line. The arcs of circles, to either side of the straight line, similarly correspond to various values of normalized line reactance $jx = jX/Z_0$. A careful look at the way in which the circles intersect shows them to be orthogonal. This means that tangents drawn to the circles at the point of intersection would be mutually perpendicular. The various circles and coordinates have been chosen so that conditions on a line with a given load (i.e., constant SWR) correspond to a circle drawn on the chart with its center at the center of the chart. Actually, this applies only to lossless lines. In the quite rare case of *lossy* RF lines, an inward spiral must be drawn instead of the circle, with the aid of the scales shown in Fig. 7-11 below the chart. This is a complicated process which will not be discussed further.

If a load is purely resistive, R/Z_0 not only represents its normalized resistance but also corresponds to the standing-wave ratio, as shown in Eq. (7-8). Thus, when a particular circle has been drawn on a Smith chart, the SWR corresponding to it may be read off the chart at the point at which the drawn circle intersects the only straight line on the chart, on the right of the chart center. This SWR is thus equal to the value of $r \pm j0$ at that point; the intersection to the left of the chart center corresponds to $1/r$. It would be of use only if it had been decided always to use values of SWR less than 1.

The greatest advantage of the Smith chart is that travel along a lossless line corresponds to movement along a correctly drawn constant SWR circle. It will be seen from close examination of the chart axes that the chart has been drawn for use with normalized impedances and admittances. As already discussed, this avoids the need to have Smith charts for every imaginable value of line characteristic impedance.[5] Also note that the chart covers a distance of only a half-wavelength, since conditions repeat exactly every half-wavelength on a lossless line. Thus the impedance at (say) 17.716 λ away from a load on a line is exactly the same as the impedance 0.216 λ from that load and can be read from the chart.

[5] On the other hand, if a particular value of Z_0 is employed widely or exclusively, it becomes worthwhile to construct a chart for that particular value of Z_0. Thus, for example, the General Radio Company makes a 50-Ω chart for use with its transmission equipment. It may also be used for any other 50-Ω situations and avoids the need for normalization.

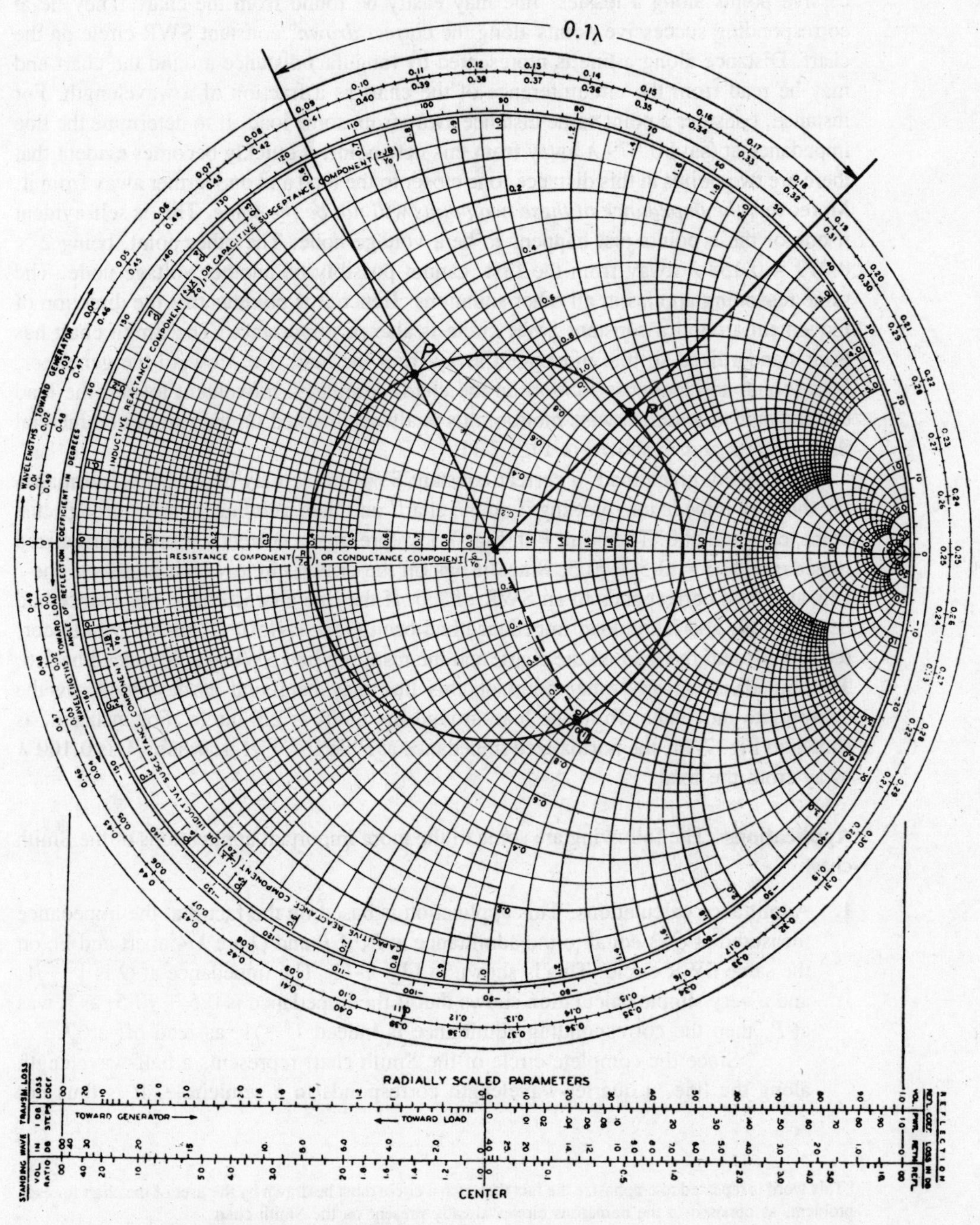

FIGURE 7-11 Smith chart.

Bearing these two points in mind, we see that impedances encountered at successive points along a lossless line may easily be found from the chart. They lie at corresponding successive points along the correct *drawn*[6] constant SWR circle on the chart. Distance along a line is represented by (angular) distance around the chart and may be read from the circumference of the chart as a fraction of a wavelength. For instance, consider a point some distance away from some load. If to determine the line impedance at (say) 0.079 λ away from this new point, it quickly becomes evident that there are two points at this distance, one closer to the load and one farther away from it. Moreover, *the impedance at these two points will not be the same*. This is self-evident if one of these points just happens to be a voltage node. The other point, being $2 \times 0.079 = 0.158\ \lambda$ away from the first, cannot possibly be another voltage node. The same reasoning applies in all other situations. Hence it is obvious that the direction of movement around a constant SWR circle is also of importance. The Smith chart has been standardized so that movement away from the load, i.e., toward the generator, corresponds to clockwise motion on the chart. Similarly, movement toward the load corresponds to anticlockwise motion; this is always marked on the rim of commercial Smith charts and is shown in Fig. 7-11.

For any given load, a correct constant SWR circle may be drawn by normalizing the load impedance, plotting it on the chart[7] and then drawing a circle through this point, centered at 0. The point P in Fig. 7-11 represents a correctly plotted normalized impedance of $z = 0.5 + j0.5$. Since it lies on the drawn circle which intersects the r axis at 2.6, it corresponds to an SWR of 2.6. If the line characteristic impedance had been 300 Ω, and if the load impedance had been $(150 + j150)$ Ω, then P would correctly represent the load on the chart, and the resulting line SWR would indeed be 2.6. The impedance at any other point on this line may be found as described, by the appropriate movement from the load around the SWR = 2.6 circle. For instance, as shown in Fig. 7-11, the normalized impedance at P' is $1.4 + j1.1$, where P' is 0.100 λ away from the load.

Applications The following are some of the more important applications of the Smith chart:

1. Admittance calculations. This application is based on the fact that the impedance measured at Q is equal to the admittance at P, if P and Q are $\lambda/4$ apart and lie on the same SWR circle. This is shown in Fig. 7-11. The impedance at Q is $1 - j1$, and a very simple calculation shows that if the impedance is $0.5 + j0.5$, as it was at P, then the corresponding admittance is indeed $1 - j1$, as read off at Q.

   Since the complete circle of the Smith chart represents a half-wavelength along the line, a quarter-wavelength corresponds to a semicircle. It is thus not

[6] This word is repeated to emphasize the fact that such a circle must be drawn by the user of the chart for each problem, as opposed to the numerous circles already present on the Smith chart.

[7] This is not *quite* as simple as it sounds. Most people are accustomed to charts consisting of straight lines only; thus the plotting and reading off of values on a Smith chart are decidedly confusing at first, and the novice requires plenty of practice.

necessary to measure $\lambda/4$ around the circle from P, but merely to project the line through P and the center of the chart until it intersects the drawn circle at Q on the other side.[8]

2. Calculation of the impedance or admittance at any point, on any transmission line, with any load, and simultaneous calculation of the SWR on the line. This may be done for lossless or lossy lines, but is much easier for the former.
3. Calculation of the length of a short-circuited piece of transmission line to give a required capacitive or inductive reactance. This is done by starting at the point $0,j0$ on the left-hand side rim of the chart, and traveling toward the generator until the correct value of reactance is reached. Alternatively, if a susceptance of known value is required, start at the right-hand rim of the chart at the point $\infty,j\infty$ and work toward the generator again. This calculation is always performed in connection with short-circuited stubs.

Example 7-6 (Students are expected to perform part of the example on their own charts.) Calculate the length of a short-circuited line required to tune out the susceptance of a load whose $Y = (0.004 - j0.002)$ S, placed on an air-dielectric transmission line of characteristic admittance $Y_0 = 0.0033$ S, at a frequency of 150 MHz.

Just as $z = Z/Z_0$, so $y = Y/Y_0$; this may be very simply checked.

Thus

$$y = \frac{0.004 - j0.002}{0.0033} = 1.21 - j0.61$$

Hence the normalized susceptance required to cancel the load's normalized susceptance is $+j0.61$. From the chart, the length of line required to give a normalized input admittance of 0.61 when the line is short-circuited is given by

$$\text{Length} = 0.250 + 0.087 = 0.337\lambda$$

Since the line has air as its dielectric, the velocity factor is 1.

Thus

$$v_c = f\lambda$$

$$\lambda = \frac{v_c}{f} = \frac{300 \times 10^6}{150 \times 10^6} = 2 \text{ m}$$

$$\text{Length} = 0.337\ \lambda = 0.337 \times 200 = 67.4 \text{ cm}$$

7-2.2 Problem Solution

In most cases, the best method of explaining problem solution with the Smith chart is to show how an actual problem of a given type is solved. In other cases, a procedure may be established without prior reference to a specific problem. Both methods of approach will be used here.

[8] Although such an application is not very important in itself, it has been found of great value in familiarizing students with the chart and with the method of converting it for use as an admittance chart, this being essential for stub calculations.

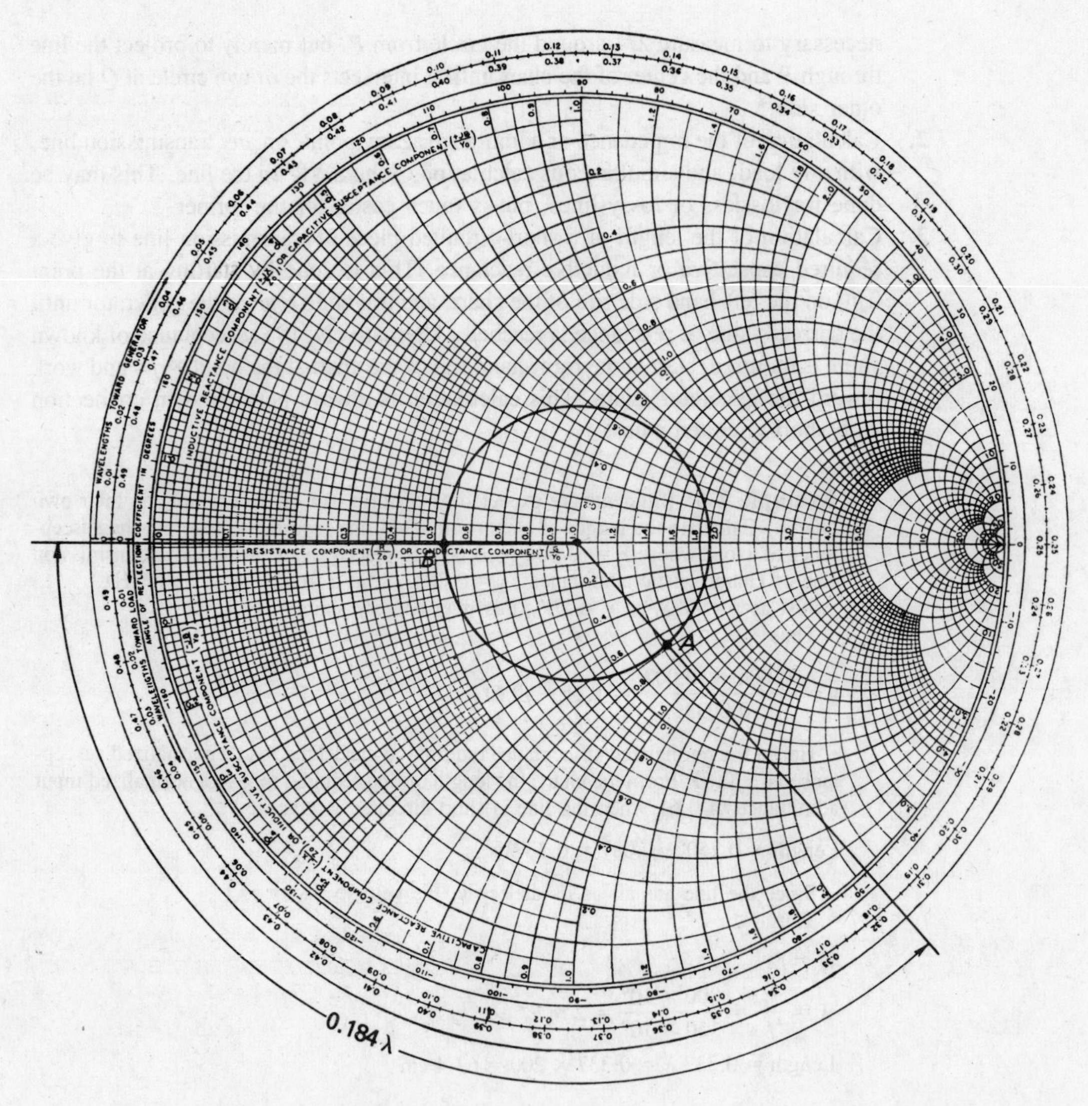

FIGURE 7-12 Smith chart solution of Example 7-7, matching with a quarter-wave transformer.

Matching of load to line with a quarter-wave transformer

Example 7-7 (Refer to Fig. 7-12.) A load $Z_L = (100 - j50)\ \Omega$ *is* connected to a line whose $Z_0 = 75\ \Omega$. Calculate

(a) The point, nearest to the load, at which a quarter-wave transformer may be inserted to provide correct matching

(b) The Z_0' of the transmission line to be used for the transformer

Solution (a) Normalize the load impedance with respect to the line; thus $(100 - j50)/75 = 1.33 - j0.67$. Plot this point *(A)* on the Smith chart. Draw a circle whose center lies at the center of the chart, passing through the plotted point. As a check, note that this circle should correspond to an SWR of just under 1.9. Moving toward the generator, i.e., clockwise, find the nearest point at which the line impedance is purely resistive (this is the intersection of the drawn circle with the only straight line on the chart). Around the rim of the chart, measure the distance from the load to this point (*B*); this distance = 0.500 − 0.316 = 0.184 λ. Read off the normalized resistance at *B*, here $r = 0.53$, and convert this normalized resistance into an actual resistance by multiplying by the Z_0 of the line. Here $R = 0.53 \times 75 = 39.8\ \Omega$.

(*b*) 39.8 Ω is the resistance which the λ/4 transformer will have to match to the 75-Ω line, and from this point the procedure is as in Example 7-4. Thus

$$Z_0' = \sqrt{Z_0 Z_R} = \sqrt{75 \times 39.8} = 54.5\ \Omega$$

Students at this point are urged to follow the same procedure to solve an example with identical requirements, but now $Z_L = (250 + j450)\ \Omega$ and $Z_0 = 300\ \Omega$. The answers are distance = 0.080 λ and $Z_0' = 656\ \Omega$.

Matching of load to line with a short-circuited stub A *stub* is a piece of transmission line which is normally short-circuited at the far end. It may very occasionally be open-circuited at the distant end, but either way its impedance is a pure reactance. To be quite precise, such a stub has an input admittance which is a pure susceptance, and it is used to tune out the susceptance component of the line admittance at some desired point. Note that short-circuited stubs are preferred because open-circuited pieces of transmission line tend to radiate from the open end.

As shown in Fig. 7-13, a stub is made of the same transmission line as the one to which it is connected. It thus has an advantage over the quarter-wave transformer, which must be constructed to suit the occasion. Furthermore, the stub may be made rigid and adjustable. This is of particular use at the higher frequencies and allows the stub to be used for a variety of loads, and/or over a range of frequencies.

Matching Procedure

1. Normalize the load with respect to the line, and plot the point on the chart.
2. Draw a circle through this point, and travel around it through a distance of λ/4 (i.e., straight through) to find the load admittance. Since the stub is placed in parallel with the main line, ***it is always necessary to work with admittances when making stub calculations.***

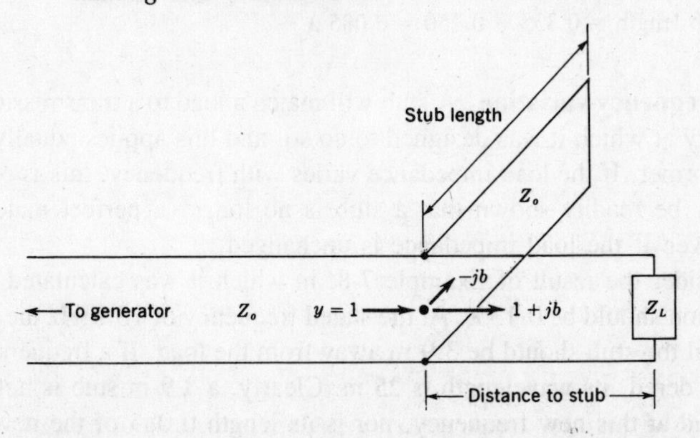

FIGURE 7-13 Stub connected to loaded transmission line.

3. *Starting from this new point (now using the Smith chart as an admittance chart),* find the point nearest to the load at which the normalized admittance is $1 \pm jb$. *This point is the intersection of the drawn circle with the* $r = 1$ *circle,* which is the only circle through the center of the chart. This is the point at which a stub designed to tune out the $\pm jb$ component will be placed. Read off the distance thus traveled around the circumference of the chart; this is the distance to the stub.
4. To find the length of the short-circuited stub, start from the point $\infty, j\infty$ on the right-hand rim of the chart, since that is the admittance of a short circuit.
5. Traveling clockwise around the circumference of the chart, find the point at which the susceptance tunes out the $\pm jb$ susceptance of the line at the point at which the stub is to be connected. For example, if the line admittance is $1 + j0.43$, the required susceptance is $-j0.43$. Ensure that the correct polarity of susceptance has been obtained; this is always marked on the chart on the left-hand rim.
6. Read off the distance in wavelengths from the starting point $\infty, j\infty$ to the new point (e.g., $b = -0.43$ as above). This is the required length of the stub.

Example 7-8 (Refer to Fig. 7-14.) A series *RC* combination, having an impedance $Z_L = (450 - j600)\ \Omega$ at 10 MHz, is connected to a 300-Ω line. Calculate the position and length of a short-circuited stub designed to match this load to the line.

In the following solution, steps are numbered as in the procedure:

1. $z_L = (450 - j600)/300 = 1.5 - j2$.
Circle plotted and has SWR = 4.6. Point plotted, *P* in Fig. 7-14.
2. $y_L = 0.24 + j0.32$, from the chart.
This, as shown in Fig. 7-14, is $\lambda/4$ away and is marked *Q*.
3. Nearest point of $y = 1 \pm jb$ is $y = 1 + j1.7$.
This is found from the chart and marked *R*. The distance of this point from the load, *Q* to *R*, is found along the rim of the chart and given by

$$\text{Distance to stub} = 0.181 - 0.051 = 0.130\ \lambda$$

Therefore the stub will be placed 0.13 λ from the load and will have to tune out $b = +1.7$; thus the stub must have a susceptance of -1.7.
4, 5, and 6. Starting from $\infty, j\infty$, and traveling clockwise around the rim of the chart, one reaches the point $0, -j1.7$; it is marked *S* on the chart of Fig. 7-14. From the chart, the distance of this point from the short-circuit admittance point is

$$\text{Stub length} = 0.335 - 0.250 = 0.085\ \lambda$$

Effects of frequency variation A stub will match a load to a transmission line only at the frequency at which it was designed to do so, and this applies equally to a quarter-wave transformer. If the load impedance varies with frequency, this is obvious. However, it may be readily shown that a stub is no longer a perfect match at the new frequency even if the load impedance is unchanged.

Consider the result of Example 7-8, in which it was calculated that the load-stub separation should be 0.13 λ. At the stated frequency of 10 MHz the wavelength is 30 m, so that the stub should be 3.9 m away from the load. If a frequency of 12 MHz is now considered, its wavelength is 25 m. Clearly, a 3.9-m stub is not 0.13 λ away from the load at this new frequency, nor is its length 0.085 of the new wavelength. Obviously the stub has neither the correct position nor the correct length at any fre-

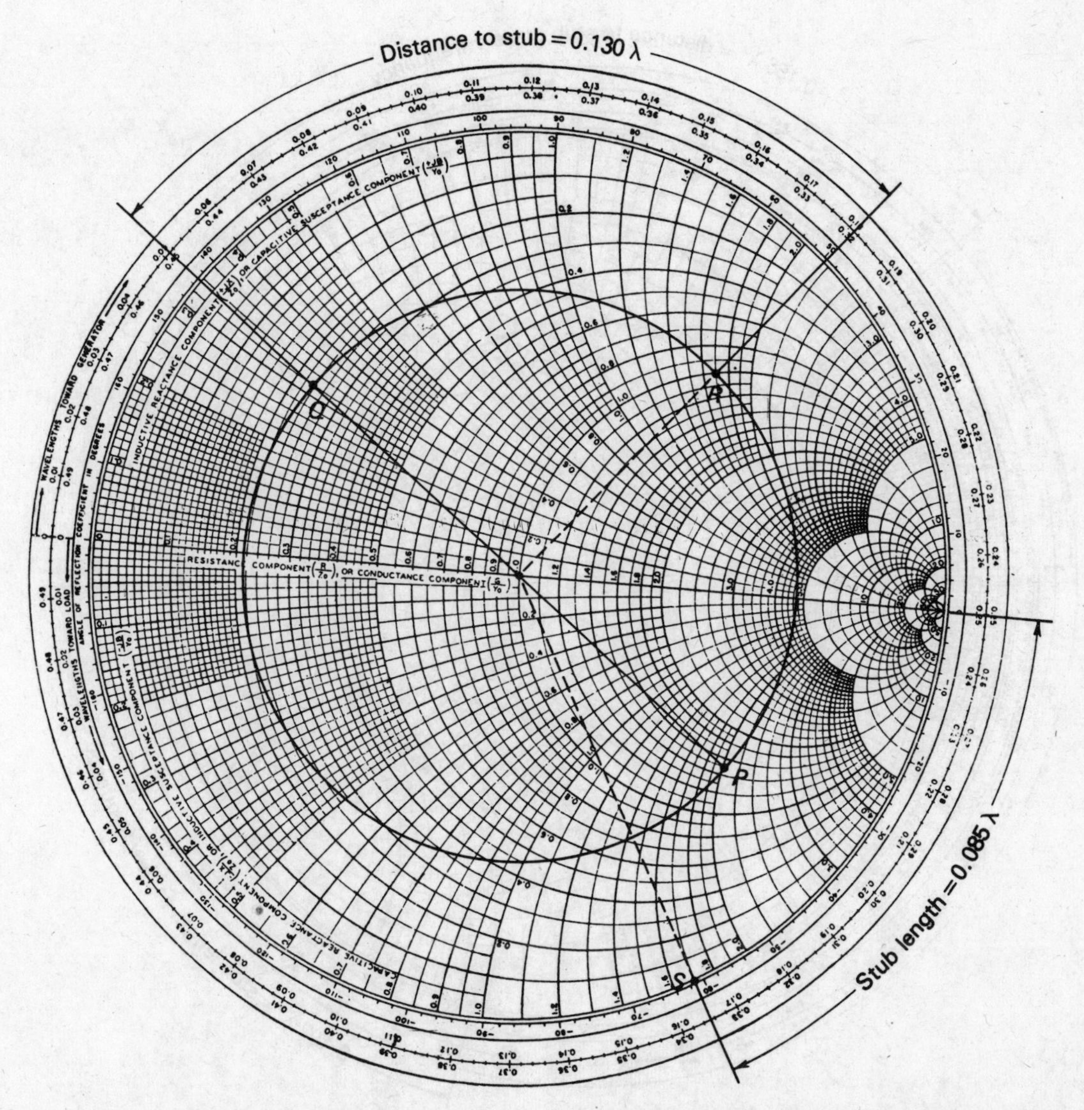

FIGURE 7-14 Smith chart solution of Example 7-8, matching with a short-circuited stub.

quency other than the one for which it was designed. Thus a mismatch will exist, although it must be said that if the frequency change is not great, neither is the mismatch.

It often occurs that a load is matched to a line at one frequency, but the setup must also be relatively lossless and efficient over a certain bandwidth. Thus some procedure must be devised for calculating the SWR on a transmission line at a frequency f'' if the load has been matched correctly to the line at a frequency f'. A procedure will now be given for a line and load matched by means of a short-circuited stub; the quarter-wave transformer situation is analogous.

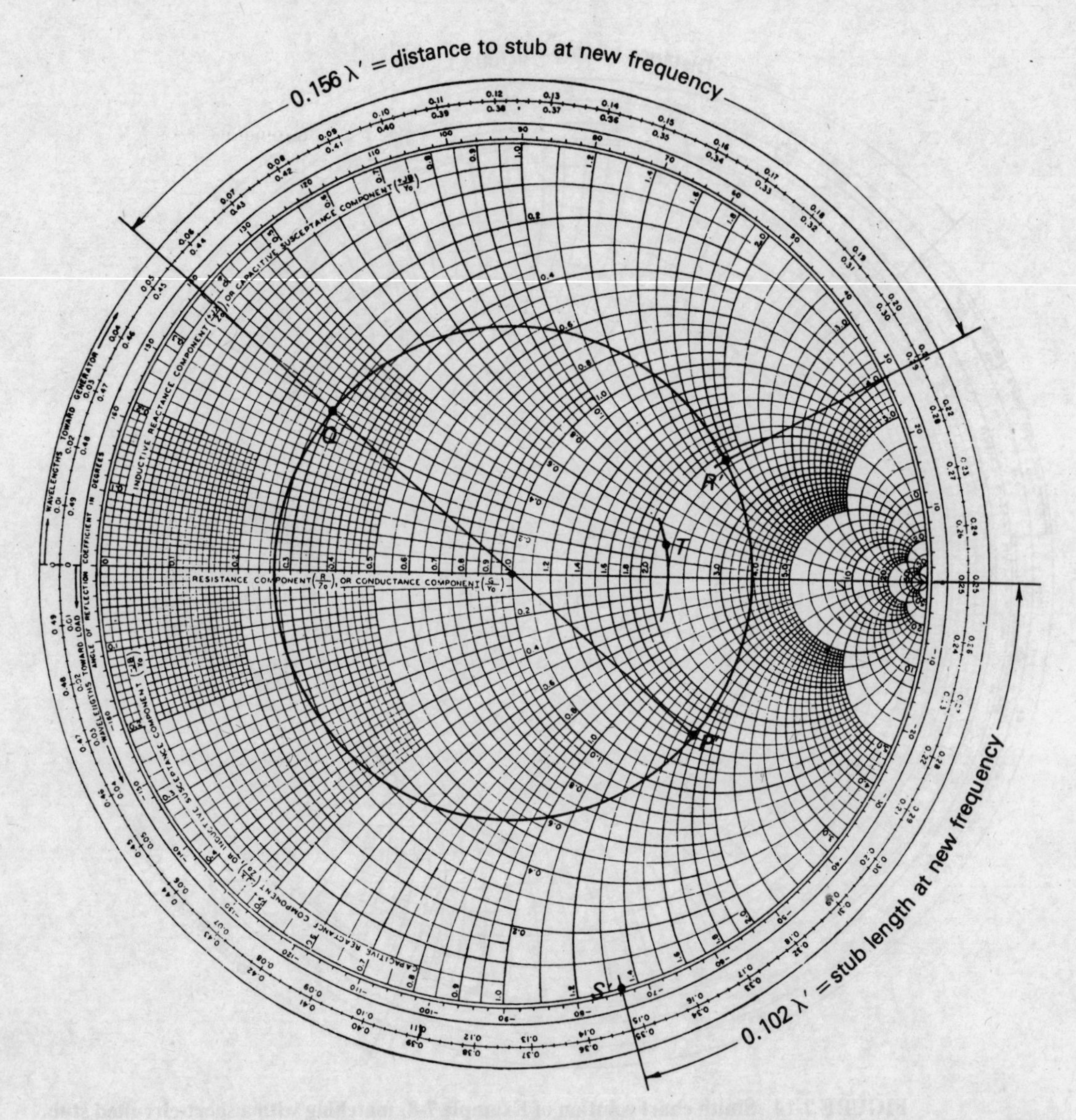

FIGURE 7-15 Smith chart solution of Example 7-9, effects of frequency change on a stub.

Example 7-9 (Refer to Fig. 7-15.) Calculate the SWR at 12 MHz for the problem of Example 7-8.

For the purpose of the procedure, it is assumed that the calculation involving the position and length of a stub has been made at a frequency f', and it is now necessary to calculate the SWR on the main line at f''. Matter referring specifically to the example will be shown [thus].

1. If data are given as to how the load impedance varies with frequency, calculate load impedance for the new frequency. [A series RC combination having $X_L = 450 - j600$ at 10 MHz will have $Z_L = 450 - j600 \times {}^{10}\!/_{12} = 450 - j500$ at 12 MHz.]

Normalize this impedance [here $z_L = (450 - j500)/300 = 1.5 - j1.67$]. Note that if the load impedance is known to be constant, this step may be omitted, since it would have been performed in the initial stub calculation.

2. Plot this point P' on the chart, draw the usual circle through it, and find Q', the normalized load admittance [here Q' is $0.30 + j0.33$] from the Smith chart.

3. Calculate the distance to the stub at the new frequency in terms of the new wavelength. [Here, since the frequency has risen, the new wavelength is shorter, and therefore a given distance is a larger fraction of it. Thus the distance to the stub is $0.130 \times 12/10 = 0.156\ \lambda'$.]

4. From the load admittance point Q', travel clockwise around the constant SWR circle through the distance just calculated [here $0.156\ \lambda'$] and read off the normalized admittance at this point R' [*here* $y_{\text{line}} = 2.1 + j1.7$]. This is the admittance at the new frequency, seen by the main line when looking toward the load at R', which is the point at which the stub was placed at the original frequency.

5. Calculate the length of the stub in terms of the new wavelength [length = $0.085 \times 12/10 = 0.102\ \lambda'$].

6. Starting at $\infty, j\infty$ as usual, this time find the susceptance of the piece of short-circuited line whose length was calculated in the preceding step. [Here the length is $0.102\ \lambda'$, and thus the susceptance from the chart of Fig. 7-15 is (at S') $y_{\text{stub}} = -j1.34$.]

7. The situation at the new frequency is that we have two admittances placed in parallel across the main line. At the original frequency, their values added so that the load was matched to the line, but at the new frequency such a match is not obtained. Having found each admittance, we may now find the total admittance at that point by addition. [Here $y_{\text{tot}} = y_{\text{stub}} + y_{\text{line}} = -j1.34 + 2.1 + j1.7 = 2.1 + j0.36$.]

8. Plot the total admittance on the chart (point T on Fig, 7-15), draw the constant SWR circle through it, and read off the SWR. This is the standing-wave ratio on the main line at f'' for a line-load-stub system that was matched at f'. [Here the SWR is 2.2. It might be noted that this is lower than the unmatched SWR of 3.9. Although a mismatch undoubtedly exists at 12 MHz, some improvement has been effected through matching at 10 MHz. This is a rule, rather than an exception, if the two frequencies are reasonably close.]

Another example is now given, covering this type of procedure from the very beginning for a situation in which the load impedance remains constant.

Example 7-10 (Refer to Fig. 7-16.) (*a*) Calculate the position and length of a short-circuited stub designed to match a 200-Ω load to a transmission line whose characteristic impedance is 300 Ω. (*b*) Calculate the SWR on the main line when the frequency is increased by 10 percent, assuming that the load and line impedances remain constant.

(*a*) $z_L = 200/300 = 0.67$. Plotting P on the chart gives an SWR = 1.5 circle; Q (admittance of load) is plotted.

Point of intersection with $r = 1$ circle, R, is plotted. Distance from load admittance, $Q - R$, is found equal to $0.11\ \lambda$; this is the distance to the stub.

At R, $y_{\text{line}} = 1 - j0.41$; hence $b_{\text{stub}} = j0.41$. Plotting S and measuring the distance of S from $\infty, j\infty$ gives stub length = $0.311\ \lambda$.

(*b*) $f'' = 110$ percent of f', so that $\lambda' = \lambda/1.1$. Thus, the distance to stub is $0.11 \times 1.1 = 0.121\ \lambda'$, and the length of stub is $0.311 \times 1.1 = 0.342\ \lambda'$.

Starting from Q and going around the drawn circle through a distance of $0.121\ \lambda'$ yields the point R', the distance to the stub attachment point at the new frequency. From the chart, the admittance looking toward the load at that point is read off as $y_{\text{line}} = 0.94 - j0.39$. Similarly, starting at $\infty, j\infty$ on the rim of the chart, and traveling around through a distance of $0.342\ \lambda'$ gives the point S'. Here the stub admittance at the new frequency is found, from the chart of Fig. 7-16, to be $y_{\text{stub}} = +j0.65$.

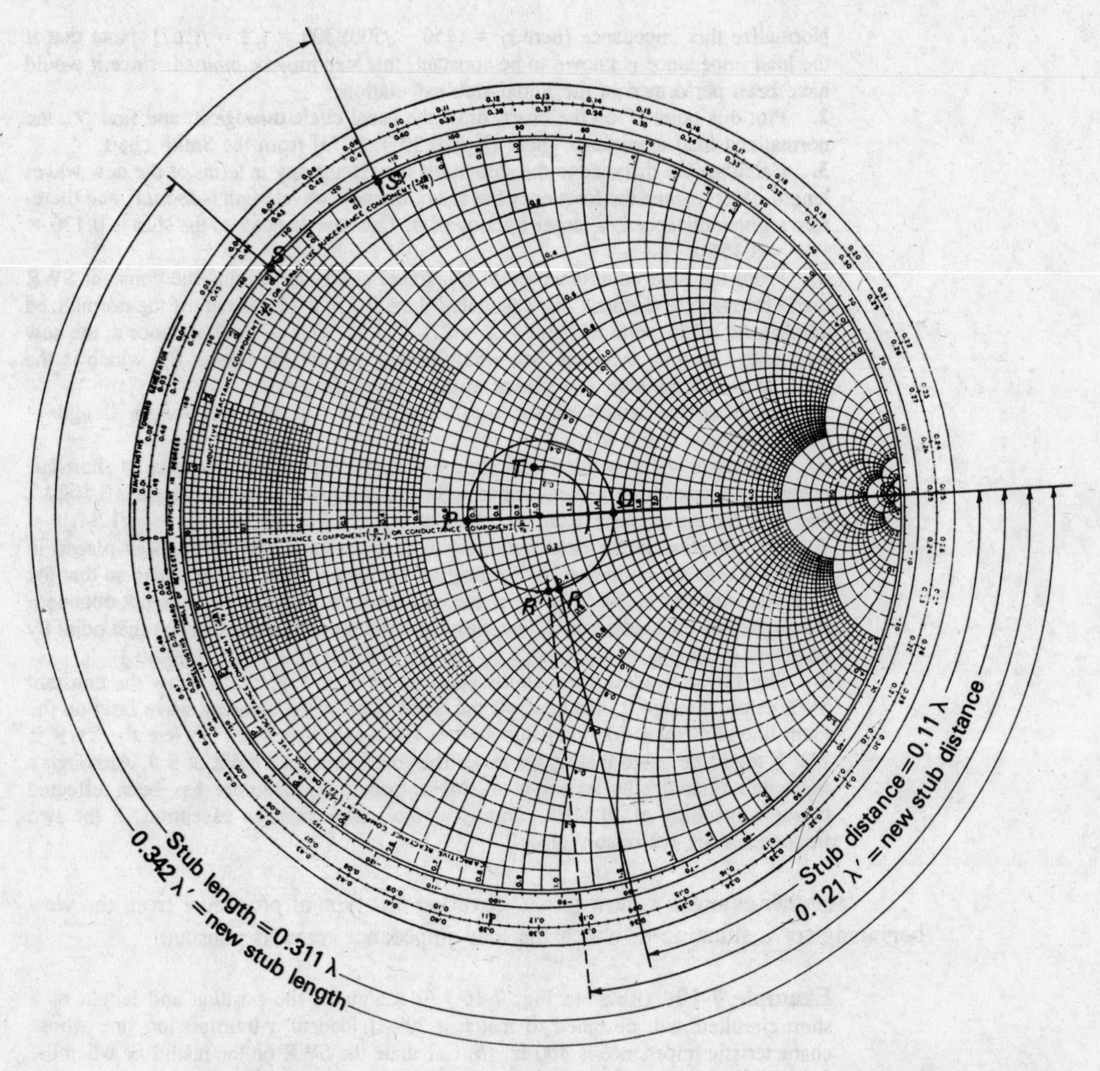

FIGURE 7-16 Smith chart solution of Example 7-10, stub matching with frequency change.

The total admittance at the stub attachment point at the new frequency is $y = y_{\text{stub}} + y_{\text{line}} = +j0.65 + 0.94 - j0.39 = 0.94 + j0.26$. Plotting this on the Smith chart, i.e., the point T of Fig. 7-16, and swinging an arc of a circle through T give SWR = 1.3. This is the desired result.

7-3 TRANSMISSION-LINE COMPONENTS

A number of situations, connected with the use of transmission lines, require components that are far easier to manufacture or purchase than to make on the spur of the moment. One very obvious requirement is for some sort of adjustable stub, which could cope with frequency or load impedance changes and still give adequate matching. Another situation often encountered is one in which it is desired to sample only the forward (or perhaps only the reverse) wave on a transmission line upon which standing waves exist. Again, it often happens that a balanced line must be connected to an unbalanced one. Finally, it would be very handy indeed to have a transmission line, for measurement purposes, on which the various quantities such as nodes, antinodes, or SWR could be measured at any point. All such eventualities are covered by special components, or "hardware," which will now be treated in turn.

7-3.1 The Double Stub

If a transmission-line matching device is to be useful in a range of different matching situations, it must have as many variable parameters, or *degrees of freedom,* as the standing-wave pattern. Since the pattern has two degrees of freedom (the SWR and the position of the first maximum), so must the stub matcher. A single stub of adjustable position and length will do the job very well at frequencies below the *microwave range*.[9] However, at such high frequencies coaxial lines are employed instead of parallel-wire lines, and difficulties with screened slots are such that stubs of adjustable position are not considered.

To provide the second degree of freedom, a second stub of adjustable position is added to the first one. This results in the double stub of Fig. 7-17 and is a commonly used matcher for coaxial microwave lines. Normally, the two stubs are placed 0.375 λ apart (λ corresponding to the center frequency of the required range), since that appears to be the optimum separation. Two variables are thus provided, and very good matching is possible. However, not all loads can be matched under all conditions, since having a second variable stub is not quite as good as having a stub of adjustable position.

Such a matcher is normally connected between the load and the main transmission line to ensure the shortest possible length of mismatched line. It naturally has the same characteristic impedance as the main line, and each stub should have a range of variation somewhat in excess of half a wavelength. The method of adjustment for matching is best described as "guesstimation," i.e., trial and error, which may or may not be preceded by a preliminary calculation. When trial and error is used, the stub nearest to the load is set at a number of points along its range, and the farther stub is

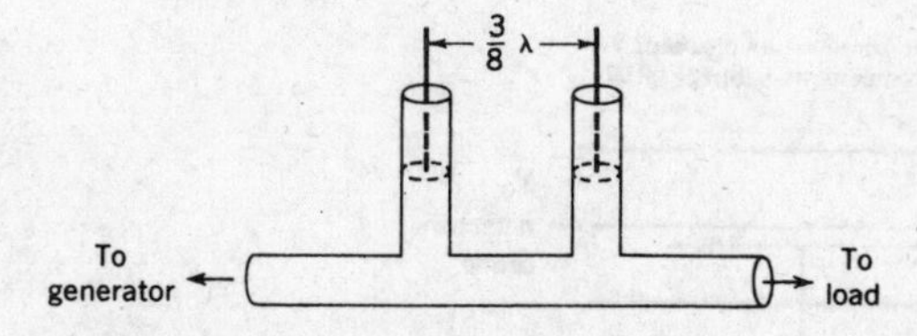

FIGURE 7-17 Double-stub matcher.

[9]This frequency range is not rigidly defined, but it is commonly understood to occupy the spectrum from 1 to 100 GHz. It is also called broadly the *centimeter range,* since the two extreme frequencies correspond to wavelengths of 30 and 0.3 cm, respectively.

racked back and forth over its entire range (at each setting point of the first stub) until the best possible match has been achieved. The SWR is, of course, monitored constantly while adjustment is taking place. Unless the load is most unusual, for example, almost entirely reactive, it should be possible to reduce the SWR on the main line to below about 1.2 with this matcher.

If almost perfect matching under all conceivable conditions of SWR and load impedance is required, a triple-stub tuner should be used. This is similar to the double-stub tuner but consists of three stubs, adjustable in length and placed 0.125 λ apart (the optimum separation in this case).

7-3.2 Directional Couplers

It is often necessary to measure the power being delivered to a load or an antenna through a transmission line. This is often done by a sampling technique, in which a known fraction of the power is measured, so that the total may be calculated. It is imperative, under these conditions, that only the forward wave in the main line is measured, not the reflected wave (if any). A number of coupling units are used for such purposes and are known as *directional couplers,* the two-hole coupler shown in Fig. 7-18 being among the most popular. This particular one is treated because it is a good illustration of transmission-line techniques and has a direct waveguide counterpart (see Sec. 10-5). Other types may be found elsewhere in the literature [6].

As indicated in Fig. 7-18, the two-hole directional coupler consists of a piece of transmission line to be connected in series with the main line, together with a piece of auxiliary line coupled to the main line via two probes through slots in the joined outer walls of the two coaxial lines. As shown, the probes do not actually touch the inner conductor of the auxiliary line. They couple sufficient energy into it simply by being near it. If they did touch, most of the energy (instead of merely a fraction) in the main line would be coupled into the auxiliary line; a fraction is all that is needed. The probes induce energy flow in the auxiliary line which is mostly in the same direction as in the main line, and provision is made to deal with energy flowing in the "wrong" direction. The distance between the probes is $\lambda/4$ but may also be any odd number of quarter-wavelengths. The auxiliary line is terminated at one end by a resistive load. This absorbs all the energy fed to it and is often termed a *nonreflecting termination.* The other end goes to a detector probe for measurement.

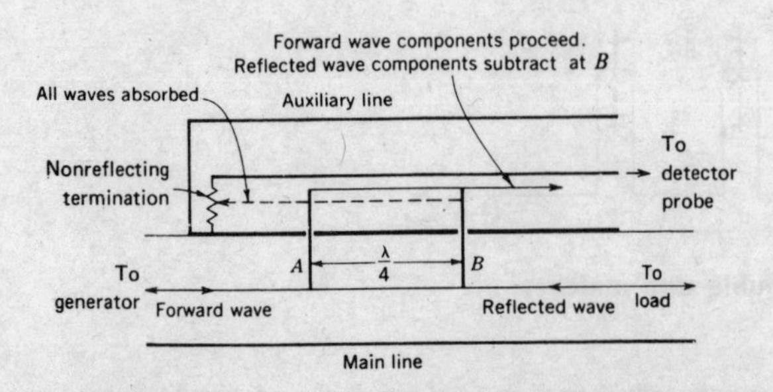

FIGURE 7-18 Coaxial two-hole directional coupler.

Any wave launched in the auxiliary line from right to left will be absorbed by the load at the left and will not, therefore, be measured. It now remains to ensure that only the forward wave of the main line can travel from left to right in the auxiliary line. The outgoing wave entering the auxiliary line at A, and proceeding toward the detector, will meet at B another sample of the forward wave. Both have traversed the same distance altogether, so that they add and travel on to the detector to be measured. There will also be a small fraction of the reverse wave entering the auxiliary line and then traveling to the right in it. However small, this wave is undesirable and is removed here by cancellation. Any of it that enters at B will be fully canceled by a portion of the reflected wave which enters the auxiliary line at A and also proceeds to the right. This is so because the reflected wave which passes B in the main line enters the auxiliary line at A and then goes to B, having traveled through a distance which is $2 \times \lambda/4 = \lambda/2$ greater than the reflected wave that entered at B. Being thus exactly 180° out of phase, the two cancel if both slots and probes are the same size and shape, and are correctly positioned.

Since various mechanical inaccuracies prevent ideal operation of this (or any other) directional coupler, some of the unwanted reflected wave will be measured in the auxiliary line. The *directivity* of a directional coupler is a standard method of measuring the extent of this unwanted wave. Consider exactly the same power of forward and reverse wave entering the auxiliary line. If the ratio of forward to reverse power measured by the detector is 30 dB, then the directional coupler is said to have a *directivity* of 30 dB. This value is common in practice.

The other important quantity in connection with a directional coupler is its *directional coupling*. This is defined as the ratio of the forward wave in the main line to the forward wave in the auxiliary line. It is measured in decibels, and 20 dB (100:1) is a typical value.

7-3.3 Baluns

A *balun*, or *balance-to-unbalance transformer*, is a circuit element used to connect a balanced line to an unbalanced line or antenna. Or, as is perhaps a little more common, it is used to connect an unbalanced (coaxial) line to a balanced antenna such as a *dipole*. At frequencies low enough for this to be possible, an ordinary tuned transformer is employed. This has an unbalanced primary and a center-tapped secondary winding, to which the balanced antenna is connected. It must also have an electrostatic shield, which is earthed to minimize the effects of stray capacitances.

For higher frequencies, several transmission-line baluns exist for differing purposes and narrowband or broadband applications. The most common balun, a narrowband one, will be described here, as shown in cross section in Fig. 7-19. It is known as the *choke*, *sleeve*, or *bazooka* balun.

As shown, a quarter-wavelength sleeve is placed around the outer conductor of the coaxial line and is connected to it at x. At the point y, therefore, $\lambda/4$ away from x, the impedance seen when looking down into the transmission line formed of the sleeve and the outer conductor of the coaxial line is infinite. In other words, the outer conductor of the coaxial line no longer has zero impedance to ground at y. Thus one of the wires of the balanced line many be connected to it without fear of being short-circuited

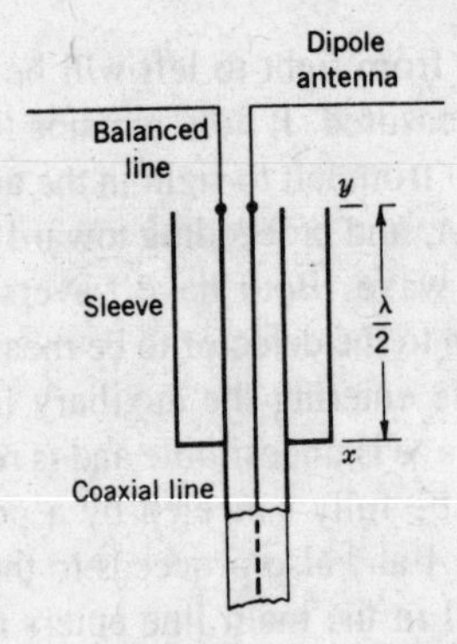

FIGURE 7-19 Choke (bazooka) balun.

to ground. The other balanced wire is connected to the inner conductor of the coaxial line. Any balanced load, such as the simple dipole antenna shown in Fig. 7-19, may now be placed upon it.

7-3.4 The Slotted Line

It can be appreciated that a piece of transmission line, so constructed that the voltage or current along it can be measured continuously over its length, would be of real use in a lot of measurement situations. At relatively low frequencies, say up to about 100 MHz, a pair of parallel-wire lines may be used, having a traveling detector connected between them. This detector is easily movable and has facilities for determining the distance of the probe from either end of the line. The *Lecher line* is the name given to this piece of equipment, whose high-frequency equivalent is the *slotted line*, as shown in Fig. 7-20.

The slotted line is a piece of coaxial line with a long narrow longitudinal slot in the outer conductor. A flat plate is mounted on the outer conductor, with a corresponding slot in it to carry the detector probe carriage. As shown, it has a rule on the side, with a vernier for microwave frequencies to indicate the exact position of the probe. The probe extends into the slot, coming quite close to the inner conductor of the line, but not touching it, as shown in Fig. 7-20*b*. In this fashion, loose coupling between line and probe is obtained which is adequate for measurements, but small enough so as not to interfere unduly. The slotted line must have the same characteristic impedance as the main line to which it is connected in series. It must also have a length somewhat in excess of a half-wavelength at the lowest frequency of operation.

Basically, the slotted line simply permits convenient and accurate measurement of the position and size of the first voltage maximum from the load, and any subsequent ones as may be desired, without interfering significantly with the quantities being measured. However, the knowledge of these quantities permits calculation of

1. Load impedance
2. Standing-wave ratio
3. Frequency of the generator being used

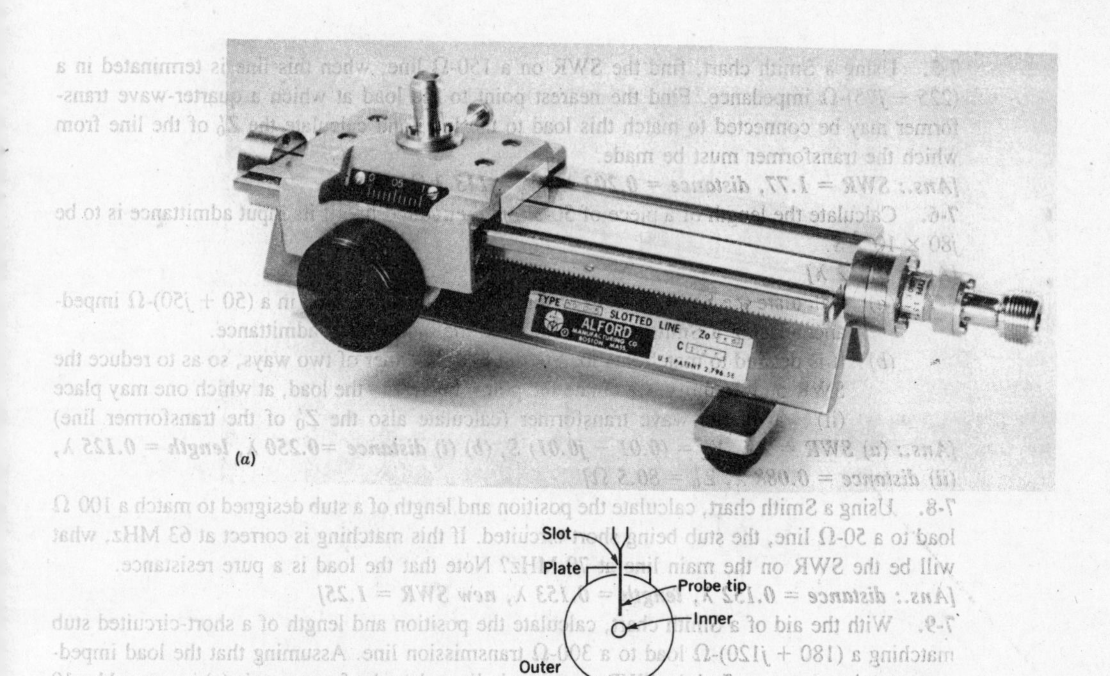

FIGURE 7-20 ***(a)*** **Commercial microwave slotted line** ***(Courtesy of the Alford Manufacturing Company);*** ***(b)*** **cross section of a slotted line.**

The practical measurement and calculations methods are normally indicated in the instructions that come with a particular slotted line. Measurement methods for these parameters that do not involve the slotted line also exist.

PROBLEMS

For self-testing questions on this chapter, see p. 686.

7-1. A lossless transmission line has a shunt capacitance of 100 pF/m and a series inductance of 4 μH/m. What is its characteristic impedance?
[Ans.: 200 Ω]

7-2. A coaxial line with an outer diameter of 6 mm has a 50-Ω characteristic impedance. If the dielectric constant of the insulation is 1.60, calculate the inner diameter.
[Ans.: 2.09 mm]

7-3. A transmission line with a characteristic impedance of 300 Ω is terminated in a purely resistive load. It is found by measurement that the minimum value of voltage upon it is 5 μV, and the maximum 7.5 μV. What is the value of the load resistance?
[Ans.: 200 or 450 Ω]

7-4. A quarter-wave transformer is connected directly to a 50-Ω load, to match this load to a transmission line whose Z_0 = 75 Ω. What must be the characteristic impedance of the matching transformer?
[Ans.: 61.2 Ω]

7-5. Using a Smith chart, find the SWR on a 150-Ω line, when this line is terminated in a (225 − j75)-Ω impedance. Find the nearest point to the load at which a quarter-wave transformer may be connected to match this load to the line, and calculate the Z_0' of the line from which the transformer must be made.
[Ans.: SWR = 1.77, distance = 0.203 λ, Z_0' = 113.3 Ω]

7-6. Calculate the length of a piece of 50-Ω *open*-circuited line if its input admittance is to be $j80 \times 10^{-3}$ *S*.
[Ans.: 0.212 λ]

7-7. (*a*) Calculate the SWR on a 50-Ω line, when it is terminated in a (50 + j50)-Ω impedance. Using a Smith chart, determine the *actual* load admittance.
(*b*) It is desired to match this load to the line, in either of two ways, so as to reduce the SWR on it to unity. Calculate the point, nearest to the load, at which one may place (ii) a quarter-wave transformer (calculate also the Z_0' of the transformer line)
[Ans.: (a) SWR = 2.6, Y_L = (0.01 − j0.01) S, (b) (i) distance =0.250 λ, length = 0.125 λ, (ii) distance = 0.088 λ, Z_0' = 80.5 Ω]

7-8. Using a Smith chart, calculate the position and length of a stub designed to match a 100 Ω load to a 50-Ω line, the stub being short-circuited. If this matching is correct at 63 MHz, what will be the SWR on the main line at 70 MHz? Note that the load is a pure resistance.
[Ans.: distance = 0.152 λ, length = 0.153 λ, new SWR = 1.25]

7-9. With the aid of a Smith chart, calculate the position and length of a short-circuited stub matching a (180 + j120)-Ω load to a 300-Ω transmission line. Assuming that the load impedance remains constant, find the SWR on the main line when the frequency is (a) increased by 10 percent; (b) doubled.
[Ans.: distance = 0.016 λ, length = 0.349 λ, (a) new SWR = 1.55, (b) new SWR = 2.7]

QUESTIONS

7-1. What different types of transmission lines are there? In what ways do their applications differ? What is it that limits the maximum power they can handle?

7-2. Draw the general equivalent circuit of a transmission line and the simplified circuit for a radio-frequency line. What permits this simplification?

7-3. Define the characteristic impedance of a transmission line. When is the input impedance of a transmission line equal to its characteristic impedance?

7-4. Discuss the types of losses that may occur with RF transmission lines. In what units are these losses normally given?

7-5. With a sketch, explain the difference between standing waves and traveling waves. Explain how standing waves occur in an imperfectly matched transmission line.

7-6. Define and explain the meaning of the term *standing-wave ratio*. What is the formula for it if the load is purely resistive? Why is a high value of SWR often undesirable?

7-7. Explain fully, with such sketches as are applicable, the concept of impedance inversion by a quarter-wave line.

7-8. For what purposes can short lengths of open- or short-circuited transmission line be used? What is a stub? Why are short-circuited stubs preferred to open-circuited ones?

7-9. When matching a load to a line by means of a stub and a quarter-wave transformer (both situated at the load), a certain procedure is followed. What is this procedure? Why are admittances used in connection with stub matching? What does a stub actually do?

7-10. What is a Smith chart? What are its applications?

7-11. Why must impedances (or admittances) be normalized before being plotted on a standard Smith chart?

7-12. Describe the double-stub matcher, the procedure used for matching with it, and the applications of the device.
7-13. What is a directional coupler? For what purposes might it be used?
7-14. Define the terms *directivity* and *directional coupling* as used with directional couplers, and explain their significance.
7-15. What is a *balun?* What is a typical application of such a device?

REFERENCES

1. See also, International Telephone and Telegraph Corporation, *Reference Data for Radio Engineers,* 6th ed., Howard W. Sams and Co. Inc., Indianapolis, 1975, pp. 24-32–24-41; this shows the Army-Navy list of preferred radio-frequency cables.
2. By the same token, when reflection is from an open circuit, it is the voltage that is unaffected, and the current is reversed in phase. Both these phenomena can be explained mathematically. See, for example, Davidson, C. W., *Transmission Lines for Communications,* The Macmillan Press Ltd., London, 1978, pp. 25–29 and 68–69.
3. *(Historical)* Smith, P. H., "Transmission Line Calculator," *Electronics,* January 1939, p. 29.
4. *(Historical)* Smith P. H., "An Improved Transmission Line Calculator," *Electronics,* January 1944, p. 130.
5. Smith, P. H., *Electronic Applications of the Smith Chart,* McGraw-Hill Book Company, New York, 1969 (copyright 1969 by Kay Electric Company, Pine Brook, N.J.).
6. Blake, Lamont, V., *Antennas,* John Wiley & Sons, Inc., New York, 1966, pp. 99–104.

8 RADIATION AND PROPAGATION OF WAVES

The very first block diagram in Chap. 1 showed a "channel" between the transmitter and receiver of a communications system, and suggested that signals (after they have been generated and processed by the transmitter) are conveyed through the channel to the receiver. In radio communications, the channel is simply the physical space between the transmitting and receiving antennas, and the behavior of signals in that channel forms the body of this chapter.

To promote the understanding of this behavior, the chapter is divided into two distinct parts. The first is electromagnetic radiation; it deals with the nature and propagation of radio waves, as well as the attenuation and absorption they may undergo along the way. Under the subheading of "effects of the environment," reflection and refraction of waves are considered, and finally interference and diffraction are explained.

The second part of the chapter treats in some detail the practical aspects of the propagation of waves. It is quickly seen that the frequency used plays a significant part in the method of propagation, as do the existence and proximity of the earth. The three main methods of propagation—around the curvature of the earth, by reflection from the ionized portions of the atmosphere, or in straight lines (depending mainly on frequency)—are discussed in detail. Then, certain aspects of microwave propagation are treated, notably so-called *superrefraction*, *tropospheric scatter* and the effects of the *ionosphere* on waves trying to travel through it. Finally, the chapter introduces extraterrestrial communications, whose applications are then treated in Section 8-2.5.

8-1 ELECTROMAGNETIC RADIATION

When electric power is applied to a circuit, a system of voltages and currents is set up in it, with certain relations governed by the properties of the circuit itself. Thus, for instance, the voltage may be high (compared to the current) if the impedance of the circuit is high, or perhaps the voltage and current are 90° out of phase because the circuit is purely reactive. In a similar manner, any power escaping into *free space* is governed by the characteristics of free space. If such power "escapes on purpose," it is said to have been *radiated,* and it then propagates in space in the shape of what is known as an *electromagnetic wave.*

Free space is space that does not interfere with the normal radiation and propagation of radio waves. Thus, it has no magnetic or gravitational fields, no solid bodies and no ionized particles. Apart from the fact that *free* space is unlikely to exist anywhere, it certainly does not exist near the earth. However, the concept of free space is used because it simplifies the approach to wave propagation, since it is possible to calculate the conditions if the space were free, and then to predict the effect of its actual properties. Also, propagating conditions sometimes do approximate those of free space, particularly at frequencies in the upper UHF region.

Since radiation and propagation of radio waves cannot be seen, all our descriptions must be based on theory which is acceptable only to the extent that it has predictive value; i.e., it may be used to forecast accurately what is going to happen next. The theory of electromagnetic radiation was propounded by the British physicist James Clerk Maxwell in 1857 and finalized in 1873. It is the fundamental mathematical explanation of the behavior of electromagnetic waves. Unfortunately, the mathematics of Maxwell's equations is too advanced to be used here. Accordingly, the emphasis will be on description and explanation of behavior, with occasional references to the mathematical background.

8-1.1 Fundamentals of Electromagnetic Waves

Electromagnetic waves are oscillations that propagate through free space with the velocity of light, which is $v_c = 299{,}792{,}500 \pm 300$ m/s (for most purposes approximated to 3×10^8 m/s). In some ways, propagation is similar to the outward travel of waves on a pond after a stone has been thrown into it, but there is a big difference. Whereas the water waves are *longitudinal* (oscillations in the direction of propagation), electromagnetic waves are *transverse* (oscillation perpendicular to the direction of propagation). Also, *the direction of the electric field, the magnetic field and propagation are mutually perpendicular in electromagnetic waves,* as Fig. 8-1 shows. This is a theoretical assumption which cannot be "checked," since the waves are invisible. However, it may be used to predict the behavior of electromagnetic waves in all circumstances, such as *reflection, refraction* and *diffraction,* to be treated later in the chapter.

Waves in free space Since no interference or obstacles are present in free space, electromagnetic waves will spread uniformly in all directions from a point source. The

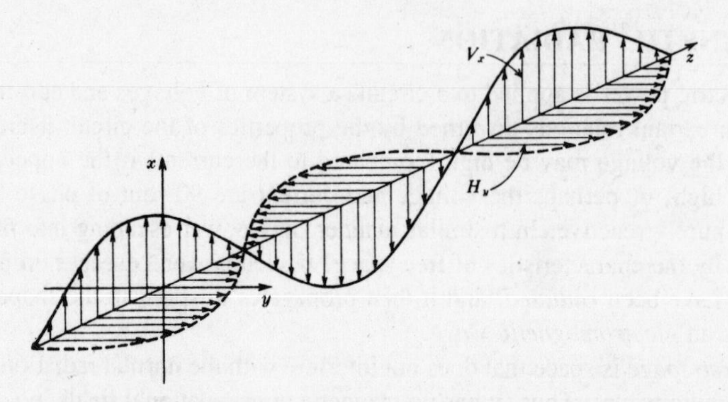

FIGURE 8-1 Transverse electromagnetic wave in free space.

wavefront is thus spherical, as shown in cross section in Fig. 8-2. To simplify the description even further, "rays" are imagined which radiate from the point source in all directions. They are everywhere perpendicular to a tangential plane of the wavefront, just like the spokes of a wheel.

At the distance corresponding to the length of ray P, the wave has a certain phase. It may have left the source at an instant when its voltage and current were maximum in the circuit feeding the source, i.e., at an instant of maximum electric and magnetic field vectors. If the distance traveled corresponds to exactly 100,000.25 wavelengths, the instantaneous electric and magnetic intensities are at that moment zero at all such points. This is virtually the definition of a wavefront; it is the plane joining all points of identical phase. Here, of course, it is spherical. If the length of ray Q is exactly twice that of ray P, then the area of the new sphere will be exactly *four times* the area of the sphere with radius P. It is seen that the total power output of the source has spread itself over four times the area when its distance from the source has doubled. Now, if *power density* is defined as radiated power per unit area, it follows that power density is reduced to one-quarter of its value when distance from the source has doubled.

It is seen that power density is inversely proportional to the square of the distance from the source. This is the *inverse-square law,* which applies universally to all forms of radiation in free space. Stating this mathematically, we have

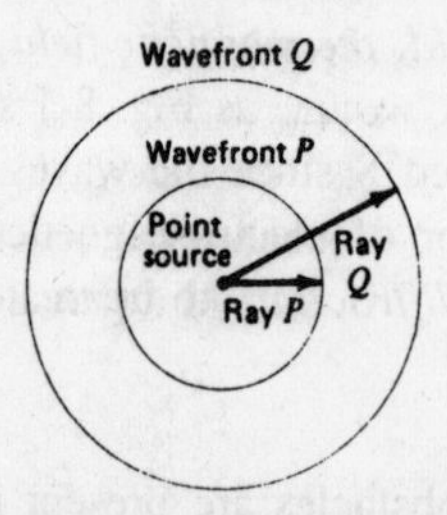

FIGURE 8-2 Spherical wavefronts.

$$\mathcal{P} = \frac{P_t}{4\pi r^2} \tag{8-1}$$

where $\mathcal{P}$ = power density at a distance r from an isotropic source
P_t = transmitted power

An *isotropic* source is one that radiates uniformly in all directions in space. Although no practical source has this property, the concept of the isotropic radiator is very useful and frequently employed. As a matter of interest, it may be shown quite simply that the inverse-square law applies also when the source is not isotropic, and students are invited to demonstrate this for themselves. However, for wavefronts to be spherical, the velocity of radiation must be constant at all points (as it is, of course, in free space). A propagation medium in which this is true is also called isotropic.

The electric and magnetic field intensities of electromagnetic waves are also important. The two quantities are the direct counterparts of voltage and current in circuits; they are measured in volts per meter and amperes per meter, respectively. Just as for electrical circuits we have $V = ZI$, so for electromagnetic waves

$$\mathcal{E} = \mathcal{Z}\mathcal{H} \tag{8-2}$$

where $\mathcal{E}$ = rms value of field strength, or intensity, V/m
$\mathcal{H}$ = rms value of magnetic field strength, or intensity, A/m
$\mathcal{Z}$ = characteristic impedance of the medium, Ω

The characteristic impedance of a medium is given by

$$\mathcal{Z} = \sqrt{\frac{\mu}{\epsilon}} \tag{8-3}$$

where μ = permeability of medium
ϵ = electric permittivity of medium

For free space,

$$\mu = 4\pi \times 10^{-7} = 1.257 \times 10^{-6}\ \text{H/m}$$
$$\epsilon = 1/36\pi \times 10^{9} = 8.854 \times 10^{-12}\ \text{F/m}$$

It will be recalled that permeability is the equivalent of inductance and permittivity is the equivalent of capacitance in electric circuits; indeed the units used above are a reminder of this. It is now possible to calculate a value for the characteristic impedance of free space. We thus have, from Eq. (8-3)

$$\mathcal{Z} = \sqrt{\frac{\mu}{\epsilon}} = \sqrt{\frac{4\pi \times 10^{-7}}{1/36\pi \times 10^{9}}} = \sqrt{144\pi^2 \times 100}$$
$$= 120\pi = 377\Omega \tag{8-4}$$

This makes it possible to calculate the field intensity at a distance r from an isotropic source. Just as $P = V^2/Z$ in electrical circuits, so $\mathcal{P} = \mathcal{E}^2/\mathcal{Z}$ for electromagnetic waves. We may now invert this relation and substitute for $\mathcal{P}$ from Eq. (8-1) and for $\mathcal{Z}$ from Eq. (8-4), thus obtaining

$$\mathscr{E} = \mathcal{P} \times \mathcal{Z}$$

$$= \frac{P_t}{4\pi r^2} \times 120\pi = \frac{30P_t}{r^2}$$

$$\therefore \mathscr{E} = \frac{\sqrt{30P_t}}{r} \tag{8-5}$$

It is seen from Eq. (8-5) that field intensity is inversely proportional to the distance from the source, since it is proportional to the square root of power density.

Finally, the wavefront must be considered once again. As discussed, it is spherical in an isotropic medium, but any small area of it at a large distance from the source may be considered to be a *plane wavefront*. This is obvious from the geometry of the situation and also from everyday experience; we know that the earth is spherical as a very close approximation, but we speak of a football field as *flat*. It represents a finite area of the earth's surface but is at a considerable distance from its center. The concept of plane waves is very useful because it greatly simplifies the treatment of the *optical* properties of electromagnetic waves, such as reflection and refraction.

Radiation and reception Antennas radiate electromagnetic waves, or, putting it differently, radiation will result from the flow of high-frequency current in a suitable circuit. Incidentally, this is predicted mathematically by the Maxwell equations, which show that current flowing in a wire is accompanied by a magnetic field around it. Furthermore, if the magnetic field is changing, as it does with *alternating* current, an electric field will be present also. As will be described in the next chapter, a proportion of the electric and magnetic field is capable of leaving the current-carrying wire. How much of it leaves the conductor depends on the relation of its length to the wavelength of the current.

Polarization

It was illustrated in Fig. 8-1 that electromagnetic waves are transverse, and the electric and magnetic fields are mutually perpendicular. Since the magnetic field surrounds the wire and is perpendicular to it, it follows that the electric field is parallel to the wire. This is a configuration that applies also after the electromagnetic wave has been radiated by the wire. The question of *polarization* thus arises.

Polarization refers to the physical orientation of the radiated waves in space. Waves are said to be *polarized* (actually *linearly polarized*) if they all have the same alignment in space. In fact, it is a characteristic of most antennas that the radiation they emit is linearly polarized. For example, a vertical antenna will radiate waves whose electric vectors will all be vertical and will remain so in free space. On the other hand, light emitted by *incoherent* sources, such as the sun or light globes, has a haphazard arrangement of field vectors and is said to be *randomly polarized*.

The wave of Fig. 8-1 is, of course, linearly polarized and is also said to be *vertically polarized,* since all the electric intensity vectors are vertical. The decision to label polarization direction after the electric intensity is not as arbitrary as it seems; this makes the direction of polarization the same as the direction of the antenna. Thus, vertical antennas radiate vertically polarized waves, and similarly horizontal antennas produce waves whose polarization is horizontal. There has been a tendency, over the years, to transfer the label to the antenna itself. Thus people often refer to antennas as vertically or horizontally polarized, whereas it is only their radiations that are so polarized.

It is also possible for antenna radiations to be circularly or even elliptically polarized, so that the polarization of the wave rotates continuously in corkscrew fashion. This will be discussed further in Sec. 9-8 in connection with helical antennas.

Reception

Just as a wire carrying HF current is surrounded by electric and magnetic fields, so a wire placed in an electromagnetic field will have a current induced in it. This is another way of saying that this wire receives some of the radiation and is therefore a receiving antenna. Since the process of reception is exactly the reverse of the process of transmission, transmitting and receiving antennas are basically interchangeable. Apart from power-handling considerations, the two types of antennas are virtually identical. In fact, a so-called principle of reciprocity exists. This states that the characteristics of antennas, such as impedance and radiation pattern, are identical regardless of use for reception or transmission, and this relation may be proved mathematically. It is of particular use for antennas employed in both functions.

Attenuation and absorption The inverse-square law shows that power density diminishes fairly rapidly with distance from the source of electromagnetic waves. Another way of looking at this is to say that electromagnetic waves are attenuated as they travel outward from their source, and this attenuation is proportional to the square of the distance traveled. Attenuation is normally measured in decibels[1] and happens to be the same numerically for both field intensity and power density. This may be shown as follows.

Let $\mathcal{P}_1$ and $\mathcal{E}_1$ be the power density and field intensity, respectively, at a distance r_1 from the source of electromagnetic waves. Let similar conditions apply to $\mathcal{P}_2$, $\mathcal{E}_2$, and r_2, with r_2 being the greater of the two distances. The attenuation of power density at the further point (compared with the nearer) will be, in decibels,

$$\alpha_P = 10 \log \frac{\mathcal{P}_1}{\mathcal{P}_2} = 10 \log \frac{P_t/4\pi r_1^2}{P_t/4\pi r_2^2} = 10 \log\left(\frac{r_2}{r_1}\right)^2$$

$$= 20 \log \frac{r_2}{r_1} \tag{8-6}$$

Similarly, for field intensity attenuation, we have

$$\alpha_E = 20 \log \frac{\sqrt{30P_t}/r_1}{\sqrt{30P_t}/r_2} = 20 \log \frac{r_2}{r_1} \tag{8-6'}$$

[1] It is also sometimes measured in ***nepers***, where the attenuation is said to be 1 neper when $\log_e (P_1/P_2) = 1$, and ***e*** is the base of the natural or ***napierian*** logarithm system (after John Napier, a sixteenth-century Scottish mathematician). The mathematical relationships are 1 neper = 8.686 dB and $e = 2.718$ approximately.

The two formulas are seen to be identical and, in fact, are used in exactly the same way. Thus, at a distance $2r$ from the source of waves, both field intensity and power density are 6 dB down from their respective values at a distance r from the source.

In free space, of course, absorption of radio waves does not occur because there is nothing there to absorb them. However, the picture is different in the atmosphere. This tends to absorb some radio waves, because some of the energy from the electromagnetic waves is transferred to the atoms and molecules of the atmosphere. This transfer causes the atoms and molecules to vibrate somewhat, and while the atmosphere is warmed only infinitesimally, the energy of the waves may be absorbed quite significantly.

Fortunately, the atmospheric absorption of electromagnetic waves of frequencies below about 10 GHz is quite insignificant. As shown in Fig. 8-3, absorption by both the oxygen and the water vapor content of the atmosphere becomes significant at that frequency and then rises gradually. Because of various molecular resonances, however, certain peaks and troughs of attenuation exist. As Fig. 8-3 shows, frequencies such as 60 and 120 GHz are not recommended for long-distance propagation in the atmosphere. It is similarly best not to use 23 or 180 GHz either, except in very dry air. On the other hand, so-called *windows* exist at which absorption is greatly reduced; frequencies such as 33 and 110 GHz fall into this category.

Figure 8-3 shows atmospheric absorption split into its two major components, with absorption due to the water vapor content of the atmosphere taken for a standard value of humidity. If humidity is increased or if there is fog, rain or snow, then this form of absorption is increased tremendously, and reflection from rainwater drops may

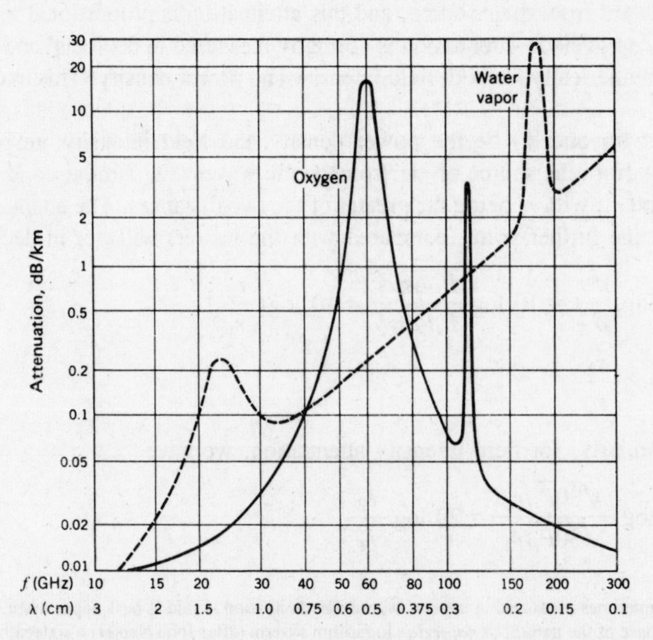

FIGURE 8-3 Atmospheric absorption of electromagnetic waves.

even take place. For example, a radar system operating at 10 GHz may have a range of 75 km in dry air, 68 km in light drizzle, 55 km in light rain, 22 km in moderate rain and 8 km in heavy rain, showing effectively how precipitation causes severe absorption at microwave frequencies. It must be repeated that such absorption is insignificant at lower frequencies, except over very long radio paths.

8-1.2 Effects of the Environment

When propagation near the earth is examined, several factors which did not exist in free space must be considered. Thus waves will be reflected by the ground, mountains and buildings. They will be refracted as they pass through layers of the atmosphere which have differing densities or differing degrees of *ionization*. Also, electromagnetic waves may be *diffracted* around tall, massive objects. They may even interfere with each other, when two waves from the same source meet after having traveled by different paths. Waves may also be absorbed by different media, but it was more convenient to consider this topic in the preceding section.

Reflection of waves There is much similarity between the reflection of light by a mirror and the reflection of electromagnetic waves by a conducting medium. In both instances the angle of reflection is equal to the angle of incidence, as illustrated in Fig. 8-4. Again, as with the reflection of light, the incident ray, the reflected ray and the normal at the point of incidence are in the one plane. Finally, the concept of *images* is used to advantage in both situations.

The proof of the equality of the angles of reflection and incidence follows the corresponding proof of what is known as *the second law of reflection* for light. Both proofs are based on the fact that the incident and reflected waves travel with the same velocity. There is yet another similarity here to the reflection of light by a mirror. Anyone who has been to a barber shop, in which there is a mirror behind as well as one in front, will have noticed not only that a huge number of images are present, but also that their brightness is progressively reduced. As expected, this is due to some absorp-

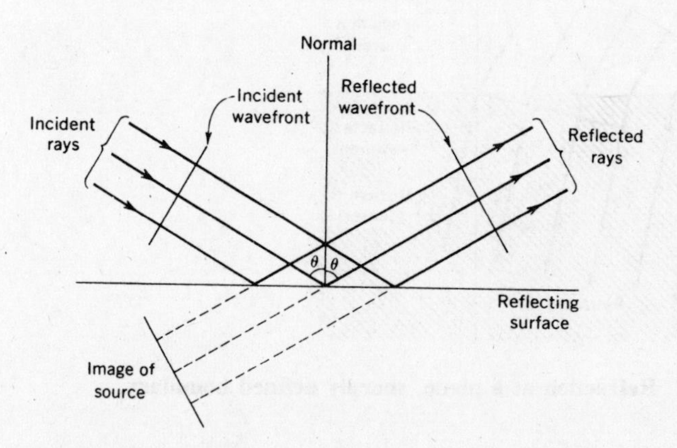

FIGURE 8-4 Reflection of waves; image formation.

tion at each reflection; this also happens with radio waves. The *reflection coéfficient* ρ is defined as the ratio of the electric intensity of the reflected wave to that of the incident wave. It is unity for a perfect conductor, and less than that for practical conducting surfaces. The difference is a result of the absorption of energy (and also its transmission) from the wave by the imperfect conductor. Transmission is a result of currents set up in the imperfect conductor, which in turn permit propagation within it, accompanied by *refraction*.

A number of other points connected with reflection must now be noted. First, it is important that the electric vector be perpendicular to the conducting surface; otherwise surface currents will be set up, and no reflection will result (this is discussed further in Sec. 10-1.2, in connection with waveguides). Second, if the conducting surface is curved, reflection will once again follow the appropriate optical laws (see also Sec. 9-7, which deals with parabolic and hyperbolic reflecting surfaces). Finally, if the reflecting surface is rough, reflection will be much the same as from a smooth surface, provided that the angle of incidence is in excess of the so-called *Rayleigh criterion*.[2]

Refraction As with light, refraction takes place when electromagnetic waves pass from one propagating medium to a medium having a different density. This situation causes the wavefront to acquire a new direction in the second medium and is brought about by a change in wave velocity. The simplest case of refraction, concerning two media with a plane, sharply defined boundary between them, is shown in Fig. 8-5.

Consider the situation in Fig. 8-5, in which a wave passes from medium A to the denser medium B, and the incident rays strike the boundary at some angle other than 90°. Wavefront P–Q is shown at the instant when it is about to penetrate the

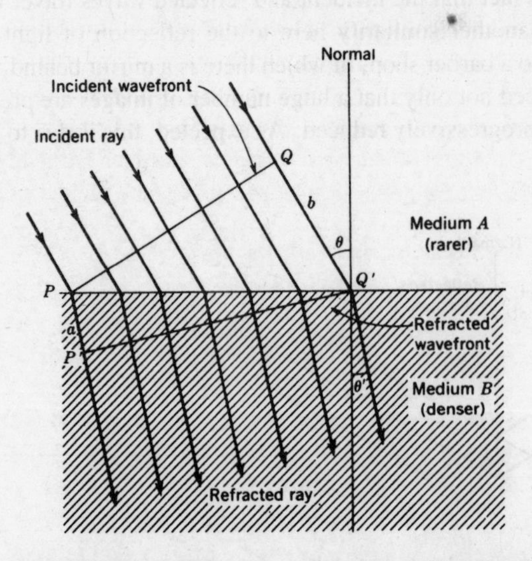

FIGURE 8-5 Refraction at a plane, sharply defined boundary.

[2] Put forward by Lord Rayleigh in 1897 and treated in most texts on optics. See also [2], pp. 65–66.

denser medium, and wavefront $P'-Q'$ is shown just as the wave has finished entering the second medium. Meanwhile, ray b has traveled entirely in the rarer medium, and has covered the distance $Q-Q'$, proportional to its velocity in this medium. In the same time ray a, which traveled entirely in the denser medium, has covered the distance $P-P'$. This is shorter than $Q-Q'$ because of the lower wave velocity in the denser medium. The in-between rays have traveled partly in each medium and covered total distances as shown; *the wavefront has been rotated.*

The relationship between the angle of incidence θ and the angle of refraction θ' may be calculated with the aid of simple trigonometry and geometry. Considering the two right-angled triangles PQQ' and $PP'Q'$, we have

$$\widehat{QPQ'} = \theta \qquad \text{and} \qquad \widehat{PQ'P'} = \theta' \tag{8-7}$$

Therefore

$$\frac{\sin\theta'}{\sin\theta} = \frac{PP'/PQ'}{QQ'/PQ'} = \frac{PP'}{QQ'} = \frac{v_B}{v_A} \tag{8-8}$$

where v_A = wave velocity in medium A
v_B = wave velocity in medium B

It will be recalled, from Eq. (7-7) and the accompanying work, that the wave velocity in a dielectric medium is inversely proportional to the square root of the dielectric constant of the medium. Substituting this into Eq. (8-8) gives

$$\frac{\sin\theta'}{\sin\theta} = \sqrt{\frac{k}{k'}} = \frac{1}{\mu} \tag{8-9}$$

where k = dielectric constant of medium A
k' = dielectric constant of medium B
μ = refractive index

Note, once again, that the dielectric constant is exactly 1 for a vacuum and very nearly 1 for air. Note also that Eq. (9-8) is an expression of the important law of optics: *Snell's law*.

When the boundary between the two media is curved, refraction still takes place, again following the optical laws (see also the discussion of dielectric lenses in Sec. 9-7). If the change in density is gradual, the situation is more complex, but refraction still takes place. Just as Fig. 8-5 showed that electromagnetic waves traveling from a rarer to a denser medium are refracted toward the normal, so we see that waves traveling the other way are bent away from the normal. However, if there is a linear change in density (rather than an abrupt change), the rays will be *curved* away from the normal rather than bent, as shown in Fig. 8-6.

The situation arises in the atmosphere just above the earth, where atmospheric density changes (very slightly, but linearly) with height. As a result of the slight refraction that takes place here, waves are bent down somewhat instead of traveling strictly in straight lines. The radio horizon is thus increased, but the effect is noticeable only for horizontal rays. Basically, what happens is that the top of the wavefront

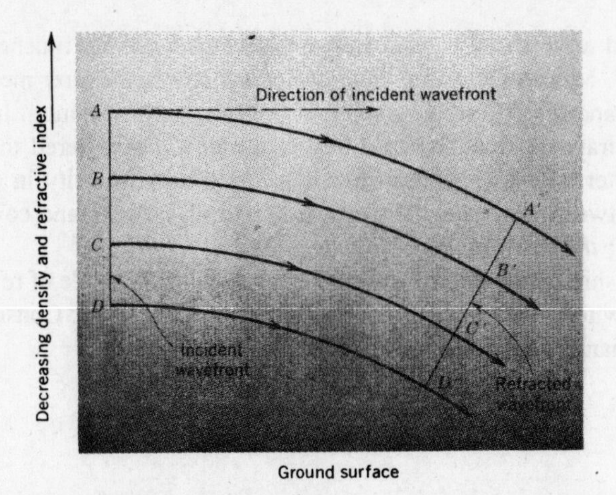

FIGURE 8-6 Refraction in a medium having linearly decreasing density (the Earth is shown flat for simplicity).

travels in rarer atmosphere than the bottom of the wavefront and therefore travels faster, so that it is bent downward. A somewhat similar situation (to be discussed later) arises when waves encounter the *ionosphere*.

Interference of electromagnetic waves Continuing with the optical properties of electromagnetic waves, we next consider interference. Interference occurs when two waves that left one source and traveled by different paths arrive at a point. This happens very often in high-frequency sky-wave propagation (see Sec. 8-2.2) and in microwave space-wave propagation (see Sec. 8-2.3). The latter case will be discussed here. It arises when a microwave antenna is located near the ground, and waves reach the receiving point not only directly but also after being reflected from the ground. This is shown in Fig. 8-7.

It is obvious that the direct path is shorter than the path with reflection. For some combination of frequency and height of antenna above the ground, the difference between paths 1 and 1′ is bound to be exactly a half-wavelength. There will thus be complete cancellation at the receiving point P if the ground is a perfect reflector and partial cancellation for an imperfect ground. Another receiving point, Q, may be lo-

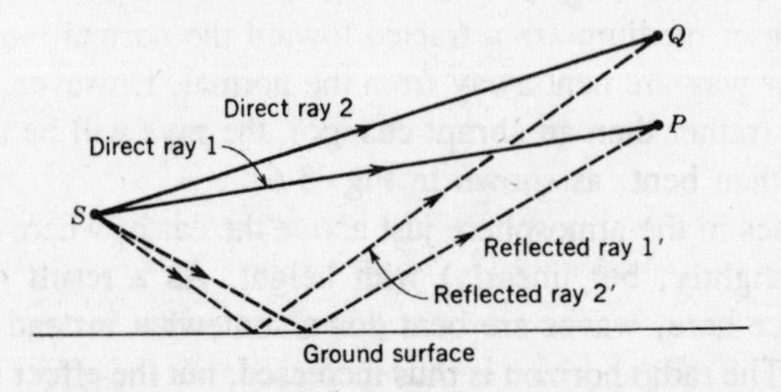

FIGURE 8-7 Interference of direct and ground-reflected rays.

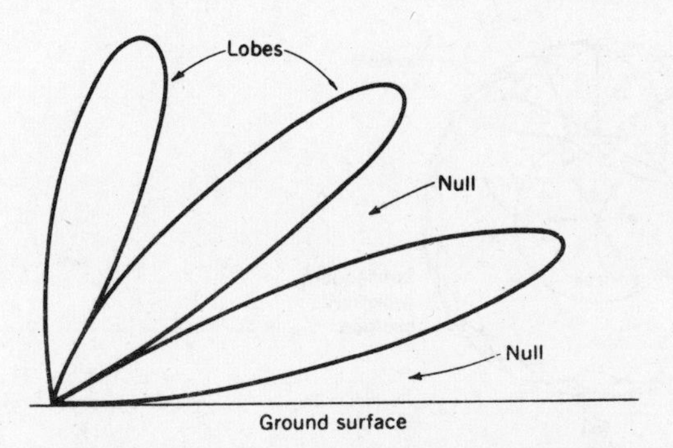

FIGURE 8-8 Radiation pattern with interference.

cated so that the path difference between 2 and 2′ is exactly one wavelength. In this case reinforcement of the received waves will take place at this point and will be partial or total, depending on the ground reflectivity. A succession of such points above one another may be found, giving an *interference pattern* consisting of alternate cancellations and reinforcements. A pattern of this form is shown in Fig. 8-8.

The curve of Fig. 8-8 joins points of equal electric intensity. The pattern is due to the presence of an antenna at a height above the ground of about a wavelength, with reflections from the ground (assumed to be plane and perfectly conducting) causing interference. A pattern such as the one shown may be calculated or plotted from actual field-strength measurements. The "flower petals" of the pattern are called *lobes*. They correspond to reinforcement points such as Q of Fig. 8-7, whereas the nulls between the lobes correspond to cancellations such as P of Fig. 8-7.

At frequencies right up to the VHF range, this interference will not be significant because of the relatively large wavelengths of such signals. In the UHF range and above, however, interference plays an increasing part in the behavior of propagating waves and must definitely be taken into account. It is certainly of great significance in radar and other microwave systems. For instance, if a target is located in the direction of one of the null zones, no increase in the transmitted radar power will make this target detectable. Again, the angle that the first lobe makes with the ground is of great significance in long-range radar. Here the transmitting antenna is horizontal and the maximum range may be limited not by the transmitted power and receiver sensitivity, but simply because the wanted direction corresponds to the first null zone. It must be mentioned that a solution to this problem consists of increasing the elevation of the antenna and pointing it downward.

Diffraction of radio waves Diffraction is yet another property shared with optics and concerns itself with the behavior of electromagnetic waves, as affected by the presence of small[3] slits in a conducting plane or sharp edges of obstacles. It was first discovered in the seventeenth century and put on a firm footing with the discovery of Huygens'

[3] That is, smaller than a wavelength, or at least of that order.

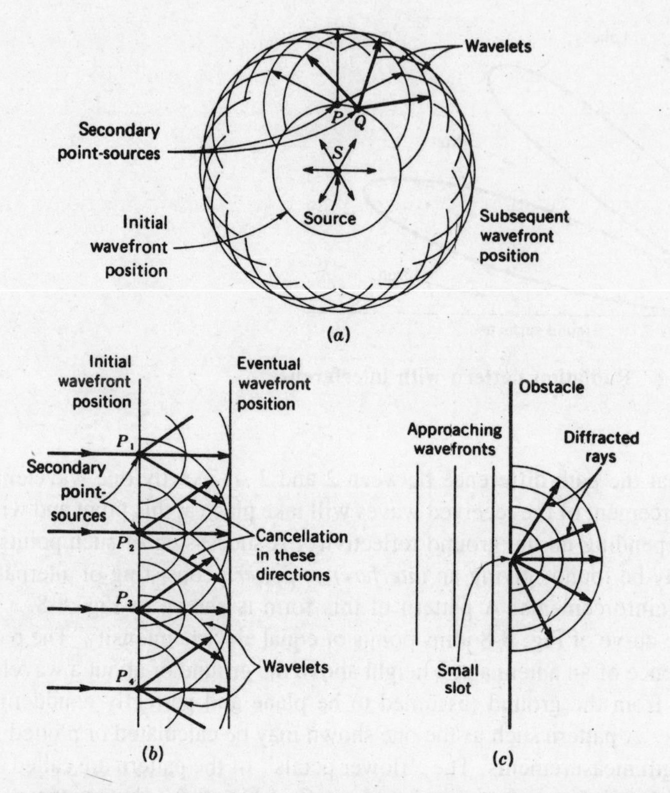

FIGURE 8-9 Diffraction. ***(a)* Of spherical wavefront; *(b)* of a plane wavefront; *(c)* through small slot.**

principle[4] fairly soon afterward. Huygens' principle states that every point on a given (spherical) wavefront may be regarded as a source of waves from which further waves are radiated outward, in a manner as illustrated in Fig. 8-9*a*. The total field at successive points away from the source is then equal to the vector sum of these secondary *wavelets*. For normal propagation, there is no need to take Huygens' principle into account, but it must be used when diffraction is to be accounted for. Huygens' principle can also be derived from Maxwell's equations.

If a plane wave is considered, as in Fig. 8-9*b*, the question that arises immediately is why the wavefront continues as a plane, instead of spreading out in all directions. The answer is that an *infinite* plane wave has been considered, and mathematics shows that cancellation of the secondary wavelets will occur in all directions other than the original direction of the wavefront; thus the wavefront does continue as a plane.

[4] Francesco Grimaldi discovered that no matter how small a slit was made in an opaque plane, light on the side opposite the source would spread out in all directions. Similarly, no matter how small a light source was constructed, a sharp shadow could not be obtained at the edge of a sharp opaque obstacle. The Dutch astronomer Christian Huygens, the founder of the wave theory of light, gave an explanation for these phenomena that was published in 1690 and is still accepted and used.

When a finite plane wave is considered, the cancellation in spurious directions is no longer complete, so that some divergence or scattering will take place. For this to be noticeable, however, the wavefront must be small, such as that obtained with the aid of the slot in a conducting plane, as in Fig. 8-9*c*. It is seen that instead of being "squeezed through" as a single ray, the wave spreads out past the slot, which now acts as Huygens' point source on a wavefront and radiates in all directions. The radiation is maximum (but not a sharp maximum if the slot is small) in front of the slot and diminishes gradually away from it.

Figure 8-10 shows what happens when a plane wave meets the edge of an obstacle. Although a sharp shadow might have been expected, diffraction takes place once again for precisely the same reasons as before. If two nearby points on the wavefront, *P* and *Q*, are again considered as sources of wavelets, it is seen that radiation at angles away from the main direction of propagation is obtained. Thus the shadow zone receives some radiation. If the obstacle edge had not been there, this side radiation would have been canceled by other point sources on the wavefront.

Radiation once again dies down away from the edge, but not so gradually as with a single slot because some interference takes place; this is the reason why two point sources on the wavefront were shown. Given a certain wavelength and point separation, it may well be that rays *a* and *a'*, coming from *P* and *Q*, respectively, have a path difference of a half-wavelength, so that their radiations cancel. Similarly, the path difference between rays *b* and *b'* may be a whole wavelength, in which case reinforcement takes place in that direction. When all the other point sources on the wavefront are taken into account, the process becomes less sharp. However, the overall result is still a succession of interference fringes (each fringe less bright than the previous) as one moves away from the edge of the obstacle.

This type of diffraction is of importance in two practical situations. First, signals propagated by means of the space wave may be received behind tall buildings, mountains and other similar obstacles as a result of diffraction. Second, in the design of microwave antennas, diffraction plays a major part in preventing the narrow pencil of radiation which is often desired, by generating unwanted side lobes; this is discussed in Sec. 9-7.1.

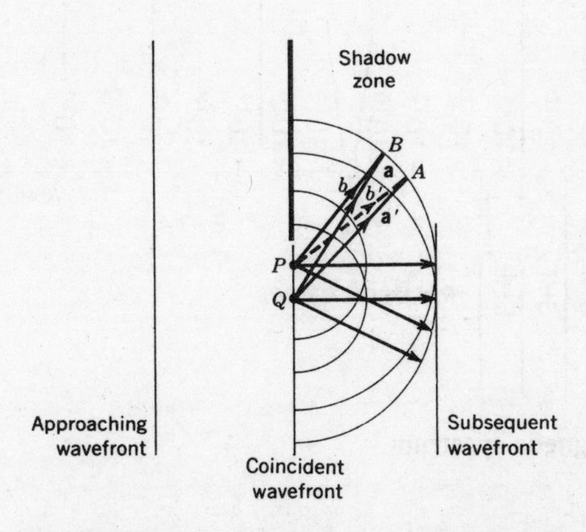

FIGURE 8-10 Diffraction around the edge of an obstacle.

8-2 PROPAGATION OF WAVES

In an earth environment, electromagnetic waves propagate in ways that depend not only on their own properties but also on those of the environment itself; some of this was seen in the preceding section. Since the various methods of propagation depend largely on frequency, the complete electromagnetic spectrum is now shown for reference in Fig. 8-11—note that the frequency scale is logarithmic.

Waves travel in straight lines, except where the earth and its atmosphere alter their path. Thus, except in unusual circumstances, frequencies above the HF generally travel in straight lines.[5] They propagate by means of so-called space waves. These are sometimes called *tropospheric waves,* since they travel in the troposphere, the portion of the atmosphere closest to the ground. Frequencies below the HF range travel around the curvature of the earth, sometimes right around the globe. The means are probably a combination of diffraction and a type of *waveguide* effect which uses the earth's surface and the lowest *ionized* layer of the atmosphere as the two waveguide walls (see Sec. 10-1.3). These *ground waves,* or *surface waves* as they are called, are one of the two original means of beyond-the-horizon propagation. For example, all broadcast radio signals received in daytime propagate by means of surface waves.

Waves in the HF range, and sometimes frequencies just above or below it, are reflected by the ionized layers of the atmosphere (to be described) and are called *sky waves*. Such signals are beamed into the sky and come down again after reflection, returning to earth well beyond the horizon. To reach receivers on the opposite side of the earth, these waves must be reflected by the ground and the ionosphere several times. It should be mentioned that neither surface waves nor sky waves are possible in space or on airless bodies such as the moon.

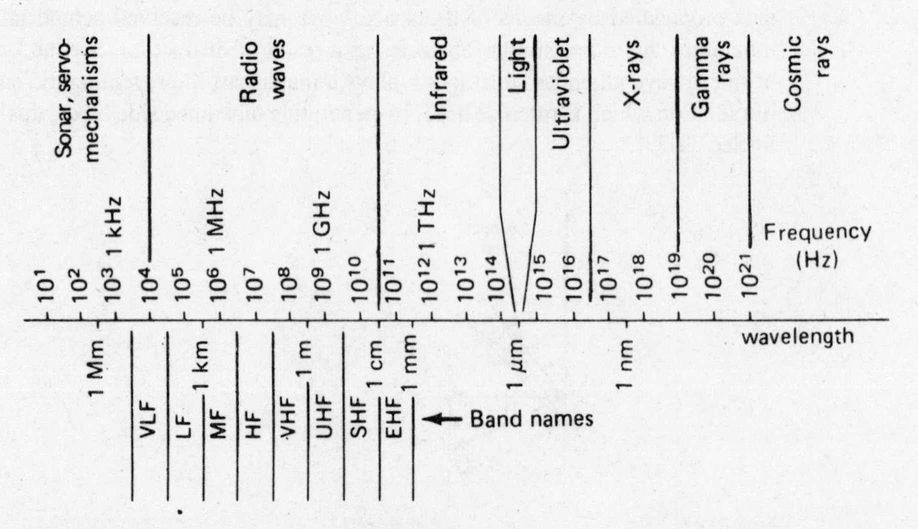

FIGURE 8-11 The electromagnetic spectrum.

[5] Except for refraction due to changing atmospheric density, as discussed in the previous section.

Two more recently developed means of beyond-the-horizon propagation are tropospheric scatter and stationary satellite communications. Each of these five methods of propagation will now be described in turn.

8-2.1 Ground (Surface) Waves

Ground waves progress along the surface of the earth and, as previously mentioned, must be vertically polarized to prevent short circuiting the electric component. A wave induces currents in the ground over which it passes and thus loses some energy by absorption. However, this is made up, to a certain extent, by energy diffracted downward from the upper portions of the wavefront.

There is another way in which the surface wave is attenuated: because of diffraction, the wavefront gradually tilts over, as shown in Fig. 8-12. As the wave propagates over the earth, it tilts over more and more, and the increasing tilt causes greater short circuiting of the electric field component of the wave and hence field strength reduction. Eventually, at some distance (in wavelengths) from the antenna, as partly determined by the type of surface over which the ground wave propagates, the wave "lies down and dies." It is important to realize this, since it shows that the maximum range of such a transmitter depends on its frequency as well as its power. Thus, in the VLF band, insufficient range of transmission can be cured by increasing the transmitting power. On the other hand, this remedy will not work near the top of the MF range, since propagation is now definitely limited by tilt.

Field strength at a distance Radiation from an antenna by means of the ground wave gives rise to a field strength at a distance, which may be calculated by use of Maxwell's equations. This field strength, in volts per meter, is given in Eq. (8-10), which differs from Eq. (8-5) by taking into account the gain of the transmitting antenna.

$$\mathscr{E} = \frac{120\pi h_t I}{\lambda d} \tag{8-10}$$

If a receiving antenna is now placed at this point, the signal it will receive will be, in volts,

$$V = \frac{120\pi h_t h_r I}{\lambda d} \tag{8-11}$$

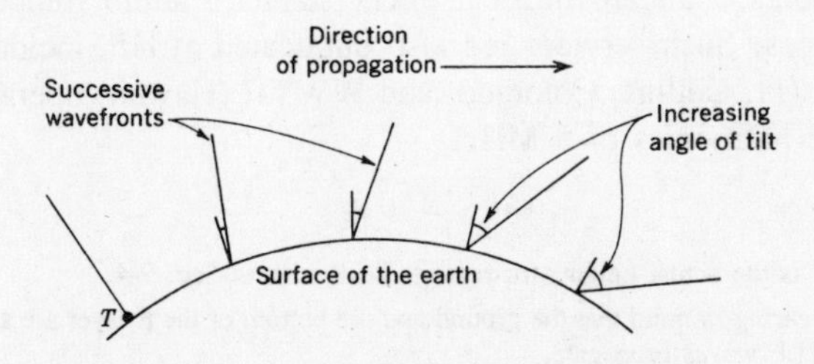

FIGURE 8-12 Ground-wave propagation.

where 120π = characteristic impedance of free space
h_t = effective[6] height of the transmitting antenna
h_r = effective height of the receiving antenna
I = antenna current
d = distance from the transmitting antenna
λ = wavelength

If the distance between the two antennas is fairly long, the reduction of field strength due to ground and atmospheric absorption reduces the value of the voltage received, making it less than shown by Eq. (8-11). Although it is possible to calculate the signal strength reduction which results, altogether too many variables are involved to make this worthwhile. Such variables include the salinity and resistivity of the ground or water over which the wave propagates and the water vapor content of the air. The normal procedure is to estimate signal strength with the aid of the tables and graphs available [1].

VLF propagation When propagation is over a good conductor like seawater, particularly at frequencies below about 100 kHz, surface absorption is small, and so is attenuation due to the atmosphere. Thus the angle of tilt is the main determining factor in the long-distance propagation of such waves. The degree of tilt depends on the distance from the antenna in wavelengths, and hence the early disappearance of the surface wave in HF propagation. Conversely, because of the large wavelengths of VLF signals, waves in this range are able to travel long distances before disappearing (right around the globe if sufficient power is transmitted).

At distances up to 1000 km, the ground wave is remarkably steady, showing little diurnal, seasonal or annual variation. Farther out, the effects of the *E* layer's contribution to propagation[7] are felt. Also, both short- and long-term signal strength variations take place, the latter including the 11-year solar cycle. The strength of low-frequency signals changes only very gradually, so that rapid fading does not occur. All in all, transmission at these wavelengths proves a very reliable means of communication over long distances.

The most frequent users of long-distance VLF transmissions are ship communications, and time and frequency transmissions. Ships use the frequencies allocated to them, from 10 to 110 kHz, for radio navigation and maritime mobile communications. The time and frequency transmissions operate at frequencies as low as 16 kHz (GBR, Rugby, United Kingdom) and 17.8 kHz (NAA, Cutler, Maine). They provide a worldwide continuous hourly transmission of stable radio frequencies, standard time intervals, time announcements, standard musical pitch, standard audio frequencies and radio propagation notices. Such services are also duplicated at HF, incidentally, by stations such as WWV (Ft. Collins, Colorado) and WWVH (Hawaii) operating at 2.5 MHz and the first five harmonics of 5 MHz.

[6] This is not quite the same as the actual height, for reasons dealt with in Sec. 9-4.

[7] See also the next section, bearing in mind that the ground and the bottom of the *E* layer are said to form a waveguide through which VLF waves propagate.

Since VLF antennas are certain to be inefficient, high powers and the tallest possible masts are used. Thus we find powers in excess of 1 MW transmitted as a rule, rather than an exception. For example, the U.S. Naval Communications Station at North-West Cape (Western Australia) has an antenna farm consisting of 13 very tall masts, the tallest 387 m high; the lowest transmitting frequency is 15 kHz.

8-2.2 Sky-Wave Propagation—The Ionosphere [11]

Even before Sir Edward Appleton's pioneering work in 1925, it had been suspected that ionization of the upper parts of the earth's atmosphere played a part in the propagation of radio waves, particularly at high frequencies. Experimental work by Appleton showed that the atmosphere receives sufficient energy from the sun for its molecules to split into positive and negative ions. They remain thus ionized for long periods of time. He also showed that there were several layers of ionization at differing heights, which (under certain conditions) reflected back to earth the high-frequency waves that would otherwise have escaped into space. The various layers, or strata, of the ionosphere have specific effects on the propagation of radio waves, and must now be studied in detail.

The ionosphere and its effects [2, 3, 4] The ionosphere is the upper portion of the atmosphere, which absorbs large quantities of radiant energy from the sun, thus becoming heated and ionized. There are variations in the physical properties of the atmosphere, such as temperature, density and composition. Because of this and the different types of radiation received, the ionosphere tends to be stratified, rather than regular, in its distribution. The most important ionizing agents are ultraviolet and α, β, and γ radiation from the sun, as well as cosmic rays and meteors. The overall result, as shown in Fig. 8-13, is a range of four main layers, D, E, F_1 and F_2, in ascending order. The last two combine at night to form one single layer.

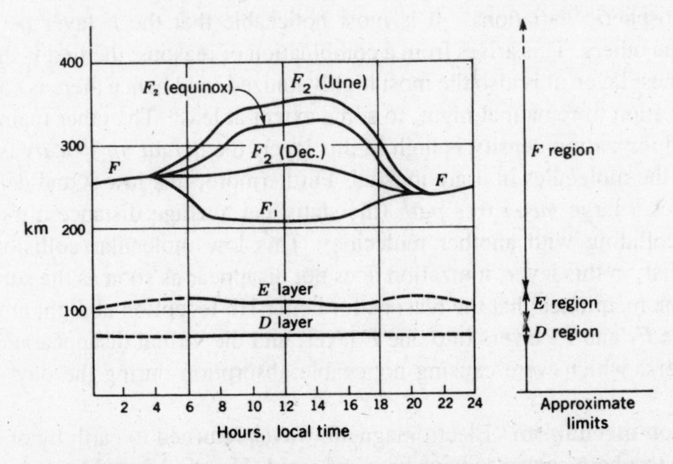

FIGURE 8-13 Ionospheric layers and their regular variations. *(By permission from F. R. East, "The Properties of the Ionosphere Which Affect HF Transmission," Point-to-Point Telecommunications, February 1965.)*

The D layer is the lowest, existing at an average height of 70 km, with an average thickness of 10 km. The degree of its ionization depends on the altitude of the sun above the horizon, and thus it disappears at night. It is the least important layer from the point of view of HF propagation. It reflects some VLF and LF waves and absorbs MF and HF waves to a certain extent.

The E layer is next in height, existing at about 100 km, with a thickness of perhaps 25 km. Like the *D* layer, it all but disappears at night; the reason for these disappearances is the recombination of the ions into molecules. This is due to the absence of the sun (at night), when radiation is consequently no longer received. The main effects of the *E layer* are to aid MF surface-wave propagation a little and to reflect some HF waves in daytime.

The E_s layer is a thin layer of very high ionization density, sometimes making an appearance with the *E layer*. It is also called the *sporadic E layer;* when it does occur, it often persists during the night also. On the whole, it does not have an important part in long-distance propagation, but it sometimes permits unexpectedly good reception. Its causes are not well understood.

The F_1 layer, as shown in Fig. 8-13, exists at a height of 180 km in daytime and combines with the F_2 layer at night; its daytime thickness is about 20 km. Although some HF waves are reflected from it, most pass through to be reflected from the F_2 layer. Thus the main effect of the F_1 layer is to provide more absorption for HF waves. Note that the absorption effect of this and any other layer is doubled, because HF waves are absorbed on the way up and also on the way down.

The F_2 layer is by far the most important reflecting medium for high-frequency radio waves. Its approximate thickness can be up to 200 km, and its height ranges from 250 to 400 km in daytime. At night it falls to a height of about 300 km, where it combines with the F_1 layer. Its height and ionization density vary tremendously, as Fig. 8-13 shows. They depend on the time of day, the average ambient temperature and the sunspot cycle (see also the following sections dealing with the normal and abnormal ionospheric variations). It is most noticeable that the *F* layer persists at night, unlike the others. This arises from a combination of reasons; the first is that since this is the topmost layer, it is also the most highly ionized, and hence there is some chance for the ionization to remain at night, to some extent at least. The other main reason is that although ionization density is high in this layer, the *actual air density* is not, and thus most of the molecules in it are ionized. Furthermore, this low actual density gives the molecules a large *mean free path* (the statistical average distance a molecule travels before colliding with another molecule). This low molecular collision rate in turn means that, in this layer, ionization does not disappear as soon as the sun sets. Finally, it must be mentioned that the reasons for better HF reception at night are the combinatin of the F_1 and F_2 layers into one *F* layer, and the virtual disappearance of the other two layers, which were causing noticeable absorption during the day.

Reflection mechanism Electromagnetic waves returned to earth by one of the layers of the ionosphere appear to have been reflected. In actual fact the mechanism involved is refraction, and the situation is identical to that described in Fig. 8-6. As the ionization density increases for a wave approaching the given layer at an angle, so the refractive index of the layer is reduced.[8] Hence the incident wave is gradually bent farther and farther away from the normal, as in Fig. 8-6.

If the rate of change of refractive index per unit height (measured in wavelengths) is sufficient, the refracted ray will eventually become parallel to the layer. It will then be bent downward, finally emerging from the ionized layer at an angle equal to the angle of incidence. Some absorption has taken place, but the wave has been returned by the ionosphere (well over the horizon if an appropriate angle of incidence was used).

Terms and definitions The terminology that has grown up around the ionosphere and sky-wave propagation includes several names and expressions whose meanings are not obvious. The most important of these terms will now be explained.

The virtual height of an ionospheric layer is best understood with the aid of Fig. 8-14. This figure shows that as the wave is refracted, it is bent down gradually rather than sharply. However, below the ionized layer, the incident and refracted rays follow paths that are exactly the same as they would have been if *reflection* had taken place from a surface located at a greater height, called the *virtual height* of this layer. If the virtual height of a layer is known, it is then quite simple to calculate the angle of incidence required for the wave to return to ground at a selected spot.

The critical frequency (f_c) for a given layer is the highest frequency that will be returned down to earth by that layer after having been beamed straight up at it. It is important to realize that there is such a maximum, and it is also necessary to know its value under a given set of conditions, since this value changes with these conditions. It was mentioned earlier that a wave will be bent downward provided that the rate of change of ionization density is sufficient, and that this rate of ionization is measured per unit wavelength. It also follows that the closer to being vertical the incident ray, the more it must be bent to be returned to earth by a layer. The result of these two effects is

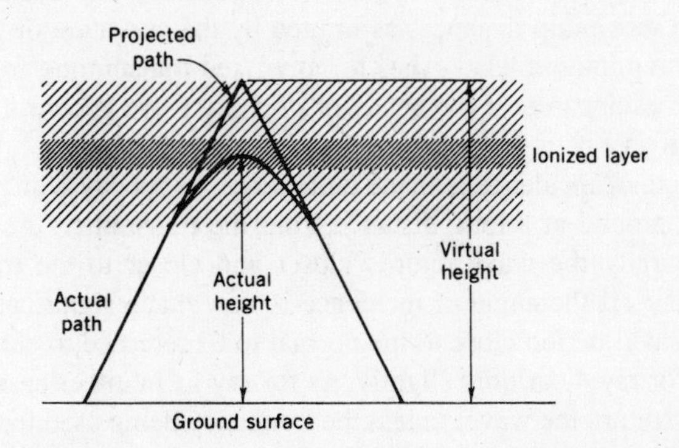

FIGURE 8-14 Actual and virtual heights of an ionized layer.

[8] Alternatively, this may be interpreted as an increase in the conductivity of the layer, and therefore a reduction in its electrical density or dielectric constant.

twofold. First, the higher the frequency, the shorter the wavelength, and the less likely it is that the change in ionization density will be sufficient for refraction. Second, the closer to vertical a given incident ray, the less likely it is to be returned to ground. Either way, this means that a maximum frequency must exist, above which rays go through the ionosphere. When the angle of incidence is normal, the name given to this maximum frequency is *critical frequency;* its value in practice ranges from 5 to 12 MHz for the F_2 layer.

The maximum usable frequency, or *MUF,* is also a limiting frequency, but this time for some specific angle of incidence other than the normal. In fact, if the angle of incidence (between the incident ray and the normal) is θ, it follows that

$$MUF = \frac{\textit{critical frequency}}{\cos\theta} = f_c \sec\theta \tag{8-12}$$

This is the so-called *secant law,* and it is very useful in making preliminary calculations for a specific MUF. Strictly speaking, it applies only to a flat earth and a flat reflecting layer. However, the angle of incidence is not of prime importance, since it is determined by the distance between the points that are to be joined by a sky-wave link. Thus MUF is defined in terms of two such points, rather than in terms of the angle of incidence at the ionosphere, it is defined as the highest frequency that can be used for sky-wave communication between two given points on earth. It follows that there is a different value of MUF for each pair of points on the globe. Normal values of MUF may range from 8 to 35 MHz, but after unusual solar activity they may rise to as high as 50 MHz. The highest working frequency between a given pair of points is naturally made less than the MUF, but it is not very much less for reasons that will be seen.

The skip distance is the shortest distance from a transmitter, measured along the surface of the earth, at which a sky wave of fixed frequency (more than f_c) will be returned to earth. That there should be a minimum distance may come as a shock. One expects there to be a maximum distance, as limited by the curvature of the earth, but nevertheless a definite minimum also exists for any fixed transmitting frequency. The reason for this becomes apparent if the behavior of a sky wave is considered with the aid of a sketch, such as Fig. 8-15.

When the angle of incidence is made quite large, as for ray 1 of Fig. 8-15, the sky wave returns to ground at a long distance from the transmitter. As this angle is slowly reduced, naturally the wave returns closer and closer to the transmitter, as shown by rays 2 and 3. If the angle of incidence is now made significantly less than that of ray 3, the ray will be too close to the normal to be returned to earth. It may be bent noticeably, as for ray 4, or only slightly, as for ray 5. In either case the bending will be insufficient to return the wave, unless the frequency being used for communication is less than the critical frequency (which is most unlikely); in that case everything is returned to earth. Finally, if the angle of incidence is only just smaller than that of ray 3, the wave may be returned, but at a distance farther than the return point of ray 3; a ray such as this is ray 6 of Fig. 8-15. This upper ray is bent back very gradually, because ion density is changing very slowly at this angle. It thus returns to earth at a considerable distance from the transmitter and is weakened by its passage.

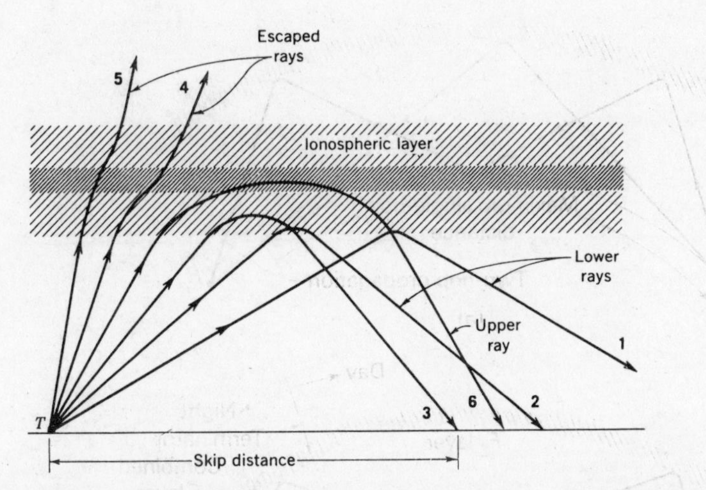

FIGURE 8-15 Effects of ionosphere on rays of varying incidence.

Ray 3 is incident at an angle which results in its being returned as close to the transmitter as a wave of this frequency can be. Accordingly, the distance is the *skip distance*. It thus follows that any higher frequency beamed up at the angle of ray 3 will not be returned to ground. It is seen that the frequency which makes a given distance correspond to the skip distance is the MUF for that pair of points.

At the skip distance, only the normal, or lower, ray can reach the destination, whereas at greater distances the upper ray can be received as well, causing interference. This is a reason why frequencies not much below the MUF are used for transmission. Another reason is the lack of directionality of high-frequency antennas, which is discussed in Sec. 9-6. If the frequency used is low enough, it is possible to receive lower rays by two different paths after either one or two hops, as shown in Fig. 8-16; the result of this is interference once again.

The *transmission path* is limited by the skip distance at one end and the curvature of the earth at the other. The longest single-hop distance is obtained when the ray is transmitted tangentially to the surface of the earth, as shown in Fig. 8-17. For the F_2

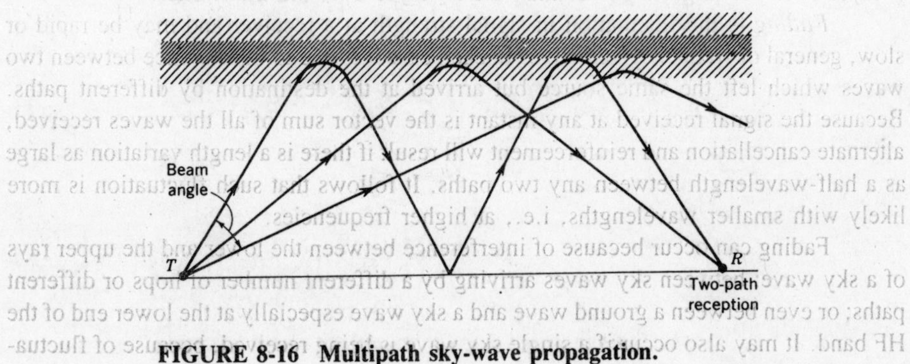

FIGURE 8-16 Multipath sky-wave propagation.

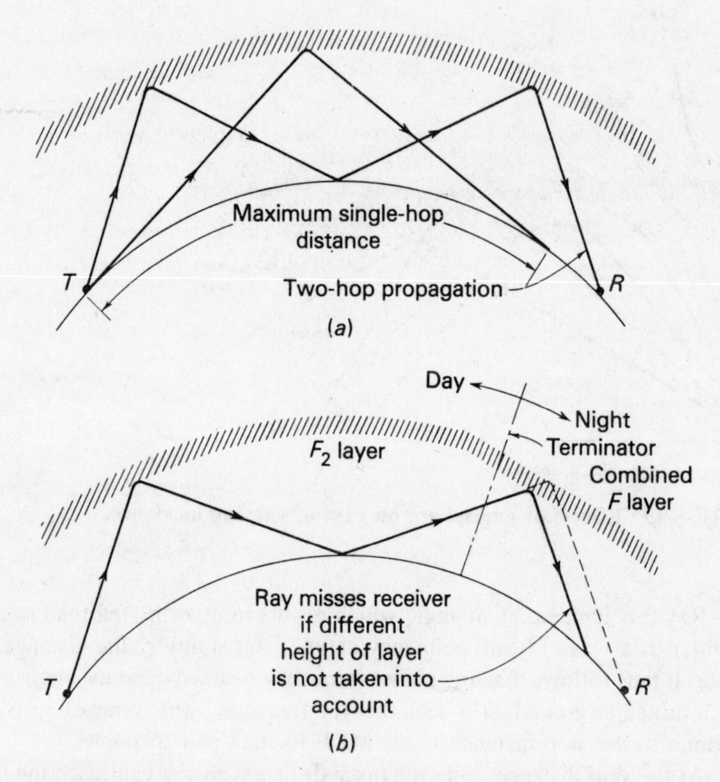

FIGURE 8-17 Long-distance sky-wave transmission paths. *(a)* North-south; *(b)* east-west.

layer, this corresponds to a maximum practical distance of about 4000 km. Since the semicircumference of the earth is just over 20,000 km, multiple-hop paths are often required, and Fig. 8-17 shows such a situation. No unusual problems arise with multihop north-south paths. However, care must be taken when planning long east-west paths to realize that although it is day "here," it is night "there," if "there" happens to be on the other side of the terminator. The result of not taking this into account is shown in Fig. 8-17*b*. A path calculated on the basis of a constant height of the F_2 layer will, if it crosses the terminator, undershoot and miss the receiving area as shown—the F layer over the target is lower than the F_2 layer over the transmitter.

Fading is the fluctuation in signal strength at a receiver and may be rapid or slow, general or frequency-selective. In each case it is due to interference between two waves which left the same source but arrived at the destination by different paths. Because the signal received at any instant is the vector sum of all the waves received, alternate cancellation and reinforcement will result if there is a length variation as large as a half-wavelength between any two paths. It follows that such fluctuation is more likely with smaller wavelengths, i.e., at higher frequencies.

Fading can occur because of interference between the lower and the upper rays of a sky wave; between sky waves arriving by a different number of hops or different paths; or even between a ground wave and a sky wave especially at the lower end of the HF band. It may also occur if a single sky wave is being received, because of fluctuations of height or density in the layer reflecting the wave. One of the more successful means of combating fading is to use space or frequency diversity (see Sec. 6-3.2).

Because fading is frequency-selective, it is quite possible for adjacent portions of a signal to fade independently, although their frequency separation is only a few dozen hertz. This is most likely to occur at the highest frequencies for which sky waves are used. It can play havoc with the reception of AM signals, which are seriously distorted by such frequency-selective fading. On the other hand, SSB signals suffer less from this fading and may remain quite intelligible under these conditions. This is because the relative amplitude of only a portion of the received signal is changing constantly. The effect of fading on radiotelegraphy is to introduce errors, and thus diversity is used here wherever possible.

Ionospheric variations The ionosphere is highly dependent upon the sun, and hence its conditions vary continuously. There are two kinds of variations. The *normal* ones have already been described as diurnal and seasonal height and thickness changes. *Abnormal* variations are due mainly to the fact that our sun is a *variable star*.

The sun has an 11-year cycle over which its output varies tremendously. Most people are unaware of this, because light variations are slight. However, the solar output of ultraviolet rays, coronae, flares, particle radiation and sunspots may vary as much as fiftyfold over that period. The extent of solar disturbance is measured by a method of sunspot counting developed by Wolf in the eighteenth century. On this basis, a pronounced 11-year (± 1 year) cycle emerges, and perhaps also a 90-year supercycle. The highest activities so far recorded were in 1778, 1871 and 1957, which was the highest ever [5].

The main sun-caused disturbances in the ionosphere are *SIDs* (*sudden ionospheric disturbances,* formerly known as *Dellinger dropouts*) and *ionospheric storms* [6]. Sudden ionospheric disturbances are caused by solar flares, which are gigantic emissions of hydrogen from the sun. Such flares are sudden and unpredictable, but more likely during peak solar activity than when the sun is "quiet." The x-ray radiation accompanying solar flares tremendously increases ionization density, right down to the *D* layer. This layer now absorbs signals that would normally go through it and be reflected from the *F* layer. Consequently, long-distance communications disappear completely, for periods up to 1 hour at a time. Studies with earth-based radioheliographs and from satellites in orbit have provided a large amount of data on solar flares, so that some short-term predictions are becoming possible. Two other points should be noted in connection with SIDs. First, only the sunlit side of the earth is affected, and second, VLF propagation is actually improved.

Ionospheric storms are caused by particle emissions from the sun, generally α and β rays. Since these take about 36 hours to reach the earth, some warning is possible after large sunspots or solar flares are noticed. The ionosphere behaves erratically during a storm, right around the globe this time, but more so in high latitudes because of the earth's magnetic field. Signal strengths drop and fluctuate quite rapidly. However, using lower frequencies often helps, since the highest ones are the most affected.

Finally, the sporadic *E* layer previously mentioned is also often included as an abnormal ionospheric disturbance. When present, it has the twin effects of preventing long-distance HF communications and permitting over-the-horizon VHF communications. The actual and virtual heights of this layer appear to be the same. This confirms the belief that the layer is thin and dense, so that actual reflection takes place.

Various national scientific bodies have ionospheric prediction programs which issue propagation notices of great value. Among them are the notices of the Central Radio Propagation Laboratory of the United States and the Ionospheric Prediction Service of the Australian Department of Science and Technology.

8-2.3 Space Waves

Space waves generally behave with merciful simplicity; they travel in (more or less) straight lines! However, since they depend on line-of-sight conditions, space waves are limited in their propagation by the curvature of the earth, except in very unusual circumstances. Thus they propagate very much like electromagnetic waves in free space, as discussed in Sec. 8-1.1. Such a mode of behavior is forced on them because their wavelengths are too short for reflection from the ionosphere, and because the ground wave disappears very close to the transmitter, owing to tilt.

Radio horizon The radio horizon for space waves is about four-thirds as far as the optical horizon. This beneficial effect is caused by the varying density of the atmosphere, and because of diffraction around the curvature of the earth. The radio horizon of an antenna is given, with good approximation, by the empirical formula

$$d_t = 4\sqrt{h_t} \tag{8-13}$$

where d_t = distance from transmitting antenna, km
h_t = height of transmitting antenna above ground, m

The same formula naturally applies to the receiving antenna. Thus the total distance will be given by addition, as shown in Fig. 8-18, and by the empirical formula

$$d = d_t + d_r = 4\sqrt{h_t} + 4\sqrt{h_r} \tag{8-14}$$

A simple calculation shows that for a transmitting antenna height of 225 m above ground level, the radio horizon is 60 km. If the receiving antenna is 16 m above ground level, the total distance is increased to 76 km. Greater distance between antennas may be obtained by locating them on tops of mountains, but links longer than 100 km are hardly ever used in commercial communications.

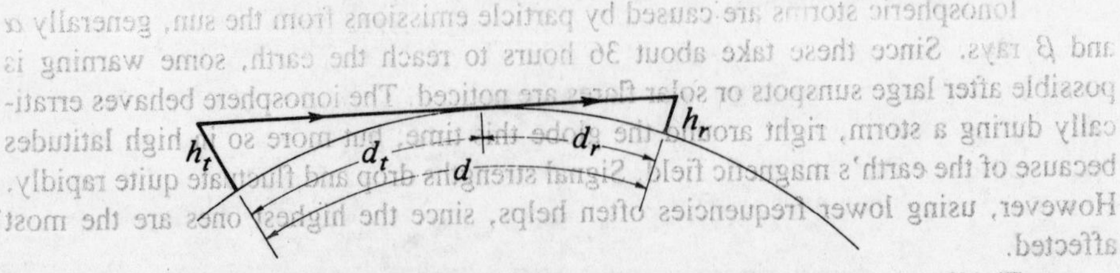

FIGURE 8-18 Radio horizon for space waves.

General considerations As treated in detail in Sec. 8-2.2, any tall or massive objects will obstruct space waves, since they travel close to the ground. Consequently, shadow zones and diffraction will result. This is the reason for the need in some areas for antennas higher than would be indicated by Eq. (8-14). On the other hand, some areas receive such signals by reflection—any object large enough to cast a radio shadow will, if it is a good conductor, cause back reflections also. Thus, in areas in front of it a form of interference known as "ghosting" may be observed on the screen of a television receiver. It is caused by the difference in path length (and therefore in phase) between the direct and the reflected rays. This situation is worse near a transmitter than at a distance, because reflected rays are stronger nearby. Finally, particularly severe interference exists at a distance far enough from the transmitter for the direct and the ground-reflected rays to be received simultaneously, as has already been shown.

Microwave space-wave propagation All the effects so far described hold true for microwave frequencies, but some are increased, and new ones are added. Atmospheric absorption and the effects of precipitation, as already discussed, must be taken into account. So must the fact that at such short wavelengths everything tends to happen very rapidly. Thus refraction, interference and absorption tend to be accentuated. One new phenomenon which occurs is *superrefraction,* also known as *ducting*.

As previously discussed, air density decreases and refractive index increases with increasing height above ground. The change in refractive index is normally linear and gradual, but under certain atmospheric conditions a layer of warm air may be trapped above cooler air, often over the surface of water. The result is that the refractive index will decrease far more rapidly with height than is usual. This happens near the ground, often within 30 m of it. The rapid reduction in refractive index (and therefore dielectric constant) will do to microwaves what the slower reduction of these quantities, in an ionized layer, does to HF waves; complete bending down takes place, as illustrated in Fig. 8-19. Microwaves are thus continuously refracted in the duct and reflected by the ground, so that they are propagated around the curvature of the earth for distances which sometimes exceed 1000 km. The main requirement for the formation of atmospheric ducts is the so-called temperature inversion. This is an increase of air temperature with height, instead of the usual decrease in temperature of 6.5°C/km in the "standard atmosphere." Superrefraction is, on the whole, more likely in subtropical than in temperate zones (see also [2], pp. 24-29).

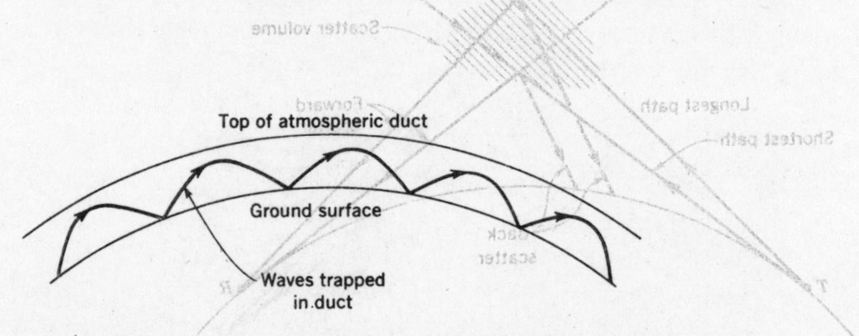

FIGURE 8-19 Superrefraction in atmospheric duct.

8-2.4 Tropospheric Scatter Propagation

Also known as *troposcatter*, or *forward scatter propagation*, tropospheric scatter propagation is a means of beyond-the-horizon propagation for UHF signals. It uses certain properties of the *troposphere*, the nearest portion of the atmosphere (within about 15 km of the ground).

Properties As shown in Fig. 8-20, two directional antennas are pointed so that their beams intersect midway between them, above the horizon. If one of these is a UHF transmitting antenna, and the other a UHF receiving one, sufficient radio energy will be directed toward the receiving antenna to make this a useful communication system. The reasons for the scattering are not fully understood, but there are two theories. One suggests reflections from "blobs" in the atmosphere, akin to the scattering of a searchlight beam by dust particles, and the other postulates reflection from atmospheric layers. Either way, this is a permanent state of affairs, not a sporadic phenomenon. The best frequencies, which are also the most often used, are centered on 900, 2000 and 5000 MHz. However, even here the actual proportion of forward scatter to signals incident on the scatter volume is very tiny—between −60 and −90 dB, or one-millionth to one-billionth of the incident power; high transmitting powers are obviously needed.

Practical considerations Although forward scatter is subject to fading, with little signal scattered forward, it nevertheless forms a very reliable method of over-the-horizon communication. It is not affected by the abnormal phenomena that afflict HF sky-wave propagation. Accordingly, this method of propagation is often used to provide long-distance telephone and other communications links, as an alternative to microwave links or coaxial cables over rough or inaccessible terrain (see also Sec. 15-2.3). Path links are typically 300 to 500 km long.

Tropospheric scatter propagation is subject to two forms of fading. The first is fast, occurring several times per minute at its worst, with maximum signal strength

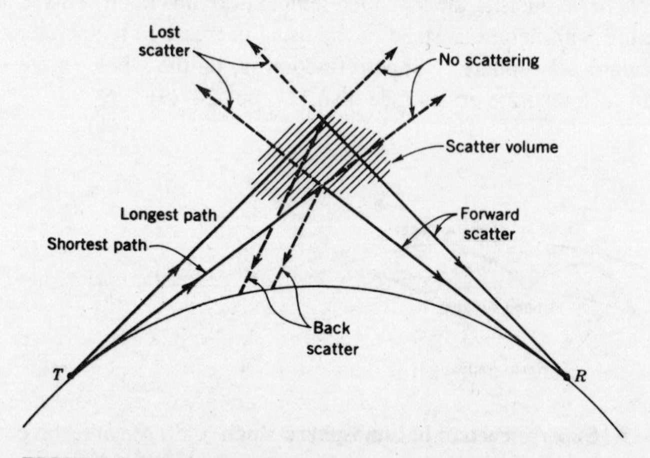

FIGURE 8-20 Tropospheric scatter propagation.

variations in excess of 20 dB. It is often called *Rayleigh fading* and is caused by multipath propagation. As Fig. 8-20 shows, scattering is from a volume, not a point, so that several paths for propagation exist within the scatter volume. The second form of fading is very much slower and is caused by variations in atmospheric conditions along the path.

It has been found in practice that the best results are obtained from troposcatter propagation if antennas are elevated and then directed down toward the horizon. Also, because of the fading problems, diversity systems are invariably employed, with space diversity more common than frequency diversity. Quadruple diversity systems are generally employed, with two antennas at either end of the link (all used for transmission and reception) separated by distances somewhat in excess of 30 wavelengths [7].

8-2.5 Extraterrestrial Communications

The most recent, and by far the fastest-growing, field of communications involves the use of various satellite relays, of which the first was launched in 1957, 12 years after the practicability and orbital positioning of stationary satellites were first described and calculated [8]. The field may be subdivided into three parts, each with somewhat differing requirements. First, there is communication with, and tracking of, fast-moving satellites in close orbits, typically 145 km in radius. Then there is communication via the *geostationary* satellites. Such satellites are placed in equatorial orbits at a height of approximately 36,000 km. This height gives a satellite the same *angular velocity* as the earth, as a result of which it appears to be stationary over a fixed spot on the equator. Such communications are discussed fully in Chap. 15. The last of the three is the communication and tracking connected with interplanetary probes.

Transionospheric space-wave propagation The ionosphere not only permits long-distance HF propagation; it also affects the propagation of waves through it. Fig. 8-15 showed that those waves which were not reflected by the ionosphere did, nevertheless, suffer varying degrees of bending away from their original paths. This suggests immediately that frequencies used here will need to be well above normal critical frequencies to minimize their refraction. If this were not done, serious tracking errors and communication difficulties would result because of the bending of radio waves. Since refraction becomes insignificant at frequencies above about 100 MHz, while atmospheric absorption is negligible up to about 14 GHz, these two considerations dictate the frequency range used in practice.

A problem encountered in transionospheric propagation is the *Faraday effect*. This causes the polarization of the radio wave to rotate as it passes through the ionosphere and is a complex process involving the presence of ionized particles and the earth's magnetic field (see also Sec. 10-5.2). As the ion density is variable, so is the Faraday effect upon any particular transmission, and so it is not practicable to calculate its extent and make appropriate allowances. If nothing at all is done, severe trouble will result. For instance, a horizontally polarized antenna will receive virtually no energy from a wave which was horizontally polarized when it left the antenna but has become vertically polarized through rotation in the ionosphere.

FIGURE 8-21 Circularly polarized antenna array for tracking satellites in low orbits. ***(Courtesy of Rohde and Schwartz, Munich.)***

Fortunately, several solutions of the problem are possible. One is to use an antenna with circular polarization (as shown in Fig. 8-21 and described in detail in Sec. 9-8), whose polarization is divided evenly between the vertical and horizontal components. Such an antenna is employed for transmission and reception at the ground station. Thus transmissions to and from the satellite may be received satisfactorily, no matter how they have been rotated in the ionosphere. It should be added that such rotation may well exceed 360° at some of the lower frequencies used. Another cure consists in using frequencies above about 1 GHz, at which Faraday rotation is negligible. The present trend for space communications is to frequencies between 1 and 14 GHz partly for the above reason.

Satellite and probe tracking The requirements for tracking and communicating with satellites in close orbits involve the use of fast-rotating, circularly polarized antennas (such as the array shown in Fig. 8-21), together with fairly low-noise receivers and medium-power transmitters. These are more or less conventional in design and operate in the allocated frequency band from 138 to 144 MHz. Most of the communication is radiotelemetry (see Sec. 13-3.2) since the function of the majority of these satellites is the gathering of scientific and other data. Because these satellites are in low orbits and therefore circumnavigate the earth in about 90 minutes, not only must the ground antennas be capable of fast rotation but also there must be strings of them around the globe because satellites disappear over the horizon in mere minutes.

Tracking interplanetary probes, such as the *Pioneer* and *Voyager* probes, is a problem of an entirely different order, especially since the distances involved may be several orders of magnitude greater. When it is considered that *Voyager 1* and 2 communicated with the earth from Saturn on a power of 30 W from about 1.5 billion km away [10], it can be seen immediately that the first requirements here are for huge directional antennas and extremely low-noise receivers. As a matter of interest, the signal power received by the antennas of NASA's Deep Space Network was only of the order of 10^{-17} W. These requirements are identical to those of radiotelescopes, which in fact are sometimes used for tracking deep-space probes.

The antennas used are of a type described in Sec. 9-7, being parabolic reflectors or horns, with diameters often in excess of 60 m. Since the relative motion of a probe is very slow, only the rotation of the earth need be taken into account. The antennas thus have *equatorial*[9] mountings with motor drives to keep them pointed at the same spot in the sky as the earth rotates. The National Aeronautics and Space Administration uses several such antennas around the world, including one at Goldstone, California, and another near Canberra, Australia.

PROBLEMS

For self-testing questions on this chapter, see p. 688.

8-1. At 20 km in free space from a point source, the power density is 200 μW/m^2. What is the power density 25 km away from this source?
[Ans.: 128μW/m^2]

8-2. Calculate the power density *(a)* 500 m from a 500-W source and *(b)* 36,000 km from a 3-kW source. Both are assumed to be omnidirectional point sources. The second value is, incidentally, the equivalent power density on the ground from a communications satellite in orbit.
[Ans.: (a) 159 μW/m^2, (b) 1.84 $\times$ 10^{-13} W/m^2]

8-3. A deep-space high-gain antenna and receiver system have a noise figure such that a minimum received power of 3.7 $\times$ 10^{-18} is required for satisfactory communication. What must be the transmitting power from a Jupiter probe, situated 800 million km from the earth? Assume that the transmitting antenna is *isotropic,* and the equivalent area of the receiving antenna has an area of 8400 m^2.
[Ans.: 3.54 kW]

8-4. A wave traveling in free space undergoes refraction after entering a denser medium, such that the original 30° angle of incidence at the boundary between the two media is changed 20°. What is the velocity of electromagnetic waves in the second medium?
[Ans.: 2.05 $\times$ 10^8 m/s]

[9] This does not mean that the antenna is located on the equator, but merely that one of its axes is made parallel to the axis of the earth. Such a mount is described and depicted in [9].

8-5. A 150-m antenna, transmitting at 1.2 MHz (and therefore by ground wave), has an antenna current of 8 A. What voltage is received by a receiving antenna 40 km away, with a height of 2 m? Note that this is a typical MF broadcasting situation.
[Ans.: 90.5 mV]

8-6. Two points on earth are 1500 km apart and are to communicate by means of HF. Given that this is to be a single-hop transmission, the critical frequency at that time is 7 MHz and conditions are idealized, calculate the MUF for those two points if the height of the ionospheric layer is 300 km.
[Ans.: 18.8 MHz]

8-7. A microwave link consists of repeaters at 40-km intervals. What must be the minimum height of transmitting and receiving antennas above ground level (given that they are the same) to ensure line-of-sight conditions?
[Ans.: 25 m]

QUESTIONS

8-1. Electromagnetic waves are said to be *transverse;* what does this mean? In what way are transverse waves different from *longitudinal* waves? Illustrate each type with a sketch.

8-2. Define the term *power density,* and explain why it is inversely proportional to the square of the distance from the source.

8-3. Explain what is meant by the terms *isotropic source* and *isotropic medium.*

8-4. Define and explain *field intensity*. Relate it to power density with the concept of *characteristic impedance of free space.*

8-5. Explain fully the concept of *linear polarization*. Can longitudinal waves be polarized? Explain.

8-6. Why does the atmosphere absorb some power from waves propagating through it? At what frequencies does this absorption become apparent? Illustrate with a sketch combining the absorption by oxygen and water vapor versus frequency, labeling the "windows."

8-7. Prove that when electromagnetic waves are reflected from a perfectly conducting medium, the angle of reflection is equal to the angle of incidence. *Hint:* Bear in mind that all parts of the wavefront travel with the same velocity, and consider what would happen if the two angles were *not* equal.

8-8. What is *refraction?* Explain under what circumstances it occurs and what causes it.

8-9. Prove, with a diagram, that electromagnetic waves passing from a denser to a rarer medium are bent away from the normal.

8-10. What is *interference* of radio waves? What are the conditions necessary for it to happen?

8-11. What is meant by the *diffraction* of radio waves? Under what conditions does it arise? Under what condition does it *not* arise?

8-12. Draw up a table showing radio-frequency ranges, the means whereby they propagate and the maximum terrestrial distances achievable under normal conditions.

8-13. Describe ground-wave propagation. What is the angle of *tilt?* How does it affect field strength at a distance from the transmitter?

8-14. Describe briefly the strata of the ionosphere and their effects on sky-wave propagation. Why is this propagation generally better at night than during the day?

8-15. Discuss the reflection mechanism whereby electromagnetic waves are bent back by a layer of the ionosphere. Include in your discussion a description of the *virtual height* of a layer. The fact that the virtual height is greater than the actual height proves something about the reflection mechanism. What is this?

8-16. Show, with the aid of a suitable sketch, what happens as the angle of incidence of a radio wave, using sky-wave propagation, is brought closer and closer to the vertical. Define the *skip distance,* and show how it is related to the maximum usable frequency.

8-17. What is fading? List its major causes.

8-18. Briefly describe the following terms connected with sky-wave propagation: *virtual height, critical frequency, maximum unsable frequency, skip distance* and *fading.*

8-19. Describe in some detail the main abnormal ionospheric variations, including a brief mention of the interference that may be caused by the *sporadic E layer.*

8-20. In connection with space-wave propagation, what is the radio horizon? How does it differ from the optical horizon?

8-21. For the three common methods of radio-wave propagation, explain briefly the mechanisms of propagation, give the approximate frequency and distance ranges of each and mention briefly the limitations of each of the modes of propagation.

8-22. Draw a sketch showing tropospheric scatter, and explain its basic principles. What are some of the frequencies which may be propagated by forward scatter? What are said to be the causes of such scattering? Where is the troposphere?

8-23. Discuss the requirements of communication with, and tracking of *(a)* satellites in close orbits and *(b)* space probes.

8-24. What factors determine the frequency range utilized for transionospheric communications?

REFERENCES

1. See, for example, International Telephone and Telegraph Corporation, *Reference Data for Radio Engineers,* 6th ed., Howard W. Sams and Co. Inc., Indianapolis, 1975, pp. 28-2–28-4.

2. Picquenard, Armel, *Radio Wave Propagation,* The Macmillan Press, Ltd., London, 1974, chap. 6.

3. *The Radio and Electronic Engineer* (whole issue), January/February 1975.

4. Reference 1, pp. 28-4–28-11.

5. Radio Society of Great Britain, *Radio Communications Handbook,* vol. 2: 5th ed., Radio Society of Great Britain, London, 1977, pp. 11.13–11.16.

6. Cook, F. E., and C. G. McCue, Solar Terrestrial Relations and Short-term Ionospheric Forecasting, pp. 11–30 in Ref. 3 (223 further references given).

7. See also Skingley, B. S., "Modern Tropospheric Scatter Systems," *Australian Electronics Engineering,* February 1977, pp. 21–25.

8. *(Historical)* Clarke, A. C., Extraterrestrial Relays, *Wireless World,* September 1945, p. 305. Arthur C. Clarke is nowadays better known as a science fiction author and recipient of the United Nations "Kalinga" Prize for the popularization of science.

9. Norton, Arthur P., and J. Gall Inglis, *Norton's Star Atlas and Reference Handbook,* 15th ed., Gall and Inglis, Edinburgh, 1964, pp. 50–51.

10. See Brejcha, A. G., "Microwave Communications from Outer Planets: The Voyager Project," *Microwave J.,* January 1980, pp. 25–35, 44, for additional details.

11. See also, for a good up-to-date summary, Rush, C. M., "HF Propagation: What We Know and What We Need to Know," *Second International Conference on Antennas and Propagation, Part 2: Propagation,* The Institution of Electrical Engineers, London, 1981, pp. 227–236.

9 ANTENNAS

The preceding chapter dealt at length with the various methods of propagation of radio waves, while only briefly mentioning how they might be launched or received. Similarly, earlier chapters took it for granted that transmitters can somehow transmit what they generate, and receivers have some means of receiving what is transmitted. Indeed, the word *antenna* was even mentioned on a number of occasions! Thus it is no secret that, in order to couple to space the output of a transmitter or the input of a receiver, some sort of interface is essential. A structure must be provided that is capable of radiating electromagnetic waves or receiving them, as the case may be. An antenna[1] is such a structure. It is generally a metallic object, often a wire or collection of wires, used to convert high-frequency current into electromagnetic waves, and vice versa. Apart from their different functions, transmitting and receiving antennas behave identically. That is, their behavior is reciprocal.

The chapter begins with fundamentals and goes on to consider simple wire antennas in free space. Then several important antenna quantities are defined and discussed, among them *antenna gain, resistance, bandwidth* and *beamwidth*. Just as the ground had a significant effect on the propagation of waves, so it modifies the properties of antennas—hence the effects of the ground are discussed next, in some detail. Then, antenna coupling and HF antenna arrays are treated. The final two major topics are microwave antennas, which are generally the most spectacular, and wideband antennas, which are generally the most complex in appearance. These last two subjects occupy more than one-third of the chapter and describe antennas such as those with *parabolic reflectors, horn antennas, lenses, helical antennas* and *log-periodic arrays*.

9-1 BASIC CONSIDERATIONS

The actual radiation mechanism may be explained quantitatively by means of Maxwell's equations. Upon examining the behavior of RF current in a wire, it is found that not all the energy applied at one end finds its way to the other; some escapes, i.e., is radiated. It is also possible to find a mathematical expression for this escaped energy, which permits the calculation not only of the quantity of energy that has been radiated

[1] According to current usage, transmitters have *antennas* and insects have *antennae;* in Australia and Great Britain the terms *antenna* and *aerial* are interchangeable.

but also the direction, or directions, in which it propagates. Because this method of dealing with radiation is too involved to be treated here, a qualitative presentation based on the behavior of traveling and standing waves on a transmission line will be used instead.

9-1.1 The Radiation Mechanism

Consider the open-circuited transmission line of Fig. 9-1. It is seen that the forward and reverse traveling waves combine to form a standing-wave pattern on the line, with a voltage antinode at the open-circuited point. This has already been discussed (in Chap. 7), but what was not mentioned at the time is that *not all* the forward energy is reflected by the open circuit. As shown, a small portion of the electromagnetic energy escapes from the system and is thus radiated. This occurs because the lines of force, traveling toward the open circuit, are required to undergo a complete phase reversal when they reach it. Not all of them are able to do this, because they possess the equivalent of mechanical inertia, and thus some do escape. It must be added that the proportion of waves escaping the system to those remaining is very small, for two reasons. First, if we consider the surrounding space as the load for the transmission line, we see that a mismatch exists, and thus very little power is dissipated in this ''load.'' Second, since the two wires are close together, it is apparent that the radiation from one tip will just about cancel that from the other. This is because they are of opposite polarities and at a distance apart that is tiny compared to a wavelength. Conversely, this is also the reason why low-frequency parallel-wire transmission lines do not radiate.

The cure for this problem seems to be an ''enlargement'' of the open circuit, i.e., spreading of the two wires, as in Fig. 9-2*a*. There is now less likelihood of cancellation of radiation from the two wire tips. By the same token, the radiating transmission line is now better coupled to the surrounding space. This is another way of saying that more power will be ''dissipated'' in the surrounding space, i.e., radiated. Moreover, because of the spreading out, waves traveling along the line find it more difficult to undergo the phase reversal at the end. Thus everything points to an increase in radiation.

The radiation efficiency of this system is improved even more when the two wires are bent so as to be in the same line, as in Fig. 9-2*b*. The electric (and also the magnetic) field is now fully coupled to the surrounding space, instead of being confined between the two wires, and the maximum possible amount of radiation results. This type of radiator is called a *dipole*. When the total length of the two wires is a

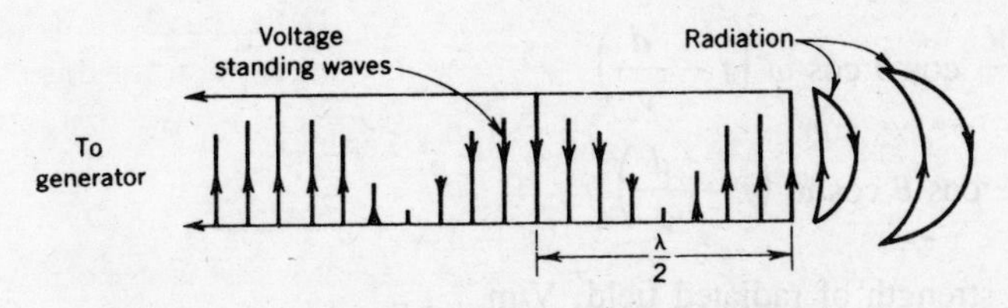

FIGURE 9-1 Radiation from transmission line.

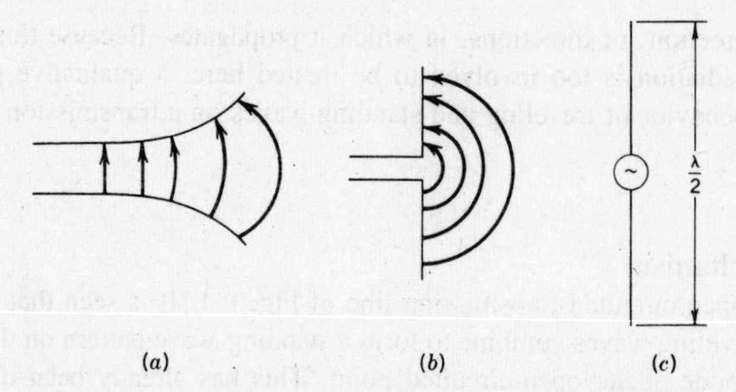

FIGURE 9-2 Evolution of the dipole. (*a*) Opened-out transmission line; (*b*) conductors in line; (*c*) half-wave dipole (center-fed).

half-wavelength, the antenna is called a *half-wave dipole*. It has the form indicated in Fig. 9-2*c*, and now even greater radiation occurs. The reason for this increase is that the half-wave dipole may be regarded as having the same basic properties (for the point of view of impedance particularly) as a similar length of transmission line. Accordingly, we have the antenna behaving as a piece of quarter-wave transmission line bent out and open-circuited at the far end. This results in the high impedance at the far ends of the antenna reflected as a low impedance at the end connected to the main transmission line. This, in turn, means that a large current will flow at the input to the half-wave dipole, and efficient radiation will take place.

The various characteristics of antennas are not normally quoted as absolute figures, but rather in comparison to those of "standard" antennas. These are theoretical simplifications, which need not necessarily exist in practice but which have properties that are easy to visualize and calculate. One such reference antenna is the *infinitesimal dipole,* which is defined as a pair of nearby conducting spheres having capacitance, with a separation and dimensions that are all negligibly small. Yet another reference antenna is the *elementary doublet,* which will now be described.

9-1.2 Elementary Doublet

The short dipole, or elementary doublet, is the simplest wire antenna. It is infinitely thin and has length l which is very short compared to the wavelength λ. Also, it is assumed that the current throughout it, I, is constant. If the RF current is $i = I \sin \omega t$, it may be shown with the aid of Maxwell's equations that the strength of the radiated field is

$$\mathscr{E} = \frac{\mathfrak{Z}(l/2)I}{d\lambda} \cos \theta \cos \omega \left(t - \frac{d}{v_c} \right)$$

$$= \frac{60\pi l I}{d\lambda} \cos \theta \cos \omega \left(t - \frac{d}{v_c} \right) \tag{9-1}$$

where $\mathscr{E}$ = strength of radiated field, V/m

$\mathfrak{Z}$ = characteristic impedance of free space = $120\pi\Omega$

d = distance of the point at which the field strength is measured from the elementary doublet
v_c = velocity of light in free space
θ = angle of inclination as shown in Fig. 9-3*a*

The first term of the equation gives the magnitude of the electric field at a distance. It shows, as did the work of Sec. 8-1.1, that this field strength depends on the power transmitted (via $\mathfrak{L}$ and I) and is inversely proportional to the distance from the radiating source (in this case, the short dipole). It is also seen that the magnitude of the electric field is proportional, for short dipoles at any rate, to their relative length, or l/λ.

The second term of Eq. (9-1) defines the radiation pattern from the elementary dipole, shown here in Fig. 9-3*b* and *c*. As might have been expected, radiation is maximum at right angles to the dipole and eventually falls away to zero in line with the antenna. This may be explained by considering that at right angles to such a short piece of wire, the distance from the remote point to any part of the wire is the same as its distance to any other point. Thus reinforcement of radiation will take place in this direction. When the distant point lies in a direction other than normal, there will be some cancellation because it is no longer true to say that its distance to all points on the elementary doublet is the same. Finally, full cancellation will take place when the angle of inclination θ is 90°. The radiation pattern cross section, as presented in Fig. 9-3*b*, is a figure eight with its axis at right angles to the dipole itself. Moreover, exactly the same radiation pattern will exist in any plane containing the elementary dipole, so that the three-dimensional pattern is the figure of revolution looking very much like a doughnut with an infinitesimally small center hole. This is also indicated by the other view of the radiation pattern, in Fig. 9-3*c*.

The last term of Eq. (9-1) is really the least important from our point of view. It merely takes into account the phase of the signal at some distant point. As compared with the phase of the signal in the antenna, it is governed by the time it takes the signal traveling at the velocity of light to reach this point.

The radiation field is not the only field surrounding the elementary doublet, or any other antenna, for that matter. Magnetic and electric fields exist also and are collectively termed the *induction field*. Such a field surrounds any current-carrying wire and, in fact, is stronger than the radiation field close to the radiator. However, the induction field diminishes very rapidly with increasing distance and becomes insignificant a few wavelengths away. The importance of the induction field does not lie in its

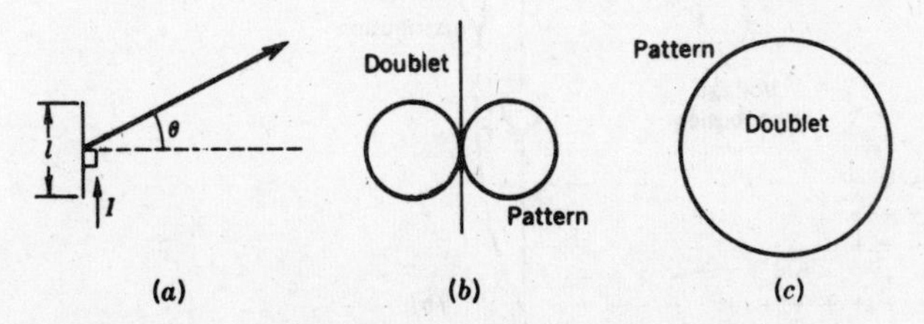

FIGURE 9-3 Elementary Doublet. (*a*) Doublet antenna; (*b*) pattern cross section in the plane of the doublet; (*c*) pattern cross section in plane perpendicular to doublet.

ability to convey information over long distances. Instead, it lies in the fact that if antennas are placed close together, the interference effects caused by the induction field must be taken into account. This interference is, in some ways, similar to the magnetic coupling of coils and will be mentioned again in connection with *antenna arrays*.

9-2 WIRE RADIATORS IN SPACE

Wire radiators are the simplest of all radiators and may be thought of as consisting of a large number of doublets connected end to end. As a result of this, their properties are similar to those of the doublet. However, it must be borne in mind that differences must exist because such antennas have finite lengths that cannot be neglected. At first, we shall consider them remote from the ground to simplify matters, just as in the preceding chapter electromagnetic waves were initially considered in free space.

9-2.1 Current and Voltage Distributions

Like a transmission line, an antenna in practice has a length that is a sizable portion of a wavelength, and sometimes even several wavelengths; it is, therefore, a circuit with distributed constants. A voltage is applied at some point, resulting in a voltage and current at that point. Traveling waves are then initiated, and possibly standing waves are set up, which means that voltage and current along the antenna generally vary from one point to the next. This antenna voltage and current distribution must have an effect on the radiated field. This field depends chiefly on the antenna length measured in wavelengths, its power losses and the terminations at its ends (if any). In addition, the thickness of the antenna wire is of importance, but for practical purposes such antennas may be assumed to be lossless and made of wire whose diameter is infinitely small when compared to a wavelength.

Figure 9-4 shows somewhat idealized voltage and current distributions along a half-wave dipole, which is the simplest practical wire antenna. One is immediately struck by their similarity to the distribution of voltage and current on a piece of quarter-

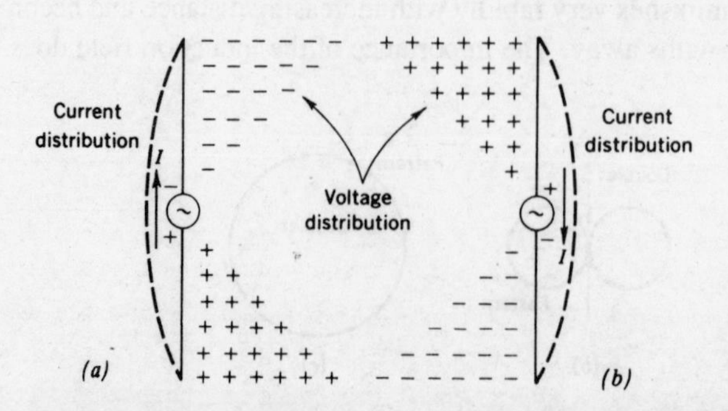

FIGURE 9-4 Voltage and current distribution on a half-wave dipole. (*a*) First half-cycle; (*b*) second half-cycle.

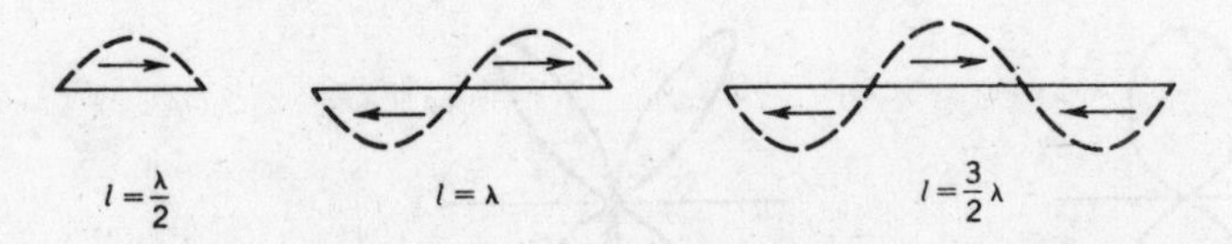

FIGURE 9-5 Current distribution on resonant dipoles.

wave transmission line open-circuited at the far end. Just as a voltage minimum and current maximum appear at the antenna feed point, so an identical situation exists $\lambda/4$ away from the open circuit on a transmission line. Again, voltage and current change polarity similarly every half-cycle, as shown here.

As indicated in Fig. 9-5, the current distributions on antennas with lengths that are multiples of $\lambda/2$ are logical extensions of those of the half-wave antenna, and similarly comparable to equivalent transmission lines. The voltage distributions are also as before and are omitted for simplicity.

9-2.2 Resonant Antennas

As might be gathered from the foregoing, a resonant antenna corresponds to a resonant transmission line, and the dipole antennas described so far have been resonant. Such an antenna can be described as an opened-out transmission line, open-circuited at the far end and of resonant length, i.e., a multiple of quarter-wavelengths so that the length of the antenna is a multiple of half-wavelengths). The reason for this last requirement is simply that the source is low-impedance and must be placed at a low-impedance point so as not to upset the standing-wave pattern. Furthermore, the nearest suitable point for this, from an open circuit, is a quarter-wavelength away.

The radiation pattern of a wire radiator in free space depends mainly on its length. For a half-wave dipole, it is like that of the elementary doublet, but somewhat flattened. The formula could be obtained by summing or by *integrating* the equation for the radiation pattern of an elementary doublet over the length of the antenna, a half-wavelength in this case; plotting would result in the pattern of Fig. 9-6*a*. The slight flattening of the pattern is due to the reinforcement, at right angles to the dipole, of the radiation in that direction from each elementary doublet. Note, once again, that the radiation pattern is a line drawn to join points in space which have equal field intensity due to this source. It is similar in meaning to an *isothermal* line drawn to join points of equal mean temperature on a weather map.

When the length of the antenna is a whole wavelength, the polarity of current on one half of the antenna is opposite to that on the other half, as shown in Fig. 9-5*b*. It is obvious, as a result, that radiation at right angles from this antenna will be zero because the field due to one half fully cancels the field due to the other half of the antenna. A direction of maximum radiation still exists, but it is no longer at right angles to the antenna; for a full-wave dipole it happens to be at 54° to the antenna. The pattern has now acquired *lobes,* and there are four in this case.

As the length of the dipole is increased to three half-wavelengths, the current distribution is changed to that of Fig. 9-5*c*. The radiation from one of the extremities of the antenna adds to that from the other, at right angles to the antenna, but both are

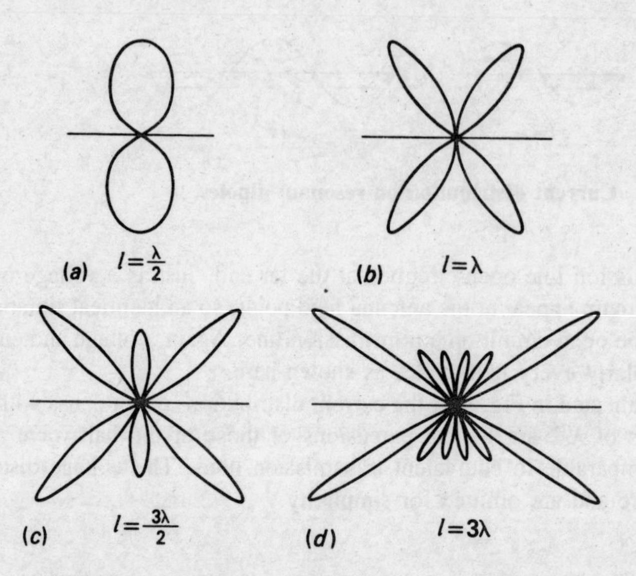

FIGURE 9-6 Radiation patterns of various resonant dipoles.

partially canceled by the radiation from the center, which carries a current of opposite polarity. There is thus radiation at right angles to the antenna (but not *maximum* radiation), and the lobe in that direction is a *minor lobe*. The direction of maximum radiation, or of the *major lobes*, is closer to the direction of the dipole itself, as shown in Fig. 9-6*c*.

If this principle is extended to a dipole of length 3 λ, it is seen that the whole process has been continued. A moment's thought shows that as the length of the resonant antenna is increased, so is the number of lobes, and the direction of the major lobes is brought closer and closer to the direction of the dipole. Continuing this trend of thought, we see that there are just as many lobes on one side of the dipole in the pattern cross section as there are sections of the antenna with currents of opposite polarity. Thus a $\frac{3}{2}$ λ antenna has three lobes on each side of the dipole itself, and a 3 λ antenna has six, the number being equal to the length of the resonant antenna expressed in half-wavelengths.

9-2.3 Nonresonant Antennas

A nonresonant antenna, like a nonresonant transmission line, is one on which there are no standing waves. These, in both cases, are suppressed by the use of a correct termination to ensure that no power is reflected, so that only a forward traveling wave will exist. In a correctly matched transmission line, all the transmitted power is dissipated in the terminating resistance. When an antenna is terminated as in Fig. 9-7*a*, about two-thirds of the forward power is radiated; the remainder is dissipated in the antenna, and none is reflected to the input.

As can be seen in Fig. 9-7*b*, the radiation pattern of the resonant antenna is similar to that of a nonresonant antenna, but they differ in one important characteristic: that of the nonresonant antenna is unidirectional. The relation between them may be

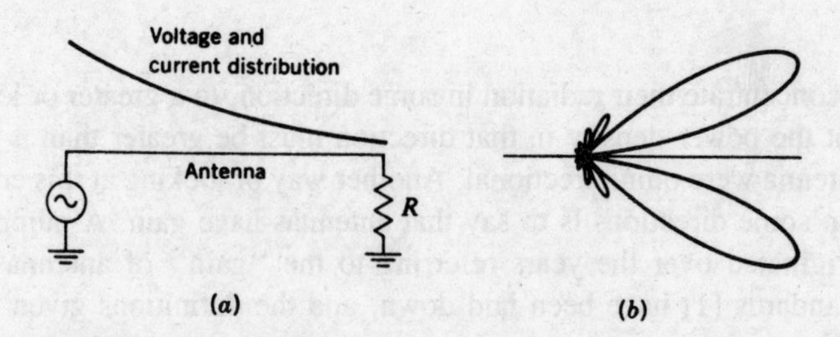

FIGURE 9-7 Nonresonant antenna. (*a*) Layout and current distribution; (*b*) radiation pattern.

deduced from standing- and traveling-wave considerations (and also confirmed mathematically). Since there are only forward traveling waves on the nonresonant antenna, its radiation pattern is, as shown, directional in the same sense as the forward traveling wave. On the other hand, standing waves exist on the resonant antenna, caused by the presence of both a reflected traveling wave and the forward wave. The radiation pattern of the resonant antenna thus consists of two parts, as shown in Fig. 9-8*a* and *b*, due to the forward and reflected waves, respectively. When the two are combined, as in Fig. 9-8*c*, the familiar bidirectional pattern results.

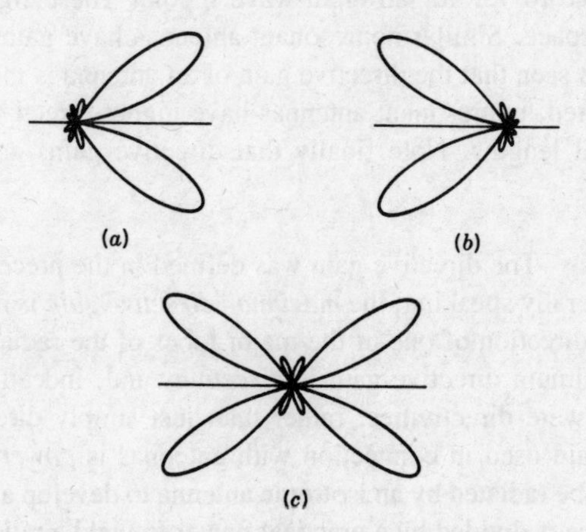

FIGURE 9-8 Synthesis of resonant antenna radiation pattern. (*a*) Due to forward wave; (*b*) due to reverse wave; (*c*) combined pattern.

9-3 TERMS AND DEFINITIONS

The preceding section showed that the radiation pattern of a wire antenna is complex, and thus some way must be found of describing and defining it. Again, something must be said about the effective resistance of antennas, their polarization and the degree to which they concentrate their radiation. Accordingly, the time has come to describe and define a number of important terms used in connection with antennas and their radiation patterns.

9-3.1 Antenna Gain

All antennas concentrate their radiation in some direction, to a greater or lesser extent. It follows that the power density in that direction must be greater than it would have been if the antenna were omnidirectional. Another way of looking at this concentration of radiation in some directions is to say that antennas have gain. A number of terms have been originated over the years referring to the "gain" of antennas. To avoid confusion, standards [1] have been laid down, and the definitions given here follow those standards.

Directive gain *Directive gain* is defined, in a particular direction, as the ratio of the power density radiated in that direction by the antenna to the power density that would be radiated by an isotropic antenna.[2] Both power densities are measured at the same distance, and both antennas radiate the same total power. Note that the directive gain is a ratio of power densities and is therefore a *power ratio*. The first step in determining the power gain of a given antenna would be the calculation or measurement of power density in the required direction, at some standard distance away. The next step would be to calculate the power density (at that distance) from an isotropic antenna radiating the same power. In the final step, the ratio of the two powers is taken. Note that the directive gain of all practical antennas is greater than unity.

The wire antennas discussed in the preceding section have gains that vary from 1.64 for a half-wave dipole to 7.1 for an eight-wave dipole. These figures are for resonant antennas in free space. Similar nonresonant antennas have gains of 3.2 and 17.4 respectively. It is thus seen that the directive gain of an antenna is increased with its length. Also, as suspected, nonresonant antennas have higher directive gains than resonant antennas of equal lengths. Note finally that directive gains are also often expressed in decibels.

Directivity and power gain The directive gain was defined in the preceding section in any direction at all. Generally speaking, the *maximum directive gain* is meant by this term, i.e., the gain in the direction of one of the major lobes of the radiation pattern. The correct name for maximum directive gain is *directivity* and, indeed, the figures quoted for wire antennas were directivities, rather than just simply directive gains.

Another form of gain used in connection with antennas is *power gain*. Once again, the power that must be radiated by an isotropic antenna to develop a certain field strength at a certain distance is divided by a practical power to yield a ratio. However, this time the practical power is that power which must be *fed* to the directive antenna to develop the same field strength at the same distance, in its direction of maximum radiation. If this definition is contrasted with the definition for directivity, only one difference is seen: for directivity the *radiated* power is considered for the directive antenna, whereas for power gain the power *fed* to the antenna is taken. Thus the two terms are identical except that power gain takes into account the antenna losses. This may be written as

[2] The isotropic antenna is a standard reference antenna, radiating equally in all directions, so that its radiation pattern is spherical. This is a very useful property and very easy to visualize, but no such antenna exists in practice.

$$A_p = \eta D \tag{9-2}$$

where A_p = power gain
D = directivity (maximum directive gain)
η = antenna efficiency = 1 for a lossless antenna

Directivity is calculated from theory, whereas power gain is of the greater practical importance. The two are almost equal for many VHF and UHF antennas, but a little more must now be said about losses in MF and LF antennas.

9-3.2 Antenna Resistance

The resistance of an antenna has two components: its radiation resistance, which it has because of the power that it converts into electromagnetic waves, and the resistance due to actual losses in the antenna. Each will now be considered.

Radiation resistance Radiation resistance is defined as the ratio of the power radiated by the antenna to the square of the current at the feed point. It is not a dc resistance, but rather an ac one, like the equivalent resistance of a parallel-tuned circuit. It is a very convenient quantity; it is part of the antenna input impedance (virtually all of it, at high frequencies), and its use greatly simplifies antenna efficiency calculations. Defining it somewhat differently, we may say that this is the resistance which, if it replaced the antenna, would dissipate precisely the same power as the antenna radiates.

Antenna losses and efficiency In addition to the energy which is radiated by an antenna, power may be dissipated as a result of antenna and ground resistance;[3] discharge or corona effects; losses in imperfect dielectrics very near to the antenna; and eddy currents induced in metallic objects within the induction field of the antenna, such as guy wires and other antennas. It is usual to represent all these by a lumped resistance R_d, the total loss resistance of the antenna. If the radiation resistance is R_r, the sum of the two is the total resistance of the antenna and also the total impedance for antennas of resonant length. The antenna efficiency then becomes

$$\eta = \frac{R_r}{R_r + R_d} \tag{9-3}$$

Low- and medium-frequency antennas are the least likely to be very efficient, because making them of resonant length often means having impossibly high structures. Even here, however, good design can ensure efficiencies of the order of 75 to 95 percent. This can be achieved by using various means (see Sec. 9-4.4) to provide as high a value of radiation resistance as possible in comparison with the loss resistance.

For short dipoles, less than a half-wavelength in *effective length*,[4] the radiation resistance is proportional to the length. It may be found from charts and tables in

[3] As can be appreciated, the ground plays an important part in the performance of antennas used for surface wave propagation. This will be described in Sec. 9-4.

[4] This is not quite the same as the actual length, for reasons discussed in Sec. 9-4.4.

antenna handbooks, measured, calculated from rather complex formulas that do not normally take the presence of the ground into consideration or found from Eq. (9-4), which holds for dipoles remote from the ground with a length l not in excess of $\lambda/8$.

$$R_r = 80\pi^2 \left(\frac{l}{\lambda}\right)^2 = 790\left(\frac{l}{\lambda}\right)^2 \qquad \textbf{(9-4)}$$

When $l/\lambda = 1/10$, Eq. (9-4) gives an accurate value of 7.9 Ω for the radiation resistance, but for a quarter-wave dipole the answer here is $R_r = 49.4\Omega$, instead of the correct value (assuming constant current) of 42 Ω.

9-3.3 Bandwidth, Beamwidth and Polarization

Bandwidth, beamwidth and *polarization* are three important terms, dealing respectively with the operating frequency range of an antenna, the degree of concentration of its radiation, and the space orientation of the waves that it radiates. They will now be introduced in turn.

Bandwidth When used in connection with antennas, the term *bandwidth* has the same meaning as in any other context. It refers to the frequency range over which operation is satisfactory and is generally taken between the half-power points. However, one complication does occur here; there are really two bandwidths, one referring to the radiation pattern and the other to the input impedance. Everything else being equal, the pattern bandwidth is equal to the difference between the frequencies at which the received power falls to one-half of maximum, in the direction of maximum radiation.

The result of all this is that when the bandwidth of an antenna is referred to, it must be qualified. Thus the criterion of "satisfactory performance" is clearly established, and it is also known which bandwidth is being referred to.

There are actually two separate requirements for large bandwidths (in excess of 10 percent) from antennas. The first is for antennas which are required to operate at a number of separate frequencies within a fairly wide frequency range. High-frequency antennas are often of this type, in which the required operation is helped by the fact that when the antenna is switched to a new frequency, compensating circuits can be switched in also. Thus matching to the feeder transmission line is maintained, with the proviso that the pattern bandwidth must not have deteriorated unduly. The other requirement, for a large operating bandwidth around a single fixed frequency, is more severe but may be solved with specially designed antennas (see Sec. 9-8).

Beamwidth The beamwidth of an antenna is the angular separation between the two half-power points on the power density radiation pattern. It is also, of course, the angular separation between the two 3-dB down points on the field strength radiation pattern of an antenna and is illustrated in Fig. 9-9. The term is used more frequently with narrow-beam antennas than with others and refers to the main lobe. Beamwidth is quoted in degrees.

Polarization As already discussed, polarization refers to the direction in space of the electric vector of the electromagnetic wave radiated from an antenna and is parallel to

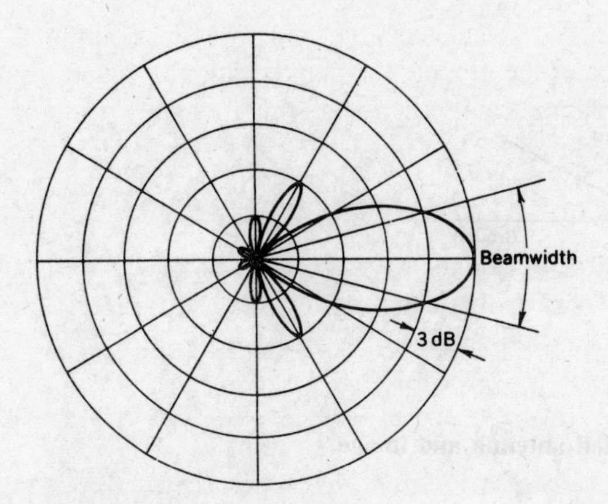

FIGURE 9-9 Beamwidth.

the antenna itself. As previously mentioned, antennas are also referred to as *vertically* or *horizontally polarized,* rather than *vertical* or *horizontal*. All VLF, LF and MF antennas, as well as many HF antennas, are made vertically polarized because of the proximity of the ground. However, there are advantages in using horizontally polarized antennas at higher frequencies, especially since most man-made noise has vertical polarization. Antennas with polarization other than linear are also sometimes used, as was mentioned in the preceding chapter; these are discussed in Sec. 9-8.

9-4 EFFECTS OF GROUND ON ANTENNAS

Since the ground is a reflecting surface, it obviously influences the radiation pattern and other properties of antennas located near it. Some of these effects have already been mentioned, but it is necessary now to delve into them a little more deeply. Because the influence of the ground depends on whether the antenna is actually grounded, or merely near the ground, these two situations will be treated separately.

9-4.1 Ungrounded Antennas

As was shown in the preceding chapter, when a source is placed near a reflecting surface, the radiation received at any distant point is the vector sum of the direct and the reflected radiations. It was also mentioned that the use of images often simplifies the situation. Here such an *image antenna* is said to exist below the ground and is a true mirror image of the actual antenna. Once the image has been established, as shown in Fig. 9-10, the resulting radiation may be treated as though it came from the antenna and its image, rather than from an antenna situated above a reflecting surface.

Matters are simplified even more if the ground can be assumed to be a perfect conductor, and therefore a perfect reflector, an assumption that is quite often justified. The currents flowing in the image now have the same magnitude as those of the actual antenna, and the overall pattern can be calculated as though caused by an *array* of two

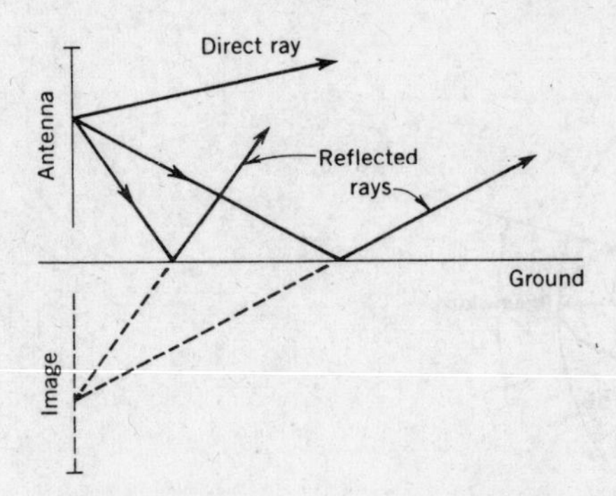

FIGURE 9-10 Ungrounded antenna and image.

nearby antennas. These are identical in length, carry currents of equal magnitudes and are separated by twice the height of the actual antenna above ground. Some typical radiation patterns are shown in Fig. 9-11.

9-4.2 Grounded Antennas

If an antenna is grounded, the earth still acts as a mirror and becomes part of the radiating system. However, whereas the ungrounded antenna with its image forms an array, the bottom of the grounded antenna is joined to the top of the image; the system acts as an antenna of double size. Thus, as shown in Fig. 9-12*a*, a grounded quarter-wave vertical radiator effectively has a quarter-wavelength added to it by its image. The voltage and current distributions on such a grounded $\lambda/4$ antenna, commonly called the *basic Marconi antenna,* are the same as those of the half-wave dipole in space and are shown in Fig. 9-12*b*.

The Marconi antenna has one important advantage over the ungrounded, or *Hertz,* antenna: to produce any given radiation pattern, it need be only half as high. On the other hand, since the ground here plays such an important role in producing the required characteristics, the ground conductivity ***must*** be good. Where it is poor, an artificial ground is used, as described in the next section.

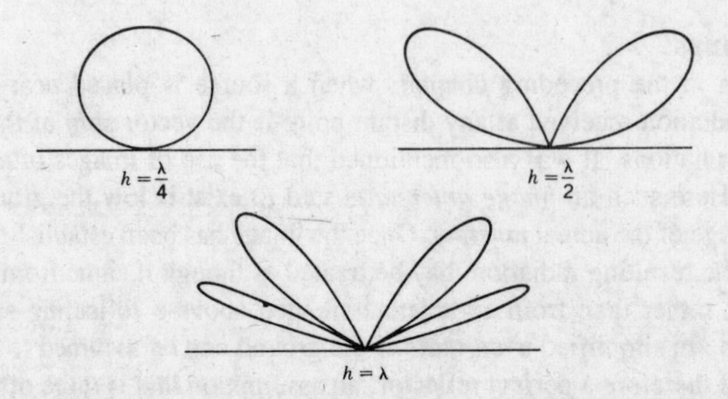

FIGURE 9-11 Radiation patterns of an ungrounded half-wave dipole located at varying heights above the ground.

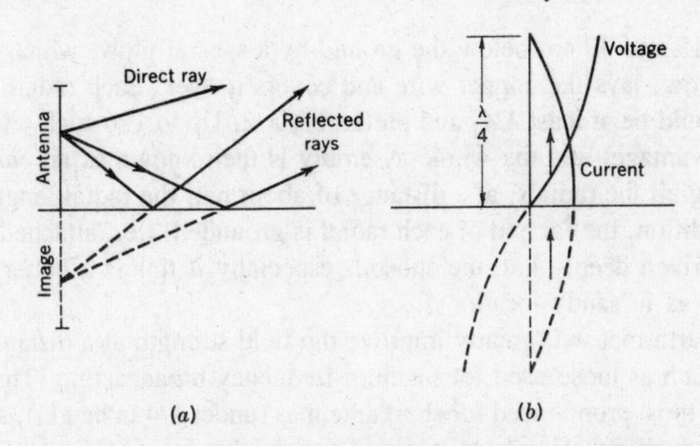

FIGURE 9-12 Grounded antennas. (*a*) Antenna and image; (*b*) voltage and current distribution on basic Marconi antenna.

The radiation pattern of a Marconi antenna depends on its height, and a selection of patterns is shown in Fig. 9-13. Each pattern is the cross section of a solid of revolution, with the antenna as the axis. It is seen that horizontal directivity improves with height up to a certain point, after which the pattern "lifts off" the ground. What has happened, of course, is that cancellation in the horizontal direction has occurred because of opposing currents in the various parts of an antenna of this effective height.

9-4.3 Grounding Systems

The earth has generally been assumed to be a perfect conductor so far, but it often stops short of this ideal. It is for this reason that the best ground system for a vertical grounded radiator is a network of buried wires directly under the antenna. This network consists of a large number of "radials" extending from the base of the tower and

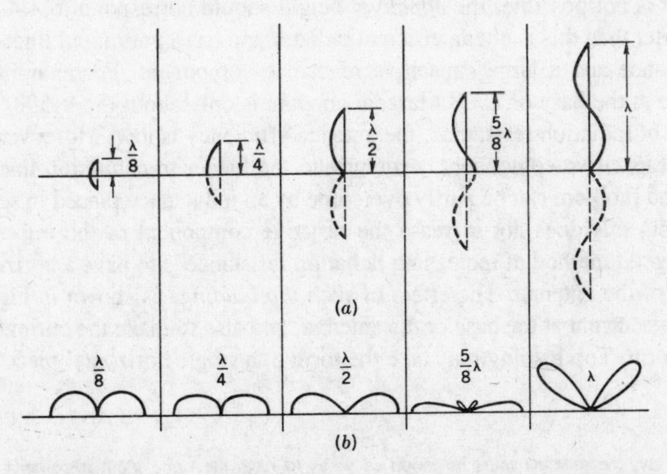

FIGURE 9-13 Characteristics of vertical grounded antennas. (*a*) Heights and current distributions; (*b*) radiation patterns.

placed between 15 and 30 cm below the ground by a special plow, which simultaneously digs a furrow, lays the copper wire and covers it over. Each radial wire has a length which should be at least $\lambda/4$, and preferably $\lambda/2$. Up to 120 such wires may be used to good advantage, and the whole assembly is then known as an *earth mat*. A conductor joining all the radials, at a distance of about half the radial length, is often employed. In addition, the far end of each radial is grounded, i.e., attached to a metal stake which is driven deeply into the subsoil (especially if this is a better conductor than the topsoil, as in sandy locations).

A good earth mat will greatly improve the field strength at a distance of Marconi antennas, such as those used for medium-frequency broadcasting. The improvement is naturally most pronounced for short antennas (under $\lambda/4$ in height), and/or with soils of poor conductivity. However, even an antenna between $\lambda/4$ and $\lambda/2$, standing on soil with good conductivity, will have its radiation pattern improved noticeably.

It sometimes happens that the ground itself is of low conductivity, and yet a ground mat cannot be laid; this arises most often in rocky areas or with antennas on top of tall buildings. In such situations a counterpoise is used, being a small edition of the earth mat, but this time above the surface level. It again consists of a system of radials, this time supported above ground and insulated from it. The supports should be few and far between, and made of a material with low dielectric losses; wood, for example, should never be used.

9-4.4 Effects of Antenna Height

At low and medium frequencies, where wavelengths are large, it often becomes impracticable to use an antenna of resonant length, as has already been mentioned. The vertical antennas used at those frequencies are, therefore, too short electrically. This has several interesting results, which are now explored.

Top loading The actual antenna height should be at least a quarter-wavelength, but where this is not possible, the *effective*[5] height should correspond to $\lambda/4$. An antenna much shorter than this is not an efficient radiator and has a poor input impedance with a low resistance and a large capacitive reactance component. For example, the input impedance at the base of a $\lambda/8$ Marconi antenna is only about $(8 - j500)\ \Omega$; with this low value of radiation resistance, the antenna efficiency is low. Moreover, because of the large capacitive component, matching to the feeder transmission line is difficult. This second problem can be partly overcome by an inductance placed in series with the antenna, but this does not increase the resistive component of the impedance.

A good method of increasing radiation resistance is to have a horizontal portion at the top of the antenna. The effect of such *top loading*, as shown in Fig. 9-14, is to increase the current at the base of the antenna, and also to make the current distribution more uniform. Top loading may take the form of a single horizontal piece, resulting in

[5]That is to say, the antenna must be modified so as to have the same input impedance and horizontal radiation field as a vertical wire of greater height, which then becomes the antenna's effective height.

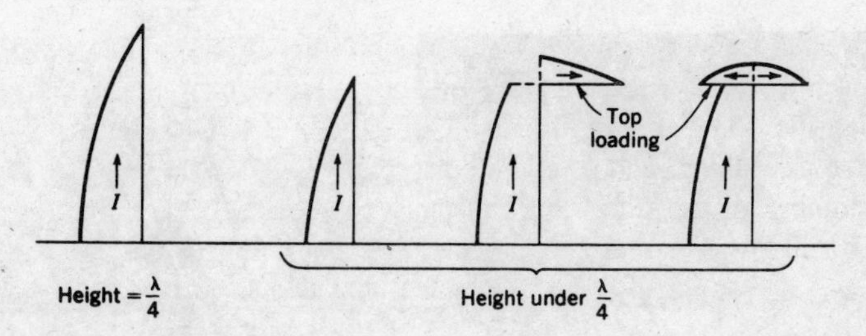

FIGURE 9-14 Top loading.

the inverted-L and T antennas of Fig. 9-14. It may also take the form of a "top hat," as shown in Fig. 9-15. The top hat also has the effect of adding capacitance in series with the antenna, thus reducing its total capacitive input reactance.

The radiation pattern of a top-loaded antenna is much the same as that of the basic Marconi, because the current distribution is also much the same, as shown in Fig. 9-14. Since the current in the horizontal portion is much smaller than in the vertical part, the antenna is still considered as a vertically polarized radiator. More often than not, the decision as to what type of top load to use and how much of it to have is dictated by the facilities available and costs, rather than by optimum design factors. One might add, incidentally, that design in this context often means inspired guesswork, especially in the case of top-loaded tapering towers.

Optimum length It is true to say for VLF and LF antennas that they should be as tall as possible, in view of the enormous wavelengths involved. When considering MF antennas, however, one should note that there is such a thing as an antenna that is too tall. A glance at Fig. 9-13 reveals this. An antenna whose height is a wavelength is useless for ground-wave propagation, because it radiates nothing along the ground. Accordingly, an optimum height must exist somewhere between "too short" and a full wavelength. A further check of Fig. 9-13 reveals that the horizontal field strength increases with height, up to about $\frac{5}{8}\lambda$. Unfortunately, when the height of the antenna exceeds $\lambda/2$, other lobes are formed. Depending on their strength and angle, their presence will cause quite objectionable sky-wave interference. This holds true for all vertical radiators taller than about 0.53λ, so that this height is not exceeded in practice for antennas used in ground-wave propagation.

Effective length The term *effective electrical length* has been used on a number of occasions and must now be explained. It refers to the fact that antennas behave as though (electrically) they were taller than their physical height. The first reason for this is the effect of top loading. The second reason is generally called *end effects,* the result of physical antennas having finite thickness, instead of being infinitely thin. In consequence, the propagation velocity within the antenna is some 2 to 8 percent less than in free space, so that the wavelength within the antenna is shorter by the same amount.

FIGURE 9-15 Antenna mast with "top hat." *(Courtesy of the Australian Telecommunications Commission.)*

The antenna thus appears longer than if wavelength had been calculated on the basis of velocity in free space. Finally, if the cross section of the antenna is nonuniform, as in tapered towers, this last situation is further complicated.

For all the preceding reasons, it is standard procedure to build these antennas slightly taller than needed and then to trim them down to size. This procedure is generally more fruitful than effective length calculation from first principles or from charts available in antenna handbooks.

9-5 ANTENNA COUPLING AT MEDIUM FREQUENCIES

Low- and medium-frequency antennas are the ones least likely to be of resonant effective height and are therefore the least likely to have purely resistive input impedances. This precludes the connection of such an antenna directly, or via transmission line, to the output tank circuit of a transmitter; some sort of matching network will have to be interposed.

9-5.1 General Considerations

A *coupling network,* or *antenna coupler,* is a network composed of reactances and transformers, which may be lumped or distributed. The coupling network is said to provide *impedance matching* and is employed for any or all of the following reasons:

1. To tune out the reactive component of the antenna impedance, making the impedance look resistive to the transmitter; otherwise detuning will take place when the antenna is connected. This function involves the provision of variable reactances.
2. To provide the transmitter (and also transmission line, if used) with the correct value of load resistance. This involves having one or more adjustable transformers.
3. To prevent the illegal radiation of spurious frequencies from the system as a whole. This function requires the presence of filtering, generally low-pass, since the spurious frequencies are most likely to be harmonics of the transmitter's frequency.

It should be noted that whereas the first two functions apply to low- and medium-frequency transmitters more than to other types, the last requirement applies equally at all frequencies. One other consideration sometimes applies, specifically to transmitters in which the output tank is series-fed and single-tuned. Here the antenna coupler must also prevent the dc supply from reaching the antenna. If this is not done, two serious problems will arise: antenna insulation difficulties and danger to operators. The danger will be caused by the fact that, whereas RF burns are serious and painful, those coming from the dc high-voltage supply to the power amplifier are *lethal* more often than not.

9-5.2 Selection of Feed Point

The half-wave dipole antennas presented so far have always been drawn with the feeding generator connected to the center. Although many practical antennas are fed in this way, the arrangement is by no means essential. The point at which a particular antenna is fed is determined by several considerations, of which perhaps the most important is the antenna impedance. This, as has been shown, varies from point to point along the antenna, so that some investigation of the options is necessary.

Voltage and current feed When a dipole has an effective length that is resonant, the impedance at its center will be purely resistive. This impedance will be high if there is a current node at the center, as with a full-wavelength antenna, or low if there is a voltage node at the center, as with a half-wave dipole. According to common jargon, an antenna is said to be *current-fed* if it is fed at a point of current maximum; thus a center-fed half-wave dipole or Marconi antenna is current-fed. Conversely, a center-fed full-wave antenna is said to be *voltage-fed.*

Both these terms are rather loose and tend to lose their meaning if an antenna is fed at some intermediate point. The definition of current feed has been extended to include all feed-point impedances below 600 Ω, and voltage feed for impedances in excess of 600 Ω. Even so, it would be better still to refer to *low-impedance feed* and *high-impedance feed.*

Feed-point impedance As has been shown, the current is maximum in the center and zero at the ends of a half-wave dipole in space, or a grounded quarter-wave Marconi, whereas the voltage is just the reverse. In a practical antenna the voltage or current values will be low (not zero) so that the antenna impedance will be finite at those points. Thus we have several thousands of ohms at the ends, and 72 Ω in the center, both values purely resistive. As a result, broadcast antennas are often center-fed in practice, 72 Ω being a useful impedance from the point of view of transmission lines. It is for this reason that antennas, although called *grounded,* are often in fact insulated from the ground electrically. Thus, the base of the antenna stands on an insulator close to the ground and is fed between base and ground, i.e., at the center of the antenna-image system.

9-5.3 Antenna Couplers

Although all antenna couplers must fulfill the three requirements outlined, there are still individual differences among them, governed by how each antenna is fed. This, in turn, depends on whether a transmission line is used, whether it is balanced or unbalanced and what value of standing-wave ratio is caused by the antenna.

Directly fed antennas These antennas are coupled to their transmitters without transmission lines, generally for lack of space. To be of use, a line connecting an antenna to its transmitter ought to be at least a half-wave in length, and at least the first quarter-wave portion of it should come away at right angles to the antenna. This may be difficult to accomplish, especially at low frequencies, for shipboard transmitters or those on tops of buildings.

Figure 9-16*a* shows the simplest method of direct coupling. The impedance seen by the tank circuit is adjusted by moving coil L_1, or by changing the number of turns with a traveling short circuit. To tune out the antenna reactance, either C_1 or L_1 is shorted out, and the other component is adjusted to suit. This is the simplest coupling network, but by no means the best, especially since it does not noticeably attenuate harmonics.

The pi (π) coupler of Fig. 9-16*b* is a much better proposition. It affords a wider reactance range and is also a low-pass filter, giving adequate harmonic suppression. However, even it will not provide satisfactory coupling if the antenna is very short, thus having a mostly capacitive input impedance. It is better, under those conditions, to increase the height of the antenna.

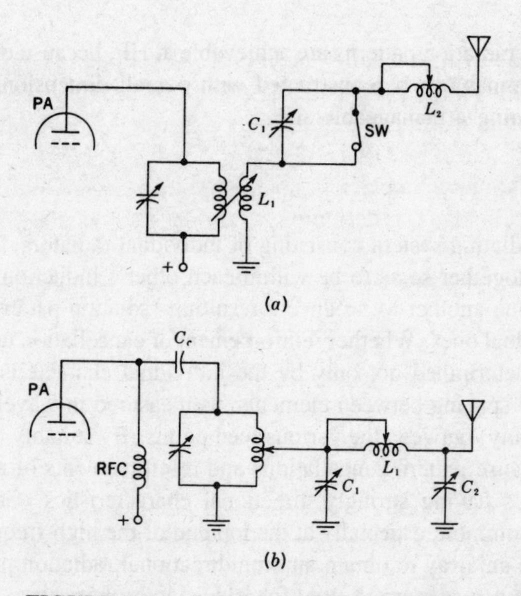

FIGURE 9-16 Antenna coupling. (*a*) Direct coupler; (*b*) π coupler.

Coupling with a transmission line The requirements are similar to those already discussed. However, balanced lines, and therefore balanced coupling networks, are often used, as shown in Fig. 9-17. The output tank is tuned accordingly, and facilities must be provided to ensure that the two legs of the coupler can be kept balanced. At higher frequencies distributed components such as quarter-wave transformers and stubs can be used.

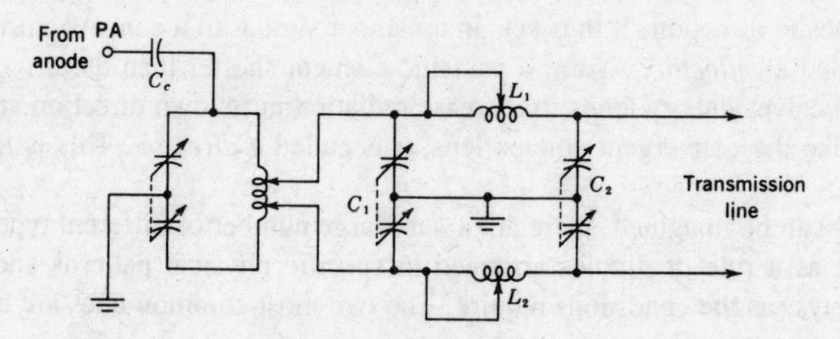

FIGURE 9-17 Symmetrical π coupler.

9-6 DIRECTIONAL HIGH-FREQUENCY ANTENNAS

HF antennas are likely to differ from lower-frequency ones for two complementary reasons. These are essentially the HF transmission/reception requirements and the ability to meet them. Since much of HF communication is likely to be point-to-point, the requirement is for fairly concentrated beams instead of omnidirectional radiation.

By the same token, such radiation patterns are achievable at HF, because of the shorter wavelengths. Thus, antennas can be constructed with overall dimensions of several wavelengths while retaining a manageable size.

9-6.1 Dipole Arrays

An antenna array is a radiating system consisting of individual radiators, or elements. These are placed close together so as to be within each other's induction field. They therefore interact with one another to produce a resulting radiation pattern that is the vector sum of the individual ones. Whether reinforcement or cancellation takes place in any given direction is determined not only by the individual characteristics of each element, but also by the spacing between elements, as measured in wavelengths, and the phase difference (if any) between the various feed points. By suitably arranging an array, it is possible to cause pattern cancellations and reinforcements of a nature that will result in the array's having strongly directional characteristics. Gains well in excess of 50 are not uncommon, especially at the top end of the high-frequency band. It is also possible to use an array to obtain an omnidirectional radiation pattern in the horizontal plane, as with *turnstile* arrays used for television broadcasting. However, it is generally true to say that HF arrays are more likely to be used to obtain directional behavior rather than to create omnidirectional patterns.

Parasitic elements It is not necessary for all the elements of an array to be connected to the output of the transmitter, although this does, in fact, happen in quite a number of arrays. An element so connected is called a *driven* element, whereas a radiator not directly connected is called a *parasitic* element. Such a parasitic radiator receives energy through the induction field of a driven element, rather than by a direct connection to the transmission line. As a generalization, a parasitic element longer than the driven one and close to it reduces signal strength in its own direction and increases it in the opposite direction. It thus acts in a manner similar to a concave mirror in optics and is called a *reflector*. Again, a parasitic element shorter than the driven one from which it receives energy tends to increase radiation in its own direction and therefore behaves like the convergent convex lens; it is called a *director*. This is illustrated in Fig. 9-21.

As can be imagined, there are a very large number of different types of arrays, consisting as a rule of dipoles arranged in specific physical patterns and excited in various ways, as the conditions require. The two most common ones are now treated.

Broadside array Possibly the simplest array consists of a number of dipoles of equal size, equally spaced along a straight line (i.e., *collinear*), with all dipoles fed in the same phase from the same source. Such an arrangement is called a *broadside array* and is shown in Fig. 9-18, together with the resulting pattern.

As indicated, the broadside array is strongly directional at right angles to the plane of the array, while radiating very little in the plane; the name comes from the naval term *broadside*. If some point is considered along the line perpendicular to the plane of the array, it is seen that this distant point is virtually equidistant from all the dipoles forming the array. Thus the individual radiations, already quite strong in that

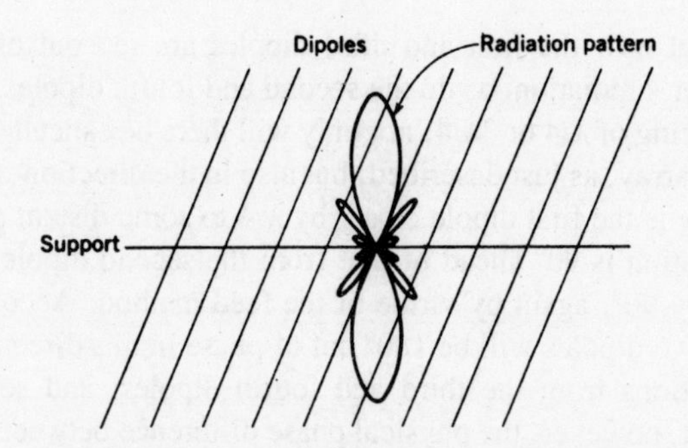

FIGURE 9-18 Broadside array and pattern.

direction, are reinforced. In the direction of the plane, however, there is little radiation, because the dipoles do not radiate in the direction in which they point, and because of cancellation in the direction of the line joining the center. This happens because any distant point along that line is no longer equidistant from all the dipoles, which will therefore cancel each other's radiation in that direction (all the more so if their separation is $\lambda/2$, which it very often is).

Typical antenna lengths in the broadside array are from 2 to 10 wavelengths, typical spacings are $\lambda/2$ or λ, and dozens of elements may be used in the one array. Note that any array that is directional at right angles to the plane of the array is said, by inference, to have *broadside action*.

End-fire array The physical arrangement of the *end-fire array* is the same as that of the broadside array. However, although the magnitude of the current in each element is still the same as in every other element, there is now a phase difference between these currents. This is progressive from left to right in Fig. 9-19, as there is a phase lag between the succeeding elements equal in hertz to their spacing in wavelengths. The pattern of the end-fire array, as shown, is quite different from that of the broadside array. It is in the plane of the array, not at right angles to it, and is unidirectional rather than bidirectional. Note that any array with that pattern arrangement is said to have *end-fire action*.

There is no radiation at right angles to the plane of the array because of cancellation. A point along the line perpendicular to the plane of the array is still equidistant

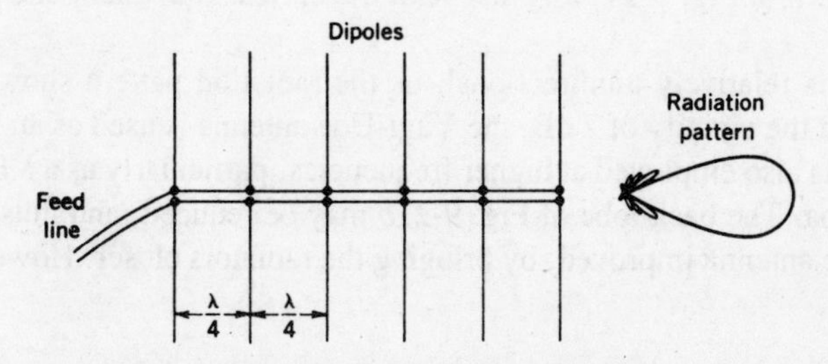

FIGURE 9-19 End-fire array and pattern.

from all the elements, but now the first and third dipoles are fed out of phase and therefore cancel each other's radiation, as do the second and fourth dipoles, and so on. With the usual dipole spacing of $\lambda/4$ or $3\lambda/4$, not only will there be cancellation at right angles to the plane of the array, as just described, but also in the direction from right to left in Fig. 9-19. Not only is the first dipole closer by $\lambda/4$ to some distant point in that direction (so that its radiation is 90° ahead of that from the second dipole) but it also leads the second dipole by 90°, again by virtue of the feed method. Accordingly, the radiations from the first two dipoles will be 180° out of phase in this direction and will cancel, as will the radiations from the third and fourth dipoles, and so on. In the direction from left to right, however, the physical phase difference between the dipoles is made up by the phase difference in feeding. Therefore addition takes place, resulting in strong unidirectional radiation.

Both the end-fire and broadside arrays are called *linear,* and both are resonant since they consist of resonant elements. As such, both arrays have a narrow bandwidth, which makes each of them particularly suitable for single-frequency transmission, but not so useful for reception, where the requirement is generally the ability to receive over a reasonable frequency range.

9-6.2 Folded Dipole and Applications

As shown in Fig. 9-20, the folded dipole is a single antenna, but it consists of two elements. The first is fed directly while the second is coupled conductively at the ends. The radiation pattern of the folded dipole is the same as that of a straight dipole, but its input impedance is greater. This may be shown by noting, as seen in Fig. 9-20, that if the total current fed in is I and the two arms have equal diameters, then the current in each arm is $I/2$. If this had been a straight dipole, the total would have flowed in the first (and only) arm. Now with the same power applied, only half the current flows in the first arm, and thus the input impedance is four times that of the straight dipole. Hence $R_r = 4 \times 72 = 288\ \Omega$ for a half-wave folded dipole with equal diameter arms.

If elements of unequal diameters are used, transformation ratios from 1.5 to 25 are practicable, and if greater ratios are required, more arms can be used. Although the folded dipole has the same radiation pattern as the ordinary dipole, it has two advantages: its higher input impedance and its greater bandwidth (as explained in Sec. 9-8).

The Yagi-Uda antenna[6] A Yagi-Uda antenna is an array consisting of a driven element and one or more parasitic elements. They are arranged collinearly and close together, as shown in Fig. 9-21, together with the optical equivalent and the radiation pattern.

Since it is relatively unidirectional, as the radiation pattern shows, and has a moderate gain in the vicinity of 7 dB, the Yagi-Uda antenna is used as an HF transmitting antenna. It is also employed at higher frequencies, particularly as a VHF television receiving antenna. The back lobe of Fig. 9-21*b* may be reduced, and thus the *front-to-back ratio* of the antenna improved, by bringing the radiators closer. However, this has

[6] More often, but less correctly, known as the *Yagi*. Invented by Prof. Uda and first described in English by Hidetsugu Yagi [2].

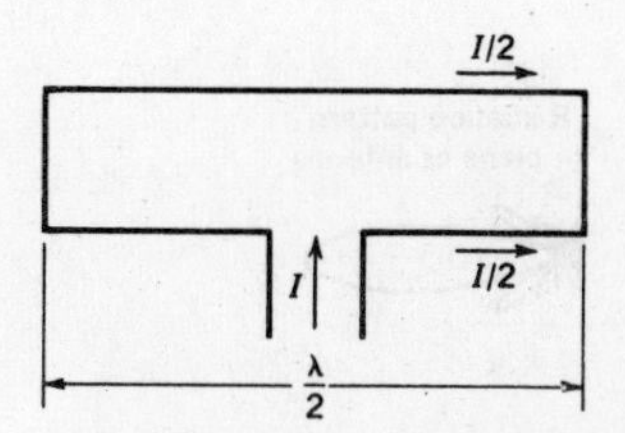

FIGURE 9-20 Folded dipole.

the adverse effect of lowering the input impedance of the array, so that the separation shown, 0.1λ, is an optimum value.

The precise effect of the parasitic element depends on its distance and tuning, i.e., on the magnitude and phase of the current induced in it. As already mentioned, a parasitic element resonant at a lower frequency than the driven element (i.e., longer) will act as a mild reflector, and a shorter parasitic will act as a mild "concentrator" of radiation. As a parasitic element is brought closer to the driven element, then regardless of its precise length, it will load the driven element more and therefore reduce its input impedance. This is perhaps the main reason for the almost invariable use of a folded dipole as the driven element of such an array.

The Yagi-Uda antenna admittedly does not have high gain, but it is very compact, relatively broadband because of the folded dipole used and has quite a good unidirectional radiation pattern. As used in practice, it has one reflector and several directors which are either of equal length or decreasing slightly away from the driven element. Finally, it must be mentioned that the folded dipole, along with one or two other antennas, is sometimes called a *supergain antenna,* because of its good gain and beamwidth per unit area of array.

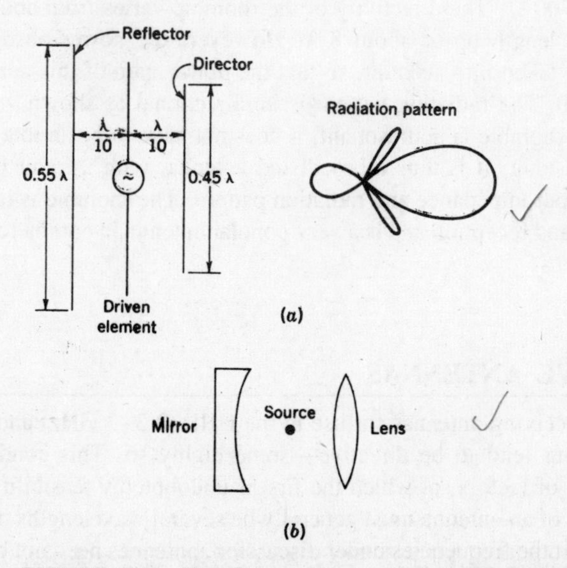

FIGURE 9-21 Yagi-Uda antenna. (*a*) Antenna and pattern; (*b*) optical equivalent.

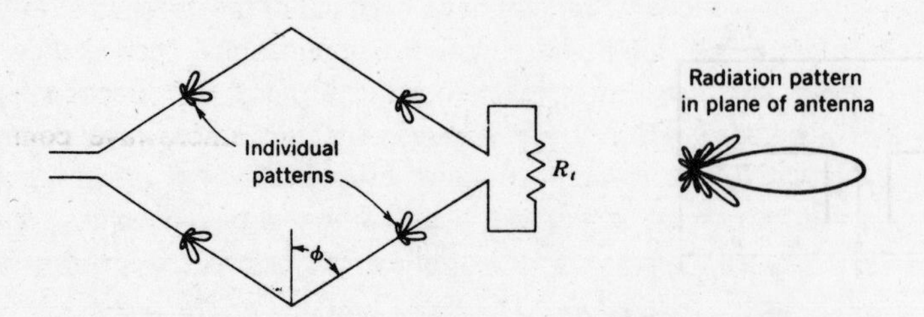

FIGURE 9-22 Rhombic antenna and radiation patterns.

9-6.3 Nonresonant Antennas—The Rhombic

A major requirement for HF is the need for a multiband antenna capable of operating satisfactorily over most or all of the 3- to 30-MHz range, for either reception or transmission. One of the obvious solutions is to employ an array of nonresonant antennas, whose characteristics will not change too drastically over this frequency range.

A very interesting and widely used antenna array, especially for point-to-point working, is shown in Fig. 9-22. This is the *rhombic antenna,* which consists of nonresonant elements arranged differently from any previous arrays. It is a planar rhombus which may be thought of as a piece of parallel-wire transmission line pinched out in the middle. The lengths of the (equal) radiators vary from 2 to 8 λ, and the angle of tilt, ϕ, varies from 40 to 75°, being closely related to the leg length.

The four legs are considered as nonresonant antennas. This is achieved by treating the two sets as a transmission line correctly terminated in its characteristic impedance at the far end; thus only forward waves are present. Since the termination absorbs some power, the rhombic antenna must be terminated by a resistor which, for transmission, is capable of absorbing about one-third of the power fed to the antenna. The terminating resistance is often in the vicinity of 800 Ω and the input impedance varies from 650 to 700 Ω. The directivity of the rhombic varies from about 20 to 90, increasing with leg length up to about 8 λ. However, the power absorbed by the termination must be taken into account, so that the *power* gain of this antenna ranges from about 15 to 60. The radiation pattern is unidirectional as shown.

Because the rhombic is nonresonant, it does not have to be an integral number of half-wavelengths long. It is thus a broadband antenna, with a frequency range at least 4:1 for both input impedance and radiation pattern. The rhombic is ideally suited to HF transmission and reception and is a very popular antenna in commercial point-to-point communications.

9-7 UHF AND MICROWAVE ANTENNAS

Transmitting and receiving antennas for use in the UHF (0.3–3 GHz) and microwave (1–100 GHz) regions tend to be directive—some highly so. This condition results from a combination of factors, of which the first is undoubtedly feasibility. That is to say, the dimensions of an antenna must generally be several wavelengths in order for it to have high gain. At the frequencies under discussion, antennas need not be physically large to have multiple-wavelength dimensions, and consequently several arrangements and concepts are possible which might have been out of the question at lower frequencies. Again, a number of UHF and microwave applications, such as radar, are in the direction-finding and measuring field, so that the need for directional antennas is self-evident. Similarly, several applications, such as microwave communications links, are essentially point-to-point services, often in areas in which interference between various services must be avoided; the use of directional antennas greatly helps in this regard. Finally, as frequencies are raised, so, ultimately, the performance of active devices deteriorates. That is to say, the maximum achievable power from output devices falls off, whereas the noise of receiving devices increases; it can be seen that having high-gain (and therefore directional) antennas helps greatly to overcome these problems.

In many ways, the VHF region, spanning the 30–300 MHz frequency range, is an "overlap" region. Thus, some of the HF techniques so far discussed can be extended into the VHF region, and some of the UHF and microwave antennas about to be discussed can also be used at VHF. Finally, it should be noted that the majority of antennas discussed in Sec. 9-8 are VHF antennas. Interestingly, one of the most commonly seen VHF antennas used around the world is the Yagi-Uda, most often used as a TV receiving antenna.

9-7.1 Antennas with Parabolic Reflectors

The parabola is a plane curve, defined as the locus of a point which moves so that its distance from another point (called the *focus*) plus its distance from a straight line (*directrix*) is constant. These geometric properties yield an excellent microwave or light reflector, as will be seen.

Geometry of the parabola Figure 9-23 shows a parabola *CAD* whose focus is at *F* and whose axis is *AB*. It follows from the definition of the parabola that

$$FP + PP' = FQ + QQ' = FR + RR' = k \tag{9-5}$$

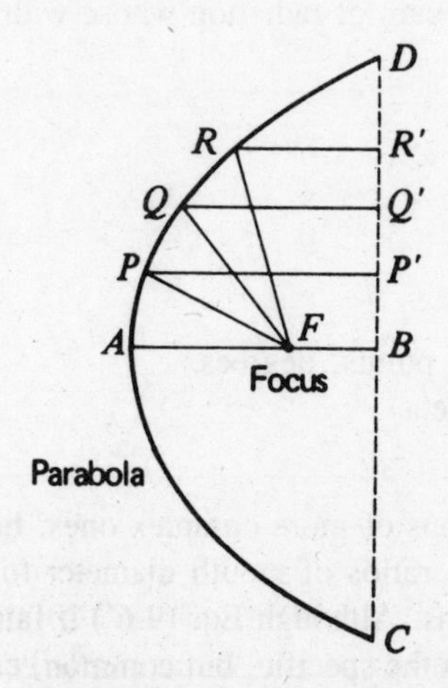

FIGURE 9-23 Geometry of the parabola.

where k = a constant, which may be changed if a different shape of parabola is required

AF = focal length of the parabola

Note that the ratio of the focal length to the mouth diameter (AF/CD) is called the *aperture* of the parabola, just as in camera lenses.

Consider a source of radiation placed at the focus. All waves coming from the source and reflected by the parabola will have traveled the same distance by the time they reach the directrix, no matter from what point on the parabola they are reflected. *All such waves will thus be in phase.* As a result, radiation is very strong and concentrated along the AB axis, but cancellation will take place in any other direction, because of path-length differences. Thus the parabola is seen to have properties that lead to the production of concentrated beams of radiation.

A practical reflector employing the properties of the parabola will be a three-dimensional surface, obtained by revolving the parabola about the axis AB. The resulting geometric surface is the *paraboloid,* often called a *parabolic reflector* or *microwave dish.* When it is used for reception, exactly the same behavior is manifested, so that this is also a high-gain receiving directional antenna reflector. Such behavior is, of course, predicted by the *principle of reciprocity,* which states that the properties of an antenna are independent of whether it is used for transmission or reception. The reflector is directional for reception because only the rays arriving from the BA direction, i.e., normal to the directrix, are brought together at the focus. On the other hand, rays from any other direction are canceled at that point, again owing to path-length differences. The reflector provides a high gain because, like the mirror of a reflecting telescope, it collects radiation from a large area and concentrates it all at the focal point.

Properties of paraboloid reflectors The directional pattern of an antenna using a paraboloid reflector has a very sharp main lobe, surrounded by a number of minor lobes which are much smaller. The three-dimensional shape of the main lobe is like that of a fat cigar, in the direction AB. If the *primary,* or *feed,* antenna is nondirectional, then the paraboloid will produce a beam of radiation whose width is given by the formulas

$$\phi = \frac{70\lambda}{D} \tag{9-6}$$

$$\phi_0 = 2\phi \tag{9-6'}$$

where λ = wavelength, m

ϕ = beamwidth between half-power points, degrees

ϕ_0 = beamwidth between nulls, degrees

D = mouth diameter, m

Both equations are simplified versions of more complex ones, but they apply accurately to large apertures, that is, large ratios of mouth diameter to wavelength. They are thus accurate for small beamwidths. Although Eq. (9-6′) if fairly universal, Eq. (9-6) contains a restriction. It applies in the specific, but common, case of illumination which falls away uniformly from the center to the edges of the paraboloid reflector. This decrease away from the center is such that power density at the edges of the reflector is 10 dB down on the power density at its center. There are two reasons for such a decrease in illumination: (1) No primary antenna can be truly isotropic, so that some reduction in power density at the edges must be accepted. (2) Such a uniform decrease in illumination has the beneficial effect of reducing the strength of minor lobes. Note, however, that the whole area of the reflector is illuminated, despite the decrease toward the edges. After all, if only half the area of the reflector were illuminated, the reflector might as well have been only half the size in the first place.

Example 9-1 Calculate the beamwidth between nulls of a 2-m paraboloid reflector used at 6 GHz. *Note:* Such reflectors are often used at that frequency as antennas in outside broadcast television microwave links.

$$\phi_0 = 2 \times \frac{70\lambda}{D} = 140 \times \frac{0.05}{2}$$
$$= 3.5°$$

The gain of an antenna using a paraboloid reflector is influenced by the aperture ratio (D/λ) and the uniformity (or otherwise) of the illumination. If the antenna is lossless, and its illumination falls away to the edges as previously discussed, then the power gain, as a good approximation, is given by

$$A_p = 6\left(\frac{D}{\lambda}\right)^2 \qquad \textbf{(9-7)}$$

where A_p = directivity (with respect to isotropic antenna)
= power gain if antenna is lossless
D = mouth diameter of reflector, m

It will be seen later in this section how this relationship is derived from a more fundamental one. Meanwhile, it is worth pointing out that, by coincidence, the power gain of an antenna with a uniformly illuminated paraboloid, *with respect to a half-wave dipole,* is given by a formula approximately the same as Eq. (9-7).

Example 9-2 Calculate the gain of the antenna of Example 9-1.

$$A_p = 6\left(\frac{D}{\lambda}\right)^2 = 6\left(\frac{200}{5}\right)^2 = 9600$$

Example 9-2 shows that the *effective radiated power* (ERP) of such an antenna would be 9600 W if the actual power fed to the primary antenna were 1 W. The ERP is the product of power fed to the antenna and its power gain. It is seen that very large gains and narrow beamwidths are obtainable with paraboloid reflectors—excessive size is the reason why they are not used at lower frequencies, such as the VHF region occupied by television broadcasting. In order to be fully effective and useful, a paraboloid must have a mouth diameter of at least 10 λ, as will be seen; at the lower end of the television band, at 63 MHz, this diameter would need to be at least 48 m. On the

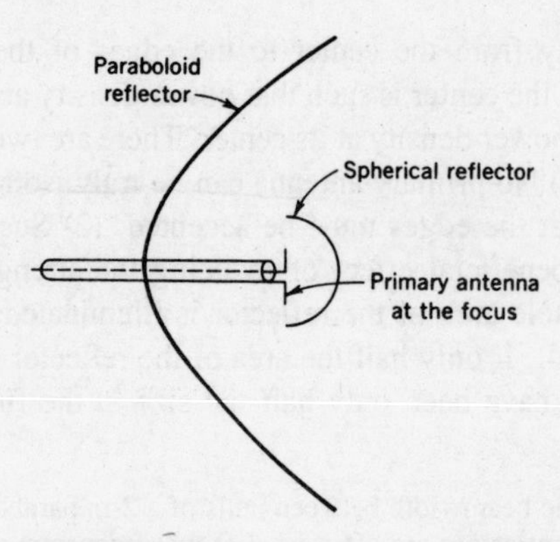

FIGURE 9-24 Center-fed paraboloid reflector with spherical shell.

other hand, of course, these figures illustrate the relative ease of obtaining high directive gains from practical microwave antennas.

Feed mechanisms As already discussed, the primary antenna is placed at the focus of the paraboloid for best results in transmission or reception. However, the direct radiation from the feed, which is not reflected by the paraboloid, tends to spread out in all directions and hence partially spoils the directivity. Several methods are used to prevent this, one of them being the provision of a small spherical reflector, as shown in Fig. 9-24, to redirect all such radiation back to the paraboloid. Another method is to use a small dipole array at the focus, such as a Yagi-Uda or an end-fire array, *pointing at the paraboloid reflector*.

Figure 9-25 shows yet another way of dealing with the problem: a *horn antenna* (to be discussed in Sec. 9-7.2) pointing at the main reflector. It has a mildly directional pattern, as will be seen, in the direction in which its mouth points; thus direct radiation from the feed antenna is once again avoided. It should be mentioned at this point that, although the feed antenna and its reflector obstruct a certain amount of reflection from the paraboloid when they are placed at its focus, this obstruction is slight indeed. For example, if a 30-cm-diameter reflector is placed at the center of a 3-m dish, simple arithmetic shows that the area obstructed is only 1 percent of the total. Similar reasoning is applied to the horn primary, which obscures an equally small proportion of the total area. Note, in conjunction with Fig. 9-25, that the actual horn is not shown here, but the bolt-holes in the waveguide flange indicate where it would be fitted.

Another feed method, the *Cassegrain feed,* is named after an early-eighteenth-century astronomer and is adopted directly from astronomical reflecting telescopes; it is illustrated in Figs. 9-26 and 9-27. It uses a hyperboloid secondary reflector, as shown. One of its foci coincides with the focus of the paraboloid, resulting in the action shown (for transmission) in Fig. 9-26. The rays emitted from the feed horn antenna are reflected from the paraboloid mirror, the effect on the main paraboloid reflector being the same as that of a feed antenna at the focus. The main reflector then *collimates* (renders parallel) the rays in the usual manner.

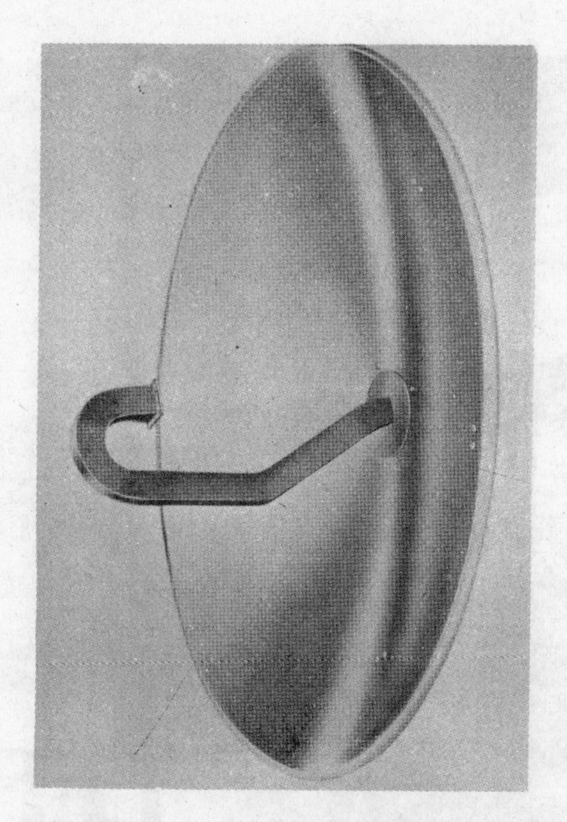

FIGURE 9-25 Paraboloid reflector with horn feed. *(Courtesy of Andrew Antennas of Australia.)*

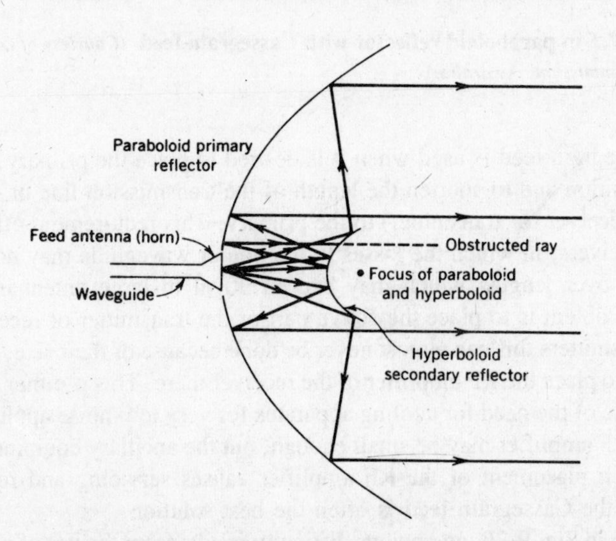

FIGURE 9-26 Geometry of the Cassegrain feed.

FIGURE 9-27 27.5-m paraboloid reflector with Cassegrain feed. *(Courtesy of Overseas Telecommunications Commission, Australia.)*

The Cassegrain feed is used when it is desired to place the primary antenna in a convenient position and to shorten the length of the transmission line or *waveguide* connecting the receiver (or transmitter) to the primary. This requirement often applies to low-noise receivers, in which the losses in the line or waveguide may not be tolerated, especially over lengths which may exceed 30 m in large antennas. Another solution to the problem is to place the active part of the transmitter or receiver at the focus. With transmitters this can almost never be done because of their size, and it may also be difficult to place the RF amplifier of the receiver there. This is either because of its size or because of the need for cooling apparatus for very low-noise applications (in which case the RF amplifier may be small enough, but the ancillary equipment is not). In any case, such placement of the RF amplifier causes servicing and replacement difficulties, and the Cassegrain feed is often the best solution.

As shown in Fig. 9-26, an obvious difficulty results from the use of a secondary reflector, namely, the obstruction of some of the radiation from the main reflector. This is a problem, especially with small reflectors, because the dimensions of the hyperboloid are determined by its distance from the horn primary feed and the mouth diameter of the horn itself, which in turn is governed by the frequency used. One of the ways of overcoming this obstruction is by means of a large primary reflector (which is not always economical or desirable), together with a horn placed as close to the subreflector as possible. This is shown in Fig. 9-27 and has the effect of reducing the required diameter of the secondary reflector. Alternatively, vertically polarized waves are emitted by the feed, are reflected back to the main mirror by a hyperboloid consisting of vertical bars and have their polarization twisted by 90° by a mechanism [3] at the surface of the paraboloid. The reflected waves are now horizontally polarized and pass freely through the vertical bars of the secondary mirror.

Other parabolic reflectors The full paraboloid is not the only practical reflector that utilizes the properties of the parabola. Several others exist, and three of the most common are illustrated in Fig. 9-28. Each of them has an advantage over the full paraboloid in that it is much smaller, but in each instance the price paid is that the beam is not as directional in one of the planes as that of the paraboloid. With the *pillbox* reflector, for example, the beam is very narrow horizontally, but not nearly so directional vertically. First appearances might indicate that this is a very serious disadvantage, but there are a number of applications where it does not matter in the least. In ship-to-ship radar, for instance, *azimuth* directivity must be excellent, but elevation selectivity is immaterial—another ship is bound, after all, to be on the surface of the ocean!

Another form of the cut paraboloid is shown in Fig. 9-29, in cross section. This is the *offset paraboloid* reflector, in which the focus is located outside the aperture (just below it, in this case). If an antenna feed is now placed at the focus, the reflected and collimated rays will pass harmlessly above it, removing any interference. This method is often used if, for some reason, the feed antenna is rather large compared with the reflector.

A relatively recent development of the offset reflector is the *torus* antenna, similar to the cut paraboloid, but parabolic along one axis and circular along the other. By placing several feeds at the focal point, it is possible to radiate or receive several beams simultaneously, to or from the (circular) geostationary satellite orbit. The first torus antenna for satellite communications came into service in Anchorage, Alaska, in

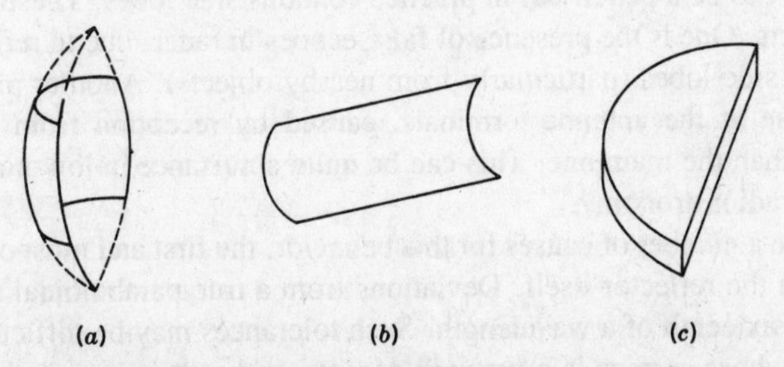

FIGURE 9-28 Parabolic reflectors. (*a*) Cut paraboloid; (*b*) parabolic cylinder; (*c*) "pillbox."

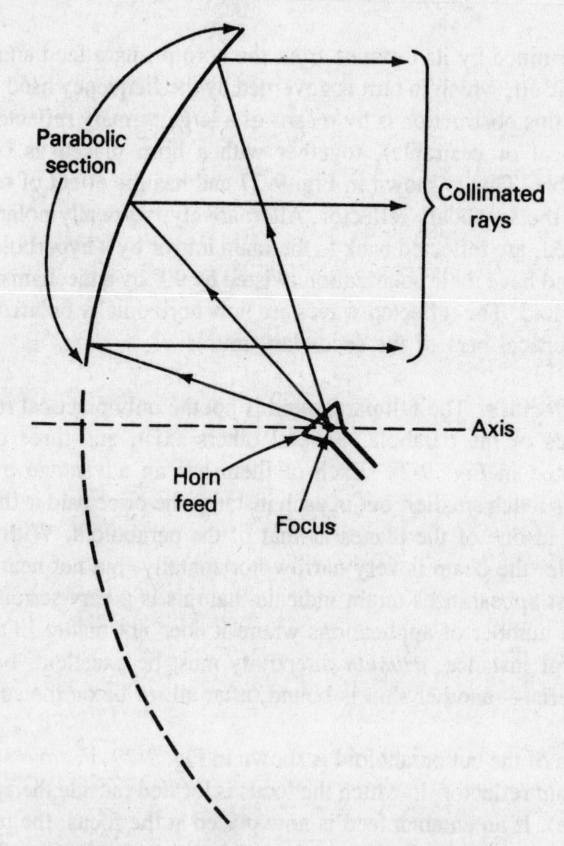

FIGURE 9-29 Offset paraboloid reflector.

1981. The 10-m antenna installed there is capable of receiving signals from up to seven satellites simultaneously. See [4] for further details.

Two other fairly common reflectors which embody the parabolic reflector exist: the *hoghorn* and the *Cass-horn;* they will both be treated with other horn antennas.

Shortcomings and difficulties The beam from an antenna with a paraboloid reflector ought, in theory, to be a pencil but in practice contains side lobes. These have several unpleasant effects. One is the presence of false echoes in radar, due to reflections from the direction of side lobes (particularly from nearby objects). Another problem is the increase in noise at the antenna terminals, caused by reception from sources in a direction other than the main one. This can be quite a nuisance in low-noise receiving systems, e.g., radioastronomy.

There are a number of causes for this behavior, the first and most obvious being imperfections in the reflector itself. Deviations from a true paraboloidal shape should not exceed one-sixteenth of a wavelength. Such tolerances may be difficult to achieve in large dishes whose surface is a network of wires rather than a smooth, continuous skin. A mesh surface is often used to reduce wind loading on the antenna and extra strain on the supports and also to reduce surface distortion caused by uneven wind force distribution over the surface. Such surface strains and distortion cannot be eliminated completely and will occur as a large dish is pointed in different directions.

Diffraction is another cause of side lobes and will occur around the edges of the paraboloid, producing interference as described in the preceding chapter. This is the reason for having reflectors with a mouth diameter preferably in excess of 10 wavelengths. Some diffraction may also be caused by the waveguide horn support, as in Fig. 9-25, or the quadripod supports of the secondary reflector in Fig. 9-27.

The finite size of the primary antenna also influences the beamwidth of antennas using paraboloid reflectors. Not being a true point source, the feed antenna cannot *all* be located at the focus. Defects known as *aberrations* are therefore produced; the main lobe is broadened and side lobes are reinforced. Increasing the aperture of the reflector, so that the focal length is about one-quarter of the mouth diameter, is of some help here. So is the use of a Cassegrain feed, which partially helps to concentrate the radiation of the feed antenna to a point.

Finally, the fact that the primary antenna does not radiate evenly at the reflector will also introduce distortion. If the primary is a dipole, it will radiate more in one plane than the other, and so the beam from the reflector will be somewhat flattened. This may be avoided by the use of a *circular horn* as the primary, but difficulties arise even here. This is because the complete surface of the paraboloid is not uniformly illuminated, since there is a gradual tapering of illumination toward the edges, which was mentioned in connection with Eq. (9-7). This has the effect of giving the antenna a virtual area that is smaller than the real area and leads, in the case of receiving antennas, to the use of the term *capture area*. This is the effective receiving area of the parabolic reflector and may be calculated from the power received and its comparison with the power density of the signal being received. The result is the area of a fully and evenly illuminated paraboloid required to produce that signal power at the primary. The capture area is simply related to the actual mouth area by the expression

$$A_0 = kA \tag{9-8}$$

where A_0 = capture area
A = actual area
k = constant depending on the antenna type and configuration = 0.65 (approximately) for a paraboloid fed by a half-wave dipole

Equation (9-8) may be used to indicate how Eq. (9-7) is derived from a more fundamental relation,

$$A_p = \frac{4\pi A_0}{\lambda^2} = \frac{4\pi kA}{\lambda^2} \tag{9-7$'$}$$

Substituting for the area of the paraboloid mouth, we have

$$A_p = \frac{4\pi k(\pi D^2/4)}{\lambda^2} = \frac{\pi^2 kD^2}{\lambda^2} = 0.65\pi^2\left(\frac{D}{\lambda}\right)^2 = 6.4\left(\frac{D}{\lambda}\right)^2$$

$$\approx 6\left(\frac{D}{\lambda}\right)^2 \tag{9-7}$$

See [9], pp. 143–163 and 174–181, for additional information on antennas with reflectors.

9-7.2 Horn Antennas

As will be seen from the next chapter, a waveguide is capable of radiating energy into open space if it is suitably excited at one end and open at the other. This radiation is much greater than that obtained from the two-wire transmission line described at the beginning of this chapter, but it suffers from similar difficulties. Only a small proportion of the forward energy in the waveguide is radiated, and much of it is reflected back by the open circuit. As with the transmission line, the open circuit is a discontinuity which matches the waveguide very poorly to space. In addition, diffraction around the edges will give the radiation a poor, nondirective pattern. To overcome these difficulties, the mouth of the waveguide may be opened out, as was done to the transmission line, but this time an electromagnetic horn results instead of the dipole.

Basic horns When a waveguide is terminated by a horn, such as any of those shown in Fig. 9-30, the abrupt discontinuity that existed is replaced by a gradual transformation. Provided that impedance matching is correct, all the energy traveling forward in the waveguide will now be radiated. Directivity will also be improved, and diffraction reduced.

There are several possible horn configurations; three of the most common are shown here. The *sectoral horn* flares out in one direction only and is the equivalent of the pillbox parabolic reflector. The *pyramidal horn* flares out in both directions and has the shape of a truncated pyramid. The *conical horn* is similar to it and is thus a logical termination for a circular waveguide. If the *flare angle* ϕ of Fig. 9-30*a* is too small, resulting in a shallow horn, the wavefront leaving the horn will be spherical rather than plane, and the radiated beam will not be directive. The same applies to the two flare angles of the pyramidal horn. On the other hand, if ϕ is too small, so will be the mouth area of the horn, and directivity will once again suffer (not to mention that diffraction is now more likely). It is therefore apparent that the flare angle has an optimum value and is, in fact, closely related to the length L of Fig. 9-30*a*, as measured in wavelengths.

In practice, ϕ varies from 40° when $L/\lambda = 6$, at which the beamwidth in the

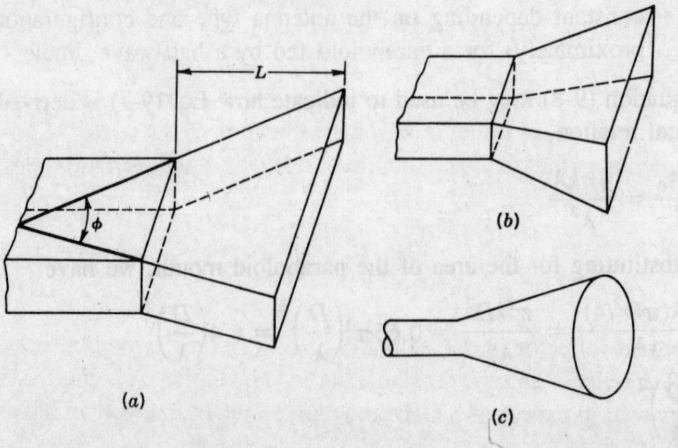

L
φ
(a)
(b)
(c)

FIGURE 9-30 Horn antennas. (*a*) Sectoral; (*b*) pyramidal; (*c*) circular.

plane of the horn in 66° and the maximum directive gain is 40, to 15° when $L/\lambda = 50$, for which beamwidth is 23° and gain is 120. Naturally, the use of a pyramidal or conical horn will improve overall directivity because flare is now in more than one direction; however, as mentioned in connection with parabolic reflectors, this is not always necessary. The horn antenna is not nearly as directive as an antenna with a parabolic reflector, but it does have quite good directivity, an adequate bandwidth (in the vicinity of 10 percent) and simple mechanical construction. In addition, it is a very convenient antenna to use with a waveguide. Simple horns such as the ones shown (or with exponential instead of straight sides) are often employed, sometimes by themselves and sometimes as primary radiators for paraboloid reflectors.

Some conditions dictate the use of a short, shallow horn, in which case the wavefront leaving it is curved, not plane as so far considered. When this is unavoidable, a *dielectric lens* may be employed to correct the curvature. Lens antennas are described in the next section. See also [9], pages 141–143 and 163–173, for additional information on basic horn antennas.

Special horns There are two antennas in use which are rather difficult to classify, since each is a cross between a horn and a parabolic reflector; they are the *Cass-horn* and the *triply folded horn reflector,* the latter more commonly called the *hoghorn antenna*.

In the Cass-horn antenna, radio waves are collected by the large bottom surface shown in Fig. 9-31, which is slightly (parabolically) curved, and are reflected upward at an angle of 45°. Upon hitting the top surface, which is a large hyperbolic cylinder, they are reflected downward to the focal point which, as indicated in Fig. 9-31*b*, is situated in the center of the bottom surface. Once there, they are collected by the conical horn placed at the focus. In the case of transmission the exact reverse happens.

This type of horn reflector antenna has a gain and beamwidth comparable to those of a paraboloid reflector of the same diameter. Like the Cassegrain feed after which is it named, it has the geometry to allow the placement of the receiver (or transmitter) at the focus, this time without any obstruction. It is therefore a low-noise antenna and is used in satellite tracking and communications stations. The one shown comes from such a station in Carnarvon (Western Australia).

The hoghorn antenna of Fig. 9-32 is another combination of paraboloid and horn. It is a low-noise microwave antenna like the Cass-horn and has similar applications. As shown, it consists of a parabolic cylinder joined to a pyramidal horn, with rays emanating from, or being received at, the apex of the horn. An advantage of the hoghorn antenna is that the receiving point does not move when the antenna is rotated about its axis [5].

9-7.3 Lens Antennas

The paraboloid reflector is one example of how optical principles may be applied to microwave antennas, and the lens antenna is yet another. It is used as a collimator at frequencies well in excess of 3 GHz and works in the same way as a glass lens used in optics.

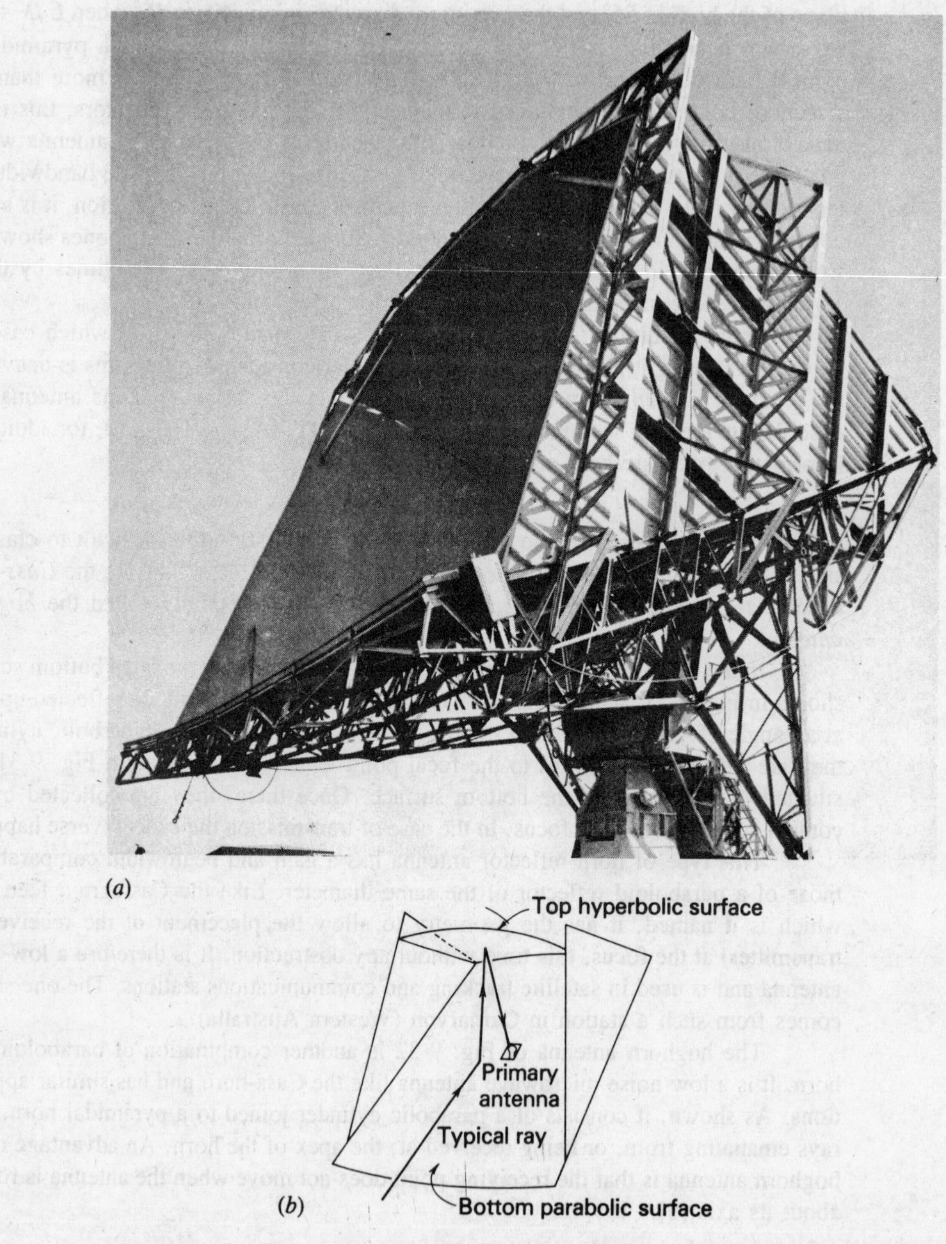

FIGURE 9-31 Cass-horn antenna. (*a*) Large Cass-horn for satellite communication *(Courtesy of Overseas Telecommunications Commission, Australia)*; (*b*) feeding the Cass-horn.

Principles Figure 9-33 illustrates the operation of a dielectric lens antenna. Looking at it from the optical point of view, as in Fig. 9-33*a*, we see that refraction takes place, and the rays at the edges are refracted more than those near the center. Thus a divergent beam is collimated, as evidenced by the fact that the rays leaving the lens are parallel. It is assumed that the source is at the focal point of the lens. The reciprocity of antennas

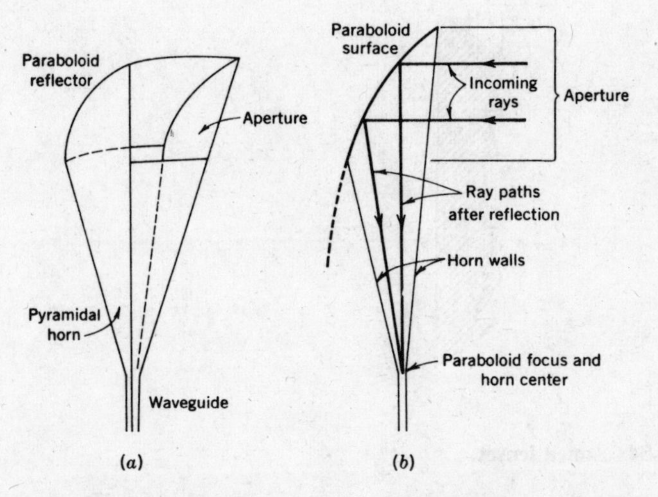

FIGURE 9-32 Hoghorn antenna. (*a*) Perspective view; (*b*) ray paths.

is nicely illustrated: if a parallel beam is received, it will be converged for reception at the focal point. Using an electromagnetic wave approach, we note that a curved wavefront is present on the source side of the lens. Also, we know that a plane wavefront is required on the opposite side of the lens, to ensure a correct phase relationship. The function of the lens must therefore be to straighten out the wavefront. The lens does this, as shown in Fig. 9-33*b*, by greatly slowing down the portion of the wave in the center. The parts of the wavefront near the edges of the lens are slowed only slightly, since those parts encounter only a small thickness of the dielectric material in which velocity is reduced. Note that, in order to have a noticeable effect on the velocity of the wave, the thickness of the lens at the center must be an appreciable number of wavelengths.

Practical considerations Lens antennas are often made of polystyrene, but other materials are also employed. However, all suffer from the same problem of excessive thickness at frequencies below about 10 GHz. Admittedly, magnifying glasses (the optical counterparts) are in everyday use, but what is not often realized is how thick they are when compared to the wavelength of the ''signal'' they pass. The thickness in

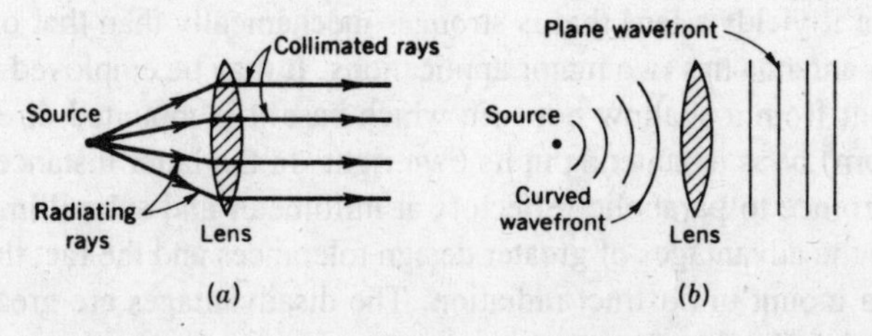

FIGURE 9-33 Operation of the lens antenna. (*a*) Optical explanation; (*b*) wavefront explanation.

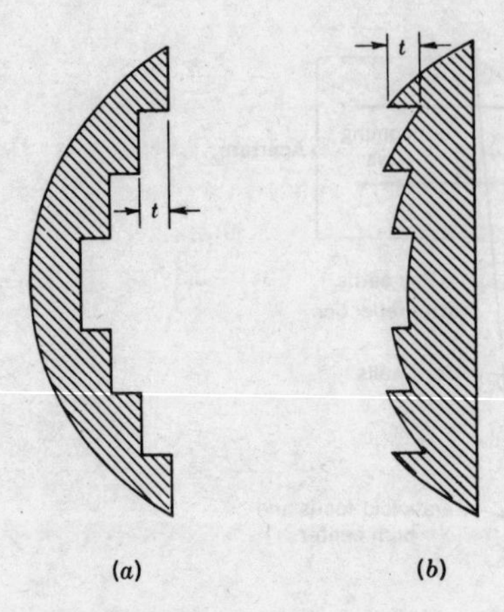

FIGURE 9-34 Zoned lenses.

the center of a typical magnifying glass may well be 6 mm, which, compared to the 0.6-μm wavelength of yellow light, is exactly 10,000 wavelengths! Dielectric antenna lenses do not have to be nearly as thick, relatively, but it is seen that problems with thickness and weight can still arise.

Figure 9-34 shows the *zoning*, or *stepping*, of dielectric lenses. This is often used to cure the problem of great thickness required of lenses used at lower microwave frequencies or for strongly curved wavefronts. Not only would the lens be thick and heavy without zoning, but it would also absorb a large proportion of the radiation passing through it. This is because any dielectric with a large enough refractive index must, for that very reason, absorb a lot of power.

The function of a lens is to ensure that signals are in phase after they have passed through it. A stepped lens will ensure this, despite appearances. What happens simply is that the phase difference between the rays passing through the center of the lens, and those passing through the adjacent sections, is 360° or a multiple of 360°—this still ensures correct phasing. Putting it differently, we see that the curved wavefront is so affected that the center portion of it is slowed down, not enough for the edges of the wavefront to catch up, but enough for the edges of the previous wavefront to catch the center portion. A disadvantage of the zoned lens is that it has only a small frequency range, unlike the unstepped lens. This is because the thickness of each step, *t*, is obviously related to the wavelength of the signal. However, since it effects a great saving in bulk, it is often used. Of the two zoning methods, the method of Fig. 9-34*b* is preferable, since it yields a lens that is stronger mechanically than that of Fig. 9-34*a*.

The lens antenna has two major applications. It may be employed to correct the curved wavefront from a shallow horn (in which case it is mounted directly over the mouth of the horn) or as an antenna in its own right. In the latter instance, lenses may be used in preference to parabolic reflectors at millimeter and submillimeter frequencies. They have the advantages of greater design tolerances and the fact that there is no primary antenna mount to obstruct radiation. The disadvantages are greater bulk, expense and design difficulties.

9-8 WIDEBAND AND SPECIAL-PURPOSE ANTENNAS

It is often desirable to have an antenna capable of operating over a wide frequency range. This may occur because a number of widely spaced channels are used, as in short-wave transmission or reception, or because only one channel is used (but it is wide), as in television transmission and reception. In TV reception, the requirement for wideband properties is magnified by the fact that it is desirable to use the same receiving antenna for a group of neighboring channels. By and large, a need exists for antennas whose radiation pattern and input impedance characteristics remain constant over a wide frequency range.

Of the antennas thus far treated, the horn (with or without paraboloid reflector), the rhombic and the folded dipole exhibit broadband properties for both impedance and radiation pattern. This was stated at the time for the first two, but the folded dipole will now be treated from this new point of view.

The special antennas to be described include the discone, helical and log-periodic antennas, as well as some of the simpler loops used for direction finding.

9-8.1 Folded Dipole (Bandwidth Compensation)[7]

A simple compensating network for increasing the bandwidth of a dipole antenna is shown in Fig. 9-35*a*. The *LC* circuit is parallel-resonant at the half-wave dipole resonant frequency. At this frequency its impedance is, therefore, a high resistance, not affecting the total impedance seen by the transmission line. Below this resonant frequency the antenna reactance becomes capacitive, while the reactance of the *LC* circuit becomes inductive. Above the resonant frequency the converse applies, the antenna becoming inductive, and the tuned circuit capacitive. Over a small frequency range near resonance, there is thus a tendency to compensate for the variations in antenna reactance, and the total impedance remains resistive in situations in which the imped-

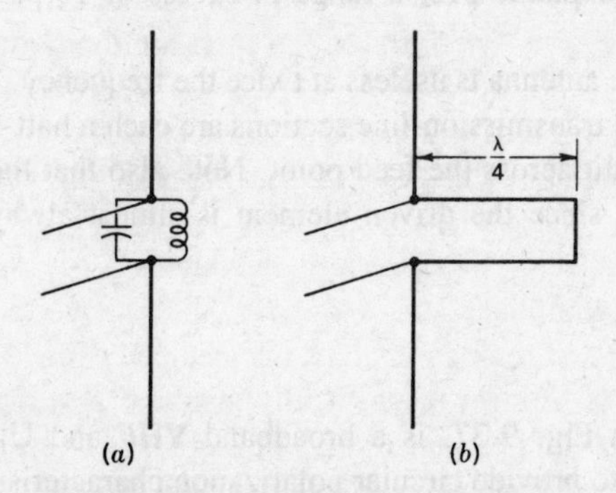

FIGURE 9-35 Impedance-bandwidth compensation for half-wave dipole. (*a*) LC circuit; (*b*) transmission line.

[7] After Everitt, William L. (ed.), *Fundamentals of Radio and Electronics,* 2d ed., Prentice-Hall, Inc., Englewood Cliffs, N.J., 1958. Figures 9-35 and 9-36 are taken from the above, by permission of the publishers.

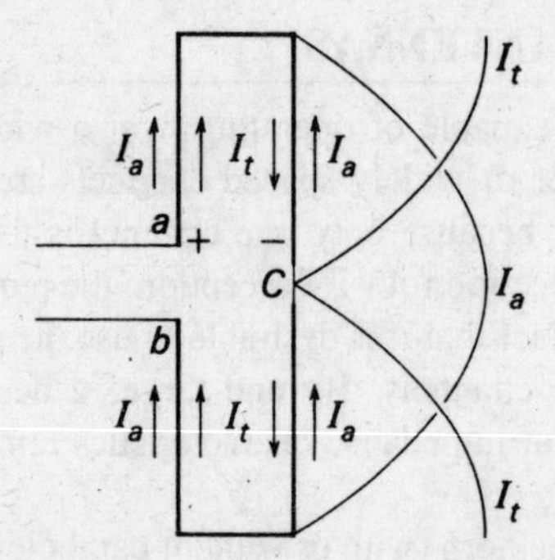

FIGURE 9-36 Folded dipole showing antenna and line currents.

ance of the antenna alone would have been heavily reactive. This compensation is both improved and widened when the Q of the resonant circuit is lowered. Moreover, it can be achieved just as easily with a short-circuited quarter-wave transmission line, as in Fig. 9-35*b*. The folded dipole provides the same type of compensation as the transmission-line version of this network.

Reference to Fig. 9-36 shows that the folded dipole may be viewed as two short-circuited, quarter-wave transmission lines, connected together at C and fed in series. The transmission line currents are labeled I_t, whereas the antenna currents are identical to those already shown for a straight half-wave dipole and are labeled I_a. When a voltage is applied at a–b, both sets of currents flow, but the antenna currents are the only ones contributing to the radiation. The transmission-line currents flow in opposite directions, and their radiations cancel. However, we do have two short-circuited quarter-wave transmission lines across a–b, and, explained in the preceding paragraph, the antenna impedance will remain resistive over a significant frequency range. Indeed, it will remain acceptable over a range in excess of 10 percent of the center frequency.

It should be noted that the antenna is useless at twice the frequency. This comes about because the short-circuited transmission-line sections are each a half-wavelength long now, throwing a short circuit across the feed point. Note also that the Yagi-Uda antenna is similarly broadband, since the driven element is almost always a folded dipole.

9-8.2 Helical Antenna

A helical antenna, illustrated in Fig. 9-37, is a broadband VHF and UHF antenna which is used when it is desired to provide circular polarization characteristics, mainly for reasons as described in Sec. 8-2.5.

As shown, the antenna consists of a loosely wound helix backed up by a *ground plane*, which is simply a screen made of "chicken" wire. There are two modes of radiation: *normal* (meaning *perpendicular*) and *axial*. In the first, radiation is in a direction at right angles to the axis of the helix. The second mode is more interesting because it produces a broadband, fairly directional radiation in the axial direction. If the helix circumference approximates a wavelength, it may be shown that a traveling

FIGURE 9-37 Helical antenna. *(Courtesy of Rohde and Schwartz, Munich.)*

wave travels around the turns of the helix, and the radiant lobe in this end-fire action is circularly polarized. Typical dimensions of the antenna are indicated in Fig. 9-38.

When the helical antenna has the proportions shown, it has typical values of directivity close to 25, beamwidth of 90° between nulls and frequency range of about 20 percent on either side of center frequency. The energy in the circularly polarized wave is divided equally between the horizontal and vertical components; the two are 90° out of phase, with either one leading, depending on construction. The transmission from a circularly polarized antenna will be acceptable to vertical or horizontal antennas, and similarly a helical antenna will accept either vertical or horizontal polarization.

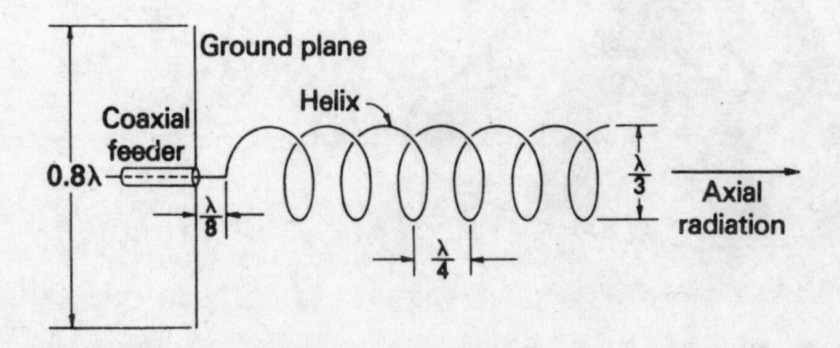

FIGURE 9-38 Dimensions of end-fire helical antenna.

The helical antenna is used either singly or in an array, as shown in Fig. 8-21, for transmission and reception of VHF signals *through* the ionosphere, as has already been pointed out. It is thus frequently used for satellite and probe communications, particularly for radiotelemetry.

When the helix circumference is very small compared to a wavelength, the radiation is a combination of that of a small dipole located along the helix axis, and that of a small loop placed at the helix turns (the ground plane is then not used). Both such antennas have identical radiation patterns, and they are here at right angles, so that the normal radiation will be circularly polarized if its two components are equal, or *elliptically* polarized if one of them predominates.

9-8.3 Discone Antenna

Pictured in Fig. 9-39, the discone antenna is, as the name aptly suggests, a combination of a *disk* and a *cone* in close proximity. It is a ground plane antenna evolved from the vertical dipole and having a very similar radiation pattern. Typical dimensions are shown in Fig. 9-40, where $D = \lambda/4$ at the lowest frequency of operation.

The discone antenna is characterized by an enormous bandwidth for both input impedance and radiation pattern. It behaves as though the disk were a reflector. As shown in Fig. 9-41, there is an inverted cone image above the disk, reflected by the disk. Now consider a line perpendicular to the disk, drawn from the bottom cone to the top image cone. If this line is moved to either side of the center of the disk, its length will vary from a minimum at the center (l_{min}) to a maximum at the edge (l_{max}) of the cone. The frequency of operation corresponds to the range of frequencies over which this imaginary line is a half-wavelength, and it can be seen that the ratio of l_{max} to l_{min}

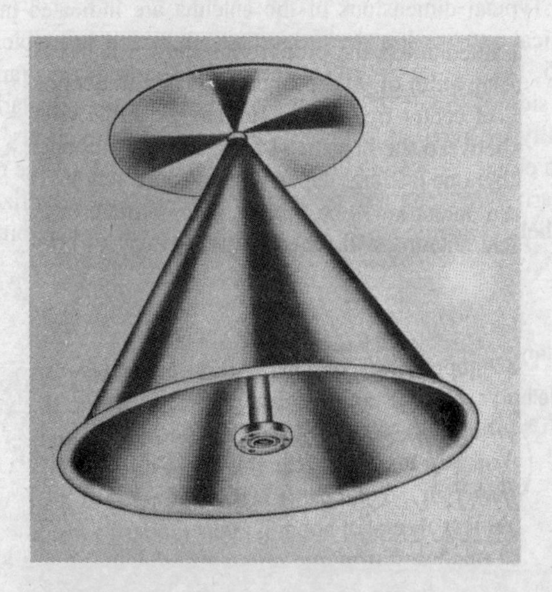

FIGURE 9-39 Discone antenna. ***(Courtesy of Andrew Antennas of Australia.)***

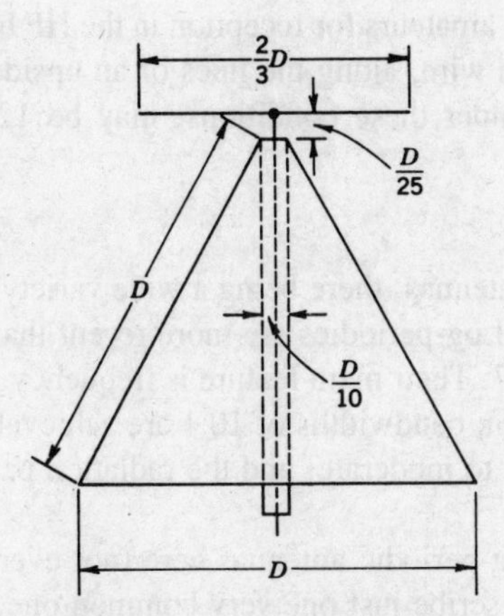

FIGURE 9-40 Dimensions of discone antenna.

is very large. The discone is thus a broadband antenna because it is a *constant-angle antenna*. For the proportions shown in Fig. 9-40, the SWR on the coaxial cable connected to the discone antenna can remain below 1.5 for a 7:1 frequency range. Overall performance is still satisfactory for a 9:1 frequency ratio.

The discone is a low-gain antenna, but it is omnidirectional. It is often employed as a VHF and UHF receiving and transmitting antenna, especially at airports, where communication must be maintained with aircraft that come from any direction.

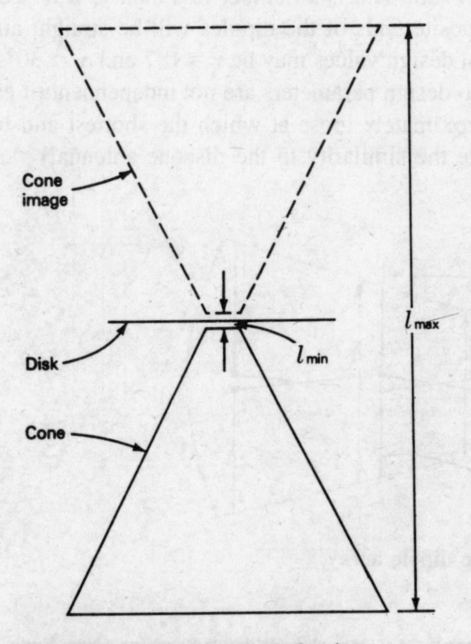

FIGURE 9-41 Discone behavior.

More recently, it has also been used by amateurs for reception in the HF band, in which case it is made of copper or aluminum wire, along the lines of an upside-down waste basket. A typical frequency range, under these conditions, may be 12 to 55 MHz.

9-8.4 Log-Periodic Antennas[8] [6,7]

Log-periodic antennas are a class of antennas, there being a wide variety of types with widely varying physical appearances. Log-periodics are more recent than most antennas, having been first proposed in 1957. Their main feature is frequency independence for both radiation resistance and pattern; bandwidths of 10:1 are achievable with ease. The directive gains obtainable are low to moderate, and the radiation patterns may be uni- or bidirectional.

It is not possible to treat all log-periodic antennas here (nor even all the common ones), so that it is proposed to describe just one very common one, the log-periodic dipole array of Fig. 9-42. This can also be used to introduce the characteristics of log-periodic antennas.

It is seen that there is a repetitiveness in the physical structure, which results in a repetitive behavior of the electrical characteristics. The array consists of a number of dipoles of different lengths and spacing, fed from a two-wire line which is transposed between each adjacent pair of dipoles. The array is fed from the narrow end, and maximum radiation is in this direction, as shown. The dipole lengths and separations are related by the formula

$$\frac{R_1}{R_2} = \frac{R_2}{R_3} = \frac{R_3}{R_4} = \tau = \frac{l_1}{l_2} = \frac{l_2}{l_3} = \frac{l_3}{l_4} \tag{9-9}$$

where τ is called the *design ratio* and is a number less than 1. It is seen that the two lines drawn to join the opposite ends of the dipoles will be straight and convergent, forming an angle α. Typical design values may be $\tau = 0.7$ and $\alpha = 30°$. As with other types of antennas, these two design parameters are not independent of each other. The cutoff frequencies are approximately those at which the shortest and longest dipoles have a length of $\lambda/2$. (Note the similarity to the discone antenna!)

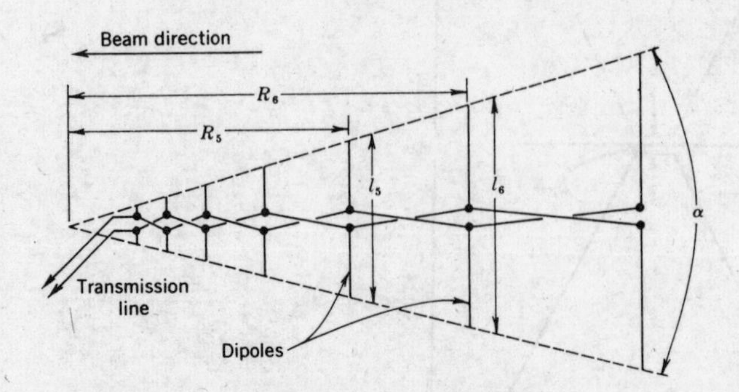

FIGURE 9-42 Log-periodic dipole array.

[8] By permission after Blake, Lamont V., *Antennas,* John Wiley & Sons, Inc., New York, 1966. Figure 9-41 is taken from the above, by permission of the publishers.

If a graph is drawn of the antenna input impedance (or SWR on the feed line) versus frequency, a repetitive variation will be noticed. If the plot is made against the *logarithm* of frequency, instead of frequency itself, this variation will be periodic, consisting of identical, but not necessarily sinusoidal, cycles. All the other properties of the antenna undergo similar variations, notably the radiation pattern. It is this behavior of the log-periodic antenna that has given rise to its name.

Like those of the rhombic, the applications of the log-periodic antenna lie mainly in the field of high-frequency communications, where such multiband steerable and fixed antennas are very often used. It has an advantage over the rhombic in that there is no terminating resistor to absorb power. Antennas of this type have also been designed for use in television reception, with one antenna for all channels including the UHF range. It must be reiterated that the log-periodic dipole array is but one of a large number of antennas of this class—there are many other exotic-looking designs, including arrays of log-periodics, as shown in [6] and [7].

9-8.5 Loop Antennas

A loop antenna is a single-turn coil carrying RF current. Since its dimensions are nearly always much smaller than a wavelength, current throughout it may be assumed to be in phase. Thus the loop is surrounded by a magnetic field everywhere perpendicular to the loop. The directional pattern is independent of the exact shape of the loop and is identical to that of an elementary doublet. The circular and square loops of Fig. 9-43 have the same radiation pattern as a short horizontal dipole, except that, unlike a horizontal dipole, a vertical loop is vertically polarized.

Because the radiation pattern of the loop antenna is the familiar doughnut pattern, no radiation is received that is normal to the plane of the loop. This, in turn, makes the loop antenna suitable for direction finding (DF) applications. For DF, it is required to have an antenna that can indicate the direction of a particular radiation. Although any of the highly directional antennas of the previous section could be used for this purpose (and are, in radar), for normal applications they have the disadvantage of being very large, which the loop is not. The DF properties of the loop are just as good at medium frequencies as those of the directional microwave antennas, except that the gain is not comparable. Also, the direction of a given radiation corresponds to

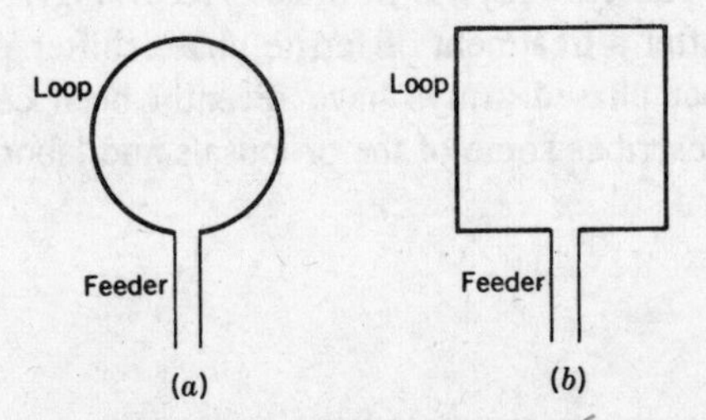

FIGURE 9-43 Loop antennas. (*a*) Circular; (*b*) square. [Note: The direction of maximum radiation is perpendicular to the plane of the loop; the shape of the radiation pattern is very similar to that in Fig. 9-6 (*a*).]

a null, rather than maximum signal. Because the loop is small, and DF equipment must often be portable, loops have direction finding as their major application.

A small loop, vertical and rotatable about a vertical axis, may be mounted on top of a portable receiver whose output is connected to a meter. This makes a very good simple direction finder. Having tuned to the desired transmission, it is then necessary to rotate the loop until the received signal is minimum; the plane of the loop is now perpendicular to the direction of the radiation. Since the loop is bidirectional, two bearings are required to determine the precise direction; if the distance between them is large enough, the distance of the source of this transmission may be found by calculation.

There are a large number of variations on the theme of the loop, far too many to consider here. They include the Alford loop, cloverleaf, Adcock antenna, and the Bellini-Tosi antenna [8].

Loops are sometimes provided with several turns and also with ferrite cores; these, being magnetic, increase the effective diameter of the loop. Such antennas are commonly found built into portable broadcast receivers. The antenna configuration explains why, if a receiver tuned to any station is rotated, a definite null will be noticed.

9-8.6 Phased Arrays

A phased array is a group of antennas, connected to the one transmitter or receiver, whose radiation beam can be adjusted electronically without any physically moving parts. Moreover, this adjustment can be very rapid indeed. More often than not, transmission or reception in several directions at once is possible. The antennas may be actual radiators, e.g., a large group of dipoles in an array (or array of arrays) pointing in the general wanted direction, or they may be the feeds for a reflector of some kind.

There are two basic types of phased arrays. In the first, a single, high-power output tube (in a transmit phased array) feeds a large number of antennas through a set of power dividers and phase shifters. The second type uses generally as many (semiconductor) generators as there are radiating elements; the phase relation between the generators is maintained through phase shifters, but this time they are low-power devices. In both types of phased arrays the direction of the beam or beams is selected by adjusting the phase difference provided by each phase shifter. This is generally done with the aid of a computer or microprocessor.

The main application of phased arrays is in radar. Accordingly, they will be further discussed in Chap. 16, after a treatment of ferrite phase shifters in Chap. 10. However, it is worth noting that phased arrays have recently been considered for satellite communications; [10] describes some of the proposals and laboratory experiments.

9-9 SYSTEM SUMMARY

An *antenna* is a structure—generally metallic and sometimes very complex—designed to provide an efficient coupling between space and the output of a transmitter or the input to a receiver. Like a transmission line, an antenna is a device with distributed constants, so that current, voltage and impedance all vary from one point to the next one along it. This factor must be taken into account when considering important antenna properties, such as impedance, gain and shape of radiation pattern.

Many antenna properties are most conveniently expressed in terms of those of *comparison antennas*. Some of these antennas are entirely fictitious but have properties that are easy to visualize. One of the important comparison antennas is the *isotropic antenna*. This cannot exist in practice. However, it is accorded the property of totally omnidirectional radiation (i.e., a perfectly spherical radiation pattern), which makes it very useful for describing the gain of practical antennas. Another useful comparison antenna is the *elementary doublet*. This is defined as a piece of infinitely thin wire, with a length that is negligible compared to the wavelength of the signal being radiated, and having a constant current along it. This antenna is very useful in that its properties assist in understanding those of practical dipoles, i.e., long, thin wires, which are often used in practice. These may be *resonant,* which effectively means that their length is a multiple of a half-wavelength of the signal, or *nonresonant,* in which case the reflected wave has been suppressed (for example, by terminating the antenna in a resistor at the point farthest from the feed point). Whereas the radiation patterns of resonant antennas are bidirectional, being due to both the forward and reflected waves, those of nonresonant antennas are unidirectional, since there is no reflected wave.

The *directive gain* of an antenna is a ratio comparing the power density generated by a practical antenna in some direction, with that due to an isotropic antenna radiating the same total power. It is thus a measure of the practical antenna's ability to concentrate its radiation. When the direction of maximum radiation of the practical antenna is taken, the directive gain becomes maximum for that antenna and is now called its *directivity*. If one now compares the input rather than radiated powers, the gain of the practical antenna drops, since some of the input power is dissipated in the antenna. The new quantity is known as the *power gain* and is equal to the directivity multiplied by the antenna efficiency.

An antenna has two *bandwidths,* both measured between half-power points: one applies to the radiation resistance and the other to radiation pattern. The *radiation resistance* is the resistive component of the antenna's ac input impedance. The *beamwidth* of an antenna is the angle between the half-power points of the main *lobe* of its radiation pattern. Because the electromagnetic waves radiated by an antenna have the electric and magnetic vectors at right angles to each other and the direction of propagation, they are said to be *polarized,* as is the antenna itself. By convention, the direction of polarization is taken to be the same as orientation of the electric vector of the radiated wave. Simple antennas may thus be *horizontally* or *vertically polarized* (i.e., themselves horizontal or vertical), respectively. More complex antennas may be *circularly polarized:* both vertically and horizontally polarized waves are radiated, with equal power in both. Where these powers are unequal, the antenna is said to be *elliptically polarized*.

Many antennas are located near the *ground,* which, to a greater or lesser extent, will reflect radio waves since it acts as a conductor. Thus, antennas which rely on the presence of the ground must be vertically polarized, or else the ground will short circuit their radiations. When the ground is a good conductor, it converts a grounded dipole into one of twice the actual height, while converting an ungrounded dipole into a two-dipole array. When its presence is relied upon, but it is a poor reflector, an *earth mat* is often laid, consisting of a network of buried wires radiating from the base of the antenna.

For grounded vertical dipoles operated at frequencies up to the MF range, the optimum *effective height* is just over a half-wavelength, although the radiation pattern of antennas with heights between a quarter- and half-wavelength is also acceptable. If the antenna is too high, objectionable side lobes which interfere with the radiated ground wave are formed. If the antenna is too low, its directivity along the ground and radiation resistance are likewise too low. A method of overcoming this is the provision of *top loading*. This is a horizontal portion atop the antenna, whose presence increases the current along the vertical portion. Together with the finite thickness of the antenna, top loading influences the *effective height* of the antenna, making it somewhat greater than the actual height.

Reactive networks known as *antenna couplers* are used to connect antennas to transmitters or receivers. Their main functions are to tune out the reactive component of the antenna impedance, to transform the resulting resistive component to a suitable value and to help tune out unwanted frequencies, particularly in a transmitting antenna. A coupler may also be used to connect a grounded antenna to a balanced transmission line or even to ensure that a transmitting antenna is isolated for dc from a transmitter output tank circuit.

Point-to-point communications are the predominant requirement in the MF range, requiring good directive antenna properties. Directional MF antennas are generally *arrays,* in which the properties of dipoles are combined to generate the wanted radiation pattern. Linear dipole arrays are often used, with *broadside* or *end-fire* radiation patterns, depending on how the individual dipoles in the array are fed. Any dipoles in an array which are not fed directly are called *parasitic elements*. These elements receive energy from the induction field surrounding the fed elements; they are known as *directors* when they are shorter than the driven element and *reflectors* when longer. The *Yagi-Uda* antenna employs a folded dipole and parasitic elements to obtain reasonable gain in the HF and VHF ranges. A much bigger antenna, the *rhombic,* is a nonresonant antenna providing excellent gain in the HF range. It consists of four wire dipoles arranged in a planar rhombus, with the transmitter or receiver located at one end; a resistor placed at the other end absorbs any power that might otherwise be reflected.

High gains and narrow beamwidths are especially required of *microwave antennas*. There are many reasons for this, with the chief ones being receiver noise, reducing power output per device as frequency is raised, and the desire to minimize the power radiated in unwanted directions. Because multiwavelength antennas are quite feasible at these frequencies, these requirements can readily be met. A large number of microwave antennas incorporate the *paraboloid reflector* in their construction. Such a reflector is made of metal and has the same properties for radio waves that an optical mirror has for light waves. That is to say, if a source is placed at the focus of the paraboloid, all the reflected rays are collimated, i.e., rendered parallel, and a very strong lobe in the axial direction is obtained. Several different methods of illuminating the paraboloid reflector are used, including the *Cassegrain feed,* in which the source is behind the reflector, and a secondary, hyperboloid reflector in front of the main one is used to provide the desired illumination. Because paraboloid reflectors can be bulky, especially at the lower end of the microwave range, cut paraboloids or parabolic cylinders are sometimes used as reflectors. Although this reduces the directivity in some directions, often this does not matter, for example, in applications such as some forms of radar.

Other microwave antennas are also in use. The chief ones are *horns* and *lenses*. A horn is an ideal antenna for terminating a waveguide and may be conical, rectangular or sectoral. More complex forms of the horns also exist, such as the hoghorn and the Cass-horn, which are really combinations of horns and paraboloid reflectors. Dielectric lenses act on microwave radiation as do ordinary lenses on light. Because of bulk, they may be stepped or zoned, but in any case they are most likely to be used at the highest frequencies. Like horns, they have good broadband properties, unless they are zoned.

Wideband antennas are required either when the transmissions themselves are wideband (e.g., television) or when working of narrow channels over a wide frequency range is the major application, as in HF communications. Horns, the folded dipole (and hence the Yagi-Uda antenna) and the rhombic all have good broadband properties. So does the *helical antenna,* which consists of a loosely wound helix backed up by a metal ground plane. This antenna has the added feature of being circularly polarized, and hence ideal for transionospheric communications. When multioctave bandwidths are required, the antennas used often have a constant-angle feature. One such antenna is the *discone,* consisting of a metal disk surmounting the apex of a metal cone. The discone is a low-gain, omnidirectional, multioctave antenna used normally in the UHF range and above, but occasionally also at HF. The *log-periodic* principle is employed to obtain very large bandwidths with quite good directivity. In a log-periodic, dipoles or other basic elements are arranged in some form of constant-angle array in which the active part of the antenna effectively moves from one end to the other as the operating frequency is changed.

Small *loop antennas* are often used for direction finding, because they do not radiate in (or receive radiation from) a direction at right angles to the plane of the loop. Accordingly, a null is obtained in this direction. Loops have many shapes and generally consist of a single turn of wire. They may also consist of several turns with a ferrite core and then make quite reasonable antennas for portable domestic receivers.

PROBLEMS

For self-testing questions on this chapter, see p. 690.

9-1. An elementary doublet is 10 cm long. If the 10-MHz current flowing through it is 2 A, what is the field strength 20 km away from the doublet, in a direction of maximum radiation?
[Ans.: 62.8 μV/m]

9-2. To produce a power density of 1 mW/m^2 in a given direction, at a distance of 2 km, an antenna radiates a total of 180 W. An isotropic antenna would have to radiate 2400 W to produce the same power density at that distance. What, in decibels, is the directive gain of the practical antenna?
[Ans.: 11.25 dB]

9-3. Calculate the radiation resistance of a $\lambda/16$ wire dipole in free space.
[Ans.: 3.09Ω]

9-4. An antenna has a radiation resistance of 72 Ω, a loss resistance of 8 Ω, and a power gain of 16. What efficiency and directivity does it have?
[Ans.: 90%, 17.8]

9-5. A 64-m-diameter paraboloid reflector, fed by a nondirectional antenna, is used at 1430 MHz. Calculate its beamwidth between half-power points and between nulls and the power gain with respect to a half-wave dipole, assuming even illumination.
[Ans.: $\phi = 0.23°$, $\phi_o = 0.46°$, $A_p = 558,000$]

9-6. A 5-m parabolic reflector, suitably illuminated, is used for 10-cm radar and is fed with 20-kW pulses. What is the effective (pulse) radiated power?
[Ans.: 300 MW]

QUESTIONS

9-1. What functions does an antenna fulfill? What does the *principle of reciprocity* say about the properties of the antenna?

9-2. What is an *elementary doublet?* How does it differ from the *infinitesimal dipole?*

9-3. Why is the maximum radiation from a half-wave dipole in a direction at right angles to the antenna?

9-4. Explain fully what is meant by the term *resonant antenna.*

9-5. What, in general, is meant by the *gain* of an antenna? What part does the *isotropic* antenna play in its calculation? How is the isotropic radiator defined?

9-6. To describe the gain of an antenna, any of the terms *directive gain, directivity* or *power gain* may be used. Define each of them, and explain how each is related to the other two.

9-7. Define the *radiation resistance* of an antenna. What is the significance of this quantity?

9-8. Discuss *bandwidth,* as applied to the two major parameters of an antenna. Also define *beamwidth.*

9-9. In what way does the effect of the ground on a nearby grounded antenna differ from that on a grounded one? What is a *basic Marconi antenna?* Show its voltage and current distribution, as well as its radiation pattern.

9-10. Describe the various factors that decide what should be the "optimum length" of a grounded medium-frequency antenna.

9-11. There are four major functions that must be fulfilled by antenna couplers (the fourth of which does not always apply). What are they?

9-12. What factors govern the selection of the feed point of a dipole antenna? How do *current feed* and *voltage feed* differ?

9-13. Draw the circuits of two typical antenna couplers, and briefly explain their operation. What extra requirements are there when coupling to parallel-wire transmission lines?

9-14. For what reasons are high-frequency antennas likely to differ from antennas used at lower frequencies? What is an *antenna array?* What specific properties does it have that make it so useful at HF?

9-15. Explain the difference between *driven* and *parasitic* elements in an antenna array. What is the difference between a *director* and a *reflector?*

9-16. Describe the end-fire array and its radiation pattern, and explain how the pattern can be made *unidirectional.*

9-17. With the aid of appropriate sketches, explain fully the operation of a *Yagi-Uda array.* List its applications. Why is it called a *supergain* antenna?

9-18. In what basic way does the *rhombic antenna* differ from arrays such as the broadside and end-fire? What are the advantages and disadvantages of this difference? What are the major applications of the rhombic?

9-19. What is a parabola? With sketches, show why its geometry makes it a suitable basis for antenna reflectors. Explain why an antenna using a *paraboloid* reflector is likely to be a highly directive *receiving* antenna.

9-20. With sketches, describe two methods of feeding a paraboloid reflector in which the primary antenna is located at the focal point. Under what conditions is this method of feed unsatisfactory?

9-21. Describe fully the *Cassegrain method* of feeding a paraboloid reflector, including a sketch of the geometry of this feeding arrangement.

9-22. Discuss in detail some shortcomings and difficulties connected with the Cassegrain feed of parabolic reflectors. How can they be overcome?

9-23. What is a horn antenna? How is it fed? What are its applications?

9-24. Explain the basic principles of operation of *dielectric lens antennas,* showing how they convert curved wavefronts into plane ones.

9-25. What is the major drawback of lens antennas, restricting their use to the highest frequencies? Show how *zoning* improves matters, while introducing a drawback of its own.

9-26. With suitable sketches, do a survey of microwave antennas, comparing their performance.

9-27. For what applications are wideband antennas required? List the various broadband antennas, giving typical percentage bandwidths for each.

9-28. Sketch a *helical antenna,* and briefly explain its operation in the *axial* mode. In what very important way does this antenna differ from the other antennas studied?

9-29. Sketch a *discone antenna,* and use the sketch to describe its operation. For what applications is it suitable? Why do its applications differ from those of a rhombic antenna?

9-30. Explain how *log-periodic antennas* acquire their name.

9-31. Describe the behavior of *loop antennas,* and show how they may be used for direction finding. What other applications do they have?

REFERENCES

1. *IEEE Std 145-1973, Definition of Terms for Antennas,* Institute of Electrical and Electronics Engineers, New York, November 1983.

2. (*Historical*) Yagi, Hidetsugu, "Beam Transmission of Ultra-short Waves," *Proc. IRE* **16,** June 1928, p. 715.

3. See Jasik, H. (ed.), *Antenna Engineering Handbook,* McGraw-Hill Book Company, New York, 1961, chap. 25, pp. 11–14.

4. Hyde, G., R. W. Kreutel, and L.V. Smith, "The Unattended Earth Terminal Multiple-Beam Torus Antenna," *COMSAT Technical Review* **4** (2):232–261 (Fall 1974).

5. This type of antenna is more fully described in Crawford, A. B., et al., "A Horn Reflector Antenna for Space Communications," *Bell System Tech. J* **40** (4):1095 (July 1961).

6. For illustrations and descriptions of several wire log-periodic antennas, see Elliott, R. S., *Antenna Theory and Design,* Prentice-Hall, Inc., Englewood Cliffs, N.J., 1981, pp. 379–385.

7. For a detailed description of a specific antenna, see Ramsdale, P. A., and P. W. Crampton, "Low Frequency Performance of Hemispherical Coverage Conical Log-Spiral Antenna, *Second International Conference on Antennas and Propagation, Part 1: Antennas,* The Institution of Electrical Engineers, London, 1981, pp. 298–302.

8. See Grob, B., and M. S. Kiver, *Applications of Electronics,* 2d ed., McGraw-Hill Book Company, New York, 1966, chap. 17.

9. Miya, K. (ed.), *Satellite Communications Technology,* KDD Engineering and Consulting, Inc., Tokyo, 1982.

10. Reudink, D. O., and Y. S. Yeh, "Scanning Spot Beams—A New Approach to Satellite Communications," *Bell Lab. Rec.,* 58(2):38–45 (February 1980).

10 WAVEGUIDES, RESONATORS AND COMPONENTS

It was seen in Chap. 8 that electromagnetic waves will travel from one point to another, if suitably radiated. Equally, Chap. 7 showed how it is possible to guide radio waves from one point to another in an enclosed system by the use of transmission lines. This chapter will, in the main, deal with *waveguides*. Any system of conductors and insulators for carrying electromagnetic waves could be called a *waveguide*, but it is customary to reserve this name for specially constructed hollow metallic pipes. They are used at microwave frequencies, for the same purposes as transmission lines were used at lower frequencies. Waveguides are preferred to transmission lines because they are much less lossy at the highest frequencies and for other reasons that will become apparent through this chapter.

The general principles of waveguide propagation are covered at first, and then rectangular, circular and odd-shaped waveguides are treated. Methods of exciting waveguides as well as basic waveguide components are then described, as are impedance matching and attenuation. *Cavity resonators* are a major topic; these are the waveguide equivalents of tuned transmission lines but are somewhat more complex because of their three-dimensional shapes. The final major section of the chapter deals with additional waveguide components, such as *directional couplers, isolators, circulators, diodes, diode mounts* and *switches*.

Having studied this chapter, students should have a very good understanding of waveguides and associated components, their physical appearance, behavior and properties. They should also have a clear grasp of means whereby microwaves may be guided, over very long distances if necessary.

10-1 RECTANGULAR WAVEGUIDES

It was seen in Chap. 7 that any system of wires may be used as a transmission line, but the simplest arrangements are invariably preferred in practice. Thus parallel-wire and coaxial lines are by far the most common. In a similar way, a pipe with *any* sort of cross section could be used as a waveguide, but the simplest cross sections are preferred. Accordingly, waveguides with constant rectangular or circular cross sections are normally employed, although other shapes may be used from time to time for special purposes. As with regular transmission lines, so in waveguides, the simplest shapes are the ones easiest to manufacture, and the ones whose properties are simplest to evaluate.

Rectangular waveguides are treated first, partly because they are very common, and partly because propagation in them is easiest to visualize and calculate. It should be noted, however, that in this context "easiest" does not, unfortunately, mean "easy"!

10-1.1 Introduction

A rectangular waveguide is shown in Fig. 10-1, as is a circular waveguide for comparison. In a typical setup, there may be an antenna at one end of a waveguide and some form of load at the other end. The antenna generates electromagnetic waves, which travel down the waveguide to be eventually received by the load. It is seen that the waves are truly *guided*.

The walls of the guide are conductors, and therefore reflections from them take place, as described in Sec. 8-1.2. It is of the utmost importance to realize, as will be explained, that *conduction of energy takes place not through the walls*, whose function is only to confine this energy, *but through the dielectric filling the waveguide*, which is usually air. *In discussing the behavior and properties of waveguides, it is necessary to speak of electric and magnetic fields, as in wave propagation, instead of voltages and currents*, as in transmission lines. This is the only possible approach, but it does make the behavior of waveguides more complex to grasp.

Applications Because the cross-sectional dimensions of a waveguide must be of the same order as those of a wavelength, use at frequencies below about 1 GHz is not

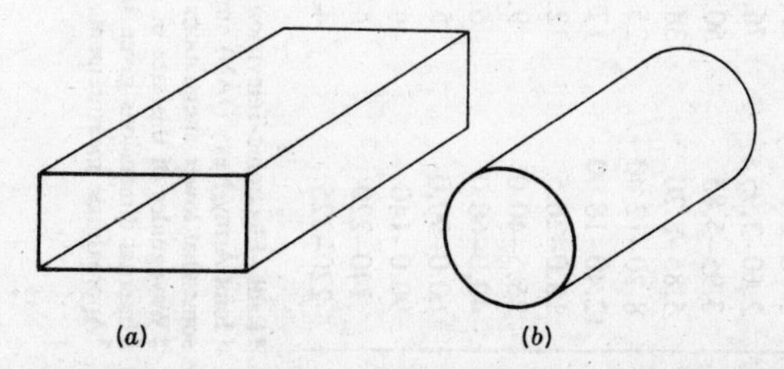

FIGURE 10-1 Waveguides. (*a*) Rectangular; (*b*) circular.

TABLE 10-1 Selected Rectangular Waveguides

USEFUL FREQUENCY RANGE, GHz	OUTSIDE DIMENSIONS, mm	WALL THICKNESS, mm	RETMA* DESIGNATION	JAN† TYPE NO.	THEORETICAL AVERAGE ATTENUATION, dB/m	THEORETICAL AVERAGE (CW) POWER RATING, kW
1.12–1.70	169 × 86.6	2.0	WR650	RG-69/U	0.0052	14,600
1.70–2.60	113 × 58.7	2.0	WR430	RG-104/U	0.0097	6400
2.60–3.95	76.2 × 38.1	2.0	WR284	RG-48/U	0.019	2700
3.95–5.85	50.8 × 25.4	1.6	WR187	RG-49/U	0.036	1700
5.85–8.20	38.1 × 19.1	1.6	WR137	RG-50/U	0.058	635
8.20–12.40	25.4 × 12.7	1.3	WR90	RG-52/U	0.110	245
12.40–18.00	17.8 × 9.9	1.0	WR62	RG-91/U	0.176	140
18.0–26.5	12.7 × 6.4	1.0	WR42	RG-53/U	0.37	51
26.5–40.0	9.1 × 5.6	1.0	WR28	RG-96/U	0.58	27
40.0–60.0	6.8 × 4.4	1.0	WR19	—	0.95¶	13
60.0–90.0	5.1 × 3.6	1.0	WR12	RG-99/U	1.50¶	5.1
90.0–140	4.0 (diam.)‡	2.0 × 1.0§	WR8	RG-138/U	2.60¶	2.2
140–220	4.0 (diam.)	1.3 × 0.64	WR5	RG-135/U	5.20¶	0.9
220–325	4.0 (diam.)	0.86 × 0.43	WR3	RG-139/U	8.80¶	0.4

* Radio-Electronic-Television Manufacturers' Association.
† Joint Army-Navy (JAN) numbers are shown for copper waveguides (there are also aluminum waveguides with identical dimensions but different US military numbers and somewhat lower attenuations), except for the last five numbers, which are for silver waveguides. Where no number is given, none exists.
‡ Waveguides of this size or smaller are circular on the outside.
§ Internal dimensions given instead of wall thickness for this waveguide and the smaller ones.
¶ Approximate measurements.

normally considered, unless special circumstances warrant it.[1] Some selected waveguide sizes, together with their frequencies of operation, are presented in Table 10-1.

The table shows how waveguide dimensions decrease as the frequency is increased (and therefore wavelength is lowered). It does not show the several waveguides larger than the WR650, nor does it show many of the overlapping sizes that are also made. Note that the reason for the rather odd dimensions is that waveguides originally were made to imperial measurements (e.g., 3.00 × 1.50 in) and have subsequently been relabeled in millimeters, not remade in round millimeter sizes. See also [1] for more extensive tables and charts. It is seen that waveguides have dimensions that are convenient in the 3- to 100-GHz range, and somewhat inconvenient much outside this range. Within the range, waveguides are generally superior to coaxial transmission lines for a whole spectrum of microwave applications, for either power or low-level signals.

Both waveguides and transmission lines can pass several signals simultaneously, but in waveguides it is sufficient for them to be propagated in different *modes* to be separated, as will be explained. They do not have to be of different frequencies. Again, a number of waveguide components are similar if not identical to their coaxial counterparts. These components include stubs, quarter-wave transformers, directional couplers and taper sections. Finally, the Smith chart may be used for waveguide calculations also. Indeed, the operation of a very large number of waveguide components may best be understood by first looking at their transmission-line equivalents.

Advantages The first thing that strikes one about the appearance of a (circular) waveguide is that it looks like a coaxial line with the insides removed. This illustrates the advantages that waveguides possess. Since it is easier to leave out the inner conductor than to put it in, waveguides are simpler to manufacture than coaxial lines. Similarly, because there is neither an inner conductor nor the supporting dielectric in a waveguide, flashover is less likely. Therefore the power-handling ability of waveguides is improved, and is about 10 times as high as for coaxial air-dielectric rigid cables of similar dimension (and much more when compared with flexible solid-dielectric cable).

Moreover, because there need be nothing but air in a waveguide, and since propagation is by reflection from the walls instead of conduction along them, power losses in waveguides are lower than in comparable transmission lines. For example, 41-mm air-dielectric cable has an attenuation of 4.0 dB/100 m at 3 GHz (which is very good for a coaxial line). This rises to 10.8 dB/100 m for a similar foam-dielectric flexible cable, whereas the figure for the copper WR284 waveguide is only 1.9 dB/100 m.

Everything else being equal, waveguides have advantages over coaxial lines in mechanical simplicity and a much higher maximum operating frequency (325 GHz as compared with 18 GHz) because of the different method of propagation.

[1] In military projects such as "Polevault" and "BMEWS," waveguides have been used at frequencies as low as 350 MHz, with cross-sectional dimensions of 53 × 26½ cm (21 × 10½ in).

10-1.2 Reflection of Waves from a Conducting Plane

In view of the way in which signals propagate in waveguides, it is now necessary to consider further what happens to electromagnetic waves when they encounter a conducting surface; therefore, this is an extension of the work in Sec. 8-1.

Basic behavior As already discussed, an electromagnetic plane wave in space is transverse-electromagnetic, or TEM; the electric field, the magnetic field and the direction of propagation are mutually perpendicular. If such a wave were sent straight down a waveguide, it would not, despite appearances, propagate in it. This is because the electric field (no matter what its direction) would be short-circuited by the walls, since the walls are assumed to be perfect conductors, and thus a potential cannot exist across them. What must be found is some method of propagation which does not require an electric field to exist near a wall and simultaneously be parallel to it. This is achieved by sending the wave down the waveguide in a zigzag fashion, bouncing it off the walls and setting up a field that is maximum at or near the center of the guide, and zero at the walls. In this case the walls have nothing to short-circuit, and therefore they do not interfere with the wave pattern set up between them; thus propagation is not hindered.

Two major consequences of the zigzag propagation are apparent. The first is that the velocity of propagation in a waveguide must be less than in free space, and the second is that waves can no longer be TEM. The second situation arises because propagation by reflection requires not only a normal component but also a component in the direction of propagation (as shown in Fig. 10-2) for either the electric or the magnetic field, depending on the way in which waves are set up in the waveguide. This extra component in the direction of propagation means that waves are no longer transverse-electromagnetic, because there is now either an electric or a magnetic additional component in the direction of propagation.

Since there are two different basic methods of propagation, names must be given to the resulting waves to distinguish them from each other. Unfortunately, nomenclature of these *modes* has always been a vexing question. The American system labels modes according to the field component that behaves as it did in free space. Thus, modes in which there is no component of electric field in the direction of propagation are called *transverse-electric* (*TE*) modes, and modes with no such component of magnetic field are called *transverse-magnetic* (*TM*). With equal logic, the

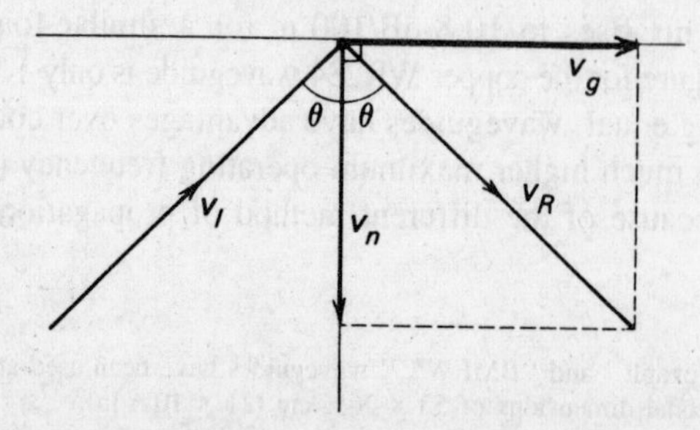

FIGURE 10-2 Reflection from a conducting surface.

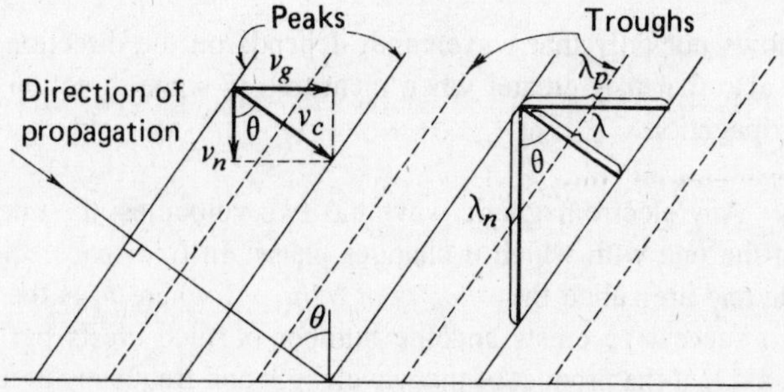

FIGURE 10-3 Plane waves at a conducting surface.

British and European systems label the modes according to the component that has behavior different from that in free space; thus modes are called H instead of TE and E instead of TM. The American system will be used here exclusively.

Plane waves at a conducting surface Consider Fig. 10-3, which shows wavefronts incident on a perfectly conducting plane (for simplicity, reflection is not shown). The waves travel diagonally from left to right, as indicated, and have an angle of incidence θ.

If the actual velocity of the waves is v_c, then simple trigonometry shows that the velocity of the wave in a direction parallel to the conducting surface, v_g, and the velocity normal to the wall, v_n, respectively, are given by

$$v_g = v_c \sin \theta \tag{10-1}$$

$$v_n = v_c \cos \theta \tag{10-2}$$

As should have been expected, Eqs. (10-1) and (10-2) show that waves travel forward more slowly in a waveguide than in free space.

Parallel and normal wavelength The concept of *wavelength* has several descriptions or definitions, all of which mean the distance between two successive identical points of the wave, such as two successive crests. To this it is now necessary to add the phrase *in the direction of measurement*, because we have so far always considered measurement in the direction of propagation (and this has been left unsaid). There is nothing to stop us from measuring wavelength in any other direction, but there has been no application for this so far. Other practical applications do exist, as in the cutting of corrugated roofing materials at an angle to meet other pieces of corrugated material.

If Fig. 10-3 is again consulted, it is seen that the wavelength in the direction of propagation of the wave is shown as λ, being the distance between two consecutive wave crests in this direction. Similarly, the distance between two consecutive crests in the direction parallel to the conducting plane, or the wavelength in that direction, is λ_p, and the wavelength at right angles to the surface is λ_n. Simple calculation again yields

$$\lambda_p = \frac{\lambda}{\sin \theta} \tag{10-3}$$

$$\lambda_n = \frac{\lambda}{\cos \theta} \tag{10-4}$$

This shows not only that wavelength depends on the direction in which it is measured, but also that it is greater when measured in some direction other than the direction of propagation.

Phase velocity Any electromagnetic wave has two velocities: the one with which it propagates and the one with which it changes phase. In free space, these are "naturally" the same and are called the *velocity of light*, ν_c, where ν_c is the product of the distance of two successive crests and the number of such crests per second. More formally, it is said that the product of the wavelength and frequency of a wave gives its velocity, and thus

$$\nu_c = f\lambda$$

$$= 3 \times 10^8 \text{ m/s} \qquad \text{in free space} \tag{10-5}$$

For Fig. 10-3 it was indicated that the velocity of propagation in a direction parallel to the conducting surface is $\nu_g = \nu_c \sin\theta$, as given by Eq. (10-1). It was also shown that the wavelength in this direction is $\lambda_p = \lambda/\sin\theta$, given by Eq. (10-3). If the frequency is f, it follows that the velocity (called the *phase velocity*) with which the wave changes phase in a direction parallel to the conducting surface is given by the product of the two. Thus

$$\nu_p = f\lambda_p$$

$$= \frac{f\lambda}{\sin\theta} \tag{10-6}$$

$$= \frac{\nu_c}{\sin\theta} \tag{10-7}$$

where ν_p = phase velocity.

A most surprising result is that there is an *apparent* velocity, associated with an electromagnetic wave at a boundary, which is greater than either its velocity of propagation in that direction, ν_g; or its velocity in space, ν_c. It should be mentioned that the theory of relativity has not been contradicted here, since neither mass, nor energy, nor signals can be sent with this velocity. It is merely the velocity with which the wave changes phase at a plane boundary, not the velocity with which it travels along the boundary. A number of other apparent velocities greater than the velocity of light can be shown to exist. For instance, consider sea waves approaching a beach at an angle, rather than straight in. The interesting phenomenon which accompanies this (it must have been noticed by most people) is that the edge of the wave appears to sweep along the beach much faster than the wave is really traveling; it is the phase velocity that provides this effect. The two velocities will be discussed again, in the next section.

10-1.3 The Parallel-Plane Waveguide

It was shown in Sec. 7-1.4, in connection with transmission lines, that reflections and standing waves are produced if a line is terminated in a short circuit, and that there is a voltage zero and a current maximum at this termination. This is illustrated again in Fig. 10-4, because it applies directly to the situation described in the previous section, involving electromagnetic waves at a conducting boundary.

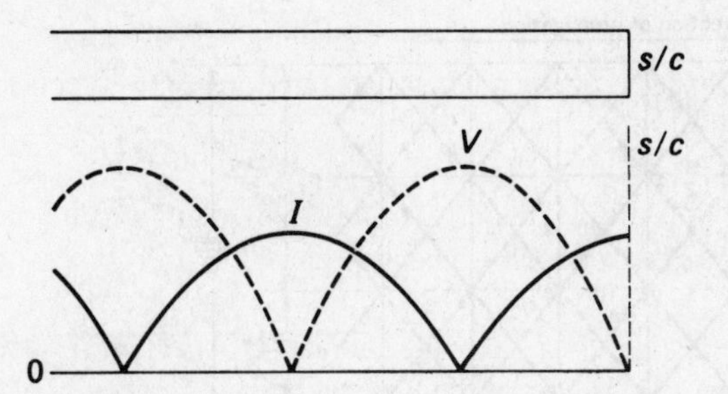

FIGURE 10-4 Short-circuited transmission line with standing waves.

A rectangular waveguide has two pairs of walls, and we shall be considering their addition one pair at a time. It is now necessary to investigate whether the second wall in a pair may be added at any distance from the first, or whether there are any preferred positions and, if so, how to determine them. Transmission-line equivalents will continue to be used, because they definitely help to explain the situation.

Addition of a second wall If a second short circuit is added to Fig. 10-4, care must be taken to ensure that it does not disturb the existing wave pattern (the feeding source must somehow be located between the two short-circuited ends). Three suitable positions for the second short circuit are indicated in Fig. 10-5. It is seen that each of them is at a point of zero voltage on the line, and each is located at a distance from the first short circuit that is a multiple of half-wavelengths.

The presence of a reflecting wall does to electromagnetic waves what a short circuit did to waves on a transmission line. A pattern is set up and will be destroyed unless the second wall is placed in a correct position. The situation is illustrated in Fig. 10-6, which shows the second wall, placed three half-wavelengths away from the first wall, and the resulting wave pattern between the two walls.

A major difference from the behavior of transmission lines is that in waveguides the wavelength normal to the walls is not the same as in free space, and thus $a = 3\lambda_n/2$ here, as indicated. Another important difference is that instead of say-

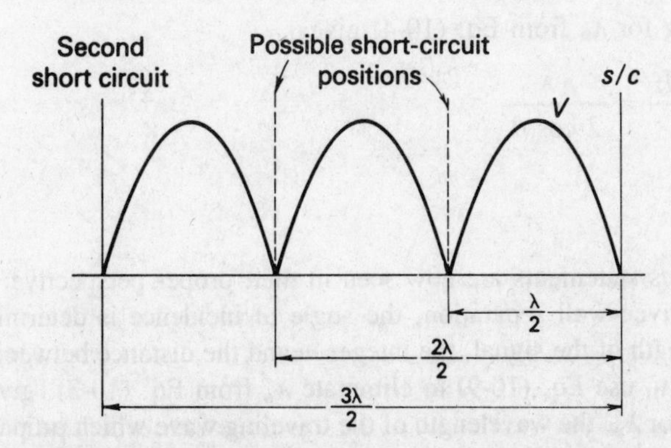

FIGURE 10-5 Placement of second short circuit on transmission line.

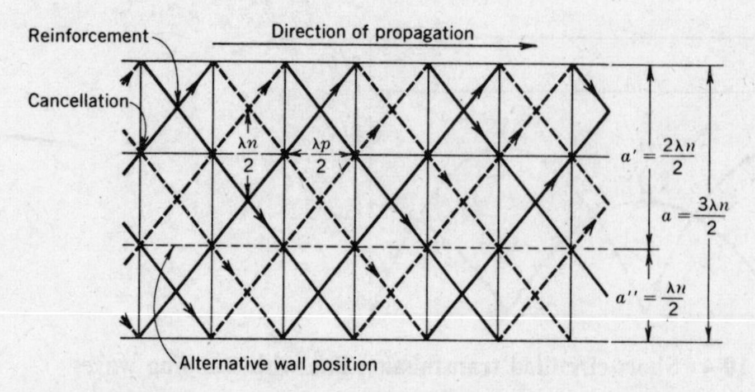

FIGURE 10-6 Reflections in a parallel-plane waveguide.

ing that "the second wall is placed at a distance that is a multiple of half-wavelengths," we should say that "the signal arranges itself so that the distance between the walls becomes an integral number of half-wavelengths, if this is possible." The arrangement is accomplished by a change in the angle of incidence, which is possible so long as this angle is not required to be "more perpendicular than 90°." Before we begin a mathematical investigation, it is important to point out that the second wall might have been placed (as indicated) so that $a' = 2\lambda_n/2$, or $a'' = \lambda_n/2$, without upsetting the pattern created by the first wall.

Cutoff wavelength If a second wall is added to the first at a distance a from it, then, as has been discussed, it must be placed at a point where the electric intensity due to the first wall is zero, i.e., at an integral number of half-wavelengths away. Putting this mathematically, we have

$$a = \frac{m\lambda_n}{2} \tag{10-8}$$

where a = distance between walls
λ_n = wavelength in a direction normal to both walls
m = number of half-wavelengths of electric intensity to be established between the walls (an integer)

Substituting for λ_n from Eq. (10-4) gives

$$a = \frac{m(\lambda/\cos\theta)}{2} = \frac{m\lambda}{2\cos\theta}$$

$$\cos\theta = \frac{m\lambda}{2a} \tag{10-9}$$

The previous statements are now seen in their proper perspective: Eq. (10-9) shows that for a given wall separation, the angle of incidence is determined by the free-space wavelength of the signal, the integer m and the distance between the walls. It is now possible to use Eq. (10-9) to eliminate λ_n from Eq. (10-3), giving a more useful expression for λ_p, the wavelength of the traveling wave which propagates down the waveguide. We then have

$$\lambda_p = \frac{\lambda}{\sin\theta} = \frac{\lambda}{\sqrt{1-\cos^2\theta}} = \frac{\lambda}{\sqrt{1-(m\lambda/2a)^2}} \tag{10-10}$$

From Eq. (10-10), it is easy to see that as the free-space wavelength is increased, there comes a point beyond which the wave can no longer propagate in a waveguide with fixed a and m. The free-space wavelength at which this takes place is called the *cutoff wavelength* and is defined as *the smallest free-space wavelength that is just unable to propagate in the waveguide under given conditions*. This implies that any larger free-space wavelength certainly cannot propagate, but that all smaller ones can. From Eq. (10-10), the cutoff wavelength is that value of λ for which λ_p becomes infinite, under which circumstance the denominator of Eq. (10-10) becomes zero, giving

$$1 - \left(\frac{m\lambda_0}{2a}\right)^2 = 0$$

$$\frac{m\lambda_0}{2a} = 1$$

$$\lambda_0 = \frac{2a}{m} \tag{10-11}$$

where λ_0 = cutoff wavelength.

The largest value of cutoff wavelength is $2a$, when $m = 1$. This means that the longest free-space wavelength that a signal may have and still be capable of propagating in a parallel-plane waveguide, is just less than twice the wall separation. *When m is made unity, the signal is said to be propagated in the dominant mode, which is the method of propagation that yields the longest cutoff wavelength of the guide.*

It follows from Eq. (10-10) that *the wavelength of a signal propagating in a waveguide is always greater than its free-space wavelength*. Furthermore, when a waveguide fails to propagate a signal, it is because its free-space wavelength is too great. If this signal must be propagated, a mode of propagation with a larger cutoff wavelength should be used; that is, m should be made smaller. If m is already equal to 1 and the signal still cannot propagate, the distance between the walls must be increased.

Finally, Eq. (10-11) may be substituted into Eq. (10-10) to give the very important universal equation for the guide wavelength, which does not depend on either waveguide geometry or the actual mode (value of m) used. The guide wavelength is obtained in terms of the free-space wavelength of the signal, and the cutoff wavelength of the waveguide, as follows:

$$\lambda_p = \frac{\lambda}{\sqrt{1-[\lambda(m/2a)]^2}} = \frac{\lambda}{\sqrt{1-[\lambda(1/\lambda_0)]^2}}$$

$$\lambda_p = \frac{\lambda}{\sqrt{1-(\lambda/\lambda_0)^2}} \tag{10-12}$$

Group and phase velocity in the waveguide As already indicated, a wave reflected from a conducting wall has two velocities in a direction parallel to the wall, namely, the *group velocity* and the *phase velocity*. The former was shown as v_g in Eq. (10-1), and the latter as v_p in Eqs. (10-6) and (10-7). These two velocities have exactly the same meanings in the parallel-plane waveguide and must now be correlated and extended further.

If Eqs. (10-1) and (10-7) are multiplied together, we get

$$v_g v_p = v_c \sin\theta \frac{v_c}{\sin\theta}$$

$$v_g v_p = v_c^2 \tag{10-13}$$

Thus the product of the group velocity and the phase velocity of a signal propagating in a waveguide is the square of the velocity of light in free space. Note that, in free space, phase and group velocities exist also, but they are then equal. It is now possible to calculate the two velocities in terms of the cutoff wavelength, again obtaining universal equations. From Eq. (10-6) we have

$$v_p = f\lambda_p$$

$$= f\frac{\lambda}{\sqrt{1-(\lambda/\lambda_0)^2}}$$

$$= \frac{v_c}{\sqrt{1-(\lambda/\lambda_0)^2}} \tag{10-14}$$

Substituting Eq. (10-14) into (10-13) gives

$$v_g = \frac{v_c^2}{v_p} = v_c^2\frac{1}{v_p} = v_c^2\frac{\sqrt{1-(\lambda/\lambda_0)^2}}{v_c}$$

$$v_g = v_c\sqrt{1-\left(\frac{\lambda}{\lambda_0}\right)^2} \tag{10-15}$$

Equation (10-15) is an important one and reaffirms that the velocity of propagation (group velocity) in a waveguide is lower than in free space. Also, as previously outlined, group velocity decreases as the free-space wavelength approaches the cutoff wavelength and eventually becomes zero when the two wavelengths are equal. The physical explanation of this is that the angle of incidence (and reflection) has become 90°, there is no traveling wave and all the energy is reflected back to the generator. There is no transmission-line equivalent of this behavior, but the waveguide may be thought of as a high-pass filter having no attenuation in the passband (for wavelengths shorter than λ_0), but very high attenuation in the stop band.

Example 10-1 A wave is propagated in a parallel-plane waveguide, under conditions as just discussed. The frequency is 6 GHz, and the plane separation is 3 cm. Calculate

(*a*) The cutoff wavelength for the dominant mode
(*b*) The wavelength in a waveguide, also for the dominant mode
(*c*) The corresponding group and phase velocities

(*a*) $\lambda_0 = \frac{2a}{m} = 2 \times \frac{3}{1} = 6$ cm

(*b*) $\lambda = \frac{v_c}{f} = \frac{3 \times 10^{10}}{6 \times 10^9} = \frac{30}{6} = 5$ cm

Since the free-space wavelength is less than the cutoff wavelength here, the wave will propagate, and all the other quantities may be calculated. Since $\sqrt{1-(\lambda/\lambda_0)^2}$ appears in all the remaining calculations, it is convenient to calculate it first. Let it be ρ; then

$$\rho = \sqrt{1-\left(\frac{\lambda}{\lambda_0}\right)^2} = \sqrt{1-\left(\frac{5}{6}\right)^2} = \sqrt{1-0.695} = 0.553$$

Then

$$\lambda_p = \frac{\lambda}{\rho} = \frac{5}{0.553} = 9.05 \text{ cm}$$

(c) $\quad \nu_g = \nu_c\rho = 3 \times 10^8 \times 0.553 = 1.66 \times 10^8 \text{m/s}$

$$\nu_p = \frac{\nu_c}{\rho} = 3 \times \frac{10^8}{0.553} = 5.43 \times 10^8 \text{m/s}$$

Example 10-2 It is necessary to propagate a 10-GHz signal in a waveguide whose wall separation is 6 cm. What is the greatest number of half-waves of electric intensity which it will be possible to establish between the two walls (i.e., what is the largest value of m)? Calculate the guide wavelength for this mode of propagation.

$$\lambda = \frac{\nu_c}{f} = \frac{3 \times 10^{10}}{10 \times 10^9} = 3 \text{ cm}$$

The wave will propagate in the waveguide as long as the waveguide's cutoff wavelength is greater than the free-space wavelength of the signal. Accordingly, we calculate the cutoff wavelengths of the guide for increasing values of m.

When $m = 1$,

$$\lambda_0 = 2 \times \frac{6}{1} = 12 \text{ cm} \qquad \text{(This mode will propagate.)}$$

When $m = 2$,

$$\lambda_0 = 2 \times \frac{6}{2} = 6 \text{ cm} \qquad \text{(This mode will propagate.)}$$

When $m = 3$,

$$\lambda_0 = 2 \times \frac{6}{3} = 4 \text{ cm} \qquad \text{(This mode will propagate.)}$$

When $m = 4$,

$$\lambda_0 = 2 \times \frac{6}{4} = 3 \text{ cm}$$

(This mode will not propagate, *because the cutoff wavelength is no longer larger than the free-space wavelength*.)

It is seen that the greatest number of half-waves of electric intensity that can be established between the walls is three. Since the cutoff wavelength for the $m = 3$ mode is 4 cm, the guide wavelength will be

$$\lambda_p = \frac{3}{\sqrt{1-(3/4)^2}} = \frac{3}{\sqrt{1-0.562}} = \frac{3}{0.661} = 4.54 \text{ cm}$$

10-1.4 Rectangular Waveguides

When the top and bottom walls are added to our parallel-plane waveguide, the result is the standard rectangular waveguide used in practice. The two new walls do not really affect any of the results thus far obtained and are not really needed in theory. In practice, however, their presence is required to confine the wave (and to keep the other two walls apart).

Modes It has already been found that a wave may travel in a waveguide in any of a number of configurations. Thus far, this has meant that for any given signal, the number of half-waves of intensity between two walls may be adjusted to suit the requirements. When two more walls exist, between which there may also be half-waves of intensity, some system must be established to ensure a universally understood description of any given propagation mode. The situation had been confused, but after the 1955 IRE (Institute of Radio Engineers) Standards were published, order gradually emerged. Modes in rectangular waveguides are now labeled $TE_{m,n}$ if they are transverse-electric, and $TM_{m,n}$ if they are transverse-magnetic. In each case m and n are integers denoting the number of half-wavelengths of intensity (electric for TE modes and magnetic for TM modes) between each pair of walls. m is measured along the X axis of the waveguide (as is the dimension a), this being the direction along the broader wall of the waveguide; n is measured in the other direction. Both are shown in Fig. 10-7.

The electric field configuration is shown for the $TE_{1,0}$ mode in Fig. 10-7; the magnetic field is left out for the sake of simplicity but will be shown in subsequent figures. It is important to realize that the electric field extends in one direction, but *changes* in this field occur at right angles to that direction. This is similar to a multilane highway with graduated speed lanes. All the cars are traveling in the same direction, but with different speeds in adjoining lanes. Although all cars in any one lane travel (say) north at high speed, along this lane no speed *change* is seen. However, a definite change in speed is noted in the east-west direction as one moves from one lane to the next. In the same way, the electric field in the $TE_{1,0}$ mode extends in the Y direction, but it is constant in that direction while undergoing a half-wave intensity change in the X direction. As a result, $m = 1$, $n = 0$, and the mode is thus $TE_{1,0}$.

The actual mode of propagation is achieved by a specific arrangement of antennas as described in Sec. 10-3.1.

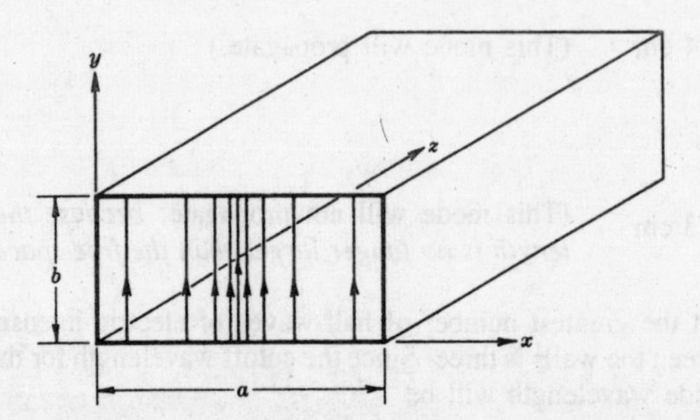

FIGURE 10-7 $TE_{1,0}$ mode in a rectangular waveguide.

The $TE_{m,0}$ modes Since the $TE_{m,0}$ modes do not actually use the broader walls of the waveguide (the reflection takes place from the narrower walls), they are not affected by the addition of the second pair of walls. Accordingly, all the equations so far derived for the parallel-plane waveguide apply to the rectangular waveguide carrying $TE_{m,0}$ modes, without any changes or reservations. The most important of these are Eqs. (10-11), (10-12), (10-14) and (10-15), of which all except the first are universal. To these equations, one other must now be added; this is the equation for the *characteristic wave impedance* of the waveguide. This is obviously related to Z, the characteristic impedance of free space, and is given by

$$Z_0 = \frac{\mathfrak{Z}}{\sqrt{1 - (\lambda/\lambda_0)^2}} \tag{10-16}$$

where Z_0 = characteristic wave impedance of the waveguide
$\mathfrak{Z} = 120\pi = 377\Omega$, as before [Eqs. (8-3) and (8-4)]

Although Eq. (10-16) cannot be derived here, it is logically related to the other waveguide equations and to the free-space propagation conditions of Chap 8. It is seen that the addition of walls has increased the characteristic impedance, as compared with that of free space, for these particular modes of propagation.

It will be seen from Eq. (10-16) that the characteristic wave impedance of a waveguide, for $TE_{m,0}$ modes, increases as the free-space wavelength approaches the cutoff wavelength for that particular mode. This is merely the electrical analog of Eq. (10-15), which states that under these conditions the group velocity decreases. In fact, it is apparent that $\nu_g = 0$ and $Z_0 = \infty$ not only occur simultaneously, when $\lambda = \lambda_0$, but are merely two different ways of stating the same thing: that the waveguide cross-sectional dimensions are now too small to allow this wave to propagate.

A glance at Eq. (10-11) will serve as a reminder that the different $TE_{m,0}$ modes all have different cutoff wavelengths and therefore encounter different characteristic wave impedances. Thus a given signal will encounter one value of Z_o when propagated in the $TE_{3,0}$ mode, and another when propagated in the $TE_{2,0}$ mode. This is the reason for the name "characteristic *wave* impedance." Clearly its value depends here on the mode of propagation as well as on the guide cross-sectional dimensions. Some of the following examples will illustrate this.

The $TE_{m,n}$ modes The $TE_{m,n}$ modes are not used in practice as often as the $TE_{m,0}$ modes (with the possible exception of the $TE_{1,1}$ mode, which does have some practical applications). All the equations so far derived apply to them except for the equation for the cutoff wavelength, which must naturally be different, since the other two walls are also used. The cutoff wavelength for $TE_{m,n}$ modes is given by

$$\lambda_0 = \frac{2}{\sqrt{(m/a)^2 + (n/b)^2}} \tag{10-17}$$

Once again the derivation of this relation is too involved to go into here, but its self-consistency can be shown when it is considered that this is actually the *universal* cutoff wavelength equation for rectangular waveguides, applying equally to all modes, including the $TE_{m,0}$. In the $TE_{m,0}$ mode, $n = 0$, so that Eq. (10-17) reduces to

$$\lambda_0 = \frac{2}{\sqrt{(m/a)^2 + (0/b)^2}} = \frac{2}{\sqrt{(m/a)^2}} = \frac{2}{m/a} = \frac{2a}{m}$$

Since this is identical to Eq. (10-11), it is seen that Eq. (10-17) is consistent. To make calculations involving $TE_{m,n}$ modes, Eq. (10-17) is used to calculate the cutoff wavelength, and then the same equations are used for the other calculations as were used for $TE_{m,0}$ modes.

The $TM_{m,n}$ modes The obvious difference between the $TM_{m,n}$ modes and those described thus far is that the magnetic field here is transverse only, and the electric field has a component in the direction of propagation. This obviously will require a different antenna arrangement for receiving or setting up such modes. Although most of the behavior of these modes is the same as for TE modes, a number of differences do exist. The first such difference is due to the fact that lines of magnetic force are closed loops. Consequently, if a magnetic field exists and is changing in the X direction, it must also exist and be changing in the Y direction. Hence $TM_{m,0}$ modes cannot exist (in rectangular waveguides). TM modes are governed by relations identical to those regulating $TE_{m,n}$ modes, except that the equation for characteristic wave impedance is reversed, because this impedance tends to zero when the free-space wavelength approaches the cutoff wavelength (it tended to infinity for TE modes). The situation is analogous to *current* and *voltage feed* in antennas. The formula for characteristic wave impedance for TM modes is

$$Z_0 = \mathfrak{X}\sqrt{1 - \left(\frac{\lambda}{\lambda_0}\right)^2} \qquad \textbf{(10-18)}$$

Equation (10-18) yields impedance values that are always less than 377 Ω, and this is the main reason why TM modes are sometimes used, especially $TM_{1,1}$. It is sometimes advantageous to feed a waveguide directly from a coaxial transmission line, in which case the waveguide input impedance must be a good deal lower than 377 Ω.

Just as the $TE_{1,1}$ is the principal $TE_{m,n}$ mode, so the main TM mode is the $TM_{1,1}$.

Example 10-3 Calculate the formula for the cutoff wavelength, in a standard rectangular waveguide, for the $TM_{1,1}$ mode.

Standard rectangular waveguides have a 2:1 aspect ratio, so that $b = a/2$. Therefore

$$\lambda_0 = \frac{2}{\sqrt{(m/a)^2 + (n/b)^2}} = \frac{2}{\sqrt{(m/a)^2 + (2n/a)^2}} = \frac{2a}{\sqrt{m^2 + 4n^2}}$$

But here $m = n = 1$. Therefore,

$$\lambda_0 = \frac{2a}{\sqrt{1+4}} = \frac{2a}{\sqrt{5}} = 0.894a$$

It is thus seen that the cutoff wavelength for the $TE_{1,1}$ and $TM_{1,1}$ modes in a rectangular waveguide is less than for the $TE_{2,0}$ mode, and, of course, for the $TE_{1,0}$

mode. Accordingly, a bigger waveguide is needed to propagate a given frequency than for the dominant mode. In all fairness, however, it should be pointed out that a square waveguide would be used for the symmetrical modes, in which case their cutoff wavelength becomes $\sqrt{2}a$, which is some improvement.

We must not lose sight of the fact that the dominant mode is the one most likely to be used in practice, with the others employed only for special applications. There are several reasons for this. For instance, it is much easier to excite modes such as the $TE_{1,0}$, $TE_{2,0}$ or $TM_{1,1}$ than modes such as the $TE_{3,7}$ or $TM_{9,5}$. The earlier modes also have the advantage that their cutoff wavelengths are larger than those of the later modes (the dominant mode is best for this, of course). Therefore smaller waveguides can be used for any given frequency. Furthermore, the dominant mode has the advantage that it can be propagated in a guide that is too small to propagate any other mode, thus ensuring that no energy loss can occur through the spurious generation of other modes. However, the higher modes do have some advantages: it may actually be more convenient to use larger waveguides at the highest frequencies (see Table 10-1), and higher modes can also be employed if the propagation of several signals through the one waveguide is contemplated. Examples are now given to illustrate the major points made so far.

Example 10-4 Calculate the characteristic wave impedance for the data of Examples 10-1 and 10-2.

In Example 10-1, ρ was calculated to be 0.553. Thus

$$Z_0 = \frac{\mathfrak{L}}{\sqrt{1-(\lambda/\lambda_0)^2}} = \frac{\mathfrak{L}}{\rho} = \frac{120\pi}{0.553} = 682\ \Omega$$

Similarly, for Example 10-2,

$$Z_0 = \frac{\mathfrak{L}}{\rho} = \frac{120\pi}{0.661} = 570\ \Omega$$

Example 10-5 A rectangular waveguide measures 3×4.5 cm internally and has a 9-GHz signal propagated in it. Calculate the cutoff wavelength, the guide wavelength, the group and phase velocities and the characteristic wave impedance for (*a*) the $TE_{1,0}$ mode and (*b*) the $TM_{1,1}$ mode.

Calculating the free-space wavelength gives

$$\lambda = \frac{v_c}{f} = \frac{3 \times 10^{10}}{9 \times 10^9} = 3.33 \text{ cm}$$

(*a*) The cutoff wavelength will be

$$\lambda_0 = \frac{2a}{m} = \frac{2 \times 4.5}{1} = 9 \text{ cm}$$

Calculating ρ, for convenience, gives

$$\rho = \sqrt{1-\left(\frac{\lambda}{\lambda_0}\right)^2} = \sqrt{1-\left(\frac{3.33}{9}\right)^2} = \sqrt{1-0.137} = 0.93$$

Thus the guide wavelength is

$$\lambda_p = \frac{\lambda}{\rho} = \frac{3.33}{0.93} = 3.58 \text{ cm}$$

The group and phase velocities are simply found from

$$v_g = v_c\rho = 3 \times 10^8 \times 0.93 = 2.79 \times 10^8 \text{ m/s}$$

$$v_p = \frac{v_c}{\rho} = \frac{3 \times 10^8}{0.93} = 3.23 \times 10^8 \text{ m/s}$$

The characteristic wave impedance is

$$Z_0 = \frac{\mathfrak{Z}}{\rho} = \frac{120\pi}{0.93} = 405 \ \Omega$$

(*b*) Running through the same routine for the $TM_{1,1}$ mode, we obtain

$$\lambda_0 = \frac{2}{\sqrt{(m/a)^2 + (n/b)^2}} = \frac{2}{\sqrt{(1/4.5)^2 + (1/3)^2}}$$

$$= \frac{2}{\sqrt{0.0494 + 0.1111}} = \frac{2}{0.4} = 5 \text{ cm}$$

$$\rho = \sqrt{1 - \left(\frac{3.33}{5}\right)^2} = \sqrt{1 - 0.444} = 0.746$$

$$\lambda_p = \frac{\lambda}{\rho} = \frac{3.33}{0.746} = 4.6 \text{ cm}$$

$$v_g = v_c\rho = 3 \times 10^8 \times 0.746 = 2.24 \times 10^8 \text{ m/s}$$

$$v_p = \frac{v_c}{\rho} = \frac{3 \times 10^8}{0.746} = 4.02 \times 10^8 \text{ m/s}$$

Because this is a TM mode, Eq. (10-18) must be used to calculate the characteristic wave impedance; hence

$$Z_0 = \mathfrak{Z}\rho = 120\pi \times 0.745 = 281 \ \Omega$$

Example 10-6 A waveguide has an internal breadth a of 3 cm, and carries the dominant mode of a signal of unknown frequency. If the characteristic wave impedance is 500 Ω, what is this frequency?

$$\lambda_0 = \frac{2a}{m} = \frac{2 \times 3}{1} = 6 \text{ cm}$$

$$\frac{\mathfrak{Z}}{Z_0} = \sqrt{1 - \left(\frac{\lambda}{\lambda_0}\right)^2}$$

$$\left(\frac{\mathfrak{Z}}{Z_0}\right)^2 = 1 - \left(\frac{\lambda}{\lambda_0}\right)^2 = 1 - \left(\frac{120\pi}{500}\right)^2 = 0.57$$

$$\left(\frac{\lambda}{\lambda_0}\right)^2 = 1 - 0.57 = 0.43$$

$$\frac{\lambda}{\lambda_0} = \sqrt{0.43} = 0.656$$

$$\lambda = 0.656\lambda_0 = 0.656 \times 6 = 3.93 \text{ cm}$$

$$f = \frac{v_c}{\lambda} = \frac{3 \times 10^{10}}{3.93} = 7.63 \times 10^9 = 7.63 \text{ GHz}$$

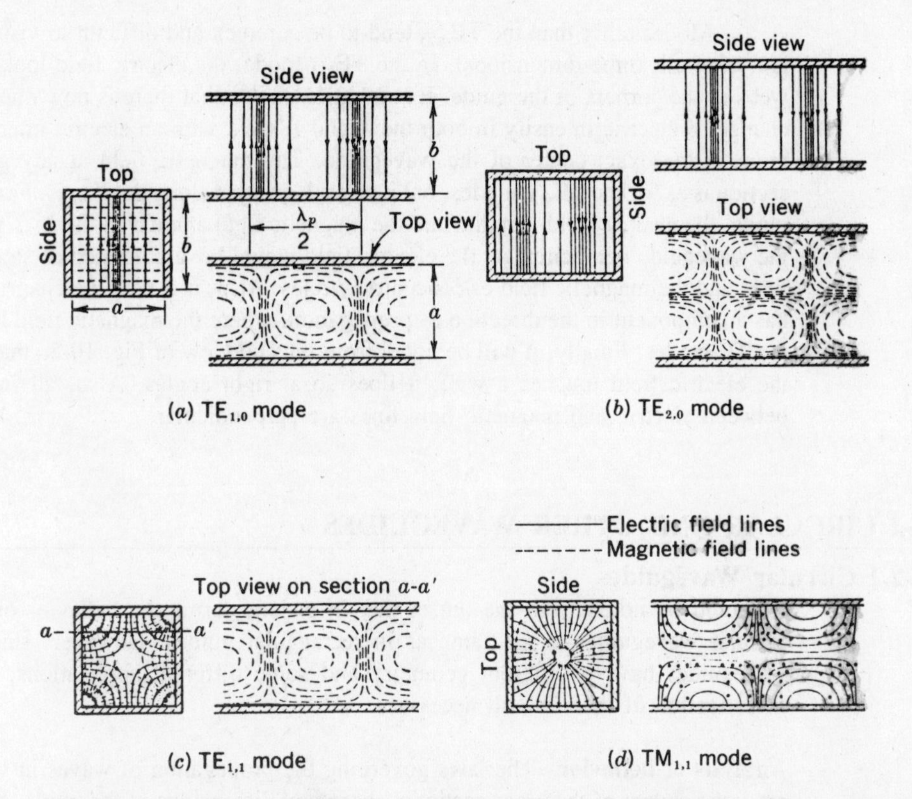

(a) $TE_{1,0}$ mode

(b) $TE_{2,0}$ mode

(c) $TE_{1,1}$ mode

(d) $TM_{1,1}$ mode

FIGURE 10-8 Field patterns of common modes in rectangular waveguides. ***(After A. B. Bronwell and R. E. Beam, Theory and Application of Microwaves, McGraw-Hill Book Company, New York, 1947. Used with permission of McGraw-Hill Book Company.)***

Field patterns The electric and magnetic field patterns for the dominant mode are shown in Fig. 10-8*a*. As was indicated, the electric field exists only at right angles to the direction of propagation, whereas the magnetic field has a component in the direction of propagation as well as a normal component. The electric field is maximum at the center of the waveguide for this mode and drops off sinusoidally to zero intensity at the walls, as shown. The magnetic field is in the form of (closed) loops, which lie in planes normal to the electric field, i.e., parallel to the top and bottom of the guide. This magnetic field is the same in all those planes, regardless of the position of such a plane along the Y axis, as evidenced by the equidistant dashed lines in the end view. This applies to all $TE_{m,0}$ modes. The whole configuration travels down the waveguide with the group velocity, but at any instant of time the whole waveguide is filled by these fields. The distance between any two identical points in the Z direction is λ_p, as implied in Fig. 10-8*a*.

The field patterns for the $TE_{2,0}$ mode, as shown in Fig. 10-8*b*, are very similar. Indeed, the only differences are that there are now two half-wave variations of the electric (and magnetic) field in the X–Y plane, as shown. The field patterns for the higher $TE_{m,0}$ modes are logical extensions of those for the first two.

Modes other than the $TE_{m,0}$ tend to be complex and difficult to visualize; they are, after all, three-dimensional. In the $TE_{1,1}$ mode, the electric field looks like cobwebs in the corners of the guide. Examination shows that there is now one half-wave change of electric intensity in both the X and Y axes, with an electric intensity maximum in the exact center of the waveguide. The magnetic field at any given cross section is as for the $TE_{m,0}$ modes, but it now also varies along the Y axis. For the $TM_{1,1}$ mode, the electric field is radial and the magnetic field annular in the X–Y plane. Had the waveguide been circular, the electric field would have consisted of straight radial lines and the magnetic field of concentric circles. Also, it is now the electric field that has a component in the direction of propagation, where the magnetic field had one for the TE modes. Finally, it will be noted from the end view of Fig. 10-8*c* that wherever the electric field touches a wall, it does so at right angles. Also, all intersections between electric and magnetic field lines are perpendicular.

10-2 CIRCULAR AND OTHER WAVEGUIDES

10-2.1 Circular Waveguides

It should be noted from the outset that in general terms the behavior of waves in circular waveguides is the same as in rectangular guides. However, since circular waveguides have a different geometry and some different applications, a separate investigation of them is still necessary.

Analysis of behavior The laws governing the propagation of waves in waveguides are independent of the cross-sectional shape and dimensions of the guide. As a result, all the parameters and definitions evolved for rectangular waveguides apply to circular waveguides, with the minor modification that modes are labeled somewhat differently. All the equations also apply here except, obviously, the formula for cutoff wavelength. This must be different because of the different geometry, and it is given by

$$\lambda_0 = \frac{2\pi r}{(kr)} \qquad \textbf{(10-19)}$$

where r = radius (internal) of waveguide
(kr) = solution of a Bessel function equation

To facilitate calculations for circular waveguides, values of (kr) are shown in Table 10-2 for the circular waveguide modes most likely to be encountered.

TABLE 10-2 Values of (*kr*) for the Principal Modes in Circular Waveguides

TE				*TM*			
MODE	(*kr*)	MODE	(*kr*)	MODE	(*kr*)	MODE	(*kr*)
$TE_{0,1}$	3.83	$TE_{0,2}$	7.02	$TM_{0,1}$	2.40	$TM_{0,2}$	5.52
$TE_{1,1}$	1.84	$TE_{1,2}$	5.33	$TM_{1,1}$	3.83	$TM_{1,2}$	7.02
$TE_{2,1}$	3.05	$TE_{2,2}$	6.71	$TM_{2,1}$	5.14	$TM_{2,2}$	8.42

Example 10-7 Calculate the cutoff wavelength, the guide wavelength and the characteristic wave impedance of a circular waveguide whose internal diameter is 4 cm, for a 10-GHz signal propagated in it in the $TE_{1,1}$ mode.

$$\lambda = \frac{v_c}{f} = \frac{3 \times 10^{10}}{10 \times 10^9} = 3 \text{ cm}$$

$$\lambda_0 = \frac{2\pi r}{(kr)} = \frac{2\pi \times 4/2}{1.84} \quad \text{(1.84 from table)}$$

$$= \frac{4\pi}{1.84} = 6.83 \text{ cm}$$

$$\lambda_p = \frac{\lambda}{\sqrt{1 - (\lambda/\lambda_0)^2}} = \frac{3}{\sqrt{1 - (3/6.83)^2}} = \frac{3}{\sqrt{1 - 0.193}}$$

$$= \frac{3}{0.898} = 3.34 \text{ cm}$$

$$Z_0 = \frac{\mathfrak{X}}{\sqrt{1 - (\lambda/\lambda_0)^2}} = \frac{120\pi}{0.898} = 420 \ \Omega$$

One of the differences in behavior between circular and rectangular waveguides is shown in Table 10-2. Since the mode with the largest cutoff wavelength is the one with the smallest value of (kr), the $TE_{1,1}$ mode is dominant in circular waveguides. The cutoff wavelength for this mode is $\lambda_0 = 2\pi r/1.84 = 3.41r = 1.7d$, where d is the diameter. Another difference lies in the different method of mode labeling, which must be used because of the circular cross section. The integer m now denotes the number of *full*-wave intensity variations around the circumference, and n represents the number of half-wave intensity changes radially out from the center to the wall. It is seen that *cylindrical coordinates* are used here.

Field patterns Figure 10-9 shows the patterns of electric and magnetic intensity in circular waveguides for the two most common modes. The same general rules apply as for rectangular guide patterns. There are the same travel down the waveguide and the same repetition rate λ_p. The same conventions have been adopted, except that now open circles are used to show lines (electric or magnetic, depending on the mode) coming out of the page, and full dots are used for lines going into the page.

Disadvantages The first drawback associated with a circular waveguide is that its cross section will be much bigger in area than that of a corresponding rectangular waveguide used to carry the same signal. This is best shown with an example.

Example 10-8 Calculate the ratio of the cross section of a circular waveguide to that of a rectangular one if each is to have the same cutoff wavelength for its dominant mode.

For the dominant ($TE_{1,1}$) mode in the circular waveguide, we have

$$\lambda_0 = \frac{2\pi r}{(kr)} = \frac{2\pi r}{1.84} = 3.41r$$

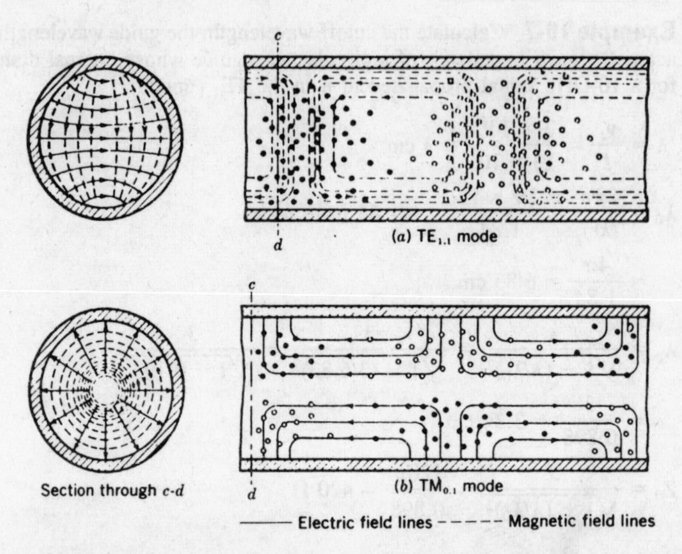

FIGURE 10-9 Field patterns of two common modes in circular waveguides. ***(From A. B. Bronwell and R. E. Beam, Theory and Application of Microwaves, McGraw-Hill Book Company, New York, 1947. Used with permission of McGraw-Hill Book Company.)***

The area of a circle with a radius r is given by

$$A_c = \pi r^2$$

In the rectangular waveguide, for the $TE_{1,0}$ mode,

$$\lambda_0 = \frac{2a}{1} = 2a$$

If the two cutoff wavelengths are to be the same, then

$$2a = 3.41r$$

$$a = \frac{3.41r}{2} = 1.705r$$

The area of a standard rectangular waveguide is

$$A_r = ab = a\frac{a}{2} = \frac{a^2}{2} = \frac{(1.705r)^2}{2} = 1.45r^2$$

The ratio of the areas will thus be

$$\frac{A_c}{A_r} = \frac{\pi r^2}{1.45r^2} = 2.17$$

It follows from Example 10-8 that (apart from any other consideration) the space occupied by a rectangular waveguide system would be considerably less than that for a circular system. This obviously weighs against the use of circular guides in some applications.

Another problem with circular waveguides is that it is possible for the plane of polarization to rotate during the wave's travel through the waveguide. This may happen because of roughness or discontinuities in the walls or departure from true circular cross section. Taking the $TE_{1,1}$ mode as an example, it is seen that the electric field usually starts out being horizontal, and thus the receiving mechanism at the other end of the guide will be arranged accordingly. If this polarization now changes unpredictably before the wave reaches the far end, as it well might, the signal will be reflected rather than received, with the obvious consequences. This mitigates against the use of the $TE_{1,1}$ mode.

Advantages and special applications Circular waveguides are easier to manufacture than rectangular ones. They are also easier to join together, in the usual plumbing fashion. In addition, rotation of polarization may be overcome by the use of modes that are rotationally symmetrical. $TM_{0,1}$ is one such mode, as seen in Fig. 10-9, and $TE_{0,1}$ (not shown) is another. The principal current application of circular waveguides is in rotational joins, as shown in Sec. 10-3.2. The $TM_{0,1}$ mode is likely to be preferred to the $TE_{0,1}$ mode, since it requires a smaller diameter for the same cutoff wavelength.

However, the $TE_{0,1}$ mode does have a practical application, albeit only an experimental one at present. It may be shown [2] that, especially at frequencies in excess of 10 GHz, this is the mode with significantly the lowest attenuation per unit length of waveguide. That is to say, there is no mode in either rectangular or circular waveguides (or any others, for that matter) for which attenuation is lower. Although that property is not of the utmost importance for short runs of up to a few meters, it becomes significant if longer-distance waveguide transmission is considered.

10-2.2 Other Waveguides

There are situations in which properties other than those possessed by rectangular or circular waveguides are desirable. For such occasions, ridged or flexible waveguides may be used, and these are now described.

Ridged waveguides Rectangular waveguides are sometimes made with single or double ridges, as shown in Fig. 10-10. The principal effect of such ridges is to lower the value of the cutoff wavelength. In turn, this allows a guide with smaller dimensions

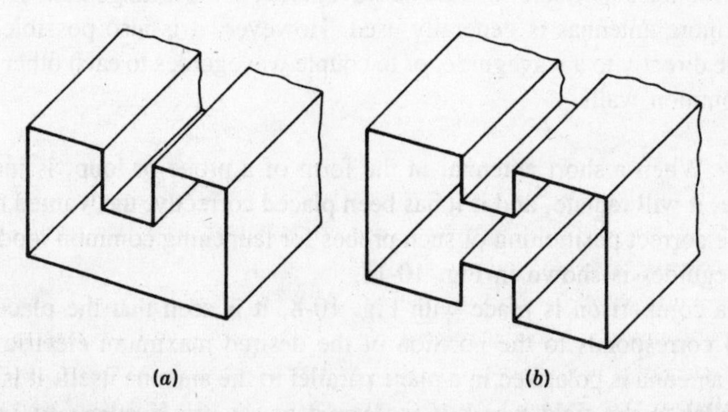

FIGURE 10-10 Ridged waveguides. (*a*) Single ridge; (*b*) double ridge.

to be used for any given frequency. Another benefit of having a ridge in a waveguide is to increase the useful frequency range of the guide; it may be shown that the dominant mode is the only one to propagate in the ridged guide over a wider frequency range than in any other waveguide. The ridged waveguide thus has a markedly greater bandwidth than an equivalent rectangular guide. However, it should be noted that ridged waveguides generally have more attenuation per unit length than rectangular waveguides and are thus not used in great lengths for standard applications.

Flexible waveguides It is sometimes required to have a waveguide section capable of movement. This may be bending, twisting, stretching or vibration, possibly continuously, and this must not cause undue deterioration in performance. Applications such as these call for flexible waveguides, of which there are several types. Among the more popular is a copper or aluminum tube having an elliptical cross section, small transverse corrugations and transitions to rectangular waveguides at the two ends. These transform the $TE_{1,1}$ mode in the flexible waveguide into the $TE_{1,0}$ mode at either end. This waveguide is of continuous construction, and thus joins and separate bends are not required. It may have a polyethylene or rubber outer cover and bends easily but cannot be readily twisted. Power-handling ability and SWR are fairly similar to those of rectangular waveguides of the same size, but attenuation in dB/m is about five times as much.

10-3 WAVEGUIDE COUPLING, MATCHING AND ATTENUATION

Having treated the theory of waveguides, it is now necessary to consider the practical aspects of their use. Methods of launching modes in waveguides will now be described in detail, as will waveguide coupling and interconnection, various junctions, accessories, methods of impedance matching and also attenuation. Auxiliary components are considered in Sec. 10-5.

10-3.1 Methods of Exciting Waveguides

In order to launch a particular mode in a waveguide, some arrangement or combination of one or more antennas is generally used. However, it is also possible to couple a coaxial line directly to a waveguide, or to couple waveguides to each other by means of slots in common walls.

Antennas When a short antenna, in the form of a probe or loop, is inserted into a waveguide, it will radiate, and if it has been placed correctly, the wanted mode will be set up. The correct positioning of such probes for launching common modes in rectangular waveguides is shown in Fig. 10-11.

If a comparison is made with Fig. 10-8, it is seen that the placement of the antenna(s) corresponds to the position of the desired maximum electric field. Since each such antenna is polarized in a plane parallel to the antenna itself, it is placed so as to be parallel to the field which it is desired to set up. Needless to say, the same arrangement may be used at the other end of the waveguide to receive each such mode.

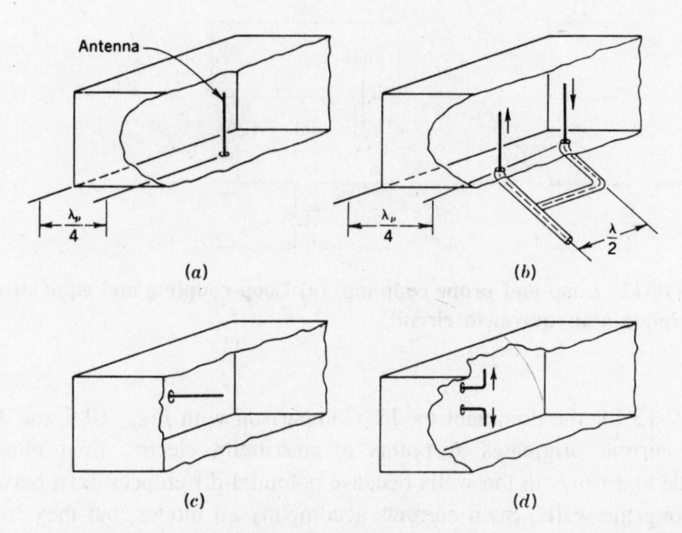

FIGURE 10-11 Methods of exciting common modes in rectangular waveguides. (*a*) $TE_{1,0}$; (*b*) $TE_{2,0}$; (*c*) $TM_{1,1}$; (*d*) $TE_{1,1}$.

When two or more antennas are employed, care must be taken to ensure that they are fed in correct phase; otherwise the desired mode will not be set up. Thus, it is seen that the two antennas used for the $TE_{1,1}$ mode are in phase (in *feed,* not in actual orientation). However, the two antennas used to excite the $TE_{2,0}$ mode are fed 180° out of phase, as required by the field pattern of Fig. 10-8*b*. Phase differences between antennas are normally achieved by means of additional pieces of transmission line, as shown here. Higher $TE_{m,0}$ modes would be launched by an extension of the principle shown, but note that their launching in a pure form might be difficult. The antenna placement for the $TE_{3,0}$ mode, requiring one antenna in the center of the guide, would almost certainly launch some $TE_{1,0}$ mode also. Again, the antenna used to launch the $TM_{1,1}$ mode is at right angles to the antennas used to launch the TE modes, because of the different orientation of the electric field. Finally, note that the depth of insertion of such a probe will determine the power it couples and the impedance it encounters. Hence adjustment of this depth may be used for impedance matching as an alternative to a stub on the coaxial line.

The $TM_{0,1}$ mode may be launched in a circular waveguide, as shown in Fig. 10-11*c*, or else by means of a loop antenna located in a plane perpendicular to the plane of the probe, so as to have its area intersected by a maximum number of magnetic field lines. It is thus seen that probes couple primarily to an electric field and loops to a magnetic field, but in each case both an electric and a magnetic field will be set up because the two are inseparable. Figure 10-12 shows equivalent circuits of probe and loop coupling and reinforces the idea of both fields being present regardless of which one is being primarily coupled to.

Slot coupling It can be appreciated that current must flow in the walls of a waveguide in which electromagnetic waves propagate. The pattern of such current flow is shown

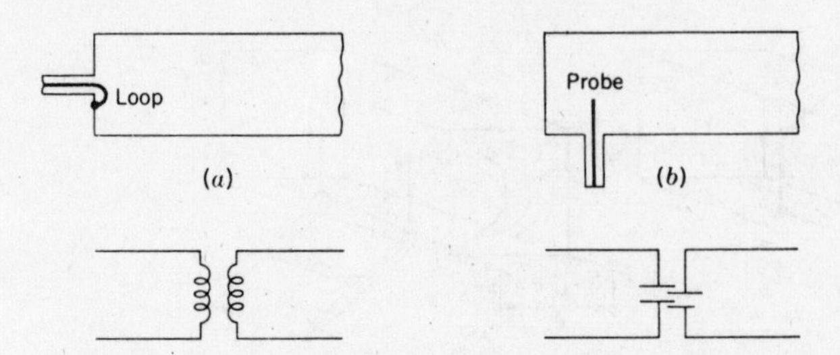

FIGURE 10-12 Loop and probe coupling. (*a*) Loop coupling and equivalent circuit; (*b*) probe coupling and equivalent circuit.

in Fig. 10-13 for the dominant mode. Comparison with Figs. 10-7 and 10-8*a* shows that the current originates at points of maximum electric field intensity in the waveguide and flows in the walls because potential differences exist between various points along the walls. Such currents accompany all modes, but they have not been shown previously, to simplify the field pattern diagrams.

If a hole or slot is made in a waveguide wall, energy will escape from the waveguide through the slot or possibly enter into the waveguide from outside. As a result, coupling by means of one or more slots seems a satisfactory method of feeding energy into a waveguide from another waveguide or cavity resonator (or, alternatively, of taking energy out).

When coupling does take place, it is either because electric field lines that would have been terminated by a wall now enter the second waveguide or because the placement of a slot interrupts the flow of wall current, and therefore a magnetic field is set up extending into the second guide. Sometimes, depending on the orientation of the slot, both effects take place. In Fig.10-13 slot 1 is situated in the center of the top wall, and therefore at a point of maximum electric intensity; thus a good deal of electric coupling takes place. On the other hand, a fair amount of wall current is interrupted, so that there will also be considerable magnetic coupling. The position of slot 2 is at a point of zero electric field, but it interrupts sizable wall current flow; thus coupling

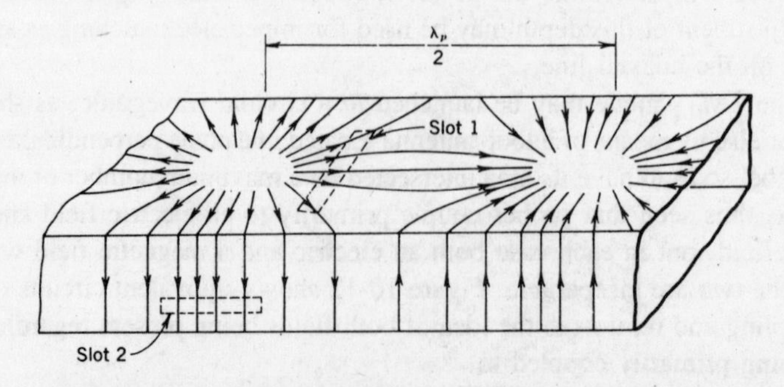

FIGURE 10-13 Slot coupling and current flow in waveguide walls for the dominant mode. ***(Adapted from M. H. Cufflin, The $H_{0,1}$ Mode and Communications, Point-to-Point Telecommunications, with permission of the publishers.)***

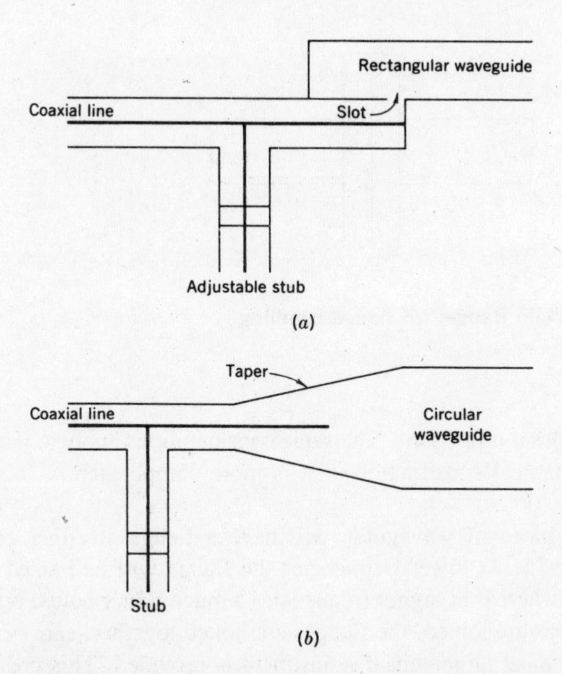

FIGURE 10-14 Coupling to waveguides from coaxial lines by means of (*a*) a slot; (*b*) a taper section.

here is primarily through the magnetic field. Slots may be situated at other points in the waveguide walls, and in each case coupling will take place. It will be determined in type and amount by the position and orientation of each slot, and also by the thickness of the walls.

Slot coupling is very often used between adjoining waveguides, as in directional couplers (see Sec. 10-5.1), or between waveguides and cavity resonators (see Sec. 10-4). Because radiation will take place from a slot, such slots may be used as antennas, and in fact they very often are.

Direct coupling to coaxial lines When a particular microwave transmission system consists of partly coaxial and partly waveguide sections, there are two standard methods of interconnection, as shown in Fig. 10-14. Diagram *a* shows a slot in a common wall, whereby energy from the coaxial line is coupled into the waveguide. In diagram *b*, coupling is by means of a taper section, in which the TEM mode in the coaxial line is transformed into the dominant mode in the waveguide. In each instance an impedance mismatch is likely to exist, and hence stub matching on the line is used as shown.

10-3.2 Waveguide Joins

When waveguide pieces or components are joined together, the coupling is generally by means of some sort of flange. The function of such a flange is to ensure a smooth mechanical junction and suitable electrical characteristics, particularly low external

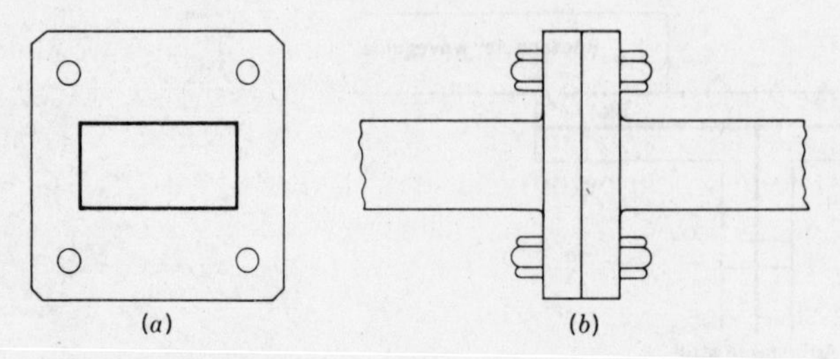

FIGURE 10-15 *(a)* **Plain flange;** *(b)* **flange coupling.**

radiation and low internal reflections. The same considerations apply to a rotating join, except that the mechanical construction of it is more complicated.

Flanges A typical piece of waveguide will have a flange at either end, such as illustrated in Fig. 10-15. At lower frequencies the flange will be brazed or soldered onto the waveguide, whereas at higher frequencies a much flatter butted plain flange is used. When two pieces are joined, the flanges are bolted together, care being taken to ensure perfect mechanical alignment if adjustment is provided. This prevents an unwanted bend or step, either of which would produce undesirable reflections. It follows that the guide ends and flanges must be smoothly finished to avoid discontinuities at the junction.

It is obviously easier to align individual pieces correctly if there is some adjustment, so that waveguides with smaller dimensions are sometimes provided with threaded flanges, which can be screwed together with ring nuts.

With waveguides naturally reduced in size when frequencies are raised, a join discontinuity becomes larger in proportion to the signal wavelength and the guide dimensions. Thus discontinuities at higher frequencies become more troublesome. To counteract this, a small gap may be purposely left between the waveguides, as shown in Fig. 10-16. The diagram shows a *choke join* consisting of an ordinary flange and a *choke flange* connected together. To compensate for the discontinuity which would otherwise be present, a circular *choke ring* of *L* cross section is used in the choke

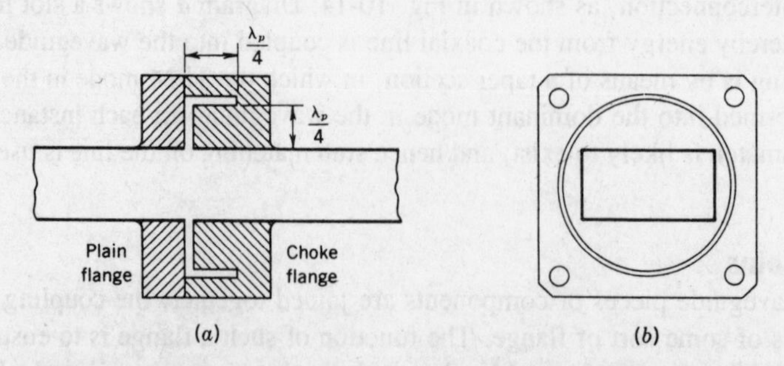

FIGURE 10-16 *(a)* **Cross section of choke join;** *(b)* **end view of choke flange.**

flange, in order to reflect a short circuit at the junction of the waveguides. This is possible because the total length of the ring cross section, as shown, is $\lambda_p/2$, and the far end is short-circuited. Thus an electrical short circuit is placed at a surface where a mechanical short circuit would be difficult to achieve.

Unlike the plain flange, the choke flange is frequency-sensitive, but optimum design can ensure a reasonable bandwidth (perhaps 10 percent of the center frequency) over which SWR does not exceed 1.05.

Rotating joins As previously mentioned, rotating joins are often used, as in radar, where a waveguide is connected to a horn antenna feeding a paraboloid reflector which must rotate for tracking. A rotating join involving circular waveguides is the most common and will be the one described here.

A typical rotary join is shown in Fig. 10-17, which (for simplicity) shows the electrical components only. The mechanical components may have varying degrees of complexity but are of subsidiary interest here. The rotating part of the waveguide is circular and carries the $TM_{0,1}$ mode, whereas the rectangular waveguide pieces leading in and out of the join carry the dominant $TE_{1,0}$ mode. The circular waveguide has a diameter which ensures that modes higher than the $TM_{0,1}$ cannot propagate. The dominant $TE_{1,1}$ mode in the circular guide is suppressed by a ring filter (as shown), which tends to short-circuit the electric field for that mode, while not affecting the electric field of the $TM_{0,1}$ mode (which is everywhere perpendicular to the ring). A choke gap is left around the circular guide join to reduce any mismatch that may occur and any

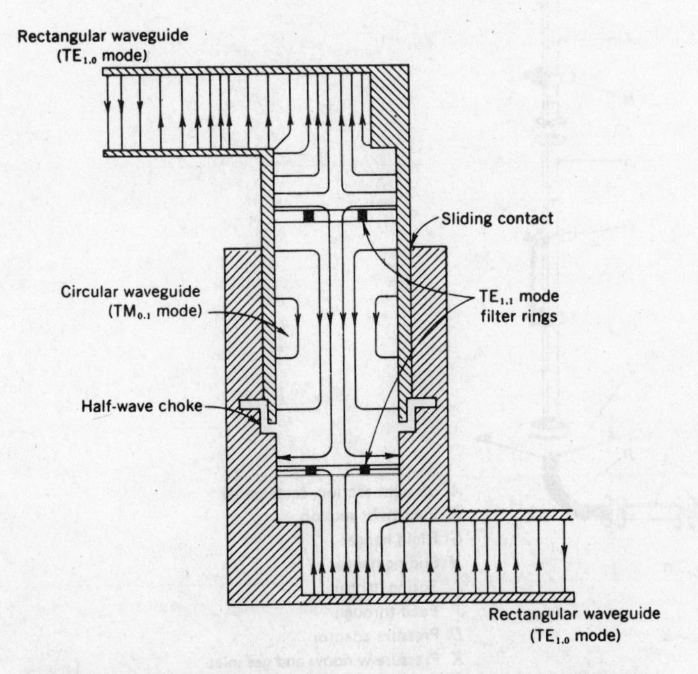

FIGURE 10-17 Rotating join, showing electric field patterns.

rubbing of the metal area during the rotation. Some sort of obstacle is often placed at each circular-rectangular waveguide junction to compensate for reflection; such obstacles are described in Sec. 10-3.5.

10-3.3 Basic Accessories

A manufacturer's catalog shows a very large number of accessories which can be obtained with waveguides for any number of purposes. Figure 10-18 shows a typical rectangular waveguide run which illustrates a number of such accessories; some of them are now described.

Bends and corners As indicated in Fig. 10-18, changes of direction are often required, in which case a bend or a corner may be used. Since these are discontinuities,

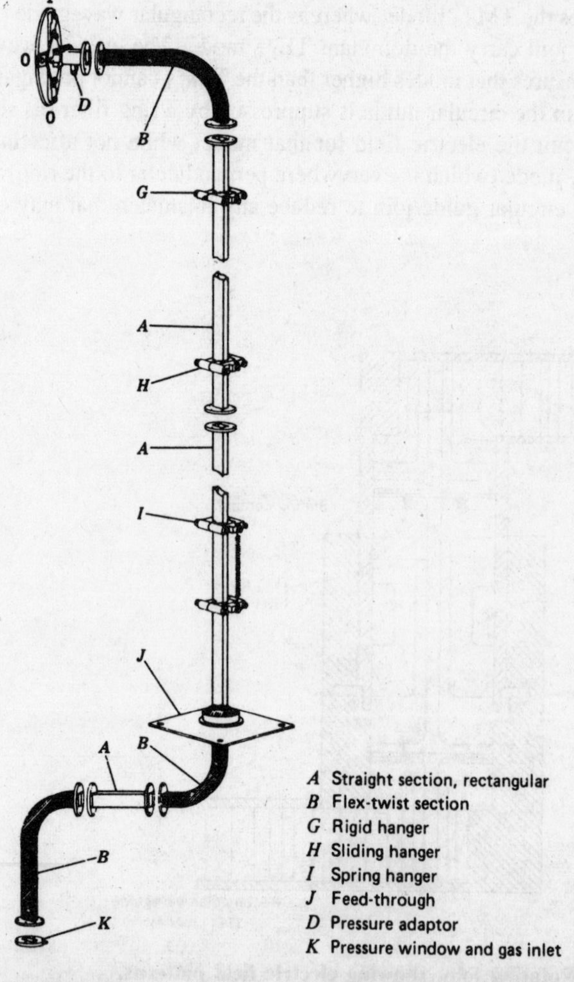

FIGURE 10-18 Rectangular waveguide run. ***(Courtesy of Andrew Antennas of Australia.)***

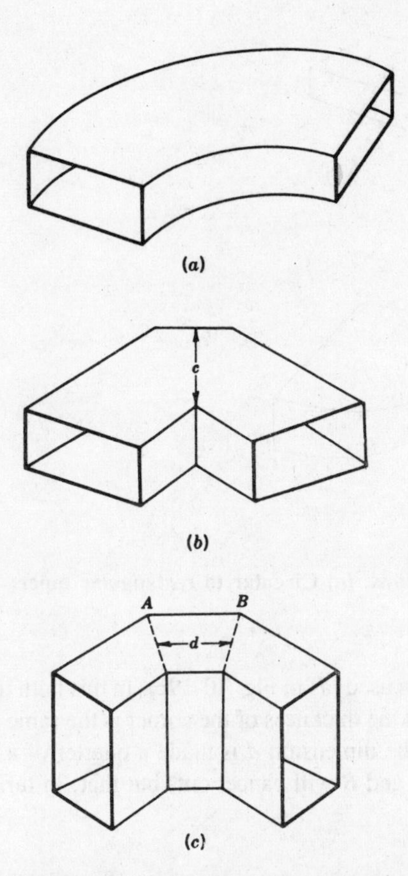

FIGURE 10-19 Waveguide bend and corners. (*a*) H-plane bend; (*b*) H-plane mitered corner; (*c*) E-plane double-mitered corner.

SWR will be increased either because of reflections from a corner, or because of a different group velocity in a piece of bent waveguide.

An H-plane bend (shown in Fig. 10-19*a*) is a piece of waveguide smoothly bent in a plane parallel to the magnetic field for the dominant mode (hence the name). In order to keep the reflections in the bend small, its length is made several wavelengths. Alternatively, if this is undesirable because of size, or if the bend must be sharp, it is possible to minimize reflections by making the mean length of the bend an integral number of guide wavelengths. In that case some cancellation of reflections takes place. It must be borne in mind, however, that the sharper the bend, the greater the mismatch introduced.

For the larger wavelengths a bend is rather clumsy, and a corner may be used instead. Because such a corner would introduce intolerable reflections if it were simply a 90° corner, a part of it is cut, and the corner is then said to be *mitered*, as in Fig. 10-19*b*. The dimension *c* depends on wavelength, but if it is correctly chosen, reflections will be almost completely eliminated. An H-plane corner is shown. With an E-plane corner, there is a risk of voltage breakdown across the distance *c*, which would naturally be fairly small in such a corner. Thus if a change of direction in the E plane is

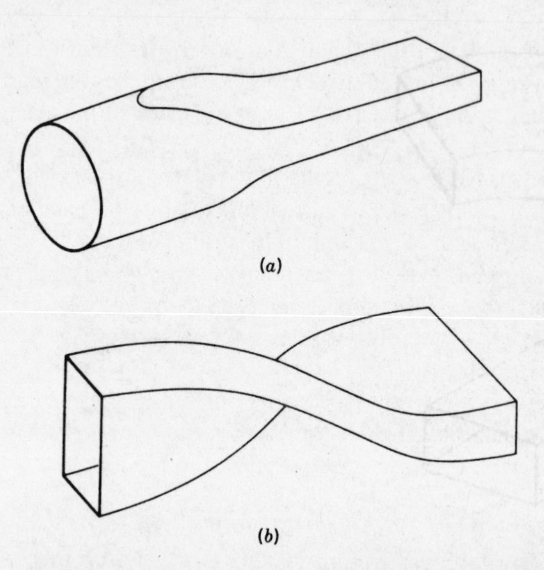

FIGURE 10-20 Waveguide transitions. (*a*) Circular to rectangular taper; (*b*) 90° twist.

required, a double-mitered corner is used (as in Fig. 10-19*c*). In this both the inside and outside corner surfaces are cut, and the thickness of the corner is the same as that of the straight portion of waveguide. If the dimension *d* is made a quarter of a guide wavelength, reflections from corners *A* and *B* will cancel out, but that, in turn, makes the corner frequency-sensitive.

Taper and twist sections When it is necessary to join waveguides having different dimensions or different cross-sectional shapes, taper sections may be used. Again, some reflections will take place, but they can be reduced if the taper section is made gradual, as shown for the circular-rectangular taper of Fig. 10-20*a*. The taper shown may have a length of two or more wavelengths, and if the rectangular section carries the dominant mode, the $TE_{1,1}$ mode will be set up in the circular section, and vice versa.

Finally, if a change of polarization direction is required, a twist section may be used (as shown in Fig. 10-20*b*), once again extending over two or more wavelengths. As an alternative, such a twist may be incorporated in a bend, such as those shown in Fig. 10-18.

10-3.4 Multiple Junctions

When it is required to combine two or more signals (or split a signal into two or more parts) in a waveguide system, some form of multiple junction must be used. For simpler interconnections T-shaped junctions are used, whereas more complex junctions may be *hybrid T* or *hybrid rings*. In addition to being junctions, these components also have other applications, and hence they will now be described in some detail.

T junctions Two examples of the T junction, or *tee,* are shown in Fig. 10-21, together with their transmission-line equivalents. Once again they are referred to as E- or H-plane trees, depending on whether they are in the plane of the electric field or the magnetic field.

All three arms of the H-plane tee lie in the plane of the magnetic field, which divides among the arms. This is thus a current junction, i.e., a parallel one, as shown by the transmission-line equivalent circuit. In a similar way, the E-plane tee is a voltage or series junction, as indicated. Each junction is symmetrical about the central arm, so that the signal to be split up is fed into it (or the signals to be combined are taken from it). However, some form of impedance matching is generally required to prevent unwanted reflections.

T junctions (particularly the E-plane tee) may themselves be used for impedance matching, in a manner identical to the short-circuited transmission-line stub. The vertical arm is then provided with a sliding piston to produce a short circuit at any desired point.

Hybrid junctions If another arm is added to either of the T junctions, then a *hybrid T junction,* or *magic tee,* is obtained; it is shown in Fig. 10-22. Such a junction is symmetrical about an imaginary plane bisecting arms 3 and 4 and has some very useful and interesting properties.

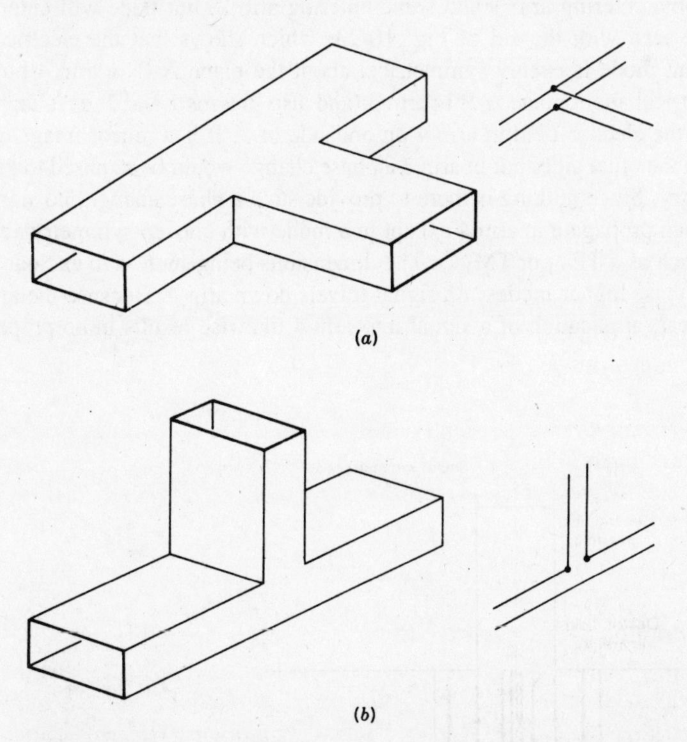

FIGURE 10-21 T junctions (tees) and their equivalent circuits. (*a*) H-plane tee; (*b*) E-plane tee.

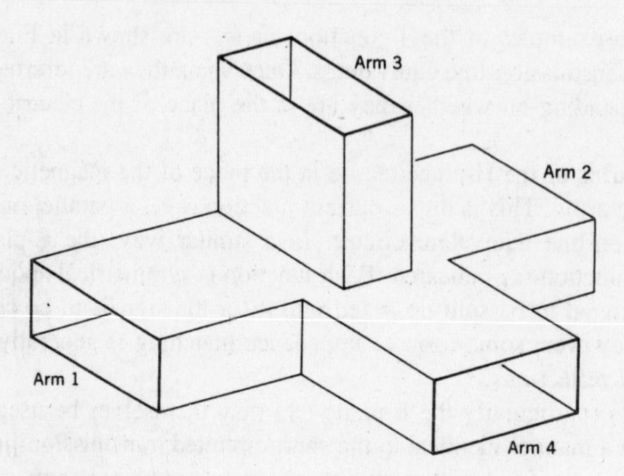

FIGURE 10-22 Hybrid T junction (magic tee).

The basic property is that arms 3 and 4 are both connected to arms 1 and 2 but not to each other. This applies for the dominant mode only, provided each arm is terminated in a correct load.

If a signal is applied to arm 3 of the magic tee, it will be divided at the junction, with some entering arm 1 and some entering arm 2, but none will enter arm 4. This may be seen with the aid of Fig. 10-23, which shows that the electric field for the dominant mode is evenly symmetrical about the plane A-B in arm 4 but is unevenly symmetrical about plane A-B in arm 3 (and also in arms 1 and 2, as it happens). That is to say, the electric field in arm 4 on one side of A-B is a mirror image of the electric field on the other side, but in arm 3 a phase change would be required to give such even symmetry. Since nothing is there to provide such a phase change, no signal applied to arm 3 can propagate in arm 4 except in a mode with uneven symmetry about the plane A-B (such as a $TE_{0,1}$ or $TM_{1,1}$). The dimensions being such as to exclude the propagation of these higher modes, no signal travels down arm 4. Because the arrangement is reciprocal, application of a signal into arm 4 likewise results in no propagation down arm 3.

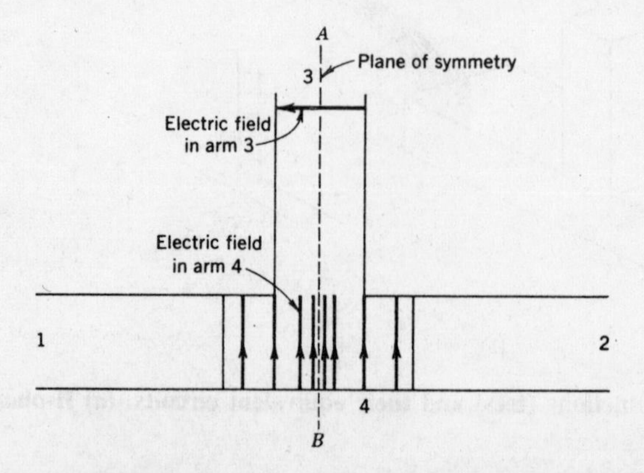

FIGURE 10-23 Cross section of magic tee, showing plane of symmetry.

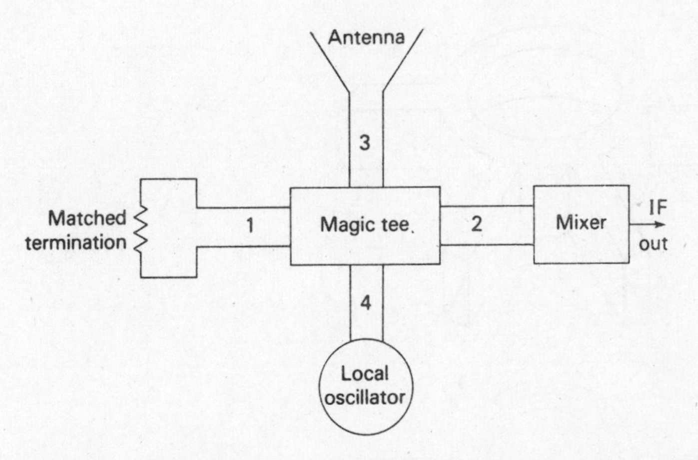

FIGURE 10-24 Magic tee application (front end of microwave receiver).

Since arms 1 and 2 are symmetrically disposed about the plane A-B, a signal entering either arm 3 or arm 4 divides evenly between these two lateral arms if they are correctly terminated. This means that it is possible to have two generators feeding signals, one into arm 3 and the other into arm 4. *Neither generator is coupled to the other, but both are coupled to the load* which, in Fig. 10-24, is in arm 2 (while arm 1 has a matched termination connected to it). The arrangement shown is but one of a number of applications of the magic tee.

It should be noted that quite bad reflections will take place at the junction unless steps are taken to prevent them. From a transmission-line viewpoint, arm 3 sees an open circuit in place of arm 4 and, across this infinite impedance, it also sees two correctly matched impedances *in parallel*. To avoid the resulting mismatch, two obstacles are normally placed at the junction, in the form of a *post* and an *iris*, each of which will be described in the next section.

Figure 10-25 shows a waveguide arrangement which looks quite different from the hybrid T and yet has very similar functions; it is the *hybrid ring*, or *rat race*. The arrangement consists of a piece of rectangular waveguide, bent in the E plane to form a complete loop whose median circumference is $1.5\lambda_p$. It has four orifices, with separation distances as shown in Fig. 10-25*b*, from each of which a waveguide emerges. If there are no reflections from the terminations in any of the arms, any one arm is coupled to two others but not to the fourth one.

If a signal is applied to (say) arm 1, it will divide evenly, with half of it traveling clockwise and the other half counterclockwise. The signal reaching arm 4 will cover the same distance, whether it has traveled clockwise or counterclockwise, and thus addition will take place at that point, resulting in some signal traveling down arm 4. Similarly, a signal reaching the input of arm 2 will have traveled a distance of $\lambda_p/4$ if traveling clockwise, and $1\frac{1}{4}\lambda_p$ if traveling counterclockwise. Thus the two portions of signal will add at that point, and propagation down arm 2 will take place. However, the signal at the mouth of arm 3 will have traveled a distance of $\lambda_p/2$ going one way and λ_p going the other, so that these two out-of-phase portions will cancel, and no signal will enter arm 3. In a similar way, it may be shown that arm 3 is

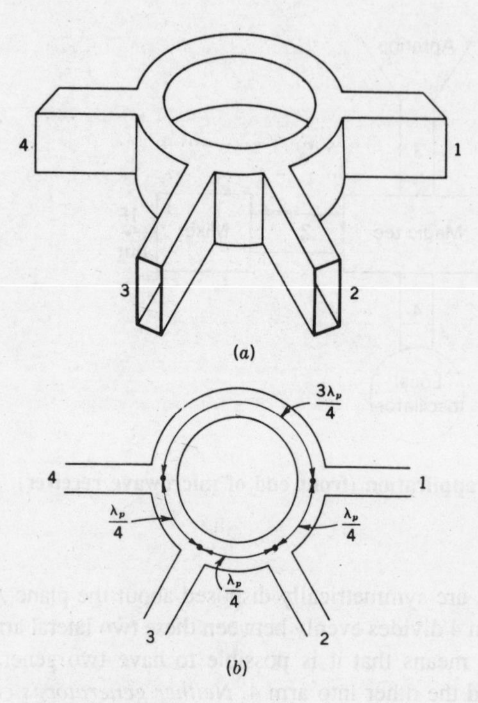

FIGURE 10-25 Hybrid ring (rat race). (*a*) Pictorial view; (*b*) plan and dimensions.

connected to arms 2 and 4, but not to arm 1. It is thus seen that behavior is very similar to that of the magic tee, although for a different reason.

The rat race and the magic tee may be used interchangeably, with the latter having the advantage of smaller bulk but the disadvantage of requiring internal matching. This the rat race does not need if the thickness of the ring is correctly chosen. Furthermore, the hybrid ring seems preferable at shorter wavelengths, since its dimensions are less critical.

10-3.5 Impedance Matching and Tuning

It was found in Secs. 7-1.5 and 7-1.6 that suitably chosen series or parallel pieces of transmission line had properties which made them useful for providing resistive or reactive impedances. It is the purpose of this section to show how the same effects are achieved in waveguides, and again transmission-line equivalents of waveguide matching devices will be used wherever applicable. Actually, some impedance matching devices have already been mentioned, and some have even been treated in detail, notably the choke ring.

Obstacles Reflections in a waveguide system cause impedance mismatches. When this happens, the cure is identical to the one that would be employed for transmission lines. That is, a lumped impedance of required value is placed at a precalculated point in the waveguide to overcome the mismatch, canceling the effects of the reflections.

Where lumped impedances or stubs were employed with transmission lines, obstacles of various shapes are used with waveguides.

The various *irises* (also called waveguide *apertures* or *diaphragms*) of Fig. 10-26 are a class of such obstacles. They may take any of the forms shown (or other similar ones) and may be capacitive, inductive or resonant. The mathematical analysis is complex, but fortunately the physical explanation is not. Consider the first capacitive iris of Fig. 10-26*a*. It is seen that potential which existed between the top and bottom walls of the waveguide (in the dominant mode) now exists between surfaces that are closer, and therefore capacitance has increased at that point. Conversely, the iris in Fig. 10-26*b* allows current to flow where none flowed before. The electric field that previously advanced now has a metal surface in its plane, which permits current flow. Energy storage in the magnetic field thus takes place, and there is an increase in inductance at that point of the waveguide.

If the iris of Fig. 10-26*c* is correctly shaped and positioned, the inductive and capacitive reactances introduced will be equal, and the aperture will be parallel-resonant. This means that the impedance will be very high for the dominant mode, and the shunting effect for this mode will be negligible. However, other modes or frequencies will be attenuated, so that the resonant iris acts as both a bandpass filter and a *mode filter*. Because irises are by their nature difficult to adjust, they are normally used to correct permanent mismatches.

A cylindrical post, extending into the waveguide from one of the broad sides, has the same effect as an iris in providing lumped reactance at that point. A post may

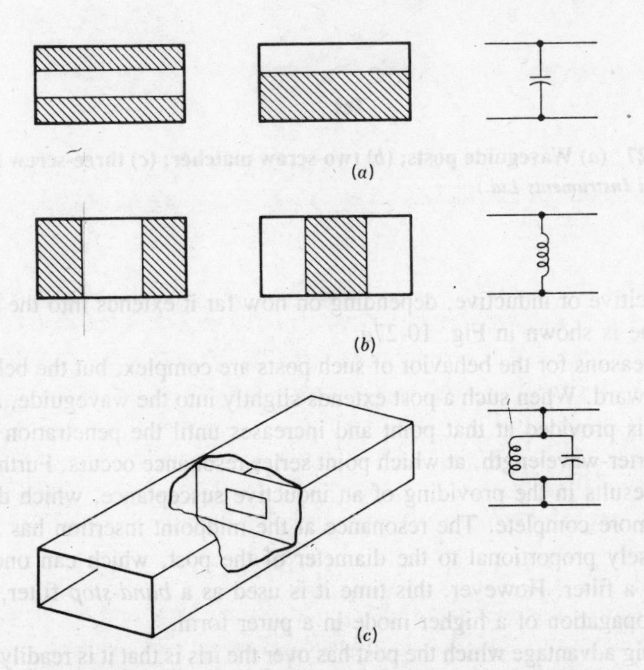

FIGURE 10-26 Waveguide irises and equivalent circuits. (*a*) Capacitive; (*b*) inductive; (*c*) resonant (perspective view).

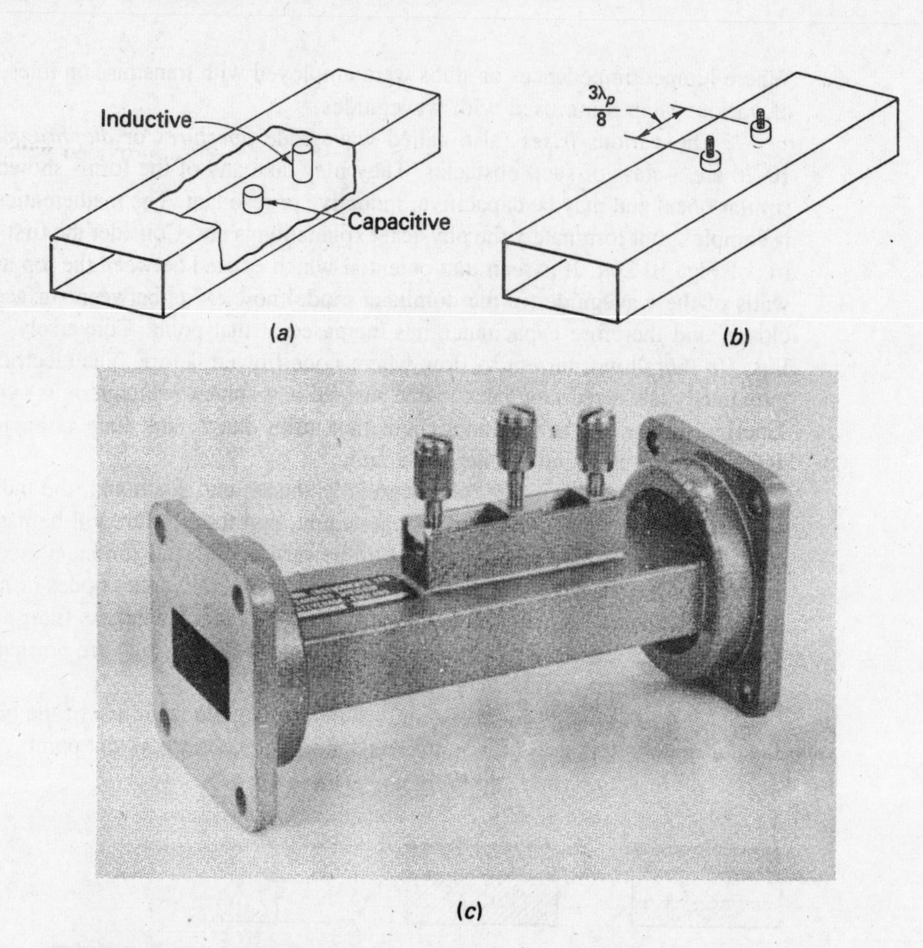

FIGURE 10-27 (*a*) Waveguide posts; (*b*) two-screw matcher; (*c*) three-screw tuner. *(Courtesy of Marconi Instruments Ltd.)*

also be capacitive or inductive, depending on how far it extends into the waveguide, and each type is shown in Fig. 10-27*a*.

The reasons for the behavior of such posts are complex, but the behavior itself is straightforward. When such a post extends slightly into the waveguide, a capacitive susceptance is provided at that point and increases until the penetration is approximately a quarter-wavelength, at which point series resonance occurs. Further insertion of the post results in the providing of an inductive susceptance, which decreases as insertion is more complete. The resonance at the midpoint insertion has a sharpness that is inversely proportional to the diameter of the post, which can once again be employed as a filter. However, this time it is used as a *band-stop* filter, perhaps to allow the propagation of a higher mode in a purer form.

The big advantage which the post has over the iris is that it is readily adjustable. A combination of two such posts in close proximity, now called *screws* and shown in Fig. 10-27*b*, is often used as a very effective waveguide matcher, similar to the double-stub tuner (Fig. 7-17). A three-screw tuner, as shown in Fig. 10-27c, may also be used, to provide even greater versatility.

Finally, it will be remembered that an E-plane tee may also be used in a manner identical to an adjustable transmission-line stub, when it is provided with a sliding, short-circuiting piston. Two such tees in close proximity are then analogous to a double-stub matcher.

Resistive loads and attenuators Waveguides, like any other transmission system, sometimes require perfectly matching loads, which absorb incoming waves completely without reflections, and which are not frequency-sensitive. One application for such terminations is in making various power measurements on a system without actually radiating any power.

The most common resistive termination is a length of lossy dielectric fitted in at the end of the waveguide and tapered very gradually (with the sharp end pointed at the incoming wave) so as not to cause reflections. Such a lossy *vane* may occupy the whole width of the waveguide, or perhaps just the center of the waveguide end, as shown in Fig. 10-28. The taper may be single or double, as illustrated, often having a length of $\lambda_p/2$, with an overall vane length of about two wavelengths. It is often made of a dielectric slab such as glass, with an outside coating of carbon film or aquadag. For high-power applications, such a termination may have radiating fins external to the waveguide, through which power applied to the termination may be dissipated or conducted away by forced-air cooling.

The vane may be made movable and used as a variable attenuator, as shown in Fig. 10-29. It will now be tapered at both ends and situated in the middle of a waveguide rather than at the end. It may be moved laterally from the center of the waveguide, where it will provide maximum attenuation, to the edges, where attenuation is considerably reduced because the electric field intensity there is much lower for the dominant mode. To minimize reflections from the mounting rods, they are made perpendicular to the electric field, as shown, and placed $\lambda_p/2$ apart so that reflections from one (if any) will tend to cancel those from the other.

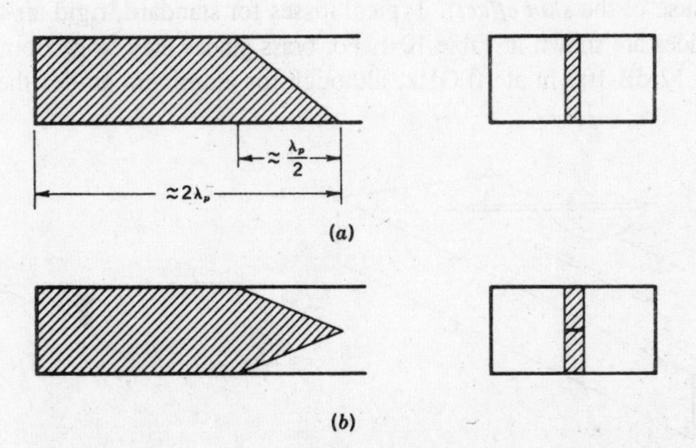

FIGURE 10-28 Waveguide resistive loads. (*a*) Single taper; (*b*) double taper.

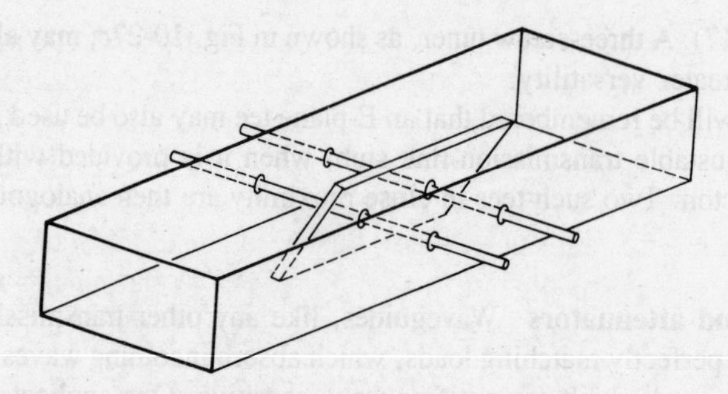

FIGURE 10-29 Movable vane attenuator.

The *flap* attenuator, shown in Fig. 10-30, is also adjustable and may be employed instead of the moving vane attenuator. A resistive element is mounted on a hinged arm, allowing it to descend into the center of the waveguide through a suitable longitudinal slot. The support for the flap attenuator is simpler than for the vane. The depth of insertion governs the attenuation, and the dielectric may be shaped to make the attenuation vary linearly with depth of insertion.

This type of attenuator is quite often used in practice, especially in situations where a little radiation from the slot is not considered significant. Both vanes and flaps are capable of attenuations in excess of 80 dB.

Attenuation in waveguides Waveguides below cutoff have attenuation for any or all of the following causes:

1. Reflections from obstacles, discontinuities or misaligned waveguide sections
2. Losses due to currents flowing in the waveguide walls
3. Losses in the dielectric filling the waveguide

The last two are similar to, but significantly less than, the corresponding losses in coaxial lines. They are lumped together and quoted in decibels per 100 meters. Such losses depend on the wall material and its roughness, the dielectric used and the frequency (because of the *skin effect*). Typical losses for standard, rigid air-filled rectangular waveguides are shown in Table 10-1. For brass guides they range from 4 dB/100 m at 5 GHz, to 12 dB/100 m at 10 GHz, although for aluminum guides they are some-

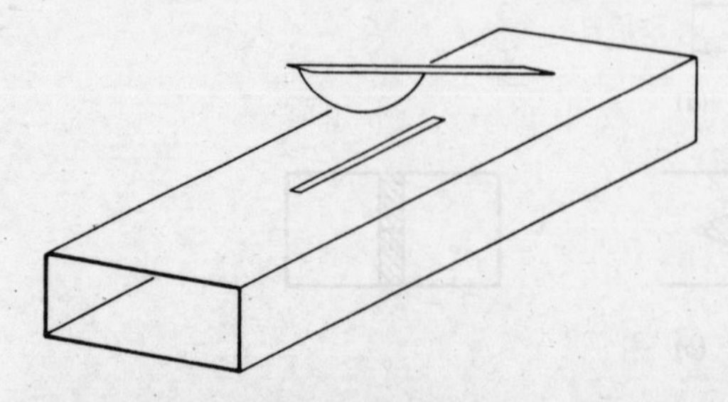

FIGURE 10-30 Flap attenuator.

what lower. For silver-plated waveguides, losses are typically 8 dB/100 m at 35 GHz, 30 dB/100 m at 70 GHz and nearly 500 dB/100 m at 200 GHz. To reduce losses, especially at the highest frequencies, waveguides are sometimes plated (on the inside, of course) with gold or platinum.

As already pointed out, the waveguide behaves as a high-pass filter. There is thus heavy attenuation for frequencies below cutoff, although the waveguide itself is virtually lossless; such attenuation is due to reflections at the mouth of the guide instead of propagation. Actually, some propagation does take place in so-called evanescent modes, but this is very slight.

For a waveguide operated well below cutoff, it may be shown that the attenuation $\mathcal{A}$ is given by

$$\mathcal{A} = e^{\alpha z} \tag{10-20}$$

and

$$\alpha = \frac{2\pi}{\lambda_o} \tag{10-21}$$

where e = base of natural logarithm system
α = attenuation factor
z = length of waveguide
λ_o = cutoff wavelength of waveguide

Under these conditions, attenuation is substantially independent of frequency and reduces to

$$\mathcal{A}_{dB} = 20 \log e^{\alpha} z = 20\alpha z \log e = \frac{40\pi z}{\lambda_o} \log e$$

$$= 40\pi \times 0.434 \times \frac{z}{\lambda_o}$$

$$= \frac{54.5z}{\lambda_o} \text{ dB} \tag{10-22}$$

where $\mathcal{A}_{dB}$ is the ratio, expressed in decibels, of the input voltage to the output voltage from a waveguide operated substantially below cutoff.

Example 10-9 Calculate the voltage attenuation provided by a 25-cm length of waveguide having $a = 1$ cm and $b = 0.5$ cm, in which a 1-GHz signal is propagated in the dominant mode.

$$\lambda_o = \frac{2a}{m} = 1 \times \frac{2}{1} = 2 \text{ cm}$$

$$\lambda = \frac{3 \times 10^{10}}{10^9} = 30 \text{ cm}$$

The waveguide is thus well below cutoff, and therefore

$$\mathcal{A}_{dB} = 54.5\frac{z}{\lambda_o} = 54.5 \times \frac{25}{2} = 681 \text{ dB}$$

Large though it is, this figure is quite realistic and is representative of the high Q possessed by a waveguide when used as a filter.

A waveguide below cutoff is often used as an adjustable, calibrated attenuator for UHF and microwave applications. Such a *piston attenuator* is a piece of waveguide to which the output of the generator is connected and within which a coaxial line may slide. The line is terminated in a probe or loop, and the distance between this coupling element and the generator end of the waveguide may be varied, adjusting the length of the waveguide and therefore its attenuation.

10-4 CAVITY RESONATORS

At its simplest, a cavity resonator is a piece of waveguide closed off at both ends with metallic planes. Where propagation in the longitudinal direction took place in the waveguide, standing waves exist in the resonator, and oscillations can take place if the resonator is suitably excited. Various aspects of cavity resonators will now be considered.

10-4.1 Fundamentals

Waveguides are used at the highest frequencies to transmit power and signals. Similarly, cavity resonators are employed as tuned circuits at such frequencies. Their operation follows directly from that of waveguides.

Operation Until now, waveguides have been considered from the point of view of standing waves between the side walls (see Figs. 10-4 to 10-6), and traveling waves in the longitudinal direction. If conducting end walls are placed in the waveguide, then standing waves, or oscillations, will take place if a source is located between the walls. This assumes that the distance between the end walls is $n\lambda_p/2$, where n is any integer. The situation is illustrated in Fig. 10-31.

As shown here and discussed in a slightly different context in Sec. 10-1.3, placement of the first wall ensures standing waves, and placement of the second wall permits oscillations, provided that the second wall is placed so that the pattern due to the first wall is left undisturbed. Thus, if the second wall is $\lambda_p/2$ away from the first, as in Fig. 10-32, oscillations between the two walls will take place. They will then

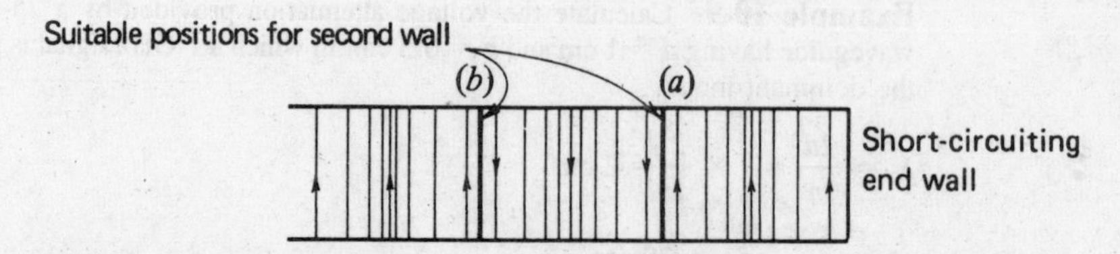

FIGURE 10-31 Transformation from rectangular waveguide propagating $TE_{1,0}$ mode to cavity resonator oscillating in (*a*) $TE_{1,0,1}$ mode; (*b*) $TE_{1,0,2}$ mode.

continue until all the applied energy is dissipated, or indefinitely if energy is constantly supplied. This is identical to the behavior of an *LC* tuned circuit.

It is thus seen that any space enclosed by conducting walls must have one (or more) frequency at which the conditions just described are fulfilled. In other words, any such enclosed space must have at least one resonant frequency. Indeed, the completely enclosed waveguide has become a cavity resonator with its own system of modes, and therefore resonant *frequencies*. The TE and TM mode-numbering system breaks down unless the cavity has a very simple shape, and it is preferable to speak of the resonant frequency rather than mode.

Each cavity resonator has an infinite number of resonant frequencies. This can be appreciated if we consider that with the resonator of Fig. 10-31 oscillations would have been obtained at twice the frequency, because every distance would now be λ_p, instead of $\lambda_p/2$. Several other resonant frequency series will also be present, based on other modes of propagation, all permitting oscillations to take place within the cavity. Naturally such behavior is not really desired in a resonator, but it need not be especially harmful. The fact that the cavity *can* oscillate at several frequencies does not mean that it *will*. Such frequencies are not generated spontaneously; they must be fed in.

Types The simplest cavity resonators may be spheres, cylinders or rectangular prisms. However, such cavities are not often used, because they all share a common defect: their various resonant frequencies are harmonically related. This is a serious drawback in all those situations in which pulses of energy are fed to a cavity. The cavity is supposed to maintain sinusoidal oscillations through the flywheel effect, but because such pulses contain harmonics and the cavity is able to oscillate at the harmonic frequencies, the output is still in the form of pulses. As a result, most practical cavities have odd shapes to ensure that the various oscillating frequencies are not harmonically related, and therefore that harmonics are attenuated.

Some typical irregularly shaped resonators are illustrated. Those of Fig. 10-32*a* might be used with *reflex klystrons,* whereas the resonator of Fig. 10-32*b* is popular for

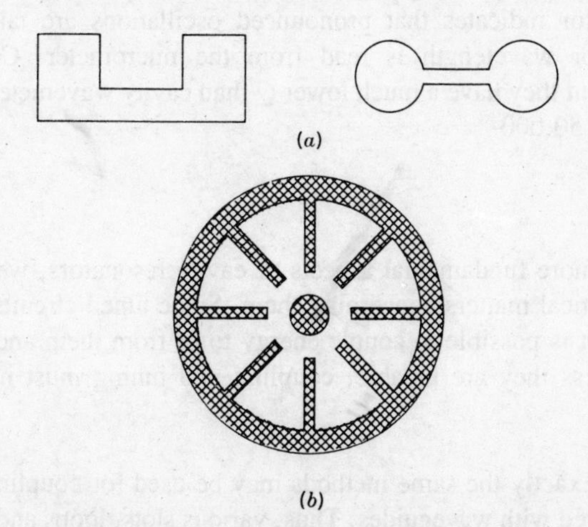

FIGURE 10-32 Reentrant cavity resonators.

use with *magnetrons* (see the next chapter). They are known as *reentrant resonators,* that is, resonators so shaped that one of the walls reenters the resonator shape. The first two are figures of revolution about a central vertical axis, and the third one is cylindrical. Apart from being useful as tuned circuits, they are also given such shapes so that they can be integral parts of the above-named microwave devices, being therefore doubly useful. However, because of their shapes, they have resonant frequencies that are not at all easy to calculate.

Note that the general size of a cavity resonator, for a given dominant mode, is similar to the cross-sectional dimensions of a waveguide carrying a dominant mode of the same signal (this is merely an approximation, not a statement of equivalence). Note further that (as with quartz crystals) the lowest frequency of oscillations of a cavity resonator is also one of most intense oscillation, as a general rule.

Applications Cavity resonators are employed for much the same purposes as tuned *LC* circuits or resonant transmission lines, but naturally at much higher frequencies since they have the same overall frequency coverage as waveguides. They may be input or output tuned circuits of amplifiers, tuned circuits of oscillators, or resonant circuits used for filtering or in conjunction with mixers. In addition, as already mentioned, they can be given shapes that make them integral parts of microwave amplifying and oscillating devices, so that almost all such devices use them, as will become plain in the course of the next chapter.

One of the many applications of the cavity resonator is as a cavity wavemeter, used as a microwave frequency-measuring device. Basically it is a simple cavity of cylindrical shape, usually with a plunger whose insertion varies the resonant frequency. Adjustment is by means of a calibrated micrometer. The plunger has absorbent material on one side of it (the back) to prevent oscillations in the back cavity, and the micrometer is calibrated directly in terms of wavelength, from which frequency may be calculated.

A signal is fed to a cavity wavemeter through an input loop, and a detector is connected to it through an output loop. The size of the cavity is adjusted with the plunger until the detector indicates that pronounced oscillations are taking place, whereupon frequency or wavelength is read from the micrometer. Coaxial line wavemeters also exist, but they have a much lower Q than cavity wavemeters, perhaps 5000 as compared with 50,000.

10-4.2 Practical Considerations

Having considered the more fundamental aspects of cavity resonators, we must now concentrate on two practical matters concerning them. Since tuned circuits cannot be used in practice unless it is possible to couple energy to or from them and are not of much practical use unless they are tunable, coupling and tuning must now be discussed.

Coupling to cavities Exactly the same methods may be used for coupling to cavity resonators as are employed with waveguides. Thus, various slots, loops and probes are used to good advantage when coupling of power into or out of a cavity is desired. It

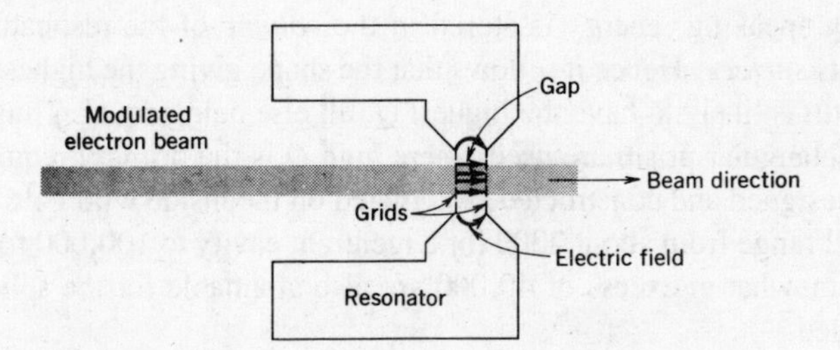

FIGURE 10-33 Coupling of cavity to electron beam.

must be realized, however, that taking an output from a cavity not only loads it but also changes its resonant frequency slightly, just as in other tuned circuits. For a cavity, this can be explained by the fact that the insertion of (say) a loop distorts the field that would otherwise have existed in the resonator. Hence a cavity may require retuning if such a loop is inserted or rotated to change the degree of coupling. It should also be mentioned that the one position of loop, probe or slot is quite capable of exciting several modes other than the desired one. This is unlikely to be a problem in practice, however, because the frequencies corresponding to these spurious modes are hardly likely to be present in the injected signal.

There is one form of coupling which is unlikely with waveguides, but quite common with cavity resonators, especially those used in conjunction with *klystrons;* this is coupling to an electron beam. The situation is illustrated in Fig. 10-33, which shows a typical klystron cavity, together with the distribution of some of the electric field.

The beam, as shown, passes through the center of the cavity. This is usually a figure of revolution about an axis coinciding with the center of the beam, with holes or mesh at its narrow gap to allow the passage of the beam. If the cavity is oscillating but the beam itself is unmodulated (having a uniform current density), then the presence of the electric field across the gap in the cavity will have an effect on the beam. This field will accelerate some electrons in it and retard others, depending on the size and polarity of the gap voltage at the time when electrons pass the gap. Alternatively, if the current of the beam is modulated and flows in pulses, as often happens in practice, the pulses will deliver energy to the cavity. This will cause oscillation if the pulse repetition rate corresponds to a resonant frequency of the cavity.

Tuning of cavities Precisely the same methods are used for tuning cavity resonators as were used for impedance matching of waveguides, with the adjustable screw, or post, perhaps the most popular. However, it is important to examine the effects of such tuning, and also loading, on the bandwidth and Q of the cavity resonator.

Q has the same meaning for cavity resonators as for any other tuned circuits and may be defined as the ratio of the resonant frequency to the bandwidth. However, it is perhaps more useful to base the definition of Q here on a more fundamental relation, i.e.,

$$Q = 2\pi \frac{\text{energy stored}}{\text{energy lost each cycle}} \tag{10-23}$$

Roughly speaking, energy is stored in the *volume* of the resonator and dissipated through its *surface*. Hence it follows that the shape giving the highest volume-to-surface-area ratio is likely to have the highest Q, all else being equal. Thus the sphere, cylinder and rectangular prism are used where high Q is the primary requirement. If a cavity is well designed and constructed, and plated on the inside with gold or silver, its unloaded Q will range from about 2000 for a reentrant cavity to 100,000 for a spherical one. Values somewhat in excess of 40,000 are also attainable for the spherical cavity when it is loaded.

When a cavity is tuned by means of a screw or sliding piston, its Q will suffer, and this should be taken into account. The Q decreases because of the extra area due to the presence of the tuning elements, in which current can flow, but this state of affairs is not always undesirable because wideband applications exist in the microwave range also.

The introduction of a solid dielectric material will have the effect of changing the resonant frequency, since the signal wavelength in the resonator is affected. Because the velocity of light in such a dielectric is less than in air, the wavelength will be reduced, and so will the size of the cavity required at any given frequency. If such a dielectric is introduced gradually, the frequency of the resonance will depend on the depth of the insertion, so that this is a useful method of tuning a cavity. However, since dielectric materials have significant losses at microwave frequencies, the Q of the cavity will be reduced by their introduction. Once again, this may or may not be desirable.

Still another method of tuning a cavity consists in having a wall that can be moved in or out slightly by means of a screw, which operates on an arm that in turn tightens or loosens small bellows. These move this wall to a certain extent. This method is sometimes used with permanent cavities built into *reflex klystrons* as a form of limited frequency shifting. Other methods of tuning include the introduction of ferrites, such as *yttrium-iron-garnet* (YIG), into the cavity. (See Sec. 10-5.2.)

Finally, it should be mentioned that it is generally difficult to calculate the frequency of oscillation of a cavity, for the dominant or any other mode, especially for a complex shape. Tuning helps, of course, because it makes design less critical. Another aid is the *principle of similitude,* which states that if two resonators have the same shape but a different size, then their resonant frequencies are inversely proportional to their linear dimensions. It is thus possible to make a scale model of a desired shape of resonator and to measure its resonant frequency. If the frequency happens to be four times too high, all linear dimensions of the resonator are increased fourfold. This also means that it may be convenient to decide on a given shape for a particular application and to keep changing dimensions for different frequencies.

10-5 AUXILIARY COMPONENTS

In addition to the various waveguide components described in Sec. 10-3, a number of others are often used, especially in measurements and similar applications. Among these are directional couplers, detector and thermistor mounts, circulators and isolators, and various switches. They differ from the previously described components in that, by and large, they are separate components, and in any case they are somewhat more specialized than the various internal elements thus far described.

10-5.1 Directional Couplers

A transmission-line directional coupler was described in Sec. 7-3.2. Its applications were stated at the time as being unidirectional power flow measurement, SWR measurement and unidirectional wave launching. Exactly the same considerations apply to waveguides. Several directional couplers for waveguides exist, and the most common ones will be described, including a direct counterpart of the transmission-line coupler, which is also commonly used with waveguides. It should also be mentioned that the hybrid T junction and hybrid ring of Sec. 10-3.4 are not normally classified as directional couplers.

Two-hole coupler The coupler of Fig. 10-34 is the waveguide analog of the transmission-line coupler of Fig. 7-18. The operation is also almost identical, the only exceptions being that the two holes are now $\lambda_p/4$ apart, and a different sort of attenuator is used to absorb backward wave components in the auxiliary guide. Students are referred to Sec. 7-3.2 for details of the operation.

This is a very popular waveguide directional coupler. It may also be used for direct SWR measurements if the absorbing attenuator is replaced by a detecting device, for measuring the components in the auxiliary guide that are proportional to the reflected wave in the main waveguide. Such a directional coupler is called a *reflectometer*, but because it is rather difficult to match two detectors, it is often preferable to use two separate directional couplers to form the reflectometer.

Other types Other directional couplers include one type that employs a single slot (with two waveguides having a different orientation). There are also a directional coupler with a single long slot so shaped that directional properties are preserved and another type which uses two slots with a capacitive coaxial loop through them. Finally, there are a series of couplers similar to the two-hole coupler, but with three or more holes in the common wall. If three holes are used, the center one generally admits

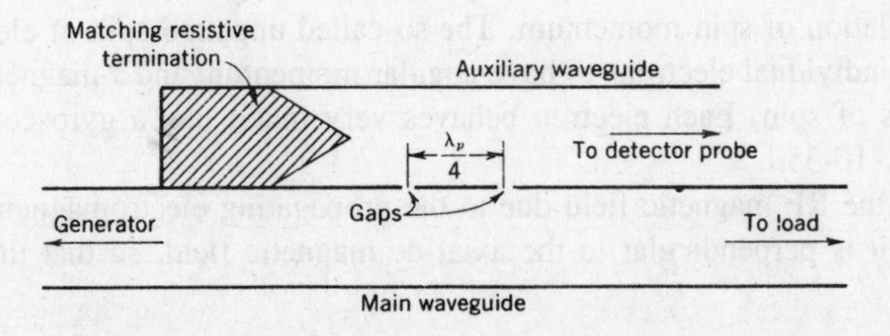

FIGURE 10-34 Two-hole directional coupler.

twice as much power as the end holes, in an attempt to extend the bandwidth of such a coupler. After all, the two-hole coupler is directional only at those frequencies at which the hole separation is $n\lambda_p/4$, where n is an odd integer.

10-5.2 Isolators and Circulators [3, 4]

It often happens at microwave frequencies that coupling must be strictly a one-way affair. This applies for most microwave generators, whose output amplitude and frequency could be affected by changes in load impedance. Consequently, some means must be found to ensure that the coupling is unidirectional from generator to load. Again, a number of semiconductor devices used for microwave amplification and oscillation are two-terminal devices, in which the input and output would interfere unless some means of isolation were found. As a result, devices such as *isolators* and *circulators* are frequently employed. They have properties much the same as directional couplers and hybrid junctions, respectively, but with different applications and construction. Since various *ferrites* are often used in isolators and circulators, these materials must be studied before the devices themselves.

Introduction to ferrites A ferrite is a nonmetallic material (though often an iron oxide compound) which is an insulator, but with magnetic properties similar to those of ferrous metals. Among the more common ferrites are *manganese ferrite* ($MnFe_2O_3$), *zinc ferrite* ($ZnFe_2O_3$) and associated ferromagnetic oxides such as *yttrium-iron-garnet*[2] [$Y_3Fe_2(FeO_4)_3$], or YIG for short. Since all these materials are insulators, electromagnetic waves can propagate in them. Because the ferrites have strong magnetic properties, external magnetic fields can be applied to them with several interesting results, including the *Faraday rotation* mentioned in connection with wave propagation in Sec. 8-2.5

When electromagnetic waves travel through a ferrite, they produce an RF magnetic field in the material, at right angles to the direction of propagation if the mode of propagation is correctly chosen. If an axial magnetic field from a permanent magnet is applied as well, a complex interaction takes place in the ferrite. The situation may be somewhat simplified if weak and strong interactions are considered separately.

With only the axial dc magnetic field present, the spin axes of the spinning electrons align themselves along the lines of magnetic force, just as a magnetized needle aligns itself with the earth's magnetic field. Electrons spin because this is a magnetic material. In other materials spin is said to take place also, but each pair of electrons has individual members spinning in opposite directions, so that there is an overall cancellation of spin momentum. The so-called unpaired spin of electrons in a ferrite causes individual electrons to have angular momentum and a magnetic moment along the axis of spin. Each electron behaves very much like a gyroscope. This is shown in Fig. 10-35*a*.

When the RF magnetic field due to the propagating electromagnetic waves is also applied, it is perpendicular to the axial dc magnetic field, so that the electrons

[2]Garnets are vitreous mineral substances of various colors and composition, several of them being quite valuable as gems.

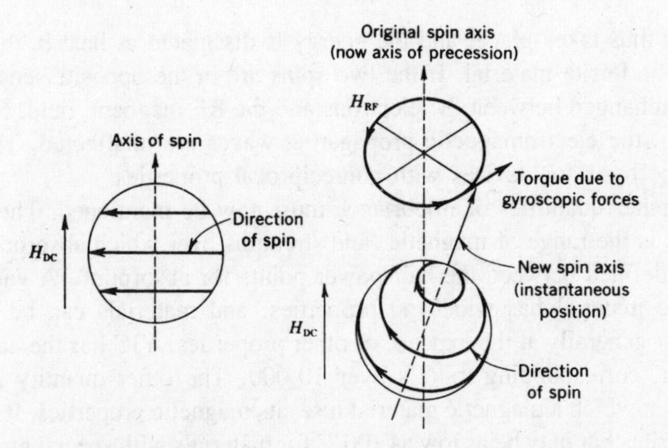

FIGURE 10-35 Effect of magnetic fields on spinning electron. ***(a)*** **dc field only;** ***(b)*** **dc and RF magnetic fields.**

precess about their original spin axis. This is due to the gyroscope forces involved and occurs at a rate that depends on the strength of the dc magnetic field. Furthermore, it is identical to the behavior that an ordinary gyroscope would exhibit under these conditions. Because of the precession, a magnetic component at right angles to the other two is produced, as shown in Fig. 10-35*b*. This has the effect of rotating the plane of polarization of the waves propagating through the ferrite and is akin to the similar behavior of light, which Michael Faraday discovered in 1845.

The amount by which the plane of polarization of the waves will be rotated depends on the length and thickness of the ferrite material, and on the strength of the dc magnetic field. The field must provide at least *saturation magnetization,* which is the minimum value required to ensure that the axes of the spinning electrons are suitably aligned. In turn, this is tied up in a rather complex fashion with the lowest usable frequency of the ferrite. This property of ferrites, whereby the plane of polarization of propagating waves is rotated, is a basis for a number of *nonreciprocal* devices. These are devices in which the properties in one direction differ from those in the other direction. Metallic magnetic materials cannot be used for such applications because they are conductors. Thus electromagnetic waves cannot propagate in them, whereas they can in ferrites, with relatively low losses.

The rate of precession is proportional to the strength of the dc magnetic field and is 3.52 MHz per ampere per meter for most ferrites. For example, if this field is 1000 A/m, the frequency of precession will be 3.52 GHz. Such a magnetic field strength is well above saturation and therefore higher than would be used if merely a rotation of the plane of polarization were required. If the dc magnetic field is made as strong as this or even stronger, the possibility of the precessional frequency being equal to the frequency of the propagated electromagnetic waves is introduced. When this happens, *gyromagnetic resonance interaction* takes place between the spinning electrons and the magnetic field of the propagating waves. If both the electrons and this magnetic field are rotating clockwise, energy is delivered to the electrons, making them rotate more violently. Absorption of energy from the magnetic field of the propagating waves thus takes place, and the energy is dissipated as heat in the crystalline structure of the ferrite material. If the two spins are in the opposite sense, energy is alternately exchanged between the electrons and the RF magnetic field. Since the net effect is zero, the electromagnetic propagating waves are unaffected. This behavior also forms the basis for devices with nonreciprocal properties.

Two other quantities of importance must now be mentioned. The first is *line width,* which is the range of magnetic field strengths over which absorption will take place and is defined between the half-power points for absorption. A wide line indicates that the material has wideband properties, and materials can be modified to possess it, but generally at the expense of other properties. YIG has the narrowest line width known, corresponding to Q's over 10,000. The other quantity is the *Curie temperature,* at which a magnetic material loses its magnetic properties. It ranges up to 600°C for ferrites but may be as low as 100°C for materials with special properties such as broad line width. It is 280°C for YIG. This places a limitation on the maximum temperature at which a ferrite may be operated, and therefore on the power dissipated. However, with external cooling, ferrite devices are available that can handle powers as high as 150 kW CW and 3 MW pulsed.

The final limitation to which ferrites may be subject is their maximum frequency of operation. For a device utilizing resonance absorption, this is dependent on the maximum magnetic field strength that can be generated and is offset somewhat by the general reduction in the size of waveguides as frequency is increased. The present upper frequency limit for commercial devices is in excess of 220 GHz.

Isolators Ferrite isolators may be based either on Faraday rotation, which is used for powers up to a few hundred watts, or on resonant absorption, used for higher powers. The Faraday rotation isolator, shown in Fig. 10-36, will be dealt with first.

The isolator consists of a piece of circular waveguide carrying the $TE_{1,1}$ mode, with transitions to a standard rectangular guide and $TE_{1,0}$ mode at both ends (the output end transition being twisted through 45°). A thin "pencil" of ferrite is located inside the circular guide, supported by polyfoam, and the waveguide is surrounded by a permanent magnet which generates a magnetic field in the ferrite that is generally about 160 A/m. A typical practical X-band (8.0 to 12.4 GHz) device may have a length of 25 mm and a weight of 100 g without the transitions.

Because the dc magnetic field (well below that required for resonance) is applied, a wave passing through the ferrite in the forward direction will have its plane of polarization shifted clockwise (through 45° in practical isolators) by the time it reaches the output end. This wave is then passed through the suitably rotated output transition, and it emerges with an *insertion loss* (attenuation in the forward direction) between 0.5 and 1 dB in practice. It has not been affected by either of the resistive vanes because they are at right angles to the plane of its electric field; this is shown in Fig. 10-36*b*.

A wave that tries to propagate through the isolator in the reverse direction *is also rotated clockwise,* because the direction of the Faraday rotation depends only on the dc magnetic field. Thus, when the wave emerges into the input transition, not only is it absorbed by the resistive vane, but also it cannot propagate in the input rectangular waveguide because of its dimensions. This situation is shown in Fig. 10-36*b*. It results

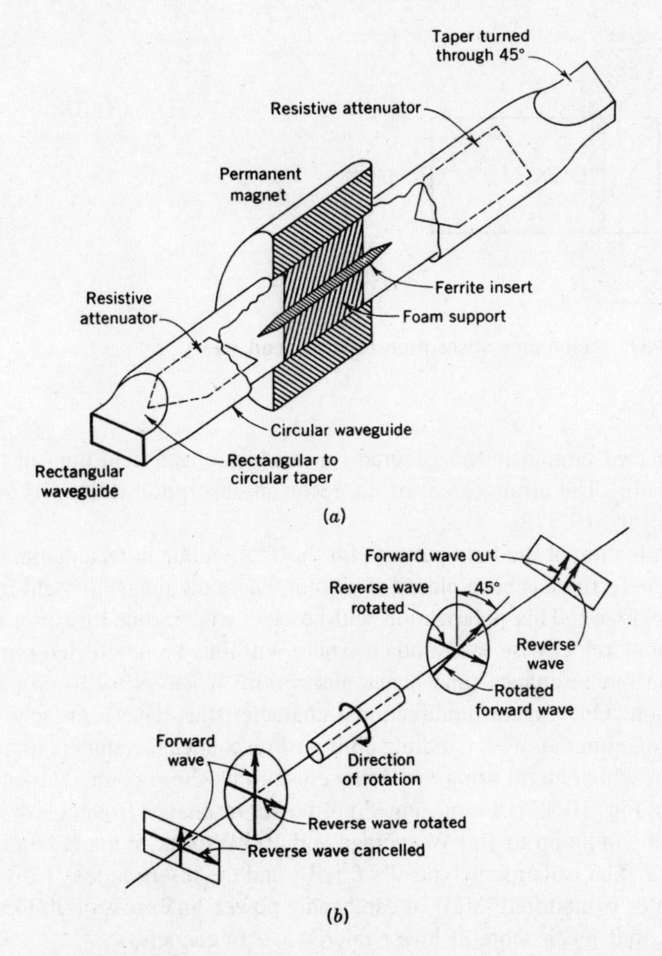

FIGURE 10-36 Faraday rotation isolator. (*a*) Cutaway view; (*b*) method of operation.

in the returned wave being attenuated by 20 to 30 dB in practice (this reverse attenuation of an isolator is called its *isolation*). Such a practical isolator will have an SWR not exceeding 1.4, with values as low as 1.1 obtainable, and a bandwidth between 5 and 30 percent of the center frequency.

This type of isolator is limited in its peak power-handling ability to about 2 kW, because of nonlinearities in the ferrite resulting in the phase shift departing from the ideal 45°. However, it has a very wide range of applications in the low-power field, since most microwave amplifiers and oscillators have output powers considerably lower than 2 kW.

The other popular type of isolator is the *resonant absorption isolator*, which is commonly used for high powers. It consists of a piece of rectangular waveguide carrying the $TE_{1,0}$ mode, with a piece of longitudinal ferrite material placed about a quarter of the way from one side of the waveguide and halfway between its ends. A permanent

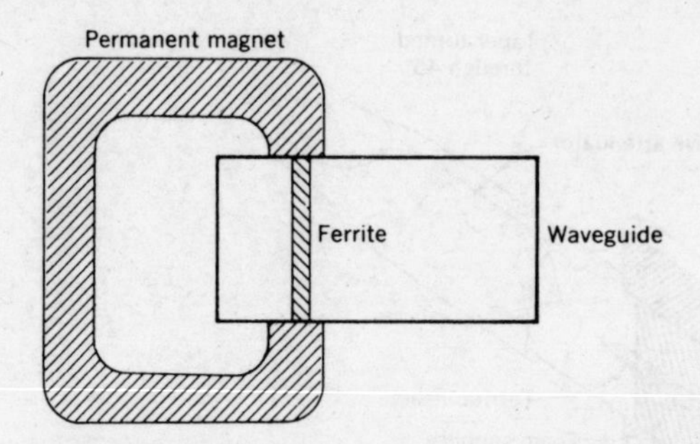

FIGURE 10-37 Resonance absorption isolator (end view).

magnet is placed around it and generates a much stronger field than in the Faraday rotation isolator. The arrangement of the resonant absorption isolator is shown schematically in Fig. 10-37.

Examination of the field patterns for the $TE_{1,0}$ mode in rectangular waveguides shows that the ferrite has been placed at a point where the magnetic field is strong and circularly polarized. This polarization will be clockwise in one direction of propagation, and counterclockwise in the other. There will thus be unaffected propagation in one direction but resonance (and hence absorption) if waves try to propagate in the other direction. Once again unidirectional characteristics have been achieved.

The maximum power-handling ability of resonance isolators is limited by temperature rise, which might bring the ferrite close to its Curie point. The one described and shown in Fig. 10-37 is a typical medium-power resonance isolator, weighing about 300 g. It can handle up to 100 W average and 10 kW peak in the X band, having an SWR of 1.15. The isolation is typically 60 dB, and the insertion loss 1 dB. When this type of isolator is modified [3(a)], it can handle powers in excess of 300 kW pulsed in the X band, and much more at lower microwave frequencies.

Circulators A circulator is a ferrite device somewhat like a rat race. It is very often a *four-port* (i.e., four-terminal) device, as shown in Fig. 10-38*a*, although other forms also exist. It has the property that *each terminal is connected only to the next clockwise terminal*. Thus port 1 is connected to port 2, but not to 3 or 4; 2 is connected to 3, but not to 4 or 1; and so on. The main applications of such circulators are either the isolation of transmitters and receivers connected to the same antenna (as in radar), or isolation of input and output in two-terminal amplifying devices such as parametric amplifiers (see Sec. 12-3).

A four-port Faraday rotation circulator is shown in Fig. 10-38. It is similar to the Faraday rotation isolator already described. Power entering port 1 is converted to the $TE_{1,1}$ mode in the circular waveguide (as before), passes port 3 unaffected because the electric field is not significantly cut, is rotated through 45° by the ferrite insert (the magnet is omitted for simplicity), continues past port 4 for the same reason that it passed port 3, and finally emerges from port 2, just as it did in the isolator. Power fed to port 2 will undergo the same fate that it did in the isolator, but now it is rotated so that although it still cannot come out of port 1, it has port 3 suitably aligned and

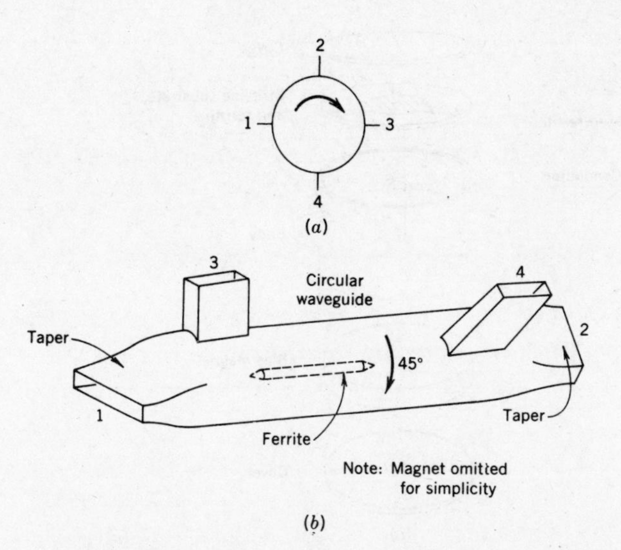

FIGURE 10-38 Ferrite circulator. (*a*) Schematic diagram; (*b*) Faraday rotation four-port circulator.

emerges from it. Similarly, port 3 is coupled only to port 4, and port 4 to port 1. This type of circulator is power-limited to the same extent as the Faraday rotation isolator, but it is eminently suitable as a low-power device. However, since it is bulkier than the Y (or wye) circulator (to be described), its use is restricted mostly to the highest frequencies, in the millimeter range and above. Its characteristics are similar to those of the isolator.

High-power circulators are fairly similar to the resonance isolator and handle powers up to 30 MW peak. (The interested student is referred to pp. 184–186 in [3(a)] for a detailed description of such a differential phase-shift circulator.)

Figure 10-39 shows a miniature Y (or wye) circulator. There are waveguide, coaxial, and stripline versions of it. The last one, as will be shown in Chap. 12, is by far the most popular. A three-port version is shown—a four-port circulator of this type is obtained by joining two wyes together. This is seen in Fig. 10-46, in a slightly different context.

With the magnet on one side of the ferrite only, and with a suitable magnetic field strength, a phase shift will be applied to any signal fed in to the circulator. If the three striplines and coaxial lines are arranged 120° apart as shown, a clockwise shift and correct terminations will ensure that each signal is rotated so as to emerge from the next clockwise port, without being coupled to the remaining port. In this fashion, circulator properties are obtained. A practical Y circulator of the type shown is typically 12 mm high and 25 mm in diameter. It handles only small powers but may have an isolation over 20 dB, an insertion loss under 0.5 dB and an SWR of 1.2, all in the X band. A similar four-port circulator, consisting of two joined wyes, will be housed in a rectangular box measuring 45 × 25 × 12 mm. It will have similar performance figures, except that the isolation is now in excess of 40 dB, and the insertion loss is about 0.9 dB.

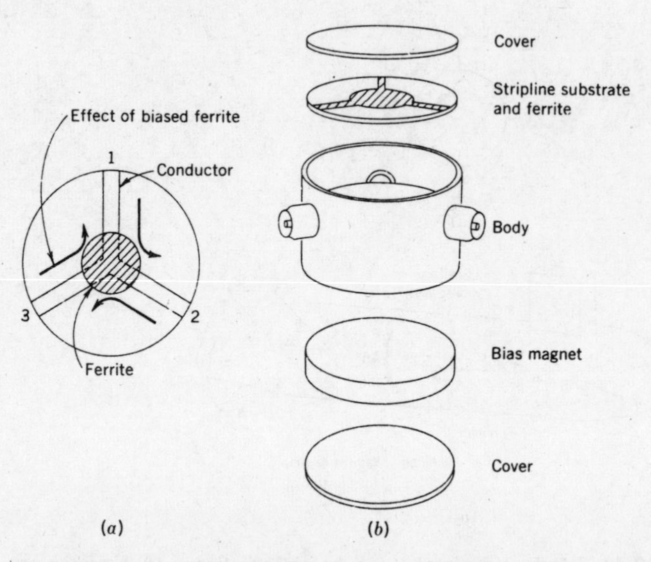

FIGURE 10-39 Y ferrite circulator. (*a*) Schematic diagram; (*b*) exploded view of stripline circulator with coaxial terminals.

10-5.3 Mixers, Detectors and Detector Mounts

As will be seen in Chaps. 11 and 12, ordinary transistor and tube RF amplifiers eventually fail at microwave frequencies, because of greatly increased noise, compared with their low-frequency performance. Unless a receiver is to be very low-noise and extremely sensitive (in which case special RF amplifiers will be used, as explained in Chap. 12), then a mixer is the first stage encountered by the incoming signal in such a receiver. Silicon point-contact diodes (called "crystal diodes") have been used as mixers since before World War II, because of their relatively low noise figures at microwave frequencies (not in excess of 6 dB at 10 GHz). *Schottky barrier* diodes have more recently been employed as microwave mixers and are described in Chap. 12. They have similar applications but even lower noise figures (below 4 dB at 10 GHz). These diodes will now be described briefly. However, what is of greater significance here is how mixer and detector diodes are mounted and used in waveguides, and the rest of this section will be devoted to that subject.

Point-contact diodes The construction of a typical point-contact silicon diode is shown in Fig. 10-40—an identical construction would be used for other semiconductor materials. It consists of a (usually) brass base on which a small pellet of silicon, germanium, gallium arsenide or indium phosphide is mounted. A fine gold-plated tungsten wire, with a diameter of 80 to 400 μm and a sharp point, makes contact with the polished top of the semiconductor pellet and is pressed down on it slightly for spring contact. This "cat's whisker," as it is known, is connected to the top brass contact, which is the cathode of the device. The semiconductor and the cat's whisker are surrounded by wax to exclude moisture and are located in a metal-ceramic housing, as shown in Fig. 10-40.

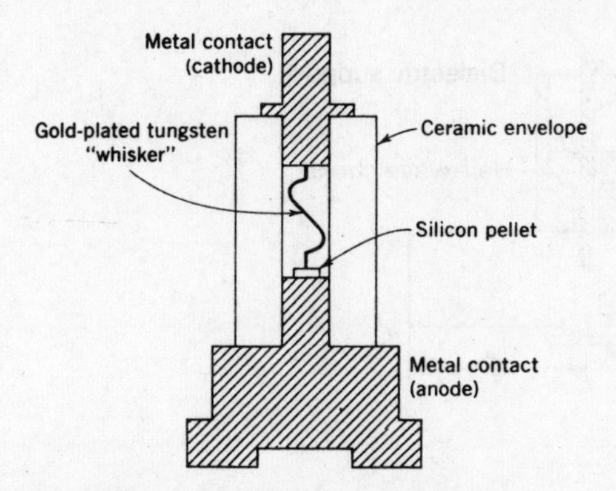

FIGURE 10-40 Diode construction.

Such diodes can be fitted into coaxial or waveguide mounts and are available at frequencies in excess of 100 GHz, although they are then noisier than at X band. As already mentioned, they are used as microwave mixers or detectors, there being some differences in diode characteristics between the two applications.

Diode mounts A diode must be mounted so that it provides a complete dc path for rectification, without unduly upsetting the RF field in the waveguide. That is to say, the mount must not constitute a mismatch which causes a high SWR. Thus, for example, the diode cannot be connected across the open end of a waveguide, or a mismatch will exist because of reflections. In fact, the diode must be connected across the waveguide for RF but not for dc (nor the IF, as the case may be). Furthermore, any reflections from it must be canceled. This suggests mounting the diode $\lambda_p/4$ from the short-circuited end of a guide and attaching it to the bottom wall of the waveguide via a half-wave choke rather than directly. This will provide an RF connection but a dc open circuit, as required. Such an arrangement is indicated in cross section in Fig. 10-41*a*, and 10-41*b* shows a more practical arrangement. Here a tuning plunger is used, instead of relying on a fixed wall $\lambda_p/4$ away to prevent mismatch—broadband operation is thus ensured. Other versions of this arrangement also exist, in which the diode is connected across the waveguide by means other than the half-wave choke. Tuning screws are also often provided on the RF input side of the diode for further matching, as shown here.

When a diode is used as a mixer, it is necessary to introduce the local oscillator signal into the cavity or waveguide, as well as the RF signal. That such a local oscillator signal was already present was assumed in Fig. 10-41; a frequently used method of introducing it is shown in Fig. 10-42.

It is sometimes important to apply automatic frequency control to the local oscillator in a microwave receiver, particularly in radar receivers. Under these circumstances, a separate AFC diode is preferred. The result is a balanced mixer, one form of which uses a magic tee junction to ensure that both diodes are coupled to the RF and local oscillator signals, but that the two signals are isolated from each other. A balanced mixer such as this is shown in Fig. 10-43.

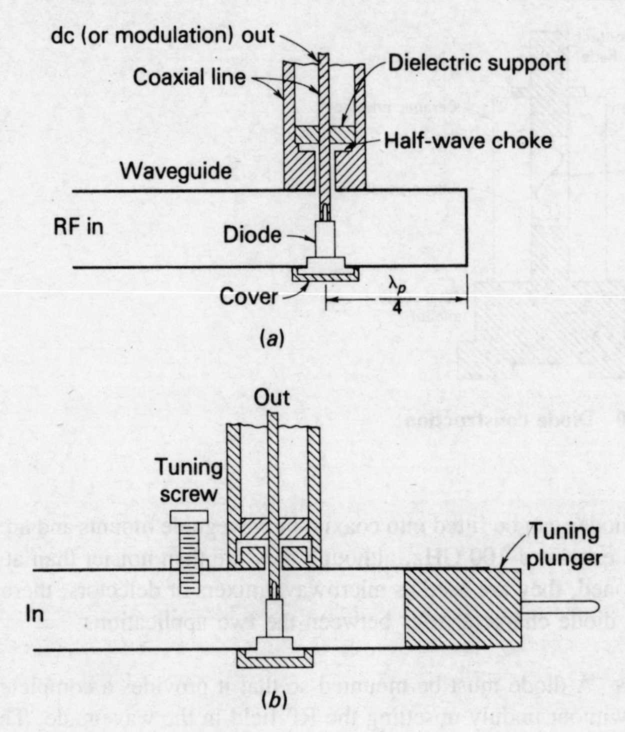

FIGURE 10-41 Diode waveguide mounts. (*a*) Simple; (*b*) tunable.

10-5.4 Switches

It is often necessary to prevent microwave power from following a particular path, or to force it to follow another path; as at lower frequencies, the component used for this purpose is called a *switch*. Waveguide (or coaxial) switches may be *mechanical* (manually operated) or *electromechanical* (solenoid-operated). They can also be *electrical,* in which case the switching action is provided by a change in the electrical properties of

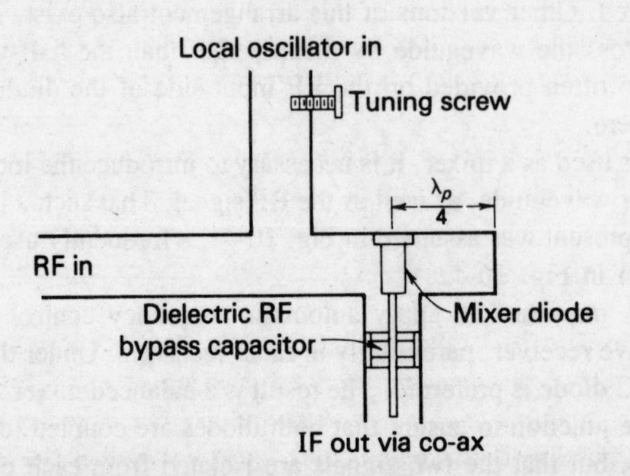

FIGURE 10-42 Method of local oscillator injection in a microwave diode mixer.

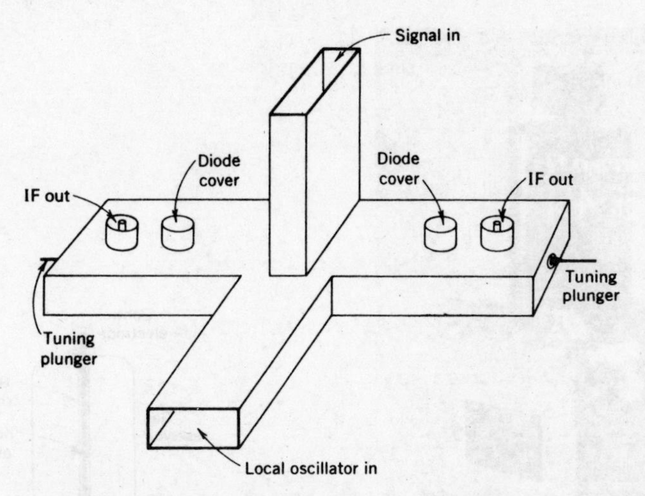

FIGURE 10-43 Hybrid T (magic tee) balanced mixer.

some device. The electrical type of switch will be the only one described here. It is conveniently categorized by the device used, which may be a gas tube, a semiconductor diode, or a piece of ferrite material, as will be seen. Finally, a very common application of such switches will be described, namely, the duplexer (as used in radar).

Gas-tube switches A typical gas-tube switch, or TR (transmit-receive) cell, is shown in Fig. 10-44. It consists basically of a piece of waveguide filled with a gas mixture, such as hydrogen, argon, water vapor and ammonia, kept at a low pressure of a few millimeters of mercury to help ionization and terminated at either end by resonant windows. These are often made of glass, which is virtually transparent to microwaves but which prevents any gas from escaping. In the center of the waveguide there is a pair of electrodes, looking faintly like a stalactite and a stalagmite and having the function of helping the ionization of the gas by virtue of being close together, thus increasing the electric field at this point.

At low applied powers, such as those coming from the antenna of a microwave receiver, the gas tube behaves much like an ordinary piece of waveguide, and the signal passes through it with an insertion loss that is typically about 0.5 dB. When a high-power pulse arrives, however, the gas in the tube ionizes and becomes an almost perfect conductor. This has the effect of placing a short circuit across the waveguide leading to the gas tube. Thus the power that passes through it does so with an attenuation that can exceed 60 dB in practice. The tube acts as a self-triggered switch, since no bias or synchronizing voltage need be applied to change it from an open circuit to a short circuit.

As will be seen in Sec. 16-1, a switch such as this must act very rapidly; from the gas tube's point of view, this means that quick ionization and deionization are required. Ionization must be quick to ensure that the initial spike of power cannot pass through the TR cell and possibly damage any equipment on the other side of it. Quick deionization is needed to ensure that the receiver connected to the other end of the tube

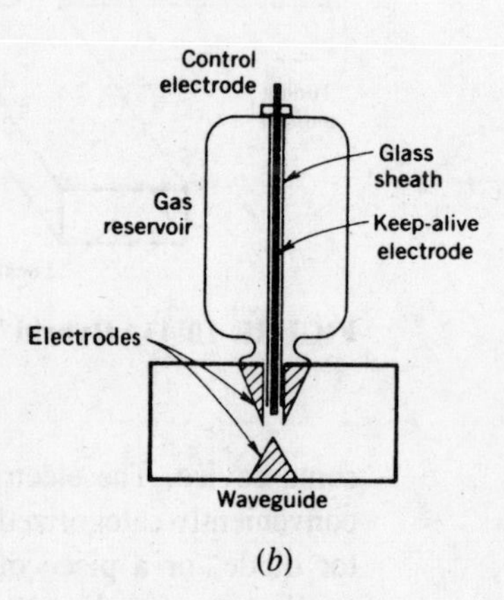

FIGURE 10-44 Gas (TR) tube. *(a)* Modern commercial tube; *(b)* simplified cross section. ***(By permission of Ferranti, Ltd.)***

does not remain disconnected from the antenna for too long. The first requirement is helped by the inclusion of a keep-alive electrode (as shown), to which a dc voltage is applied to ensure that ionization occurs as soon as any significant microwave power is applied. The second requirement may be helped by a suitable choice of gas. Finally, present-day gas tubes are capable of switching very high powers indeed (in excess of 10 MW pulsed if required).

Semiconductor diode switches A number of semiconductor diodes may be used as switches, by virtue of the fact that their resistance may be changed quickly, by a change in bias, from forward to reverse and back again. Point-contact diodes have been used for this purpose, but their power-handling ability is very low, and the most popular switching diode is the PIN diode.[3] Not only does its resistance change significantly with the applied bias, but also it is capable of handling appreciable amounts of power. Several diodes may be used in parallel to increase the power-handling ability even further.

A PIN (or any other) diode switch may be mounted as shown in Fig. 10-41 except that there is now no wall on the right-hand side of the waveguide. Instead, the

[3]This diode is described fully in Chap. 12, where its switching applications are mentioned, and further references are given.

guide continues and is eventually connected to some device such as a receiver. Such a diode switch may be *passive* or *active*. The passive type is simpler because it just has the diode connected across the waveguide. It then relies on the incidence of high microwave power to cause the diode to conduct and therefore to become a short circuit which reflects the power so as to prevent its further passage down the waveguide. An active diode switch, on the other hand, has a reverse bias applied to it in the absence of incident power. Simultaneously with the application of high power, the bias is changed to forward, and the diode once again short-circuits that portion of waveguide. Back bias is then applied at the same time as the pulse ends. The advantage of this somewhat more complex arrangement is a reduction in the forward and reverse loss (so that they are both comparable to those of the TR tube), and a very significant increase in the maximum power handled.

A practical PIN diode switch is shown in Fig. 10-45 and is seen to consist of a number of diodes in parallel. Such an arrangement allows peak powers of several hundred kilowatts to be switched. The advantages of the PIN diode switch, compared with the TR tube, are its greater life and reliability, as well as smaller size and the removal of the initial spike of power coinciding with the beginning of the pulse. It handles less power, however, and is slower in high-power applications, although in low-power switching and pulse modulation PIN diodes are capable of switching times under 10 ns.

Ferrite switches The properties of ferrites, as described in Sec. 10-5.2, make them suitable also for switching operations. A typical switch is the pair of Y circulators shown in Fig. 10-46, in which the direction of the magnetic field can be reversed for the second circulator. This is accomplished by providing bias changes in the form of current reversals through the solenoid which is used to generate the magnetic field for this circulator. It can be seen, from the previous treatment of circulators and the signal paths shown in Fig. 10-46, that in the "transmit" condition very little power from the transmitter will enter the second circulator, and most of the power that does will be dissipated in the matched load. In the "receive" state, the magnetic bias will be (externally) reversed for the second circulator, so that the signal from the antenna will be coupled to the receiver. The action of the first circulator will prevent this signal from entering the transmitter. Ferrite switches are capable of switching hundreds of kilowatts peak, with low losses, long life and high reliability, but they are not yet as fast as gas tubes.

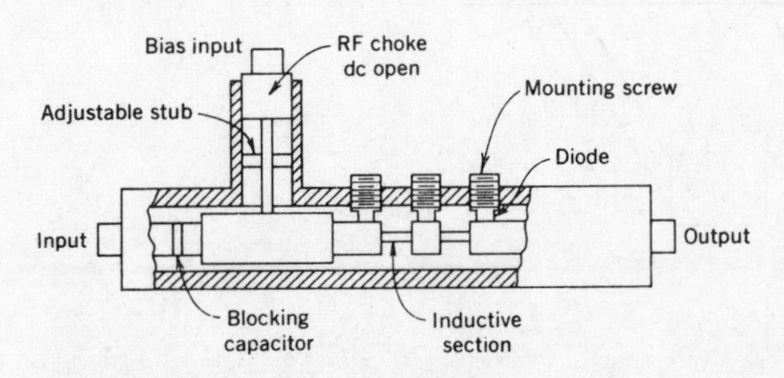

FIGURE 10-45 PIN diode switch.

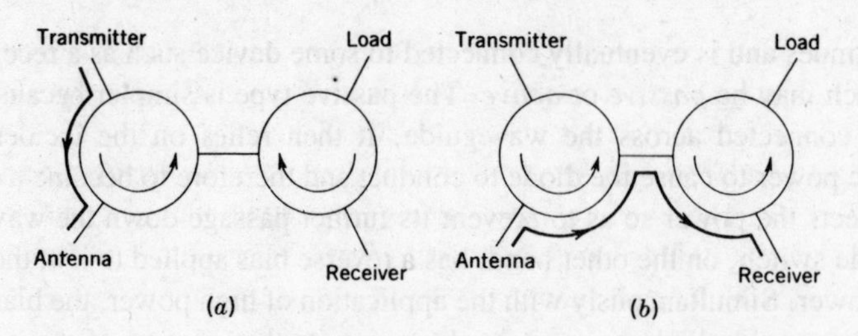

FIGURE 10-46 Schematic diagram of ferrite switch. (*a*) Transmission; (*b*) reception.

Duplexers A *duplexer* is a circuit designed to allow the use of the same antenna for both transmission and reception, with minimal interference between the transmitter and the receiver. From this description it follows that an ordinary circulator is a duplexer, but the emphasis here is on a circuit using switching for pulsed (not CW) transmission (see Sec. 16-1).

The branch-type duplexer shown in Fig. 10-47 is a type often used in radar. It has two switches, the TR and the ATR (anti-TR), arranged in such a manner that the receiver and the transmitter are alternately connected to the antenna, without ever being connected to each other. The operation is as follows.

When the transmitter produces an RF impulse, both switches become short-circuited either because of the presence of the pulse, as in TR cells, or because of an external synchronized bias change. The ATR switch reflects an open circuit across the main waveguide, through the quarter-wave section connected to it, and so does the TR switch, for the same reason. Therefore, neither of them affects the transmission, but the short-circuiting of the TR switch prevents RF power from entering the receiver or at least reduces any such power down to a tolerable level. At the termination of the transmitted pulse, both switches open-circuit by a reversal of the initial short-circuiting process. The ATR switch now throws a short circuit across the waveguide leading to the transmitter. If this were not done, a significant loss of the received signal would be incurred. At the input to the guide joining the TR branch to the main waveguide, this

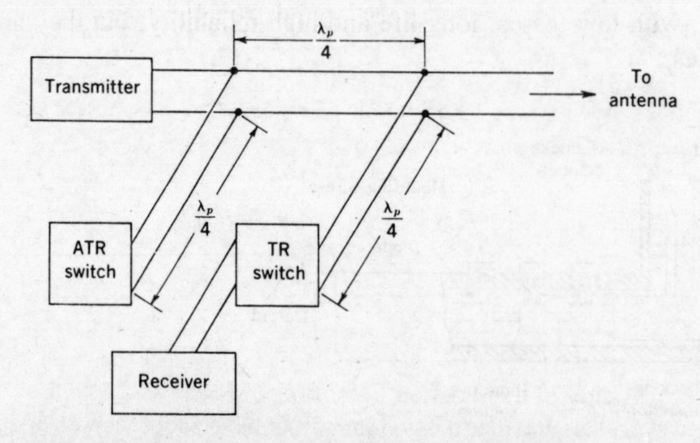

FIGURE 10-47 Branch-type duplexer for radar.

short circuit has now become an open circuit and hence has no effect. Meanwhile, the guide leading through the TR switch is now continuous and correctly matched. The signal from the antenna can thus go directly to the receiver.

As can be appreciated, the branch-type duplexer is a narrowband device, because it relies on the length of the guides connecting the switches to the main waveguide. However, single-frequency operation is very often sufficient, so that the branch-type duplexer is very common.

PROBLEMS

For self-testing questions on this chapter, ***see p. 693.***

10-1. What will be the cutoff wavelength, for the dominant mode, in a rectangular waveguide whose breadth is 10 cm? For a 2.5-GHz signal propagated in this waveguide in the dominant mode, calculate the guide wavelength, the group and phase velocities, and the characteristic wave impedance.
[Ans: 20 cm, 15 cm, 2.4×10^8 m/s, 3.75×10^8 m/s, 471 Ω]

10-2. A 6-GHz signal is to be propagated in a waveguide whose breadth is 7.5 cm. Calculate the characteristic wave impedance of this rectangular waveguide for the first three $TE_{m,0}$ modes and, if $b = 3.75$ cm, for the $TM_{1,1}$ mode.
[Ans.: 400 Ω, 506 Ω, ∞, 251 Ω]

10-3. A 6-GHz signal is to be propagated in the dominant mode in a rectangular waveguide. If its group velocity is to be 90 percent of the free-space velocity of light, what must be the breadth of the waveguide? What impedance will it offer to this signal, if it is correctly matched?
[Ans.: 5.74 cm, 419 Ω]

10-4. It is required to propagate a 12-GHz signal in a rectangular waveguide in such a manner that Z_0 is 450 Ω. If the $TE_{1,0}$ mode is used, what must be the corresponding cross-sectional waveguide dimension? If the guide is 30 cm long, how many wavelengths does that represent for the signal propagating in it? How long will this signal take to travel from one end of the waveguide to the other?
[Ans.: $a = 2.29$ cm, $l = 10\lambda_p$, $t = 1.19$ ns]

10-5. Calculate the bandwidth of the WR28 waveguide, i.e., the frequency range over which *only* the $TE_{1,0}$ mode will propagate.
[Ans.: 21.13–38.86 GHz]

10-6. A circular waveguide has an internal diameter of 5 cm. Calculate the cutoff *frequencies* in it for the following modes: *(a)* $TE_{1,1}$, *(b)* $TM_{0,1}$, *(c)* $TE_{0,1}$.
[Ans.: (a) 3.51 GHz, (b) 4.58 GHz, (c) 7.31 GHz]

10-7. A 4-GHz signal, propagating in the dominant mode, is fed to a WR28 waveguide. What length of this guide will be required to produce an attenuation of 120 dB?
[Ans.: 35.7 mm]

QUESTIONS

10-1. What are waveguides? What is the fundamental difference between propagation in waveguides and propagation in transmission lines or free space?

10-2. Compare waveguides and transmission lines from the point of view of frequency limitations, attenuation, spurious radiation and power-handling capacity.

10-3. Draw a sketch of electromagnetic wavefronts incident at an angle on a perfectly conducting plane surface. Use this sketch to derive the concept of *parallel* and *normal* wavelengths.
10-4. Define, and fully explain the meaning and consequences of, the cutoff wavelength of a waveguide. Apart from the wall separation, what else determines the actual value of the cutoff wavelength for a signal of a given frequency?
10-5. Differentiate between the concepts of *group velocity* and *phase velocity* as applied to waveguides. Derive the universal formula for the group velocity.
10-6. The $TE_{1,0}$ mode is described as the *dominant* mode in rectangular waveguides. What property does it have which makes it dominant? Show the electric field distribution at the mouth of a rectangular waveguide carrying this mode, and explain how the designation $TE_{1,0}$ comes about.
10-7. Why is the Z_0 of waveguides called the characteristic *wave* impedance, and not just simply the characteristic impedance?
10-8. What takes place in a waveguide if the wavelength of the applied signal is greater than the cutoff wavelength? Why?
10-9. What are the differences, in the propagation and general behavior, between TE and TM modes in rectangular waveguides?
10-10. Compare the practical advantages and disadvantages of circular waveguides with those of rectangular waveguides. What is a particular advantage of the former, with broadband communications applications?
10-11. Describe ridged and flexible waveguides briefly, and outline their applications. Why are they not used more often than rectangular waveguides?
10-12. With the aid of appropriate sketches, show how probes may be used to launch various modes in waveguides. What determines the number and placement of the probes?
10-13. Sketch the paths of current flow in a rectangular waveguide carrying the dominant mode, and use the sketch to explain how a slot in a common wall may be used to couple the signals in two waveguides.
10-14. Describe briefly the various methods of exciting waveguides, and explain under what circumstances each is most likely to be used.
10-15. Explain the operation of a choke join. Under what circumstances would this join be preferred to a plain flange coupling?
10-16. Draw the cross section of a waveguide rotating join, and describe it and its operation.
10-17. When would a waveguide bend be preferred to a corner? Why is an E-plane corner likely to be double-mitered? Illustrate your answer with appropriate sketches.
10-18. With the aid of a suitable diagram, explain the operation of the *hybrid T* junction (magic tee). What are its applications? What is done to avoid reflections within such a junction?
10-19. Show a pictorial view of a *hybrid ring*, and label it to show the various dimensions. Explain the operation of this *rat race*. When might it be preferred to the magic tee?
10-20. How do the methods of impedance matching in waveguides compare with those used with transmission lines? List some of the ones in waveguides.
10-21. Show a waveguide with a cylindrical post, and briefly describe the behavior of this obstacle. What can it be used for when it is inserted halfway into a waveguide? What advantage does such a post have over an iris?
10-22. Draw and title the diagram of the waveguide tuner which is the analog of a transmission-line stub matcher. What else might be used with waveguides for this purpose?
10-23. With sketches, describe waveguide matching terminations and attenuators. Include one sketch of a variable attenuator.
10-24. Discuss the applications of waveguides operated beyond cutoff.
10-25. What are cavity resonators? What applications do they have? Why do they normally have odd shapes?

10-26. Starting with a rectangular waveguide carrying the $TE_{1,0}$ mode, evolve the concept of a cavity resonator oscillating in the $TE_{1,0,2}$ mode.
10-27. Describe briefly various methods of coupling to cavity resonators. With the aid of a sketch, explain specifically how a cavity may be coupled to an electron beam.
10-28. By what methods may cavity resonators be tuned? Explain the effect of tuning on cavity Q.
10-29. With the aid of a diagram, explain *fully* the operation of a two-hole waveguide directional coupler; also state its uses.
10-30. What are ferrites? What properties do they have which distinguish them from ordinary conductors or insulators? What is YIG?
10-31. Explain the results of an interaction of dc and RF magnetic fields in a ferrite; what is the *gyromagnetic resonance interaction* that may occur?
10-32. What are the three main limitations of ferrites?
10-33. With the aid of a suitable diagram, explain the operation of a *Faraday rotation* ferrite isolator. List its applications and typical performance figures.
10-34. Use sketches to help explain the operation of a Faraday rotation circulator and a Y circulator. What applications and typical performance figures do these devices have?
10-35. List the requirements that a diode mount must fulfill if the diode is to be used as a detector or mixer mounted in a waveguide. Show a typical practical diode mount, and explain how it satisfies these requirements.
10-36. When might a *balanced* mixer be required in a waveguide system? Show, by means of a sketch, one of the ways of achieving such balanced mixing.
10-37. Briefly describe the operation of PIN diode switches. What is the difference between *passive* and *active* switching?
10-38. Compare the properties of three types of waveguide switches.
10-39. What is a duplexer? Explain the operation of a branch-type duplexer, with the help of a circuit diagram. What is the main limitation of this duplexer?

REFERENCES

1. Saad, T. S. (compiler and ed.), *Microwave Engineers' Handbook,* vol. 1, Artech House, Inc., Dedham, Mass., 1971, pp. 28–40. (Note that the attenuation figures shown on p. 36 are significantly higher than those currently recommended by the International Electrotechnical Commission.)
2. Staniforth, J. A., *Microwave Transmission,* The English Universities Press Ltd., London, 1972, pp. 95–101.
3. For ferrites in general, see also
(*a*) Hancock, K. E., "An Introduction to Microwave Ferrite Devices," *Wireless World,* part I, February 1967, pp. 85–87; part II, April 1967, pp. 185–188.
(*b*) Whicker, L. R., and C. W. Young, Jr., "The Evolution of Ferrite Control Components," *Microwave J.* November 1978, pp. 33–37 (38 additional references given).
4. For YIG in particular, see also
(*a*) Roschmann, P., "YIG Filters," *Philips Tech. Rev.* **32**(9–12): 322–327, 1971.
(*b*) Hernday, P. R., and C. J. Enlow, "A High-Performance 2-to-18 GHz Sweeper," *Hewlett-Packard Journal,* March 1975, pp. 2–14.

11 MICROWAVE TUBES AND CIRCUITS

The preceding chapter treated passive microwave devices; it is now necessary to study active ones. Consequently, this chapter deals with microwave tubes and circuits, and the next one discusses microwave semiconductor devices and associated circuitry. The order of selection is mainly historical, in that tubes preceded semiconductors by some 20 years. However, it is interesting to consider that, if one measures age in terms of worker-years of research, then actually the semiconductor devices are very much older!

By and large, the limitation for tubes on the one hand, and transistors and diodes on the other, is one of size at microwave frequencies. As frequency is raised, so, eventually, devices must become smaller. Consequently, the powers handled fall, and noise rises. The overall result at microwave frequencies is that tubes have the higher output powers, while semiconductor devices are smaller, require simpler power supplies and, more often than not, have lower noise and greater reliabilities.

There are, broadly speaking, three general types of microwave tubes. The first is the ordinary gridded tube, invariably a *triode* at the highest frequencies, which has been evolved and refined to its utmost. Then there is the class of devices in which brief, though sometimes repeated, interaction takes place between an electron beam and an RF voltage. The *klystron* exemplifies this type of device.

The third category of device is one in which the interaction between an electron beam and an RF field is continuous. This is divided into two subgroups. In the first, an electric field is used to ensure that the interaction between the electron beam and the RF field is continuous. The *traveling-wave tube (TWT)* is the prime example of this interaction. It is an amplifier, whose oscillator counterpart is called a *backward-wave oscillator (BWO)*. The second subgroup consists of tubes in which a magnetic field ensures a constant electron beam–RF field interaction.

The *magnetron,* an oscillator, uses this interaction and is complemented by the *cross-field amplifier (CFA),* which was evolved from it.

Each type of microwave tube will now be discussed in turn, and in each case state-of-the-art performance figures will be given. Also, comparisons will be drawn showing the relative advantages and applications of competing devices.

11-1 MICROWAVE TRIODES

Ordinary vacuum tubes are useless at microwave frequencies, because of a number of limitations which will now be explained. *It should be noted that such limitations also afflict transistors at UHF and above,* and they, too, are exotic versions of the lower-frequency devices. Neither in tubes nor in transistors can these limitations be completely overcome. However, it is possible to extend the useful range of both to well over 10 GHz, as will be seen.

11-1.1 Frequency Limitations of Gridded Tubes

As frequency is raised, vacuum tubes suffer from two general kinds of problems. The first is concerned with interelectrode capacitances and inductances; and the second is caused by the finite time that electrons take to travel from one electrode to another in a tube. Noise tends to increase with frequency, and thus microwave tubes are invariably triodes, these being the least noisy tubes (see Sec. 2-2.4).

"Standard" limitations At UHF and above, interelectrode capacitances cannot be ignored, nor can series electrode inductances due to internal connecting leads. Figure 11-1*a* shows the equivalent circuit of any triode. However, if this is an ordinary RF tube, the capacitances will all be of the order of 2 pF, and the inductances are typically 0.02 μH. At 500 MHz, input and output shunt impedances will thus be of the order of

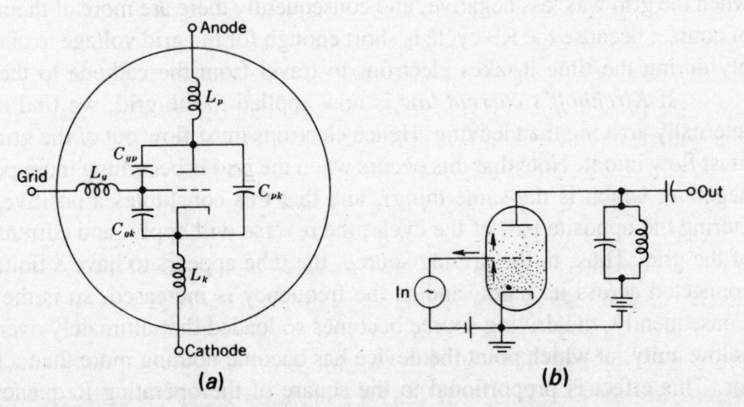

FIGURE 11-1 Triode frequency limitations. (*a*) Equivalent circuits at UHF; (*b*) noninstantaneous electron flow at UHF (arrows show direction of electron flow, not current flow).

160 Ω (capacitive), and series impedances are 60 Ω (inductive). The maximum possible inductances of external tuned circuits will be severely limited, as will their Q. Neither short circuits nor connections directly to electrodes will be possible—even internal self-oscillation will occur. Trying to use such tubes would be like employing AF devices each of whose electrodes has a 20-mH coil connected to it, as well as a 2-μF bypass capacitor! Needless to say, the difficulties increase with frequency, and would be 10 times as bad at 5 GHz.

The *skin effect* causes very significant increases in series resistance and inductance at UHF, unless tubes have been designed to minimize the effect. Also, *dielectric losses* increase with frequency. Accordingly, unless tubes and their bases are made of the lowest-loss dielectrics, efficiencies are reduced so much that proper amplification cannot be provided.

Transit time effects Transit time effects, it should be noted, apply not only to tubes but also to transistors and any other devices which depend on short times between electrodes.

At low frequencies, it is possible to assume that electrons leave the cathode and arrive at the anode of a tube *instantaneously*. This can most certainly not be assumed at microwave frequencies. That is to say, *the transit time becomes an appreciable fraction of the RF cycle*. Several awkward effects result from this situation. One of them is that the grid and anode signals are no longer 180° out of phase, thus causing design problems, especially with feedback in oscillators. Another important effect—possibly the most important—is that the grid begins to take power from the driving source. The *power is absorbed* (and dissipated) *even when the grid is negatively biased*.

Figure 11-1*b* shows the electron distribution in a triode at an instant of time when the voltage on the grid is becoming less negative. The tube is biased and driven so that the grid always remains negative, but its instantaneous voltage of course varies with the RF drive. It is seen that there are more electrons on the cathode side of the grid than on the plate side. The reason is that the electrons about to arrive at the anode left the cathode when the grid voltage was quite negative, and thus the electron flow rate was low. The electrons now in the cathode-grid space left the cathode subsequently, when the grid was less negative, and consequently there are more of them. This occurs, of course, because the RF cycle is short enough for the grid voltage to change appreciably during the time it takes electrons to travel from the cathode to the plate.

If *Kirchhoff's current law* is now applied to the grid, we find more electrons internally arriving than leaving. Hence electrons must flow out of the grid: i.e., current must flow into it. Note that this occurs when the grid is becoming more positive (or less negative, which is the same thing), and that this constitutes a positive conductance. During the opposite part of the cycle, the reverse will apply, and current will flow out of the grid. Thus, to the driving source, the tube appears to have a finite conductance connected across its input, and as the frequency is increased, so is the conductance. Consequently, the driving source becomes so loaded that ultimately overall gain drops below unity, at which point the device has become nothing more than a fancy attenuator. This effect is proportional to the square of the operating frequency.

Finally, the increased input conductance increases input noise, as was discussed in Sec. 2-2.3. Long before 1 GHz is reached, ordinary RF tubes have a noise figure very much in excess of 25 dB. As a conclusion, it is true to say that when any tube (or transistor) eventually fails at high frequencies, *transit time is the "killer,"* in one way or another.

11-1.2 UHF Triodes and Circuits [1]

Tube requirements A short transit time is essential if grid loss and noise are to be minimized, as is the reduction of internal capacitances and inductances. Also, since coaxial resonant lines and cavities are used as tuned circuits at these frequencies, tubes should be built so as to connect to such lines or resonators directly.

To reduce transit time, the various electrodes are brought as close to each other as possible, while it is ensured that they cannot touch, even when a tube is jarred. Increased anode voltage will also help to reduce the transit time, and so will increased anode current, for ratner complex reasons. Interelectrode capacitance is proportional to the electrode area and inversely proportional to the distance between electrodes. Thus, if both are reduced by the same so-called *scaling factor,* the nearness of electrodes will not increase the capacitance between them. Finally, lead inductances may be reduced by having leads in the form of rings.

The reduced area of the electrodes will undoubtedly reduce the maximum allowable dissipation; it thus follows that as operating frequency is raised, less output power is available from tubes (and any other devices). The frequency range is extended by the use of improved constructions and forced cooling, so that external anodes with cooling fins are generally used.

As a result of all these endeavors, tubes are capable of continuous powers from over 100 W at 1 GHz to about 25 mW at 10 GHz. The pulsed powers (duty cycle 0.1 percent) over that range are 15 kW to 350 W. In the middle of this range, at 3 GHz, typical gains are 10 to 20 dB, and efficiencies 30 percent CW to 55 percent pulsed. See also [2].

Of the various tubes that have been used for microwaves, only the *disk-seal triode* is still in use, because it is capable of high powers—the figures quoted above are for this type. It comes in various versions, all of which have in common extreme closeness of electrodes, high voltages, small electrode areas but large volumes (for dissipation), ceramic bodies, and electrodes in the form of disks to reduce inductance and facilitate connection to resonant lines or cavities.

One type of this triode is shown in Fig. 11-2. The enlargement of the active region shows why this is often called a *parallel-plane tube*. It will be seen that the cathode and plate have fairly large volumes but low active areas. Because these areas are plane, a planar grid is mounted between them, with very close spacing as indicated. The close spacing is possible because all the electrodes are solid and firmly attached to the body of the tube. They are thus unlikely to touch each other if the tube is subjected to mechanical vibration or shock. As shown, the electrode connections are cylinders or rings; they are often made of copper-tungsten. Together with the ceramic body, this permits high operating temperatures and thus large power dissipation. As previously mentioned, the connections to the electrodes are physically such as to permit direct working with coaxial lines. A typical tube of the family shown may have an overall

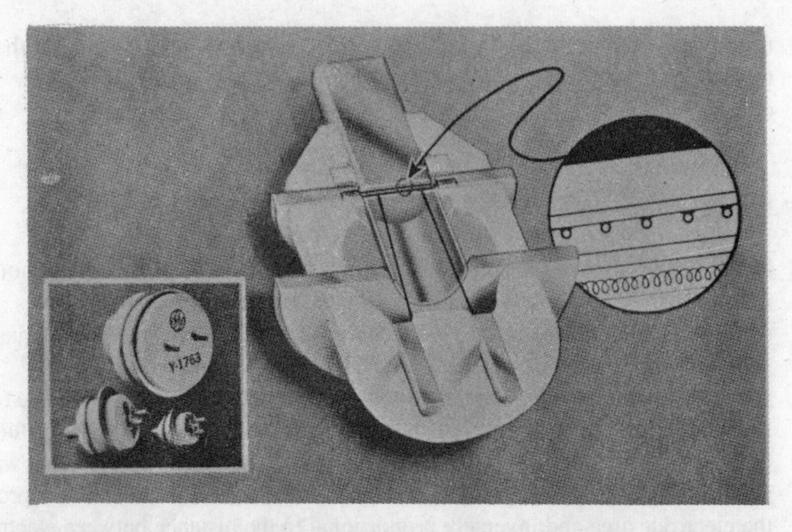

FIGURE 11-2 Modern planar triode construction, showing an enlarged active area and complete tubes (inset). ***(Courtesy of General Electric Co., Owensboro, Kentucky.)***

height of 24 mm, a diameter of 19 mm and a pulsed power output of 2 kW at 4.1 GHz. A somewhat different version of the planar triode is the *lighthouse tube,* which features radiating fins for the anode and is thus capable of higher powers.

As might well have been expected, amplifiers at UHF and above use the grounded-grid configuration, and the oscillators use the *ultra-audion* version of the Colpitts, as shown in Fig. 11-3. RF chokes and feedthrough capacitors are used for isolation and decoupling, while transmission lines are the tuned circuits.

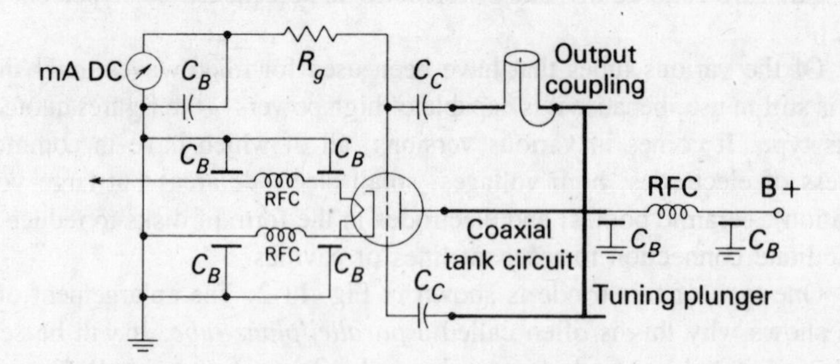

FIGURE 11-3 Microwave triode ultra-audion oscillator using coaxial resonators.

11-2 MULTICAVITY KLYSTRON

The multicavity klystron, together with all the remaining devices described in this chapter, relies on the principle of "if you can't beat them, join them." That is to say, the designs of all these tubes recognize the fact that transit time will sooner or later

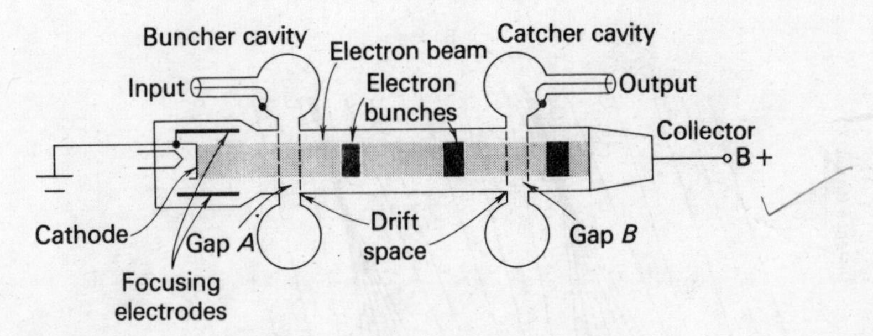

FIGURE 11-4 Klystron amplifier schematic diagram.

terminate the usefulness of any orthodox tube. They therefore use the transit time, instead of fighting it. The klystron was invented just before World War II by the Varian brothers [3], as a source and amplifier of microwaves. It provided much higher powers than had previously been obtainable at these frequencies.

11-2.1 Operation

Figure 11-4 schematically shows the principal features of a two-cavity amplifier klystron. It is seen that a high-velocity electron beam is formed, focused (external magnetic focusing is omitted for simplicity) and sent down a long glass tube to a collector electrode which is at a high positive potential with respect to the cathode. The beam passes gap A in the *buncher* cavity, to which the RF signal to be amplified is applied, and it is then allowed to drift freely, without any influence from RF fields, until it reaches gap B in the output or *catcher* cavity. If all goes well, oscillations will be excited in the second cavity which are of a power much higher than those in the buncher cavity, so that a large output can be taken. The beam is then collected by the collector electrode.

As can be seen, the cavities are reentrant and are also tunable (although this is not shown). Futhermore, they may be integral or demountable. In the latter case, the wire grid meshes are connected to rings external to the glass envelope, and cavities may be attached to the rings. The *drift space* is quite long, and the transit time in it is put to use. However, the gaps must be so short that the voltage across them does not change significantly during the passage of a particular bunch of electrons; having a high collector voltage helps in this regard.

It is apparent that the electron beam, which has a constant velocity as it approaches gap A, will be affected by the presence of an RF voltage across the gap, as was outlined in Sec. 11-4.2. However, the extent of this effect on any one electron will depend on the voltage across the gap when the electron passes this gap. It is thus necessary to investigate the effect of the gap voltage upon individual electrons.

Consider the situation when there is no voltage across the gap. Electrons passing it are unaffected and continue on the collector with the same constant velocities they had before approaching the gap (this is shown at the left of Fig. 11-5). Sometime later, after an input has been fed to the buncher cavity, an electron will pass gap A at the time when the voltage across this gap is zero and going positive. Let this be the

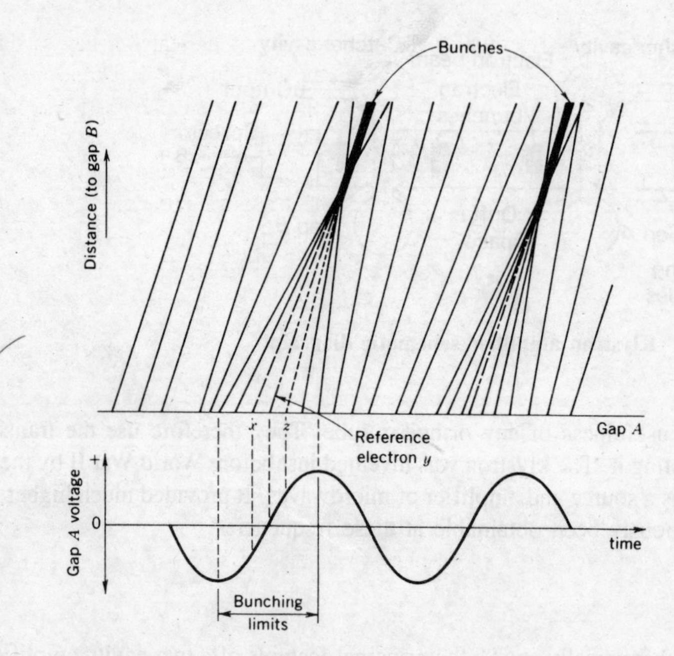

FIGURE 11-5 Applegate diagram for klystron amplifier.

reference electron y. It is of course unaffected by the gap, and thus it is shown with the same slope on the *Applegate diagram* of Fig. 11-5 as electrons passing the gap before any signal was applied.

Another electron, z, passes gap A slightly later than y, as shown. Had there been no gap voltage, both electrons would have continued past the gap with unchanged velocity, and therefore neither could have caught up with the other. Here, however, electron z is slightly accelerated by the now positive voltage across gap A, and given enough time, it will catch up with the reference electron. As shown on the Applegate diagram, it has enough time to catch electron y easily before gap B is approached. Similarly, electron x passes gap A slightly before the reference electron. However, although it passed gap A before electron y, it was retarded somewhat by the negative voltage then present across the gap. Since electron y was not so retarded, it has an excellent chance of catching electron x before gap B (this it does, as shown on the Applegate diagram).

As electrons pass the buncher gap, they are *velocity-modulated* by the RF voltage existing across this gap. Such velocity modulation would not be sufficient, in itself, to allow amplification by the klystron. However, as explained and shown on the Applegate diagram, electrons have the opportunity of catching up with other electrons in the drift space. When an electron catches up with another one, it may simply pass it and forge ahead. On the other hand, it may exchange energy with the slower electron, giving it some of its excess velocity, and the two bunch together and move on with the average velocity of the beam. As the beam progresses farther down the drift tube, so the bunching becomes more complete, as more and more of the faster electrons catch up with bunches ahead. Eventually, the current passes the catcher gap in quite pronounced bunches and therefore varies cyclically with time. This variation in current density is known as *current modulation,* and this is what enables the klystron to have significant gain.

It will be noted from the Applegate diagram that bunching can occur once per cycle, centering on the reference electron. The limits of bunching are also shown. Electrons arriving slightly after the second limit clearly are not accelerated sufficiently to catch the reference electron, and the reference electron cannot catch any electron passing gap *A* just before the first limit. Bunches thus also arrive at the catcher gap once per cycle and deliver energy to this cavity. In ordinary vacuum tubes, a little RF power applied to the grid can cause large variations in the anode current, thus controlling large amounts of dc anode power. Similarly in the klystron, a little RF power applied to the buncher cavity results in large beam current pulses being applied to the catcher cavity, with a considerable power gain as the result. Needless to say, the catcher cavity is excited into oscillations at its resonant frequency (which is equal to the input frequency), and a large sinusoidal output can be obtained because of the *flywheel effect* of the output resonator.

11-2.2 Practical Considerations

The construction of the klystron lends itself to two practical microwave applications: as a multicavity power amplifier or as a two-cavity power oscillator.

Multicavity klystron amplifier The bunching process in a two-cavity klystron is by no means complete, since there are large numbers of out-of-phase electrons arriving at the catcher cavity between bunches. Consequently, more than two cavities are always employed in practical klystron amplifiers. Four cavities are shown in the klystron amplifier schematic diagram of Fig. 11-6 and up to seven cavities have been used in practice. Partially bunched current pulses will now also excite oscillations in the intermediate cavities, and these cavities in turn set up gap voltages which help to produce more complete bunching. Having the extra cavities helps to improve the efficiency and power gain considerably. The cavities may all be tuned to the same frequency, such *synchronous tuning* being employed for narrowband operation. For broadband work, for example with UHF klystrons used as TV transmitter output tubes, or 6-GHz tubes used as power amplifiers in some satellite station transmitters, *stagger tuning* is used. Here, the intermediate cavities are tuned to either side of the center frequency, improving the bandwidth very significantly. It should be noted that cavity *Q* is so high that stagger tuning is a ''must'' for bandwidths much over 1 percent.

Two-cavity klystron oscillator If a portion of the signal in the catcher cavity is coupled back to the buncher cavity, oscillations will take place. As with other oscillators, the feedback must have the correct polarity and sufficient amplitude. The schematic diagram of such an oscillator is as shown in Fig. 11-4, except for the addition of a (permanent) feedback loop. Oscillations in the two-cavity klystron behave as in any other feedback oscillator; having been started by a switching transient or noise impulse, they continue as long as dc power is present.

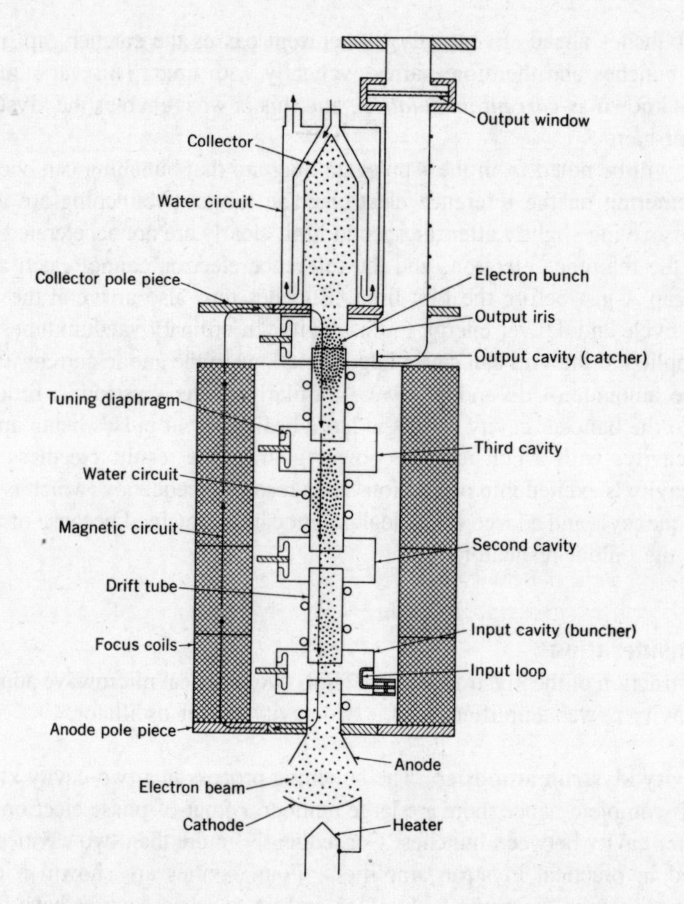

FIGURE 11-6 Four-cavity klystron amplifier schematic diagram. ***(Courtesy of Varian Associates, Inc.)***

Performance and applications The multicavity klystron is used as a medium-, high- and very high-power amplifier in the UHF and microwave ranges, for either continuous or pulsed operation. The frequency range covered is from about 250 MHz to over 95 GHz, and the power available has been described as "adequate," i.e., much higher than currently needed. Table 11-1 illustrates this situation.

This table summarizes the power requirements of the major applications for klystron amplifiers and shows how the devices are able to meet them. The gain of klystrons is also adequate. It ranges from 30–35 dB at UHF to 60–65 dB in the microwave range. Such high gain figures mean that the klystron is generally the only nonsemiconductor device in high-power amplifiers. For example, the Varian VKX-7809 X-band klystron has a CW power output of 2.2 kW with a gain of 46 dB. Hence the driving power required is only 50 mW, which is well within the capability of any semiconductor power device.

Current klystron developments are aimed at improving efficiency, providing longer lives, and reducing size; typical efficiency is 35 to 50 percent. To improve

TABLE 11-1 Klystron Amplifier Performance and Applications

APPLICATION AND TYPE OF REQUIREMENT	FREQ. RANGE, GHz	NEEDED POWER, max.	AVAILABLE POWER, max.
UHF TV transmitters (CW)	0.5–0.9	55 kW	100 kW
Long-range radar (pulsed)	1.0–12	10 MW	20 MW
Linear particle accelerator (pulsed)	2.0–3.0	25 MW	40 MW
Troposcatter links (CW)	1.5–12	250 kW	1000 kW
Earth station transmitter (CW)	5.9–14	8 kW	25 kW

reliability and MTBF (mean time between failures), tungsten-iridium cathodes are now being used to reduce cathode temperature and thus provide longer life. As regards size, a typical 50-kW UHF klystron, as shown in Fig. 11-6, may be over 2 m long, with a weight of nearly 250 kg. As may be gathered from Fig. 11-6, a large proportion of the bulk is due to the magnet, often as much as two-thirds. A 100-kW peak (2.5-kW average) X-band klystron may be 50 cm long and may weigh about 30 kg, if it uses permanent-magnet focusing as so far treated. It is possible to reduce this weight to one-third by using *periodic permanent-magnet (PPM)* focusing. In this system (see also Sec. 11-5.2 and [4]), the beam is focused by so-called magnetic lenses, which are small, strong magnets along the beam path. In between them, the beam is allowed to defocus a little. Further advancements include the distributed *interaction klystron* (see [5]), which makes this device fairly similar to a traveling-wave tube. The use of grids for modulation purposes (see Fig. 11-7 and [6]) has been rediscovered and evolved further.

The two-cavity klystron oscillator has fallen out of favor, having been displaced by CW magnetrons, semiconductor devices and the high gain of klystron and TWT amplifiers.

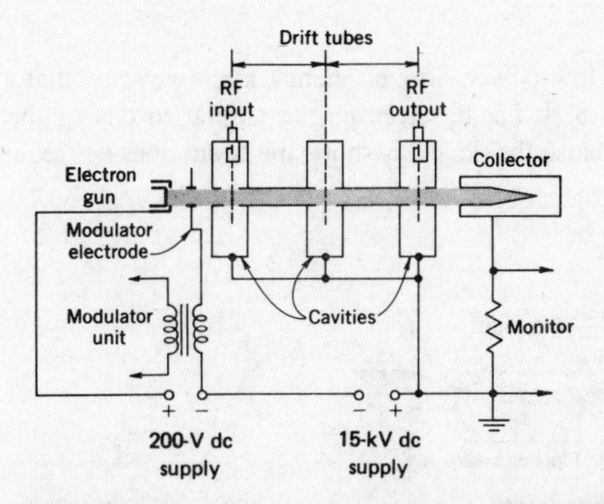

FIGURE 11-7 Three-cavity klystron pulsed amplifier with modulating grid. ***[By permission from Beck and Deering, "A Three-cavity L-band Pulsed Klystron Amplifier," Proc. IEE (London), vol. 105B, 1958.]***

Further practical aspects Multicavity klystron amplifiers suffer from the noise caused because bunching is never complete, and so electrons arrive at random at the catcher cavity. This makes them too noisy for use in receivers, but their typically 35-dB noise figures are more than adequate for transmitters.

Since the time taken by a given electron bunch to pass through the drift tube of a klystron is obviously influenced by the collector voltage, this voltage must be regulated. Indeed, the power supplies for klystrons are quite elaborate, with a regulated 9 kV at 750-mA collector current required for a typical communications klystron. Similarly, when a klystron amplifier is pulsed, such pulses are often applied to the collector. They should be flat, or else frequency drift (within limits imposed by cavity bandwidth) will take place during the pulse. As an alternative to this, and also because collector pulsing takes a lot of power, modulation of a special grid has been developed, as shown in Fig. 11-7. A typical "gain" of 20 is available between this electrode and the collector, thus reducing the modulating power requirements twentyfold. Amplitude modulation of the klystron can also be applied via this grid. However, if amplitude linearity is required, it should be noted that the klystron amplifier begins to saturate at about 70 percent of maximum power output. Beyond this point, output still increases with input but no longer linearly. This saturation is not a significant problem, all in all, because most of the CW applications of the multicavity klystron involve frequency modulation. Under such conditions, e.g., in a troposcatter link, the klystron merely amplifies a signal that is already frequency-modulated and at a constant amplitude.

11-3 REFLEX KLYSTRON [7]

It is possible to produce oscillations in a klystron device which has only one cavity, through which electrons pass twice. This is the reflex klystron, which will now be described.

11-3.1 Fundamentals

The reflex klystron is a low-power, low-efficiency microwave oscillator, illustrated schematically in Fig. 11-8. It has an electron gun similar to that of the multicavity klystron but smaller. Because the device is short, the beam does not require focusing.

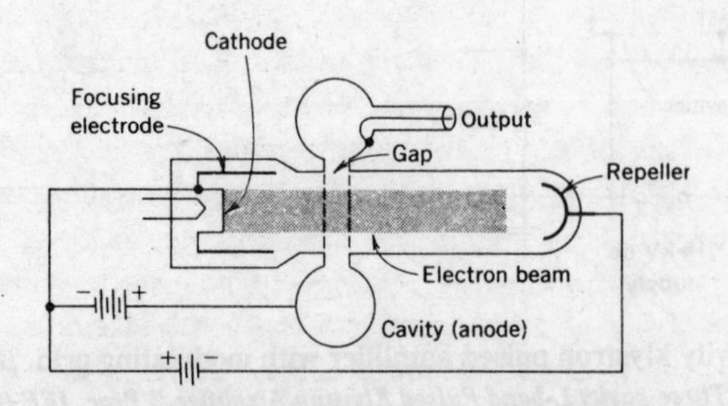

FIGURE 11-8 Reflex klystron schematic.

Having been formed, the beam is accelerated toward the cavity, which has a high positive voltage applied to it and, as shown, acts as the anode. The electrons overshoot the gap in this cavity and continue on to the next electrode, *which they never reach*. This *repeller* electrode has a fairly high negative voltage applied to it, and precautions are taken to ensure that it is not bombarded by the electrons. Accordingly, electrons in the beam reach some point in the repeller space and are then turned back, eventually to be dissipated in the anode cavity. If the voltages are properly adjusted, the returning electrons give more energy to the gap than they took from it on the outward journey, and continuing oscillations take place.

Operation As with the multicavity klystron, the operating mechanism is best understood by considering the behavior of individual electrons. This time, however, the reference electron is taken as one that passes the gap on its way to the repeller at the time when the gap voltage is zero and going negative. This electron is of course unaffected, overshoots the gap, and is ultimately returned to it, having penetrated some distance into the repeller space. An electron passing the gap slightly earlier would have encountered a slightly positive voltage at the gap. The resulting acceleration would have propelled this electron slightly farther into the repeller space, and the electron would thus have taken a slightly longer time than the reference electron to return to the gap. Similarly, an electron passing the gap a little after the reference electron will encounter a slightly negative voltage. The resulting retardation will shorten its stay in the repeller space. It is seen that, around the reference electron, earlier electrons take longer to return to the gap than later electrons, and so the conditions are right for bunching to take place. The situation can be verified experimentally by throwing a series of stones upward. If the earlier stones are thrown harder (i.e., accelerated more) than the later ones, it is possible for all of them to come back to earth simultaneously (i.e., in a bunch).

It is thus seen that, as in the multicavity klystron, velocity modulation is converted to current modulation in the repeller space, and one bunch is formed per cycle of oscillations. It should be mentioned that bunching is not nearly as complete in this case, and so the reflex klystron is much less efficient than the multicavity klystron.

Transit time As usual with oscillators, it is assumed that oscillations are started by noise or switching transients. Accordingly, what must now be shown is that the operation of the reflex klystron is such as to maintain these oscillations. For oscillations to be maintained, the transit time in the repeller space, or the time taken for the reference electron from the instant it leaves the gap to the instant of its return, must have the correct value. This is determined by investigating the best possible time for electrons to leave the gap and the best possible time for them to return.

The most suitable departure time is obviously centered on the reference electron, at the 180° point of the sine-wave voltage across the resonator gap. It is also interesting to note that, ideally, no energy at all goes into velocity-modulating the electron beam. It admittedly takes some energy to accelerate electrons, but just as much energy is gained from retarding electrons. Since just as many electrons are retarded as accelerated by the gap voltage, the total energy outlay is nil. This actually raises a most important point: *energy is spent in accelerating bodies* (*electrons in this case), but energy is gained from retarding them.* The first part of the point is obvious, and the second may be observed by means of a very simple experiment, for which the apparatus consists of a swing and a small member of the family. Once the child is swinging freely, retard the swing with some part of the body and measure the amount of energy absorbed (if still standing!).

It is thus evident that the best possible time for electrons to return to the gap is when the voltage then existing across the gap will apply maximum retardation to them. This is the time when the gap voltage is maximum positive (on the right side of the gap in Fig. 11-8). Electrons then fall through the maximum negative voltage between the gap grids, thus giving the maximum amount of energy to the gap. Therefore, the best time for electrons to return to the gap is at the 90° point of the sine-wave gap voltage. Returning after 1¾ cycles obviously satisfies these requirements; more generally it may be stated that

$$T = n + 3/4$$

where T = transit time of electrons in repeller space, cycles
n = any integer

Modes The transit time obviously depends on the repeller and anode voltages, so that both must be carefully adjusted and regulated. Once the cavity has been tuned to the correct frequency, both the anode and repeller voltages are adjusted to give the correct value of T from data supplied by the manufacturer. Each combination of acceptable anode-repeller voltages will provide conditions permitting oscillations for a particular value of n. In turn, each value of n is said to correspond to a different reflex klystron *mode*, practical transit times corresponding to the range from 1¾ to 6¾ cycles of gap voltage. Modes corresponding to $n = 2$ or $n = 3$ are the ones used most often in practical klystron oscillators.

11-3.2 Practical Considerations

Performance Reflex klystrons with integral cavities are available for frequencies ranging from under 4 to over 200 GHz. A typical power output is 100 mW, but overall maximum powers range from 3 W in the X band to 10 mW at 220 GHz. Typical efficiencies are under 10 percent, restricting the oscillator to low-power applications.

A typical X-band reflex klystron is shown in Fig. 11-9, approximately half life-size. The cutaway illustration reveals the internal construction of the electron gun. Part of the cavity is also visible, showing the smallness of the gap and the closeness of the repeller, The mechanical tuning arrangements are seen on the left and the waveguide output window on the right.

Tuning The klystron shown in Fig. 11-9 has an integral cavity, but demountable cavities are also possible, as with amplifier klystrons. The frequency of resonance is mechanically adjustable by any of the methods discussed in Sec. 10-4.2, with the adjustable screw, bellows or dielectric insert the most popular. Such *mechanical tuning* of reflex klystrons may give a frequency variation which ranges in practice from

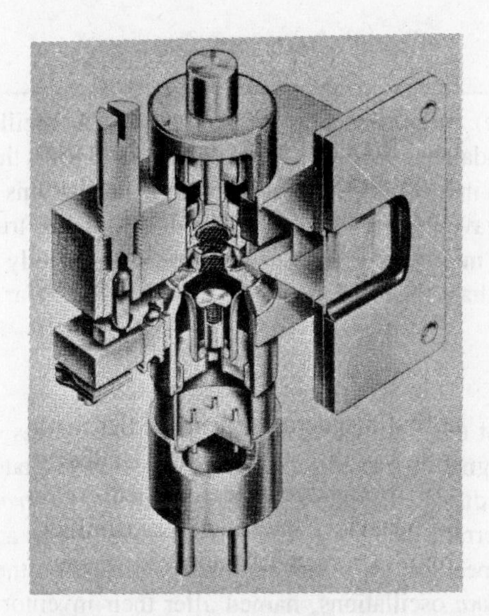

FIGURE 11-9 Reflex klystron cutaway photograph, showing construction. *(Courtesy of Varian Associates, Inc.)*

±20 MHz at X band to ±4 GHz at 200 GHz. *Electronic tuning* is also possible, by adjustment of the repeller voltage. The tuning range is about ±8 MHz at X band and ±80 MHz for submillimeter klystrons. The device is also very easy to frequency-modulate, simply by the application of the modulating voltage to the repeller.

Repeller protection It is essential to make sure that the repeller of a klystron never draws current by becoming positive with respect to the cathode. Otherwise, it will very rapidly be destroyed by the impact of high-velocity electrons as well as overheating. A cathode resistor is often used to ensure that the repeller cannot be more positive than the cathode, even if the repeller voltage fails. Other precautions may include a protective diode across the klystron or an arrangement in which the repeller voltage is always applied before the cathode voltage. Manufacturers' specifications generally list the appropriate precautions.

Applications The klystron oscillator has been replaced by various semiconductor oscillators in a large number of its previous applications, in new equipment. However, it will be found in a lot of existing equipment, as a:

1. Signal source in microwave generators
2. Local oscillator in microwave receivers
3. Frequency-modulated oscillator in portable microwave links
4. Pump oscillator for parametric amplifiers

Additionally, the reflex klystron is still a very useful millimeter and submillimeter oscillator, producing more power at the highest frequencies than most semiconductor devices, with very low AM and FM noise.

11-4 MAGNETRON

The *cavity* (or *traveling wave*) *magnetron* high-power microwave oscillator was invented in Great Britain by Randall and Boot [8]. It is a diode which uses the interaction of magnetic and electric fields in a complex cavity to provide oscillations of very high peak power (the original one gave in excess of 100 kW at 3 GHz). It is true to say that without the cavity magnetron microwave radar would have been greatly delayed and would have come too late to have been the factor it was in World War II.

11-4.1 Introduction

The cavity magnetron was not the first magnetron invented, but it was certainly the first useful one. The first magnetron was discovered in the United States by E. W. Hull, but it quickly fell into disuse. It employed the so-called *cyclotron resonance* principle but suffered from erratic behavior, low power capabilities and very low efficiency in the microwave spectrum. A somewhat improved version then appeared, employing the so-called *Habann* oscillations, named after their inventor. The cavity magnetron of Randall and Boot used the *traveling-wave* principle, on which the modern magnetron is based (as are the remaining devices discussed in this chapter, such as the TWT).

Description of the cavity magnetron The cavity magnetron, which will be referred to as the *magnetron* henceforth, is a diode, usually of cylindrical construction. It employs a *radial electric field,* an *axial magnetic field* and an anode structure with permanent cavities. As shown in Fig. 11-10, the cylindrical cathode is surrounded by the anode with cavities, and thus a radial dc electric field will exist. The magnetic field, because of a magnet like the one in Fig. 11-11, is axial, i.e., has lines of magnetic force passing through the cathode and the surrounding interaction space. The lines are thus at right angles to the structure cross section of Fig. 11-10. The magnetic field is also dc, and since it is perpendicular to the plane of the radial electric field, the magnetron is called a *crossed-field* device.

The output is taken from one of the cavities, by means of a coaxial line as indicated in both Figs. 11-10 and 11-11, or through a waveguide, depending on the power and frequency. With regard to Fig. 11-11, note that the anode surround has been

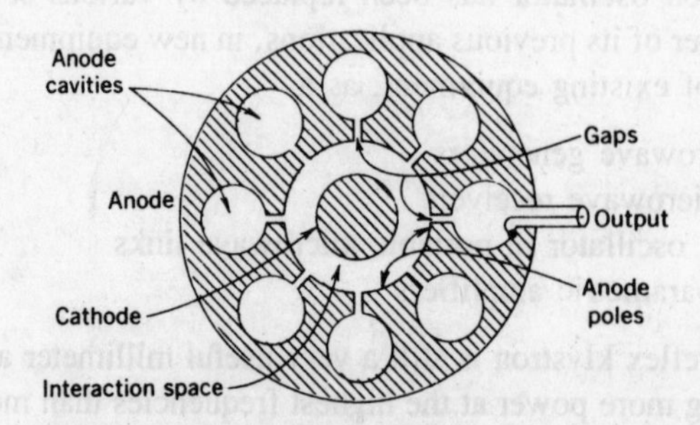

FIGURE 11-10 Cross section of hole-and-slot magnetron.

FIGURE 11-11 Sectioned magnetron (vane type) showing internal construction and magnet. ***(Courtesy of English Electric Valve Company Limited.)***

removed to make the inside visible. Note also that the cavity is of a somewhat different type, being identical to the resonator shown in Fig. 10-33*b*. Also, the output coupling loop leads to a cavity resonator to which a waveguide is connected, and thus the overall output from this magnetron is via waveguide. The rings interconnecting the anode poles are used for *strapping*, and the reason for their presence will be explained. Finally, the anode is normally made of copper, regardless of its actual shape.

The magnetron has a number of resonant cavities and must therefore have a number of resonant frequencies and/or modes of oscillation. Whatever mode is used, it must be self-consistent. For example, it is not possible for the eight-cavity magnetron (which is often used in practice) to employ a mode in which the phase difference between the adjacent anode pieces is 30°. If this were done, the total phase shift around the anode would be $8 \times 30° = 240°$, which means that the first pole piece would be 120° out of phase with itself! Simple investigation shows that the smallest practical phase difference that can exist here between adjoining anode poles, is 45° or $\pi/4$ rad, giving a self-consistent overall phase shift of 360° or 2π rad. This $\pi/4$ mode is seldom used in practice because it does not yield suitable characteristics, and the π mode is preferred for rather complex reasons. In this mode of operation, the phase difference between adjacent anode poles is π rad or 180°; it will be dealt with in detail in Sec. 11-4.2.

Effect of magnetic field Since any electrons emitted by the magnetron cathode will be under the influence of the dc magnetic field, as well as an electric field, the behavior of electrons in a magnetic field must first be investigated. Note that the behavior is not the same as was discussed in Sec. 10-5.2 in connection with ferrites. There, electrons were in the crystalline structure of a magnetic material, whereas here they are moving in the vacuum of the anode interaction space.

A moving electron represents a current, and therefore a magnetic field exerts a force upon it, just as it exerts a force on a wire carrying a current. The force thus exerted has a magnitude proportional to the product Bev, where e and v are the charge and velocity of the electron, respectively, and *B is the component of the magnetic field in a plane perpendicular to the direction of travel of the electron*. This force exerted on the electron is perpendicular to the other two directions. Thus, if the electron is moving (say) forward horizontally, and the magnetic field acts vertically downward, the path of the electron will be curved to the left. Since the magnetic field in the magnetron is constant, the force of the magnetic field on the electron (and therefore the radius of curvature) will depend solely on the forward (radial) velocity of the electron.

Effect of magnetic and electric fields When magnetic and electric fields act simultaneously upon the electron, its path can have any of a number of shapes dictated by the relative strengths of the mutually perpendicular electric and magnetic fields. Some of these electron paths are shown in Fig. 11-12 *in the absence of oscillations* in a magnetron, in which the electric field is constant and radial, and the axial magnetic field can have any number of values.

When the magnetic field is zero, the electron goes straight from the anode to the cathode, accelerating all the time under the force of the radial electric field; this is indicated by path x in Fig. 11-12. When the magnetic field has a small but definite strength, it will exert a lateral force on the electron, bending its path to the left (here). Note, as shown by path y of Fig. 11-12, that the electron's motion is no longer rectilinear. As the electron approaches the anode, its velocity continues to increase radially as it is accelerating. Therefore, the effect of the magnetic field upon it increases also, so that the path curvature becomes sharper as the electron approaches the anode.

It is possible to make the magnetic field so strong that electrons will not reach the anode at all. The magnetic field required to return electrons to the cathode after they have *just grazed* the anode is called the *cutoff field*. The resulting path is z in Fig. 11-12. Knowing the value of the required magnetic field strength is important because

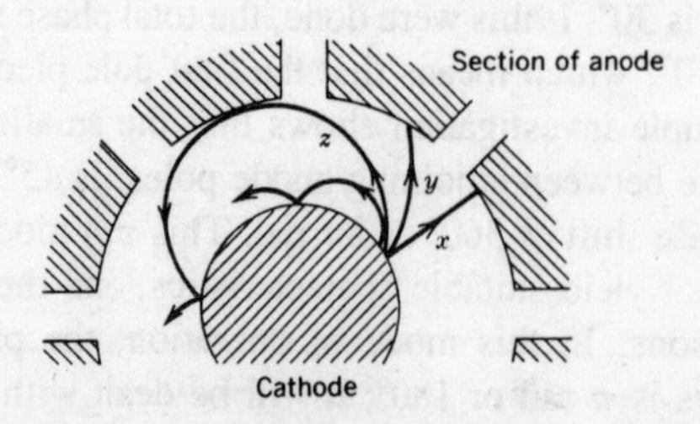

FIGURE 11-12 Electron paths in magnetron without oscillations, showing effect of increasing magnetic field.

this cutoff field just reduces the anode current to zero in the absence of oscillations. If the magnetic field is stronger still, the electron paths as shown will be more curved still, and the electrons will return to the cathode even sooner (only to be reemitted). All these paths are naturally changed by the presence of any RF field due to oscillations, but the state of affairs without the RF field must still be appreciated, for two reasons. First, it leads to the understanding of the oscillating magnetron. Second, it draws attention to the fact that unless a magnetron is oscillating, all the electrons will be returned to the cathode, which will overheat and ruin the tube. This happens because in practice the applied magnetic field is greatly in excess of the cutoff field.

11-4.2 Operation

Once again it will be assumed that oscillations are capable of starting in a device having high-Q cavity resonators, and the mechanism whereby these oscillations are maintained will be explained.

π-mode oscillations As explained in the preceding section, self-consistent oscillations can exist only if the phase difference between adjoining anode poles is $n\pi/4$, where n is an integer. For best results, $n = 4$ is used in practice. The resulting π-mode oscillations are shown in Fig. 11-13 at an instant of time when the RF voltage on the top left-hand anode pole is maximum positive. It must be realized that these are oscillations. A time will thus come, later in the cycle, when this pole is instantaneously maximum negative, while at another instant the RF voltage between that pole and the next will be zero.

In the absence of the RF electric field, electrons a and b would have followed the paths shown by the dotted lines a and b, respectively, but the RF field naturally modifies these paths. This RF field, incidentally, exists inside the individual resonators

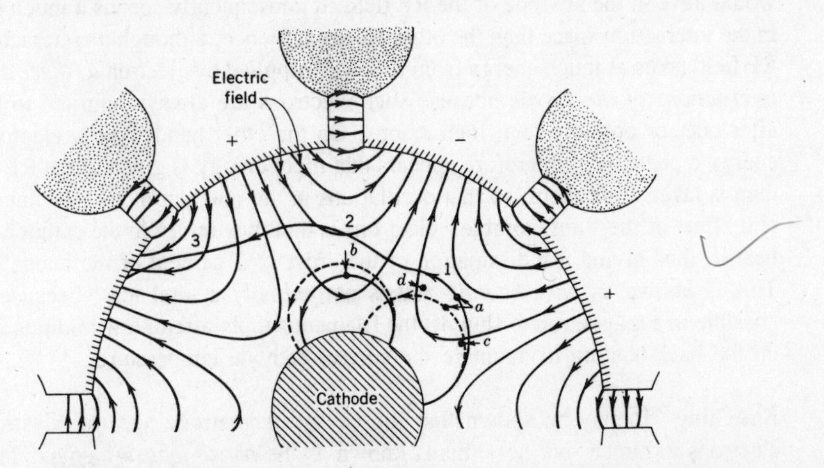

FIGURE 11-13 Paths traversed by electrons in a magnetron under π-mode oscillations. ***(From F. E. Terman, E'ectronic and Radio Engineering, McGraw-Hill Book Company, New York, 1955. Used with permission of McGraw-Hill Book Company.)***

also, but it is omitted here for simplicity. The important fact is that each cavity acts in the same way as a short-circuited quarter-wave transmission line. Each gap corresponds to a maximum voltage point in the resulting standing-wave pattern, with the electric field extending into the anode interaction space, as shown in Fig. 11-13.

Effect of combined fields on electrons The presence of oscillations in the magnetron brings in a *tangential* (RF) component of electric field. As shown, when electron *a* is situated (at this instant of time) at point 1, the tangential component of the RF electric field opposes the tangential velocity of the electron. Thus the electron is retarded by the field and gives energy to it (as happened in the reflex klystron). On the other hand, electron *b* is so placed as to extract an equal amount of energy from the RF field, by virtue of being accelerated by it. For oscillations to be maintained, more energy must be given to the electric field than is taken from it. Yet, on the face of it, this is unlikely to be the case here because there are just as many electrons of type *a* as of type *b*. However, it will be noted that electron *a* spends much more time in the RF field than electron *b*. The former is retarded, and therefore the force of the dc magnetic field on it is diminished; as a result, it can now move closer to the anode. If conditions are arranged so that by the time electron *a* arrives at point 2, the field has reversed polarity, this electron will once again be in a position to give energy to the RF field (though being retarded by it). The magnetic force on electron *a* diminishes once more, and another interaction of this type occurs (this time at point 3). This assumes that at all times the electric field has reversed polarity each time this electron arrives at a suitable interaction position. In this manner, "favored" electrons spend a considerable time in the interaction space and are capable of orbiting the cathode several times before eventually arriving at the anode.

However, an electron of type *b* undergoes a totally different experience. It is immediately accelerated by the RF field, and therefore the force exerted on it by the dc magnetic field increases. This electron thus returns to the cathode even sooner than it would have in the absence of the RF field. It consequently spends a much shorter time in the interaction space than the other electron. Hence, although its interaction with the RF field takes as much energy from it as was supplied by electron *a*, *there are far fewer interactions of the* b *type* because such electrons are always returned to the cathode after one, or possibly two, interactions. On the other hand, type *a* electrons give up energy repeatedly. It therefore appears that more energy is given to the RF oscillations than is taken from them, so that oscillations in the magnetron are sustained. The only real effect of the "unfavorable" electrons is that they return to the cathode and tend to heat it, thus giving it a dissipation of the order of 5 percent of the anode dissipation. This is known as *back-heating* and is not actually a total loss, because it is often possible in a magnetron to shut off the filament supply after a few minutes and just rely on the back-heating to maintain the correct cathode temperature.

Bunching It may be shown that the cavity magnetron, like the klystrons, causes electrons to bunch, but here this is known as the *phase-focusing effect*. This effect is rather important. Without it, favored electrons would fall behind the phase change of the electric field across the gaps, since such electrons are retarded at each interaction with the RF field. To see how this effect operates, it is most convenient to consider another electron, such as *c* of Fig. 11-13.

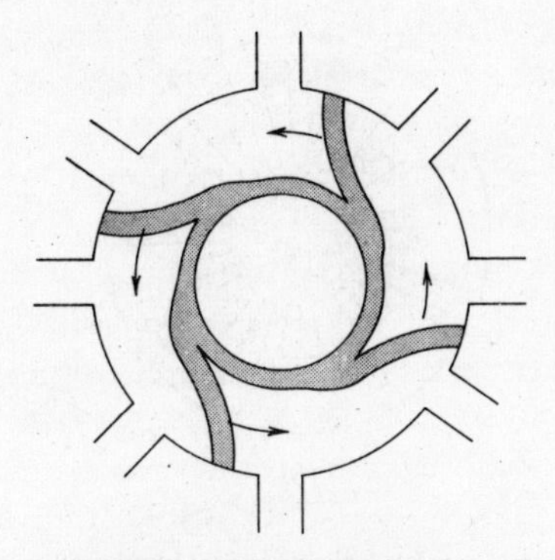

FIGURE 11-14 Bunched electron clouds rotating around magnetron cathode (individual electron paths not shown).

Electron *c* contributes some energy to the RF field. However, it does not give up as much as electron *a*, because the tangential component of the field is not as strong at this point. As a result, this electron appears to be somewhat less useful than electron *a*, but this is so only at first. Electron *c* encounters not only a diminished tangential RF field but also a component of the *radial* RF field, as shown. This has the effect of accelerating the electron radially outward. As soon as this happens, the dc magnetic field exerts a stronger force on electron *c*, tending to bend it back to the cathode but also accelerating it somewhat in a counterclockwise direction. This, in turn, gives this electron a very good chance of catching up with electron *a*. In a similar manner, electron *d* (shown in Fig. 11-13) will be retarded tangentially by the dc magnetic field. It will therefore be caught up by the favored electron; thus, a bunch takes shape. In fact, it is seen that being in the favored position means (to the electron) being in a position of equilibrium. If an electron slips back or forward, it will quickly be returned to the correct position with respect to the RF field, by the phase-focusing effect just described.

Figure 11-14 shows the wheel-spoke bunches in the cavity magnetron. These bunches rotate counterclockwise with the correct velocity to keep up with RF phase changes between adjoining anode poles. In this way a continued interchange of energy takes place, with the RF field receiving much more than it gives. As has already been discussed, the RF field changes polarity. Thus each favored electron, by the time it arrives opposite the next gap, meets the same situation of there being a positive anode pole above it and to the left, and a negative anode pole above it and to the right. It is not difficult to imagine that the electric field itself is rotating counterclockwise at the same speed as the electron bunches. As a matter of fact, the cavity magnetron is called the *traveling-wave magnetron* precisely because of these rotating fields.

11-4.3 Practical Considerations

The operating principles of a device are important but do not give the entire picture of that particular device. Accordingly, a number of other significant aspects of magnetron operation must now be considered.

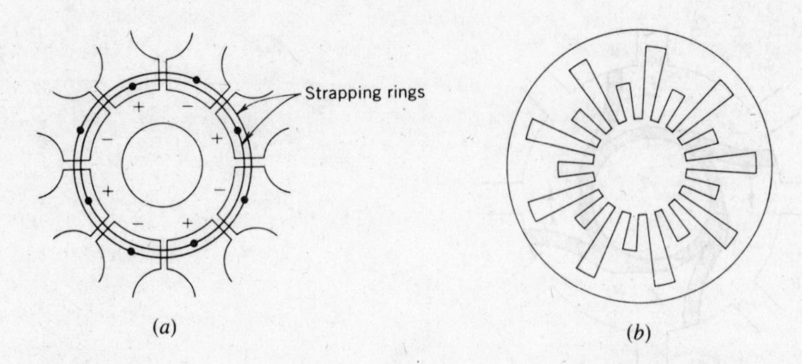

FIGURE 11-15 **(*a*) Hole-and-slot magnetron with strapping; (*b*) rising-sun magnetron anode block.**

Strapping Because the magnetron has eight (or more) coupled cavity resonators, several different modes of oscillation are possible. Unfortunately, the oscillating frequencies corresponding to the different modes are not the same. As it happens, some are quite close to one another, so that, through *mode jumping*, a 3-cm π-mode oscillation which is normal for a particular magnetron could, spuriously, become a 3.05-cm 3/4 π-mode oscillation. The dc electric and magnetic fields, adjusted to be correct for the π mode, would still support the spurious mode to a certain extent, since its frequency is not too far distant. The result might well be oscillations of reduced power, at the wrong frequency.

Magnetrons using identical cavities in the anode block normally employ *strapping* to prevent mode jumping. Such strapping was shown in Fig. 11-11 for the vane cavity system, and it is now seen in Fig. 11-15*a* for the hole-and-slot cavity arrangement. Strapping consists of two rings of heavy-gauge wire connecting alternate anode poles. These are the poles that should be in phase with each other for the π mode. The reason for the effectiveness of strapping in preventing mode jumping may be simplified by pointing out that, since the phase difference between alternate anode poles is other than 2π rad in other modes, these modes will quite obviously be prevented. The actual situation is somewhat more complex.

Strapping may become unsatisfactory because of losses in the straps in very high-power magnetrons or because of strapping difficulties at very high frequencies. In the latter case, the cavities are small, and there are generally a lot of them (16 and 32 are common numbers), to ensure that a suitable RF field is maintained in the interaction space. This being so, so many modes are possible that even strapping may not prevent mode jumping. A very good cure consists in having an anode block with a pair of cavity systems of quite dissimilar shape and resonant frequency. Such a *rising-sun* anode structure is shown in Fig. 11-15*b* and has the effect of isolating the π-mode frequency from the others. Consequently the magnetron is now unlikely to oscillate at any of the other modes, because the dc fields would not support them. Note that strapping is not required with the rising-sun magnetron.

Frequency pulling and pushing It should be recognized that the resonant frequency of magnetron can be altered somewhat by changing the anode voltage. Such *frequency pushing*, as it is called, is due to the fact that the change in anode voltage has the effect of altering the orbital velocity of the electron clouds of Fig. 11-14. This in turn alters the rate at which energy is given up to the anode resonators and therefore changes the oscillating frequency, cavity bandwidth permitting. The effect of all this is that power changes will result from inadvertent changes of anode voltage, but *voltage tuning* of magnetrons is quite feasible.

Like any other oscillator, the magnetron is susceptible to frequency variations due to changes of load impedance. This will happen regardless of whether such load variations are purely resistive or involve load reactance variations, but it is naturally more severe for the latter. The frequency variations, known as *frequency pulling,* are caused by changes in the load impedance reflected into the cavity resonators. They must be prevented, all the more so because the magnetron is a power oscillator. Unlike most other oscillators, therefore, it is not followed by a buffer.

The various characteristics of a magnetron, including the optimum combinations of anode voltage and magnetic flux, are normally plotted on *performance charts* and *Rieke diagrams* [9a]. From these the best operating conditions are selected.

11-4.4 Types, Performance and Applications

Magnetron types The magnetron, perhaps more than any other microwave tube, lends itself to a variety of types, designs and arrangements. Magnetrons using hole-and-slot, vane and rising-sun cavities have already been discussed. Figure 11-16*a*

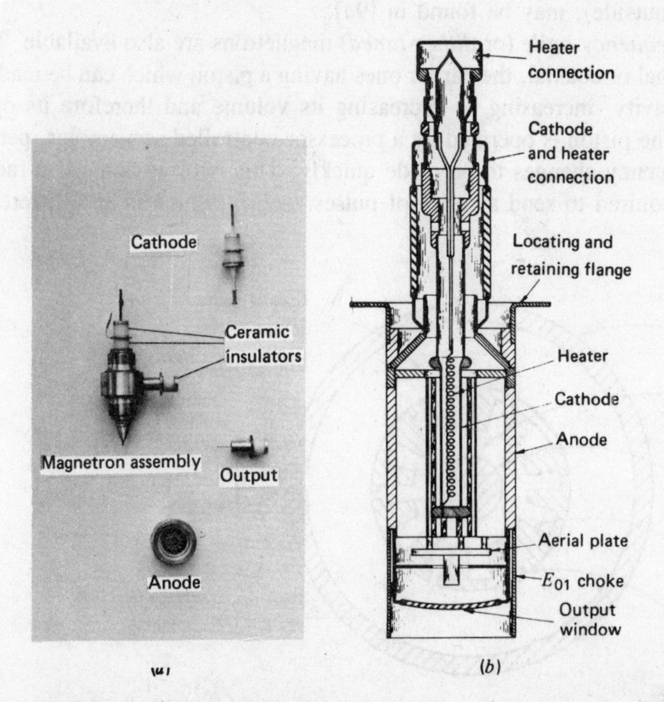

FIGURE 11-16 Pulsed magnetron construction (magnets omitted); (*a*) 20-kW conventional metal-ceramic magnetron; (*b*) 5-MW "long-anode" coaxial magnetron. *(Courtesy of English Electric Valve Co. Ltd.)*

shows the construction of another vane-type magnetron, this time using an all-metal-ceramic structure. This type is in current use in marine radar, producing about 25 kW pulsed in the X band. The all-up weight of such a magnetron is typically 2 kg. A very high-power (5 MW pulsed at 3 GHz) magnetron is shown in Fig. 11-16*b*. It features an anode that is about three times normal length and thus has the required volume and external area to allow high dissipation and therefore output power. A magnetron such as this may stand over 2 m high, and have a weight in excess of 60 kg without the magnet.

A most interesting feature of Fig. 11-16*b* is that it shows a *coaxial* magnetron. The cross section of a coaxial magnetron structure, similar to the one of Fig. 11-16*b*, is shown in Fig. 11-17. It is seen that there is an integral coaxial cavity present in this magnetron. The tube is built so that the Q of this cavity is much higher than the Q's of the various resonators, so that it is the coaxial cavity which determines the operating frequency. Oscillations in this cavity are in the coaxial $TE_{0,1}$ mode, in which the electric field is circular, as is the magnetic field in Fig. 10-9*b* ($TM_{0,1}$ mode in circular waveguides). Furthermore, it is possible to attenuate the resonator modes without interfering with the coaxial mode, so that mode jumping is all but eliminated. Frequency pushing and pulling are both significantly reduced, while the enlarged anode area, as compared with a conventional magnetron, permits better dissipation of heat and consequently smaller size for a given output power. The MTBF of coaxial magnetrons is also considerably longer than that of conventional ones. Further information on coaxial magnetrons, including *inside-out* coaxial magnetrons (in which the cathode is on the outside), may be found in [9*a*].

Frequency-agile (or *dither-tuned*) magnetrons are also available. They may be conventional or coaxial, the earlier ones having a piston which can be made to descend into the cavity, increasing or decreasing its volume and therefore its operating frequency. The piston is operated by a processor-controlled servomotor, permitting very large frequency changes to be made quickly. This is of advantage in radar, where it may be required to send a series of pulses each of which is at a different radio fre-

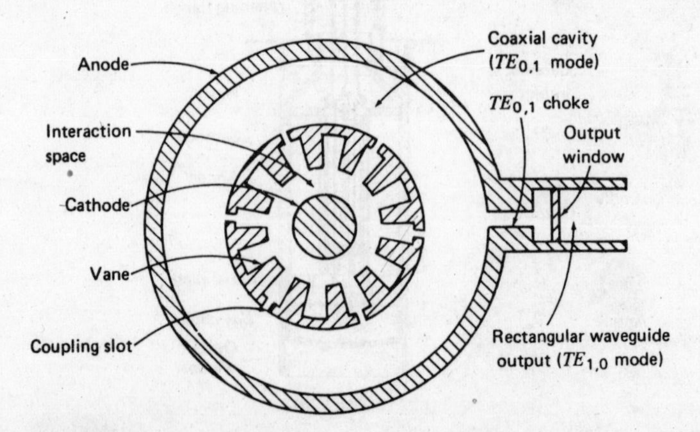

FIGURE 11-17 Cross section of coaxial magnetron; the magnetic field (not shown) is perpendicular to the page.

quency. The benefits of doing this are improved resolution and more difficulty (for the enemy) in trying to jam the radar. See also Chap. 16 and [9*b*]. Dither tuning by electronic methods has recently been produced, yielding very rapid frequency changes—during the transmission of one pulse, if required—with a range typically 1 percent of the center frequency. The methods used have included extra cathodes, electron injection and the placing of PIN diodes inside the cavity. Such magnetrons are described in [10].

Voltage-tunable magnetrons (*VTMs*) are also available for CW operation, though they are not very efficient. For this and other reasons they are not suited to pulsed radar work. These use low-Q cavities, cold cathodes (and hence back-heating) and an extra *injection electrode* to help bunching. The result is a magnetron whose operating frequency may be varied over an octave range by adjusting the anode voltage. Very fast sweep rates, and indeed frequency modulation, are possible. See [11] for an excellent description of the VTM.

Performance and applications The traditional applications of the magnetron have been for pulse work in radar and linear particle accelerators. The *duty cycle* (fraction of total time during which the magnetron is actually ON) is typically 0.1 percent; the methods of pulsing are explained in Chap. 16. The powers required range from 10 kW to 5 MW, depending on the application and the operating frequency. The maximum available powers range from 10 MW in the UHF band, through 2 MW in the X band, to 10 kW at 100 GHz. Current efficiencies are of the order of 50 percent; a significant size reduction is being achieved, especially for larger tubes, with the aid of two advancements. One is the development of modern permanent magnet materials, which has resulted in reduced electromagnet bulk. The other advance is in cathode materials. By the use of such substances as thoriated tungsten, much higher cathode temperatures (1800°C compared with 1000°C) are being achieved. This helps greatly in overcoming the limitation set by cathode heating from back bombardment.

VTMs are available for the frequency range from 200 MHz to X band, with CW powers up to 1000W (10 W is typical). Efficiencies are higher, up to 75 percent. Such tubes are used in sweep oscillators, in *telemetry* and in missile applications.

Fixed-frequency CW magnetrons are also available; they are used extensively for industrial heating and microwave ovens. The operating frequencies are around 900 MHz and 2.5 GHz, although typical powers range from 300 W to 10 kW. Efficiencies are typically in excess of 70 percent.

11-5 TRAVELING-WAVE TUBE (TWT)

Like the multicavity klystron, the TWT is a *linear-beam* tube used as a microwave amplifier. Unlike the klystron, however, it is a device in which *the interaction between the beam and the RF field is continuous*. The TWT was invented independently by Kompfner [12] in Britain and then Pierce [13] in the United States, shortly after World War II. Each of them was dissatisfied with the very brief interaction in the multicavity klystron, and each invented a *slow-wave structure* in which *extended interaction* took place. Because of its construction and operating principles, as will be seen, the TWT is capable of enormous bandwidths. Its main application is as a medium- or high-power amplifier, either CW or pulsed.

11-5.1 TWT Fundamentals

In order to prolong the interaction between an electron beam and an RF field, it is necessary to ensure that both are moving in the same direction with approximately the same velocity. This relation is quite different from the multicavity klystron, in which the electron beam travels but the RF field is stationary. The problem that must be solved is that an RF field travels with the velocity of light, while the electron beam's velocity is unlikely to exceed 10 percent of that, even with a very high anode voltage. The solution is to retard the RF field with a slow-wave structure. Several such structures are in use, as will be seen, the helix and a waveguide coupled-cavity arrangement being the most common.

Description A typical TWT using a helix is shown in Fig. 11-18. An electron gun is employed to produce a very narrow electron beam, which is then sent through the center of a long axial helix. The helix is made positive with respect to the cathode, and the collector even more so. Thus the beam is attracted to the collector and acquires a high velocity. It is kept from spreading, as in the multicavity klystron, by a dc axial magnetic field, whose presence is indicated in Fig. 11-18 though the magnet itself is not shown. The beam must be narrow and correctly focused, so that it will pass through the center of the helix without touching the helix itself.

Signal is applied to the input end of the helix, via a waveguide as indicated, or through a coaxial line. This field propagates around the helix with a speed that is hardly different from the velocity of light in free space. However, the speed with which the electric field advances *axially* is equal to the velocity of light multiplied by the ratio of helix pitch to helix circumference. This can be made (relatively) quite slow and approximately equal to the electron beam velocity. The axial RF field and the beam can now interact continuously, with the beam bunching and giving energy to the field. Almost complete bunching is the result, and so is high gain.

Operation The TWT may be considered as the limiting case of the multicavity klystron, one that has a very large number of closely spaced gaps, with a phase change that

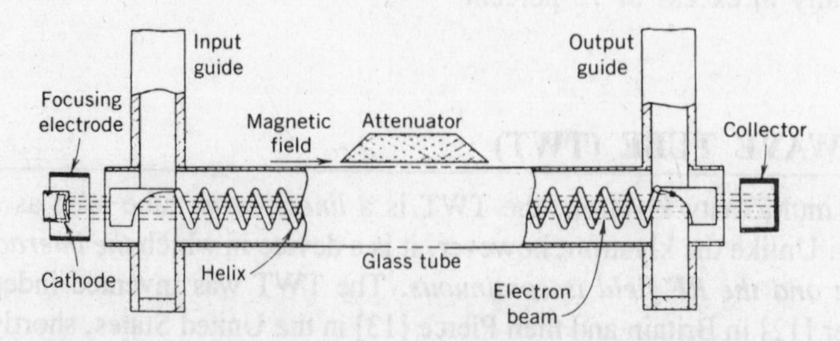

FIGURE 11-18 Helix-type traveling-wave tube; propagation along the helix is from left to right. *[By permission from F. Harvey, Microwave Engineering, Academic Press Inc. (London) Ltd., 1963.]*

progresses at approximately the velocity of the electron beam. This also means that there is a lot of similarity here to the magnetron, in which much the same process takes place, but around a closed circular path rather than in a straight line.

Bunching takes place in the TWT through a process that is a cross between those of the multicavity klystron and the magnetron.

Electrons leaving the cathode at random quickly encounter the weak axial RF field at the input end of the helix, which is due to the input signal. As with the passage of electrons across a gap, velocity modulation takes place and with it, between adjacent turns, some bunching. Once again it takes theoretically no power to provide velocity modulation, since there are equal numbers of accelerated and retarded electrons. By the time this initial bunch arrives at the next turn of the helix, the signal there is of such phase as to retard the bunch slightly and also to help the bunching process a little more. Thus, the next bunch to arrive at this point will encounter a somewhat higher RF electric field than would have existed if the first bunch had not made its mark.

The process continues as the wave and electron beam both travel toward the output end of the helix. Bunching becomes more and more pronounced until it is almost complete at the output end. Simultaneously the RF wave on the helix grows (exponentially, as it happens) and also reaches its maximum at the output end; this situation is shown in Fig. 11-19.

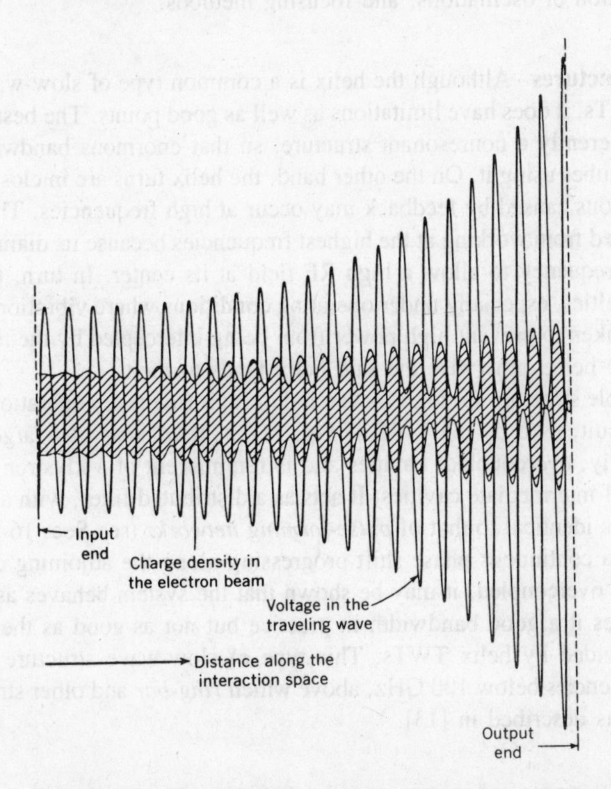

FIGURE 11-19 Growth of signal and bunching along traveling-wave tube. ***(By permission from Reich, Skalnik, Ordung, and Krauss, Microwave Principles, D. Van Nostrand Company, Inc., Princeton, N.J., 1957.)***

The interaction between the beam and the RF field is very similar to that of the magnetron. In both devices electrons are made to give some of their energy to the RF field, through being slowed down by the field, and in both devices a phase-focusing mechanism operates. It will be recalled that this tends to ensure that electrons bunch and that the bunches tend to keep arriving in the most favored position for giving up energy. There is at least one significant difference between the devices, and it deals with the methods of keeping the velocity of the beam much the same as that of the RF field, even though electrons in the beam are continually retarded. In the magnetron this is done by the dc magnetic field, but since there is no such field here (no component of it at right angles to the direction of motion of the electrons, at any rate), the axial dc electric field must provide the energy. A method of doing this is to give the electron beam an initial velocity that is slightly greater than that of the axial RF field. The extra initial velocity of electrons in the beam balances the retardation due to energy being given to the RF field.

11-5.2 Practical Considerations

Among the points to be considered now are the various types of slow-wave structures in use, prevention of oscillations, and focusing methods.

Slow-wave structures Although the helix is a common type of slow-wave structure in use with TWTs, it does have limitations as well as good points. The best of the latter is that it is inherently a nonresonant structure, so that enormous bandwidths can be obtained from tubes using it. On the other hand, the helix turns are in close proximity, and so oscillations caused by feedback may occur at high frequencies. The helix may also be prevented from working at the highest frequencies because its diameter must be reduced with frequency to allow a high RF field at its center. In turn, this presents focusing difficulties, especially under operating conditions where vibration is possible. Care must be taken to prevent high power from being intercepted by the (by now very small-diameter) helix; otherwise the said helix tends to melt.

A suitable structure for high-power and/or high-frequency operation is the *coupled-cavity* circuit, used by the TWT of Fig. 11-20. It consists of a large number of coupled (actually, *over*coupled) cavities and is reminiscent of a klystron with a very large number of intermediate cavities. It acts as a distributed filter, with a principle of operation that is identical to that of *pulse-forming networks* (see Sec. 16-2.1). Essentially, there is a continuous phase shift progressing along the adjoining cavities. Because these are overcoupled, it may be shown that the system behaves as a bandpass filter. This gives it a good bandwidth in practice but not as good as the exceptional bandwidth provided by helix TWTs. This type of slow-wave structure tends to be limited to frequencies below 100 GHz, above which *ring-bar* and other structures may be employed, as described in [13].

Prevention of oscillations Figure 11-19 shows the exponential signal growth along the traveling-wave tube, but it is not to scale—the actual gain could easily exceed 80

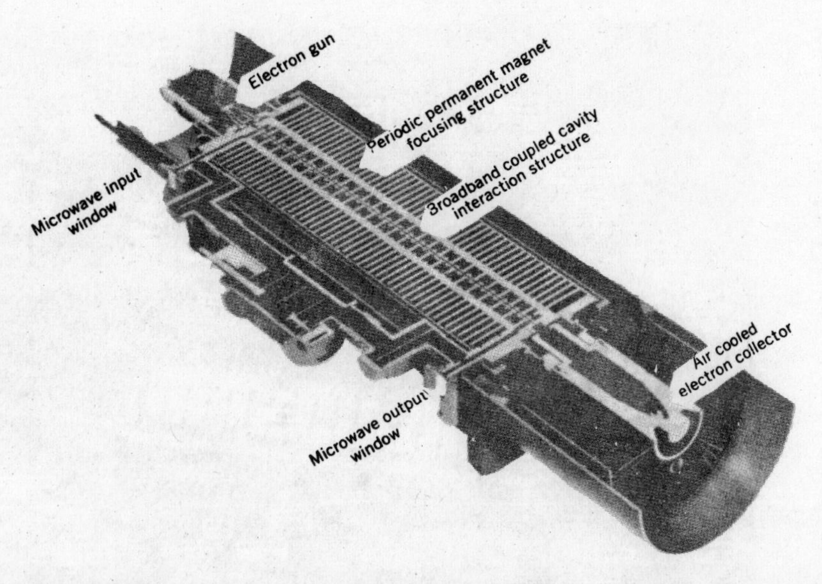

FIGURE 11-20 Cross section of high-power traveling-wave tube, using a coupled-cavity slow-wave structure and periodic permanent-magnet focusing. ***(Courtesy of Electron Dynamics Division, Hughes Aircraft Company.)***

dB. Oscillations are thus possible in such a high-gain device, especially if poor load matching causes significant reflections along the slow-wave structure. The problem is aggravated by the very close coupling of the slow-wave circuits. Thus all practical tubes use some form of attenuator (which has the subsidiary effect of somewhat reducing gain). Both forward and reverse waves are attenuated, but the forward wave is able to continue and to grow past the attenuator, because bunching is unaffected. With helix tubes, the attenuator may be a lossy metallic coating (such as aquadag or Kanthal) on the surface of the glass tube. As shown in Fig. 11-20, with a coupled-cavity slow-wave structure there are really several (three, in this case) loosely coupled, self-contained structures, between which attenuation takes place. It should be noted that Fig. 11-19 shows a simplified picture of signal and bunching growth, corresponding to a TWT without an attenuator.

Focusing Because of the length of the TWT, focusing by means of a permanent magnet is somewhat awkward, and focusing with an electromagnet is bulky and wasteful of power. On the other hand, the solenoid does provide an excellent focusing magnetic field, so that it is often employed in high-power (ground-based) radars. The latest technique in this field is the integral solenoid, a development that makes the assembly light enough for airborne use. Figure 11-21 shows the cross section of a TWT with this type of focusing.

To reduce bulk, *periodic permanent-magnet* focusing is very often used. This PPM focusing was mentioned in connection with klystron amplifiers and is now illustrated in Fig. 11-20. PPM is seen to be a system in which a series of small magnets are

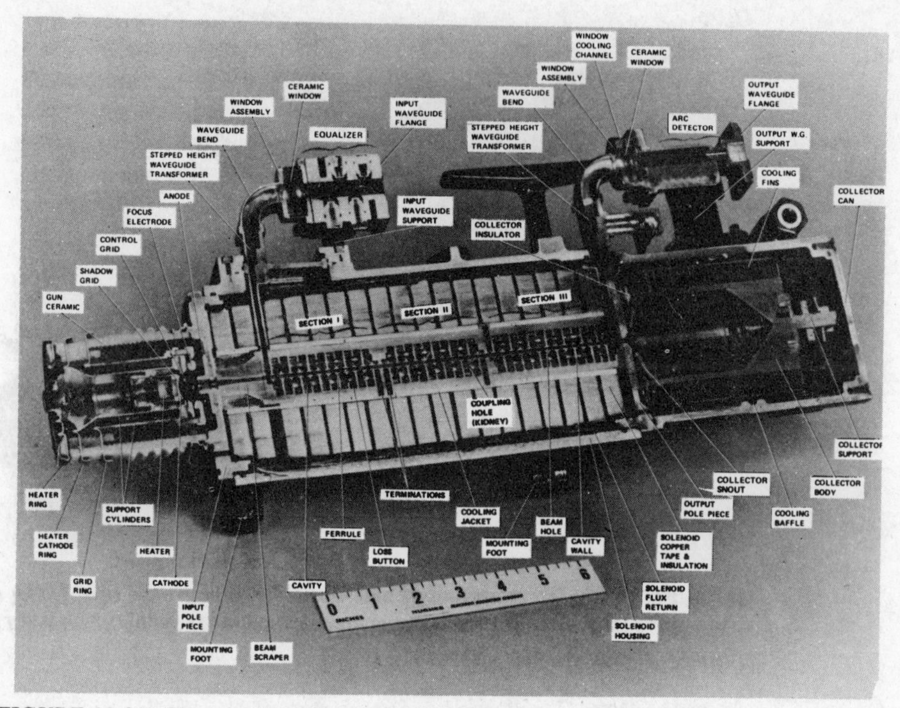

FIGURE 11-21 Cross section of complete 9-kW pulsed X-band traveling-wave tube, with a three-section coupled-cavity slow-wave structure and integral solenoid focusing. *(Courtesy of Electron Dynamics Division, Hughes Aircraft Company.)*

located right along the tube, with spaces between adjoining magnets. The beam defocuses slightly past each pole piece but is refocused by the next magnet; note that the individual magnets are interconnected. The system illustrated is the so-called radial magnet (as opposed to axial magnet) PPM. A good discussion of the various focusing methods and their merits is provided by [14].

11-5.3 Types, Performance and Applications

The TWT is the most versatile and most frequently used microwave tube. There are broadly four types, each with particular applications and performance requirements. These are now described.

TWT types The most fruitful method of categorizing traveling-wave tubes seems to be according to size, power levels and type of operation. Within each category, of course, various slow-wave structures and focusing methods may be used.

The first TWTs were broadband, low-noise low-level amplifiers used mainly for receivers. That is now a much-diminished application, because transistor amplifiers have much better noise figures, much lower bulk and comparable bandwidths. However, they are not as radiation-immune as the TWT and hence not as suitable for hazardous environments. The TWX19, whose performance is given in Table 11-2, is typical of such tubes. It comes all enclosed with its power supply and draws just a few watts from the mains. The package measures about 30 × 5 × 5 cm and weighs about 1½ kg.

The second type is the CW power traveling-wave tube. It is represented by several of the entries in Table 11-2 (all those that produce watts or kilowatts of CW). The 677H is typical, weighing just under 2¾ kg and measuring 7 × 7 × 41 cm. The major application for this type of TWT is in satellite communications, either in satellite earth stations (types 614H and 870H in Table 11-2) or aboard the satellites themselves (type 677H). However, this type is also increasingly used in CW radar and electronic countermeasures (ECM) (see Chap. 16); indeed, tubes such as type 819H in Table 11-2 are designed for this application.

Pulsed TWTs are representative of the third category, and several are shown in Table 11-2; they are considerably bigger and more powerful than the preceding two types. A representative tube is the Hughes 797H, illustrated in Fig. 11-21. This TWT produces 9 kW in the X band, with a duty cycle of 50 percent. It weighs just over 20 kg, draws 2.5 A at 8 kV dc and measures 53 × 15 × 20 cm.

The fourth type is the newest, still under active development; it comprises *dual-mode* TWTs. These are tubes with military applications, capable of being used as either CW or pulsed amplifiers. They are comparable in size, power, weight and mains requirements to the medium-power communications TWTs. As an example, the type 562H tube in Table 11-2 weighs 4.5 kg and is 45 cm long. Although the TWT in general represents a fairly mature technology, the dual-mode tube does not. In [14, 16 and 17] the performance and difficulties encountered with dual-mode TWTs, including whether grids should be used and if so, what kind, are discussed.

Performance Low-level, low-noise TWTs are available in the 2- to 40-GHz range, and three are shown in Table 11-2. Such tubes generally use helixes and have octave bandwidths or sometimes even more. Their gains range from 25 to 45 dB and noise figures from 4 to 17 dB, while typical power output is 1 to 100 mW. They tend now to be used mostly for replacement purposes, having been displaced by transistor (FET or bipolar) amplifiers in most new equipment except in specialized applications.

By virtue of their applications, CW power tubes are made essentially in two power ranges: up to about 100 W and over about 500 W. Several of them are featured in Table 11-2. The frequency range covered is from under 1 to over 100 GHz, with typically 2 to 15 percent bandwidths. Available output powers exceed 10 kW with gains that may be over 50 kB, and efficiencies are in the 25 to 35 percent range with normal techniques. However, with a so-called *depressed collector* (see also [14, 15]) efficiencies can exceed 50 percent. This is a system in which the collector potential is made lower than the cathode potential to reduce dissipation and thus improve efficiency. The tube of Fig. 11-21 uses the depressed collector technique. TWTs of this type employ the helix when octave bandwidths are required and the coupled-cavity

TABLE 11-2 Typical Traveling-Wave Tubes

MAKE AND MODEL	FREQUENCY RANGE, GHz	POWER OUT, max.	DUTY CYCLE	NOISE FIGURE, dB	POWER GAIN, dB	FOCUSING
EEV* N1047M	2.7–3.2	1.5 mW	CW	4.0	24	Solenoid
M-OV† TWX19	7–12	1 mW	CW	11.0	38	PPM
TMEC‡ M9346	26.5–40	5 mW	CW	17.0	40	?
EEV* N1073	3.55–5	16 W	CW		41	PPM
Hughes 677H	5.9–6.4	125 W	CW		45	PPM
Hughes 551H	2–4	1 kW	CW		30	Solenoid
Hughes 614H	5.9–6.4	8 kW	CW		40	Solenoid
Hughes 876H	14.0–14.5	700 W	CW		43	PPM
Hughes 870H	14.0–14.5	5 kW	CW		35	Integral solenoid
Hughes 819H	54.5–55.5	5 kW	CW		20	Solenoid
Hughes 985H	84–86	200 W	CW		47	PPM
Ferranti LY70	2.7–3.7	10 kW	2.5%		48	PPM
Hughes 8754H	9–18	1.5 kW	8.0%		45	PPM
Hughes 835H	16–16.5	200 kW	1.0%		60	PPM
EEV* N1061	9–9.45	900 kW	0.5%		33	Solenoid
Hughes 562H§	2–4	200 W/1 kW	CW/5%		30/30	PPM
EEV* N10011§	9–10.5	210/820W	CW/50%		29/49	PPM

*English Electric Valve Company.
†M-O Valve Company.
‡Teledyne MEC.
§Dual-mode tubes.

structure for narrower bandwidths. Focusing is PPM most often, and a noise figure of 30 dB is typical. For space applications, reliabilities of the order of 50,000 hours (nearly 6 years) mean time between failures are now available.

Over the frequency range of approximately 2 to 100 GHz, pulsed TWTs are available with peak outputs from 1 to about 250 kW typically. However, powers in the megawatt range are also possible. Bandwidths range from narrow (5 percent) to three octaves with helix tubes at the lower end of the power range. All manners of focusing and slow-wave structures are employed. Duty cycles can be much higher than for magnetrons or klystrons, 10 percent or higher being not uncommon. All other performance figures are as for CW power TWTs.

Dual-mode TWTs are currently available for the 2- to 18-GHz spectrum. Power outputs range up to 3 kW pulsed and 600 W CW, with a maximum 10:1 *pulse-up ratio* (peak pulse power to CW ratio for the same tube), which should be raised even more in the near future. The remaining data are as for single-mode pulsed TWTs, and two dual-mode tubes are shown in Table 11-2.

Applications As has been stated, traveling-wave tubes are very versatile indeed. The low-power, low-noise ones have been used in radar and other microwave receivers, in laboratory instruments and as drivers for more powerful tubes. Their hold on these applications is much more tenuous than it was, because of semiconductor advances. As will be seen in the next chapter, transistor amplifiers, *tunnel* diodes and *Schottky* diodes can handle a lot of this work, while the TWT never could challenge *parametric amplifiers* and *masers* for the lowest-noise applications.

Medium- and high-power CW TWTs are used for communications and radar, including *ECM*. The vast majority of space-borne power output amplifiers ever employed have been TWTs because of the high reliability, high gain, large bandwidths and constant performance in space. The majority of satellite earth stations use TWTs as output tubes, and so do quite a number of tropospheric scatter links. Broadband microwave links also use TWTs, generally employing tubes in the under 100-W range. Note that all the foregoing applications are discussed in Chap. 15. CW traveling-wave tubes are also used in some kinds of radar, and also in radar jamming, which is a form of ECM. In this application, the TWT is fed from a broadband noise source, and its output is transmitted to confuse enemy radar.

CW tubes will of course handle FM and may be used either to amplify AM signals or to generate them. For AM generation, the modulating signal is fed to the previously mentioned special grid. However, it must be noted that the TWT, like the klystron amplifier, begins to saturate at about 70 percent of maximum output and ceases to be linear thereafter. Although this does not matter when amplifying FM signals, it most certainly does matter when AM signals are being amplified or generated, and in this case the tube cannot be used for power outputs exceeding 70 percent of maximum.

Pulsed tubes find applications in airborne and ship-borne radar, as well as in high-power ground-based radars. They are capable of much higher duty cycles than klystrons or magnetrons and are thus used in applications where this feature is required. Some of these are covered in Chap. 16, and others, including requirements of dual-mode operation, are described in [14] and [18].

11-6 OTHER MICROWAVE TUBES

Various other microwave tubes will now be introduced and briefly treated. They are the *crossed-field amplifier* (CFA), *backward-wave oscillator*, and some miscellaneous high-power amplifiers.

11-6.1 Crossed-Field Amplifier

The CFA is a microwave power amplifier based on the magnetron and looking very much like it. It is a cross between the TWT and the magnetron in its operation. That is, it uses an essentially magnetron structure to provide an interaction between crossed dc electric and magnetic fields on the one hand, and an RF field on the other hand. However, it uses a slow-wave structure akin to that of the TWT to provide a continuous interaction between the electron beam and a *moving RF field*. (It will be noted that in the magnetron interaction is with a *stationary RF field*.) The crossed-field amplifier is more recent than most other microwave tubes, having been first proposed in the early 1960s.

Operation The cross section of a typical CFA is shown in Fig. 11-22; the similarity to a coaxial magnetron is striking in its appearance. It would have been even more striking if, as used in practice, a vane slow-wave structure had been shown, with waveguide connections. The helix is illustrated here purely to simplify the explanation. Indeed, practical CFAs and magnetrons are very difficult to tell apart by mere looks, except for one unmistakable giveaway: unlike magnetrons, CFAs have RF *input* connections.

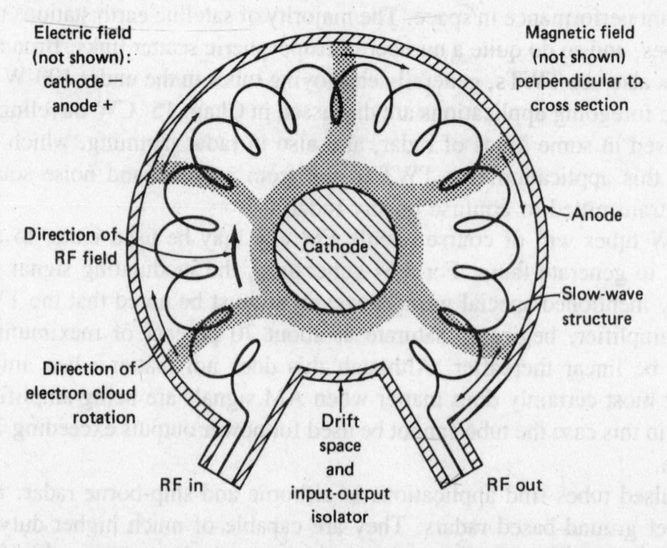

FIGURE 11-22 Simplified cross section of continuous-cathode, forward-wave crossed-field amplifier.

As in the magnetron, the interaction of the various fields results in the formation of bunched electron clouds. However, in this case an input signal is supplied and receives energy from electron clouds traveling in the same direction as the RF field. Consequently, as in the TWT, signal strength grows along the slow-wave structure, and gain results. It will be seen in Fig. 11-22 that there is an area free of the slow-wave structure. This provides a space in which electrons drift freely, isolating the input from the output to prevent feedback and hence oscillations. An attenuator is sometimes used also, akin to the TWT arrangement.

In the tube shown, the direction of the RF field and the electron bunches is the same; this is thus a *forward-wave* CFA. *Backward-wave* CFAs also exist, in which the two directions are opposed. There are also CFAs which have a grid located near the cathode in the drift-space area, with an accelerating anode nearby. They are known as *injected-beam* CFAs. The various CFA types are discussed in great detail in [9*c*] and [19].

Practical considerations The majority of crossed-field amplifiers are pulsed devices. CW and dual-mode CFAs are also available, although their performance and other details tend to be shrouded in military secrecy. However, dual-mode operation is easier for CFAs than for TWTs, because here both the electric and the magnetic fields can be switched to alter power output. Thus 10:1 or higher power ratios for dual-mode operation are feasible.

Pulsed CFAs are available for the frequency range from 1 to 50 GHz, but the upper frequency is a limit of existing requirements rather than tube design. CFAs are quite small for the power they produce (like magnetrons), and that is a significant advantage for airborne radars. The maximum powers available are well over 10 MW in the UHF range (with an excellent efficiency of up to 70 percent), 1 MW at 10 GHz (efficiency up to 55 percent) and 400 kW CW in the S-band. The excellent efficiency contributes to the small relative size of this device and of course to its use. Duty cycles are up to about 5 percent, better than magnetrons but not as high as TWTs. Bandwidths are quite good at up to 25 percent of center frequency (and one octave for some injected-beam CFAs). However, the relatively low gains available, typically 10 to 20 dB, are a disadvantage, in that the small size of the tube is offset by the size of the driver, which the klystron or TWT, with their much higher gains, would not have required.

A typical forward-wave CFA is the Varian SFD257. It operates over the range 5.4 to 5.9 GHz, producing a peak power of 1 MW with a duty cycle of 0.1 percent. The efficiency is 50 percent, gain 13 dB, and noise figure approximately 36 dB, a little higher than for a corresponding klystron. The anode voltage is 30 kV dc, and the peak anode current is 70 A. The tube, like a number of magnetrons, uses back-heating for the cathode, and indeed both it and the anode are liquid-cooled. The whole package, with magnet, weighs 95 kg and looks just like a high-power magnetron with an extra set of RF terminals.

Crossed-field amplifiers are used almost entirely for radar and electronic countermeasures. Further information on their applications and methods of use may be found in [18, 19, 20 and 9*c*].

11-6.2 Backward-Wave Oscillator

A backward-wave oscillator (BWO) is a microwave CW oscillator with an enormous tuning and overall frequency coverage range. It operates on TWT principles of electron beam-RF field interaction, generally using a helix slow-wave structure. In general appearance the BWO looks like a shorter, thicker TWT.

Operation [21] If the presence of starting oscillations may be assumed, the operation of the BWO becomes very similar to that of the TWT. Electrons are ejected from the electron-gun cathode, focused by an axial magnetic field and collected at the far end of the glass tube. They have meanwhile traveled through a helix slow-wave structure, and bunching has taken place, with bunches increasing in completeness from the cathode to the collector. An interchange of energy occurs, exactly as in the TWT, with RF along the helix growing as signal progresses toward the collector end of the helix. Now, unlike the TWT, the BWO does not have an attenuator along the tube. As a gross simplification, oscillations may be thought of as occurring simply because of reflections from an imperfectly terminated collector end of the helix. There is thus feedback, and the output is collected from the *cathode end* of the helix, toward which reflection took place. Because the helix is essentially a nonresonant structure, bandwidth (if one may use such a term with an oscillator) is very high, and the operating frequency is determined by the collector voltage together with the associated cavity system.

Bandwidth is limited by the interaction between the beam and the slow-wave structure. To increase this interaction, the BWO has a ring cathode which sends out a hollow beam, with maximum intensity near the helix.

Practical aspects [22] Backward-wave oscillators are used as signal sources in instruments and transmitters. They can also be made broadband noise sources, whose output, amplified by an equally wideband TWT, is transmitted as a means of enemy radar confusion. The frequency spectrum over which BWOs can be made to operate is vast, stretching from 1 to well over 1000 GHz. For example, the Thomson-CSF CO 08 provides about 50 mW CW over the range 320 to 400 GHz, while 0.8 mW CW has been reported, from another BWO, at 2000 GHz. The normal output range of BWOs is 10 to 100 mW CW, but tubes with outputs over 20 W, at quite high frequencies, have also been produced. The tuning range of a BWO is an octave typically, up to about 40 GHz. At higher frequencies multiple helixes or coupled cavities are used, with a consequent bandwidth reduction to typically a half-octave. However, at the lower end of the spectrum, frequency ranges over 3:1 are possible from the one tube. For example, the ITT F-2513 produces an average of 25 mW over the range 1.3 to 4.0 GHz. Furthermore, the rate at which the BWO frequency may be changed is very high, being measured in gigahertz per microsecond.

Permanent magnets are normally used for focusing, since this results in simplest magnets and smallest tubes. However, solenoids are used at the highest frequencies, since it has been found that they give the best penetration and distribution for the axial magnetic field. A recent development in this respect has been the use of samarium-cobalt permanent magnets to reduce weight and size.

The Siemens RWO 170 is a typical BWO and produces an average power output of 10 mW. It is electronically tunable over the range from 60 GHz (at which the collector voltage is 500 V) to 110 GHz (collector voltage 2500 V). The average collector current is 12 to 15 mA and dissipation about 30 W. Together with its power supply and magnet, it weighs 2 kg.

11-6.3 Miscellaneous Tubes

A number of less common microwave tubes have been proposed and built, mainly for very high powers. The most successful of these appear to be:

1. The Twystron,[1] which is a hybrid combination of klystron driver and TWT output section in tandem in the same envelope.
2. The extended interaction amplifier (EIA), which is a multicavity klystron with interconnected multigap cavities.

Pulsed powers in excess of 10 MW have been obtained from both types of tube, neither of which is in common use.

Multiple-beam klystrons (within the one envelope) have also been proposed, as have Ubitron (undulating beam interaction) amplifiers, using a so-called fast-wave structure.

Detailed information on the appearance, operation and performance of the foregoing tubes is provided by [5].

The latest newcomer to the ranks of high-power microwave tubes is the gyrotron, first described in the U.S.S.R. in the late 1970s. It can be used as an amplifier or oscillator and is an adaptation of either the klystron (gyroklystron) or TWT (gyro-TWT), such that a cyclotron resonance takes place in its cavity. It is potentially capable of very high powers at the highest frequencies, powers in excess of 1 kW CW having been reported experimentally at frequencies over 100 GHz. In [23] and [24] are details of experimental gyroklystrons and gyro-TWTs, respectively.

PROBLEMS

For self-testing questions on this chapter, see p. 697.

QUESTIONS

11-1. Vacuum tubes suffer from limitations at high frequencies. Describe these failings, and indicate the steps that may be taken to extend the usefulness of vacuum tubes to higher frequencies.

11-2. Explain the transit-time effect as it affects high-frequency amplifying devices (hot-cathode or semiconductor) of orthodox construction.

11-3. Draw and describe the circuit of a typical microwave triode amplifier or oscillator, and indicate the performance figures of which it may be capable.

11-4. Describe the two-cavity klystron amplifier, with the aid of a schematic diagram which shows the essential components of this tube as well as the voltages applied to the electrodes.

[1] Twystron® is a registered trademark of Varian Associates.

11-5. Explain how bunching takes place in the klystron amplifier around the electron which passes the buncher cavity gap when the gap voltage is zero and becoming positive.
11-6. Make a clear distinction between *velocity modulation* and *current modulation*. Show how each occurs in the klystron amplifier, and explain how current modulation is necessary if the tube is to have significant power gain.
11-7. Why do practical klystron amplifiers generally have more than two cavities? How can broadband operation be achieved in multicavity klystrons?
11-8. Discuss the applications and performance of the multicavity klystron amplifier, and draw up a performance table. Why should the collector voltage be kept constant for this tube?
11-9. Describe the reflex klystron oscillator with the aid of a suitable schematic diagram; indicate the polarity of the voltages applied to the various electrodes.
11-10. Explain the operation of the reflex klystron oscillator. Why is the transit time so important in this device?
11-11. List and discuss the applications and limitations of the reflex klystron and two-cavity klystron oscillators.
11-12. Describe fully the effect of a dc axial field on the electrons traveling from the cathode to the anode of a magnetron, and then describe the combined effect of the axial magnetic field and the radial dc electric field. Define the *cutoff field*.
11-13. Explain how oscillations are sustained in the cavity magnetron, with suitable sketches, assuming that the π-mode oscillations already exist. Make clear why more energy is given to the RF field than is taken from it.
11-14. With the aid of Fig. 11-13, explain the *phase-focusing* effect in the cavity magnetron, and show how it allows electron bunching to take place and prevents favored electrons from slipping away from their relative position.
11-15. What is the purpose of *strapping* in a magnetron? What are the disadvantages of strapping under certain conditions? Show the cross section of a magnetron anode cavity system that does not require strapping.
11-16. With the aid of a cross-sectional sketch of a coaxial magnetron, explain the operation of this device. What are its advantages over the standard magnetron? What is done to ensure that the coaxial cavity is the one that determines the frequency of operation?
11-17. Describe briefly what is meant by *coaxial, frequency-agile* and *voltage-tunable* magnetrons.
11-18. Discuss the performance of magnetrons and the applications to which this performance suits them.
11-19. With the aid of a schematic diagram, describe the traveling-wave tube. What is a slow-wave structure? Why does the TWT need such a structure?
11-20. How does the function of the magnetic field in a TWT differ from its function in a magnetron? What is the fundamental difference between the beam-RF field interaction in the two devices?
11-21. Discuss briefly the three methods of beam focusing in TWTs.
11-22. What are the power capabilities and practical applications of the various types of traveling-wave tubes? What are the major advantages of CW and pulsed TWTs?
11-23. With the aid of a schematic sketch, briefly describe the operation of the crossed-field amplifier.
11-24. Compare the multicavity klystron, traveling-wave tube and crossed-field amplifier from the point of view of basic construction, performance and applications.
11-25. Briefly compare the applications of the multicavity klystron, TWT, magnetron and CFA. What are the most significant advantages and disadvantages of each tube?
11-26. Describe the practical aspects of BWO performance, explaining why such wide tuning ranges are possible from this device.

REFERENCES

1. See also G. E. Application Note, *Performance and Application of Microwave Gridded Vacuum Tubes and Microwave Circuit Modules,* General Electric Company, Owensboro, Ky., 1973, 13 pp.

2. "Special Report: Microwave Power Sources," *Microwave J.*, April 1975, pp. 18–28.

3. (*Historical*) Varian, R. H., and S. F. Varian, "A High-Frequency Amplifier and Oscillator," *J. Appl. Phys.* **10:**140, 321 (1939).

4. Faillon, G., J. Hervier and E. D. Maloney, "Magnetic-Cell Focused Klystron," *Microwave J.*, August 1975, pp. 39–41.

5. Staprans, Armand, E. W. McCune and J. A. Ruetz, High-Power Linear-Beam Tubes, *Proc. IEEE* **61:**299–329 (March 1973).

6. Lien, E. L, "Advances in Klystron Amplifiers," *Microwave J.*, December 1973, pp. 33–39.

7. For a detailed treatment of the reflex klystron, see Kennedy, G., *Electronic Communication Systems,* 2d ed., McGraw-Hill Book Company, New York, 1977.

8. (*Historical*) Boot, H. A. H., and J. T. Randall, "The Cavity Magnetron," *J. Inst. Elec. Eng.*, vol. 93, pt. IIIA, pp. 928–938, 1946.

9. Applications Bulletins, Varian, Beverly, Mass. 1974, as follows:

(a) "Introduction to Coaxial Magnetrons."

(b) "The Frequency-Agile Magnetron Story."

(c) "Introduction to Pulsed Crossed-Field Amplifiers."

10. Pickering, A. H., "Electronic Tuning of Magnetrons," *Microwave J.*, July 1979, pp. 73–78.

11. Weinstein, M., "Voltage Tunable Magnetron," *Microwave J.*, November 1978, pp. 64–65.

12. (*Historical*) Kompfner, R., "The Travelling-Wave Valve," *Wireless World* **52:**369 (1946).

13. (*Historical*) Pierce, J. R., and L. M. Field, "Traveling-Wave Tubes," *Proc. IRE* **35:**108 (February 1947).

14. Hughes Aircraft Company, Electron Dynamics Division, *Hughes TWT and TWTA Handbook,* Torrance, Calif., 1981, 56 pp.

15. Leborgne, R. H., "A 700 W Ku-Band TWT for Satcom Ground Terminals," *Microwave J.*, July 1981, pp. 65–68.

16. Hamilton, J. J., "Dual-Mode TWT's: The Nitty Gritty," *Microwaves,* May 1974, pp. 38–43.

17. Pallakoff, Owen E., "To Grid, or Not to Grid?" in *ibid,* pp. 52–56.

18. Moats, Robert R. (Dr.), "CFA's and TWTA's for ECM Systems," *Microwave J.*, September 1973, pp. 33–37.

19. Skowron, John F., "The Continuous-Cathode (Emitting-Sole) Crossed-Field Amplifier," *Proc. IEEE* **61** (3):330–356 (March 1973).

20. Kaisel, S. F., "Microwave Tube Technology Review," *Microwave J.*, July 1977, pp. 23–26.

21. See also Adam, Stephen F., *Microwave Theory and Applications,* Prentice-Hall, Inc., Englewood Cliffs, N.J., 1969, pp. 125–134.

22. Kantorowicz, G., P. Palluel, and J. Pontvianne, "New Developments in Submillimeter-Wave BWOs," *Microwave J.*, February 1979, pp. 57–59.

23. Jory, H., S. Hegji, J. Shively and R. Symons, "Gyrotron Developments," *Microwave J.*, August 1978, pp. 30–32.

24. Ferguson, P. E., G. Valier and R. S. Symons, "Gyrotron-TWT Operating Characteristics," *IEEE Trans. Microwave Theory and Tech.*, **MTT-29** (8):794–799 (August 1981).

12 SEMICONDUCTOR MICROWAVE DEVICES AND CIRCUITS

No segment of the microwave field has had more research devoted to it, over the past three decades, than the field of solid-state devices and circuits. This has resulted, and continues to result, in a tremendous proliferation of, and improvements in, semiconductor devices for microwave amplification, oscillation, switching, limiting, frequency multiplication and other functions. For the systems designer, the result of these continuing endeavors has been greater flexibility, improved performance, generally greater reliability, reduced sizes and power requirements, and importantly the ability to produce some systems that would not otherwise have been possible.

It would be entirely feasible to write a large book on each of the major sections of this chapter. Anyone doing this, however, would be well advised to produce a series of loose-leaf books, for easier updating. In this, one of the longest chapters of the text, an attempt will be made to explain the basic principles of each type of device, to discuss its practical aspects and applications, to describe and show its appearance, and to indicate its state-of-the-art performance figures. Different devices that may be used for similar purposes will be compared from a practical point of view. A number of explanations will be deliberately simplified very significantly, because, for the most part, we shall be dealing with extremely complex processes, despite the simple outward appearance of some of the devices themselves.

The chapter begins with a treatment of certain passive microwave circuits, notably *microstrip*, *stripline* and *surface acoustic wave (SAW)* components. They are not semiconductor devices themselves, but since they are often used in conjunction with solid-state microwave devices, this is a convenient place to treat them.

We then continue with a presentation of microwave transistors, both bipolar and field-effect. As with microwave triodes in the preceding chapter, it will be assumed that students already understand how transistors work in general. The stress will thus be laid on their high-frequency limitations and what makes microwave transistors different in construction and behavior from lower-frequency ones. The section concludes with an introduction to microwave integrated circuits.

The next section is devoted to varactor diodes. These are diodes whose capacitance is linearly variable with the change in applied bias. This property makes the diodes ideal for electronic tuning of oscillators and for low-loss frequency multiplication, as will be seen. Another important application of varactors is in *parametric amplifiers,* which form the next major portion of the chapter. As will be explained, extremely low-noise amplification of (microwave) signals can be obtained by a suitable variation of a reactive parameter of an RLC circuit. Varactor diodes fit the bill, since their capacitance parameter is easily variable.

***Tunnel diodes* and their applications are the next topic studied. They are diodes which, under certain circumstances, exhibit a negative resistance. It will be shown that this results in their use as amplifiers and oscillators. Indeed, tunnel diodes will be employed as a vehicle in the general explanation of how amplification is possible with a device that has negative resistance.**

The *Gunn effect* and *Gunn diodes*—so-called after their inventor—are treated next. These are devices in which negative resistance is obtained as a *bulk* property of the material used, rather than a junction property. Gunn diodes are now very common medium-power oscillators for microwave frequencies, with a host of applications that will be enumerated.

Another class of power devices depends on *controlled avalanche* to produce microwave oscillations or amplification. The *IMPATT* and *TRAPATT* diodes are the most commonly encountered members of the family, and both are treated in the next section of the chapter. They are then followed in the second-last section by a treatment of the previously mentioned *Schottky barrier* and *PIN diodes,* used for mixing/detection and limiting/switching, respectively.

The final topic covered is the amplification of microwaves or light by means of the quantum-mechanical effect of stimulated emission of radiation. In other words, the topic covers masers, lasers and a number of other optoelectronic devices. By no means do all the versions of these devices use semiconductors to achieve their aims, but some do, and this is a very convenient place to treat them.

All in all, this is a very wide-ranging chapter.

12-1 PASSIVE MICROWAVE CIRCUITS

Transmission lines and waveguides were invented at the time of, and used in conjunction with, microwave electron tubes such as those treated in the preceding chapter. They are still so used at medium and high powers. Again, being low-loss, they are used at low powers where significant distances are traversed, as in connecting antennas to receivers. However, transmission lines are considerably bulkier than semiconductor microwave devices, and consequently their use would prevent the reduction in circuit size and weight which would otherwise be obtainable. Consequently, *stripline* and *microstrip* have been developed and are used for circuit interconnections with solid-state devices. In addition, they may also be used for passive components, as can another class of devices, using SAW principles. Finally, *microwave integrated circuits* (MICs) are by no means uncommon, and indeed great benefits in many applications result from their use. All these developments are now discussed in turn.

12-1.1 Stripline and Microstrip Circuits

Stripline and *microstrip* are physically related to transmission lines but are treated here because they are microwave circuits used in conjunction with semiconductor microwave devices. As illustrated in Fig. 12-1 *stripline* consists of flat *metallic ground planes,* separated by a thickness of dielectric in the middle of which a thin metallic strip has been buried. As shown, the conducting strip in *microstrip* is on top of a layer of dielectric resting on a single ground plane. Typical dielectric thicknesses vary from 0.1 to 1.5 mm, although the metallic strip may be as thin as 10 μm.

Stripline and microstrip were developed as an alternative conducting medium to waveguides and are now used very frequently in a host of microwave applications in which miniaturization has been found advantageous. Such applications include receiver front ends, low-power stages of transmitters and low-power microwave circuitry in general.

Stripline is evolved from the coaxial transmission line. It may be thought of as flattened-out coaxial line in which the edges have been cut away. Propagation is similarly by means of the TEM (transverse electromagnetic) mode as a reasonable approximation. Microstrip is analogous to a parallel-wire line, consisting of the top strip and its image below the ground plane. The dielectric is often Teflon, alumina or silicon. It is possible to use several independent strips with the same ground planes and dielectric, for both types of circuits. Semiconductor microwave devices are often packaged for direct connection to stripline or microstrip.

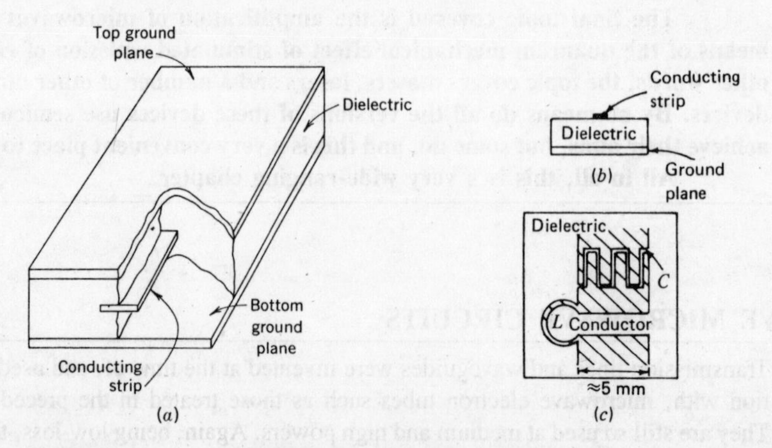

FIGURE 12-1 (*a*) Stripline; (*b*) microstrip cross section; (*c*) microstrip LC circuit.

As was shown in Chap. 10, waveguides are used not only for interconnection but also as circuit components. The same applies to stripline and microstrip (and indeed to coaxial lines). Figure 12-1*c* shows a microstrip *LC* circuit—typical capacitances possible are up to 1 pF, and typical inductances up to 5 nH. The stripline version would be very similar, with just a covering of dielectric and a second ground plane. Transformers can be made similar to the single-turn coil shown, and passive filters and couplers [1] may also be fabricated. Resistances are obtained by using a patch of high-resistance metal such as Nichrome, instead of the copper conductor. Ferrite may be readily blended into such circuits; and so isolators, circulators and duplexers (all described in Chap. 10) are quite feasible. Indeed, Fig. 10-39 shows the construction of a ferrite stripline circulator.

Microstrip has the advantage over stripline in being of simpler construction and easier integration with semiconductor devices, lending itself well to printed-circuit and thin-film techniques. On the other hand, there is a far greater tendency with microstrip to radiate from irregularities and sharp corners. Thus there is a lower isolation between adjoining circuits in microstrip than in stripline. Finally, both Q and power-handling ability are lower with microstrip.

In comparison with waveguides (and coaxial lines), stripline has two significant advantages: reduced bulk and greater bandwidth. The first of these goes without saying, while the second is due to a restriction in waveguides. In practice, these are used over the 1.5:1 frequency range, limited by cutoff wavelength at the lower end and the frequency at which higher modes may propagate at the upper end. There is no such restriction with stripline, and so bandwidths greater than 2:1 are entirely practicable. A further advantage of stripline, as compared with waveguides, is greater compatibility for integration with microwave devices, especially semiconductor ones. On the debit side, stripline has greater losses, lower Q and much lower power-handling capacity than waveguides. Circuit isolation, although quite good, is not in the waveguide class. The final disadvantage of stripline (and consequently of microstrip) is that components made of it are not readily adjustable, unlike their waveguide counterparts.

Above about 100 GHz, stripline and microstrip costs and losses rise significantly. However, at frequencies lower than that, these circuits are very widely used, particularly at low and medium powers. Two extra references [2, 3] on this still developing field are given.

12-1.2 SAW Devices

Surface acoustic waves may be propagated on the surfaces of solid piezoelectric materials, at frequencies in the VHF and UHF regions. Devices employing SAW principles were first discussed in the late 1960s, then moved out of the laboratory in about 1974, and since about 1978 have found many applications as passive components in the low microwave range.

As is well known, the application of an ac voltage to a plate of quartz crystal will cause it to vibrate and, if the frequency of the applied voltage is equal to a mechanical resonance frequency of the crystal, the vibrations will be intense. Because quartz is piezoelectric, all mechanical vibrations will be accompanied by electric oscillations at the same frequency. The mechanical vibrations can be made very stable in frequency, and consequently piezoelectric crystals find many applications in stable oscillators and filters. However, as the desired frequency of operation is raised, so quartz plates must be made thinner and thus more fragile, so that crystal oscillators are not normally likely to operate at fundamental frequencies much in excess of 50 MHz. It is, of course, possible to multiply the output frequency of an oscillator almost indefinitely (see also Sec. 12-3.3), but inconvenience would be avoided if multiplication were unnecessary. This may be done with SAW resonators, which employ thin lines etched on a metallic surface electrodeposited on a piezoelectric substrate. The etching is performed by using photolithography or electron beam techniques, while the most commonly used piezoelectric materials are quartz and lithium niobate.

A simplified sketch of a typical interdigitated SAW resonator is shown in Fig. 12-2. As shown, traveling waves in both directions result from the application of an RF voltage between the two electrodes, but the resulting standing wave is maintained adequately only at the frequency at which the distance between adjoining "fingers" is equal to an (acoustic) wavelength, or a multiple of a wavelength along the surface of the material. As with other piezoelectric processes, an electric oscillation accompanies the mechanical surface oscillation.

If the device is used as a filter, only those frequencies that are close to the resonant frequency of the SAW resonator will be passed. Because the mechanical Q is high (though not quite as high as that of a quartz crystal being used as a standard resonator), the SAW device is a narrowband bandpass filter. To use the SAW resonator to produce oscillations, one need merely place it, in series with a phase-shift network, between the input and output of an amplifier. The phase shift is then adjusted so as to provide positive feedback, and the amplifier will produce oscillations at the frequency permitted by the SAW resonator.

There is no obvious lower limit to the operating frequency of a SAW resonator,

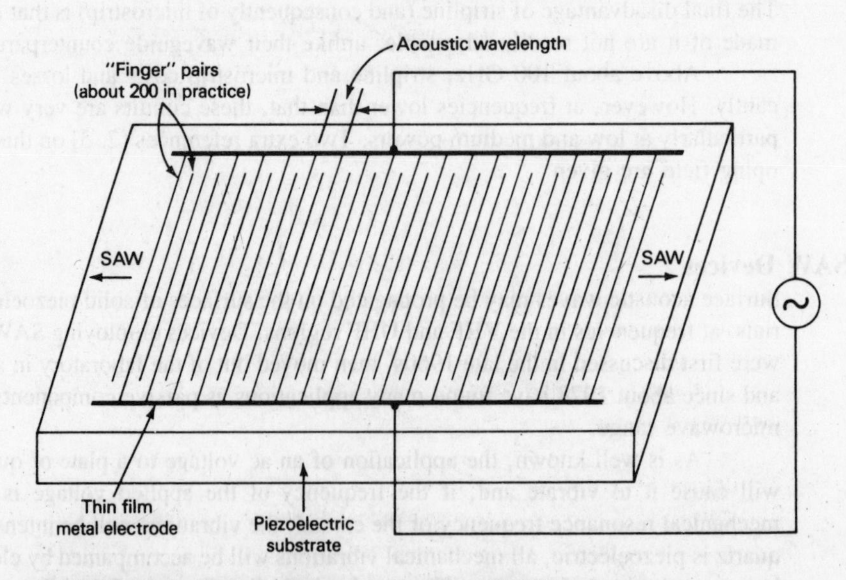

FIGURE 12-2 Basic surface acoustic wave (SAW) resonator.

except that it is unlikely to be used below about 50 MHz, because at such frequencies straightforward crystal oscillators can be used. The upper frequency limit is governed by photoetching accuracy. Noting that, as usual, wavelength = v/f and the velocity of the acoustic wave is approximately 3000 m/s, it is easy to calculate that the finger separation at 5 GHz should be 0.6 μm, and of course the fingers themselves must be thinner still. In consequence, 5 GHz represents the current upper limit of SAW resonator operation.

By necessity, this has been an abbreviated treatment of a wide and developing field. For those who wish to delve more deeply into the various aspects of surface acoustic waves and the resulting devices, [4] and [5] are provided.

12-2 TRANSISTORS AND INTEGRATED CIRCUITS

All devices that try to overcome the limitations of transit time eventually lose. This was stated in connection with vacuum tubes near the beginning of Chap. 11 and applies equally here. However, as with tubes, so with transistors, useful operation can be extended well into the microwave region.

12-2.1 High-Frequency Limitations

As stated, transistors suffer from high-frequency limitations. These are of a twofold nature. On the one hand, there are the same difficulties as those encountered with tubes (Sec. 11-1.1). On the other hand, there is some difficulty in specifying accurately the performance of microwave transistors in a manner which would make it relatively easy for the equipment designer to use them.

Limitations If one has become familiar with the limitations of vacuum tubes at high frequencies, it is relatively easy to predict the limitations of transistors. Thus, capacitances between electrodes play an important part in determining high-frequency response. As is well known, both current gains, α and β, eventually acquire reactive components which make both complex at first and eventually unusable. However, interelectrode capacitances in bipolar transistors depend also on the width of the depletion layers at the junctions, which in turn depend on bias. Thus the situation is somewhat more complex than with tubes, whose interelectrode capacitances are not so bias-dependent. The difficulty here is not that the transistor has a poorer high-frequency response; quite the opposite. It is simply a greater difficulty in finding parameters with which to describe the behavior so as to give a meaningful picture to the circuit designer.

Electrode inductances have more or less the same nuisance value as with tubes, but since transistors are smaller, electrode leads are shorter. Thus suitable geometry and the use of low-inductance packages go a long way toward reducing the effects of lead inductance.

The effect of transit time is identical to that in tubes, although its actual operation is somewhat different. The smaller distances traveled in transistors are counterbalanced by the slower velocities of current carriers, but overall the maximum attainable frequencies are somewhat higher than for tubes. In traveling across a bipolar transistor, the holes or electrons drift across with velocities determined by the ion mobility (basically higher for Ge and GaAs than Si[1]), the bias voltages and the transistor construction. We first find majority carriers suffering an emitter delay time, and then the injected carriers encounter the base transit time, which is governed by the base thickness and impurity distribution. The collector depletion-layer transit time comes next. This is governed mainly by the limiting drift velocity of the carriers (if a higher voltage were applied, damage might result) and the width of the depletion layer (which is heavily dependent on the collector voltage). Finally, electrons or holes take some finite time to cross the collector, as they did with the emitter.

Specification of performance Several methods are used to describe and specify the overall high-frequency behavior of RF transistors. Older specifications showed the alpha and beta cutoff frequencies, respectively $f_{\alpha b}$ and $f_{\alpha e}$. The first is the frequency at which α, the common-base current gain, falls by 3 dB, and the second applies similarly to β, the common-emitter current gain.The two figures are simply interconnected. Since we know that

$$\beta = \frac{\alpha}{1 - \alpha} \tag{12-1}$$

it follows that, for the usual values of β,

$$f_{\alpha e} = \frac{f_{\alpha b}}{\beta} \tag{12-2}$$

These frequencies are no longer commonly in use. They have been replaced by f_T, the (current) gain-bandwidth frequency. This may simply be used as a gain-bandwidth product at low frequencies or, alternatively, as the frequency at which β falls to unity, i.e., the highest frequency at which *current* gain may be obtained. It is very nearly equal to $f_{\alpha b}$ in most cases, although it is differently defined.

Up to a point, f_T is proportional to both collector voltage and collector current and reaches its maximum for typical bipolar RF transistors at V_{ce} = 15 to 30 V and I_c in excess of about 20 mA. This situation is brought about by the higher drift velocities and therefore shorter transit times corresponding to the higher collector voltage and current.

Finally, there is one last frequency of interest to the user of microwave transistors. This is the maximum possible frequency of oscillation, $f_{\max}$. Strangely enough, it is higher than f_T because, although β has fallen to unity at this frequency, power gain has not. In other words, at $\beta = 1$ output impedance is higher than input impedance, voltage gain exists, and thus both regeneration and oscillation are possible. However, it should be mentioned that although the use of transistors above the beta cutoff frequency is certainly possible and very often used in practice, the various calculations are not as easy as at lower frequencies. The transistor behaves as both an amplifier and a low-pass filter, with a 6 dB per octave gain drop above a frequency whose precise value depends on the bias conditions.

[1] Germanium, gallium arsenide and silicon, respectively.

To help with design of transistor circuits at microwave frequencies, scattering-(S) parameters have been evolved [6]. These consider the transistor as a two-port, four-terminal network under matched conditions. The parameters themselves are the forward and reverse transmission gains, and the forward and reverse reflection coefficients. Their advantage is relatively easy measurement and plotting on the Smith chart, but further treatment is outside the scope of this chapter. (See [6].)

12-2.2 Microwave Transistors and Integrated Circuits

Silicon bipolar transistors were first on the microwave scene, followed by GaAs field-effect transistors. Indeed, FETs now have noticeably lower noise figures, and in the C band and above they yield noticeably higher powers. A description of microwave transistor constructions and a discussion of their performance now follow.

Transistor construction The various factors that contribute to a maximum high-frequency performance of microwave transistors are complex. They include the already mentioned requirement for high voltages and currents, and two other conditions. The first of these is a small electrode area to reduce interelectrode capacitance. The second is very narrow active regions to reduce transit time.

For bipolar transistors, these requirements translate themselves into the need for a very small emitter junction and a very thin base. Silicon planar transistors offer the best bipolar microwave performance; fabrication difficulties, together with the excellent performance of GaAs FETs, have prevented the manufacture of GaAs bipolars. Epitaxial diffused structures are used, giving a combination of small emitter area and large emitter edge. The first property gives a short transit time through the emitter, and the second a large current capacity. The *interdigitated* transistor, shown in Fig. 12-3, is by far the most common bipolar in production. The transistor shown has a base and emitter layout that is similar to two hands with interlocking fingers; hence its

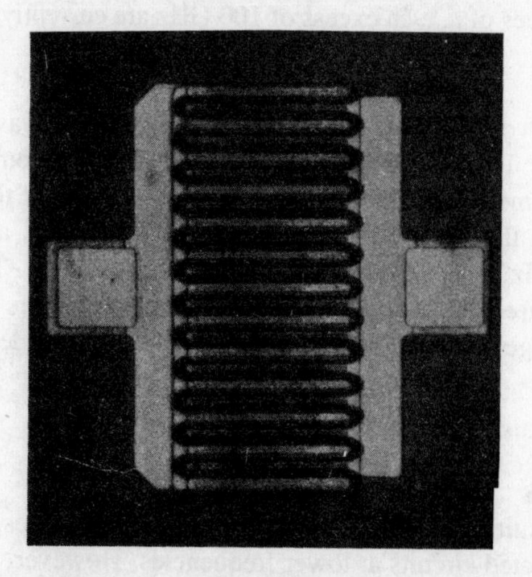

FIGURE 12-3 Geometry of an interdigitated planar microwave transistor. ***(Courtesy of Texas Instruments, Inc.)***

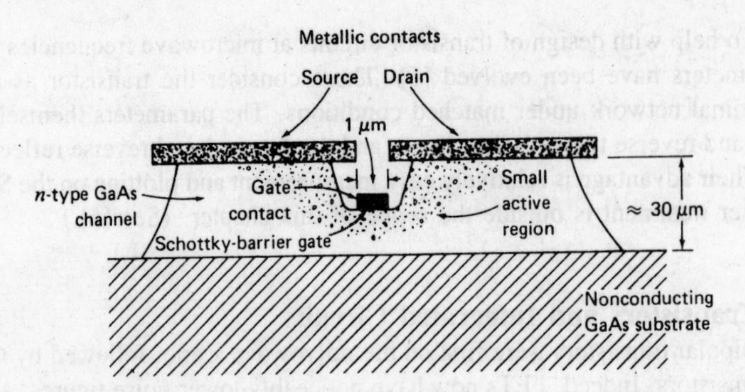

FIGURE 12-4 Construction of microwave mesa field-effect transistor (MESFET) chip, with a single Schottky-barrier gate.

name. The chip illustrated has overall dimensions (less contacts) of about 70 × 70 μm; the emitter contact is on the left, the base on the right and the collector underneath. The thickness of each emitter (and base) "finger" in the transistor shown is 0.5 μm. This yields values of f_{max} in excess of 20 GHz; 0.25-μm geometries have been proposed.

The most common microwave FET uses a *Schottky-barrier gate* (i.e., a metal-semiconductor one; see also Sec. 12-8.2). Figure 12-4 demonstrates why this device is also known as a *MESFET*. The cross section shows it to be of mesa construction. The top metallic layer has been etched away, as has a portion of the n-type GaAs semiconductor underneath. The metallic Schottky-barrier gate stripe is deposited in the resulting groove. It has a typical length of 1 μm (the normal range is 0.5–3μm). The width of the gate is not shown in the cross section; 300–2400 μm is a typical range. Dual-gate GaAs FETs are also available, in which the second gate may be used for the application of AGC in receiver RF amplifiers. For photographs and further details of fabrication, construction and performance, students are referred to Taylor, et al. [7]. It should be mentioned that values of f_{max} in excess of 100 GHz are currently achievable; see also [8].

Packaging and circuits Two typical methods of packaging microwave transistor chips are shown in Fig. 12-5. The Avantek stripline package at the bottom has a body thickness of 1 mm and a diameter just under 5 mm. The TO-72 can at the top has a 7½-mm diameter and much the same height. The TO-72 package is available for frequencies up to about 2 GHz, especially for silicon bipolar transistors. The stripline packages are used for higher frequencies, up to about 30 GHz, for bipolars or FETs; for still-higher frequencies or large bandwidths, the transistor chips are bonded directly to the associated circuitry.

12-2.3 Microwave Integrated Circuits

Because of the inherent difficulties of operation at the highest frequencies, MICs took longer to develop than integrated circuits at lower frequencies. However, by the mid-1970s, *hybrid* MICs had become commercially available, at first with sapphire sub-

FIGURE 12-5 Microwave transistor package types. Top: T0-72 can; bottom: Stripline (beam-lead) package. ***(Courtesy of Avantek, Inc.)***

strates and subsequently with (insulator) gallium arsenide substrates. In these circuits, thick or thin metallic film was deposited onto the substrate, and the passive components were etched onto the film, while the active components, such as transistors and diodes, were subsequently soldered or bonded onto each chip. In the early 1980s, however, *monolithic* MICs became commercially available. In these circuits, all the components are fabricated on each chip, using metallic films as appropriate for passive components and injection doping of the GaAs substrate to produce the requisite diodes and FETs. In view of the size reduction initially available from monolithic MICs, it appeared at first that they would completely take over the field, but significant improvements were made in hybrid circuits, with a consequent resurgence of their use. It would appear that the two types will be used side by side for the foreseeable future, and interested students are referred to Yamasaki and Maki [9] for a discussion of the relative merits of the two types of MICs.

A typical hybrid MIC amplifier is illustrated in Fig. 12-6. This is an Avantek miniature GaAs FET hybrid MIC, with overall dimensions (including connectors and dc power feedthrough) of about 40 × 20 × 4 mm—its volume is thus under 0.2 in^3 The two-stage amplifier produces an output of 10 mW, with a gain of 9 dB and a noise figure of 8 dB, over the very wide frequency range of 6 to 18 GHz. It is seen that the two modules on either side of center are identical balanced amplifiers, with the two transistors located above each other in the middle of each module as indicated. In a working amplifier, a lid is welded on, dry nitrogen is pumped in, and the amplifier is hermetically sealed.

A Texas Instruments monolithic MIC chip is shown in Fig. 12-7. This is a high-gain four-stage GaAs FET power amplifier developed for satellite communications. Although the chip measures only 1 × 5.25 × 0.15 mm, it produces an output of

FIGURE 12-6 Hybrid GaAs FET MIC amplifier. ***Note:*** **Hermetically sealed cover removed.** ***(Courtesy of Avantek, Inc.)***

1.3 W at 7.5 GHz, with a good frequency response from 6.5 to 8 GHz and an efficiency of 30 percent; the gain is 32 dB. The gate widths range from 300 μm for the input FET to 2400 μm for the output FET. Silicon nitride capacitors are used, and a fair amount of gold plating is used to reduce resistance. This amplifier is further described in [10], and [11] is a good overall reference for monolithic MICs.

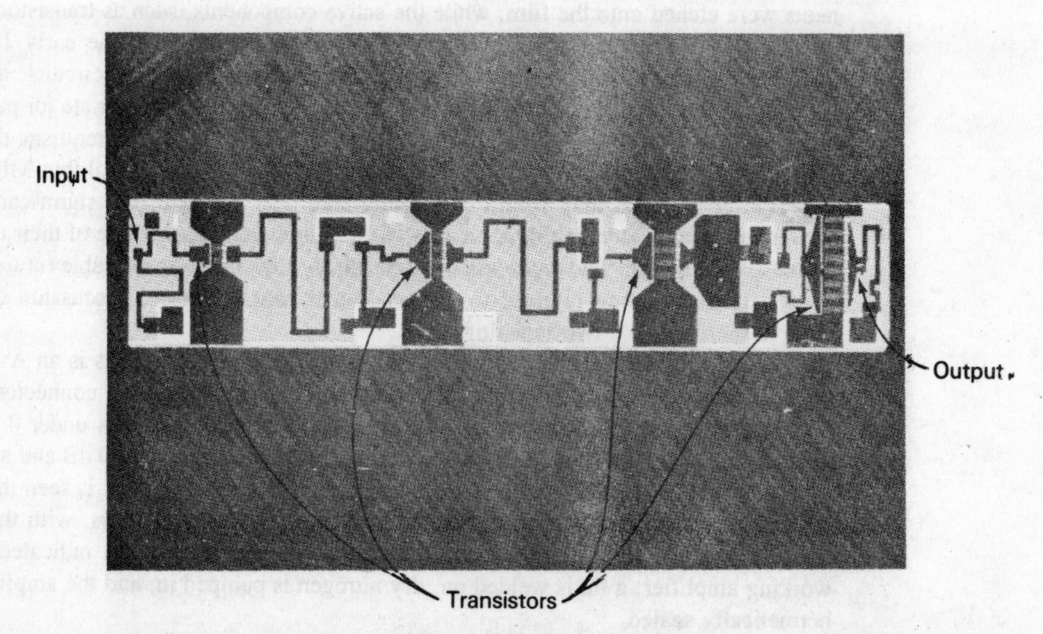

FIGURE 12-7 GaAs FET monolithic MIC four-stage high-gain power amplifier. ***(Courtesy of Texas Instruments, Inc.)***

12-2.4 Performance and Applications of Microwave Transistors and MICs

The power and noise capabilities of microwave transistors and MICs have been improving spectacularly over more than a decade, with good improvements in bandwidth and efficiency over the same period.

Bipolar transistors are available for frequencies up to about 8 GHz, where power devices produce up to about 150 mW output, while low-noise transistors have noise figures of the order of 14 dB. Neither is as good as the corresponding figure for GaAs FETs. However, bipolars do very well at lower microwave frequencies: transistors such as the Avantek ones shown in Fig. 12-5 produce noise figures as low as 2.8 dB at 4 GHz and 1.8 dB at 2 GHz, and power bipolars can produce over 1 W per transistor at 4 GHz.

GaAs FETs are available, as discrete transistors and/or MICs, right through the Ka band (26.5 to 40 GHz) and are becoming available for higher frequencies. Powers of several watts per transistor are available up to 15 GHz, and hundreds of milliwatts to 30 GHz. Noise figures below 1 dB are attainable at 4 GHz and are still only about 2 dB at 20 GHz. However, it should be noted that the noise figures of amplifiers, be they bipolar or FET, are not as good as those of individual transistors. The major reason for this is the low gain per stage, typically 5 to 8 dB at X band (8 to 12.5 GHz). Thus, as explained in Sec. 2-3.2, the noise of transistors and components beyond the input stage makes a significant contribution to the total noise.

As has been mentioned, FETs have the advantage over bipolars at the highest frequencies because they are able to use GaAs, which has a higher ion mobility than silicon. They also have higher peak electron velocities, the two advantages providing a faster transit time and lower dissipation. FETs are thus able to work at higher frequencies, with higher gain, lower noise and better efficiency. Other semiconductor materials currently being investigated as potentially useful at microwave frequencies, because of possible advantages in electron mobility and drift velocity over gallium arsenide, include gallium-indium arsenide (GaInAs). Interested students are referred to Gardner et al. [12].

With such excellent performance, transistor amplifiers (and oscillators) have found many microwave applications, especially as their prices have fallen. The advantages of transistors over other microwave devices include long shelf and working lives, small size and electrode voltages, and low power dissipation together with good efficiencies, of the order of 40 percent. The noise figures and bandwidths are also excellent. Computer control of design and manufacture has resulted in good reliability and repeatability of characteristics for both field-effect and bipolar transistors.

Low-noise transistor amplifiers are employed in the front ends of all kinds of microwave receivers, for both radar and communications. That is, unless the requirement is for extremely low noise, in which case transistors are used to amplify the output of more exotic RF amplifiers (treated later in this chapter). The application for microwave power transistors is as power amplifiers or oscillators in a variety of situations. For example, they serve as output stages in microwave links, driver amplifiers in a wide range of high-power transmitters (including radar ones), and as output stages in broadband generators and phased array radars (see Sec. 16-3). See also [47].

12-3 VARACTOR AND STEP-RECOVERY DIODES AND MULTIPLIERS

Step-recovery diodes are junction diodes which can store energy in their capacitance and then generate harmonics by releasing a pulse of current. They are very useful as microwave frequency multipliers, sometimes by very high factors. The *varactor,* or variable capacitance diode, is also a junction diode. It has the very useful property that its junction capacitance is easily varied electronically. This is done simply by changing the reverse bias on the diode. This single property makes this diode one of the most useful and widely employed of all microwave semiconductor devices.

12-3.1 Varactor Diodes

Varactor diodes were first used in the early 1950s as simple voltage-variable capacitances and later for frequency modulation of oscillators. They thus represent a very mature semiconductor microwave art. As materials and construction improved, so did the maximum operating frequencies, until the stage has now been reached where the most common applications are in tuning,[2] in microwave frequency multipliers (treated next) and in the very low-noise microwave parametric amplifiers (see Sec. 12-4).

Operation When reverse-biased, almost any semiconductor diode has a junction capacitance which varies with the applied back bias. If such a diode is manufactured so as to have suitable microwave characteristics, it is then usually called a *varactor diode*; Fig. 12-8 shows its essential characteristics. Apart from the fact that the capacitance variation must be appreciable in a varactor diode, it must be capable of being varied at a microwave rate, so that high-frequency losses must be kept low. The basic way in which such losses are reduced is the reduction in the size of the active parts of the diode itself.

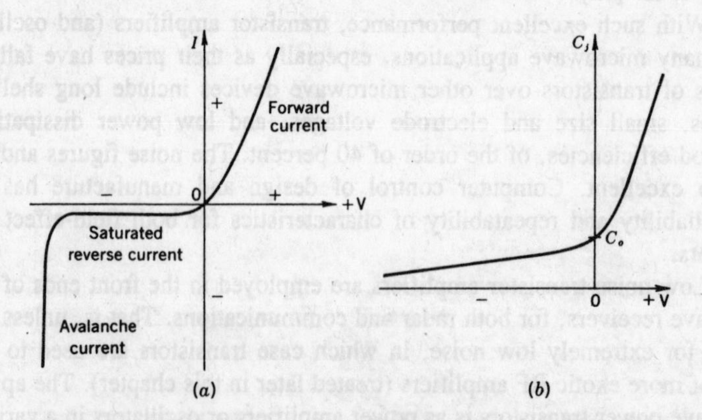

FIGURE 12-8 Varactor diode characteristics. (*a*) Current vs. voltage; (*b*) junction (depletion layer) capacitance vs. voltage.

[2]For an illustration of this application, see Fig. 12-32.

In a diffused-junction diode, the junction is depleted when reverse bias is applied, and the diode then behaves as a capacitance, with the junction itself acting as a dielectric between the two conducting materials. The width of the depletion layer depends on the applied bias, and the capacitance is naturally inversely proportional to the width of this layer; it may thus be varied with changes in the bias. This is shown in Fig. 12-8*b*, where C_0 represents the junction capacitance for zero bias voltage. Finally, as with all other diodes, avalanche occurs with very high reverse bias. Since this is likely to be destructive, it forms a natural limit for the useful operating range of the diode.

Materials and construction Diffused-junction mesa silicon diodes were used originally at microwave frequencies, but by now they have to a large extent been supplanted by gallium arsenide varactors. Figure 12-9 shows a varactor diode made of gallium arsenide. GaAs has such advantages as a higher maximum operating frequency (up to nearly 1000 GHz) and better functioning at the lowest temperatures (of the order of −269° C, as in parametric amplifier applications). Both advantages are due mainly to the higher mobility of charge carriers exhibited by gallium arsenide.

Characteristics and requirements Above all, the varactor diode (no matter how it is made or what it is made from) is a diode, i.e., a rectifier. The diode conducts normally in the forward direction, but the reverse current saturates at a relatively low voltage (as Fig. 12-8*a* shows) and then remains constant, eventually rising rapidly at the avalanche point. For varactor applications, the region of interest lies between the reverse saturation point, which gives the maximum junction capacitance, and a point just above avalanche, at which the minimum diode capacitance is obtained. Conduction and avalanche are thus seen to be the two conditions which limit the reverse voltage swing and therefore the capacitance variation.

Within the useful operating region, the varactor diode at high frequencies behaves as a capacitance in series with a resistance. At higher frequencies still, the stray

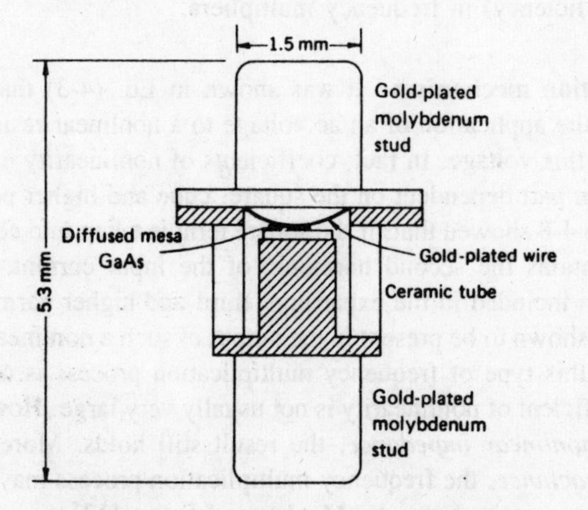

FIGURE 12-9 Varactor diode construction.

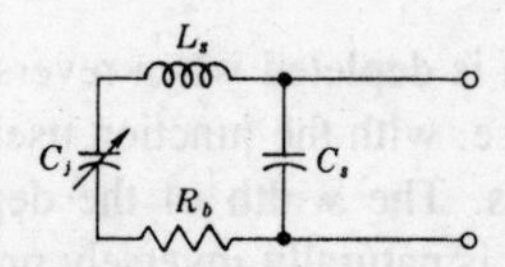

FIGURE 12-10 Varactor diode equivalent circuit.

lead inductance becomes noticeable, and so does the stray fixed capacitance between the cathode and anode connections. The equivalent circuit diagram of Fig. 12-10 then applies. For a typical silicon varactor, C_0 = 25 pF, C_{min} = 5 pF, R_b = 1.3 Ω, C_s = 1.4 pF, and L_s = 0.013 μH.

To be suitable for parametric amplifier service, as will be seen in Sec. 12-4, a varactor diode should have a large capacitance variation, a small value of minimum junction capacitance and the lowest possible value of series resistance R_b (to give low noise). For harmonic generation, much the same requirements apply (although possibly the low value of R_b is a little less important), but now power-handling ability assumes a greater significance. Unfortunately, base resistance and minimum junction capacitance are largely tied to each other, so that these two requirements can be satisfied only in a compromise fashion. The *resistive cutoff frequency* is often used as a figure of merit; it is given by

$$f_c = \frac{1}{2\pi R_b C_{min}} \qquad \textbf{(12-3)}$$

Values of f_c well over 1000 GHz are available from gallium arsenide varactors. However, this does not mean that varactors may be operated at such high frequencies. f_c is measured at a relatively low frequency (e.g., 50 or 500 MHz). It is a figure of merit, a convenient way of relating base resistance and minimum junction capacitance. Operation at frequencies much above $f_c/10$ is inadvisable, because at such frequencies there is a gradual increase in base resistance, partly through the skin effect. Consequently the diode Q drops, and the result is increased noise in parametric amplifiers or increased dissipation (lowered efficiency) in frequency multipliers.

Frequency multiplication mechanism It was shown in Eq. (4-3) that the output current resulting from the application of an ac voltage to a nonlinear resistance is not merely proportional to this voltage. In fact, coefficients of nonlinearity exist, and the output current is thus in part dependent on the square, cube and higher powers of the input voltage. Equation 4-8 showed that, if the square term is taken into consideration, the output voltage contains the second harmonic of the input current. Had higher nonlinearity terms been included in the expansion, third and higher harmonics of the input would have been shown to be present in the output of such a nonlinear resistance.

Unfortunately, this type of frequency multiplication process is not very efficient, because the coefficient of nonlinearity is not usually very large. However, if Eq. (4-3) is applied to a *nonlinear impedance*, the result still holds. Moreover, if this impedance is a *pure reactance*, the frequency multiplication process may be 100 percent efficient in theory, as was shown by Manley and Rowe [13].

Since the capacitance of a varactor diode varies with the applied reverse bias, the diode acts as a nonlinear capacitance (i.e., a nonlinear capacitive reactance). The varactor diode is consequently a very useful device, especially since it will operate at frequencies much higher than the highest operating frequencies of transistor oscillators.

12-3.2 Step-Recovery Diodes

A step-recovery diode, also known as a *snap-off varactor,* is a silicon or gallium arsenide *p-n* junction diode, of a construction similar to that of the varactor diode. It is an epitaxial diffused junction diode, designed to store charge when it is conducting with a forward bias. When reverse bias is applied, the diode very briefly discharges this stored energy, in the form of a sharp pulse very rich in harmonics. The duration of this pulse is typically 100 to 1000 ps, depending on the diode design. This snap time must in practice be shorter than the reciprocal of the output frequency; for example, for an output frequency of 8 GHz, snap time should be less than $T = \frac{1}{8} \times 10^{-9} = 1.25 \times 10^{-10} = 125$ ps.

As will be shown in the next section, a step-recovery diode is biased so that it conducts for a portion of the input cycle. The depletion layer of the junction is charged during this period. When the input signal changes polarity and the diode is biased off, it then produces this sharp pulse, which is very rich in harmonics. All that is then needed in the output is a tuned circuit operating at the wanted harmonic, be it the second or the twentieth. If the circuit is correctly designed, efficiencies well in excess of $1/n$ are possible, where n is the frequency multiplication factor. For example, this means that feeding (say) 12 W at 0.5 GHz to a snap-off varactor may result in decidedly more than 1.2 W out at 5 GHz.

It is also possible to use these diodes without a tuned output circuit, to produce multiple harmonics in so-called "comb generators." Also possible is the stacking of two or more step-recovery (or varactor) diodes in the one package, to provide a higher power-handling capacity.

12-3.3 Frequency Multipliers

Practical circuits A typical multiplier chain is shown in Fig. 12-11. The first stage is a transistor crystal oscillator, operating in the VHF region, and this is the only circuit in the chain to which dc power is applied. The next stage is a step-recovery multiplier by 10, bringing the output into the low-GHz range. This multiplier is likely to have lumped input circuitry and stripline or coaxial output. With 10 × multiplication, the efficiency will be of the order of 20 percent, as shown in Fig. 12-11. Another snap-off 5 × multiplier now brings the output into the X band, with comparable efficiency. Normal varactors are used from this point onward. The reason is an increasing difficulty, beyond the X band, in constructing step-recovery diodes with snap-times sufficiently short to meet the $1/f_{out}$ criterion.

The circuit of Fig. 12-12 shows a simple frequency tripler, which could be varactor or step-recovery. It can also be taken as the equivalent of a higher frequency

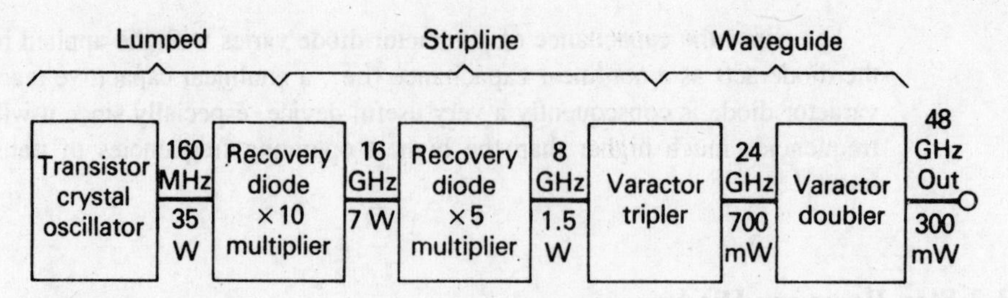

FIGURE 12-11 Step-recovery/varactor diode frequency multiplier with typical powers and frequencies shown.

stripline or cavity tripler. Note that the diode bias is provided by resistor R_B, in a leak-type arrangement. For correct operation of a snap-off varactor multiplier, the value of the resistance is normally between 100 and 500 $k\Omega$. No circulator is necessary to isolate input from output, because the two operate at different frequencies, and the filters provide all the isolation required. Note finally that the tripler is provided with an idler circuit, which is a tuned circuit operating at the frequency of $f_{out} - f_{in}$. Idlers are further discussed in conjunction with parametric amplifiers, where the need for them is explained in Sec. 12-4.1.

Performance, comparison and applications Snap-off varactors multiply by high factors with better efficiency than ordinary varactor chains, and so they are used by preference where possible. However, as mentioned previously, varactors produce higher output powers from about 10 GHz, and step-recovery diodes are not available for frequencies above 20 GHz, while varactors can be used well above 100 GHz. Again, as already mentioned, snap-off devices are suitable for comb generators, whereas the others are not. Finally, it has been found that varactor diodes are preferable to step-recovery diodes for broadband frequency multipliers. These are circuits in which the input frequency may occur anywhere within a bandwidth of up to 20 percent, and any such frequency must be multiplied by a given factor.

Step-recovery diodes are available for power outputs in excess of 50 W at 300 MHz, through 10 W at 2 GHz to 1 W at 10 GHz. Multiplication ratios up to 12 are commonly available, and figures as high as 32 have been reported. Efficiency can be in excess of 80 percent for triplers at frequencies up to 1 GHz. With an output frequency of 12 GHz, 5 × multiplier efficiency drops to 15 percent.

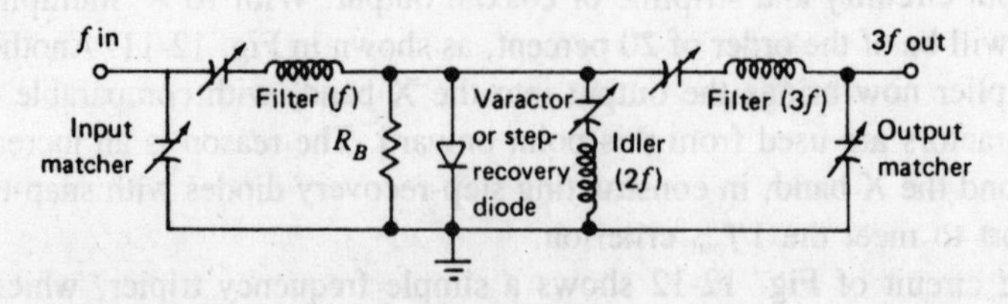

FIGURE 12-12 Diode tripler circuit.

For varactor diodes, the maximum power output ranges from more than 10 W at 2 GHz to about 25 mW at 100 GHz; most varactors at frequencies above 10 GHz are gallium arsenide. Tripler efficiencies range from 70 percent at 2 GHz to just under 40 percent at 36 GHz, and a GaAs varactor doubler efficiency of 54 percent at 60 GHz has also been reported.

For many years, frequency multiplier chains provided the highest microwave powers available from semiconductors, but other developments have overtaken them. At the lower end of the microwave spectrum, GaAs FETs are capable of higher powers, as are Gunn and IMPATT diodes (see following sections) from about 20 to at least 100 GHz. Unless the highest frequency stabilities are required (note that it is the output of a *crystal* oscillator that is multiplied), it is more likely that a transistor Gunn or IMPATT oscillator will be used up to about 100 GHz. One of the current applications of multiplier chains is to provide a low-power signal used to phase-lock a Gunn or IMPATT oscillator.

Varactors are used widely for tuning (see Fig. 12-32, for example), for frequency-modulating microwave oscillators, and as the active devices in parametric amplifiers, as will be shown in the next section. They are produced by a mature, well-established manufacturing technique, with consequent good reliability and comparatively low prices.

12-4 PARAMETRIC AMPLIFIERS

Although the use of parametric amplifiers for high-frequency, low-noise amplification is relatively recent, the principles themselves are not at all new. They were first propounded by Lord Rayleigh in the 1880s for mechanical systems, and by R. Hartley in the 1930s for electrical applications. The need for extremely low-noise amplification, as in radiotelescopes, space probe tracking and communications, and tropospheric scatter receivers, spurred the design of useful parametric microwave amplifiers in the late 1950s. The low-noise properties of such amplifiers and the advent of suitable varactor diodes were the greatest stimuli.

12-4.1 Basic Principles

The parametric amplifier uses a device whose reactance is varied in such a manner that amplification results. It is low-noise because no resistance need be involved in the amplifying process. A varactor diode is now always used as the active element. Amplification is obtained when the reactance (capacitive here) is varied electronically in some predetermined fashion at some frequency *higher* than the frequency of the signal being amplified. The name of the amplifier stems from the fact that capacitance is a *parameter* of a tuned circuit.

Fundamentals To understand the operation of one of the forms of the parametric amplifier, consider an *LC* circuit oscillating at its natural frequency. If the capacitor plates are physically pulled apart at the instant of time when the voltage between them is at its positive maximum, then work is done on the capacitor since a force must be applied to separate the plates. This work, or energy addition, appears as an increase in the voltage across the capacitor. Since $V = q/C$ and the charge q remains constant, voltage is inversely proportional to capacitance. Since the capacitance has been reduced by the pulling apart of the plates, voltage across them has increased proportionately. The plates are now returned to their initial separation just as the voltage between them passes through zero, which involves no work. As the voltage passes through the negative maximum, the plates are pushed apart, and voltage increases once again. The process is repeated regularly, so that energy is taken from the "pump" source and added to the signal, *at the signal frequency;* amplification will take place if an input circuit and a load are connected. In practice, the capacitance is varied electronically (as could be the inductance). Thus the reactance variation can be made at a much faster rate than by mechanical means, and it is also sinusoidal rather than a square wave.

Comparing the principles of the parametric amplifier with those of more conventional amplifiers we see that the basic difference lies in use of a variable reactance (and an ac power-supply) by the former, and a variable resistance[3] (and a dc power supply) by the latter.

The basic parametric amplifier just described requires the capacitance variation to occur at a *pump* frequency that is exactly twice the resonant frequency of the tuned circuit, and hence twice the signal frequency. It is thus phase-sensitive; this is a property that sometimes limits its usefulness. This mode of operation is called the *degenerate mode,* and it may also be shown that the amplifier is a negative-resistance one (see also Sec. 12-5.3).

Amplification mechanism The introduction laid down the basis of parametric amplification, and Fig. 12-13 illustrates the process graphically. It will be seen that (as outlined) the voltage across the capacitor is increased by pumping at each signal voltage peak. Furthermore, the energy thus given to the circuit is not removed when the plates are restored to their initial position (i.e., when the capacitance of the diode is restored to its original value) because this is done when the voltage across the capacitance is instantaneously zero.

The process of signal buildup is shown in Fig. 12-13*c*. Note that it requires more energy in each successive step to increase the voltage across the capacitance, because the peak charge is greater each time. The capacitor voltage would tend to increase indefinitely, except that the driving power is finite. Thus in practice the buildup progresses until the energy added at each peak equals the maximum energy available from the pump source.

If the pump frequency is other than twice the signal frequency, beating between the two will occur, and a difference signal, called the *idler* frequency, will appear. The amplitude of this idler signal is equal to the amplitude of the output signal, and its presence is an automatic consequence of using a pump frequency such that $f_p \neq 2f_s$. This means that *if the idler signal is suppressed, the amplifier will have no gain.*

[3] In an ordinary transistor amplifier, for example, changes in base current cause changes in collector current when the collector supply voltage is constant; it may thus be said that the collector resistance is being changed.

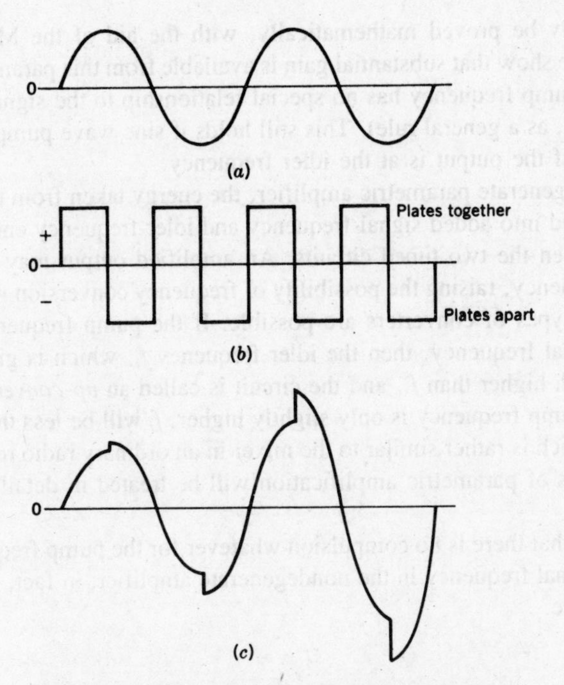

FIGURE 12-13 Parametric amplification with square-wave pumping in degenerate mode. (*a*) Signal input voltage; (*b*) pumping voltage, (*c*) output voltage buildup.

Figure 12-14 shows two simple parametric amplifier circuits. In the basic diagram (Fig. 12-14*a*) degenerate operation takes place, whereas for Fig. 12-14*b* $f_p \neq 2f_s$, and the pumping is called *nondegenerate*. Thus an idler circuit is necessary for amplification to take place, and one is provided. The pump frequency tuned circuit has been left out in each case for the sake of simplicity. Note that nothing prevents us from taking the output at the idler frequency, and in fact there are a number of advantages in doing this, as will be shown.

The nondegenerate parametric amplifier, like the degenerate one, produces gain, with the pump source being a net supplier of energy to the tank circuit. Unfortu-

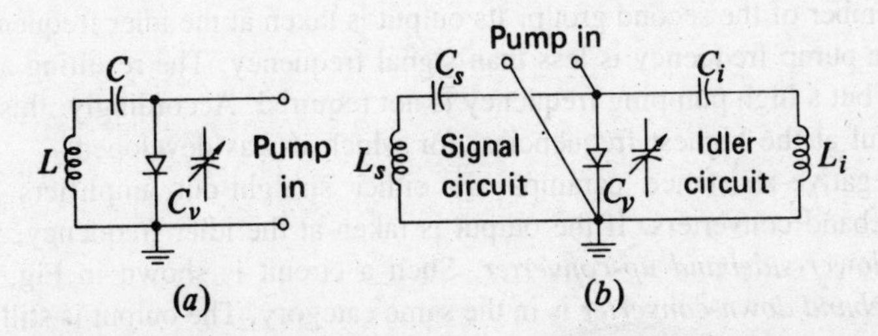

FIGURE 12-14 Basic parametric amplifiers. (*a*) Degenerate; (*b*) nondegenerative, showing idler circuit.

nately, this can only be proved mathematically, with the aid of the Manley-Rowe relations [13]. These show that substantial gain is available from this parametric amplifier, in which the pump frequency has no special relationship to the signal frequency (except to be higher, as a general rule). This still holds if sine-wave pumping is used, and it also applies if the output is at the idler frequency.

In the nondegenerate parametric amplifier, the energy taken from the pumping source is transformed into added signal-frequency and idler-frequency energy and divides equally between the two tuned circuits. An amplified output may thus be obtained at either frequency, raising the possibility of frequency conversion *with gain*. In fact, two different types of converters are possible. If the pump frequency is much higher than the signal frequency, then the idler frequency f_i, which is given by $f_i = f_p - f_s$, will be much higher than f_s, and the circuit is called an *up-converter*. On the other hand, if the pump frequency is only slightly higher, f_i will be less than f_s, and a *down-converter*, which is rather similar to the mixer in an ordinary radio receiver, will result. These aspects of parametric amplification will be treated in detail in the next section.

Note finally that there is no compulsion whatever for the pump frequency to be a multiple of the signal frequency in the nondegenerate amplifier; in fact, it seldom is a multiple in practice.

12-4.2 Amplifier Circuits

The basic types of parametric amplifiers have already been treated in detail, but several others also exist. They differ from one another in the variable reactance used, the bandwidth required and the output frequency (signal or idler). Various other characteristics of parametric amplifiers must also now be discussed, such as practical circuits, their performance and advantages, and lastly the important noise performance.

Amplifier types When classifying parametric amplifiers, the first thing to decide is the device whose parameter will be varied. This is now always a varactor, whose capacitance is varied, but a variable inductance can also be used. Indeed, the first parametric amplifiers were of this type, using an RF magnetic field to pump a small ferrite disk. Such amplifiers are no longer used, mainly because their noise figures do not compare with those available from varactor amplifiers.

Parametric amplifiers (or *paramps*) may be divided into two main groups: negative-resistance and positive-resistance. The *upper-sideband up-converter* is the only useful member of the second group. Its output is taken at the idler frequency $f_i = f_p + f_s$, and the pump frequency is less than signal frequency. The resulting amplifier has low gain, but a high pumping frequency is not required. Accordingly, this amplifier is most useful at the highest frequencies, for which it was developed.

Negative-resistance paramps are either straight-out amplifiers ($f_0 = f_i$) or lower-sideband converters. If the output is taken at the idler frequency, we have the two-port *lower-sideband up-converter*. Such a circuit is shown in Fig. 12-15. The *lower-sideband down-converter* is in the same category. The output is still taken at the idler frequency, but this is now lower than the signal frequency. Both these amplifiers are nondegenerate.

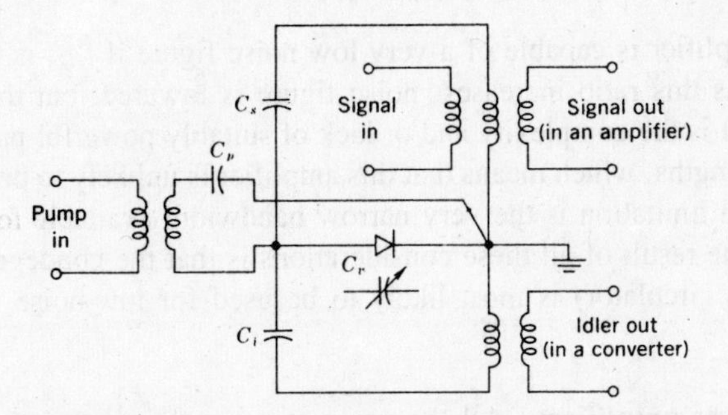

FIGURE 12-15 Parametric amplifier or converter.

The (straight-out) amplifier may be degenerate or not, depending on whether pump frequency is twice signal frequency. The two types share the disadvantage of being one-port (two-terminal) amplifiers. The nondegenerate amplifier is the one in which the pump frequency is (much) higher than the signal frequency but is quite unrelated to it. The circuit of Fig. 12-15 also applies here.

Any paramp can belong to one of two broad classes. First there are narrowband amplifiers using a varactor diode that is part of a tuned circuit, as so far treated. Alternatively, paramps can be wideband, in which case a number of diodes are used as part of a traveling-wave structure, as will be shown.

Narrowband amplifiers The negative-resistance parametric amplifier is the type almost always used in practice. The most commonly used types are the nondegenerate one-port amplifier and the two-port lower-sideband up-converter, in that order. The circuit of Fig. 12-15 could be either type, depending on where the output is taken. The one-port amplifier may suffer from a lack of stability and low gain due mainly to the fact that the output is taken at the input frequency. On the other hand, the pump power is low and so is noise, and the amplifier can be made small, rugged and inexpensive.

Undoubtedly the fundamental drawback of this amplifier, as it stands, is that the input and output terminals are in parallel, as shown in Fig. 12-15. This, of course, applies to all two-terminal amplifiers. If such an amplifier is followed by a relatively noisy stage such as a mixer, then the noise from the mixer, present at the output of the parametric amplifier, will find its way to the amplifier's input. It will therefore be reamplified, and the noise performance will suffer.

In order to overcome this difficulty, a circulator is used (see Sec. 10-5.2). The four-terminal version of the Y stripline circulator of Fig. 11-39 is particularly suitable. The arrangement is then identical to that of Fig. 12-23, with the paramp replacing the tunnel diode amplifier shown there. The output of the antenna feeds the parametric amplifier, whose output can go only to the mixer. Any noise present at the input of the mixer can be coupled neither to the paramp nor to the antenna; as shown, it goes only to the matched termination. The circulator itself can generate some noise, but this may be reduced with proper techniques (such as cooling, as will be seen).

If the output is taken at the idler frequency (in Fig. 12-15), a two-port lower-sideband up-converter results, for which a circulator is not required. It has been shown that this type of amplifier is capable of a very low noise figure if f_i/f_s is in excess of about 10. In fact, as this ratio increases, noise figure is lowered, but there are two limitations. The first is the complexity and/or lack of suitably powerful pump sources at millimeter wavelengths, which means that this amplifier is unlikely to be used above X band. The second limitation is the very narrow bandwidth available for minimum noise conditions. The result of all these considerations is that the nondegenerate one-port amplifier (with circulator) is most likely to be used for low-noise narrowband applications.

Traveling-wave diode amplifiers All the parametric amplifiers thus far described use cavity or coaxial resonators as tuned circuits. Since such resonators have high Q's and therefore narrow bandwidths, parametric amplifiers using them are anything but broadband; the available literature does not describe any such amplifier exceeding a bandwidth of 10 percent. However, it is possible to use traveling-wave structures for parametric amplifiers to provide bandwidths as large as 50 percent of the center frequency, with other properties comparable to those of narrowband amplifiers.

As shown in Fig. 12-16, a typical traveling-wave amplifier employs a multistage low-pass filter, consisting of either a transmission line or lumped inductances, with suitably pumped shunt varactor diodes providing the shunt capacitances. The signal and pump frequencies are applied at the input end of the circuit, and the required output is taken from the other end. If the filter is correctly terminated at the desired output frequency, this will not be reflected back to the input, and thus unilateral operation is obtained, even for a negative-resistance amplifier without a circulator. The only real disadvantage is a lower gain than with narrowband amplifiers.

In order to obtain useful amplification, the pump, signal and idler frequencies must all fall within the passband of the filter, whereas the sum of the signal and the pump frequencies must fall outside the passband. This suggests that the pump frequency must not be very much higher than the signal frequency, or filtering will be difficult. As the wave progresses along the filter (lumped or transmission-line), the signal and idler voltages grow at the expense of the pumping signal. Although this power conversion becomes more complete as the length of the line is increased, the growth rate reduces. Hence maximum gain is achieved for a certain optimum length of line (or number of lumped sections), particularly as ohmic losses increase with the length.

Noise—cooling The noise figures of practical parametric amplifiers are extremely low, a very close second only to those of cooled multilevel masers (Sec. 12-9.2). The reason for such low noise is that the variable transconductance used in the amplifying process is reactive, rather than resistive as in the more orthodox amplifiers. Once noise

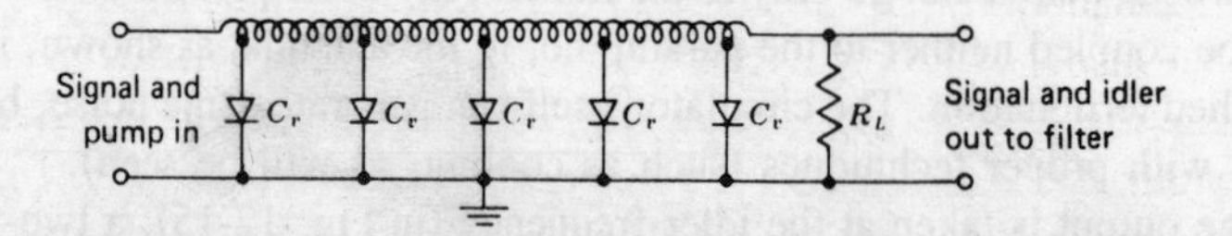

FIGURE 12-16 Basic traveling-wave parametric amplifier.

contributions due to associated circuitry (such as the circulator) have been minimized, the only noise source in the parametric amplifier is the base resistance, sometimes called the *spreading resistance,* of Fig. 12-10. This being the case, it seems that cooling the paramp and associated circuitry should have the effect of lowering its noise considerably.

Those paramps that are not operated at room temperature (290 K, or 17° C, is considered standard) may be cooled to about 230 K by using *Peltier thermoelectric cooling*. The next step is to use *cryogenic* cooling with liquid nitrogen (to 77 K) or with liquid helium (4.2 K). Cryogenic apparatus is outside the scope of this book, although one system is shown in outline in Fig. 12-28. *Dewar flasks, cryodyne* and other cryogenic systems are described in [14], and Peltier cooling is explained in [15].

It must be emphasized that cooling is used with some parametric amplifiers in an attempt to improve their performance; it is neither compulsory nor always employed. As a matter of fact, although the noise temperature improvement which results from cooling is significant, it is not as great as might be expected. It would appear that the spreading resistance is increased as temperature is lowered, perhaps because of a decrease in the mobility of the varactor's charge carriers. The point is uncertain, however, because measurements at extremely low temperatures are rather difficult to make.

Cryogenic cooling tends to be bulky and expensive, and consequently the current trend is away from cryogenically cooled amplifiers, except for the most exacting applications, as in radiotelescopes, some satellite earth stations, and space communication terminals. Thus applications requiring very good but not critical noise figures, including portable earth stations, are likely to use Peltier cooled or uncooled paramps. The other current design feature is the use of solid-state (especially Gunn) oscillators for pumps, although a lot of existing parametric amplifiers still use klystrons or even varactor chains.

The Ferranti parametric amplifier of Fig. 12-17 is uncooled. In the figure, terminal 1 is the RF input, 2 is the pump input, 3 the RF output and 4 the connection point for temperature stabilizing equipment. The circulator is behind the paramp box, which measures approximately 70 × 45 × 25 mm. The Gunn diode pump source is located next to the circulator, together with its cavity and thermal stabilizer. The panel has its own dc supply and simply plugs into the mains. Two diodes are connected back-to-back in a coaxial circuit, using a system pioneered by the firm. The varactors in the paramp are high-quality, high-frequency GaAs ones. A similar parametric amplifier is described in detail in [16].

Performance comparisons There are so many different types of parametric amplifiers and temperatures at which they may be used that tabular comparison is considered the most convenient. Accordingly, Table 12-1 compares a number of typical paramps; note the degradation in noise figure with increased temperature and/or operating frequency. Note also the lower bandwidth of converters as compared with nondegenerate one-port amplifiers, while the traveling-wave amplifier has by far the greatest percentage bandwidth.

The comparison in Table 12-2 is between paramps and other low-noise amplifiers. Note that the best, rather than typical, performances are included in Table 12-2.

FIGURE 12-17 Uncooled 5-GHz parametric amplifier with ancillary equipment. ***(Courtesy of Ferranti Ltd., Solid State Microwave Group.)***

As already mentioned, parametric amplifiers find use in microwave receivers which require extremely low-noise temperatures. At the lowest point, in radiotelescopes and satellite and space probe tracking stations, they compete with masers. They are used in earth stations, sometimes in communications satellites and, increasingly, in radar receivers.

TABLE 12-1 Performance Comparison of Various Parametric Amplifier Types

AMPLIFIER TYPE	WORKING TEMPERATURE, K	f_{in}, GHz	f_p, GHz	f_{out}, GHz	POWER GAIN, dB	BAND WIDTH, MHz	NOISE FIGURE, dB	NOISE TEMPERATURE, K
Degenerate*	4.2	6.00	12.0	6.00	14	10	0.3	21
Degenerate*	290	5.85	11.7	5.85	18	8	3.0	300
Nondegenerate*	4.2	4.2	23.0	4.2	22	40	0.2	14
Nondegenerate*	77	4.1	23.0	4.1	20	60	0.6	45
Nondegenerate*	290	3.95	61.0	3.95	60	500	1.0	80
(Not known)*	235	3.95	?	3.95	60	500	0.75	55
Nondegenerate*	290	60.0	105.0	60.0	14	670	6.0	865
LSB up-converter	290	0.9	26.5	25.6	16	2.5	1.0	80
USB up-converter	77	1.0	20.0	21.0	10	0.1	0.4	29
Traveling-wave	290	3.4	8.5	3.4	10	720	3.5	370

*All these amplifiers are one-port and hence require circulators.

TABLE 12-2 Comparison of Various Low-Noise Amplifiers*

TYPE	f_{out}, GHz	POWER GAIN, dB	BAND WIDTH, MHz	NOISE TEMPERATURE, K	COOLING
Parametric amplifier	4.00	19	40	8	Very helpful
Traveling-wave paramp	4.10	12	500	16	
Three-level ruby maser	8.00	10	5	6	Compulsory (with
Traveling-wave maser	5.80	20	25	11	liquid helium)
Tunnel-diode amplifier	4.00	30	75	400	Helps (but de-
Tunnel-diode amplifier	3.00	10	2,000	500	stroys simplicity)
GaAs FET amplifier	3.00	32	2,000	200	As above
Low-noise TWT	3.00	25	2,000	600	Not practicable

*The figures shown are for the best available commercial amplifiers, of which the paramps and masers are cooled down to 4.2 K. Typical noise temperatures for mixers, which may be used instead, are approximately 700 K.

12-5 TUNNEL DIODES AND NEGATIVE-RESISTANCE AMPLIFIERS

The tunnel, or Esaki [17], diode is a thin-junction diode which, under low forward-bias conditions, exhibits negative resistance. This makes the tunnel diode, invented in the late 1950s, useful for oscillation or amplification. Because of the thin junction and short transit time, it lends itself well to microwave applications.

12-5.1 Principles of Tunnel Diodes

The equivalent circuit of the tunnel diode, when biased in the negative-resistance region, is shown in Fig. 12-18. At all except the highest frequencies, the series resistance and inductance can be ignored. The resulting diode equivalent circuit is thus reduced to the parallel combination of the junction capacitance C_j and the negative resistance $-R$. Typical values of the circuit components of Fig. 12-18 are $r_s = 6\ \Omega$, $L_s = 0.1$ nH, $C_j = 0.6$ pF and $R = -75\ \Omega$.

The tunnel diode was the first, and for several years the only, solid-state device which merely required the application of a small dc voltage for negative resistance to manifest itself. However, after the initial exclamations of joy at the remarkable invention died down, tunnel diode oscillators were found to be unstable in their frequency of operation and were all but discarded. The reasons for the instability have subsequently been found and cured, and so tunnel diodes are again in use, but now mainly for amplifiers, as will be seen.

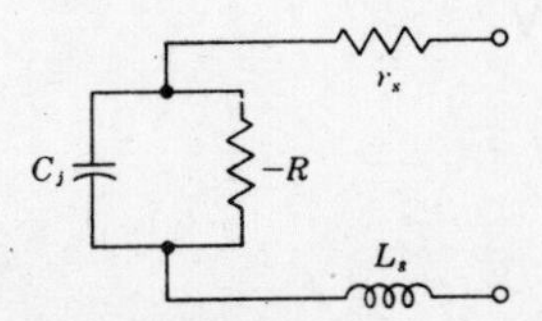

FIGURE 12-18 Tunnel-diode equivalent circuit.

The junction capacitance of the tunnel diode is highly dependent on the bias voltage and temperature. Connecting a tuned circuit directly across it will undoubtedly yield an unstable oscillator, particularly since the effective Q of the circuit is relatively low. However, if a high-Q cavity is *loosely coupled* to the diode, a highly stable oscillator is obtained, with a relative independence of temperature, bias voltage or diode parameter variation.

Description of behavior The tunnel diode is a semiconductor *p-n* junction diode. It differs from the usual rectifier-type diodes in that the semiconductor materials are very heavily doped, perhaps as much as 1000 times more than in ordinary diodes. This heavy doping results in a junction which has a depletion layer that (with a typical thickness of 0.01 μm) is so thin as to permit *tunneling* to occur. In addition, the thinness of the junction allows microwave operation of the diode because it considerably shortens the time taken by the carriers to cross the junction. A current-voltage characteristic for a typical germanium tunnel diode is shown in Fig. 12-19. It is seen that at first forward current rises sharply as voltage is applied, where it would have risen slowly for an ordinary diode (whose characteristic is shown for comparison). Also, reverse current is much larger for comparable back bias than in other diodes, owing to the thinness of the junction.

The interesting portion of the characteristic begins at the point A on the curve of Fig. 12-19; this is the *voltage peak*. As the forward bias is increased past this point, the forward current drops and continues to drop until point B is reached; this is the *valley voltage*. At B the current starts to increase once again and does so very rapidly as bias is increased further. From this point the characteristic resembles that of an ordinary diode. Apart from the voltage peak and valley, the other two parameters normally used to specify the diode behavior are the peak current and the peak-to-valley current ratio, which here are 2 mA and 10, respectively, as shown.

The diode voltage-current characteristic illustrates two important properties of the tunnel diode. First it shows that the diode exhibits *dynamic negative resistance*

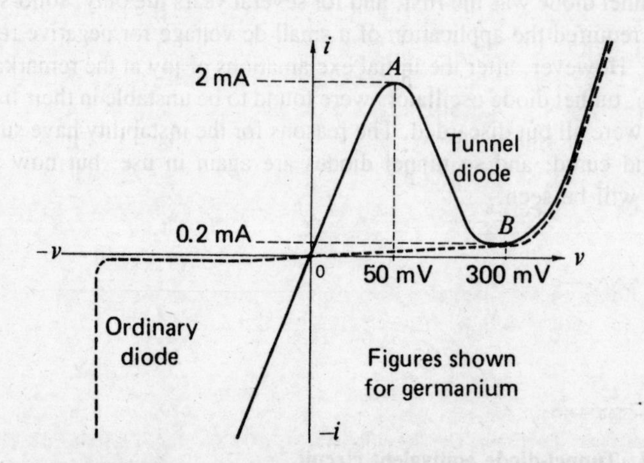

FIGURE 12-19 Tunnel-diode voltage-current characteristic.

between A and B and is therefore useful for oscillator (and amplifier, as will be seen) applications. Second since this negative resistance occurs when both the applied voltage and the resulting current are low, the tunnel diode is a relatively low-power device. A quick calculation shows that in order to stay within the negative-resistance region, the voltage variation must be restricted to 300 − 50 = 250 mV (peak-to-peak) = 88.4 mV rms, whereas the current range is similarly 1.8 mA (peak-to-peak) = 0.63 mA. Thus the load power is very roughly 88.4 × 0.635 = 56 μW. Other factors have been neglected, but the figure is of the right order.

Diode theory Unless energy is imparted to electrons from some external source, the energy possessed by the electrons on the *n* side of the junction is insufficient to permit them to *climb over* the junction barrier to reach the *p* side. However, *quantum mechanics* shows that there is a small but finite probability that an electron which has insufficient energy to climb the barrier can, nevertheless, find itself on the other side of it if this barrier is thin enough, without any loss of energy on the part of the electron.[4] This is the tunneling phenomenon which is responsible for the behavior of the diode over the region of interest.

Figure 12-20 shows energy-level diagrams for the tunnel diode for three interesting bias levels. The cross-hatched regions represent energy states in the conduction band occupied by electrons, whereas the shaded areas show the energy states occupied by electrons in the valence bands. The levels to which energy states are occupied by

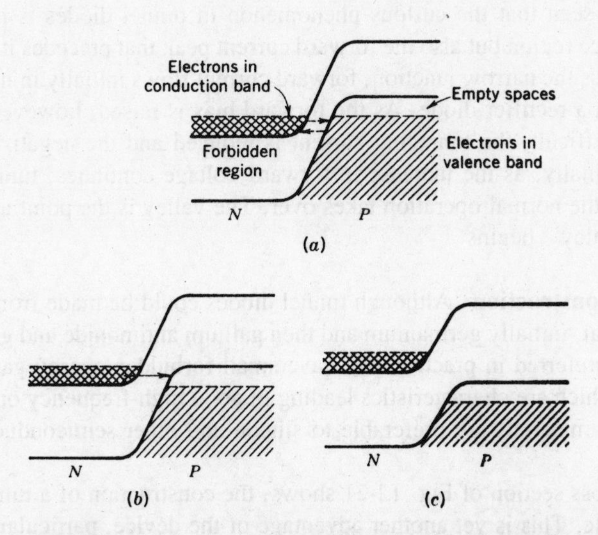

FIGURE 12-20 Energy-level diagrams for tunnel-diode junction at (*a*) zero bias voltage; (*b*) peak voltage; (*c*) valley voltage. *(Courtesy of RCA.)*

[4] Students may check the highly complex mathematical proof of this in Ronald F. Soohoo, *Microwave Electronics*, Addison-Wesley Publishing Co., Reading, Mass., 1971, pp. 184–196. Alternatively, and with a great deal of justification, they may assume that "quantum mechanics" and "magic" are synonymous in this instance!

electrons on either side of the junction are shown by dotted lines. When the bias voltage is zero, these lines are at the same height. Electrons can now tunnel from one side of the junction to the other because of its thinness, but the tunneling currents in the two directions are the same; thus no effective overall current flows. This is shown in Fig. 12-20*a*.

When a small forward bias is applied to the junction, the energy level of the *p* side is lowered (as compared with the *n* side). As shown in Fig. 12-20*b*, electrons are able to tunnel through from the *n* side. This is possible because the electrons in the conduction band there find themselves opposite vacant states on the *p* side. Tunneling in the other direction is not possible, because the valence-band electrons on the *p* side are now opposite the forbidden energy gap on the *n* side. This gap, shown here at its maximum, represents the peak of the diode characteristic.

When the forward bias is raised beyond this point, tunneling will decrease, as may be seen with the aid of Fig. 12-20*c*. The energy level on the *p* side is now depressed further, with the result that fewer *n*-side free electrons are opposite unoccupied *p*-side energy levels. As the bias is raised, forward current drops; this corresponds to the negative-resistance region of the diode characteristic. Finally, as Fig. 12-20*c* shows, a forward bias is reached at which there are no conduction-band electrons opposite valence-band vacant states, and tunneling stops altogether. The point at which this happens is the valley of Fig. 12-19, to which the energy-level diagram of Fig. 12-20*c* corresponds. When forward voltage is increased even further, "normal" forward current flows and increases, as with ordinary rectifier diodes.

It is thus seen that the curious phenomenon in tunnel diodes is not only the negative-resistance region but also the forward current peak that precedes it. As a result of tunneling across the narrow junction, forward current flows initially in much greater quantities than in a rectifier diode. As the forward bias is raised, however, tunneling becomes more difficult, the tunneling current is reduced and the negative-resistance region results. Finally, as the increase in forward voltage continues, tunneling stops completely, and the normal operation takes over. The valley is the point at which this "return to normalcy" begins.

Materials and construction Although tunnel diodes could be made from any semiconductor material, initially germanium and then gallium antimonide and gallium arsenide have been preferred in practice. All have small forbidden energy gaps and high ion mobilities, which are characteristics leading to good high-frequency or high-speed operation. These materials are preferable to silicon and other semiconductors in this regard.

As the cross section of Fig. 12-21 shows, the construction of a tunnel diode is remarkably simple. This is yet another advantage of the device, particularly since the fabrication is also simple. A very small tin dot, about 50 μm in diameter, is soldered or alloyed to a heavily doped pellet (about 0.5 mm square) of *n*-type Ge, GaSb or GaAs. The pellet is then soldered to a Kovar pedestal, used for heat dissipation, which forms the anode contact. The cathode contact is also Kovar, being connected to the tin dot via a mesh screen used to reduce inductance. The diode has a ceramic body and a hermetically sealing lid on top. Note the tiny dimensions of the pill package.

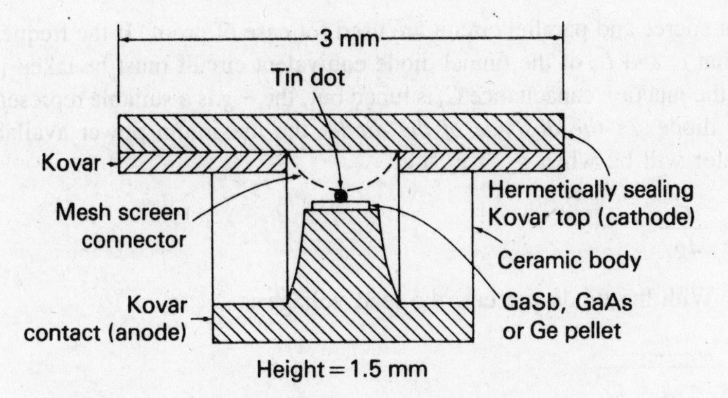

FIGURE 12-21 Construction of typical tunnel diode.

12-5.2 Negative-Resistance Amplifiers

The classical application of the tunnel diode was in microwave oscillators, especially after it was realized that the secret of stable oscillations lay in loosely coupling the diode to its tuned circuit. However, other semiconductor devices have subsequently appeared, producing far more microwave power than the tunnel diode ever could. Consequently, the tunnel diode has been superseded in some of its traditional oscillator applications. Nonetheless, it is important to realize that the tunnel diode is a fully fledged active device, like the transistor, so that amplification may be performed with it. *It will now be used as a vehicle to introduce negative-resistance amplifiers in general!* These are common at microwaves, and indeed negative-resistance parametric amplifiers have already been met.

Theory of negative-resistance amplifiers It can be shown that a circuit incorporating a negative resistance is capable of significant power gain. This is obvious, actually; since negative-resistance oscillators are able to oscillate, it is clear that the negative resistance must be making up all the circuit losses. That is, it feeds power into the circuit, which dissipates some and puts out the rest. This is similar to the feedback oscillator situation, in which βA must at least equal unity, and therefore gain certainly exists. The proof for the tunnel diode now follows, but it is really independent of the particular device used to provide the negative resistance.

Consider the basic negative-resistance amplifier of Fig. 12-22. It consists of an input current source i_s, together with the source conductance g_s, connected to a negative conductance $-g$. Across this the load conductance g_L is also connected. The

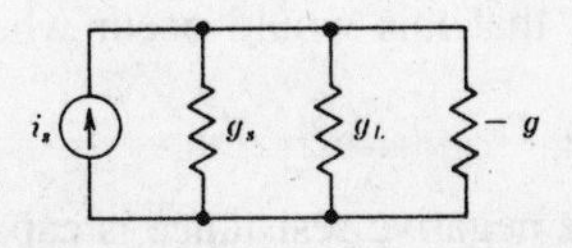

FIGURE 12-22 Basic negative-resistance amplifier.

current source and parallel circuit are used for ease of proof. If the frequency is not so high that r_s and L_s of the tunnel-diode equivalent circuit must be taken into account, and if the junction capacitance C_j is tuned out, the $-g$ is a suitable representation of the tunnel diode. *In the absence of the diode,* the maximum power available from the generator will be when $g_L = g_s$, i.e.,

$$P_{\max} = \frac{i_s^2}{4g_s} \tag{12-4}$$

With the diode present, the load voltage is

$$v_L = \frac{i_s}{g_s - g + g_L} \tag{12-5}$$

Thus the power delivered to the load is

$$P_L = v_L^2 g_L = \frac{g_L i_s^2}{(g_s - g + g_L)^2} \tag{12-6}$$

If the presence of the diode has permitted power gain, the ratio of Eq. (12-5) to Eq. (12-4) is greater than unity. Thus

$$A_P = \frac{P_L}{P_{\max}} = \frac{i_s^2 g_L/(g_s - g + g_L)^2}{i_s^2/4g_s}$$
$$= \frac{4g_s g_L}{(g_s - g + g_L)^2} \tag{12-7}$$

For maximum power transfer, the load and generator conductances are made equal as before. With this new condition we have

$$A_P = \frac{4g_L^2}{(2g_L - g)^2}$$
$$= \frac{4g_L^2}{4g_L^2 - 4g_L g + g^2}$$
$$= \frac{4g_L^2}{4g_L^2 + g(g - 4g_L)} \tag{12-8}$$

Equation (12-8) can obviously be greater than 1, provided that the second term in its denominator is negative, i.e., *provided that* $4g_L$ *is greater than* g. If this applies, A_P exceeds unity, real power gain is available, and the circuit may be used as an amplifier. However, care must be taken to ensure that the denominator of Eq. (12-8) is not reduced to zero, which would happen for a value of g such that the last term of Eq. (12-8) is equal to -1. Simple algebra shows that this would occur when

$$g = 2g_L \qquad (\text{if } g_L = g_s \text{ as before}) \tag{12-9}$$

It is seen that an amplifier containing a negative resistance is capable not only of power gain but also of *infinite gain* (and therefore oscillation). This occurs when Eq. (12-9) holds, and it gives the lower limit for the value of g, and hence the upper limit for the value of the negative resistance. (Note that the lower limit of the negative resistance is governed by the requirement that $4g_L$ must be greater than g.) We have thus proved that the negative-resistance amplifier is capable of power gain if the negative resistance has a value between the limits just described. If it strays outside these limits, either Eq. (12-8) exceeds unity, and therefore power gain is less than 1, or else it becomes negative, and oscillations take place.

Tunnel-diode amplifier theory For frequencies below self-resonance, Eq. (12-7) must be enlarged to include the junction capacitance of the diode. This capacitance is tuned out in an amplifier, but including it yields a useful result. Thus

$$A_P = \frac{4g_s g_L}{(g_s + g_L - g + j\omega C_j)^2} \tag{12-10}$$

This, in turn, gives a resistive cutoff frequency, or figure of merit, for such a diode, which corresponds to the frequency at which the magnitude of ωC_j equals the magnitude of $-g$. Past this frequency, the negative resistance of the tunnel diode disappears. This frequency is given by

$$g = \omega_r C_j$$

$$R = \frac{1}{\omega_r C_j}$$

$$\omega_r = \frac{1}{RC_j}$$

$$f_r = \frac{1}{2\pi RC_j} \tag{12-11}$$

The series diode loss resistance r_s of Fig. 12-18 has been neglected in this derivation, because it is much smaller than the negative resistance (generally being no more than one-tenth of the negative resistance) and thus its effect is very small. An alternative interpretation of Eq. (12-11) is that it represents the gain-bandwidth product of a tunnel-diode amplifier.

12-5.3 Tunnel-Diode Applications

In all its applications, the tunnel diode should be loosely coupled to its tuned circuit. With lumped components, this is done by means of a capacitive divider, with the diode connected to a tapping point, while the divider is across the tuned circuit itself. In a cavity, the diode is placed at a point of significant, but not maximum, coupling. The other point of significance is the application of dc bias. This must be connected to the diode without interfering with the tuned circuit. The simplest way of doing this is with a filter, as shown in Fig. 12-23. Basically, this filter prevents the diode from being short-circuited by the supply source, while ensuring that no positive resistance is added to interfere with the negative resistance of the diode. Also, the addition of capacitance across the diode is avoided. Care must be taken to ensure that the bias inductance does not introduce spurious frequencies in the passband.

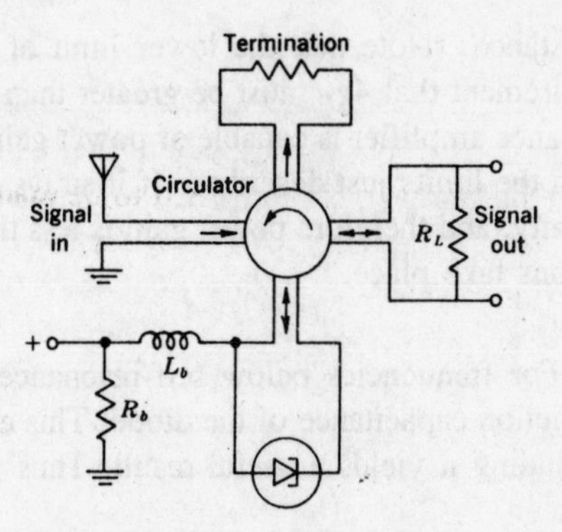

FIGURE 12-23 Tunnel-diode amplifier with circulator. ***(Based on a figure from "Tunnel Diodes," by courtesy of RCA.)***

Amplifiers As shown in Fig. 12-23, the tunnel-diode amplifier (TDA), like the parametric amplifier, requires a circulator to separate the input from the output. Indeed, their layouts are very similar, with the very significant difference that no pump source is required for the TDA.

Tables 12-1 and 12-2 show a number of low-noise microwave amplifier performance figures, including those of tunnel-diode amplifiers. It is seen that the tunnel diode is a low-noise device. The twin reasons for this are the low value of the parasitic resistance r_s (producing low thermal noise) and the low operating current (producing low shot noise). In such low-noise company, TDAs are as broadband as any, are very small and simple and have output levels on a par with paramps and masers. The available gains are high, and operating frequencies in excess of 50 GHz have been reported.

Amplifier applications Tunnel-diode amplifiers may be used throughout the microwave range as moderate-to-low-noise preamplifiers in all kinds of receivers. However, GaAs FET amplifiers are more likely to be used in current equipment up to 18 GHz. Large bandwidths and high gains are available from multistage amplifiers, the circuits and power requirements are very simple (typically a few milliamperes at 10 V dc), and noise figures below 5 dB are possible well above X band. Finally, it is worth noting that TDAs are immune to the ambient radiation encountered in interplanetary space, and so are practicable for space work.

Other applications Because tunnel diodes are *diodes*, they may be used as mixers. Being also capable of active oscillation, they may be used as self-excited mixers, in a manner similar to the transistor mixer of Sec. 6-2.2. Being high-speed devices, tunnel diodes also lend themselves to high-speed switching and logic operations, as flip-flops and gates. Finally, they are used as low-power oscillators up to about 100 GHz, because of their simplicity, frequency stability and immunity to radiation [18].

12-6 GUNN EFFECT AND DIODES

It was being prophetically said in the early 1960s that as long as the important properties of semiconductors depended on junctions—and these had to be made thinner and thinner as frequency was raised—then high powers from such semiconductor devices were simply not possible at microwave frequencies, certainly not above X band. Fortunately, since then other devices have been invented or adapted which do not depend on junction behavior and which are capable of producing adequate microwave powers. One of these classes of devices depends on controlled avalanche and takes into account the transit time in its operation. This *Avalanche and Transit Time* family is treated in Sec. 12-7. The other class of devices exhibits microwave power properties that depend on the behavior of *bulk* semiconductors, rather than junctions. The *Gunn effect* is the main representative of this class of devices and is now treated.

12-6.1 Gunn Effect

In 1963, Gunn [19] discovered the *transferred electron* effect which now bears his name. This effect is instrumental in the generation of microwave oscillations in bulk semiconductor materials. The effect was found by Gunn to be exhibited by *gallium arsenide* and *indium phosphide,* but *cadmium telluride* and *indium arsenide* have also subsequently been found to possess it. Gunn's discovery was a breakthrough of great importance. It marked the first instance of useful semiconductor device operation depending on the *bulk properties* of a material.

Introduction If a relatively small dc voltage is placed across a thin slice of gallium arsenide, such as the one shown in Fig. 12-24, then negative resistance will manifest itself under certain conditions. Basically, these consist merely of ensuring that the voltage gradient across the slice is in excess of about 3300 V/cm. Oscillations will then occur if the slice is connected to a suitably tuned circuit. It is seen that the voltage gradient across the slice of GaAs is very high. Hence the electron velocity is also high, so that oscillations will occur at microwave frequencies.

It must be reiterated that the Gunn effect is a *bulk* property of semiconductors and does not depend, as do other semiconductor effects, on either junction or contact properties. As established painstakingly by Gunn, the effect is independent of *total*

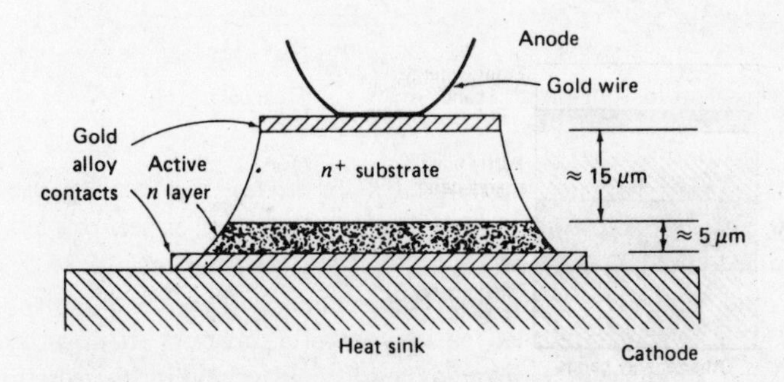

FIGURE 12-24 Epitaxial GaAs Gunn slice.

voltage or current and is not affected by magnetic fields or different types of contacts. Furthermore, it occurs in *n-type* materials *only,* so that it must be associated with electrons rather than holes. Having determined that the voltage required was proportional to the sample length, the inventor concluded that the electric field, in volts per centimeter, was the factor determining the presence or absence of oscillations. He also found that a threshold value of 3.3 kV/cm must be exceeded if oscillations are to take place. Finally, he found that the frequency of the oscillations produced corresponded closely to the time that electrons would take to traverse such a slice of *n*-type material as a result of the voltage applied. This suggests that a bunch of electrons, here called a *domain,* is formed somehow, occurs once per cycle and arrives at the positive end of the slice to excite oscillations in the associated tuned circuit.

Negative resistance Although the device itself is very simple, its operation (as might be suspected) is not quite so simple. Gallium arsenide is one of a fairly small number of semiconductor materials which, in an *n*-doped sample, have an empty energy band higher in energy than the highest filled (or partly filled) band. Furthermore, the size of the forbidden gap between these two is relatively small. This does not apply to some other semiconductor materials, such as silicon and germanium. The situation for gallium arsenide is illustrated in Fig. 12-25, in which the highest levels shown also have the highest energies.

When a voltage is applied across a slice of GaAs which is doped so as to have excess electrons (i.e., *n*-type), these electrons flow as a current toward the positive end of the slice. The greater the potential across the slice, the higher the velocity with which the electrons move toward the positive end, and therefore the greater the current. Thus far, the device is behaving as a normal positive resistance. In other diodes, the component of velocity toward the positive end, imparted to the electrons by the applied voltage, is quite small compared to the random thermal velocity that these electrons possess. In this case, however, so much energy is imparted to the electrons by the extremely high voltage gradient that instead of traveling faster and therefore constituting a larger current, their flow actually slows down. This is because such electrons have acquired enough energy to be transferred to the higher energy band, which is normally empty, as shown in Fig. 12-25. This gives rise to the name *transferred-electron* effect, which is often given to this phenomenon. *Electrons have thus been trans-*

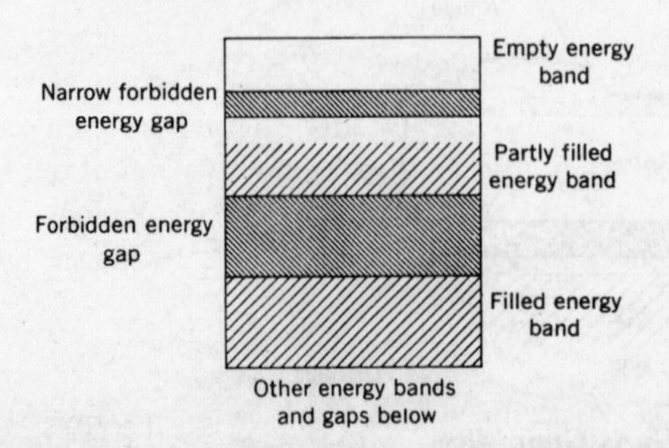

FIGURE 12-25 Important energy levels in gallium arsenide.

ferred from the conduction band to a higher energy band in which they are much less mobile, and thus the current has been reduced as a result of a voltage rise. Note that, in a sense, gallium arsenide is a member of a group of unusual semiconductor substances. In a lot of others, the energy required for this transfer of electrons would be so high, because of a higher forbidden energy gap, that the complete crystal structure might be distorted or even destroyed by the high potential gradient before any transfer of electrons could take place.

It is thus seen that as the applied voltage rises past the *threshold negative-resistance value,* current falls, and thus the classical case of negative resistance is exhibited. All good things must come to an end, unfortunately. Eventually the voltage across the slice becomes sufficient to remove electrons from the higher-energy, lower-mobility band, so that current will increase with voltage once again. The voltage-current characteristic of such a slice of gallium arsenide is seen to be very similar to that of a tunnel diode, but for vastly different reasons.

Gunn domains It was stated in the preceding section that the oscillations observed in the initial GaAs slice were compatible with the formation and transit time of electron bunches. It follows, therefore, that the negative resistance just described is not the only effect taking place. The other phenomenon is the formation of *domains,* the reasons for which may now be considered.

It is reasonable to expect that the density of the doping material is not completely uniform throughout our sample of gallium arsenide. Hence it is entirely possible that there will be a region, perhaps somewhere near the negative end, where the impurity concentration is less than average. In such an area there are fewer free electrons than in other areas, and therefore this region is less conductive than the others. As a result of this, there will be a greater than average potential across it. Thus, as the total applied voltage is increased, this region will be the first to have a voltage across it large enough to induce transfer of electrons to the higher energy band. In fact, such a region will have become a *negative-resistance domain.*

A domain like this is obviously unstable. Electrons are being taken out of circulation at a fast rate within it, the ones behind bunch up and the ones in front travel forward rapidly. In fact, the whole domain moves across the slice toward the positive end with the same average velocity as the electrons before and after it, about 10^7 cm/s in practice. Note that such a domain is self-perpetuating. As soon as some electrons in a region have been transferred to the less conductive energy band, fewer free electrons are left behind. Thus this particular region becomes less conductive, and therefore the potential gradient across it increases. The domain is quite capable of traveling and may be thought of as a low-conductivity, high-electron-transfer region, corresponding to a negative pulse of voltage. When it arrives at the positive end of the slice, a pulse is received by the associated tank circuit and shocks it into oscillations. It is actually this arrival of pulses at the anode, rather than the negative resistance proper, which is responsible for oscillations in Gunn diodes.[5]

[5] The term *diode* is a misnomer for Gunn devices since there is no junction, nor is rectification involved. The device is called a *diode* because it has two terminals, and the name is also convenient because it allows the use of *anode* for the "positive end of the slice."

With the usual applied voltages, once a domain forms, insufficient potential is left across the rest of the slice to permit another domain to form. This assumes that the sample is fairly short; otherwise the situation can become very complex, with the possibility that other domains may form.[6] When the domain in a short sample arrives at the anode, there is once again sufficient potential to permit the formation of another domain somewhere near the cathode. It is seen that only one domain, or pulse, is formed per cycle of RF oscillations, and so energy is received by the tank circuit in correct phase to permit the oscillations to continue.

Miscellaneous considerations The following brief notes and observations are now made:

1. The Gunn diode/oscillator has received various names. Although *Gunn* (diode or oscillator) will always be used here, students should be aware of names such as *transferred-electron devices* (or *TED*) and *transferred-electron oscillator* (or *TEO*), which are also frequently used.
2. The Gunn effect is eminently viable, and Gunn diodes are very widely used.
3. Although only gallium arsenide diodes have thus far been considered, indium phosphide (InP) Gunn diodes are becoming widely used, especially at the highest frequencies. InP has properties quite similar to those of GaAs, while also offering the advantages of a higher peak-to-valley ratio in its negative resistance characteristic and lower noise. Students are referred to Fank, et al. [20] for a comparison of GaAs and InP as microwave semiconductor materials.
4. Students must surely be wondering by now why such ''oddball'' operating principles are used by the majority of solid-state microwave devices—and they have not even encountered the IMPATT diode as yet! The only explanation the author can put forward is that all the simple devices got themselves invented a long time ago, and *they* didn't work at microwave frequencies!

12-6.2 Gunn Diodes and Applications

Gunn diodes A practical Gunn diode consists of a slice like the one shown in Fig. 12-24, sometimes with a buffer layer between the active layer and the substrate, mounted in any of a number of packages, depending on the manufacturer, the frequency and the power level. Encapsulation identical to that shown for varactor diodes in Fig. 12-9 is common. As it happens, the power that must be dissipated is quite comparable.

Gunn diodes are grown epitaxially out of GaAs or InP doped with silicon, tellurium or selenium. The substrate, used here as an ohmic contact, is highly doped for good conductivity, while the thin active layer is less heavily doped. The gold alloy

[6] The domain described is sometimes called a *dipole domain*. An *accumulation domain* may also occur (particularly in a longer sample), where a more highly doped region is involved, and a current accumulation travels toward the anode.

contacts are electrodeposited and used for good ohmic contact and heat transfer for subsequent dissipation. Diodes have been made with active layers varying in thickness from 40 to about 1 μm at the highest frequencies. The actual structure is normally square, and so far GaAs diodes predominate commercially. For further descriptions of the Gunn diode and its manufacture, students are referred to [21] and [22].

Diode performance As a good approximation, the equivalent circuit of a GaAs X-band Gunn diode consists of a negative resistance of about 100 ohms (100 Ω) in parallel with a capacitance of about 0.6 pF. Such a commercial diode will require a 9-V dc bias, and, with an operating current of 950 mA, the dissipation in its (cathode) heat sink will be 8.55 W. Given that the output (anywhere in the range 8 to 12.4 GHz) is 300 mW, the efficiency is seen to be 3.5 percent. A higher-frequency Gunn diode, operating over the range of 26.5 to 40 GHz, might produce an output of 250 mW with an efficiency of 2.5 percent.

Overall, GaAs Gunn diodes are available commercially for frequencies from 4 GHz (1 to 2 W CW maximum) to about 100 GHz (50 mW CW maximum). Over that range, the maximum claimed efficiencies drop from 20 to about 1 percent, but for most commercial diodes 2.5 to 5 percent is normal. InP diodes, not yet as advanced commercially, have a performance that ranges from 500 mW CW at 45 GHz (efficiency of 6 percent) to 100 mW CW at 90 GHz (efficiency of 4.5 percent); higher powers and operating frequencies are expected to be forthcoming [23]. Other options available include two or more diodes in one oscillator package for higher CW outputs, and diodes for pulsed outputs. In the latter case, commercial diodes produce up to a few dozen watts pulsed, with 1 percent duty cycles and efficiencies somewhat better than for CW diodes. See also [23].

Gunn oscillators Since the Gunn diode consists basically of a negative resistance, all that is required in principle to make it into an oscillator is an inductance to tune out the capacitance, and a shunt load resistance not greater than the negative resistance. This has already been discussed in conjunction with the tunnel diode. In practice, however, a coaxial cavity operating in the TEM mode has been found the most convenient for fixed frequency (but with some mechanical tuning) operation. A typical coaxial Gunn oscillator is shown in Fig. 12-26. If some electrical tuning is required as well, a varactor may be placed in the cavity, at the opposite end to the Gunn diode. The dimensions shown in Fig. 12-26 are selected to provide suitable diode mounting and dissipation, as well as freedom from spurious mode oscillations. Further information on the factors involved is available in [24].

YIG-tuned (see Sec. 10-5.2) Gunn VCOs are available for instrument applications, featuring frequency ranges as large as 2 octaves, much greater than is possible with varactors. A selection of such oscillators is shown in Fig. 12-27. The 300-g, 50 × 50 mm package contains a Gunn slice on a heat sink, and a cavity with a small YIG sphere. There is a heater for the YIG sphere, to keep it at a constant temperature, and a coil for altering the magnetic field. The instantaneous frequency of oscillation is governed by the cavity frequency, which in turn depends on the YIG sphere and the magnetic field by which it is surrounded. It is the Gunn diode, rather than the tuning

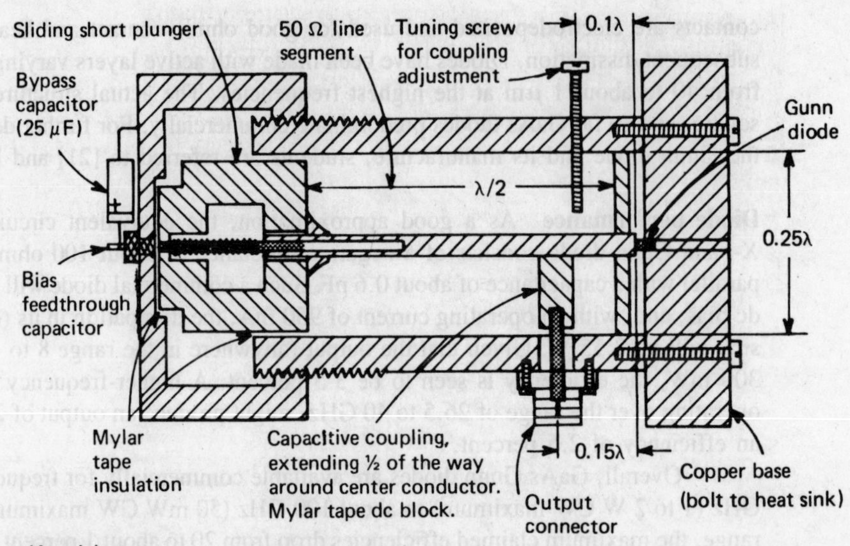

FIGURE 12-26 Cross section of typical Gunn coaxial oscillator cavity. ***(Courtesy of Microwave Associates International.)***

mechanism, that determines the frequency limits.[7] Avantek oscillators of the type shown in Fig. 12-27 cover the range from 1 to 12.4 GHz, with typically octave tuning (or sweeping) ranges for each. Frequency modulation is also possible, via the terminals provided, and in all very rapid frequency changes can be made. Such VCOs are designed as backward-wave oscillator replacements, certainly at the lower end of the BWO's operating spectrum. Typical power outputs range up to 50 mW, and total power consumption may be 5 W, including power for the YIG sphere. See also [24] for details of a YIG-tuned sweeper.

Finally, it should be mentioned that the noise performance of Gunn oscillators is quite acceptable. Spurious AM noise is on par with that of the klystron (which itself is very good), while spurious FM noise is worse, but not too high for normal applications. Injection locking with a low-amplitude, high-stability signal helps to reduce FM noise quite significantly.

Gunn diode amplifiers As was shown in connection with the tunnel diode, a device exhibiting negative resistance may be used as an amplifier, and of course the Gunn diode qualifies in this respect. However, Gunn diode amplifiers are not used nearly as much as Gunn oscillators. The reasons are many. On the one hand, Gunn diode ampli-

[7] When the frequency of the resonator is changed, the diode itself responds by generating its domain at a distance from the anode such that the transit time of the domain corresponds to a cycle of oscillations. Thus, as frequency is raised, the formation point of the domain moves closer to the anode. The oscillations eventually stop when this point is more than halfway across the slice.

FIGURE 12-27 YIG-tuned voltage-controlled oscillators. *(Courtesy of Avantek, Inc.)*

fiers cannot compete for power output and low noise with GaAs FET amplifiers at frequencies below about 30 GHz, and at higher frequencies they cannot compete with the power output or efficiency of electron tube or IMPATT (see next section) amplifiers. Accordingly, the niche which is left for them is as low- to medium-power medium-noise amplifiers in the 30- to 100-GHz frequency range. Over that range, they are capable of amplifying with noise figures of the order of 20 to 30 dB, relatively low efficiency and power gain per stage, and an output power that is perhaps two to four times as high would be expected from a single-diode oscillator (this is achieved by combining the output of several diodes in the final stage). One avenue of approach for improvement is to use a hybrid tunnel diode-Gunn diode amplifier, in which the tunnel diode input stages significantly reduce the noise figure. Noting that the foregoing applies to gallium arsenide diodes, another avenue of approach is to use indium phosphide devices. The early results with InP Gunn diodes are most encouraging, with noise figures as low as 12 dB reported for amplifiers in the 50- to 60-GHz range [23].

Finally, it should be said that, for reasons identical to those applying to YIG-tuned Gunn oscillators, Gunn amplifiers, be they GaAs or InP, are capable of broadband operation, 2:1 bandwidth ranges being not unusual. In this respect, they are greatly superior to IMPATT amplifiers.

Gunn diode applications Having taken the microwave world more or less by storm, Gunn diode oscillators are widely used and also intensely researched and developed. They are employed frequently as low- and medium-power oscillators in microwave receivers and instruments. The majority of parametric amplifiers now use Gunn diodes as pump sources. They have the advantage over IMPATT diodes (as will be seen in the next section) of having much lower noise, this being an important criterion in the selection of a pump oscillator. Where very high pump frequencies are required, the technique of using a lower-frequency Gunn oscillator and doubling the frequency with a varactor multiplier is often used.

The higher-power Gunn oscillators (250 to 2000 mW) are used as power output oscillators, generally frequency-modulated, in a wide variety of low-power transmitter applications. These currently include police radar, CW Doppler radar (see Chap. 16), burglar alarms and aircraft rate-of-climb indicators. For information on a host of other applications, students are also referred to [2] of Chap. 11.

12-7 AVALANCHE EFFECTS AND DIODES

In 1958, Read [26] at Bell Telephone Laboratories proposed that the delay between voltage and current in an avalanche, together with transit time through the material, could make a microwave diode exhibit negative resistance. Because of fabrication difficulties and the large amounts of heat that would have to be dissipated, such a diode was not produced until 1965, by Johnston and associates [27] at the same laboratories. The diode was subsequently called the *IMP*act *A*valanche and *T*ransit *T*ime (IMPATT) diode. Two years later, at RCA Laboratories this time, a method of operating the IMPATT diode that seemed anomalous at the time was discovered by Prager et al. [28]. This device, now called the *TRA*pped *P*lasma *A*valanche *T*riggered *T*ransit (TRAPATT) diode, also exhibits negative resistance and holds out a promise of high pulsed powers at the lower microwave frequencies. Both diodes will now be studied.

12-7.1 IMPATT Diodes[8]

Introduction It was shown in Sec. 12-5.1 that the tunnel diode has a *dynamic dc negative resistance*. This meant that, over a certain range, current decreased with an increase in voltage, and vice versa.[9] This particular point was pursued no further, it being taken for granted that any device which exhibits a dynamic negative resistance *for direct current* will also exhibit it *for alternating current*. That is to say, if an alternating voltage is applied, current will rise when voltage falls, at an ac rate. We may thus now redefine negative resistance as *that property of a device which causes the current through it to be* 180° *out of phase with the voltage across it*. The point is important here, because this is the only kind of negative resistance exhibited by the IMPATT diode. One hastens to add that such a negative resistance is quite sufficient. It would not have mattered if the tunnel diode, for instance, had only this kind of negative resistance (without exhibiting it for dc voltage or current variations)—after all, the oscillations are ac. To summarize: if it can be shown that the voltage current in the IMPATT diode are 180° out of phase, negative resistance in this device will have been proved.

[8] The author is indebted to Dr. Ir. Marinus T. Vlaardingerbroek of the Philips Research Laboratories, Eindhoven, the Netherlands, for the wealth of material provided on the IMPATT diode and also other solid-state microwave devices.

[9] No device known to the author has a *static* negative resistance, i.e., voltage is applied one way, and current flows the other way.

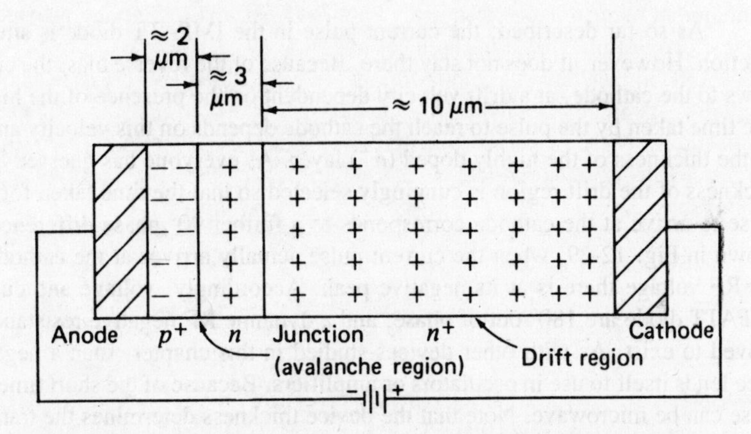

FIGURE 12-28 IMPATT diode (single-drift) schematic diagram.

IMPATT diode A combination of delay involved in generating avalanche current multiplication, together with delay due to transit time through a drift space, provides the necessary 180° phase difference between applied voltage and the resulting current in an IMPATT diode. The cross section of the active region of this device is shown in Fig. 12-28. Note that it *is* a diode, the junction being between the p^+ and the *n* layers.

An extremely high voltage gradient is applied to the IMPATT diode, of the order of 400 kV/cm, eventually resulting in a very high current. A normal diode would very quickly break down under these conditions, but the IMPATT diode is constructed so as to be able to withstand such conditions repeatedly.[10] Such a high potential gradient, back-biasing the diode, causes a flow of minority carriers across the junction. If it is now assumed that oscillations exist, we may consider the effect of a positive swing of the RF voltage superimposed on top of the high dc voltage. Electron and hole velocity has now become so high that these carriers form additional holes and electrons by knocking them out of the crystal structure, by so-called *impact ionization*. These additional carriers continue the process at the junction, and it now snowballs into an avalanche. If the original dc field was just at the threshold of allowing this situation to develop, this voltage will be exceeded during the whole of the positive RF cycle, and avalanche current multiplication will be taking place during this entire time. However, *since it is a multiplication process, avalanche is not instantaneous*. Indeed, as shown in Fig. 12-28, the process takes a time such that the current pulse maximum, at the junction, occurs at the instant when the RF voltage across the diode is zero and going negative. A 90° phase difference between voltage and current has thus been obtained.

[10] A normal diode breaks down under avalanche conditions because of the enormous powers generated. Consider that the thickness of an IMPATT diode's active region is a few micrometers, to ensure the correct transit time for microwave operation. Its cross-sectional area is similarly tiny, to ensure a small capacitance. With the high-voltage gradient and resulting high current, the power being generated is of the order of *100 MW/cm³*. The delay between the proposal for the IMPATT diode and its first realization was due in no small measure to the problems involved in dissipating such vast amounts of heat. This had to be done, to ensure a satisfactorily low operating temperature for the IMPATT diode, so that it would not be destroyed by melting. Typical operating temperatures of commercial diodes are of the order of 250°C.

As so far described, the current pulse in the IMPATT diode is situated at the junction. However, it does not stay there. Because of the reverse bias, the current pulse flows to the cathode, at a drift velocity dependent on the presence of the high dc field. The time taken by the pulse to reach the cathode depends on this velocity and of course on the thickness of the highly doped (n^+) layer. As everyone has guessed by now, the thickness of the drift region is cunningly selected so that the time taken for the current pulse to arrive at the cathode corresponds to a further 90° phase difference. Thus, as shown in Fig. 12-29, when the current pulse actually arrives at the cathode terminal, the RF voltage there is at its negative peak. Accordingly, voltage and current in the IMPATT diode are 180° out of phase, and a dynamic RF negative resistance has been proved to exist. As with other devices studied in this chapter, such a negative resistance lends itself to use in oscillators or amplifiers. Because of the short times involved, these can be microwave. Note that the device thickness determines the transit time, to which the IMPATT diode is very sensitive. Accordingly, and unlike the Gunn diode, the IMPATT diode is essentially a narrowband device (especially when used in an amplifier). For further information on the operation and fabrication of the IMPATT diode, [29] and [30] are recommended.

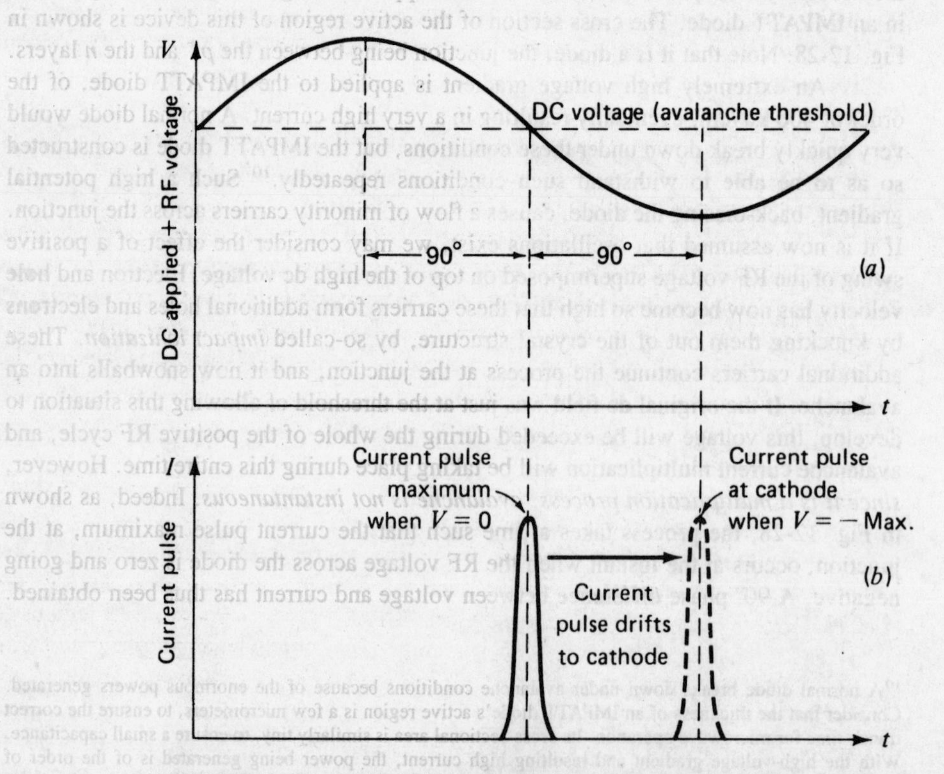

FIGURE 12-29 IMPATT diode behavior. (*a*) Applied and RF voltage; (*b*) resulting current pulse and its drift across diode. (*Note:* Relative size of RF voltage exaggerated.)

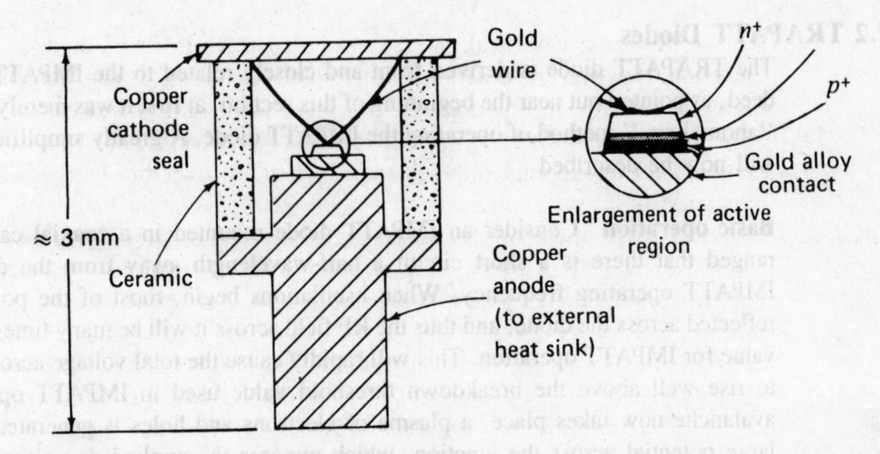

FIGURE 12-30 Typical IMPATT diode.

Practical considerations Commercial IMPATT diodes have been available for quite some time. They are made of either silicon, gallium arsenide or even indium phosphide. The diodes are mostly mesa, and epitaxial growth is used for at least part of the chip; some have Schottky barrier junctions. Gallium arsenide is theoretically preferable [30] and should give lower noise, higher efficiencies and higher maximum operating frequencies. However, silicon is cheaper and easier to fabricate. Accordingly, silicon IMPATT diodes, which came first, are even now preferred for many applications; indeed, it is silicon diodes that currently provide the highest output powers at the highest operating frequencies (in excess of 200 GHz).

The IMPATT diode shown in Fig. 12-30 is a typical commercial diode for use below about 50 GHz and could house either a GaAs or an Si chip. At higher frequencies, beam-lead packages almost identical in appearance to those shown in Fig. 12-5 tend to be preferred. The construction is deceptively simple. However, a lot of thought and development has gone into its manufacture, particularly the contacts, which must have extremely low ohmic and thermal resistance. Additionally, in a practical circuit, the IMPATT diode is generally embedded in the wall of a cavity, which then acts as an external heat sink.

Until a few years ago, practical IMPATT diodes were unlike Read's original proposal. This called for a double-drift region, whereas Figs. 12-28 and 12-30 show diodes with single- (n^+) drift regions. The reason for the initial departure from what was theoretically a higher-efficiency structure was difficulty in fabrication, but this problem has now been solved. For some years IMPATT diodes with two drift regions (one n^+ and the other p^+) have been made commercially. In the manufacturing process, an n layer is epitaxially grown on an n^+ substrate. The p layer is then grown epitaxially or by ion implantation, and finally the p^+ layer is formed by diffusion. These p^+-p-n-n^+ devices were at first known as RIMPATT (Read-IMPATT) diodes, but they are now commonly known as double-drift IMPATT diodes. They are undoubtedly the versions used at the highest frequencies and for the highest output powers, as will be seen.

12-7.2 TRAPATT Diodes

The TRAPATT diode is derived from and closely related to the IMPATT diode. Indeed, as pointed out near the beginning of this section, at first it was merely a different, "anomalous," method of operating the IMPATT diode. A greatly simplified operation will now be described.

Basic operation Consider an IMPATT diode mounted in a coaxial cavity, so arranged that there is a short circuit a half-wavelength away from the diode at the IMPATT operating frequency. When oscillations begin, most of the power will be reflected across the diode, and thus the RF field across it will be many times the normal value for IMPATT operation. This will rapidly cause the total voltage across the diode to rise well above the breakdown threshold value used in IMPATT operation. As avalanche now takes place, a plasma of electrons and holes is generated, placing a large potential across the junction, which opposes the applied dc voltage. The total voltage is thereby reduced, and the current pulse is trapped behind it. When this pulse travels across the n^+ drift region of the semiconductor chip, the voltage across it is thus much lower than in IMPATT operation. This has two effects. The first is a much slower drift velocity, and consequently longer transit time, so that for a given thickness the operating frequency is several times lower than for corresponding IMPATT operation. The second point of great interest is that, when the current pulse does arrive at the cathode, the diode voltage is much lower than in an IMPATT diode. Thus dissipation is also much lower, and efficiency much higher. The operation is akin to class C, and indeed the TRAPATT diode lends itself to pulsed instead of CW operation. It must be reiterated that the foregoing is a much simplified description of the operating mechanism. A more detailed explanation may be obtained from [31].

Practical considerations Although they were first proposed in the early 1970s, commercial TRAPATT diodes are only now becoming commercially available. They tend to be planar silicon diodes, with structures corresponding to those of IMPATT diodes but with gradual, rather than abrupt, changes in doping level between the junction and the anode. Furthermore, they are likely to use complementary n^+-p-p^+ structures as shown in Fig. 12-31, instead of the p^+-n-n^+ IMPATT chip of Fig. 12-28, for reasons

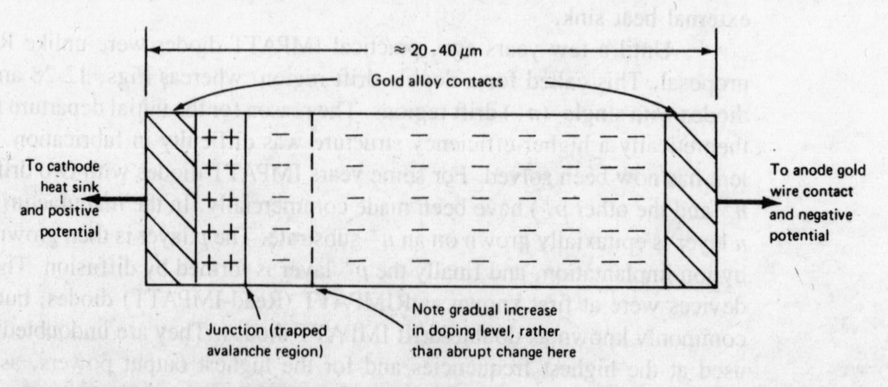

FIGURE 12-31 TRAPATT diode schematic.

of better dissipation. The two figures should be examined in conjunction with each other.

Because the drift velocity in a TRAPATT diode is much less than in an IMPATT diode, either operating frequencies must be lower or the active regions must be made thinner. In fact, both these considerations are borne out by results thus far obtained. On the one hand, most good experimental TRAPATT results have been for frequencies under 10 GHz, and on the other hand, it has been found that by the time 5 GHz is reached, the width of the depletion layer is only 2 μm. However, since the TRAPATT pulse is rich in harmonics, amplifiers or oscillators can be designed to tune to these harmonics, and operation above X band in this manner is possible. For further details of TRAPATT operation, see [31].

12-7.3 Performance and Applications of Avalanche Diodes

IMPATT diode performance Commercial diodes are currently produced over the frequency range from 4 to about 200 GHz, over which range the maximum output power per diode varies from nearly 20 W to about 50 mW. This means that, above about 20 GHz, the IMPATT diode produces a higher CW power output per unit than any other semiconductor device. Typical efficiency is about 10–20 percent up to 40 GHz, reducing to 1 percent as frequency is raised to 200 GHz. Several diodes' outputs may be combined (as indicated in [32]), giving a significantly greater output. Pulsed powers are generally one magnitude higher. Note that the above figures, for the most part, are for single-drift diodes.

Laboratory devices have produced as much as 30 W CW at 12 GHz, 300 mW at 140 GHz and 75 mW at 220 GHz, with one laboratory reporting 1 mW CW at over 300 GHz. Pulsed powers similarly range from about 50 W at 10 GHz to 3 W at 140 GHz. However, experimental results should be taken with a grain of salt. What is often reported is the best result obtained from several specially made diodes. What is often not reported is that maximum efficiency need not coincide with maximum output power or that a diode died of thermal runaway soon after the experiment. On the other hand, it should be noted that results being currently obtained from double-drift IMPATT diodes augur well for the device, especially as regards efficiency, for which figures in excess of 20 percent are being consistently reported, together with higher powers at the highest frequencies.

The biggest problem of IMPATT operation is noise. Avalanche is a very noisy process, and the high operating current helps the generation of shot noise. Thus IMPATT diode oscillators are not as good as either klystrons or Gunn diodes for spurious AM or FM noise, by quite a significant margin. When used as amplifiers, IMPATT diodes produce noise figures of the order of 30 dB, not as good as TWT amplifiers.

IMPATT oscillators and amplifiers The dynamic impedance of an IMPATT diode is $-10\ \Omega$ in parallel with 1 pF, as a good approximation. Like the Gunn diode, therefore, it has a negative resistance which must be placed in a low-impedance environment. Figure 12-32 shows a suitable arrangement. The IMPATT diode is located at

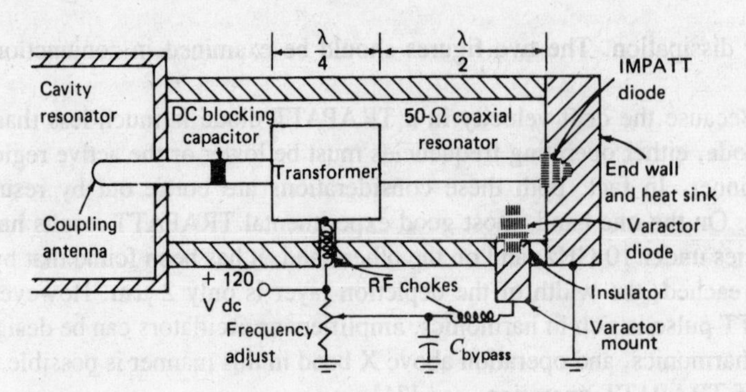

FIGURE 12-32 IMPATT diode oscillator with varactor electronic tuning.

the end of the center conductor in a low-impedance coaxial resonator, and a quarter-wave transformer is used to step up the impedance seen at its point of connection. Oscillations are basically at the frequency at which the length of the coaxial resonator is a half-wave, but this is influenced by the capacitance of the varactor diode. This diode is used for tuning, with its capacitance varied by a change in the applied bias. Frequency modulation could be achieved in exactly the same manner. Typical frequency variation is a few hundred megahertz at 10 GHz. Because of their close dependence on transit time through the entire drift space, IMPATT diodes do not lend themselves to tuning over nearly as wide a frequency range as Gunn diodes. Consequently YIG tuning is not used, since varactors match IMPATTs in that regard.

IMPATT diode amplifiers are available with outputs similar to those of oscillators at about the same frequency range. They are comparable to Gunn diode amplifiers in that they also require circulators, but efficiencies for Gunn amplifiers (up to 10 percent) and power outputs are much higher. Gain is similarly 6 to 10 dB per stage, and bandwidths are up to about 10 percent of the center frequency. Higher frequencies of operation, to over 100 GHz, are another attraction, but noise is still a problem. Additional information on various aspects of IMPATT oscillators and amplifiers is available from [29] to [32].

Performance of TRAPATT oscillators and amplifiers As was explained in a preceding section, TRAPATT operation requires a large RF voltage swing, the kind unlikely to be obtained from switching transients. It seems that TRAPATT oscillators most probably start in the IMPATT mode, then switch over when oscillations have built up sufficiently. The circuit must thus be arranged to permit this to happen. However, no such difficulties are encountered with TRAPATT amplifiers, where an adequately large signal is present, being the input. Another practical point which must be taken into account is the extreme TRAPATT sensitivity to harmonics. Thus, when operating in the fundamental mode, care must be taken to ensure that the second, third and even fourth harmonics cannot be maintained in the tuned circuit. In [31] several TRAPATT circuits and their operational aspects are discussed.

Most TRAPATT oscillators and amplifiers are still in the laboratory stage. However, such impressive results have been obtained that it should not be long before units are available commercially, now that the initial difficulties in establishing and maintaining coherent oscillations seem to have been overcome [31]. With typical duty cycles of the order of 0.1 percent, pulsed powers as high as 200 W at 3 GHz have been produced, with efficiencies in excess of 30 percent. At about 1 GHz, pulsed output powers of 600 W and (independently) an efficiency of 75 percent have been obtained. These figures are for oscillators, but amplifier figures should be comparable. As previously discussed, performance above X band is not very impressive, because of the mode of operation. However, it should be mentioned that many microwave operations take place at X band or below.

Applications of avalanche diodes IMPATT diodes are more efficient and more powerful than Gunn diodes. However, they have not replaced Gunn diodes, and the reason is mainly their noise and the higher supply voltages needed. It also happens that the majority of low-power microwave oscillator applications can be adequately covered by Gunn diodes, except at the highest frequencies, where they are no match for IMPATTS. However, with the current development in IMPATT and TRAPATT diodes proceeding apace, their use in practical systems is wide and increasing, but they are taking over from low- and medium-power tubes, rather than Gunn diodes. For example, most parametric amplifier designers do not want IMPATTS, because of noise. However, long-distance communications carriers are replacing many of their TWT transmitters with IMPATT ones in microwave links in the large field covered by powers under 10 W. IMPATTs can also eventually replace BWOs and low-power CW magnetrons in several types of CW radar and electronic countermeasures. Finally, when commercial TRAPATT oscillators and amplifiers can produce several hundred watts pulsed, with efficiencies in excess of 30 percent and duty cycles close to 1 percent, a very wide pulsed radar field will be open to them. The first applications here are likely to be in airborne and marine radars.

12-8 OTHER MICROWAVE DIODES

Having treated in detail the microwave "active" diodes, we are now left with some "passive" microwave diodes to consider. They are passive only to the extent that they are not used in power generation or amplification; apart from that, they are very active indeed in mixers, detectors and power control. The devices in question are the *PIN*, *Schottky-barrier* and *backward* diodes, which will now be looked at in turn.

12-8.1 PIN Diodes

The PIN diode consists of a narrow layer of *p*-type semiconductor separated from an equally narrow layer of *n*-type material by a somewhat thicker region of *intrinsic* material. In actual fact, the intrinsic layer is a lightly doped *n*-type semiconductor. The name of the diode is derived from the construction (*p*-intrinsic-*n*). Although gallium arsenide is used in the construction of PIN diodes, silicon tends to be the main material. The reasons for this are easier fabrication, higher powers handled and higher resistivity of intrinsic region. The PIN diode is used for microwave power switching, limiting and modulation. It was first proposed by R. N. Hall in 1952, and its potential as a microwave switch was first recognized by Uhlir [33] in 1958.

Construction The construction of the PIN diode is shown in Fig. 12-33. The advantage of the planar construction is the lower series resistance while conducting. Encapsulation for such a chip takes any of the forms already shown for other microwave diodes. The in-line construction has a number of advantages, including reduced diode shunt capacitance. Also, as shown in Fig. 12-33*c* and *d*, it lends itself ideally to beam-lead encapsulation, thus interworking excellently with stripline circuits. This construction is often preferred in practice, except perhaps for the highest powers. When fairly large dissipations are involved, the planar construction is better adapted to mounting on a heat sink.

Operation The PIN diode acts as a more or less ordinary diode at frequencies up to about 100 MHz. However, above this frequency it ceases to be a rectifier, because of the carrier storage in, and the transit time across, the intrinsic region. At microwave frequencies the diode acts as a variable resistance, with a simplified equivalent circuit as in Fig. 12-34*a* and a resistance-voltage characteristic as in Fig. 12-34*b*.

When the bias is varied on a PIN diode, its microwave resistance changes from a typical value of 5 to 10 kΩ under negative bias to the vicinity of 1 to 10 Ω when the bias is positive. Thus, if the diode is mounted across a 50-Ω coaxial line, it will not significantly load the line when it is back-biased, so that power flow will be unaffected. When the diode is forward-biased, however, its resistance becomes very low, so that most of the power is reflected and hardly any is transmitted: the diode is acting

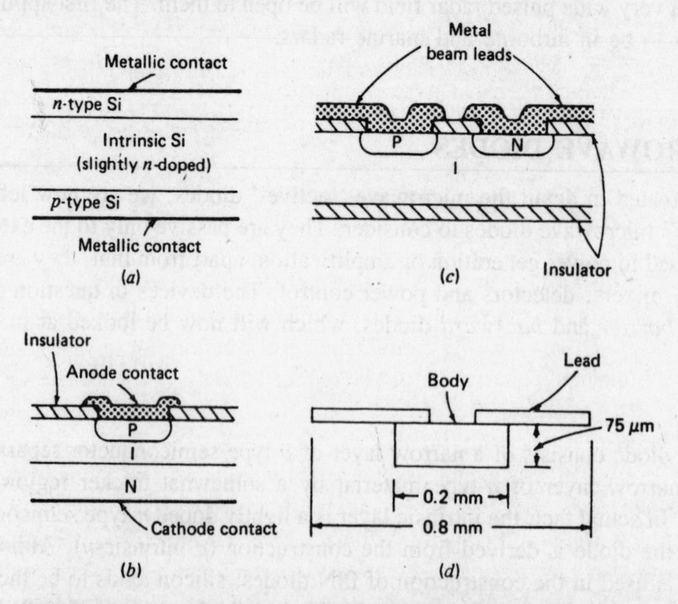

FIGURE 12-33 PIN diode. (*a*) Schematic diagram; (*b*) planar diode; (*c*) planar diode with in-line orientation; (*d*) beam-lead mounting of in-line diode.

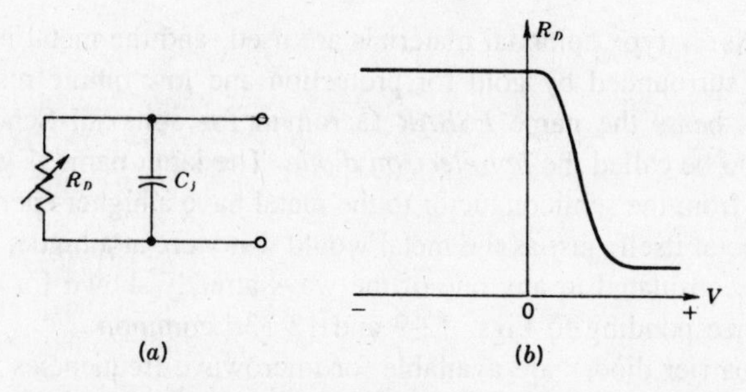

FIGURE 12-34 PIN diode high-frequency behavior. (*a*) Equivalent circuit; (*b*) resistance variation with bias.

as a switch. In a similar fashion, it may be used as a (pulse) modulator. Several diodes may be used in series or in parallel in a waveguide or coaxial line, to increase the power handled or to reduce the transmitted power in the *OFF* condition.

Performance and applications The applications of the PIN diode are as outlined; they have already been treated in Sec. 10-5.4, and Fig. 10-45 showed a coaxial PIN diode switch. Diodes are available with resistive cutoff frequencies up to about 700 GHz. As for varactor diodes (which see), the operating frequencies do not exceed one-tenth of the above figure. However, at least one instance of operation at 150 GHz, with specially constructed diodes, has been reported. Individual diodes may handle up to about 200 kW peak (or 200 W average), although typical levels are one magnitude lower. On the other hand, several diodes may be combined to handle as much as 1 MW peak. Actual switching times vary from approximately 40 ns for high-power limiters to as little as 1 ns at lower powers. Further information on the construction and performance of PIN diodes may be found in [34] while [35] describes in detail a high-power PIN limiter.

12-8.2 Schottky-Barrier Diode

Schottky junctions have been shown and described throughout this chapter, in conjunction with various devices that use them in their construction (for instance, see Fig. 12-4 and its description). Accordingly it will be realized that the Schottky-barrier diode is an extension of the oldest semiconductor device of them all—the point-contact diode. Here the metal-semiconductor interface is a surface—the Schottky barrier—rather than a point contact. It shares the advantage of the point-contact diode in that there are no minority carriers in the reverse-bias condition; that is, there is no significant current from the metal to the semiconductor with back bias. Thus the delay present in junction diodes, due to hole-electron recombination time, is absent here. However, because of a larger contact area (barrier) between the metal and semiconductor than in the point contact diode, the forward resistance is lower, and so is noise.

The most commonly used semiconductors are "the old faithfuls," silicon and gallium arsenide. As usual, GaAs has the lower noise and higher operating frequency limits; silicon is easier to fabricate and is consequently used at X band and below, in preference to GaAs. *N*-type epitaxial materials are used, and the metal is often a thin layer of titanium surrounded by gold for protection and low ohmic resistance. The device sometimes bears the name *ESBAR* (acronym for epitaxial Schottky-barrier) diode and may also be called the *hot-electron diode*. The latter name is given because electrons flowing from the semiconductor to the metal have a higher energy level than electrons in the metal itself, just as the metal would if it were at a higher temperature. The diodes are encapsulated in any one of the ways already shown for other diodes, with packages corresponding to Figs. 12-9 and 12-33*d* common.

Schottky-barrier diodes are available for microwave frequencies up to at least 100 GHz. Like point-contact diodes, they are used as detectors and mixers, mounted as shown previously in Figs. 10-40 to 10-42 and discussed in Sec. 10-5.3. The noise figures of mixers using Schottky-barrier diodes are excellent, rising for as low as 4 dB at 2 GHz to 15 dB near 100 GHz. At frequencies much above X band, GaAs diodes are preferred, since they have lower noise, as discussed. At the highest frequencies, point-contact diodes are preferred, since they have lower shunt capacitances. [36] and [37] are articles on recent developments in Schottky-barrier diodes, showing constructions and describing their behavior and performance. For a comparison of Schottky-barrier diode performance with that of other low-noise front ends, see Table 12-2.

12-8.3 Backward Diodes

It is possible to remove the negative-resistance peak and valley region from the tunnel diode of Sec. 12-5.1, by suitable doping and etching during manufacture. When this is done, the voltage-current characteristic of Fig. 12-35 results. This shows the rather unusual situation in which, for small applied voltages, the forward current is actually much smaller than the reverse current. The reverse current is large, it will be recalled, because of the very high doping. On the other hand, forward current is low at first because tunneling has been stopped. This diode can therefore be used as a small-signal

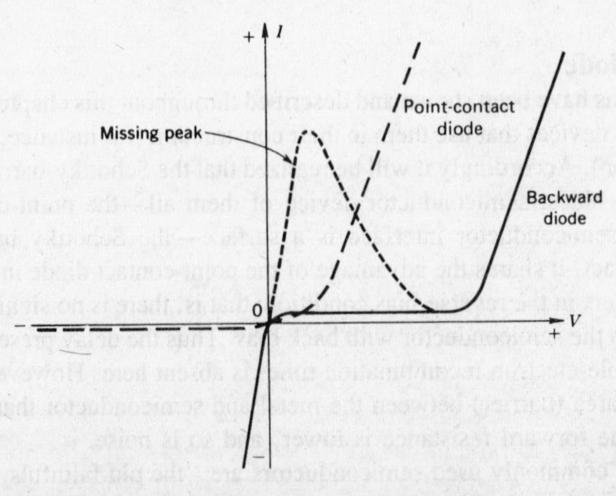

FIGURE 12-35 Backward diode voltage-current characteristic.

rectifier. It has the advantage not only of a narrow junction, and therefore a high operating speed and frequency, but also of a current ratio (reverse to forward!) which is much higher than in conventional rectifiers.

When GaAs is used, a maximum signal of about 0.9 V may be applied to the diode before it begins to conduct heavily in the forward direction. This value, although higher than for germanium (silicon is an unsuitable material, it will be recalled), is nevertheless quite low. This naturally means that the backward diode is limited, just like the tunnel diode, to lower operating levels. Despite this, the backward diode, or *tunnel rectifier* as it is sometimes called, is in quite common use. Aside from having a high current ratio in the two directions, the backward diode is a low-noise device. Consequently, it is used in such applications as video detection and low-level mixing, as in Doppler radar (see Sec. 16-3), for instance. Another of its attractions is that it requires a local oscillator signal up to 10 dB lower than that needed by a point-contact diode.

12-9 STIMULATED-EMISSION (QUANTUM-MECHANICAL) AND ASSOCIATED DEVICES

The first *really* low-noise microwave amplifier produced *Microwave Amplification by Stimulated Emission of Radiation;* hence the acronym *maser*. This brand new principle was developed to fruition by Townes and his colleagues [38] in 1954 and provided extremely low-noise amplification of microwave signals by a *quantum-mechanical* process. The *laser,* or optical maser (*l* stands for light), is a development of this idea, which permits the generation or amplification of *coherent* light. In this instance, coherent means single-frequency, in-phase, polarized and directional—just like microwave radio waves. This was also put forward by Professor Townes, in 1958 [39]. The overall work was of sufficient importance to make him the 1964 corecipient of the Nobel Prize for physics. The first practical laser was demonstrated by Maiman [40] in 1960.

12-9.1 Fundamentals of Masers

As was found with ferrites in Sec. 10-5.2, certain materials have atomic systems that can be made to resonate magnetically at frequencies dependent on the atomic structure of the material and the strength of the applied magnetic field. When such a resonance is stimulated by the application of a signal at that frequency, absorption will take place, as in the *resonant absorption* ferrite isolator. Alternatively emission will occur, if the material is suitably excited, or pumped, from another source. It is upon this behavior that the maser is based.

The material itself may be gaseous, such as ammonia, or solid-state, such as ruby. Ammonia was the original material used, and it is still used for some applications, notably in the so-called *atomic clock* frequency standards. *Extreme* is the correct word to use in describing the stability of such an oscillator. The atomic clock built at Harvard University in 1960 has a cumulative error which would cause it to be incorrect by only 1 second after more than 30,000 years! However, from the point of view of microwave amplification, ammonia gas suffered from the disadvantage of yielding amplifiers that worked at only one frequency and whose bandwidth was very narrow. This description will therefore be aimed mainly at the ruby maser.

Fundamentals of operation As is well known, the electrons belonging to the atoms of a substance can exist in various energy levels, corresponding to different orbit shells for the individual atoms. At a very low temperature, most of the electrons exist in the lowest energy level, but they may be raised by the addition of *specific amounts of energy*. *Quantum theory* shows that a quantum, or bundle of energy, may provide the required energy to raise the level of an electron, provided that

$$E = hf \tag{12-12}$$

where E = energy difference, joules
f = photon frequency, Hz
h = Planck's constant = 6.626×10^{-34} joule · s

Having been excited by the absorption of a quantum, the atom may remain in the excited state, but this is most unlikely to last for more than perhaps a microsecond. It is far more likely that the photon of energy will be reemitted, at the same frequency at which it was received, and the atom will thus return to its original, or *ground*, state. The foregoing assumes, incidentally, that the reemission of energy has been *stimulated* at the expense of absorption. This may be done by such measures as the provision of a structure resonant at the desired frequency and the removal of absorbing atoms, as was done in the original gas maser.

It is also possible to supply energy to these atoms in such quantities and at such a frequency that they are raised to an energy level which is much higher than the ground state, rather than immediately above it. This being the case, it is then possible to make the atoms emit energy at a frequency corresponding to the difference between the top level and a level intermediate between the top level and the ground state. This is achieved by the combination of the previously mentioned techniques (the cavity now resonates at this new frequency) and the application of an input signal at the desired frequency. Pumping thus occurs at the frequency corresponding to the energy difference between the ground and the top energy levels. Reemission of energy is stimulated at the desired frequency, and the signal at this frequency is thus amplified. *Practically no noise is added to the amplified signal*. This is because there is no resistance involved and no electron stream to produce shot noise. In addition, the material that is being stimulated has been cooled to a temperature only a few degrees above absolute zero. It now only remains to find a substance capable of being stimulated into radiating at the frequency which it is required to amplify, and low-noise amplification will be obtained.

The original substance was the gas ammonia, while hydrogen and cesium featured prominently among the materials used subsequently. The gaseous substance had the advantage of allowing absorbing atoms to be removed easily. Unfortunately, since the operating frequency was determined very rigidly by the energy levels in ammonia, the range of frequencies over which the system operated, i.e., its bandwidth, was extremely narrow (of the order of 3 kHz at a frequency of approximately 24 GHz). Moreover, there was no method whatsoever of tuning the maser, so that signals at other

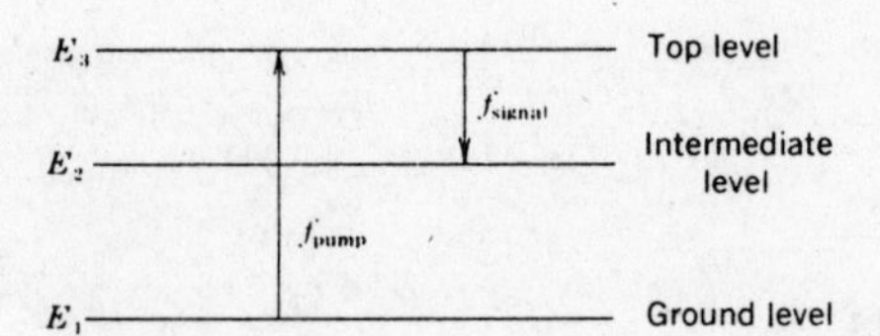

FIGURE 12-36 Energy levels in ruby relevant to maser operation.

frequencies just could not be amplified. To overcome these difficulties, the traveling-wave ruby maser was invented.[11]

The ruby maser A gaseous material is inconvenient in a maser amplifier, as can be appreciated. The search for more suitable materials revealed ruby, which is a crystalline form of silica (Al_2O_3) with a slight natural doping of chromium. Ruby has the advantages of being solid, having suitably arranged energy levels, and being *paramagnetic*, which virtually means "slightly magnetic." This last property is due to the presence of chromium atoms, which have *unpaired electron spins* (like the ferrites of Sec. 10-5.2). These are capable of being aligned with a dc magnetic field, and this permits not only reradiation of energy from atoms in the desired direction but also some tuning facilities.

Figure 12-36 shows the energy-level situation in a three-level maser, introduced in the previous section. Energy at the correct pump frequency is added to the atoms in the crystal lattice of ruby, raising them to the uppermost of the levels shown (there are many other levels, but they are of no interest here). Normally, the number of electrons in the third energy level is smaller than the number in the ground level. However, as pumping is continued, the number of electrons in level 3 increases until it is about equal to the number in the first level. At this point the crystal saturates, and so-called *population inversion* has been accomplished.

Since conditions have been made suitable for reradiation (rather than absorption) of this excess energy, electrons in the third level may give off energy at the original pump frequency and thus return to the ground level. On the other hand, they may give off smaller energy quanta at the frequency corresponding to the difference between the third and second levels and thus return to the intermediate level. A large number of them take the latter course, which is stimulated by the presence of the cavity surrounding the ruby, which is resonant at this frequency. This course is further aided by the presence of the input signal at this frequency. Since the amount of energy radiated or emitted by the excited ruby atoms at the signal frequency exceeds the energy applied at the input (it does not, of course, exceed the pumping energy), amplification results.

The presence of the strong magnetic field (typically about 4 kA/m) has the effect of providing a difference between the three desired energy levels that corre-

[11] Note that the foregoing was greatly simplified, with a view especially to simplifying the explanation of the solid-state maser. Also, some slight liberties with the truth had to be taken in order to present an overall picture that is essentially correct and *understandable*.

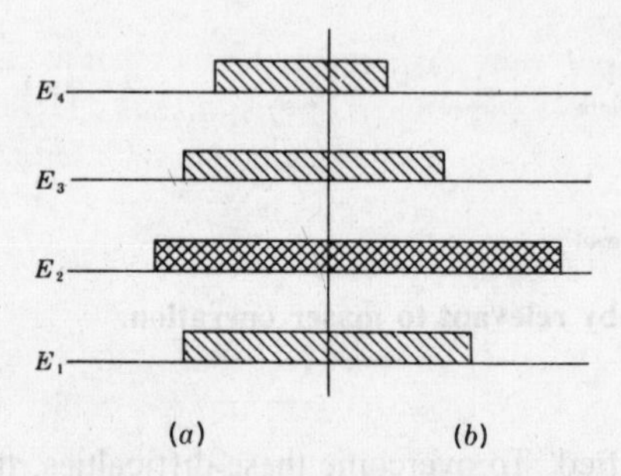

FIGURE 12-37 Energy level populations in suitably pumped ruby. (*a*) At room temperature; (*b*) at liquid helium temperature. (Note the reduction in the fourth-level population in the latter case and the accompanying significant population inversion in levels 2 and 1.)

sponds to the required output frequency. Any adjustment of this magnetic field will alter the energy levels of the ferrous chromium atoms and therefore provide a form of tuning. This is similar to the situation in ferrites, where it was found that a change in the dc magnetic field changed the frequency of *paramagnetic resonance*. This field strength can be altered to permit the ruby maser to be operated over a frequency range from below 1 to above 6 GHz. For frequencies as high as 10 GHz and above, other materials are often used. *Rutile* is a very common alternative; this is titanium oxide (TiO_2) with a light doping by iron. At the higher frequencies, the required magnetic fields tend to be rather strong, so that the magnet is very often cooled also, to take advantage of *superconductivity* and therefore to give a reduction in the power required to maintain the magnetic field.

In order to consider the effect of cooling the ruby with liquid helium (which is almost invariably done[12]) it is helpful to consider Fig. 12-37. Figure 12-37*a* shows the situation at room temperature. It is seen that because of the relatively high energy possessed by the electrons at this temperature, quite a number of electrons normally exist in the fourth level, apart from the three so far mentioned. This has the undesirable effect of reducing the number of electrons in the ground level. There are thus fewer electrons whose energy level can be raised from the first to the third, and consequently fewer electrons that can reradiate their excess energy at the correct frequency. The high temperature is thus said to *mask* the maser effect. However, if cooling is applied, the overall energy possessed by the electrons is reduced, as is the number of electrons at the fourth level. As seen in Fig. 12-37*b* there are now an adequate number of electrons that can be jumped from the ground to the third level and then down again to the intermediate level. Maser action is thus maintained. Note that no maser has (yet) operated satisfactorily at room temperature. Even if such operation were possible, the noise level would be raised sufficiently to make the noise figure of the maser a very poor second to that of the parametric amplifier.

The noise figure of the cooled ruby maser is governed by the same factors as that of the ammonia maser and is therefore equally low. There is the slight noise due to

[12] Cooling with liquid nitrogen down to only 77 K can also be used, but it results in an increase in noise and a reduction in gain.

the random motion of electrons in the ruby (caused by the fact that the temperature of the crystal is above absolute zero). However, most of the noise is due to the associated components, such as the waveguide leading from the antenna, and the noise created at the input to the following amplifier. The first of these problems may be reduced by making the waveguide run as short as possible. This involves mounting the maser at the prime focus of the antenna. Such a solution is practicable only if a Cassegrain or folded horn antenna is used (see Chap. 9), and in fact that is done in practice. The problem of noise from succeeding stages is alleviated in a number of ways. One involves cooling the circulator (which must sometimes be used, as will be seen), in the same way as in a parametric amplifier. It is also possible to increase the gain of the maser, thereby reducing noise reflected from succeeding stages, by making it a two-stage amplifier. Finally, the amplifier following the maser can be made a relatively low-noise one, by the use of tunnel diodes or FETs.

12-9.2 Practical Masers and Their Applications

Practical solid-state masers The term *solid-state* is used deliberately here; it does not mean "semiconductor." In terms of the somewhat older maser parlance, it means the opposite of gaseous, i.e., ruby.

The cross section of a ruby cavity maser is shown in Fig. 12-38. It is seen to be a single-port amplifier, so that a circulator is needed, as shown, just as in so many other microwave amplifiers. Again, just as in the parametric amplifier, a tuned circuit must be provided for the pump signal as well as for the signal to be amplified. This is not difficult to achieve, but it should be realized that the cavity must be able to oscillate at both frequencies.

From a communications point of view, a disadvantage of the cavity maser is that its bandwidth is very narrow, being governed to a large extent by the cavity itself.

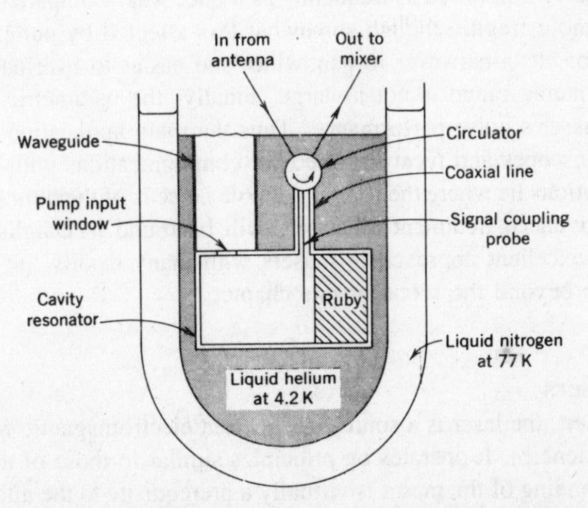

FIGURE 12-38 Schematic diagram of cryogenically cooled ruby maser cavity amplifier (magnet not shown).

It may be typically 1.5 MHz at 1.5 GHz, but some compromise at the expense of gain is possible, noting that the gain-bandwidth product is about 35 MHz. Increasing the bandwidth to even 25 MHz is not practicable, however, since gain by then would not be much in excess of unity.

The solution to the problem is one that has already been encountered a number of times in this chapter: the use of a traveling-wave structure. The resulting operating system is then virtually identical to the one used in the TW paramp. The signal to be amplified now travels along the ruby via a slow-wave structure and grows at the expense of the pump signal. The traveling-wave maser has not only an increased bandwidth but also effectively four terminals, so that a circulator is no longer needed. Such TW masers are used in some older satellite earth stations, built before the subsequent paramp developments.

Performance and applications A typical TW maser operating at 1.6 GHz may have a 25-dB gain, a bandwidth of 25 MHz and a 48-GHz pump requiring 140 mW of CW power. The last two figures are also applicable to the cavity maser, and both types are capable of a noise temperature better than 20 K, i.e., a noise figure better than 0.3 dB. A glance at Table 12-2 will serve as a reminder that the noise performance of masers is unsurpassed.

A disadvantage of the maser is that it is a very low-level amplifier and may saturate for input levels well over 1 μW. While this makes it suitable for radioastronomy and other forms of extraterrestrial communications, radar is a typical application in which a maser could not be used. Not only can much larger radar signals be received in the course of duty, but so can jamming; this would certainly overload a maser RF amplifier, though fortunately without permanent damage. However, the maser would take about 1 s to recover, during which it would be unusable. Also, care must be taken not to point the antenna at the ground when a maser amplifier is used, or the ground temperature will create sufficient noise to overload the maser once again.

Because of the improvements that the parametric amplifier has undergone in the last decade, the maser is not used as frequently as it once was. Compared to the paramp it is bulkier and more fragile, though somewhat less affected by pump noise or frequency fluctuations. It is narrower in bandwidth and easier to overload, which also means that its dynamic range is not as large. Finally, the parametric amplifier has approached the maser's noise performance. Thus the main application for the maser now is in radiotelescopes and receivers used for communications with space probes. That is, its applications lie where the lowest possible noise is of the utmost importance.

A more advanced treatment of masers will be found in Dalglish et al. [41], chap. 4. It is an excellent approach to masers with many details, including photographs, which are beyond the scope of this chapter.

12-9.3 Fundamentals of Lasers

As already indicated, the laser is a source of coherent electromagnetic waves at infrared and light frequencies. It operates on principles similar to those of the maser, and indeed an understanding of the maser is virtually a prerequisite to the understanding of its more spectacular stablemate. However, the frequencies are *much* higher; for visible light, these range from 430 to 750 terahertz (THz) (i.e., 430,000 to 750,000 GHz!). It can thus be seen that the scope and information-carrying capacity of lasers is immense, to say the least.

Introduction The first laser using ruby was proposed in 1958 and demonstrated in 1960, while the first continuously operating laser followed in 1961 and used a mixture of helium and neon gases. Since then, a very large number of other materials have been found suitable, including the other inert gases, argon and krypton, as well as the gallium arsenide and other semiconductor diodes.

Ruby laser The ruby laser is similar to the ruby cavity maser, to some extent, in that stimulation is applied to raise the chromium atoms to a higher energy level to secure a population inversion once again. However, this time pumping is with light, rather than with microwave, energy. Also, no magnetic field is required to modify the existing energy levels because these are already suitable for laser action. The cavity is also different, as can be seen from Fig. 12-39. This shows that two parallel mirrors are used, one fully silvered and the other partly so, to enable the coherent light radiation to be emitted through that end. The mirrors must be parallel to a high degree of accuracy and must be separated by a distance that is an exact number of half-wavelengths apart (in the ruby, at the desired frequency). Such an arrangement is called a *Fabry-Perot resonator*. The spiral flash tube, as shown, pumps light energy into the ruby in pulses, which are generated by the charge and discharge of a capacitor. Finally, cooling is used to keep the ruby at a constant temperature, since quite a lot of the energy pumped into it is dissipated into heat, instead of being radiated as coherent light. Although this cooling also helps laser action, as it did with the maser, room temperature operation is normal.

Pumping raises the electrons to a high energy level, different from that which operated in the maser, since the photon energy is now much higher, because of the higher frequency [this is in accord with Eq. (12/2)]. Electrons so raised in energy may fall back either to the ground state, emitting uncoordinated radiation, or else to an intermediate level, as a large number of them do. The energy they lose in the process appears in the form of heat and/or fluorescence. The intermediate level is quasi-stable;

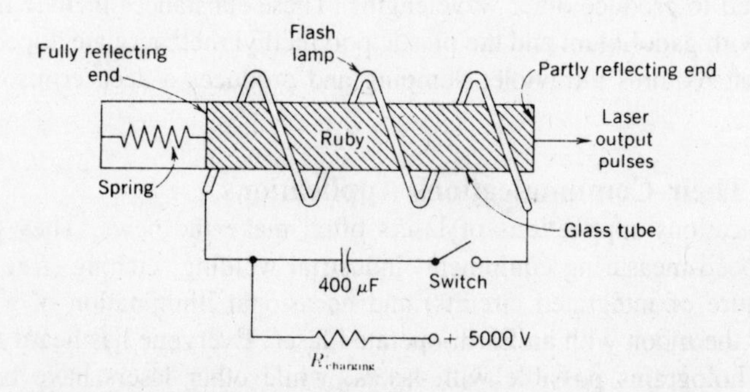

FIGURE 12-39 Basic ruby pulsed laser.

electrons remain at it for a few milliseconds, which corresponds to the pumping period. Then their energy rapidly falls to the ground level, with ensuing radiation at the desired frequency. The energy discharge from some of the chromium atoms triggers and coordinates the discharge from the others, with a resulting *correct phase relationship* of all the photons radiated. A large number of these may not escape through the cylindrical sidewalls of the ruby. However, the photons traveling longitudinally are reflected from the silvered end walls and travel back and forth, triggering off other atoms. In this fashion energy builds up, until it is sufficient to escape through the partly silvered end wall, in the form of a very intense short pulse of coherent light that is almost completely *monochromatic* (i.e., single-frequency). The ruby crystal is now in its original state, ready for the next pumping pulse from the flash tube.

The beam of light leaving the ruby crystal is very narrow and almost parallel, with a divergence of less than 0.1°. The frequency spread, or line width, is also very small, of the order of about 1 GHz at a center frequency that is roughly 500,000 GHz (or 500 THz). However, the efficiency is poor (in the vicinity of 1 percent), so that pulsed operation is preferable, in order to permit the dissipated heat to be removed before the next pulse. As previously mentioned, cooling also helps, and liquid nitrogen is sometimes used for this. If the chromium doping of the ruby is increased, CW operation becomes possible. However, the output level is then only milliwatts instead of the megawatts of peak power available with pulsed operation.

It is possible to shorten the pulse duration, without altering the *average* power output of the ruby laser, by the process of *Q-spoiling,* whose effect is to intensify the peak radiated pulse power. In this process, also known as *Q-switching,* one of the ends of the ruby rod is made transparent, and the other is left partly silvered. A mirror is situated behind the unsilvered end, with a shutter placed in front of it. The shutter is closed during pumping, thus preventing laser action and "spoiling" the Q of the Fabry-Perot resonator. This has the effect of greatly helping the population inversion and permits an even larger number of electrons to be situated at the intermediate level. The shutter is opened at the end of the pumping period. With the second mirror now in place, oscillations build up extremely quickly and produce a most intense flash of very short duration; peak powers in excess of 1000 MW are possible.

Two other points should now be raised in connection with solid-state lasers. The first is simply that, as so far described, the laser is an oscillator, unlike the maser. The second is that solid-state lasers are not restricted to using ruby, and other materials have been used to produce other wavelengths. These substances include neodymium, glass doped with gadolinium and the plastic polymethyl methacrylate doped with europium. The last requires ultraviolet pumping and produces a deep crimson light.

12-9.4 CW Lasers and Their Communications Applications

Noncommunications applications of lasers often make the news. They include distance- and speed-measuring equipment, industrial welding, etching (fine enough for the manufacture of integrated circuits) and occasional illumination of a ludicrously small area of the moon with an Earth-operated laser. Everyone has heard of the three-dimensional holograms possible with lasers, while other lasers have been used in optical and other surgery, and still others suggested for military uses (the oft-promised death ray!). Each of these is a valid laser application, but none of them falls into the category of *communications* applications.

We shall concentrate in this section on those applications of lasers which involve conveying information at a distance. Although it is not essential to have a continuous-wave laser for such work, it does help, and so CW lasers will be the only ones now treated. Before they are, together with a mention of modulation and detection, it is worth suggesting where they are likely to be used. In fact, it is unlikely that laser links will ever be used in the same way as microwave links or satellite links. As has often been pointed out, too many things interfere with light in the atmosphere: fog, dust, rain and clouds can all interfere, and so can flying pigeons. It seems that the most spectacular application of laser communications will be in space, while the most frequent workaday one is to send information along optical fibers. This application will be treated in Sec. 15-2.2—it will be seen that lasers are used along the coaxial cable principles, rather than radio link ones.

Gas lasers The first CW laser, in 1961, was a gas laser using a mixture of helium and neon gases. These are still used, and a simplified He-Ne laser is shown in Fig. 12-40. It operates in a manner similar to that of the ruby laser, with the following differences.

1. The mirrors must be as close as possible to being ideally parallel; hence the bellows of Fig. 12-40 which are used for fine adjustment.
2. The mirrors must be optically flat, to better than a wavelength,[13] if proper laser action is to take place.
3. RF pumping is now required, at a frequency of about 28 MHz for helium-neon. Energy is discharged into the gas mixture via the ring contacts shown.
4. Emission is not at one frequency but at several so-called *lines*. This behavior is due in part to the atomic structure of the gases.
5. Each of the emission lines is extremely pure, having a line width of only a few hertz[14]; that is, each emitted frequency is extremely close to being monochromatic.

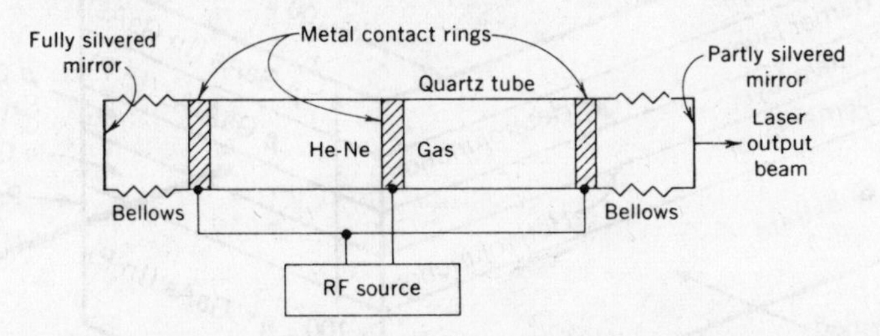

FIGURE 12-40 Schematic diagram of simple CW gas laser. (Note bellows for mirror adjustment; this is the equivalent of cavity tuning.)

[13] This is not as exacting as might at first appear—amateur reflector telescope mirrors are normally ground to an accuracy of one-eighth of a wavelength or better.

[14] In practical lasers, gas mixtures provide the narrowest lines. those of solid-state lasers are one magnitude wider and the lines of semiconductor lasers are one magnitude wider still.

6. The beam divergence from parallel is similarly less than in a ruby laser.
7. Such multifrequency oscillation is possible because the dimensions of the resonator (i.e., the distance between the mirrors) are very much greater than a wavelength. The behavior is exactly the same as in a simple oversized cavity resonator, capable of supporting a large number of modes.

Because pumping is continuous, unlike in the solid-state laser, continuous operation is possible. The early gas lasers operated in the infrared region and produced a few milliwatts with low efficiency. Subsequent improvements have included the use of much shorter tubes to give single rather than multiple lines, laser action with greater efficiency and in the visible spectrum and, more recently, the use of a mixture of carbon dioxide, nitrogen and helium gases. This last device operates in the far infrared spectrum at a wavelength of 10.6 μm, corresponding to a frequency of 28,300 GHz. The process has an efficiency of the order of 20 percent or more, and CW powers as high as 1000 W are possible.

Semiconductor lasers It was discovered in 1962 that a gallium arsenide diode, such as the one shown in Fig. 12-41, is capable of producing laser action. This occurs when the diode is forward-biased, so that effective dc pumping is needed (a very convenient state of affairs). Depending on its precise chemical composition, the GaAs laser is capable of producing an output within the range of 0.75 to 0.9 μm, i.e., in the near infrared region (light occupies the 0.39 to 0.77 μm range).

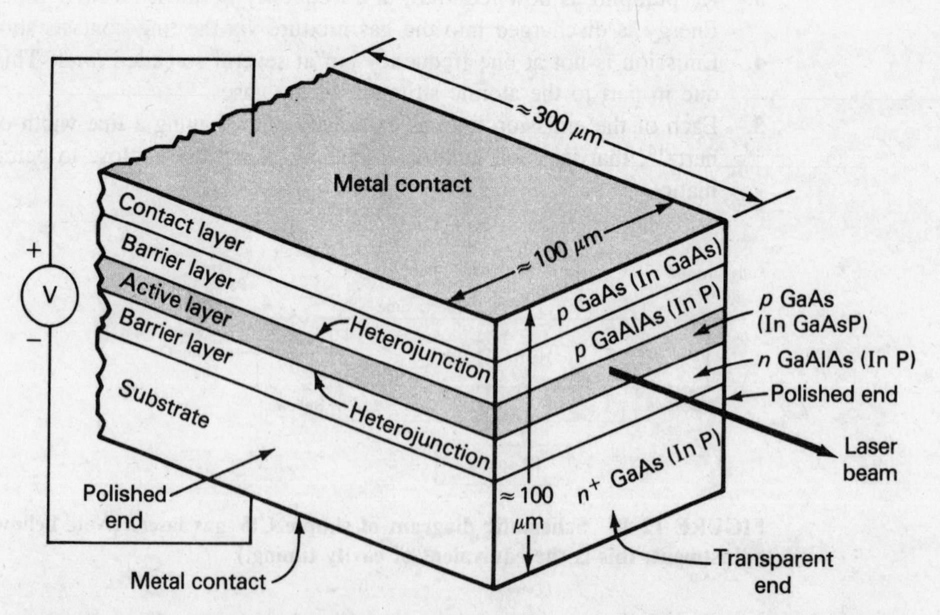

FIGURE 12-41 Double heterojunction semiconductor laser. The materials outside the parentheses are for a gallium arsenide laser operating in the 0.75- to 0.9-μm wavelength range; those inside parentheses are for an indium gallium arsenide phosphide laser operating over the range of 1.2 to 1.6 μm.

Briefly, the device is an *injection laser,* in which electrons and holes originating in the GaAlAs layers cross the *heterojunctions* (between dissimilar semiconductor materials, GaAlAs and GaAs in this case) and give off their excess recombination energy in the form of light. The heterojunctions are opaque, and the active region is constrained by them to the *p*-layer of GaAs, which is a few micrometers thick, as shown. The two ends of the slice are very highly polished, so that reinforcing reflection takes place between them as in other lasers, and a continuous beam is emitted in the direction shown. The laser is capable of powers in excess of 1 W, which is far higher than the 1 mW, or so, necessary to send along optic fibers, as will be seen in Chap. 15.

The indium gallium arsenide phosphide laser, also illustrated in Fig. 12-41, is a much more recent development than the GaAs device, having been evolved during the late 1970s. The motive force was a desire to produce laser outputs at wavelengths longer than those which the GaAs laser is capable of producing, to take advantage of "windows" in the transmission spectrum of optic fibers—these are discussed in more detail in Chap. 15. Consequently, the InGaAsP lasers are less well developed at the time of writing, and so many of the world's optic fiber communications systems still operate at wavelengths of about 0.85 μm, whereas, as will be shown in Chap. 15, transmissions at wavelengths of 1.3 or 1.55 μm incur significantly less attenuation than at 0.85 μm in optic fibers. By the early to mid-1980s, the teething problems with the new laser materials were being solved, and all new lightwave systems were being designed for wavelengths of 1.3 μm or greater.

For additional information on the construction, manufacture and operation of semiconductor lasers, [42] and [43(*a*)] are highly recommended.

12-9.5 Other Optoelectronic Devices

Although *light-emitting diodes* and *photodiodes* are not quantum-mechanical devices, they are semiconductor devices closely associated with lasers. Accordingly, it is most convenient to treat them here, and they will now be briefly discussed in turn.

Light-emitting diodes (LEDs) The construction of an LED is similar to that of a laser diode, as indeed is the operational mechanism. Once again electrons and holes are injected across heterojunctions, and light energy is given off during recombination. The materials used are the same as for the corresponding laser diodes, but the structure is simpler, there are no polished ends and laser action does not take place. Consequently, power output is lower (perhaps one-twentieth) than for the laser, a much wider beam of light results and the light itself is no longer monochromatic. A small lens is often used to couple the output of the LED to the optic fiber.

Despite the foregoing, the LED does have a number of advantages over the laser. For example, it is a good deal cheaper and tends to be more reliable. Moreover, the LED, unlike the laser, is not temperature-sensitive, so that it can operate over a large temperature range without the need for elaborate temperature control circuits which the laser may require. In practice, lasers tend to be used in a fairly large proportion of practical systems, especially the more exacting ones, noting that pulse modulation is normally used, and the light output of lasers can be pulsed at much higher rates than that of LEDs. Additional information on LEDs is provided in [43(*g*)] and [44].

Photodiodes A PIN diode, such as any of the ones shown in Fig. 12-33, is capable of acting as a photodiode. If a large reverse bias, of the order of 20 V or more, is applied to such a diode, no current will flow. However, if the diode absorbs light quanta through a window on the *p* side, each quantum will cause an electron-hole pair to be created in the intrinsic depletion layer, and a corresponding current will flow in the external circuit. Within limits, this current will be proportional to the intensity of the impinging light, so that photodetection is taking place.

The original photodiode semiconductor was germanium, and it is still used for wavelengths in excess of about 1.1 μm; for shorter wavelengths silicon is preferred. Because of the well-known sensitivity of germanium to temperature, research is currently taking place among the newer semiconductor materials, such as GaAlAs and InGaAs, to find a replacement for the germanium PIN photodetector.

Avalanche photodiodes (APDs) A problem with the PIN photodiode is that it is not overly sensitive; as can be seen, no gain takes place in the device, in that a single photon cannot create more than one hole-electron pair. This problem is overcome by the use of the avalanche photodiode, which, in some respects, operates in a manner similar to the IMPATT diode.

An APD, such as the one shown in Fig. 12-42, is operated with a reverse voltage close to break-down. Like the IMPATT, the APD is capable of withstanding

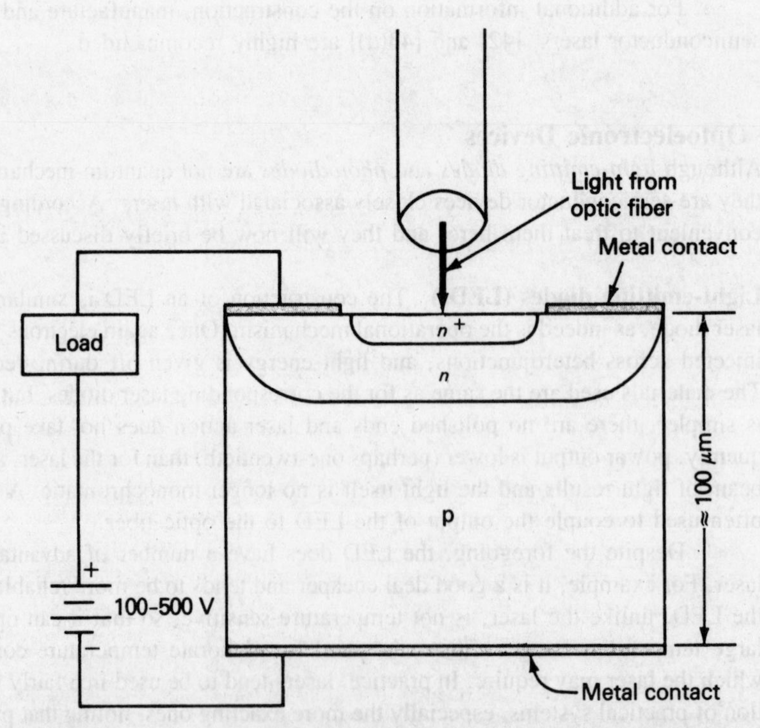

FIGURE 12-42 Avalanche photodiode construction and schematic. (Note similarity to IMPATT diode schematic in Fig. 12-28.)

sustained break-down. As in the PIN photodetector, a light quantum impinging on the diode will cause a hole-electron pair to be created, but this time avalanche multiplication can take place, as in the IMPATT, so that the initial electron-hole pair will cause several others to be created, with consequently increased current flowing through the eternal circuit. The extent of avalanche multiplication can be gauged from the fact that a typical APD is 10 to 150 times more sensitive than a PIN photodetector.

The materials used for APDs are the same as for the corresponding PIN diodes. Because the voltage gradient across the APD is so high, electron and hole drift is higher than for the PIN diode, and the response time is similarly faster, typically 2 nS compared with 5 nS for the PIN diode. It follows that the APD can be used for higher pulse modulation rates than the PIN. As a matter of fact, there is a fairly close correlation between light transmitters and receivers in fiber-optic systems. That is to say, those less exacting systems which use LEDs for transmission are also likely to use PIN photodiodes for reception. Similarly, the systems requiring higher sensitivities and higher modulation bit rates are likely to use lasers for transmission and avalanche photodiodes for reception.

Additional information on photodetectors—both PIN and APD—will be found in [42], [43(*b*)], [46] and [48].

PROBLEMS

For self-testing questions on this chapter, see pp. 699.

12-1. A microwave signal has a purely resistive output impedance of 500 Ω, and its load is matched for maximum power transfer. A negative resistance is now placed across the circuit, turning it into an amplifier. If the value of this negative resistance is −200 Ω what will be the power gain of the amplifier?

[Ans.: 16 (12 dB)]

12-2. If, in Prob. 12-1, the load and source resistance are now both 1000 Ω, what must be the value of the negative resistance to give a power gain of 23 dB?

[Ans.: 528 Ω]

QUESTIONS

12-1. With the aid of appropriate sketches, describe basic stripline and microstrip circuits. From what previously studied transmission media are they derived?

12-2. What are the advantages and disadvantages of stripline and microstrip with respect to waveguides and coaxial transmission lines? What are the conditions under which waveguides and coax would be preferred?

12-3. What are the applications of microstrip and stripline circuits? Which is the more convenient to use in hybrid MICs? Why?

12-4. Discuss the construction and applications of surface acoustic wave devices, illustrating the answer with a sketch of a typical SAW component.

12-5. Discuss the high-frequency limitations of transistors, comparing and contrasting them with those of vacuum tubes.

12-6. Illustrating your answer with sketches, describe the construction of microwave bipolar and field-effect transistors.

12-7. Compare the performance and general construction of hybrid and monolithic MICs.
12-8. Discuss the performance and applications of microwave transistors and MICs, illustrating your answer with graphs of power output and noise versus frequency.
12-9. With the aid of suitable sketches, discuss the materials, construction and characteristics of microwave varactors,
12-10. Discuss briefly the basic theory of varactor frequency multipliers. Define the term *nonlinear capacitance*.
12-11. Discuss the capabilities and applications of varactor and snap-recovery diode frequency multipliers.
12-12. What is a parametric amplifier? Discuss its fundamentals *in full,* and state the ways in which it differs from an orthodox amplifier.
12-13. Describe the *nondegenerate* negative-resistance parametric amplifier. Show a simple circuit of this device, and explain the function of the *idler* circuit.
12-14. What is the most common type of very low-noise parametric amplifier? Show the block diagram of such a device, explaining carefully the function of the circulator. Does the use of the circulator have any drawbacks? Can its use be avoided?
12-15. Draw the circuit diagram of a representative TW parametric amplifier, and briefly explain how it works. Why must the pump frequency be not *too* much higher than the signal frequency in this type of amplifier?
12-16. Discuss the noise performance of parametric amplifiers and the factors influencing it. Why is *cryogenic* cooling sometimes used? Is it compulsory? What are the advantages of *not* cooling cryogenically?
12-17. Discuss the advantages and list the applications of parametric amplifiers. Contrast the applications of paramps cooled by various means with those of uncooled ones.
12-18. Using energy-band (Fermi level) diagrams, explain the tunnel-diode characteristic (voltage-current curve) point by point. Take it for granted that quantum-mechanical tunneling will take place under favorable conditions.
12-19. Discuss the problems connected with the biasing of a tunnel diode and their solution. Illustrate the discussion with a practical tunnel-diode circuit.
12-20. Explain why it is possible to obtain amplification by using a device which exhibits negative resistance.
12-21. Discuss the performance, advantages and applications of tunnel-diode amplifiers, and then compare them, in turn, with each of the other microwave low-noise amplifiers.
12-22. What is the significant and very important difference between the *Gunn effect* and all the other properties of semiconductors?
12-23. Explain fully the Gunn effect, whereby negative resistance, and therefore oscillations, are obtainable under certain conditions from bulk gallium arsenide and similar semiconductors. Why are Gunn devices called *diodes?*
12-24. Sketch a Gunn diode construction, and describe it briefly. What are some of the performance figures of which Gunn diodes are capable?
12-25. What are Gunn *domains?* How are they formed? What do they do?
12-26. How does the domain formation in a Gunn diode respond to the tuning of the cavity to which the diode is connected? Sketch a cavity Gunn oscillator.
12-27. Describe the construction, fabrication and encapsulation of Gunn diodes.
12-28. Discuss the performance and operation of (*a*) YIG-tuned Gunn oscillators; (*b*) Gunn diode amplifiers.
12-29. What do the acronyms *IMPATT* and *TRAPATT* stand for?
12-30. What are the applications of Gunn oscillators and amplifiers?
12-31. Draw the schematic diagram of an IMPATT diode, and fully explain the two effects that combine to produce a 180° phase difference between the applied voltage and the resulting current pulse.

12-32. Show an encapsulated IMPATT diode, and discuss some of the practical considerations involved. What is a double-drift IMPATT diode?

12-33. Briefly describe the basic operating mechanism of TRAPATT diodes, using a suitable sketch. Why is the drift through this diode much slower than through a comparable IMPATT diode? What implications does this have for the operational frequency range of the TRAPATT diode?

12-34. Compare the performance of IMPATT and TRAPATT oscillators with that of Gunn oscillators and amplifiers. Consider also their relative applications.

12-35. What is the major drawback of avalanche devices? What limitations does this place on their applications?

12-36. With the aid of a suitable sketch, describe the construction of a PIN diode. What does PIN stand for? Briefly explain the operation of this diode.

12-37. Discuss the performance and applications of Schottky-barrier diodes, and list the competitors for those applications.

12-38. Write a survey of semiconductor diode and bulk effect microwave generators, describing briefly the construction, operation, performance and applications of each.

12-39. How does the backward diode differ from the tunnel diode? What is this device used for?

12-40. What is a maser? What does its name signify? What applications does it have?

12-41. Discuss the fundamentals of the maser, and explain the various levels at which electrons may be found, the connection between the pumping frequency and these levels and finally what is done to make the electrons reemit the energy they receive from the pump source, instead of absorbing it. Why is the maser such a low-noise device?

12-42. Show the energy levels in a ruby crystal relevant to maser operation. What is meant by the terms *population inversion* and *saturation?* How does the presence of the magnetic field affect the situation? What else can the magnetic field be used for?

12-43. From what point of view is cooling of a ruby maser with liquid helium preferable to cooling with liquid nitrogen? Discuss the causes of noise in a maser amplifier, and describe some of the steps taken in practice to reduce it.

12-44. What are the capabilities and performance of the maser?

12-45. Discuss fully the operation of the ruby laser. Show a basic sketch of one.

12-46. What are the outstanding characteristics of the ruby laser? Describe the process of *Q-spoiling* and its function. What is the big disadvantage of this laser from a communications point of view?

12-47. Compare and contrast the operation and applications of the gas laser with those of the ruby laser.

12-48. Briefly explain the operation of a semiconductor laser, using a sketch showing the construction of this device

12-49. What is the major application of semiconductor lasers? How do GaAs and InGaAsP devices compare in this regard?

12-50. How does the performance of light-emitting diodes compare with that of semiconductor lasers? What are their respective applications?

12-51. Briefly discuss the operation and relative merits of PIN and avalanche photodiodes.

REFERENCES

1. See Rosen, A., et al., "Silicon as a Millimeter-Wave Monolithically Integrated Substrate—A New Look," *RCA Rev.* **42**(4):633–660 (December 1981), especially pp. 636–645.

2. Itoh, T., "Open Guiding Structures for mmW Integrated Circuits," *Microwave J.*, September 1982, pp. 113–126 (*42 additional references given*).

3. Faulkner, J., et al., ''Soft Microstrip with Integral Ground Plane Aids in Supercomponent Integration,'' *Microwave J.*, November 1983, pp. 105–115.

4. Kansy, R., ''Understanding SAW Resonators,'' *Microwaves & RF*, November 1983, pp. 99–106.

5. Hartmann, C. S., et al., ''SAW Devices for Military Communications, Radar and EW Systems,'' *Microwave J.*, July 1982, pp. 73–86.

6. Kurokawa, K., ''Power Waves and the Scattering Matrix,'' *IEEE Trans. Microwave Theory Tech.* **MTT-13**(3):194 (March 1965) *Original reference*.

7. Taylor, G. C., et al., ''GaAs Power Field-Effect Transistors for K-Band Operation,'' *RCA Rev.* **42**(4):508–521 (December 1981).

8. Berentz, J. J., G. C. Dalman and C. A. Lee, ''Improved FET Design Reaches Millimeter Waves,'' *Microwaves*, February 1982, pp. 67–71.

9. Yamasaki, H., and D. Maki, ''Hybrid vs. Monolithic—Is More Monolithic Better?'' *Microwave J.*, November 1982, pp. 95–100.

10. Wisseman, W. R., ''GaAs MMICs—A Technology Assessment,'' *Microwave J.*, November 1982, pp. 20–34 (especially p. 22).

11. Ch'en, D. R., and D. R. Decker, ''MMICs—The Next Generation of Microwave Components,'' *Microwave J.*, May 1980, pp. 67–78.

12. Gardner, P. D., et al., ''$Ga_{0.47}In_{0.53}As$ Metal Insulator Field-Effect Transistors (MISFETs) for Microwave Frequency Applications,'' *RCA Rev.* **42**(4):542–556 (December 1981).

13. Manley, J. M., and H. E. Rowe, ''Some General Properties of Non-linear Elements,'' pt. 1, ''General Energy Relations,'' *Proc. IRE*, July 1956, pp. 904–913 (*Original and classical reference*).

14. Miya, K. (ed.), *Satellite Communications Engineering*, Lattice Co., Ltd., Tokyo, 1975, pp. 297–302.

15. Considine, D. M. (ed.), *Van Nostrand's Scientific Encyclopaedia*, 5th ed., Van Nostrand, New York, 1976, pp. 2188–2189.

16. Fowler, L., and F. N. Savage, ''Rugged Miniaturised Parametric Amplifiers (for Military Applications),'' *Eur. Electron.*, issue 1, 1981.

17. (*Historical*) Esaki, L., ''New Phenomenon in Ge *p-n* Junctions,'' *Phys. Rev.*, vol. 109, 1958, pp. 603–604.

18. Blaine, C., ''Tunnel Diode Oscillators,'' *Microwave J.*, September 1982, pp. 156–157.

19. (*Historical*) Gunn, J. B., ''Microwave Oscillations of Current of III-IV Semiconductors,'' *Solid State Commun.*, vol. 1, September 1963, pp. 88–91.

20. Fank, F. B., J. D. Crowley and J. J. Berentz, ''InP Material & Device Development for mm-Waves,'' *Microwave J.*, June 1979, pp. 86–91.

21. Acket, G. A., R. Tijburg and P. J. de Waard, ''The Gunn Diode,'' *Philips Tech. Rev.* **32** (9–12):370–379 (1971, published in 1973).

22. Wandinger, L., ''mm-Wave InP Gunn Devices: Status and Trends,'' *Microwave J.*, March 1981, pp. 71–78.

23. Fank, F. B., and J. D. Crowley, ''Gunn Effect Devices Move Up in Frequency and Become More Versatile,'' *Microwave J.*, September 1982, pp. 143–147.

24. Magarshack, J., ''Gunn-Effect Oscillators and Amplifiers,'' *Philips Tech. Rev.* **32**(9–12):397–404 (1971, published in January 1973).

25. Hernday, P. R., and C. J. Enlow, ''A High-Performance 2- to 18-GHz Sweeper,'' *Hewlett-Packard J.*, March 1975, pp. 2–14.

26. (*Historical*) Read, W. T., ''A Proposed High-Frequency Negative Resistance Diode,'' *Bell Syst. Tech. J*, vol 37, March 1958, pp. 401–446.

27. (*Historical*) Johnston, R. L., B. C. DeLoach, Jr. and B. G. Cohen, ''A Silicon Diode Microwave Oscillator,'' *Bell System Tech. J.*, vol. 44, February 1965, pp. 369–372.

28. (*Historical*) Prager, H. J., K. K. N. Chang and S. Weisbrod, "High-Power, High-Efficiency Silicon Avalanche Diodes at Ultra High Frequencies," *Proc. IEEE (Corr.)*, vol. 55, April 1967, pp. 586–587.
29. Kuno, H. J., and T. T. Fong, "Solid State mm-Wave Sources and Combiners," *Microwave J.*, June 1979, p. 47–48, 73–75 & 85.
30. Masse, D., et al., "GaAs IMPATT diodes for Satellite Communications," *Microwave J.*, March 1982, pp. 71–78.
31. Davies, R., B. H. Newton and J. G. Summers, "The TRAPATT Oscillator," *Philips Tech. Rev.* **40**(4):99–111 (1982).
32. Chang, K., G. M. Hayashibara and F. Thrower, "140 GHz Silicon IMPATT Power Combiner Development," *Microwave J.*, June 1981, pp. 65–71 & 77.
33. Uhlir, A., Jr., "The Potential in Semiconductor Diodes in High-Frequency Communications," *Proc. IRE*, vol. 46, June 1958, pp. 1099–1115. (*Original and classical reference*)
34. Microwave Associates, Inc., PIN Diode Applications Bulletins 4301–4303, 4304A, 4305A and 4306B, Burlington, Mass., December 1972–January 1974.
35. Patel, S. D., and H. Goldie, "A 100 kW Solid-State Coaxial Limiter for L-Band," *Microwave J.*, pt. I, December 1981, pp. 61–65; pt. II, January 1982, pp. 91–97.
36. Anand, Y., "X-Band High Burnout Silicon Schottky Barrier Diodes," *Microwave J.*, March 1979, pp. 56–61.
37. Kerr, A. R., and Y. Anand, "Schottky Diode mm Detectors," *Microwave J.*, December 1981, pp. 67–71.
38. (*Historical*) Gordon, J. P., J. J. Zeigler and C. H. Townes, "The Maser—New Type of Microwave Amplifier, Frequency Standard, and Spectrometer," *Phys. Rev.*, vol. 99, August 1955, pp. 1264–1274.
39. (*Historical*) Shawlow, A. L., and C. H. Townes, "Infrared and Optical Masers," *Phys. Rev.*, vol. 112, December 1958, p. 1950.
40. (*Historical*) Maiman, T. H., "Optical and Microwave-Optical Experiments in Ruby," *Phys. Rev. Letters*, vol. 4, June 1960, p. 564.
41. Dalglish, H. N., et al., *Low-Noise Microwave Amplifiers*, Syndics of the Cambridge University Press, London, 1968.
42. Bar-Chaim, N., I. Ury and A. Yariv, "Integreated Optoelectronics," *IEEE Spectrum*, May 1982, pp. 38–45.
43. (*ITU*) *Telecommun. J.*, vol. 48, no. 11, November 1981, ("Special Issue on Optical Fibers" pt. 1, "Optical Components"): (*a*) Newman, D. H., M. R. Matthews and I. Garrett, "Sources for Optical Fibre Communications," pp. 673–679. (*b*) Smith, D. R., M. C. Brain and R. C. Hooper, "Receivers for Optical Fibre Communication Systems," pp. 680–685.
44. Bergh, A. A., J. A. Copeland and J. H. Wernick, "The Growing Role of LEDs," *Bell Lab. Rec.*, January 1982, pp. 7–9.
45. Lacy, E. A., *Fiber Optics*, Prentice-Hall, Englewood Cliffs, N. J., 1982, pp. 39–52.
46. Ibid., pp. 131–143.
47. *Microwave J.*, November 1983 (whole issue).

13 PULSE COMMUNICATIONS

Apart from telegraphy, there was hardly any pulse-type communication until just before World War II. Then the advent of television and radar quickly changed the picture. Now, a lot of telecommunication is in pulse (*digital*) form, and the proportion is increasing rapidly. This sharp increase in digital communication, increasingly at the expense of *analog* communication, is caused by two interworking factors. The first is the fact that a lot of information to be transmitted is in pulse form to start with, and so sending it in that form is clearly the simplest technique. The second factor has been the advent of large-scale integration, which has permitted the use of complex coding systems that take the best advantage of channel capacities. Accordingly, it is very important to have a good working knowledge of the fundamentals of pulse and data communications, the former covered in this chapter and the latter in Chap. 14.

To achieve the above aim, this chapter is divided into three major parts. The first deals with *information theory,* touched upon in Chap. 1. This is a treatment of what is sent through a communications system, rather than the system itself. Until the excellent pioneering efforts of Shannon [1] and his colleagues, which culminated in the late 1940s, hardly any such work had been carried out, but now it is commonplace to talk about binary systems, bits, dits, and channel capacities. These topics will be covered to familiarize students with the measurement of information rates and capacities.

The second section describes various *pulse modulation systems.* Here, we are concerned with the transmission of continuous (or analog) signals in the form of pulses, by use of various sampling techniques. Among the systems discussed will be *pulse-code modulation* (PCM) and *delta modulation,* both much in demand for the transmission of telephony and a host of other communications. Methods of generating and decoding each of the pulse modulation systems will be introduced. Time-division multiplexing, i.e., the sequential sending of pulses from many sources through the one channel, will be covered in Sec. 15-1.2.

Finally, we look at two classes of pulse communications. The first is the traditional one: *telegraphy,* which also encompasses *telex*. This, of course, is a case of sending words rather than speech, generally in one direction only. To that extent, it may be thought of as not being real time, in the sense that a two-way conversation is real time. There is thus a fair degree of latitude in encoding and

modulating telegraphy—fundamentals of both functions will be treated. Finally *telemetry* is introduced as an important and interesting application of pulse communications.

13-1 INFORMATION THEORY

Information theory is a quantitative body of knowledge which has been established about "information," to enable systems designers and users to use the channels allocated to them as efficiently as possible. It is necessary to assign "information" a precise value if one is to deal scientifically with it. For transmission systems, "information" means exactly the same as it does in other situations, as long as it is realized that "meaning" is quite unimportant when it comes to measuring the quantity of information. This may come as a shock, until one considers the fact that "information" here is a physical quantity, such as mass. Accordingly, one determines the mass of a given object in kilograms, and such mass is not in the least determined by the type of material weighed. One pound of platinum does, after all, have the same mass as one pound of horseradish!

Information theory is thus seen to be the scientific study of information and of the communications systems designed to handle it. These systems include telegraphy (which just about gave birth to information theory), radio communications, computers and many other systems concerning themselves with the processing or storage of signals, including even molecular biology. The theory is used to establish, precisely and mathematically, the rate of information issuing from any source, the information capacity of any channel, system or storage device, and the efficiency of codes by means of which this information is sent. The type of code used in any one case will depend on the form and type of information sent and also, most importantly, on the noise prevailing in the communications system.

13-1.1 Information in a Communications System

Communications system The general communications system has already been described in detail in the first chapter; it is Shannon's familiar *information source—transmitter—channel—receiver—destination* system of Fig. 1-1. However, the subject was at the time treated as an introduction to communication systems in general, rather than from the point of view of information theory, as in the present case.

The most fundamental idea of information theory is that information is a measurable physical quantity, such as mass, heat or any other form of energy. This may be made quite clear with an analogy.[1]

> for example, we can imagine an information source to be like a lumber mill producing lumber at a certain point. The channel . . . might correspond to a conveyor system for transporting the lumber to a second point. In such a situation there are two important

[1] Courtesy of Encyclopaedia Britannica, Inc.

quantities: the rate R (in cubic feet per second) at which lumber is produced at the mill, and the capacity C (in cubic feet per second) of the conveyor. These two quantities determine whether or not the conveyor system will be adequate for the lumber mill. If the rate of production R is greater than the conveyor capacity C, it will certainly be impossible to transport the full output of the mill; there will not be sufficient space available. If R is less than or equal to C, it may or may not be possible, depending on whether the lumber can be packed efficiently in the conveyor. Suppose, however, that we allow ourselves a sawmill at the source. This corresponds in our analogy to the encoder or transmitter. Then the lumber can be cut up into small pieces in such a way as to fill out the available capacity of the conveyor with 100% efficiency. Naturally in this case we should provide a carpenter shop at the receiving point to fasten the pieces together in their original form before passing them on to the consumer.

The analogy is very apt and sound; both the rate of production of information by the source and the carrying capacity of a channel can be measured to determine compatibility. The fact that information *can* be measured was one of the earliest and most important results of information theory, and on this important basis most of the other work is established.

Measurement of information Having said what information *is not* (it is not *meaning*), we now state specifically what information *is*. Accordingly, *information is defined as the choice of one message out of a finite set of messages*. Meaning is immaterial; in this sense, a table of random numbers may well contain as much information as a table of world track-and-field records. Indeed, it may well be that a cheap fiction book contains more information than this textbook, if it happens to contain a larger number of choices from a set of possible messages (the set being the complete English language in this case). Also, when measuring information, it must be taken into account that some choices are more likely than others and therefore contain less information. Any choice that has a probability of 1, i.e., is completely unavoidable, is fully redundant and, therefore, contains no information. An example is the letter "u" in English when it follows the letter "q."

The binary digit (bit) The simplest possible choice is the one involving two equally likely events, called *equiprobable events:* each has the probability[2] of exactly 0.5. This choice occurs in a large number of everyday events, such as tossing a coin or predicting the color of the next card from a pack held face down. This being the case, it is convenient to use the information content of this kind of choice as the basic unit of information. Thus, *the bit is defined as the quantity of information required to permit the correct selection of one out of a pair of equiprobable events*. It therefore follows that

[2]Normally, the words "possibility" and "probability" are synonyms, but here they are used with their mathematical (or statistical) meanings. Hence "possibility" retains its everyday meaning, as something that may happen, in the sense of not being impossible, whereas "probability" is the mathematical chance or likelihood of the event taking place. Thus the probability of "certain" is 1, that of "impossible" is 0, and that of "fifty-fifty" is 0.5.

$$\left[\begin{array}{l}\text{Number of bits of information which must} \\ \text{be given to enable the correct selection of one} \\ \text{event from a set of } N \text{ equiprobable events}\end{array}\right] = \log_2 N \qquad \textbf{(13-1)}$$

Equation (13-1) may be derived by logic. Returning to the prediction of the color of the next card, we see that this is a simple choice of one out of two equiprobable events. One bit of information is required for its correct predictions; i.e., one must be told whether the next card is black or red (let black = 1, red = 0). If the correct prediction of the *suit* of the next card is required, that is a slightly more complicated selection sequence, but it may be indicated as follows. Let club or heart = 1, and spade or diamond = 0. Instead of saying "club," we can say "11," and in a similar fashion spade = 10, heart = 01, and diamond = 00. Thus, to indicate any of the suits, we provide *two* bits of information. As can be seen, there are two successive choices out of pairs of equiprobable events, to indicate 0 or 1 each time. Of course, $\log_2 4 = 2$, so that Eq. (13-1) is vindicated.

Using this successive choice of one out of two, it is possible to extend the system indefinitely. Figure 13-1 shows diagrammatically a switching system designed to select any one event out of 16 equally probable ones, by means of bit-by-bit switching. It is seen that a total of 4 bits are required to allow correct selection of any one of the 16 numbers, since four successive choices of one out of two equiprobable events must be made. To select the number 6, for instance, the information or orders given would be up, down, down, up, or UDDU. Perhaps they would be 0110 or even OPPO, where 0, meaning "up," stands for a binary zero or no pulse, whereas 1, meaning "down," is the other binary digit (replaced by P, or pulse, in a signaling system). Once again it will be seen that the number of bits required is exactly as predicted by Eq. (13-1) ($\log_2 16 = 4$), so that the equation has been justified. It is apparent that the quantity of information required, as measured in bits, depends solely on the probability of predicting the information. The lower the probability, the more information is given. It is equally apparent that since choice between alternatives plays an

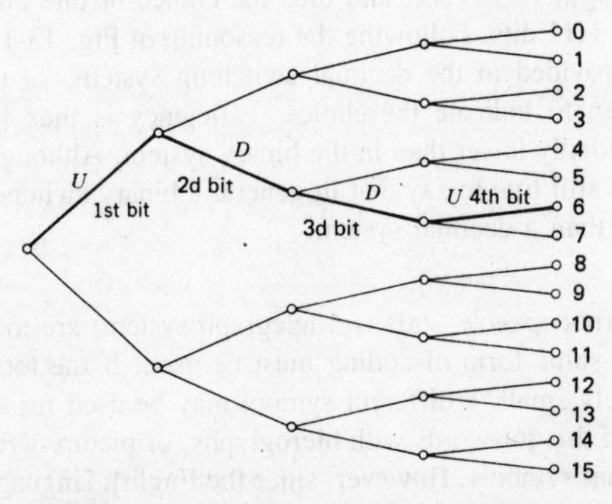

FIGURE 13-1 Selection of 1 out of 16 equiprobable events using binary switching; 4 bits are required to select No. 6.

important part in information theory, students should have some familiarity with the binary number system; hence the insistence on this in the Preface and the limited review provided in the following chapter.

13-1.2 Coding

In measuring the amount of information, we have so far concentrated on a choice of one from 2^n equiprobable events, using the binary system; thus the number of bits involved has always been an integer. In fact, *if we do use the binary system for signaling, the number of bits required will always be an integer*. For example, it is not possible to choose one from a set of 13 equiprobable events in the binary system by giving 3.7 bits ($\log_2 13 = 3.7$). It is necessary to give 4 bits of information, which corresponds to having the switching system of Fig. 13-1 with the last three places never used. The efficiency of using a binary system for the selection of one of 13 equiprobable events is

$$\eta = \frac{3.7}{4} \times 100 = 92.5 \textit{ percent}$$

which is considered a high efficiency. The situation is that a choice of one from 13 conveys 3.7 bits of information, but if we are going to use a binary system of selection or signaling, 4 bits must be given and the resulting inefficiency accepted.

At this point, it is worth noting that the binary system is used widely but not exclusively. The decimal system is also used, and here the unit of information is the decimal digit, or *dit*. A choice of one from a set of 10 equiprobable events involves 1 dit of information and may be made, in the decimal system, with a rotary switch. It is simple to calculate that since we have $\log_2 10 = 3.32$,

$$1 \text{ dit} = 3.32 \text{ bits} \tag{13-2}$$

Just as a matter of interest, it is possible to compare the efficiency of the two systems by noting that $\log_{10} 13 = 1.11$, and thus the choice of one out of 13 equiprobable events involves 1.11 dits. Following the reasoning of Fig. 13-1, 100 switching positions must be provided in the decimal switching system, so that 2 dits of information will be given to indicate the choice. Efficiency is thus $\eta = 1.11/2 \times 100 = 55.5$ percent, decidedly lower than in the binary system. Although this is only an isolated instance, it is still true to say that in general a binary switching or coding system is more efficient than a decimal system.

Baudot code If *words* (not *speech*—this is a telegraph system) are to be sent by a communications system, some form of coding must be used. If the total number of words or ideas is relatively small, a different symbol may be used for each word or object. The Egyptians did this for words with hieroglyphs, or picture writing, and we do it for objects with circuit symbols. However, since the English language contains at least 800,000 words and is still growing (who had heard of "quasarology," or "ecodoom" 10 years ago?), this method is out of the question. Alternatively, a different pulse, perhaps having a different width or amplitude, may be used for each letter and symbol. Since here are 26 letters in English and roughly the same number of other symbols, this gives a total of about 50 different pulses. Such a system could be used, but it never is, because it would be very vulnerable to distortion by noise.

If we consider pulse-amplitude variation and amplitude modulation, then each symbol in such a system would differ by 2 percent of modulation from the previous, this being only one-fiftieth of the total amplitude range. Thus the word ''stop'' might be transmitted as /38/40/30/32/, each figure being the appropriate percentage modulation. Suppose a very small noise pulse, having an amplitude of only one-fiftieth of the peak modulation amplitude, happens to superimpose itself on the transmitted signal at that instant. This signal will be transformed into /40/42/32/34/, which reads ''tupq'' in this system and is quite meaningless. It is obvious that a better system must be found. As a result of this, almost all the systems in use are binary systems, in which the sending device sends fully modulated pulses (''marks'') or no-pulses (''spaces''). Noise now has to compete with the full power of the transmitter, and it will be a very large noise pulse indeed that will convert a transmitted mark into a space, or vice versa.

Since information in English is drawn from 26 choices (letters), there must be on the average more than 1 bit per letter. In fact, since $\log_2 26 = 4.7$ and a binary sending system is to be used, each letter must be represented by 5 bits. If all symbols are included, the total number of different signals nears 60. The system is in use with tele-typewriters, whose keyboards are similar to those of ordinary typewriters. It is thus convenient to retain 5 bits per symbol and to have carriage-shift signals for changing over from letters to numerals, or vice versa.

The CCITT No. 2 code shown in Fig. 13-2*a* is an example of how a series of five binary signals can indicate any one from up to 60 letters and other symbols. The code is based on an earlier one proposed by J. M. E. Baudot, the only difference being an altered allocation of code symbols to various letters. In the middle of a message, a word of n letters is indicated by $n + 1$ bits; the last bit is used for the space. For example, the center portion of the message ''I have caught 25 fish today'' would read as in Fig. 13-3. The use of codes for the transmission of data will be covered more fully in the next chapter.

The ARQ[3] code was developed from the Baudot code by H. C. A. Van Duuren in the late 1940s, and is an example of an error-detecting code widely used in radio telegraphy. As shown, 7 bits are used for each symbol, but of the 128 possible combinations that exist, only those containing 3 marks and 4 spaces are used. There are 35 of these, and 32 of them are used as shown in Fig. 13-2*b*. The advantage of this system is that it offers protection against single errors. If a signal arrives so mutilated that some of the code groups contain a mark-to-space proportion other than 3:4, an ARQ signal is sent, and the mutilated information is retransmitted. There is no such provision for the detection of errors in the Baudot-based codes, but they do have the advantage of requiring only 5 bits per symbol, as opposed to 7 here.

The Hartley law The Baudot code was shown as an example of a simple and widely used binary code, but it may also be employed as a vehicle for providing a very

[3]This is a telegraphic code group and means *automatic request for repetition*.

Figures	Letters	CCITT-2 code					ARQ code						
		1	2	3	4	5	1	2	3	4	5	6	7
—	*A*	●	●						●	●		●	
?	*B*	●			●	●			●	●			●
:	*C*		●	●	●		●			●	●		
Who are you?	*D*	●			●				●	●	●		
3	*E*	●						●	●	●			
%	*F*	●		●	●				●			●	●
@	*G*		●		●	●	●	●					●
£	*H*			●		●	●		●			●	
8	*I*		●	●			●	●	●				
Bell	*J*	●	●		●			●				●	●
(	*K*	●	●	●	●					●		●	●
)	*L*		●			●	●	●				●	
.	*M*			●	●	●	●		●				●
,	*N*			●	●		●		●		●		
9	*O*				●	●	●				●	●	
Ø	*P*		●	●		●	●			●		●	
1	*Q*	●	●	●		●				●	●		●
4	*R*		●		●		●	●			●		
'	*S*	●		●				●		●		●	
5	*T*					●	●				●		●
7	*U*	●	●	●				●	●			●	
=	*V*		●	●	●	●	●			●			●
2	*W*	●	●			●		●			●		●
/	*X*	●		●	●	●			●		●	●	
6	*Y*	●		●		●			●		●		●
+	*Z*	●				●		●	●				●
Carriage return					●		●					●	●
Line feed			●				●		●	●			
Figures shift		●	●		●	●		●			●	●	
Letters shift		●	●	●	●	●				●	●	●	
Space				●			●	●		●			
Unperforated tape											●	●	●
	(*a*)						(*b*)						

FIGURE 13-2 Telegraphic codes. (*a*) CCITT-2; (*b*) ARQ.

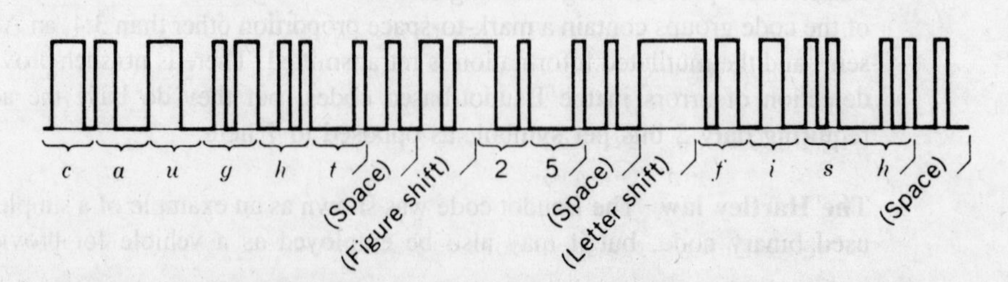

FIGURE 13-3 Example of use of CCITT No. 2 code.

fundamental and important law of information theory. This is the Hartley law [2] and may be demonstrated by logic.

A quick glance at the CCITT-2 code of Fig. 13-2*a* reveals that, on the average, just as many bits of information are indicated by pulses as by no-pulses. This means, of course, that the signaling rate in pulses per second depends on the information rate in bits per second at that instant. Now, the pulse rate is by no means constant. If the letters "Y" and "R" are sent one after the other, the pulse rate will be at its maximum and exactly equal to half the bit rate. At the other end of the scale, the letter "E," followed by "T," would provide a period of time during which no pulses are sent. Accordingly, it is seen that when information is sent in a binary code at a rate of b bits per second, the instantaneous pulse rate varies randomly between $b/2$ pulses per second and zero. It follows that a band of frequencies, rather than just a single frequency, is required to transmit information at a certain rate with a particular system. It will be recalled from Chap. 1 that pulses consist of the fundamental frequency and harmonics, in certain proportions. However, if the harmonics are filtered out at the source, and only fundamentals are sent, the original pulses can be re-created at the destination (with multivibrators). This being the case, the highest frequency required to pass b bits per second in this system is $b/2$ Hz (the lowest frequency is still 0). It may thus be said that, if a binary coding system is used, the channel capacity in bits per second is equal to twice the bandwidth in hertz. This is a special case of the Hartley law and is expanded in Sec. 14-2.2. The general case states that, *in the total absence of noise,*

$$C = 2\ \delta f \log_2 N \tag{13-3}$$

where C = channel capacity, bits per second
δf = channel bandwidth, Hz
N = number of coding levels

When the binary coding system is used, the above general case is reduced to $C = 2\delta f$, since $\log_2 2 = 1$. The Hartley law shows that the bandwidth required to transmit information at a given rate is proportional to the information rate. Also, in the absence of noise, the Hartley law shows that the greater the number of levels in the coding system, the greater the information rate that may be sent through a channel. What happens when noise is present was indicated in the preceding section (i.e., "tupq" for "stop") and will be enlarged upon in the next section. Meanwhile, extending the Hartley law to its logical conclusion, as was done by the originator, we have

$$\begin{aligned} H &= Ct \\ &= 2\ \delta f t \log_2 N \end{aligned} \tag{13-4}$$

where H = total information sent in a time t, bits
t = time, seconds

The foregoing assumes, of course, that an information source of sufficient capacity is connected to the channel.

13-1.3 Noise in an Information-Carrying Channel

As has already been indicated, noise has an influence on the information-carrying capacity of a channel. This idea will now be explored further, as will means of combating noise.

Effects of noise That noise has some harmful effect has already been demonstrated. To quantify the effect, consider again the earlier suggestion that each letter in the alphabet could be represented by a different signal amplitude, using 32-scale code. If this were done, the information flow would be greatly speeded (according to the Hartley law), since each letter would now be represented by one symbol instead of five. However, unless transmitting power were raised tremendously, noise would cause so many errors as to make the multilevel system useless. The truth of this may be shown by considering the power required for the binary coding system and for any other system under the same noise conditions.

As can be appreciated, for a given transmission and coding system, there is such a thing as a threshold noise level; as long as noise does not exceed it, practically no errors occur. When a binary code is used, noise must compete with the full power of the transmitter to affect the signal, and practical results show that a signal-to-noise ratio of 30 dB ensures virtually error-free reception. This corresponds to a noise power of 1/1000 of signal power, i.e., an rms noise voltage of 1/31.6 of the rms signal voltage maximum. Let us take this S/N ratio as a practical requirement and consider the effect of this condition on increased signaling levels.

If it is now decided to double signaling speed by doubling the number of amplitude levels to four, the transmitted power will have to be increased to retain the 30-dB S/N ratio at the receiver. In terms of the maximum permitted amplitude, the new levels will be 0, 1/3, 2/3 and 1, where they were 0 and 1 in the binary system. This means that the difference in voltage levels is now one-third of what it was, the difference in power levels is one-ninth, and therefore *transmitted power must be multiplied ninefold when the signaling speed is doubled*. Similarly, if an eight-level code is used, each amplitude level difference is one-seventh of the original, necessitating a 49-fold increase in transmitting power to return to the original 30-dB S/N ratio. Finally, if the proposed 32-level code were used, the power transmitted would have to be increased by a factor of $31^2 = 961$. It is easy to deduce that this power increase is logarithmic and is given by

$$\frac{P_n}{P_2} = (n - 1)^2 \tag{13-5}$$

where n = number of levels in the code
P_n = power required in the n-level code
P_2 = power level required in the binary code

It can be seen that, in noise-limited conditions, the advantage of a binary system is such as to outweigh almost all other considerations.

Capacity of a noisy channel The preceding section showed that transmitted power must be raised considerably, if a constant signal-to-noise ratio is to be kept when the number of coding levels is increased to raise the signaling speed. The Shannon-Hartley theorem gives a formula for the capacity of a channel when its bandwidth and noise level are known. This capacity is

$$C = \delta f \log_2(1 + S/N) \tag{13-6}$$

where C = channel capacity, bits per second
δf = bandwidth, Hz
S/N = ratio of total signal power to total random noise power at the input to the receiver, within the frequency limits of this channel, i.e., over the bandwidth δf

The Shannon-Hartley theorem is too complex to derive here. However, its validity is demonstrated in [3], pp. 421–423.

Example 13-1 Calculate the capacity of a standard 4-kHz telephone channel with a 32-dB signal-to-noise ratio.

Standard telephone channels occupy the frequency range of 300 to 3400 Hz. Also, the actual signal-to-noise ratio is antilog (32/10) = antilog (3.2) = 1585. We thus have

$$\begin{aligned} C &= \delta f \log_2 (1 + S/N) = 3100 \times \log_2 (1 + 1585) \\ &= 3100 \times \log_2 1586 = 3100 \times 10.63 \\ &= 32{,}953 \text{ bits per second} \end{aligned}$$

The Shannon-Hartley theorem shows a limit that cannot be exceeded by the signaling speed in a channel in which the noise is purely random. It may be used as a very good approximation for the ultimate channel capacity of most transmission channels, although practical noise distributions are never perfectly random. Example 13-1 shows the limiting channel speed for a typical telephone channel to be approximately 33 kilobits per second. Speeds used in practice over such channels do not normally exceed 10.8 kilobits per second (10.8 kbps), as discussed in Sec. 14–2.2. If the answer to Example 13-1 is equated with Eq. (13-3), it will be seen that 39.8 code levels would be required to reach the Shannon speed limit for this channel, resulting in a system that is too complex in practice.

It would be incorrect to assume that doubling the bandwidth of a noise-limited channel will automatically double its capacity—that would be misinterpreting Eq. (13-6). Consider the following example.

Example 13-2 A system has a bandwidth of 4 kHz and a signal-to-noise ratio of 28 dB at the input to the receiver. Calculate

(*a*) its information-carrying capacity

(*b*) the capacity of the channel if its bandwidth is doubled, while the transmitted signal power remains constant.

(*a*) S/N = antilog (28/10) = antilog (2.8) = 631

$$\begin{aligned} C_1 &= 4000 \times \log_2 (1 + 631) = 4000 \times 9.304 \\ &= 37{,}216 \text{ bits per second} \end{aligned}$$

(*b*) If the signal-to-noise ratio in the 4-kHz channel is 631:1, this can be interpreted as a noise power of 1 mW at some point in the channel where the signal power is 631 mW. The signal power is unchanged here when the bandwidth is doubled, but Eq.

(2-1) showed that the noise power in a system is doubled when the bandwidth of the system is doubled. We thus have

$$C_2 = 8000 \times \log_2 (1 + 631/2) = 8000 \times \log_2 (1 + 315.5)$$
$$= 8000 \times 8.306 = 66{,}448 \text{ bits per second}$$

As a matter of interest, taking a ratio of the two capacities gives

$$C_2/C_1 = 66{,}488/37{,}216 = 1.785$$

It is seen from the above example that capacity was increased, but certainly not doubled, when the bandwidth was doubled. This implies that useful possibilities of trading bandwidth for signal-to-noise ratio exist. Indeed, such tradeoffs are often made in system design, especially in power-limited situations. It also should be realized that, if channel capacity seems low in a given situation, this does not mean that a wanted amount of information cannot be sent over a given channel. As Eq. (13-4) amply shows, it merely means that sending this amount of information takes longer.

Finally, it must be emphasized that the Shannon-Hartley theorem represents a fundamental limitation. *The only consequence of trying to exceed the Shannon limit would be an unacceptable error rate*. In practical transmission systems, error rates greater than 1 error in 10^5 are generally considered not good enough.

Redundancy The preceding has assumed, although this was not stated explicitly at the time, that all messages sent through the noise-limited channel were *unpredictable*. That is, they were assumed to be random, without any redundancy whatever. If redundant messages are sent, it is generally possible to work out from context the correct version of an erroneous message. In this fashion, error rates can be very significantly reduced.

Redundancy is that which is not essential—it can be removed from a signal and yet leave the remainder intelligible. All those who have sent telegrams which contain only the key words, leaving out all the articles and simple verbs, for instance, will have taken advantage of the redundancy in the language to save money. The letter "u" always follows the letter "q" in English, and so it is fully redundant. Anyone with an ounce of imagination could work out the correct spelling of long words if they were transmitted with a couple of non-key letters missing. By sending a message over a noise-limited channel, from which most redundancy had been eliminated, it would be possible to increase the effective signaling speed quite substantially.

It is also possible to go the other way, deliberately introducing redundancy because the error rate of a channel is too high. The ARQ 7-bit code of Sec. 13-1.2 can obviously, because of its deliberate redundancy, be used in noise conditions where the CCITT-2 5-bit code would be useless. The following chapter will discuss several data transmission codes which deliberately introduce redundant bits to permit their use. Similarly, when sending numbers over a noisy channel, it would be possible to introduce redundancy by sending each number as a triplet. For example, the number 195 could be sent as 111999555, in the hopes that in marginal noise conditions such redundancy would be sufficient to cancel out any errors.

Redundancy is seen as a means of reducing error rates, sometimes very greatly, in noisy conditions. However, the *TANSTAAFL*[4] principle regrettably applies to this

[4] *There ain't no such thing as a free lunch!*

situation. Because more information is being sent, either it will take longer to send, or it will require a greater bandwidth to send in a given time. If the two telegraphic codes are taken as examples, it is seen that, with a given bandwidth, a message in ARQ (7 bits per letter) will take 1.4 times[5] as long to send as the same message in CCITT-2 (5 bits per letter). However, if the difference is between a slower, intelligible message, and a faster, useless one, the price is worth paying.

For students seeking additional information on the topics of Sec. 13-1, [3], chap. 13, and [4], chaps. 11 and 12, are highly recommended. The former is more mathematical and the latter more descriptive.

13-2 PULSE MODULATION

13-2.1 Introduction—Types

Pulse modulation may be used to transmit *analog* information, such as continuous speech or data. It is a system in which continuous waveforms are *sampled* at regular intervals. Information regarding the signal is transmitted only at the sampling times, together with any synchronizing pulses that may be required. At the receiving end, the original waveforms may be reconstituted from the information regarding the samples, if these are taken frequently enough. Despite the fact that information about the signal is not supplied continuously, as in AM and FM, the resulting receiver output can have negligible distortion.

Pulse modulation may be subdivided broadly into two categories, *analog* and *digital*. In the former, the indication of sample amplitude may be infinitely variable, while in the latter a code which indicates the sample amplitude to the nearest predetermined level is sent. *Pulse-amplitude* and *pulse-time* modulation, to be treated next, are both analog, while the *pulse-code* and *delta* modulation systems of the later sections are both digital. All the modulation systems to be discussed have sampling in common, but they differ from each other in the manner of indicating the sample amplitude.

The two types of analog pulse modulation, pulse-amplitude and pulse-time modulation, correspond roughly to amplitude and frequency modulation. The digital systems are quite unlike anything that we have so far studied. The reasons why pulse modulation is increasingly used instead of the more familiar continuous modulation systems will become apparent as the chapter progresses.

Pulse-amplitude modulation (PAM) Pulse-amplitude modulation, the simplest form of pulse modulation, is illustrated in Fig. 13-4. It forms an excellent introduction to pulse modulation in general. PAM is a pulse modulation system in which the signal is sampled at regular intervals, and each sample is made proportional to the amplitude of the signal at the instant of sampling. The pulses are then sent by either wire or cable, or else are used to modulate a carrier. As shown in Fig. 13-4, the two types are double-polarity PAM, which is self-explanatory, and single-polarity PAM, in which a fixed dc level is added to the signal, to ensure that the pulses are always positive. As will be seen shortly, the ability to use constant-amplitude pulses is a major advantage

[5] But see also how letters are separated from each other, in Sec. 13-3.1.

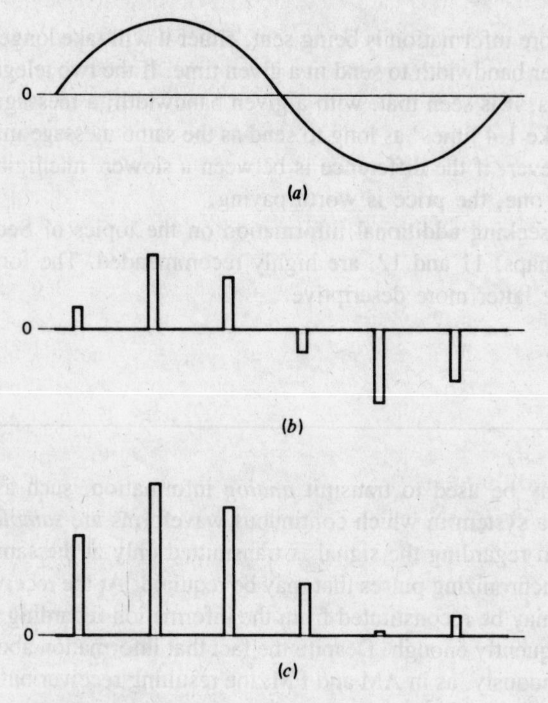

FIGURE 13-4 Pulse-amplitude modulation. (*a*) Signal; (*b*) double-polarity PAM; (*c*) single-polarity PAM.

of pulse modulation, and since PAM does not utilize constant-amplitude pulses, it is infrequently used. When it is used, the pulses *frequency-modulate* the carrier.

It is very easy to generate and demodulate PAM. In a generator, the signal to be converted to PAM is fed to one input of an AND gate. Pulses at the sampling frequency are applied to the other input of the AND gate to open it during the wanted time intervals. The output of the gate then consists of pulses at the sampling rate, equal in amplitude to the signal voltage at each instant. The pulses are then passed through a pulse-shaping network, which gives them flat tops. As mentioned above, frequency modulation is then employed, so that the system becomes PAM-FM. In the receiver, the pulses are first recovered with a standard FM demodulator. They are then fed to an ordinary diode detector, which is followed by a low-pass filter. If the cutoff frequency of this filter is high enough to pass the highest signal frequency, but low enough to remove the sampling frequency ripple, an undistorted replica of the original signal is reproduced.

Pulse-time modulation (PTM) In PTM the signal is sampled as before, but the pulses indicating instantaneous sample amplitudes themselves all have a constant amplitude. However, one of their timing characteristics is varied, being made proportional to the sampled signal amplitude at that instant. The variable characteristic may be the width, position or frequency of the pulses, so that three different types of PTM are possible. Pulse-frequency modulation has no significant practical applications and will be omitted. The other two forms of PTM will be discussed below. It should be noted that all forms of PTM have the same advantage over PAM as frequency modulation has over amplitude modulation. In all of them the pulse amplitude remains constant, so that amplitude limiters can be used to provide a good degree of noise immunity.

Sampling theorem The sampling theorem states that, *if the sampling rate in any pulse modulation system exceeds twice the maximum signal frequency, the original signal can be reconstructed in the receiver with minimal distortion.* The proof is beyond the scope of this book, but it may be found in [3], pp. 157–162. The sampling theorem is used in practice to determine minimum sampling speeds. Consider pulse modulation used for speech. Transmission is generally over standard telephone channels, so that the audio frequency range is 300 to 3400 Hz. For this application, a sampling rate of 8000 samples per second is almost a worldwide standard. This pulse rate is, as can be seen, comfortably more than twice the highest audio frequency. The sampling theorem is satisfied, and the resulting system is free from sampling error.

13-2.2 Pulse-Width Modulation

Introduction The pulse-width modulation of PTM is also often called PDM (pulse-duration modulation) and, less often, PLM (pulse-length modulation). In this system, as shown in Fig. 13-5, we have a fixed amplitude and starting time of each pulse, but the width of each pulse is made proportional to the amplitude of the signal at that instant. In Fig. 13-5, there may be a sequence of signal sample amplitudes of (say) 0.9, 0.5, 0 and −0.4 V. These can be represented by pulse widths of 1.9, 1.5, 1.0 and 0.6 μs, respectively. The width corresponding to zero amplitude was chosen in this system to be 1.0 μs, and it has been assumed that signal amplitude at this point will vary between the limits of +1 V (width = 2 μs) and −1 V (width = 0 μs). Zero amplitude is thus the average signal level, and the average pulse width of 1 μs has been made to correspond to it. In this context, a negative pulse width is not possible. It would make the pulse end before it began, as it were, and thus throw out the timing in

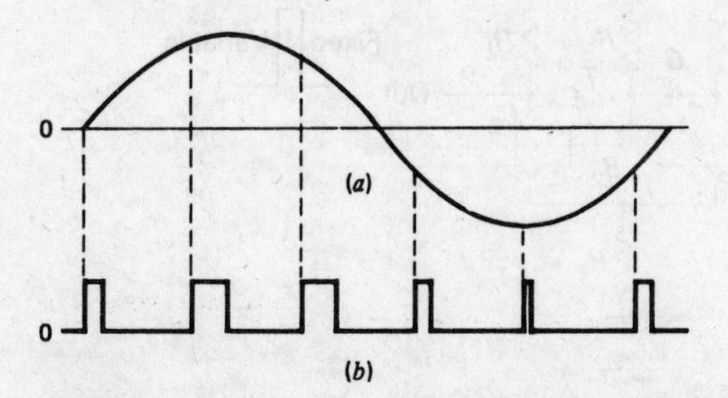

FIGURE 13-5 Pulse-width modulation. (*a*) Signal; (*b*) PWM (width variations exaggerated).

the receiver. If the pulses in a practical system have a recurrence rate of 8000 pulses per second, the time between the commencement of adjoining pulses is $10^6/8000 = 125\ \mu s$. This is adequate not only to accommodate the varying widths but also to permit *time-division multiplexing,* as explained in Chap. 15.

Pulse-width modulation has the disadvantage, when compared with *pulse-position modulation* (PPM), which will be treated next, that its pulses are of varying width and therefore of varying power content. This means that the transmitter must be powerful enough to handle the maximum-width pulses, although the average power transmitted is perhaps only half of the peak power. On the other hand, PWM still works if synchronization between transmitter and receiver fails, whereas pulse-position modulation does not, as will be seen.

Generation and demodulation of PWM Pulse-width modulation may be generated by applying trigger pulses (at the sampling rate) to control the starting time of pulses from a monostable multivibrator, and feeding in the signal to be sampled to control the duration of these pulses. The circuit diagram for such an arrangement is shown in Fig. 13-6.

The emitter-coupled monostable multivibrator of Fig. 13-6 makes an excellent voltage-to-time converter, since its gate width is dependent on the voltage to which the capacitor C is charged. If this voltage is varied in accordance with a signal voltage, a series of rectangular pulses will be obtained, with widths varying as required. Note that the circuit does the twin jobs of sampling and converting the samples into PWM.

It will be recalled that the stable state for this type of multivibrator is with T_1 OFF and T_2 ON. The applied trigger pulse switches T_1 ON, whereupon the voltage at C_1 falls as T_1 now begins to draw collector current, the voltage at B_2 follows suit and T_2 is switched OFF by regenerative action. As soon as this happens, however, C begins to charge up to the collector supply potential through R. After a time determined by the supply voltage and the RC time constant of the charging network, B_2 becomes sufficiently positive to switch T_2 ON. T_1 is simultaneously switched OFF by regenerative action and stays OFF until the arrival of the next trigger pulse. The voltage that the base of T_2 must reach to allow T_2 to turn on is slightly more positive than the voltage across the common emitter resistor R_k. This voltage depends on the current flowing through

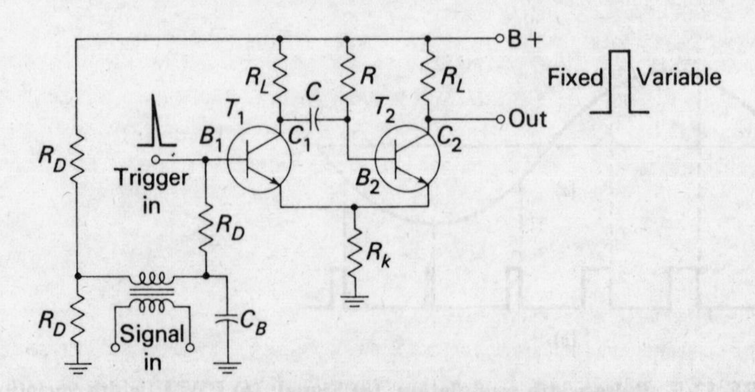

FIGURE 13-6 Monostable multivibrator generating pulse-width modulation.

the circuit, which at the time is the collector current of T_1 (which is then ON). The collector current, in turn, depends on the base bias, which is governed by the instantaneous changes in the applied signal voltage. Thus, admittedly in a roundabout fashion, the applied modulation voltage controls the voltage to which B_2 must rise to switch T_2 ON. Since this voltage rise is linear, the modulation voltage is seen to control the period of time during which T_2 is OFF, that is, the pulse duration. It should be noted that this pulse duration is very short compared to even the highest signal frequencies, so that no real distortion arises through changes in signal amplitude while T_2 is OFF.

The demodulation of pulse-width modulation is quite a simple process. PWM is merely fed to an integrating circuit from which a signal emerges whose amplitude at any time is proportional to the pulse width at that time.[6]

13-2.3 Pulse-Position Modulation (PPM)

The amplitude and width of the pulses is kept constant in this system, while the position of each pulse, in relation to the position of a recurrent reference pulse,[7] is varied by each instantaneous sampled value of the modulating wave. As mentioned in connection with PWM, pulse-position modulation has the advantage of requiring constant transmitter power output, but the disadvantage of depending on transmitter-receiver synchronization.

Generation and demodulation of PPM Pulse-position modulation may be obtained very simply from PWM, as shown in Fig. 13-7. Considering PWM and its generation again, it is seen that each such pulse has a leading edge and a trailing edge (like any other pulse, of course). However, in PWM the locations of the leading edges are fixed, whereas those of the trailing edges are not. Their position depends on pulse width, which is determined by the signal amplitude at that instant. Thus, it may be said that *the trailing edges of PWM pulses are, in fact, position-modulated*. The method of obtaining PPM from PWM is thus accomplished by "getting rid of" the leading edges and bodies of the PWM pulses. This is surprisingly easy to achieve.

Figure 13-7*a* and *b* shows, once again, PWM corresponding to a given signal. If the train of pulses thus obtained is differentiated, then, as shown in Fig. 13-7*c*, another pulse train results. This has positive-going narrow pulses corresponding to leading edges and negative-going pulses corresponding to trailing edges. If the position corresponding to the trailing edge of an unmodulated[8] pulse is counted as zero displacement, then the other trailing edges will arrive earlier or later. They will therefore have a time displacement other than zero; this time displacement is proportional to the instantaneous value of the signal voltage. The differentiated pulses corresponding to the leading edges are removed with a diode clipper or rectifier, and the remaining pulses, as shown in Fig. 13-7*d*, are position-modulated.

[6] This principle is also employed in the very efficient so-called *class D* amplifiers. The integrating circuit most often used there is the loudspeaker itself.

[7] This means that the transmitter must send synchronizing pulses, to operate timing circuits in the receiver.

[8] An unmodulated PWM pulse is one that is obtained when the instantaneous signal value is zero. These pulses are appropriately labeled in Fig. 13-7*b*.

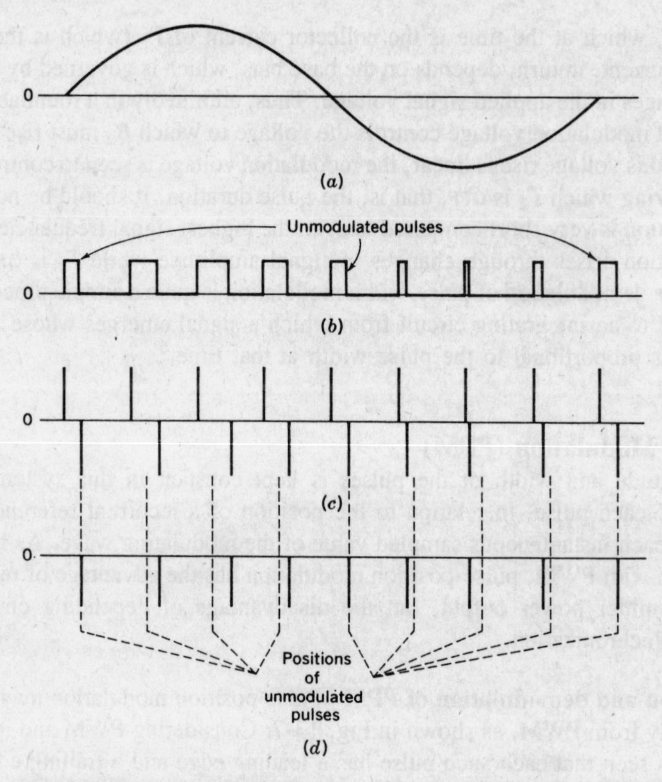

FIGURE 13-7 Generation of pulse-position modulation. (*a*) Signal; (*b*) PWM; (*c*) differentiated; (*d*) clipped (PPM).

When PPM is demodulated in the receiver, it is again first converted into PWM; this is done with a flip-flop, or bistable multivibrator. One input of the multivibrator receives trigger pulses from a local generator which is synchronized by trigger pulses received from the transmitter, and these triggers are used to switch OFF one of the stages of the flip-flop. The PPM pulses are fed to the other base of the flip-flop and switch that stage ON (actually by switching the other one OFF). The period of time during which this particular stage is OFF depends on the time difference between the two triggers, so that the resulting pulse has a width that depends on the time displacement of each individual PPM pulse. The resulting PWM pulse train is then demodulated as already described.

13-2.4 Pulse-Code Modulation

Pulse-code modulation is just as different from the forms of pulse modulation so far studied as they were from AM or FM. PAM and PTM differed from AM and FM because, unlike in those two continuous forms of modulation, the signal was sampled and sent in pulse form. However, like AM and FM, they were forms of *analog* communication—in all these forms a signal is sent which has a characteristic that is infinitely variable and proportional to the modulating voltage. In common with the other forms of pulse modulation, PCM also uses the sampling technique, but it differs from the others in that it is a *digital* process. That is, instead of sending a pulse train capable of continuously varying one of the parameters, the PCM generator produces a series of numbers, or digits (hence the name *digital* process). Each one of these digits, almost always in binary code, represents the *approximate amplitude* of the signal sample at that instant. The approximation can be made as close as desired, but it is always just that–an *approximation*.

Principles of PCM In PCM, the total amplitude range which the signal may occupy is divided into a number of standard levels, as shown in Fig. 13-8. Since these levels are transmitted in a binary code, the actual number of levels is a power of 2; 16 levels are shown here for simplicity, but practical systems use as many as 128. By a process called *quantizing*, the level actually sent at any sampling time is the nearest standard (*or quantum*) level. As shown in Fig. 13-8, should the signal amplitude be 6.8 V at any time, it is not sent as a 6.8-V pulse, as it might have been in PAM, nor as a 6.8-μs-wide pulse as in PWM, but simply as the digit 7, because 7 V is the standard amplitude nearest to 6.8 V. Furthermore, the digit 7 is sent at that instant of time as a series of pulses corresponding to the number 7. Since there are 16 levels (2^4), 4 binary places are required; the number becomes 0111, and could be sent as 0PPP, where

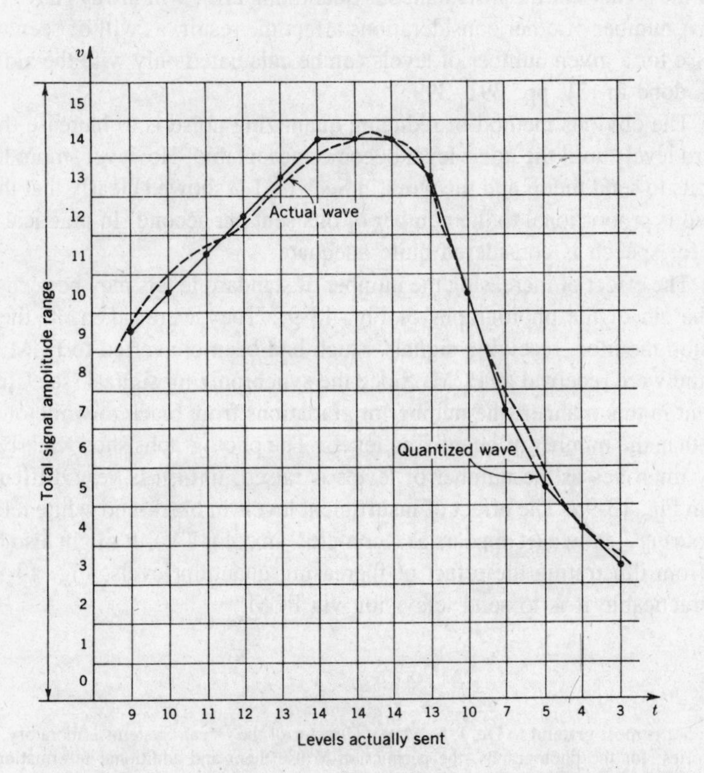

FIGURE 13-8 Quantization of signal for pulse-code modulation.

P = pulse and 0 = no-pulse. Actually, it is often sent as a binary number back-to-front, i.e., as 1110, or PPP0, to make demodulation easier (as will be seen).

As shown in Fig. 13-8, the signal is continuously sampled, quantized, coded and sent, as each sample amplitude is converted to the nearest standard amplitude and into the corresponding back-to-front binary number. Provided sufficient quantizing levels are used, the result cannot be distinguished from that of analog transmission. A supervisory or signaling bit is generally added to each code group representing a quantized sample. Hence each group of pulses denoting a sample, here called a *word*, is expressed by means of $n + 1$ bits, where 2^n is the chosen number of standard levels.

Quantizing noise If the actual signal is compared with the signal in Fig. 13-8, it is seen immediately that the quantizing process introduces some distortion. This is *quantizing noise*, called *noise* because errors are random in character. The randomness occurs simply because the difference between the digit sent and the actual signal at that instant is completely unpredictable, i.e., random.

It will be noted that the biggest error that can occur is equal to half the size of the sampling interval. Thus, in the system of Fig. 13-8, this maximum error is 1/32 of the total signal amplitude range. However, it would be wrong to assume that the signal-to-(quantizing) noise ratio of this system is necessarily 32:1. This is because neither the signal nor the instantaneous quantizing error will always have its maximum value. A number of other considerations affect the result, as will be seen, and quantizing noise for a given number of levels can be calculated only with the aid of statistics. This is done in [3], pp. 391–399.

The obvious method of reducing quantizing noise is to increase the number of standard levels until the noise level becomes acceptable. However, more levels require more bits to send them, and the work in Sec. 13-1.2 showed clearly that the bandwidth required is proportional to the number of bits sent per second. In practical systems 128 levels for speech is considered quite adequate.

The effect of increasing the number of standard levels may be seen qualitatively with the aid of the photographs of Fig. 13-9.[9] They were taken off the screen of a television monitor, receiving signals which had been converted to PCM before being sent, and were received as PCM. Since the synchronizing signals (see Chap. 17) were also sent in this manner, the number of gradations from black to white on each picture is less than the number of quantizing levels. The photographs show clearly how picture quality improves as the number of levels is raised, until it is very difficult to see any noise in Fig. 13-9*d*. The effect of insufficient levels in black-and-white television is the "contouring" shown (it appears as "confetti" in color TV). It might also be added that apart from illustrating the effect of increasing quantum levels, Fig. 13-9 also shows how practicable it is to send television via PCM.

[9] The author is most grateful to Dr. J. S. Mayo, Director of the Ocean Systems Laboratory, Bell Telephone Laboratories, for the photographs, the permission to use them and additional information received. The photographs first appeared in *Scientific American*, March 1968, in an article by Dr. Mayo.

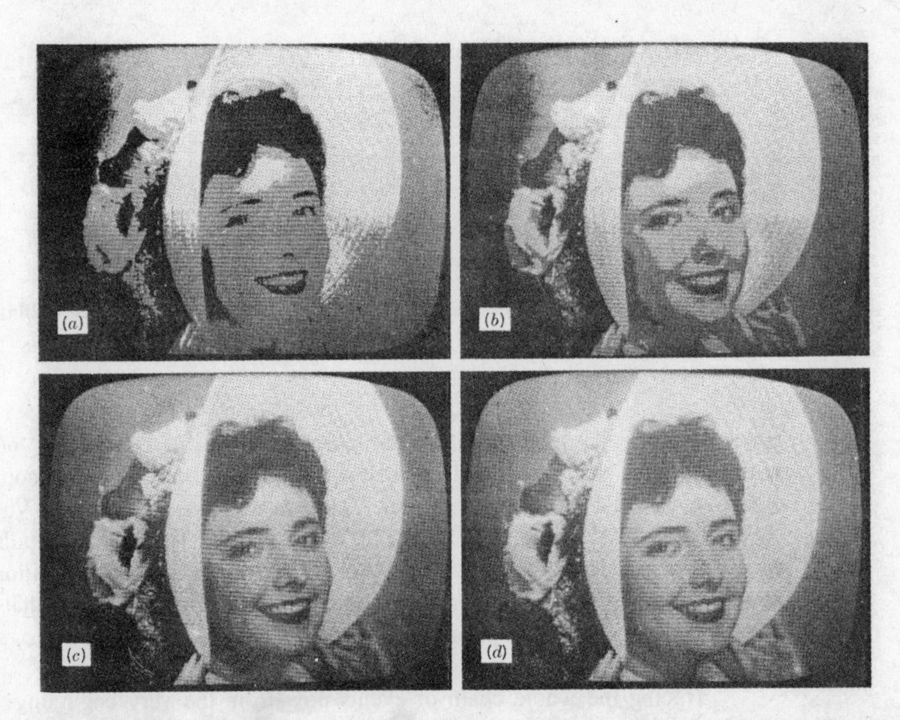

FIGURE 13-9 Television received via PCM. (*a*) 4 standard levels; (*b*) 8 standard levels; (*c*) 16 standard levels; (*d*) 32 standard levels. *(By courtesy of Dr. J. S. Mayo, Bell Telephone Laboratories.)*

Generation and Demodulation of PCM Generation of PCM is a complex business. Essentially, the signal is sampled and converted to PAM, the PAM is quantized and encoded, and supervisory signals are added. The signal is then sent directly via cable, or modulated and transmitted. Because PCM is highly immune to noise, as will be shown, amplitude modulation may be used, so that PCM–AM is quite common. The actual details of PCM quantizing and encoding may be found in [5].

At the receiver, the signaling information is extracted, and PCM is translated into corresponding PAM pulses which are then demodulated in the usual way. In fact, the ''quantized wave'' of Fig. 13-8 would be the output for that signal from an ideal PCM receiver. One of the methods of reconverting PCM is most ingenious and simple. It is worth investigating and may be explained with the aid of Fig. 13-10. It depends on the fact, mentioned previously and shown again here, that binary numbers are sent back-to-front to indicate each sample. Thus, for example, 3 is sent as 1100 (in Fig. 13-10) instead of 0011, and 13 is sent as 1011 instead of 1101.

An integrating *RC* circuit is used for reconversion in the receiver. The applied pulses are fed to the circuit, and it is then sampled and discharged immediately after the arrival of the last pulse. Also, this circuit has a time constant such that the charge due to one pulse decays to exactly one-half of its value by the time the next pulse arrives. If it is assumed, for simplicity, that each pulse has an amplitude of 1 V and that each pulse or space has a duration of 1 μs, it is then seen that each ''last'' pulse contributes

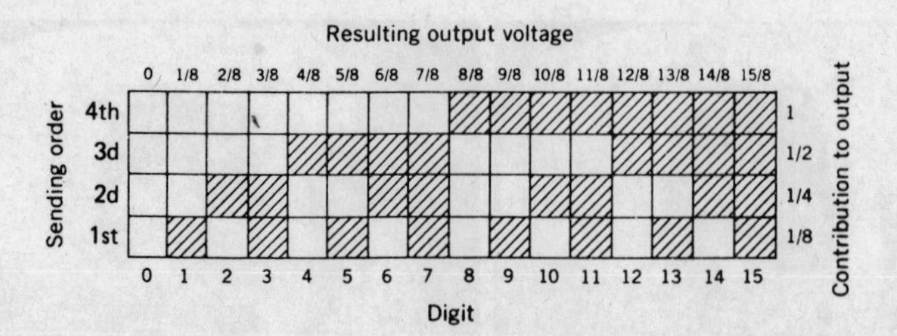

FIGURE 13-10 Reconversion of PCM; shaded areas represent pulses, blank areas no pulses.

precisely 1 V to the total output. In the intervening time of 1 μs, each "next-to-last" pulse will have decayed to half its original value, so that its output contribution is 0.5 V. Again, the contribution of each "third-last" pulse will be 0.25 V, and that of each "fourth-last" pulse will be 0.125 V and so on. The indicated pulses will be provided at the output only if a pulse is received in the indicated positions. If there is no pulse, then obviously there is no contribution to the output. All that happens is that the voltage due to previously arrived pulses decays to half its value, as it is supposed to.

Tracing the whole chain of events now from the very beginning, we have a signal wave sampled at regular intervals. At any instant of sampling, its amplitude might be 10.1 V; the sampling circuit notes this and converts it to the digit 10, which is the nearest standard level in the 4-bit code of Fig. 13-8. The coding circuits convert 10 into 0101, its back-to-front binary code equivalent, and the pulse train OPOP is sent. This is received, demodulated if necessary, and fed as OPOP to the *RC* circuit. At the time this circuit is sampled (and also discharged to permit the next digit to be reconverted), the last pulse has only just arrived, and so its contribution to the output is 1 V. There is no previous pulse, but the one before that has placed a charge on the capacitor which has decayed to 0.25 V at the sampling instant. The total voltage taken from the capacitor is thus 1.25 V, which is exactly $^{10}/_{8}$ V; this is the original amplitude of 10 expressed in $^{1}/_{8}$-V units in this example. Similarly, 4 becomes 0.5 V, or $^{4}/_{8}$ V, and 15 is 1.875 V, that is, $^{15}/_{8}$ V. As shown fully in Fig. 13-10, each binary set of pulses has been reconverted into a pulse whose amplitude is proportional to the original digit sent. A final advantage of this method, apart from its simplicity, is that it is capable of almost indefinite extension.

Effects of noise Those forms of pulse modulation which, like FM, transmit constant-amplitude signals, are equally amenable to signal-to-noise-ratio improvement with amplitude limiters of one form or another. Thus PWM, PPM and PCM have all the advantages of frequency modulation when it comes to noise performance; this is best illustrated by means of Fig. 13-11. Figure 13-11*a* shows the effect of noise being superimposed on pulses with vertical sides. It is seen that noise will have no effect at all unless its peaks are so large that they can be mistaken for pulses, or so large negatively that they can mask legitimate pulses. This is ensured by the slicer, or double clipper, which selects the amplitude range between *x* and *y* in Fig. 13-11 for further transmission, thus removing *all* the effects of noise.

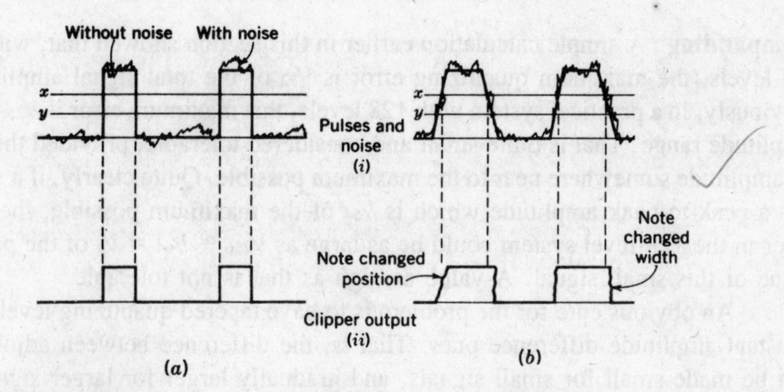

FIGURE 13-11 Effects of noise on pulses. (*a*) Vertical pulses; (*b*) sloping pulses.

The transmitted pulses in a practical system cannot have sides with perfectly vertical slopes. These must "lean," as shown in an exaggerated fashion in Fig. 13-11*b*. Noise will now superimpose itself on the pulses' sides, and the result may well be a change in width or position as shown. This will affect PWM and PPM, but not nearly as much as an amplitude change would have affected PAM. The situation is virtually identical to that discussed in connection with the noise triangle in Chap. 5. Furthermore, the obvious method of reducing the effects of noise in PWM or PPM is to send pulses with steeper sides. This will increase the bandwidth required, so that these two forms of pulse modulation share with frequency modulation the ability to *trade bandwidth for improved noise performance*.

Good as these systems are, *PCM is much better* for noise immunity. As shown in Fig. 13-11*b*, sloping pulses are affected by noise, but this does not matter in PCM at all. *Provided that the signal-to-noise ratio is not so poor that noise pulses can be mistaken for normal pulses or can obliterate them, the effect of noise on PCM will be nil*. This is because PCM depends only on the presence or absence of pulses at any given time, not on any characteristic of the pulses which could be distorted. It is possible to predict statistically the error rate in PCM due to random noise. Consider a channel signal-to-noise ratio of 17 dB, which would be considered pitifully low for any other modulation system so far studied. For PCM in a 4-kHz channel, this would yield an error rate close to 1 error in 10,000 characters sent; when S/N = 23 dB, the error rate falls to 1 in about 8×10^8. Such an error rate is negligible, corresponding to about one error every 30 hours in a system sampling 8000 times per second, using an 8-bit code, and operating 24 hours per day.

Since errors will occur in PCM only when noise pulses are large enough, it is seen that the digital modulation system does not suffer from a gradual, subtle deterioration. Indeed, *pulse-code modulation can be relayed without degradation when the signal-to-noise ratio exceeds about 21 dB*. This gives PCM an enormous advantage over analog modulation in relay systems, since even in the best of analog systems some degradation will occur along every link and through every repeater. Since the process is cumulative, any such system will have to start with a much higher S/N ratio and also use more low-noise equipment en route than would have been needed by PCM. The ability to be relayed without *any* distortion and to use poor-quality transmission paths, is a very significant incentive to use digital, rather than analog, modulation systems.

Companding A simple calculation earlier in this section showed that, with 16 standard levels, the maximum quantizing error is 1/32 of the total signal amplitude range. Obviously, in a practical system with 128 levels, this maximum error is 1/256 of the total amplitude range. That is quite small and considered tolerable, provided the signal has an amplitude somewhere near to the maximum possible. Quite clearly, if a small signal has a peak-to-peak amplitude which is 1/64 of the maximum possible, the quantizing error in the 128-level system could be as large as $1/256 \div 1/64 = 1/4$ of the peak-to-peak value of this small signal. A value as high as that is not tolerable.

An obvious cure for the problem is to have tapered quantizing levels instead of constant-amplitude difference ones. That is, the difference between adjoining levels can be made small for small signals, and gradually larger for larger signals. In this way, the quantizing noise could be "distributed" so as to affect small signals somewhat less and large signals somewhat more. In practice, this kind of tapered system is difficult to implement, because it would significantly complicate the (already complex) quantizer design. However, there is a suitable alternative.

It is possible to predistort the signal before it is modulated, and "un-distort" it after demodulation. This is always done in practice, and is shown in Fig. 13-12. The process is known as "companding," since it consists of compressing the signal at the transmitter and expanding it at the receiver. With companding, exactly the same results are obtained as with tapered quantizing, but much more easily. The signal to be transmitted is passed through an amplifier which has a correctly adjusted nonlinear transfer characteristic, favoring small-amplitude signals. These are then artificially large when they are quantized, and so the effect of quantizing noise upon them is reduced. The correct amplitude relations are restored by the expander in the receiver. It should be noted that disagreement about the companding law (between the United States and Europe) has been a real impediment in establishing worldwide PCM standards for

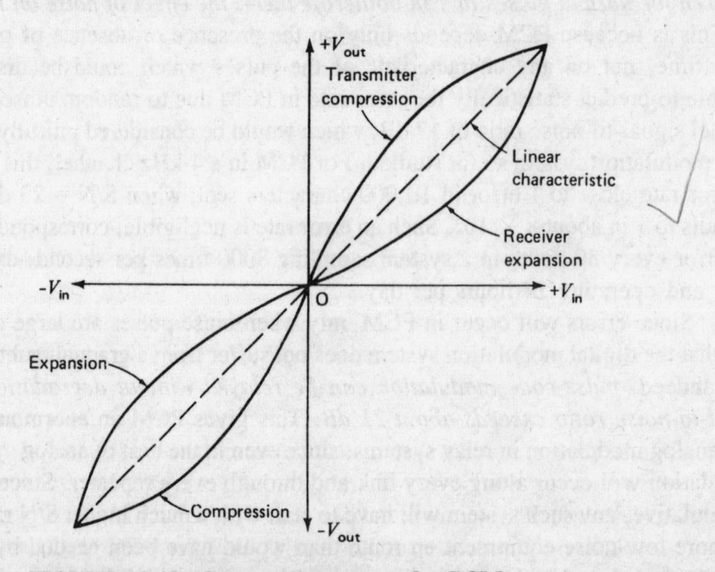

FIGURE 13-12 Companding curves for PCM.

telephony. Fortunately, there has at least been agreement that companding *should be* used, and standard converters are available.

Advantages and applications of PCM A person may well ask, at this stage, "If PCM is so marvelous, why are any other modulation systems used?" There are three answers to this question, namely:

1. The other systems came first.
2. PCM requires very complex encoding and quantizing circuitry.
3. PCM requires a large bandwidth compared to analog systems.

PCM was invented by Alex H. Reeves in Great Britain, in 1937. When he patented it the following year, it was an astonishingly detailed and complete system. However, its very complexity prevented its immediate use—there were no really suitable electronic devices to implement it. By the end of World War II it was being used in a microwave relay system designed and operated by the U.S. Army Signal Corps. Soon after the war the system was adapted and evolved for commercial use by the Bell System. PCM then received a real boost from a very important paper of Oliver et al. [6], in 1948, but it was some 10 years before it was actually used for telephony. Its first practical application in commercial telephony was in short-distance, medium-density work, in Great Britain and the United States in the early 1960s. Semiconductors and integration (it was not yet "large-scale" then) made its use practicable. Subsequently, quite a number of new communication facilities built around the world have used PCM, and its use has grown very markedly during the 1980s.

As regards the second point, it is perfectly true that PCM requires much more complex modulating procedures than analog systems. However, multiplexing equipment is very much cheaper, and repeaters do not have to be placed so close together because PCM tolerates much worse signal-to-noise ratios. Especially because of very large-scale integration, the complexity of PCM is no longer a significant cost penalty.

Although the large bandwidth requirements still represent a problem, it is no longer as serious as it had earlier been, because of the advent of large-bandwidth fiber-optic systems. However, the large bandwidth requirements should be recognized. A typical first-level PCM system is the Bell T1 digital transmission system in use in North America. As described in greater detail in Sec. 15-1.2, it provides 24 PCM channels with time-division multiplexing. Each channel, as so far discussed, requires 8 bits per sample and thus 24 channels will need $24 \times 8 + 1 = 193$ bits—the extra 1 bit is an additional sync signal. With a sampling rate of 8000 per second, a total of $8000 \times 193 = 1{,}544{,}000$ bps will be sent by using this system. Work earlier in this chapter showed that the bandwidth in hertz would have to be at least half that figure, but the practical system in fact uses a bandwidth of 1.5 MHz as an optimum figure. It will be shown in Chap. 15 that 24 channels correspond to two groups, requiring a bandwidth of 96 kHz if frequency-division multiplex is used. PCM is seen to require 16 times as much bandwidth for the same number of channels. However, the situation in practice is not quite so bad, because economies of scale begin to appear when higher levels of digital multiplexing are used (see also Sec. 15-1.2).

The following considerations ensured that, at first, the main application of PCM for telephony was in 24-channel frames over wire pairs which previously had carried only one telephone conversation each. Their performance was not good enough to provide 24 FDM channels, but after a little modification 24 PCM channels could be carried over the one pair of wires. Since the mid-1970s, however, the picture has changed dramatically. First, very large-scale integration reduced costs significantly. Then came the proliferation of digital systems, such as data transmissions, which were clearly advantaged by not having to be converted into analog prior to transmission and reconverted to digital after reception. Finally, fiber-optic systems became practical, with two effects. On the one hand, the current state of development of lasers and receiving diodes is such that digital operation is preferable to analog because of nonlinearities (see also Sec. 15-2.2). On the other hand, huge bandwidths, e.g., 565 Mbps per pair of fibers, have become available without attendant huge costs. Consequently, and as further discussed in Sec. 15-1.2, the use of PCM in the broadband networks of advanced countries is increasing by leaps and bounds.

PCM also finds use in space communications. Indeed, the *Mariner IV* probe was an excellent example of the noise immunity of PCM, when, back in 1965, it transmitted the first pictures of Mars. Admittedly, each picture took 30 minutes to transmit, whereas it takes only 1/30 s in TV broadcasting. However, the *Mariner IV* transmitter was, at the time, just over 200,000,000 km away, and the transmitting power was only 10 W. PCM was used; no other system would have done the job.

Other digital pulse modulation systems PCM was the first digital system, but by now several others have been proposed. The major ones will now be mentioned, but it should be noted that none of them is in widespread use.

Differential PCM is quite similar to ordinary PCM. However, each word in this system indicates the difference in amplitude, positive or negative, between this sample and the previous sample. Thus the relative value of each sample is indicated, rather than the absolute value as in normal PCM. The rationale behind this system is that speech is redundant, to the extent that each amplitude is related to the previous amplitude, so that large variations from one sample to the next are unlikely. This being the case, it would take fewer bits to indicate the size of the amplitude change than the absolute amplitude, and so a smaller bandwidth would be required for the transmission. The differential PCM system has not found wide acceptance because complications in the encoding and decoding process appear to outweigh any advantages gained.

Delta modulation is a digital modulation system which has many forms, but at its simplest it may be equated with the basic form of differential PCM. In the simple form of delta modulation, there is just 1 bit sent per sample, to indicate whether the signal is larger or smaller than the previous sample. This system has the attraction of extremely simple coding and decoding procedures, and the quantizing process is also very simple. However, delta modulation cannot readily handle rapid amplitude variations, and so quantizing noise tends to be very high. Even with companding and more complex versions of delta modulation, it has been found that the transmission rate must be close to 100 kbits per second to give the same performance for a telephone channel as PCM gives with 64 kbits/s (8000 samples per second × 8 bits per sample).

Other digital systems also exist. Together with differential PCM and various forms of delta modulation, they are described in [5], pp. 198–218, and [3], pp. 216–222. In [7] is a performance comparison, including computer simulation, of the various digital systems.

13-3 PULSE SYSTEMS

This section deals with two examples or applications of pulse communications. The first is *telegraphy*, which is by far the oldest form of telecommunications,[10] indeed the one that gave birth to the pulse modulation systems just discussed, and also to information theory. The second system is an application of pulse modulation. It is *telemetry*, or the making of measurements and automatic transmission of results to some distant point.

13-3.1 Telegraphy (and Telex)

Telegraphy is a form of communication that employs typewriterlike machines operating at a maximum speed of about 60 words per minute (wpm) to send written messages from one point to another. In telegraphy, a user lodges a written message for transmission at a telegraph or post office. The message is subsequently transmitted to the office nearest to the addressee, and delivered in typewritten form, perhaps after having been first telephoned through, if that was part of the request. Telex combines the above system with subscriber dialing techniques, similar to those discussed in Sec. 15-4. The originator of the message composes the address of the message and then its text and sends the message, which is automatically routed to the wanted subscriber, and printed out by machine. Thus the only difference between telex and telegraphy is in the signaling method and the actual sending procedure. All other aspects of the two systems are identical and will be considered together under the general title of telegraphy in this section. Note that telegraphy and telex, together with facsimile and similar systems, are often collectively referred to as *record services*. This is because they provide a printed record, unlike telephony.

General system description In either telegraphy or telex, the transmitting teletypewriter produces a set of coded pulses when a given key is tapped. In the receiving mode, the same machine prints out the appropriate letter when a given code is received. A message is typed out at one end of a link and is printed out at the other end, either simultaneously or very soon afterward. The transmitting typewriter also prints a copy of what it sends, for checking and filing. Teletypewriters may be of the page-printer or tape-printer form. The former, perhaps the more common of the two, prints on a roll of paper having much the same width as typing paper.

[10] The first proposal for an electric telegraph was made by an anonymous writer to a Scottish newspaper in 1753 who suggested the sending of electric sparks over as many insulated wires as there are letters in the alphabet.

To avoid limitations due to the maximum typing speed and efficiency of the operator, there is an alternative system in which the message is typed out in advance on perforated tape (see also Fig. 14-21) or in a word processor, at any convenient speed. It can then be transmitted at maximum speed, so that the capacity of the channel and the machine are not wasted. Teletypewriters are pictured, and further described, in [4], pp. 112–119; keyboard layouts are also shown.

In the code used for teletype, each character has five binary elements, so that the code either resembles or (what is more likely) is identical to the CCITT-2 code. The only exception to this is that most HF radio links use the ARQ code. In practice, each element, i.e., each mark or space, occupies 22 ms. In addition, each letter is preceded by a 22-ms space and followed by a 31-ms mark, so that each letter occupies approximately 7½ elements, or 163 ms. Since the average telegraphic word is considered as consisting of six letters, it is seen that each word takes 1 s to transmit on the average; hence the 60-wpm limit mentioned earlier.

From the examples so far given, it would be excusable to think that simple AM is used for all radiotelegraphy; as it happens, nothing could be further from the truth. AM is a good system to use for illustrations and examples, but it is not a very good system for transmitting telegraphy over long distances.

Frequency-shift keying (FSK) It would be quite possible to transmit teletype by the ordinary ON-OFF keying of the transmitter. In other words, we could use amplitude modulation with pulses, ON corresponding to mark and OFF to space. However, such a system has the inherent disadvantage that there is no real indication for the space. In addition, a system such as this would suffer from all the usual ailments of amplitude modulation, as a result of which it is never used for automatic telegraphy (it is, of course, widely used for manual Morse code CW operation). A system known as *frequency-shift keying* is generally used instead.

FSK is a system of frequency modulation. In it, the nominal unmodulated carrier frequency corresponds to the *mark* condition, and a *space* is represented by a downward frequency shift. The amount was 850 Hz in the original wideband FSK system designed for HF radio. For transmission by line or broadband systems, the current shift is 60 Hz, as laid down in CCITT Rec. R35. This is known as narrowband FSK, or frequency-modulated voice-frequency telegraph (FMVFT). FSK is still often used for HF radio transmissions, with a frequency shift that is commonly 170 Hz. As with other forms of FM, the main advantage of the wideband system is greater noise immunity, while the narrowband systems are used to conserve the allocated frequency spectrum. Note that FSK may be thought of as an FM system in which the carrier frequency is midway between the mark and space frequencies, and modulation is by a square wave. In practice, of course, only the fundamental frequency of the square wave is transmitted, and regeneration takes place in the receiver.

In the FSK generator, the frequency shift may be obtained by applying the varying dc output of the telegraph machine to a varactor diode in a crystal oscillator. At the receiving end, the signal is demultiplexed (if, as is common, FDM was used to send a number of telegraph or telex transmissions together) and applied to a standard phase discriminator. From the discriminator, signals of either polarity will be available. After some pulse shaping, they are applied to the receiving teletypewriter. If the telegraph transmission is by HF radio, the phase discriminator works at a (fairly low) intermediate frequency, although other methods are also possible for demodulation. An amplitude limiter is always used in the receiver, to take full advantage of the noise immunity of FSK.

Other transmission methods *ON-OFF keying* of the transmitter is sometimes used at MF but generally only for Morse code. In this system, carrier is present for a mark, and absent for a space. Some HF transmissions may still use *two-tone modulation*. This is an AM system, in which the carrier is modulated with one tone for a mark, and another audio for a space. In such HF telegraph transmissions, the two tones are generally 170 Hz apart. Neither system has the noise immunity of FSK.

Phase-shift keying (PSK) is a relatively new system, in which the carrier may be phase-shifted by +90° for a mark, and by −90° for a space. In *four-phase,* or *quadrature PSK,* system, the possible phase shifts are +135°, +45°, −45° and −135°, so that 2 bits of information can be indicated instead of one as in the other systems. PSK has a number of similarities to FSK and is further described in [3], pp. 222–227, and [4], pp. 257–264.

Multiplexing Where a telegraph transmission is sent over a single wire pair, or as the only modulation on a radio transmitter, no multiplexing is necessary. However, it is far more common to send telegraphy (and telex) over channels designed for telephony, where the available bandwidth is 300 to 3400 Hz. This is far more than an FMVFT channel requires, and so frequency-division multiplex, akin to the systems described in Sec. 15-1.1, is often used. For broadband systems, the subdivision of the telephone channel is governed by the previously mentioned CCITT Rec. R35. This allocates 120 Hz to each telegraph channel, so that 24 channels can be fitted into the telephone bandwidth. The nominal carrier frequency is 420 Hz for the lowest-frequency channel, while channel 24 is allocated a nominal carrier frequency of 3180 Hz. The 120-Hz bandwidth allocated to each channel is quite adequate for the 60-Hz frequency shift and leaves a further 60 Hz as a guard band.

Time-division multiplex systems for telegraphy are now becoming widespread. As explained in Sec. 15-1.2, pulses from various transmissions are interleaved in time, and here they happen to be pulses for telegraph transmitters. TDM systems naturally lend themselves to being used with other pulse-code transmissions, in wholly digital systems. It is expected that as many as 128 simultaneous telegraph transmissions, time-division multiplexed, could eventually be sent over one telephone channel.

Telegraph speeds The maximum rate at which intelligence can be transmitted over a telegraph circuit is naturally proportional to the bandwidth of the circuit. The bandwidth, in turn, is governed by the duration of the shortest signal element and is inversely proportional to it. This comes about because bandwidth is dependent on the pulse repetition rate, and obviously the shorter each pulse, the greater the repetition rate per second. As a result of all these considerations, telegraph speed is expressed numerically as the reciprocal of the duration in seconds of the shortest signaling element. The *baud* is the name given to the unit of telegraphic speed. Thus the speed in bauds expresses the number of the shortest elements of signal that may occur per second. For the system so far described, in which each short element occupies 22 ms, the speed in bauds is

$$\text{Speed} = \frac{1}{L}$$

$$= \frac{1}{22 \times 10^{-3}} = 45 \text{ bauds} \tag{13-7}$$

where L is the duration of the shortest signal element.

13-3.2 Telemetry [8]

Telemetry, as the name suggests, consists of performing measurements on distant objects. Although hydraulic or wire circuits may be used for this purpose, this section will deal with radiotelemetry only, particularly as this form is now prevalent. The alternatives to radiotelemetry are either having observers on the spot or having some form of multitrack recorders on the site from which data may be collected when convenient. However, there are situations in which neither of these alternatives is possible, and in fact radiotelemetry arose in the early 1940s from the need to obtain in-flight information from fighter aircraft and then early missiles. These had such excellent chances of crashing out of control that placing observers or even recorders on board was considered too risky. Since it was even then required to monitor several quantities simultaneously, the demand for multichannel radiotelemetry arose immediately. The systems were quickly extended to include remote control of such flights, followed by the remote control and light testing of guided missiles, satellite launching and control in orbit, and finally control of extraterrestrial probes. Many other applications of radiotelemetry also exist, including remote-location gathering of meteorological data and control of petroleum pipeline systems.

Since radiotelemetry can become quite complex, with problems associated with transmitters, antenna location and orientation, and noise and interference, the other two forms of data acquisition are always investigated first. Because of space and weight limitations, there is a limit on the total data that can be stored, however, and for obvious reasons human observers are sometimes unacceptable. As a result, radiotelemetry is used widely but not indiscriminately.

Methods Since several channels must always be sent simultaneously in radiotelemetry, and it is impracticable to use a different radio link for each, either frequency-division or time-division multiplexing is always used, or (as shown in Fig. 13-13) both.

If, for example, telemetry is used with a pilotless aircraft, not all the telemetered variables change at the same rate. Airspeed and altitude change relatively slowly, whereas wing flutter and attitude may change quite rapidly. As a result, there is need for both narrowband channels for handling the slow variations and (relatively) wideband channels for the faster variations. The mixture is achieved by a process known as *subcommutating*, of which a mechanical version is illustrated in Fig. 13-13.

Subcommutating consists of taking one of the wideband channels and subdividing it into several narrowband ones. As it stands, the system uses FDM in general and

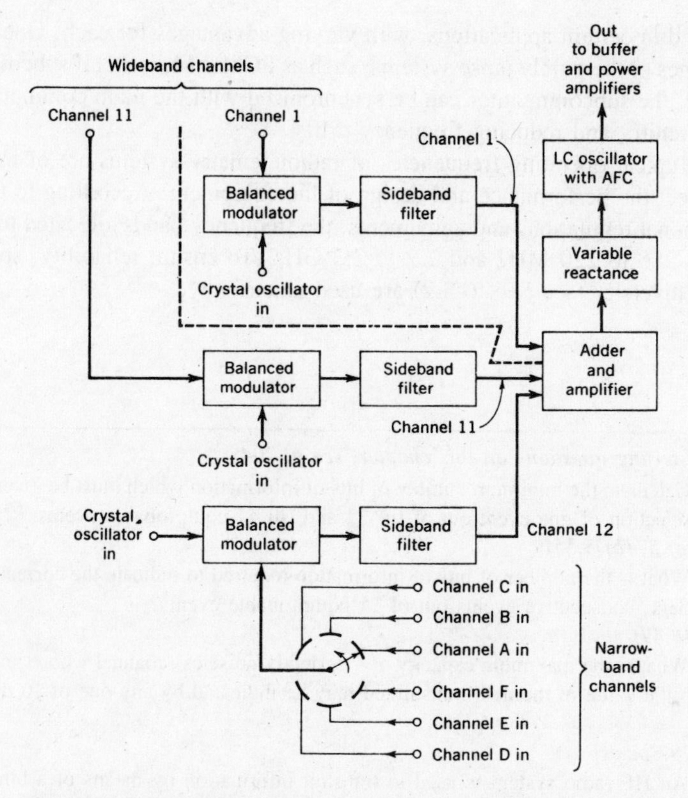

FIGURE 13-13 Radiotelemetry transmitter using frequency-division multiplex with TDM for subcommutation.

TDM for the subcommutated channels, and mechanical commutation can be used for these because the variation is slow. However, it is equally possible to use FDM for the subcommutated channels or else TDM for both.

Since complex modulation arrangements are in force, some sort of standardization is required when describing a given system, and by custom one works from the individual channel. For example, SSB/FM is a system in which each of the channels modulates it subcarrier using SSB, and then all the modulated subcarriers frequency-modulate the main carrier; this is in fact the FDM system described in Sec. 15-1.1. PCM/AM means time-division multiplex of pulse-code modulation, with the pulse train amplitude-modulating the main carrier. Finally, PWM/AM/FM would be a system in which the subcarriers are amplitude-modulated by their signals, and in turn they frequency-modulate the main carrier. Also, one of the channels has been subcommutated, and each of the subchannels is fed PWM.

A large number of telemetering systems are used in practice, since rigid standardization has not fully taken place in this field. However, the FM/FM frequency-division multiplexing system is used quite widely in the United States and is extended to PAM/FM/FM when subcommutating is needed. It has the advantages of reliability and flexibility, but it requires greater bandwidth and carrier strength than the purely pulse systems. Systems such as PPM/AM, PWM/FM, PAM/FM and even PCM/FM have all been used in certain applications, with varying advantages for each. One of the main advantages of the purely pulse systems, such as PCM/AM, is that if subcommutation is required, the subcommutator can be synchronized with the main commutator, simplifying circuitry and reducing frequency drifts.

Typical operating frequencies of radiotelemetry systems are of interest, since they affect the performance and design of the equipment. According to the FCC and international regulations and agreements, the frequency bands allocated to radiotelemetry are 216 to 220 MHz and 2.2 to 2.3 GHz. To ensure reliability, space and frequency diversity (see Sec. 6-3.2) are used extensively.

PROBLEMS

For self-testing questions on this chapter, see p. 704.

13-1. Calculate the minimum number of bits of information which must be given to permit the correct selection of one event out of (*a*) 32 and (*b*) 47 equiprobable events.
[Ans.: (a) 5, (b) 5.55]

13-2. What is the number of bits of information required to indicate the correct selection of 3 independent, consecutive events out of 75 equiprobable events?
[Ans.: 18.69]

13-3. What is the maximum capacity of a perfectly noiseless channel whose width is 120 Hz, in which the value of the data transmitted may be indicated by any one of 10 different amplitudes?
[Ans.: 797 bits/s]

13-4. An HF radio system is used to transmit information by means of a binary code. The transmitting power is 50 W, and the noise level at the receiver input is such that the consequent error rate is just acceptable. The operator now decides to double the information flow rate by using a four-level code instead of the binary code. To what level must the transmitting power be raised to retain the same error rate?
[Ans.: 450 W]

13-5. At the input to the receiver of a standard telephone channel, the noise power is 50 μW and the signal to power is 20 mW. Calculate the Shannon limit for the capacity of the above channel under these conditions, and then when the signal power is halved.
[Ans.: 26,807 bits/s, 23,718 bits/s]

13-6. A 2-kHz channel has a signal-to-noise ratio of 24 dB. (*a*) Calculate the maximum capacity of this channel. (*b*) Assuming constant transmitting power, calculate the maximum capacity when the channel bandwidth is (*i*) halved, (*ii*) reduced to a quarter of the original value.
[Ans.: (a) 15,957 bits/s, (b) (i) 8,975 bits/s, (ii) 4,987 bits/s]

QUESTIONS

13-1. Define and explain *information* and *information theory*. What are the aims of information theory? Why is *meaning* divorced from *information?*

13-2. What is the mathematical definition of *information?* What is the difference between *possibility* and *probability?*

13-3. Define the *bit* of information. What are equiprobable events? Give in full the formula used to calculate the number of bits of information required in a given situation.

13-4. Why must a code of the Baudot form be used to send *words* by telegraph? Why cannot a different symbol be used or each separate word or perhaps each letter?

13-5. Derive the Hartley law (verbally) for binary codes, using the CCITT-2 code to prove the relation.

13-6. Explain why any binary-type code is noise-resistant, and explain why an enormous power increase is required when a more complex code is used.

13-7. Quote the Shannon-Hartley theorem, defining each term in the formula. What is the fundamental importance of this theorem?

13-8. With the aid of the Shannon-Hartley theorem, explain why doubling the bandwidth of a channel, while keeping a constant transmitting power, will not automatically double the channel capacity.

13-9. What is the fundamental difference between pulse modulation, on the one hand, and frequency and amplitude modulation on the other?

13-10. What is pulse-width modulation? What other names does it have? How is it *de*modulated?

13-11. Define and describe pulse-position modulation, and explain with waveforms how it is derived from PWM.

13-12. In what way is pulse-code modulation different from other modulation systems, pulse or otherwise, previously studied? What makes it a *digital* system?

13-13. Explain fully what pulse-code modulation is. Draw one complete cycle of some irregular waveform, and show how it is *quantized,* using eight standard levels.

13-14. Explain what is meant by the *ability to trade bandwidth for improved noise performance,* as applied to pulse modulation. Explain why PCM is more noise-resistant than the other forms of pulse modulation.

13-15. Explain why quantizing noise could affect small-amplitude signals in a PCM system far more than large signals. With the aid of sketches, show how *tapered* quantizing levels could be used to counteract this effect.

13-16. What is *companding?* Why is it used? Why is it preferable to quantizing with tapered steps? Illustrate your answer with a sketch of typical companding curves.

13-17. What are the advantages and applications of pulse-code modulation?

13-18. Briefly describe some *digital* modulation systems other than PCM.

13-19. What is telegraphy? Describe briefly the system and machines used for transmitting and receiving it.

13-20. How does the 60-wpm limit come about for standard telegraphy? How can it be reached even with a slow teletypist?

13-21. Explain fully what is meant by frequency-shift keying.

13-22. Why is multiplexing almost always used with telegraphy transmissions over broadband systems? Explain briefly the operating arrangements in a standard 24-channel FMVFT system.

13-23. What is *telemetry?* What transmission or storage systems can be used for telemetry? Under what conditions is *radio*telemetry preferred?

13-24. With the aid of a block diagram, explain the operation of a PPM/AM/FM radiotelemetry transmitter.

REFERENCES

1. Shannon, C. E., and W. Weaver, *The Mathematical Theory of Communication,* University of Illinois Press, Urbana, Ill., 1949. (*Standard reference*)
2. (*Historical*) Hartley, R. V. L., "Transmission of Information." *Bell System Tech. J.,* vol. 7, pp. 535–564, 1928.

3. Taub, Herbert, and D. L. Schilling, *Principles of Communication Systems*, McGraw-Hill Book Company, New York, 1971.
4. Martin, James, *Telecommunications and the Computer*, Prentice-Hall, Inc., Englewood Cliffs, N.J., 1969.
5. Cattermole, K. W., *Principles of Pulse Code Modulation*, Iliffe Books Ltd., London, 1969.
6. Oliver, B. M., J. R. Pierce and C. E. Shannon, ''The Philosophy of P.C.M.,'' *Proc. IRE*, vol. 36, November 1948, pp. 1324–1331. (*Classical reference*)
7. Kikkert, C. J., ''A Comparison of Code Modulation Systems,'' *Proc, IREE* (*Aust.*), March 1975, pp. 44–48.
8. This is meant only as an introduction to telemetry, especially from the point of view of its being an application of pulse modulation and multiplexing. Students interested specifically in this very wide field are referred to Gruenberg, Elliot L. (editor-in-chief), *Handbook of Telemetry and Remote Control*, McGraw-Hill Book Company, New York, 1967. Chapters 1 and 4 are the most relevant to the work covered here.

14 DIGITAL COMMUNICATIONS

Data communications became important with the expansion of the use of computers and data processing, and have continued to develop into a major industry providing the interconnection of computer peripherals and transmission of data between distant sites. The terminology, equipment and procedures for data communications comprise the subject of this chapter.

Data communications depend on digital electronics, and so a brief consideration of their fundamentals will set the stage for a discussion of other aspects of data transmission. Integral to the discussion of digital electronics is the subject of the binary number system because, as will be seen, it is virtually synonymous with digital electronics.

A brief history of computer technology demonstrates the interdependence of this field with that of data communications. Data communications are at once a support industry for data processing and an independent industry enabling other fields to take advantage of the advances in transmission it develops.

At the heart of data communications is the transmission channel, the medium of data transfer. The channel has inherent limitations which determine its suitability for data communications. This chapter will discuss channel limitations and characteristics, and it will demonstrate the impact which these have on data transmission. Bandwidth, frequency, noise, distortion, transmission speed and other channel considerations are the daily fare of the data communications engineer.

Data is coded prior to being transferred, and it can be appreciated that numerous codes are in use. Some are specific to particular applications, such as the *Hollerith code* for punched cards, and others are more universal, such as the *ASCII* code for general data processing. The chapter does not attempt to discuss all of the data codes, but a large and representative sample is presented, illustrating the major codes and their strengths and limitations.

Data communication using digital data must be very accurate because the redundancy available with analog signals is not present with digital signals. Errors can therefore be catastrophic. To limit the extent of the deterioration which errors impose, much has been done to develop error detection and correction mechanisms, and several of these will be discussed.

The data set is the basic equipment of data communication, since it transforms the digital data into signals compatible with transmission circuits. The various types and capabilities of data sets will be illustrated.

The chapter concludes with a discussion of network techniques. *Networking*, using point-to-point or fixed circuits for transmission, has become important as a method of improving data communication efficiency and economy. Accordingly the various network systems and the popular *protocols* are covered in some detail.

14-1 DIGITAL TECHNOLOGY

Digital technology is a branch of electronics and communications which utilizes *discontinuous signals*, i.e., signals which appear in discrete steps rather than having the continuous variations characteristic of analog signals. The value of digital techniques derives from the ability to construct unique codes to represent different items of information. These codes are the language of computers and the other types of digital electronic equipment which have revolutionized modern society.

Even though digital technology has been on the scene for only a relatively short period of time, it is difficult to remember how life was conducted prior to computers and their peripherals. Indeed, digital technology has become so pervasive today that few fields of endeavor remain for which digital processing is not important or even essential. Digital technology is of particular importance when information is to be gathered, stored, retrieved and/or evaluated. Digital processing is used so widely because it provides economical and rapid manipulation of data.

14-1.1 Digital Fundamentals

Comparison of analog and digital signals The meaning of the term "digital" is fundamental to an understanding of the benefits of digital technology. Digital data must be compared with analog data in order to understand the distinction between the two. An analog signal is best illustrated by a sine wave as shown in Fig. 14-1. Notice that the sine wave is continuous, and the value of the analog signal at any given instant can be anywhere within the range of the signal's extremes.

The digital signal does not provide this continuous representation of the original signal. Instead, the digital signal represents data as a series of digits such as a number. This digital representation can be considered as a code which approximates the actual value. Various digital systems are available. For example, a digital system which is decimal in nature can be used to represent the sine wave. To do this, the range of values

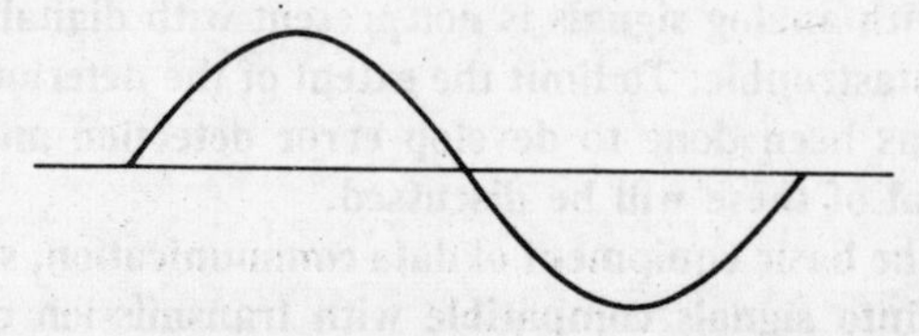

FIGURE 14-1 Sine wave; example of an analog signal.

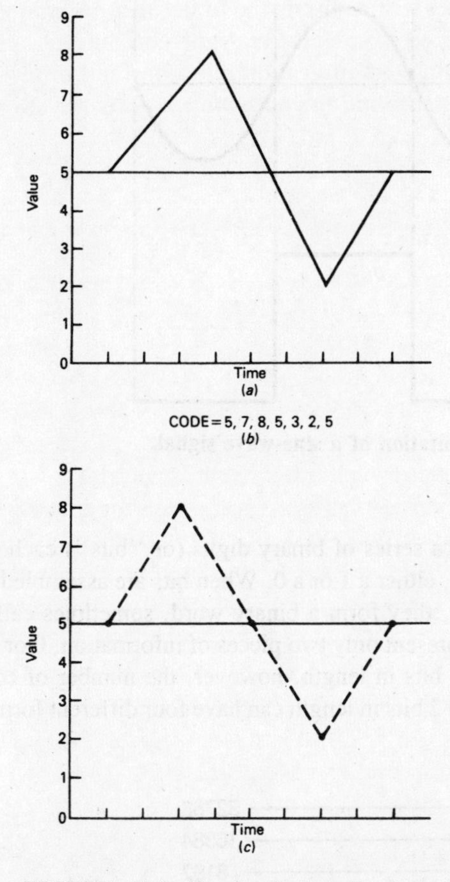

FIGURE 14-2 Ten-level digitalization. (*a*) Analog signal is sampled and a code (*b*) is derived; (*c*) code is used to reconstruct original waveform.

would be divided into 10 levels. The signal would then be sampled at set intervals and the appropriate level at that instant would be determined. Since only 10 levels would be available, however, the level which is closest would be used for a given sample if the signal were not to fall exactly on one of the levels. As shown in Fig. 14-2, the code could later be used to reconstruct the signal, producing a waveform similar to the original, with only minor discrepancies.

The binary digital system The binary digital system is based on the binary numbering system. This system has only two numbers, 0 and 1. If the sine wave is represented by the most basic binary digital system, a 1 will be generated if the signal value is above the halfway point. A 0 will result from signals below the halfway point. Figure 14-3 shows the binary representation of the sine-wave signal. A series of the sine-wave cycles would generate a series of pulses with spaces between them as shown.

The word "digital," as it is commonly used in information systems, refers to information coded in the form of binary states. In the preceding chapter, it was shown

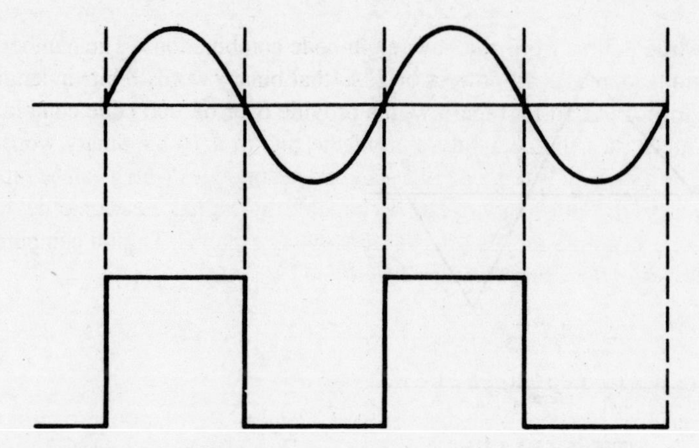

FIGURE 14-3 Binary representation of a sine-wave signal.

that a binary code consists of a series of binary digits (or "bits") each of which can have either of only two values, either a 1 or a 0. When bits are assembled together into a group of a particular length, they form a binary word, sometimes called a byte. A single binary character can represent only two pieces of information, 0 or 1. As bits are assembled into words several bits in length, however, the number of code combinations increases. A binary word 2 bits in length can have four different forms, 00, 01, 10

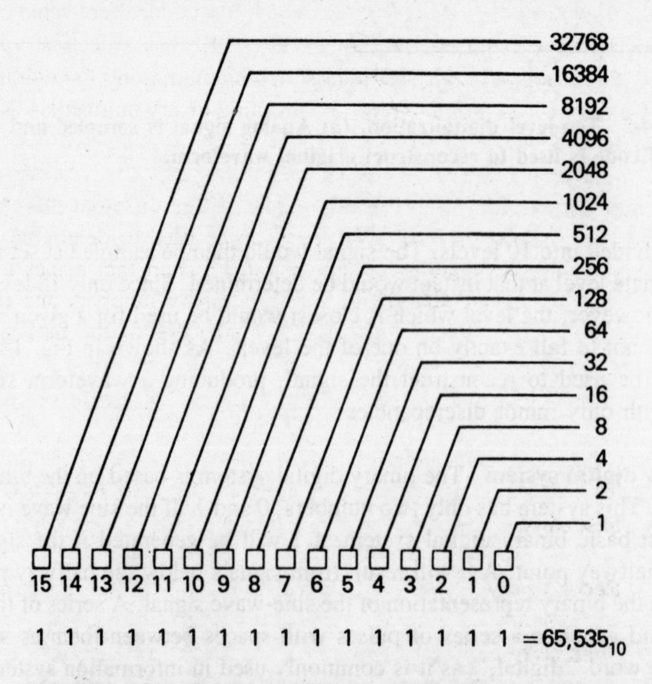

FIGURE 14-4 Decimal value of a 16-bit binary word.

and 11, while a 3-bit word can have eight code combinations. The number of possible combinations increases by powers of 2, so that binary words 8 bits in length have 256 different forms and 16-bit binary words provide over 65,000 code combinations. Figure 14-4 illustrates the decimal value of the bits in a 16-bit binary word.

If information is reduced to a standard set of codes which can be represented by binary words, the information can be handled by digital electronic devices. This is what is done in computers and digital processing systems. Digital computer codes are often based on the binary number system.

14-1.2 The Binary Number System

The decimal number system, using the base or *radix* 10, is familiar to most people. The digits for the system are defined as the non-negative integers which are smaller than the radix. For the decimal system, therefore, the digits are 0 to 9. This means that when one column of numbers increases beyond 9, the highest single digit, a carry is made to the next column while the first column returns to zero. The number 10 results. When the first two columns reach 99, carries are made from the first column to the second column and from the second column to the third column and the number 100 results. This familiar procedure is illustrated in Fig. 14-5.

The radix of the binary number system is the number 2, so that the digits of the binary system are limited to only two, 0 and 1. Therefore, when 1 is added to a binary number column, the first count changes the column from 0 to 1. When one more is added, the radix has been reached, so a carry is accomplished to the next column as shown in Fig. 14-6. The binary number 10, which has the decimal value of 2, results. If 1 is added to the 10_2, the binary number 11, which has a decimal value of 3, is obtained. Adding one to 11_2 requires a carry from column one to column two and a carry from column two to column three. The resulting binary number is 100, which has a decimal value of 4.

Binary/decimal conversion The decimal values of the different places in a 16-bit binary number were shown in Fig. 14-4; they can be used to determine the decimal value of binary numbers. Figure 14-7 demonstrates a chart that makes it possible to determine the decimal equivalent of each 1 in a binary number, and to add these decimal numbers together to determine the decimal value of the binary number. A

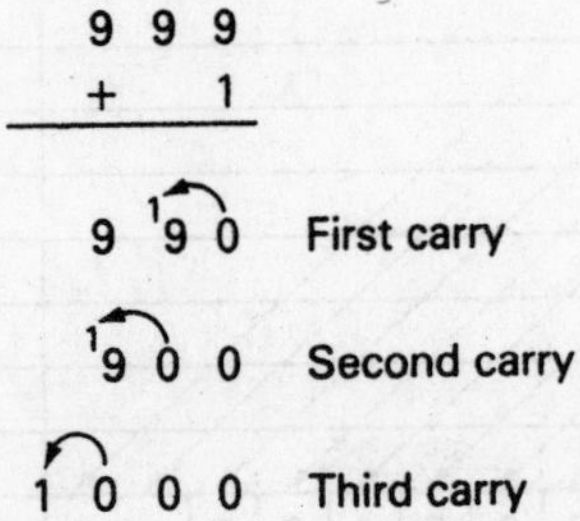

```
   9 9 9
 +     1
 -------
   9 ¹9 0    First carry
  ¹9 0 0     Second carry
 1 0 0 0     Third carry
```

FIGURE 14-5 Decimal addition.

$$\begin{array}{r} 0\ 0\ 1_2 \\ +\ 1_2 \\ \hline 0\ 1\ 0_2 \end{array}$$

$$\begin{array}{r} 0\ 1\ 0_2 \\ +\ 1_2 \\ \hline 0\ 1\ 1_2 \end{array}$$

$$\begin{array}{r} 0\ 1\ 1_2 \\ +\ 1_2 \\ \hline 1\ 0\ 0_2 \end{array}$$

FIGURE 14-6 Binary addition.

binary word is entered beneath the binary position line. For each "1" in the binary word, the diagonal is followed to the value in column A. These values are entered in the blanks under column B and when added provide the decimal value of the binary word.

Conversion from decimal to binary uses a different process, whereby the decimal number is divided by two and the remainder recorded each time, as shown in Fig. 14-8. Since each division is by 2, a remainder of either 1 or 0 must be obtained. The conversion is complete when the division by two cannot continue.

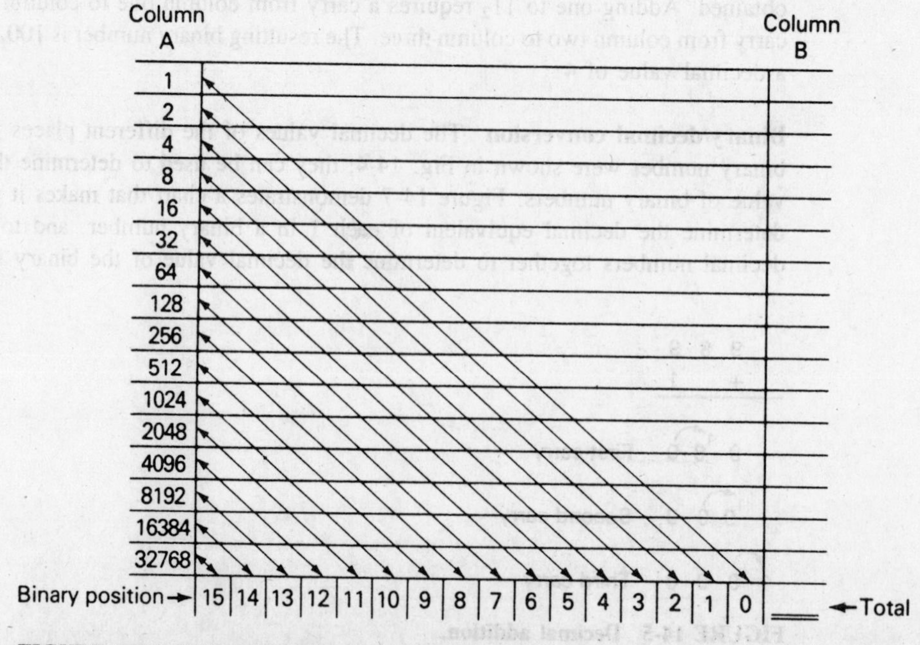

FIGURE 14-7 Binary to decimal conversion chart.

207_{10}

	Remainder
2 0 7 / 2 ⟶	1
1 0 3 / 2 ⟶	1
5 1 / 2 ⟶	1
2 5 / 2 ⟶	1
1 2 / 2 ⟶	1
6 / 2 ⟶	0
3 / 2 ⟶	1
1 / 2 ⟶	1

Binary equivalent = 01101111

FIGURE 14-8 Decimal to binary conversion process.

Binary addition and subtraction Addition of binary numbers follows the four simple rules shown below:

$0_2 + 0_2 = 0_2$
$0_2 + 1_2 = 1_2$
$1_2 + 0_2 = 1_2$
$1_2 + 1_2 = 10_2$

Examples of binary addition are given in Fig. 14-9.

Subtraction of binary numbers can be accomplished by using standard subtraction procedures, but this is not usually done in computer and data transmission systems. Instead, subtraction is accomplished by adding the *2s complement* of the subtrahend to the minuend. The *2s complement* is defined as the number which when added to the original number will result in a sum of unity. The 2s complement can be obtained in a two-step process. First, the *1s complement* of the subtrahend is obtained by converting each digit of the subtrahend to the opposite value. The 2s complement is then obtained by adding 1 to the 1s complement. The following example demonstrates the process.

$$\begin{array}{r} 1 \\ +\ 1 \\ \hline 1\ 0 \end{array} \qquad \begin{array}{r} 1\ 0 \\ +\ 1 \\ \hline 1\ 1 \end{array}$$

$$\begin{array}{r} 1\ 1 \\ +\ 1 \\ \hline 1\ 0\ 0 \end{array} \qquad \begin{array}{r} 1\ 0\ 0 \\ +\ 1 \\ \hline 1\ 0\ 1 \end{array}$$

$$\begin{array}{r} 1\ 1\ 1\ 1 \\ +\ \ \ \ \ \ \ 1 \\ \hline 1\ 0\ 0\ 0\ 0 \end{array} \qquad \begin{array}{r} 1\ 1\ 1\ 1 \\ +\ \ \ 1\ 1\ 1 \\ \hline 1\ 0\ 1\ 1\ 0 \end{array}$$

FIGURE 14-9 Binary addition.

Example 14-1 If the binary number 0101 is to be subtracted from the binary number 1110, the process is as follows:

$$\begin{array}{r} 1110 \\ -\underline{0101} \end{array}$$

1. Convert to 1s complement:

0101 = 1010

2. Convert to 2s complement:

1010 + 1 = 1011

3. Add:

$$\begin{array}{r} 1110 \\ +\ \underline{1011} \\ 1\ 1001 \end{array}$$

4. Discard the carry:

1001 = result

This process seems involved, but it is simple to accomplish with digital circuitry and is the common method of binary subtraction.

The important point about the binary number system is to recognize that it can be used to represent any decimal number. As indicated, all binary numbers will consist entirely of 0s and 1s. This is advantageous for use in digital electronic systems which are designed to utilize binary codes. When codes are based on the binary number system, they are computable, that is, they can be added and subtracted or evaluated to determine which is larger. The value of this will become clear later in the chapter.

14-1.3 Digital Electronics

Digital devices are based on electronic circuits which can represent two states, ON or OFF. The ability to design integrated circuits consisting of thousands of transistors on wafer-thin, postage-stamp-sized chips has made digital electronics both feasible and extraordinarily powerful. The circuits are used to store, evaluate, manipulate and modify digital code. Integrated circuitry has made it possible to design circuits that can handle large quantities of digitally encoded data at enormous speeds.

The transistor switch The fundamental circuit used with digital electronics is the transistor switch, illustrated in Fig. 14-10 in its simple form. If the base is held at a ground potential, the transistor will be cut off; i.e., no current will flow and the voltage at the collector will be equal to the value of the positive supply. When, however, the base is made positive, the transistor is turned on and current is conducted between the collector and emitter. The collector potential now assumes a voltage close to ground since the low resistance of the transistor forms a voltage divider with the collector resistance. The output, taken from the collector, will therefore be almost equal to the supply voltage when the base is grounded and almost at ground potential when a positive voltage is applied to the base. In typical digital terminology, the transistor

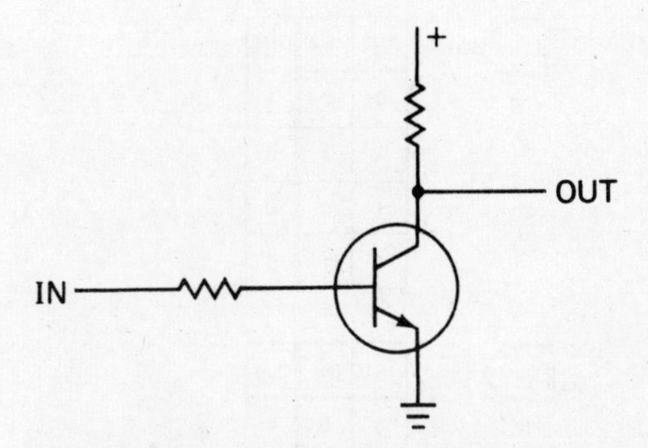

FIGURE 14-10 Transistor switch or inverter.

forms an inverter, since its output is high when a low voltage is on its input, and vice versa.

Digital gates If a second input is added to the inverter, another digital building block is formed, the two-input *NAND gate*. This gate will provide a low output only when both inputs are high. Figure 14-11 is the schematic of a NAND gate from the digital family called transistor-transistor-logic (TTL). The NAND gate in the figure uses a "totem-pole" output stage which helps to speed up the transition from low to high. Other digital circuits are created from combinations of NAND gates and inverters. Some of the common digital circuits are shown in Fig. 14-12. The *truth tables* next to each of the logic diagrams indicate the actions of the gates. These basic gates are combined into more complex circuits needed for data processing.

The flip-flop One of the intermediate building blocks of data circuits is the flip-flop, or latch. The latch simply accepts an input consisting of either a high or a low, and then

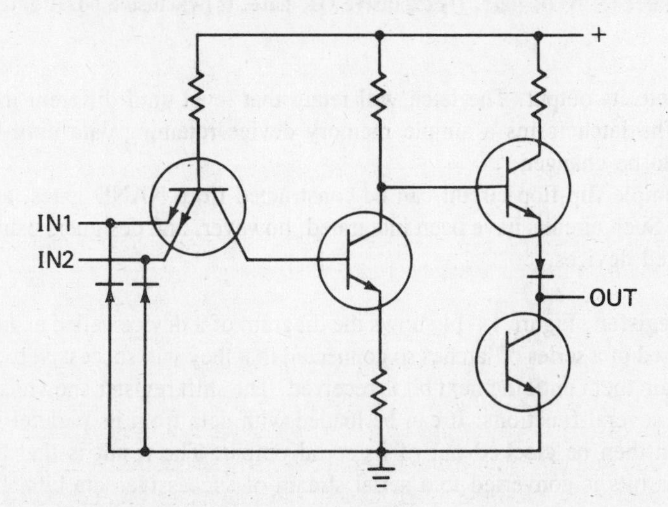

FIGURE 14-11 Two-input NAND gate.

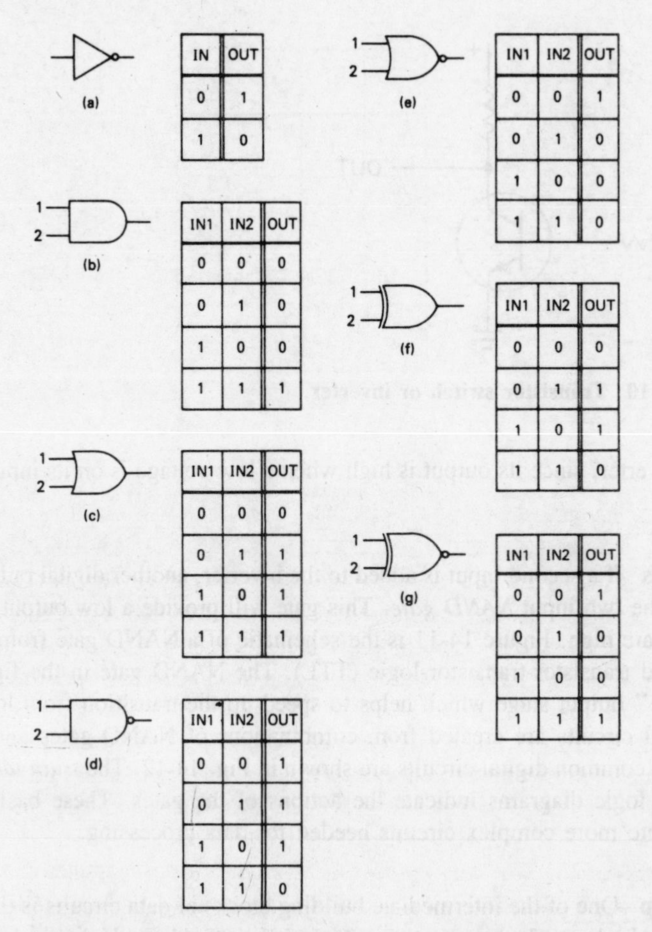

(a)

IN	OUT
0	1
1	0

(b)

IN1	IN2	OUT
0	0	0
0	1	0
1	0	0
1	1	1

(c)

IN1	IN2	OUT
0	0	0
0	1	1
1	0	1
1	1	1

(d)

IN1	IN2	OUT
0	0	1
0	1	1
1	0	1
1	1	0

(e)

IN1	IN2	OUT
0	0	1
0	1	0
1	0	0
1	1	0

(f)

IN1	IN2	OUT
0	0	0
0	1	1
1	0	1
1	1	0

(g)

IN1	IN2	OUT
0	0	1
0	1	0
1	0	0
1	1	1

FIGURE 14-12 Common digital circuits. *(a)* Inverter; *(b)* AND gate; *(c)* OR gate; *(d)* NAND gate; *(e)* NOR gate; *(f)* exclusive OR gate; *(g)* exclusive NOR gate.

latches it onto its output. The latch will retain that level until different input data is received. The latch forms a simple memory device retaining data until the system requires it to be changed.

A simple flip-flop circuit can be constructed from NAND gates, as shown in Fig. 14-13. Such circuits have been integrated, however, and designers usually choose the integrated devices.

The shift register Figure 14-14 shows the diagram of a device called a *shift register*. It is composed of a series of latches so connected that they will successively accept data bits and retain them until the next bit is received. The shift register shown is capable of performing several functions. It can be loaded with data from its parallel inputs, and the data can then be clocked out of its serial output. The result is that a stream of parallel data bits is converted to a serial stream of successive data bits. In a similar

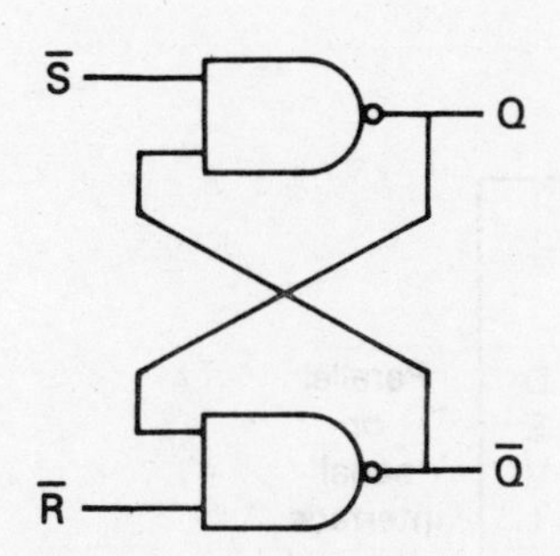

$\overline{S}$	$\overline{R}$	Q	$\overline{Q}$
0	1	1	0
1	0	0	1
1	1	NO CHANGE	
0	0	DISALLOWED	

FIGURE 14-13 Flip-flop circuit.

manner, it is possible for the device to clock in a string of serial data bits through its serial input and then to make these bits available at its parallel outputs. These serial-to-parallel and parallel-to-serial conversions are important for data systems and data transmission. It should be noted that computer systems deal with data in a parallel form, although transmission systems most often require data to be in a serial form. Thus, in order for the computer data to be transmitted and then provided to other computer peripherals, it is normally necessary for serial-to-parallel and parallel-to-serial conversions to be made. Shift registers and similar devices will be noted as forming part of the data sets and other data transmission devices to be discussed later in the chapter.

The microprocessor The microprocessor is a large-scale integrated circuit that has revolutionized computer technology. The microprocessor performs all the decision making and control functions necessary for computer operations. At the most elemental level, it is simply a combination of the gates and latches previously described. The power of the microprocessor results from the large number of gates which make up the device and the digital coding designed into it, which allow many decision options. Figure 14-15 is a block diagram of a typical microprocessor. The figure illustrates that

Q0	+V
T/C	Q1
$\overline{K}$	Q2
J	Q3
RST	DP3
CLK	DP2
P/S	DP1
GND	DP0

T/C	True output or complement
$\overline{K}$&J	Inputs
RST	Reset
CLK	Clock
P/S	Parallel or serial
Q0-Q3	Output
DP0-DP3	Inputs

FIGURE 14-14 Shift register capable of parallel in, parallel out, serial in, or serial out.

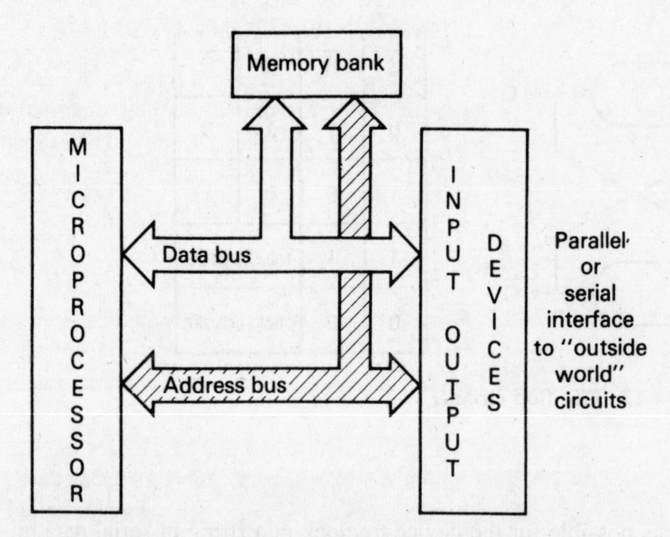

FIGURE 14-15 Typical microprocessor system.

the microprocessor is a *bus-oriented* device, i.e., it receives and sends out data as parallel binary words over common interconnections. All of the devices to which the microprocessor is connected share the data bus. Each device is permitted to use the data bus only when it is sending or receiving data.

The microprocessor has some latch circuits, called registers, built into it. Code words and data words are latched into these registers. Thousands of gates built into the microprocessor then cause different actions to take place, depending on the code and data words in the registers. The microprocessor performs its actions one at a time. It obtains an instruction from its program and places this in the control register. On the basis of the action that the instruction requires, the microprocessor will obtain data from, or will route data to, the other devices which make up the system. The peripheral devices are responsible for converting the incoming data into a form the microprocessor can use. When the appropriate data word is recognized, the microprocessor will carry out some action, e.g., addition, subtraction, storage and so on. This action is one step of data processing. Since the microprocessor can perform data manipulations very rapidly, processing of large quantities of data can be done even with small systems.

The microprocessor communicates with devices which "speak its language." That is, devices used in a microprocessor system must be bus-oriented and must send and receive data as binary words. The use of binary words signifies that the microprocessor system requires parallel data. In an 8-bit system, all 8 bits would arrive at the data port of the system device at the same time. If data arrives at the data port 1 bit at a time, i.e., in serial form, it will be converted before delivery to the microprocessor. The serial data must be converted into parallel data words by the peripheral device. These devices will be discussed later in this chapter; one typical communications application of microprocessors, radio receiver control, is discussed in Sec. 6-3.3.

14-2 FUNDAMENTALS OF DATA COMMUNICATIONS SYSTEMS

Data communications became important when the rapid transfer of data became both necessary and feasible. In other words, data communications emerged as a natural result of the development of sophisticated computer systems. The milestones in this development are now outlined.

14-2.1 The Emergence of Data Communications Systems

Computer systems history The early history of the development of computing machines is replete with impressive names. The French scientist Blaise Pascal is credited with the invention of the first adding machine in 1642. His machine was mechanical in nature, using gears to store numbers.

The mechanical model was followed up in 1822 by Charles Babbage, professor of mathematics at Cambridge University in England. Babbage used gears and punched cards to produce the first general purpose digital computer, which he called the analytic engine, but it was never completed or put into use.

Census taking provided the incentive for Herman Hollerith to use punched cards in the first data processing operation. Their successful application to the 1890 U.S. National Census demonstrated the value to be realized from automatic data processing systems. The laborious, time-consuming task of sorting census data by hand was reduced in both time required and effort expended, because punched cards were put into the machine which automatically sorted them.

Howard Aiken of Harvard University combined the mechanical processes of Babbage with the punched-card techniques of Hollerith to develop an electromechanical computer. The Harvard Mark 1, as it was called, was capable of multiplying and dividing at rates significantly faster than previously possible. The electromechanical nature of the device, which used punched cards and punched tape for data and control, limited its speed and capability, however.

The first fully electronic computer was developed at the University of Pennsylvania by Dr. John Mauchly and J. Presper Eckert, Jr. The computer used 18,000 electron tubes to make and store its calculations. Called the Electronic Numerical Integrator and Calculator (ENIAC), this device could, in 1946, multiply 300 numbers per second (approximately 1000 times as fast as Aiken's computer). As fast as ENIAC was, however, the lack of external control and the bulk and power consumption resulting from the use of vacuum tubes precluded large-scale production.

The milestone which marked the beginning of the modern age of computers was the development of the transistor. This device was significantly smaller than the electron tube, required much less electrical power to operate, and generated very much less heat. With the subsequent development of integrated circuits, it became possible to design equipment consisting of hundreds and thousands of transistors but requiring minimal space. This advance has made computers with amazing speed and impressive capability commonplace. Concurrently with the development of smaller, faster, and more sophisticated computers, developments in storage devices were also made.

Computer systems have been classed into three generations. The first generation consisted of vacuum-tube-based machines. They used magnetic drums for internal storage and magnetic tape for external storage. These computers were slow compared to modern machines and, owing to their bulk, they required data to be brought to them.

Second-generation computers using transistors began to appear in 1959. The internal storage used magnetic cores, with small doughnuts of magnetic material wired into frames that were stacked into large cores. This form of storage represented a tremendous increase in speed and reduction in bulk over previous storage methods. The external storage in second-generation computers used magnetic disks. This form of storage also added to increased speed and greater ''online'' storage capability as compared to magnetic tape systems.

Beginning in 1964, a third generation of computers began to emerge. These computers utilized integrated circuits to increase capability and decrease size, while integrated technology also provided improved internal storage capability. Solid-state memory, being totally electronic, greatly increased the speed and capacity of the internal memory while decreasing its cost and complexity. External memory continued to use magnetic disks, which became larger and faster.

It was stated that early computers required data to be brought to them. This data was usually prepared by using punch cards or magnetic tape. The cards or tapes would then be carried to the computer where they would be processed. The transfer of data in this fashion was called *batch processing*. Transport might be no farther than from the next room, or again, it might be from the other side of the world. As each batch of data was received, it was placed into line with other batches of data which were processed one after another. Reports were generated, files were updated, new tapes were made, and the revised data was routed to appropriate locations in the form of punched cards or magnetic tape. The inefficiency of such a system is easily seen in retrospect.

Later-model computers are provided with the capability of handling numerous input devices directly. These multitask computers treat the incoming data in much the same way as the earlier computers did. Incoming data is received from the various input devices and is lined up, or ''queued,'' by the computer. The computer will then process the incoming data according to internal procedures. If the computer reaches a place with one batch of data where it can link the data to storage, printers or other devices, the computer will begin to process another batch. The modern computers are so fast in their operation that they can handle many users without the users even being aware that others are on the system. This capability has made it necessary for computer data to be transported in ways other than by punch cards or magnetic tape. *The ability of the computer to service many input-output devices simultaneously has made data communications essential.*

The rise of data systems It was the ability to handle multiple tasks and numerous remote terminals which promoted the rise of the data transmission industry. Initially, standardization was sought for the interconnections needed between the computer and the various peripheral devices. This standardization took the form of standard connectors, signaling formats and signal levels. As these standards became recognized by the industry, it became desirable to extend them to the transmission media used for medium- and long-haul transmission of data.

The need for transmission standards became really acute when computer facilities began to use the telephone system for their transmission requirements. The pervasiveness of the telephone system made it ideal for interconnection of computers with remote sites, but one major problem was encountered; because the telephone system was designed for voice communication, some modifications were required for data transmission. Indeed, much of the current body of data transmission engineering information is the product of telephone system engineers. Initially, data utilized dedicated circuits which could be specifically adapted for data transmission. As the need for data transmission increased, however, it became advantageous for data uses to be accommodated over standard voice-grade channels. Modifications to telephone circuit equipment were made, and new devices such as acoustic couplers, which made the telephone system accessible for widespread data transmission, were designed. Data communication now has its own language, equipment and standards. It is an industry in itself and is certainly an integral part of the current computerized society.

14-2.2 Characteristics of Data Transmission Circuits

Bandwidth requirements Data in most instances consists of pulse-type energy. The data stream is similar to a square-wave signal with rapid transitions from one voltage level to another, with the repetition rate depending on the binary representation of the data word. For instance, if an 8-bit word has the value 01010101, the resulting voltage graph would appear as a series of four square waves with each negative half-cycle equal to each positive half-cycle. If, however, the data word has the form 00001111, the voltage graph would appear as a single square wave with negative and positive half-cycles equal but longer than the first example. Figure 14-16 shows the voltage graphs for these and other binary words. It can be seen that data circuits must provide a bandwidth for the data transmissions they carry. This will be governed by the pulse rate variations just explained, and by the fact, indicated in Chap. 1, that even a single square wave occupies a frequency range because of the harmonics present.

Since many data transmissions utilize telephone channels, the bandwidth of the telephone is an appropriate consideration. As shown in Secs. 4-3.1 and 15-1.1, the internationally accepted standard telephone channel occupies the frequency range of

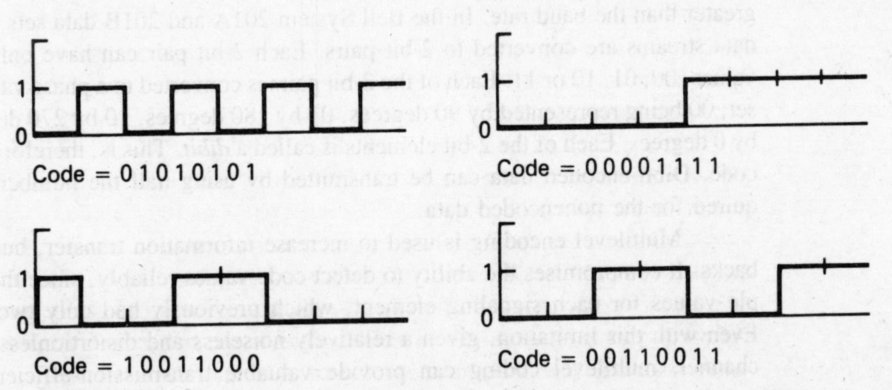

FIGURE 14-16 Digital code waveforms showing frequency variations for different codes.

300 to 3400 Hz, this referred to within the industry as a 4-kHz channel. In certain difficult or expensive applications, such as HF radio or some submarine cables, 3-kHz circuits, in which the frequency range is 300 to 2800 Hz, are used. Neither channel will encompass all of the audible spectrum, but each will cover the range into which speech falls and convey enough of the components of speech to ensure intelligibility and voice recognition. The signals which fall outside the channel bandwidth are attenuated by filters so that they will not interfere with other signals.

When data is sent over telephone channels, the speed must be limited to ensure that the bandwidth required by the data transmission will not exceed the telephone channel bandwidth. The faster the data is transmitted, the greater the bandwidth will need to be to accommodate it.

Data transmission speeds The rate of data transfer depends on several aspects of the transmission channel, of which signaling speed is very important. Transmission engineers often refer to the transmission speed of a communications channel as the channel's baud rate. The baud (defined in Sec. 13-3.1) is an important unit of signaling speed. In a system in which all pulses have equal duration, the speed in bauds is equal to the maximum rate at which signal pulses are transmitted. This should be recognized as different from information bit rate. In a system which uses only one information bit per signaling pulse, i.e., a binary system, the baud rate and the bit rate happen to be the same. In systems which encode the data in such a way that more than one information bit can be placed on each signaling pulse, the information bit rate will exceed the baud rate.

To relate baud rate to bandwidth, the observations of the twentieth-century electrical engineer Nyquist are used. Nyquist determined that one cycle of a transmission can contain a maximum of two bauds. This relation was derived in Sec. 13-1.2, in a slightly different context. The result is that the maximum signaling speed in bauds is equal to twice the bandwidth of the channel. As previously indicated, this is theoretical, however, and could be achieved only in an ideal channel which had no noise or distortion.

As indicated above, the baud is a unit of signaling speed, but information transfer can occur at a rate equal to or different from the baud rate. Multilevel and encoded data elements can be used to provide information transfer rates at speeds greater than the baud rate. In the Bell System 201A and 201B data sets, for example, data streams are converted to 2-bit pairs. Each 2-bit pair can have only one of four values, 00, 01, 10 or 11. Each of the 2-bit pairs is converted to a phase value in the data set, 00 being represented by 90 degrees, 01 by 180 degrees, 10 by 270 degrees, and 11 by 0 degrees. Each of the 2-bit elements is called a *dibit*. This is, therefore, a four-level code. Dibit-encoded data can be transmitted by using half the number of bauds required for the nonencoded data.

Multilevel encoding is used to increase information transfer, but it has drawbacks. It compromises the ability to detect code values reliably, since there are multiple values for each signaling element, which previously had only two: ON or OFF. Even with this limitation, given a relatively noiseless and distortionless transmission channel, multilevel coding can provide valuable transmission-efficiency improvements.

Equation (13-6) gave the formula for the maximum capacity for a noisy channel with a given noise level. This formula provides the ideal expectations, which are not realizable in practice. Nonetheless, the Shannon-Hartley law does set the upper limit for a channel and encourages continued coding improvements to increase channel capacity. For instance, if Example 13-2 is recalculated for a voice-grade channel with a 3100-Hz bandwidth and a signal-to-noise ratio of 30 dB, the Shannon-Hartley maximum bit rate of 30,880 bps is obtained for this standard channel. The data rates of common systems are limited to a maximum rate of about 10,800 bps for a voice-grade channel. Faster data rates are prevented by noise other than random in the channel and other channel limitations. The advantages of faster data rates over voice-grade channels must be weighed against the design and implementation cost of advanced data communications systems.

Noise The Shannon-Hartley law is related to random noise, but impulse noise can also be harmful to signals. The sampling theorem (see Sec. 13-2.1) shows that all values of a signal can be determined by sampling the signal at a rate equal to at least twice the bandwidth. Noise affects this sampling process because the noise pulse will be interpreted as a data bit (see Fig. 14-17), if the noise impulse occurs at the time a sample is taken, and has an amplitude equal to or exceeding the minimum level recognized by the system as a mark. The potential for impulse noise to become a source of errors increases with the number of levels of each code element. To achieve the 30,880-bps rate mentioned in the above example, it may be shown that five levels would be required for each code element. A noise-free channel would be necessary to preclude noise-induced data errors, but noise-free channels do not exist in practice. It is noise, among other impairments, which tends to limit the actual 4-kHz channel data speeds to 10,800 bps or less.

The effect of noise on the data channel can be reduced by increasing the signal-to-noise ratio. For an ideal 3-kHz channel, the Nyquist rate (twice the bandwidth, as discussed) would be 6000 bps. A binary system using this channel would require a minimum signal-to-noise ratio of 3:1, or 4.8 dB. This is calculated by using Eq. (14-1), as follows:

$$S/N = 2^{NR/\delta f} - 1 \tag{14-1}$$

where S/N = Signal-to-noise ratio
NR = Nyquist rate
δf = Channel bandwidth

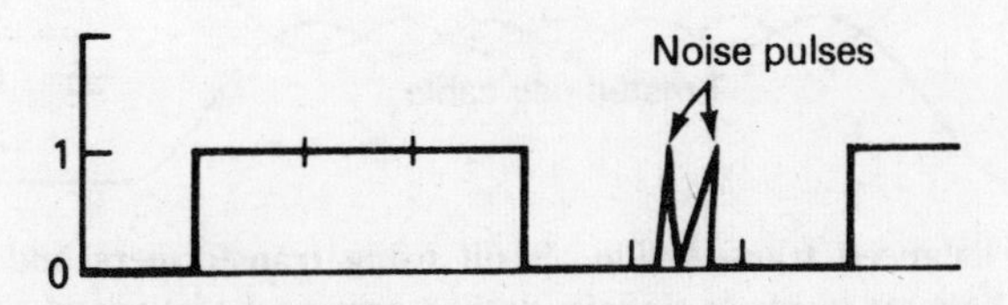

FIGURE 14-17 Data stream with noise pulse.

For the ideal 3-kHz channel:

$$\begin{aligned} S/N &= 2^{6000/3000} - 1 \\ &= 3 \quad \text{or } 3{:}1 \end{aligned}$$

To obtain the decibel value:

$$\begin{aligned} dB &= 10 \log S/N \\ &= 10 \log 3 \\ &= 4.8 \end{aligned}$$

It can be shown that a system using a three-level code must have a signal-to-noise ratio of 8.5 dB, or 3.7 dB greater, for equal performance in the same channel. A four-level code requires a signal-to-noise performance of 11.7 dB. Improvement in the signal-to-noise ratio makes use of multilevel encoding feasible.

Crosstalk Any transmission system which conveys more than one signal simultaneously can experience *crosstalk,* which is the reception of portions of a signal from one channel in another channel. This is common in multiplexed systems in which inadequate procedures are employed to ensure that overmodulation of the various carriers of the multiplexed groups is prevented. In modern transmission systems which convey many channels of voice and data simultaneously, the systems will become ''loaded,'' or heavily utilized, so that the control of levels of the individual channels and the group levels becomes very important in order to preclude crosstalk. Data transmission engineers have developed specific level-setting parameters to ensure that as the circuit loading increases, crosstalk will not become a problem.

Crosstalk can also occur through electromagnetic interaction between adjacent wires. If the wires of two signal-carrying circuits run parallel with each other, it is possible for the signal from one circuit to be induced by electromagnetic radiation into the second circuit. This phenomenon becomes more pronounced when the length of parallel circuits is extensive. This type of crosstalk is reduced by using twisted pair cables and balanced circuits along with shielding.

In a balanced circuit, a transformer is placed at each end of the circuit. The transformers are carefully constructed to provide a center tap which is at the exact electrical center of the winding which connects to the transmission circuit. The center taps at each end are grounded. As shown in Fig. 14-18, if twisted pair cables are used for the transmission circuit, noise or signals from other circuits will be induced into

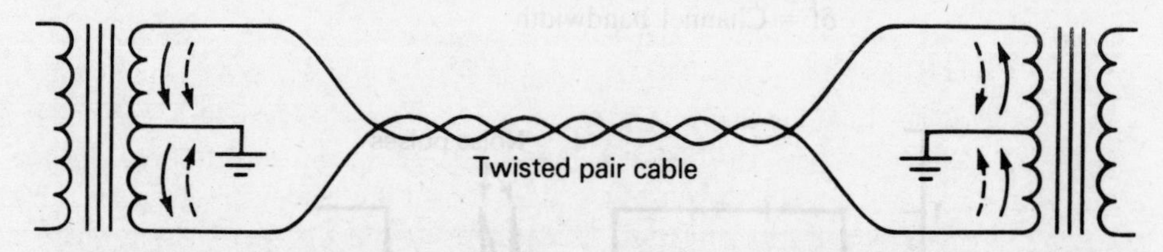

FIGURE 14-18 A balanced transmission circuit using transformers and twisted-pair cable. Solid arrows indicate in-phase signals; dashed arrows depict out-of-phase noise or crosstalk.

both wires at equal levels. When the crosstalk or noise reaches the transformer, it enters as out-of-phase signals from the two wires and cancels out in the transformer windings. The circuit signal, however, enters the transformer in phase. Each side of the transformer forms a circuit with ground and the signal transfers through the transformer intact. The crosstalk and noise are reduced, but the signal is unaffected.

Another way to reduce crosstalk is to use shielded cables. If the twisted pairs are placed inside a braided or metal foil shield, the induction between pairs cannot take place as easily. The shields are grounded to drain off the induced signals and noise.

Echo suppressors As discussed in Sec. 15-4.3, echo suppressors or echo cancellers are used on long-distance circuits, in an effort to overcome echoes caused by circuit imbalances. This is of significance to data transmission because a lot of it occurs over the public switched telephone network, nationally and internationally.

Although the use of echo suppressors improves voice communications, it is incompatible with data transmission. Because a lot of data transmissions are bothway, or quickly alternating from one direction to the other, they require the capability of bidirectional transmission at standard levels, or at least rapid response and interrupt capability. For this type of operation to be accomplished, it is necessary to disable the echo suppressor. In fact, so-called ''tone-disablable'' echo suppressors have been designed to accommodate the needs of data users. If a 2025-Hz tone is applied to the line for approximately 300 ms prior to the start of transmission, such an echo suppressor will be disabled and bidirectional communication can proceed. If a gap in the transmission greater than 100 ms occurs, the echo suppressor will be reactivated.

Distortion Communication channels tend to react to signals of different speeds within their passband in different ways. Specifically, signals of different frequencies can be passed by the channel with different values of amplitude attenuation and at different propagation speeds. The result is distortion.

Of great importance to systems using phase modulation is phase delay (or envelope delay) distortion. *Phase delay distortion* occurs in a channel when signals of one frequency are passed through the circuit at a different speed than other signals. The resulting distortion can take the form of intersymbol interference. Since characters which have lower-frequency components pass at a different speed than data characters with high-frequency components, it is possible in higher-speed circuits for portions of one character to enter or remain in the time slot allocated to other characters.

Equalizers Phase delay distortion can be reduced to acceptable levels by using equalization on the channel. As shown in Fig. 14-19, it is possible to plot the delay characteristics of the channel and insert an equalizer which can be adjusted to compensate for the delay abnormalities. The result is a channel relatively free of phase delay.

Equalizers can be obtained which are automatic in nature. These equalizers precede data transmission with a short ''training period'' during which test pulses are used to determine the delay characteristics of the channel. The equalizer automatically varies its delay characteristics while sampling the return signal to determine when the channel delay plus equalizer delay reach proper tolerances. At that time, data transmis-

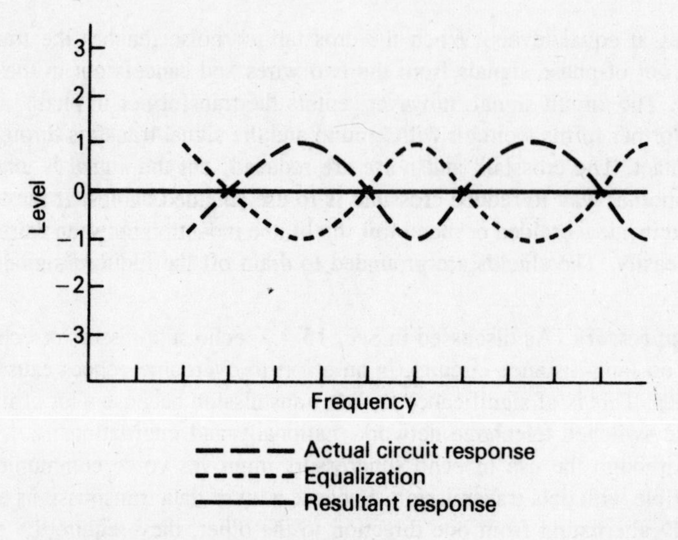

FIGURE 14-19 Circuit equalization.

sion commences. The data is thereafter sampled during transmission to ensure that equalization settings are appropriate, with modifications made as required. This type of equalization is called *adaptive equalization*.

Preset equalization or conditioning follows the same processes as adaptive equalization except that the equalization is set prior to transmission and then updated only during breaks in transmission, using special test sequences. This is not as flexible as adaptive equalization, since the transmission must be interrupted to permit transmission of test data sequences whenever the channel characteristics alter. However, it is quite acceptable for dedicated circuits with fixed terminations. It is possible to lease national or international circuits that have been conditioned to domestic or international standards. Understandably, though, such circuits are more expensive than unequalized circuits.

14-2.3 Digital Codes

Various types of equipment are used in computer systems to send and receive data: keyboards, video terminals, printers, paper tape punches and readers, paper card punches and readers, and magnetic storage devices. Each of these types of equipment generates and receives data in the form of codes. The fact that all use encoded data, however, does not mean that all use the same code. Indeed, several codes exist and are common among digital data systems. The reasons for more than one encoding system are several.

Codes evolved during the development of data systems. Some of these codes replaced existing codes, but as new encoding systems developed, the previous systems continued alongside the new codes.

Standardization is not easy to accomplish. It is difficult to convert all users to a single coding scheme, since some codes are advantageous for one use although others are better for different applications. Adopting nationwide and especially worldwide standards is normally a lengthy and sometimes frustrating process. As in many other areas, the marketplace and politics make the ultimate decision.

The capability of modern data systems has reduced the necessity of establishing a single encoding scheme. Modern computers can easily deal with different codes by simply converting them to the code used by the computer. With speeds of several million operations per second for many current computers, the time invested in code translation is negligible. The result is that several encoding systems are in use within data systems and can be expected to continue in use for some time. It is necessary, therefore, that these major encoding systems be given due consideration.

The Baudot code Named for the telegraph pioneer, J. M. E. Baudot, the Baudot code is a 5-bit code which has been used in telegraphy and paper-tape systems. With only 5 bits available, the basic code is limited to 32 different code combinations ($5^2 = 32$). Shift codes have been incorporated into the Baudot code to indicate whether a code is upper- or lowercase. This increases the number of code combinations to 64, of which 6 are used for function codes, leaving only 58 available codes. The alphabet, numbers and functions require 42 of these 58. This limits the ability of the Baudot code to provide extra punctuation and computing codes. Figure 14-20 shows the Baudot encoding scheme. Another limitation of the Baudot code is evident in the figure: the code is not sequential, limiting its ability to be used for computation.

Early teletypewriter machines used the Baudot code for intercommunications. Many of these machines incorporated a paper-tape punch and reader mechanism in their systems. Figure 14-21 illustrates the use of a Baudot code with paper tape. The use of shift characters to indicate that succeeding characters are letters or figures is also shown.

																											Blank	Letters	Figures	Space	Carriage Return	Line Feed
A	↑	⊕	○	↗	3	→	↘	↓	8	↙	←	↖	•	⦶	9	0	1	4	Bell	5	7	⦶	2	/	6	+	—					
B	—	5/8	1/8	$	3	1/4	&		8	1	1/2	3/4	•	7/8	9	O	1	4	Bell	5	7	3/8	2	/	6	"						
C	—	?	:	$	3	!	&	#	8	1	(	)	•	,	9	O	1	4	Bell	5	7	;	2	/	6	"	≀	↓	↑	■	<	≡
	A	B	C	D	E	F	G	H	I	J	K	L	M	N	O	P	Q	R	S	T	U	V	W	X	Y	Z		TAPE SYMBOLS ONLY				
1	X	X		X	X	X				X	X						X		X		X		X	X	X	X		X	X			
2	X		X				X		X	X	X	X				X	X	X			X	X	X					X	X			X
3			X			X		X	X		X		X	X		X	X		X		X	X		X	X			X		X		
4		X	X	X		X	X			X	X		X	X	X			X				X		X				X	X		X	
5		X					X	X				X	X		X	X	X			X		X	X	X	X	X		X	X			

FIGURE 14-20 The Baudot code. This 5-element code uses letter shift and figure shift symbols to expand the number of combinations it can provide. Line A, weather symbols; line B, used for fractions; line C, used for communications.

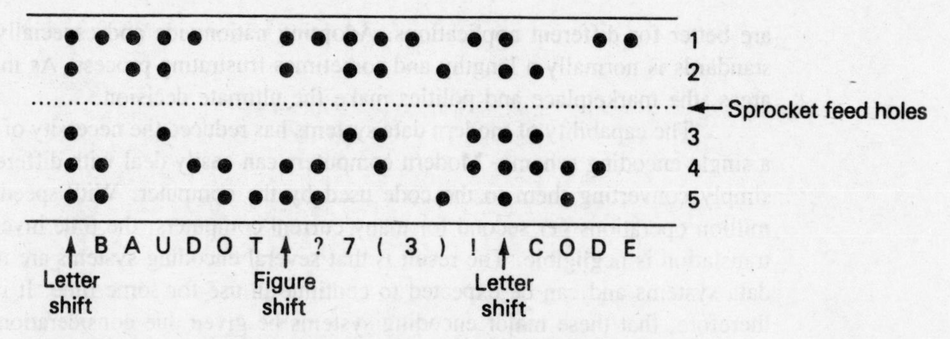

FIGURE 14-21 Baudot code as punched into paper tape.

The binary code Binary encoding forms the basis of several coding schemes. If straight binary encoding is used, 256 different combinations are possible for an 8-bit character. Binary encoding is not used unmodified in many situations, however, for several reasons. Although 256 combinations are available, this is inadequate for representation of large numbers. Also, it was learned early that errors can occur during transmission of data, but the use of an unmodified 8-bit binary code did not permit any means of error detection. The most useful code would incorporate an error-detecting bit, called a *parity bit*. For use with numbers, the binary code was modified so that only the lower 4 bits were needed. This system, called *binary-coded-decimal* (BCD), counts binarily from 0 to 9, as shown in Fig. 14-22. The sequence uses a second 8-bit word to represent each successive decimal column. As one binary word reaches decimal 10, it returns to zeros and a carry is added into the next binary word. The use of BCD encoding to represent a four-digit decimal number is shown in Fig. 14-23.

One of the uses for BCD encoding is for data representation on magnetic tape. Data is recorded on magnetic tape in much the same way as audio; a recording head creates a magnetic pattern on the tape which represents the information. For data recording, the recording is made on several tracks. A 1 results in a magnetized spot

0 0 0 0	0 0 0 0	0 0 0 0	0 0 0 0	=0
0 0 0 0	0 0 0 0	0 0 0 0	0 0 0 1	=1
0 0 0 0	0 0 0 0	0 0 0 0	0 0 1 0	=2
0 0 0 0	0 0 0 0	0 0 0 0	0 0 1 1	=3
0 0 0 0	0 0 0 0	0 0 0 0	0 1 0 0	=4
0 0 0 0	0 0 0 0	0 0 0 0	0 1 0 1	=5
0 0 0 0	0 0 0 0	0 0 0 0	0 1 1 0	=6
0 0 0 0	0 0 0 0	0 0 0 0	0 1 1 1	=7
0 0 0 0	0 0 0 0	0 0 0 0	1 0 0 0	=8
0 0 0 0	0 0 0 0	0 0 0 0	1 0 0 1	=9
0 0 0 0	0 0 0 0	0 0 0 1	0 0 0 0	=10

FIGURE 14-22 Binary coded decimal.

```
00000100    00001001    00001000    00000011
    4           9           8           3
```

FIGURE 14-23 Decimal 4983 represented in BCD.

```
1              1              8
  1    1  1  1     1          4 Binary
  1 1        1 1   1 1        2 weights
1    1   1   1 1      1       1
9 6 2 1 0 5 8 4 3 7 0 0 9 6 2
        1 = Magnetic domain
```

FIGURE 14-24 BCD code recorded on magnetic tape.

being recorded, while a 0 leaves the spot unmagnetized. For recording BCD, four tracks are used, with each character being represented by a pattern of magnetized spots on the track, similar to the holes in a punched tape (see Fig. 14-24).

An extension of the BCD code is the 7-bit *alphanumeric* code. This code uses BCD for representation of numbers, but adds two extra bits to represent letters and punctuation marks (see Fig. 14-25). A seventh bit is used to provide parity for error detection. These 7 bits are recorded on seven parallel tracks on the magnetic tape.

ASCII code One of the more universal codes is the American Standard Code for Information Interchange (ASCII). ASCII is based on a binary progression, as demonstrated in Fig. 14-26. It should be noted that the code is arranged so that the numbers are represented by a standard BCD progression within the last bits shown on the left of the chart, while the preceding 3 bits, shown at the top of the chart, specify whether a number, letter or character is being represented by the last 4 bits. For example, the table shows that an ASCII code of 0110001 represents the number 1, while 1000001 represents a capital "A," and the code 1100001 represents a lowercase "a." By using a standard binary progression, ASCII makes possible mathematical operations with numbers. Since the letters are also in a binary progression, alphabetizing can be accomplished by using simple binary mathematical procedures.

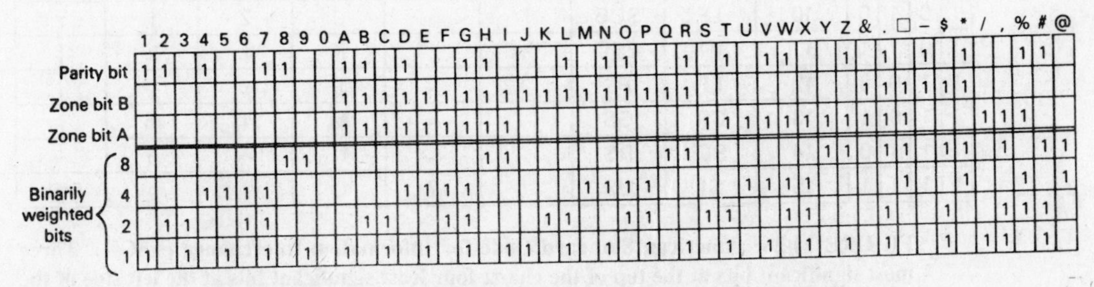

FIGURE 14-25 The alphanumeric code; even-bit parity is used.

Most modern computers use *hexadecimal* notation internally. Hexadecimal notation represents a 4-bit binary word with one of 16 symbols (0,1,2,3,4,5,6,7,8,9, A,B,C,D,E,F). An 8-bit word is easily accommodated in these computers. Since ASCII is a 7-bit code, it is normally converted into 8-bit words by using the most significant bit as a parity bit for error detection. Typically, the parity bit is given the value (1 or 0) which will result in the sum of the 1s in the ASCII data word being even. When checked after transmission, if the parity bit does not result in an even sum, an error is assumed and the data is retransmitted. Error detection is covered in more detail in Sec. 14-2.4.

EBCDIC Another popular code is called the Extended Binary Coded Decimal Interchange Code (EBCDIC). EBCDIC is also based on the binary-coded decimal format, as its name implies, but it differs from the ASCII code in several respects. As shown in Fig. 14-27, EBCDIC uses all 8 bits for information, so that no parity bit is available. Also, although EBCDIC follows a BCD progression for the numbers, the numbers follow the letters rather than preceding them as they do in ASCII. Approved by the International Telephone and Telegraph Consultative Committee (CCITT), EBCDIC has similarities to the Baudot code; it was mentioned in Chap. 13 under the name "CCITT No. 2."

Bits				b_7 →	0	0	0	0	1	1	1	1
				b_6 →	0	0	1	1	0	0	1	1
				b_5 →	0	1	0	1	0	1	0	1
b_4 ↓	b_3 ↓	b_2 ↓	b_1 ↓	Column → / Row ↓	0	1	2	3	4	5	6	7
0	0	0	0	0	NUL	DLE	SP	0	@	P	`	p
0	0	0	1	1	SOH	DC1	!	1	A	Q	a	q
0	0	1	0	2	STX	DC2	"	2	B	R	b	r
0	0	1	1	3	ETX	DC3	#	3	C	S	c	s
0	1	0	0	4	EOT	DC4	$	4	D	T	d	t
0	1	0	1	5	ENQ	NAK	%	5	E	U	e	u
0	1	1	0	6	ACK	SYN	&	6	F	V	f	v
0	1	1	1	7	BEL	ETB	'	7	G	W	g	w
1	0	0	0	8	BS	CAN	(	8	H	X	h	x
1	0	0	1	9	HT	EM	)	9	I	Y	i	y
1	0	1	0	10	LF	SUB	*	:	J	Z	j	z
1	0	1	1	11	VT	ESC	+	;	K	[	k	{
1	1	0	0	12	FF	FS	,	<	L	\	l	|
1	1	0	1	13	CR	GS	–	=	M	]	m	}
1	1	1	0	14	SO	RS	.	>	N	^	n	~
1	1	1	1	15	SI	US	/	?	O	_	o	DEL

FIGURE 14-26 American Standard Code for Information Interchange (ASCII). Three most significant bits at the top of the chart; four least significant bits at the left side of the chart. The number 6 would have 011 from the top of the chart and 0110 from the side: 6 = 011 0110.

$b_7\ b_6\ b_5\ b_4$ \\ b_0	0	0	0	0	0	0	0	0	1	1	1	1	1	1	1	1
b_1	0	0	0	0	1	1	1	1	0	0	0	0	1	1	1	1
b_2	0	0	1	1	0	0	1	1	0	0	1	1	0	0	1	1
b_3	0	1	0	1	0	1	0	1	0	1	0	1	0	1	0	1
0 0 0 0	NUL	SOH	STX	ETX	PF	HT	LC	DEL			SMM	VT	FF	CR	SO	SI
0 0 0 1	DLE	DC1	DC2	DC3	RES	NL	BS	IL	CAN	EM	CC	CU1	FS	GS	RS	US
0 0 1 0	DS	SOS	FS		BYP	LF	EOB	PRE			SM	CU2		ENQ	ACK	BEL
0 0 1 1			SYN		PN	RS	UC	EOT				CU3	DC4	NAK		SUB
0 1 0 0	SPACE										¢		<	(	+	/
0 1 0 1	&										!	$	*	)	;	¬
0 1 1 0											^	,	%	_	>	?
0 1 1 1											:	#	@	'	=	"
1 0 0 0		a	b	c	d	e	f	g	h	i						
1 0 0 1		j	k	l	m	n	o	p	q	r						
1 0 1 0			s	t	u	v	w	x	y	z						
1 0 1 1																
1 1 0 0		A	B	C	D	E	F	G	H	I						
1 1 0 1		J	K	L	M	N	O	P	Q	R						
1 1 1 0			S	T	U	V	W	X	Y	Z						
1 1 1 1	0	1	2	3	4	5	6	7	8	9						¤

FIGURE 14-27 Extended Binary Coded Decimal Interchange Code (EBCDIC).

Hollerith code Several codes are in use for punched cards, many of them specific to particular manufacturers. One of the more universal punched-card codes is the Hollerith code. This code is used with an 80-column card, as shown in Fig. 14-28. It is seen that the code for a number, letter, punctuation or control character is punched into the card as a pattern of rectangular slots using variations of 12 horizontal rows. The logical arrangement of the Hollerith code makes it convenient for sorting and computing applications.

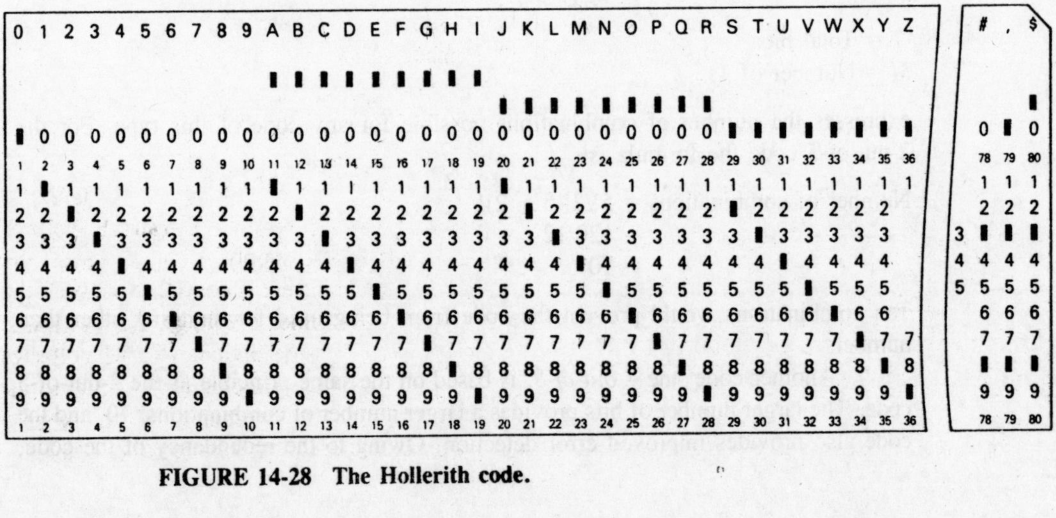

FIGURE 14-28 The Hollerith code.

14-2.4 Error Detection and Correction

Errors enter the data stream during transmission and are caused by noise and transmission system impairments. Because errors compromise the data and in some cases render it useless, procedures have been developed to detect and correct transmission errors. The processes involved with error correction normally result in an increase in the number of bits per second which are transmitted, and naturally this increases the cost of transmission. Procedures which permit error correction at the receiver location are complicated, and so it is necessary for data users to determine the importance of the transmitted data and to decide what level of error detection and correction is suitable for that data. The tolerance the data user has for errors will decide which error control system is appropriate for the transmission circuit being used for the user's data.

Error detection The 5-bit Baudot code provides no error detection at all, because it uses all 5 bits to represent characters. If only 1 bit is translated (by error) to its opposite value, a totally different character will be received and the change will not be apparent to the receiver. The inability of such codes to detect errors has led to the development of other codes which provide for error control.

Constant-Ratio Codes

A few codes have been developed which provide inherent error detection when used in ARQ (automatic request for repeat) systems. The *2-out-of-5 code* follows a pattern which results in every code group having two *1s* and three *0s*. When the group is received, the receiver will be able to determine that an error has occurred if the ratio of *1s* to *0s* has been altered. If an error is detected, a NAK (do not acknowledge) response is sent and the data word is repeated. This testing procedure continues word for word.

This code has some limitations, however. An odd number of errors will always be detected, but an even number of errors may go undetected. Even more limiting is the problem that this code will severely reduce the number of available code combinations. The formula

$$\text{Number of combinations} = \frac{T!}{M!(T-M)!} \tag{14-2}$$

T = Total bits
M = Number of 1s

expresses the number of combinations possible for any code of this type. For the 2-out-of-5 code the formula is:

$$\begin{aligned}\text{Number of combinations} &= 5!/2!(5-2)! \\ &= 120/12 \\ &= 10\end{aligned}$$

Ten combinations would prevent the code from being used for anything other than numbers.

Another code, the *4-out-of-8*, is based on the same principle as the 2-out-of-5 code. The larger number of bits provides a larger number of combinations, 70, and the code also provides improved error detection. Owing to the redundancy of the code, however, its efficiency for transmission is reduced.[1] Codes such as the 2-out-of-5 and 4-out-of-8, which depend on the ratio of *1s* to *0s* in each code group to indicate that errors have occurred, are called *constant-ratio codes*.

Redundant Codes

Most error-detection systems use some form of redundancy to check whether the received data contains errors. This means that information additional to the basic data is sent. In the simplest system to visualize, the redundancy takes the form of transmitting the information twice and comparing the two sets of data to see that they are the same. Statistically, it is very unlikely that a random error will occur a second time at the same place in the data. If a discrepancy is noted between the two sets of data, an error is assumed and the data is caused to be retransmitted. When two sets of data agree, error-free transmission is assumed.

Retransmission of the entire message is very inefficient, because the second transmission of a message is 100 percent redundant. In this case as in all cases, redundant bits of information are unnecessary to the meaning of the original message. It is possible to determine transmission efficiency by using the following formula:

$$\text{Efficiency} = \text{Information bits/total bits} \tag{14-3}$$

In the above case of complete retransmission, the number of information bits is equal to one-half the number of total bits. The transmission efficiency is therefore equal to 0.5, or 50 percent. In a system with no redundancy, information bits equal total bits and the transmission efficiency is 100 percent. Most systems of error detection fall between these two extremes; efficiency is sacrificed to obtain varying degrees of security against errors which would otherwise be undetected.

Parity-Check Codes

A popular form of error detection employing redundancy is the use of a *parity-check bit* added to each character code group. Codes of this type are called *parity-check codes*. The parity bit is added to the end of the character code block according to some logical process. The most common parity-check codes add the *1s* in each character block code and append a *1* or *0* as required to obtain an odd or even total, depending on the code system. Odd parity systems will add a *1* if addition of the *1s* in the block sum is odd. At the receiver, the block addition is accomplished with the parity bit intact, and appropriate addition is made. If the sum provides the wrong parity, an error during transmission will be assumed and the data will be retransmitted.

Parity bits added to each character block provide what is called *vertical parity*, which is illustrated in Fig. 14-29. The designation vertical parity is explained by the figure which shows the parity bit at the top of each column on the punched tape.

Parity bits can also be added to rows of code bits. This is called *horizontal parity* and is also illustrated by Fig. 14-29. The code bits are associated into blocks of specific length with the horizontal parity bits following each block. By using the two

[1] The application of Eq. (14-4) shows that, if there were no restriction of the number of 1s in a code group, 8 bits would provide 40,320 combinations, 576 times as many as are provided by the 4-out-of-8 code.

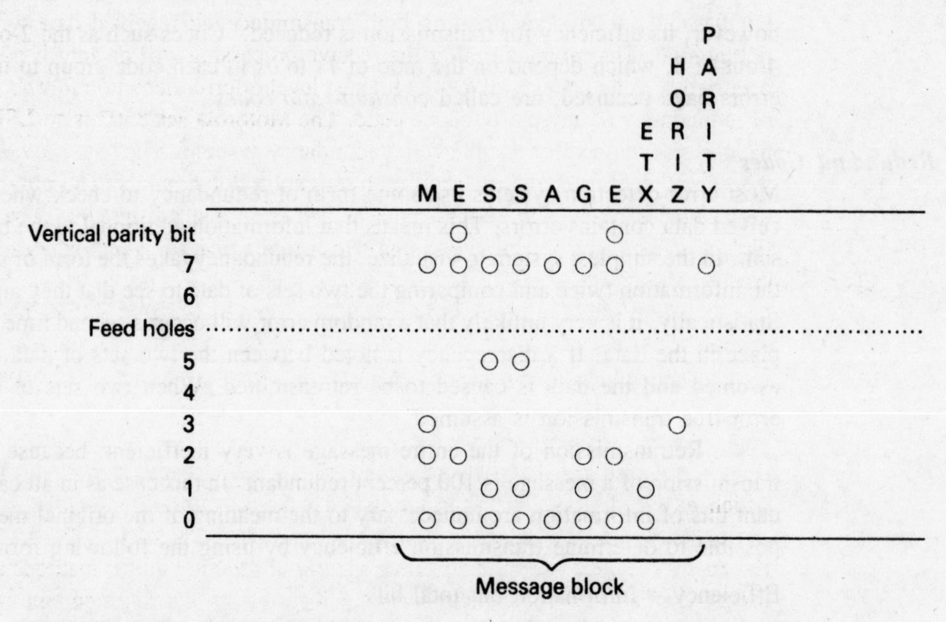

FIGURE 14-29 Vertical and horizontal parity used with a paper tape code.

parity schemes concurrently, it becomes possible to determine which bit is in error. This is explained in Fig. 14-30, where even parity is expected for both horizontal and vertical parity. Note that here one column and one row each display improper parity. By finding the intersection of the row and column, the bit in error can be identified. Simply changing the bit to the opposite value will restore proper parity both horizontally and vertically. These types of parity arrangements are sometimes called *geometric codes*.

Another group of parity-check codes are referred to as *cyclic codes*. These use shift registers with feedback to create parity bits based on polynomial representations of the data bits. The process is somewhat involved and will not be fully described here,

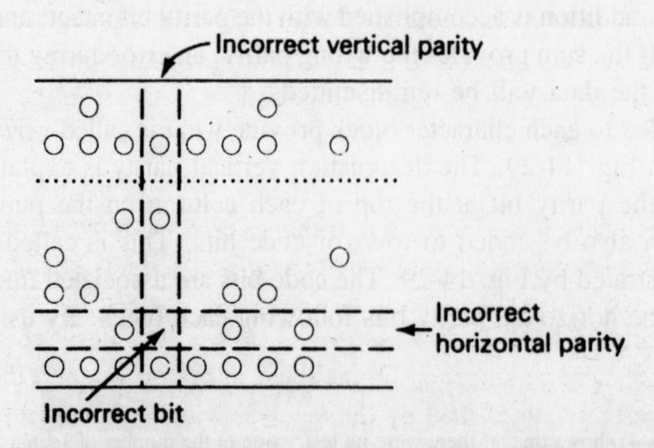

FIGURE 14-30 Error detection using vertical and horizontal parity.

but basically it involves treating both transmitted and received data with the same polynomial. The remainder after the receive processing will be zero if no errors have occurred. Cyclic codes provide the highest level of error detection for the same degree of redundancy of any parity-check code. The Motorola MC8503 is an LSI chip which has been developed for use in cyclic redundancy systems. The chip provides for use in systems which utilize any of four more common polynomials. The polynomial to be used is selected by a three-digit code which is applied to the chip. The MC8503 is typical of the error-detection and correction sophistication which is possible with microchip technology.

One additional type of parity-check encoding scheme differs from those described previously in that it does not require the data to be grouped into blocks. Instead, the data is treated as a stream of information bits into which parity bits are interspersed according to standard rules of encoding. The process is more involved than some of the other schemes and is typically reserved for higher-data-speed applications. *Convolutional codes,* as these are called, are particularly well suited to systems which utilize forward error-correcting procedures as described below.

Error correction Detecting errors is clearly of little use unless methods are available for the correction of the detected errors. Correction is thus an important aspect of data transmission.

Retransmission

The most popular method of error correction is retransmission of the erroneous information. For the retransmission to occur in the most expeditious manner, some form of automatic system is needed. A system which has been developed and is in use is called the automatic request for repeat (ARQ), also called the positive acknowledgment/negative acknowledgment (ACK/NAK) method. The request for repeat system transmits data as blocks. The parity for each block is checked upon receipt, and if no parity discrepancy is noted, a positive acknowledgment (ACK) is sent to the transmit station and the next block is transmitted. If, however, a parity error is detected, a negative acknowledgment (NAK) is made to the transmit station which will repeat the block of data. The parity check is again made and transmission continues according to the result of the parity check. The value of this kind of system stems from its ability to detect errors after a small amount of data has been sent. If retransmission is needed, the redundant transmission time is held to a minimum. This is much more efficient than retransmission of the total message if only one or two data errors have occurred.

Forward Error-Correcting Codes

For transmission efficiency, error correction at the receiver without retransmission of erroneous data is naturally preferred, and a number of methods of accomplishing this are available. Codes which permit correction of errors by the receive station without retransmission are called *forward error-correcting codes*. The basic requirement of such codes is that sufficient redundancy be included in the transmitted data for error correction to be properly accomplished by the receiver without further input from the transmitter.

A	B	C	D	E	F	G	H	I	J	K	L	M	N	O	P	Q	R	S	T	U	V	W	X	Y	Z
1	2	3	4	5	6	7	8	9	10	11	12	13	14	15	16	17	18	19	20	21	22	23	24	25	26

a.

A	D	D
K	I	D
B	E	G

b.

1	4	4	9 (I)
11	9	4	24 (X)
2	5	7	14 (N)
14	18	15	47 (−26 = 11 = K)
(N)	(R)	(O)	

A D D K I D B E G N R O I X N K = DATA STRING TRANSMITTED

Check letters

A D D K I N B E G N R O I X N K = DATA STRING RECEIVED

c.

A	D	D	I
K	I	N	X
B	E	G	N
N	R	O	K

d.

1	4	4	9
11	9	14	24
2	5	7	14
14	18	15	11 (+26 = 47)

Row total incorrect

Incorrect letter

Column total incorrect

FIGURE 14-31 Three-level matrix sum forward error correcting code. (*a*) Message in triplets; (*b*) triplets as numbers with check sums; (*c*) received data with error; (*d*) error check and correction.

One forward error-correcting code is the *matrix sum,* shown in Fig. 14-31, which illustrates the use of a three-level matrix sum system. Note that the sum of the rows is equal to the sum of the columns; this is important for the encoding scheme's ability to find and correct errors. The transmitted message consists of the information bits plus the letters representing the sum of each column and row and the total. When received, the matrix is reconstructed and the sums are checked to determine whether they agree with the original sums. If they agree, error-free transmission is assumed, but if they disagree, errors must be present. The value of using this method is that it makes it possible for the receiver not only to determine which sums are incorrect but also to correct the erroneous values. In Fig. 14-31*a*, note that the row and column discrepancies identify the matrix cell that is incorrect. By replacing the incorrect number with the value which agrees with the check sums, the message can be restored to the correct form. Such error correction requires intervention by a computer or by a smart terminal of some kind. The transmission efficiency also suffers when this kind of code is used; however, if retransmission is used instead, the redundancy it requires can easily offset the inefficiency of the matrix sum code. Forward error correction is particularly well suited to applications which place a high value on the timeliness of data reception.

A three-level matrix sum code will provide for approximately 90 percent error-correction confidence. Larger matrices will increase this confidence level significantly, and it may be shown that a nine-level matrix will provide a 99.9 percent confidence level. The larger matrix has the additional benefit of increasing the ratio of information bits to error check bits. The result of this is increased transmission efficiency, 81 percent for the nine-level matrix versus 56 percent for the three-level matrix.

An interesting error-detecting code is the *hamming code,* named for R. W. Hamming, an error-correction pioneer. This code adds several parity-check bits to a data word. Consider the data word 1101. The hamming code adds three parity bits to the data bits as shown below:

P_1	P_2	1	P_3	1	0	1	
1	2	3	4	5	6	7	Bit Location

The first parity bit, P_1, provides even parity from a check of bit locations 3, 5, and 7, which are 1, 1, and 1, respectively. P_1 will therefore be 1 to achieve even parity. P_2 checks locations 3, 6, and 7 and is therefore a 0 in this case. Finally, P_3 checks locations 5, 6, and 7 and is a 0 here. The resulting 7-bit word is:

1	0	1	0	1	0	1
P	P	D	P	D	D	D

If the data word is altered during transmission, so that location five changes from a 1 to a 0, the parity will no longer be correct. The hamming encoding permits evaluation of the parity bits to determine where errors occur. This is accomplished by assigning a 1 to any parity bit which is incorrect and a 0 to one which is correct. If the three parity bits are all correct, 0 0 0 results and no errors can be assumed. In the case of the above described error, the code has the form:

1 0 1 0 0 0 1

P_1 (which checks location 3, 5, and 7) should now be a 1 and is therefore incorrect. It will be given a 1. P_2 checks 3, 6, and 7 and is therefore still correct. It receives a value of 0. P_3 checks 5, 6, and 7 and should be a 1, but it is wrong here, and so it receives a value of 1. The three values result in the binary word 1 0 1, which has a decimal value of 5. This means that the location containing the error is five, and the receiver has been able to pinpoint the error without retransmission of data.

The hamming code is therefore capable of locating a single error, but it fails if multiple errors occur in the one data block.

Codes such as the *hagelbarger* and *bose-chaudhuri* are capable of detecting and correcting multiple errors, by increasing the number of parity bits to accomplish their error correction. In the case of the hagelbarger code, one parity bit is sent after each data bit. This represents 100 percent redundancy. It may be shown that the code can correct up to six consecutive errors, but error bursts must be separated by large blocks of correct data bits. The bose-chaudhuri code can be implemented in several forms with different ratios of parity bits to data bits. The code was first implemented with 10 parity bits per 21 data bits. Redundancy again approaches 100 percent.

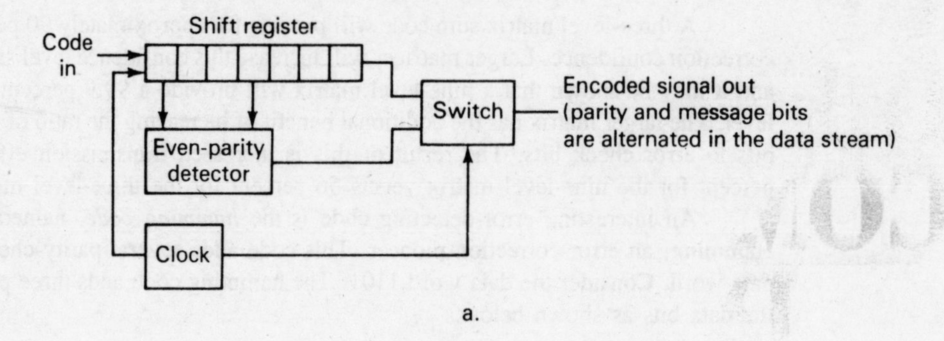

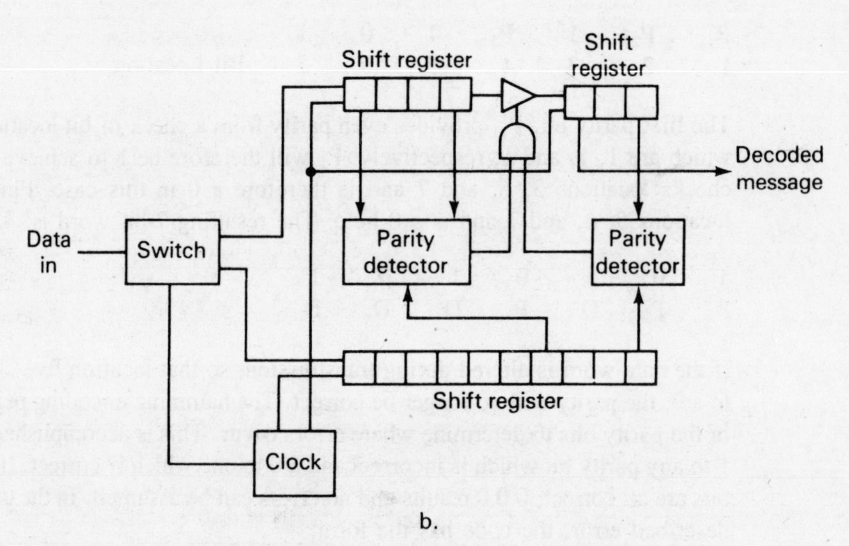

FIGURE 14-32 Hagelbarger code. (*a*) Encoder; (*b*) decoder.

Figure 14-32 illustrates the use of shift registers and logic devices to implement the hagelbarger code. The increased complexity and decreased transmission efficiency are offset by improved immunity to transmission errors for data requiring high degrees of accuracy. More information on these complex codes may be found in [3] and [4].

14-3 DATA SETS AND INTERCONNECTION REQUIREMENTS

Data sets or *modems* are used to interface digital source and sink equipment to interconnecting circuits. The modem at the transmitting station changes the digital output from a computer or business machine to a form which can be easily sent via a communication circuit, while the receiving modem reverses the process. Modems differ in rate of data transmission, modulation methods and bandwidth, and standards have been developed to provide compatibility between various manufacturers' equipment and systems.

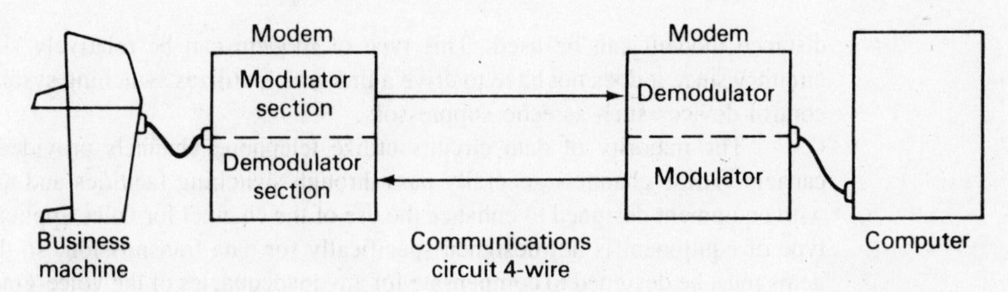

FIGURE 14-33 Communications circuit using modems.

14-3.1 Modem Classification

The name *modem* is a contraction of the terms *MO*dulator and *DEM*odulator; as the name implies, both functions are included in a modem. When used in the transmitting mode, the modem accepts digital data and converts it to analog signals for use in modulating a carrier signal. At the receive end of the system, the carrier is demodulated to recover the data.

Modems are placed at both ends of the communications circuit, as shown in Fig. 14-33.

Modes of modem operation Modems are described in several ways, one distinction between modems being the mode of operation. A data set which provides transmission in only one direction is referred to as operating in the *simplex* mode. This type of data set uses only one transmission channel, so that no signaling is available in the direction from the receiver to the transmitter. This is an economical method of data transfer, but it is very limited in its application; it clearly does not accommodate error correction and requests for retransmission.

Some modems provide for data transfer in both directions, but the data flow takes turns, with flow in one direction at one time and in the opposite direction at a second time. This type of modem operation is referred to as *half-duplex;* it requires only one transmission channel, but the channel must be bidirectional. Some economies result from half-duplex operation, but speed of transmission is reduced because of the necessity of sharing the same circuit and waiting while the transmission circuit components accomplish turnaround.

Full-duplex operation permits transmission in both directions at the same time. Two circuits[2] are required, one for each direction of transmission. Modems are placed at each end of the circuits to provide modulation and demodulation.

Modem interconnection Modems differ according to the method of interfacing with the communications circuits. If the circuit is a short and dedicated line, a limited

[2]That is, two 2-wire circuits or one 4-wire circuit.

distance modem can be used. This type of modem can be relatively simple in its circuitry since it does not have to drive a line which utilizes switching systems and line control devices such as echo suppressors.

The majority of data circuits utilize telephone channels provided by public carriers. These channels generally pass through switching facilities and are provided with equipment designed to enhance the use of the channel for voice applications. This type of equipment is not designed specifically for data transmission, so that the modems must be designed to compensate for any inadequacies of the voice-grade channel. Two broad types of modems are available for this type of service, the hard-wired modem and the acoustically coupled data set.

A hard-wired modem connects directly to the communication circuit in a semipermanent way. Such modems may be self-contained devices which connect to terminals and business machines, or they may be incorporated in the business machine. Connected to the communications circuit at all times, the hard-wired units can be *polled* (automatically contacted by the computer) and interrogated at any time. If associated with proper business machines and computers, these modems can send and receive data without human intervention. The one limitation of the hard-wired modem is that it precludes mobility since, being hard-wired, the equipment must remain connected to the circuit terminals.

The acoustically coupled modem solves the mobility problem. A standard telephone handset can be placed in the foam cups of an acoustic coupler, and the transmitter and receiver sounds will be conveyed to and from the telephone channel by transmit and receive elements of the acoustic coupler. The modem components of the acoustic coupler form an interface with the business machine. Using this device, a person is able to interconnect with any computer system which has dial-up interconnect capability. Acoustic couplers are often built into briefcase-sized units which include a typewriterlike terminal and a printer, providing the ability to access and manipulate data from any telephone. The portability and ease of connection afforded by the acoustic coupler are obtained at the expense of other capabilities. Since standard telephone circuits are typically used, speed of transmission is limited. The ability to have the system "on line" continuously is obviously not possible.

Modem data transmission speed Modems are generally classified according to the important characteristic of transmission speed as follows:

MODEM CLASSIFICATION	DATA RATE HANDLED (BPS)
Low-speed	Up to 600
Medium-speed	600 to 2400
High-speed	2400 to about 10,800

All of the above modems can operate within a single 300- to 3400-Hz (4-kHz) telephone channel. As speed increases beyond approximately 19,000 bps, a wideband modem is needed, as is a wideband channel. Wideband circuits are available, generally in multiples of 4-kHz circuits, but the cost is significantly greater than for voice-grade circuits.

TABLE 14-1 Modem Specifications

MODEM TYPE	DATA TRANSFER RATE	MODULATION TYPE
103A	300 bps	FSK*
113A	300 bps	FSK
202C	1200/1800 bps	FSK
202D	1800 bps	FSK
202E	1200/1800 bps	FSK
203A/B/C	3600/7200 bps	VSB†
208A/B	4800 bps	8-phase PSK‡
209A	9600 bps	QAM§
301B	40.8 kbps	PSK
303B	19.2 kbps	VSB
303C	50.0 kbps	VSB
303D	230.4 kbps	VSB

*FSK = frequency-shift keying.
†VSB = vestigial sideband.
‡PSK = phase-shift keying.
§QAM = quadrature amplitude modulation.
Note: This is not an exhaustive list.

Modem modulation methods Modems utilize various types of modulation methods, the most common being frequency-shift keying (FSK), which shifts a carrier frequency to indicate a mark or a space. Encoded data can be transferred through communication systems designed for voice transmission because the frequency shifting is limited to the 4-kHz bandwidth of the voice-grade channel. The FSK signal is also analog in nature, enhancing its compatibility with communications circuits.

Other types of modulation schemes are used, such as phase-shift-keying (PSK), four-phase PSK and eight-phase PSK, quadrature AM (QAM) and vestigial sideband AM. Table 14-1 lists some of the various types of modems in use in the United States, according to their Bell System designations, showing data transfer rates and modulation methods.

14-3.2 Modem Interfacing

RS-232 interface In the United States, a standard interconnection between business machine and modem is supplied by the RS-232 interface. The RS-232 interface has been defined by the Electronic Industries Association (EIA) to ensure compatibility between data sets and terminal equipment. The interface uses a 25-pin Cannon or Cinch plug, where each of the 25 pins has been given a specific function by EIA, as shown in Table 14-2. The United States military data communications system uses a similar interface designated as MIL-188C, and an international interface similar to the RS-232 is also available.

The RS-232 interface specifications limit the interconnecting cable to a length of 50 ft (15 m) or, if this length is exceeded, the load capacitance at the interface point

TABLE 14-2 RS-232 Pin Assignment

PIN	ASSIGNMENT	EIA DESIGNATIONS
01	Frame ground	AA
02	Transmitted data	BA
03	Received data	BB
04	Request to send (RTS)	CA
05	Clear to send (CTS)	CB
06	Data set ready (DSR)	CC
07	Signal ground	AB
08	Received line signal detector (LSD)	CF
09	Test	
10	Test	
11	Not assigned	
12	Secondary LSD	SCF
13	Secondary CTS	SCB
14	Secondary transmitted data	SCA
15	Transmitter signal element timing (modem to terminal)	DB
16	Secondary received data	SBB
17	Receiver signal element timing	DD
18	Not assigned	
19	Secondary RTS	SCA
20	Data terminal ready	CD
21	Signal quality detector	CG
22	Ring indicator (R)	CE
23	Data signal rate selector	CA/CI
24	Transmit signal element timing (terminal to modem)	DA
25	Not assigned	

must not be greater than 2500 pF. This limitation insures that signals will operate at appropriate standards of quality.

The interface also specifies the voltage levels with which data and control signals are exchanged between data sets and business machines. Each pin in the 25-pin connector will carry either a binary 0 or a 1 to indicate activation or deactivation of control functions or data values. A binary 1 is used for marking and signifies OFF, while the 0 is used for spacing and signifies ON.

The RS-232 interface can accommodate several different types of data circuit operation, using different combinations of circuit lines. For example, point-to-point dedicated system will require a minimum number of control lines in the interface, while for circuits which operate in a half-duplex mode, line-turnaround must be provided since the same pair of wires is used for both send and receive. Control circuits which will accomplish these functions are included in the RS-232 interface. In the case of another type of operation, systems which involve several remote terminals connected to a data circuit follow particular sequences of operation. The terminal wishing to send data will signal with a request-to-send (circuit CA, designated in Table 14-2, will change state), and the data set responds to the request-to-send by conducting procedures which will inform the receive station modem of the request-to-send and will conduct such tests and system set-up sequences as may be required. When the start-up procedures are completed, the receive modem will send a clear-to-send to the transmit modem, whereupon the transmit modem will cause circuit BA to change states, and transmission of data will begin. Data will be sent as alternating binary states of circuit BA, and thus data will be transferred in a serial mode. At the receive station, circuit BB will reflect the binary status of the data and will be interpreted by the business machine for processing.

Other interfaces Several new interface standards have also been developed. Listed as RS-422, RS-423 and RS-449, these interfaces expand the flexibility of the RS-232. Two connectors replace the 25-pin connector of RS-232 with a 37-pin connector providing all interchange circuits except secondary channel circuits, which are provided by a separate 9-pin connector. The new standards extend the 15-m (50-ft) range of RS-232 to 60 m (200 ft). The maximum signaling rate increases under the new standards from the 20,000 bps of RS-232 to 2.048 Mbps. Ten additional exchange circuits not included under RS-232C are provided in RS-449, while three circuits provided by RS-232 have been deleted. Balanced and unbalanced circuits have been provided by the new standards, and integrated circuit technology has been considered in the definition of the electrical characteristics of the interface. The new standards have been devised to facilitate interconnection with RS-232 equipment with minimal modification.

14-3.3 Interconnection of Data Circuits to Telephone Loops

In the United States, a recent FCC ruling, in part 68 of the Rules and Regulations, permits for the first time non-telephone company interconnection to telephone company circuits. This ruling has placed the responsibility for much of the necessary interconnection circuitry on the manufacturer of data equipment, which must be registered with the FCC. Three types of customer equipment have been identified by the new rules: the *permissive data set*, the *fixed-loss loop data set*, and the *programmed data set*. Each of these data sets interfaces with telephone company supplied jacks, whose type is determined by the type of data set to be connected.

The permissive data set The permissive data set provides a maximum output level of −9 dBm, while the guideline is that the circuit signal level must not exceed −12 dBm. Since the standard line loss of a business loop is 3 dB, the permissive data set can be used with any of three jacks supplied by U.S. telephone companies, including the standard voice jack, RJ11C, which includes no provision for signal attenuation.

The fixed-loss-loop data set The fixed-loss-loop data set can have a maximum of −4 dBm signal level. This type of data set requires connection to a universal jack, RJ41S, which includes an adjustable resistive pad to limit output to the required −12 dBm as measured at the time of installation. Measurement of signal level will include loop losses.

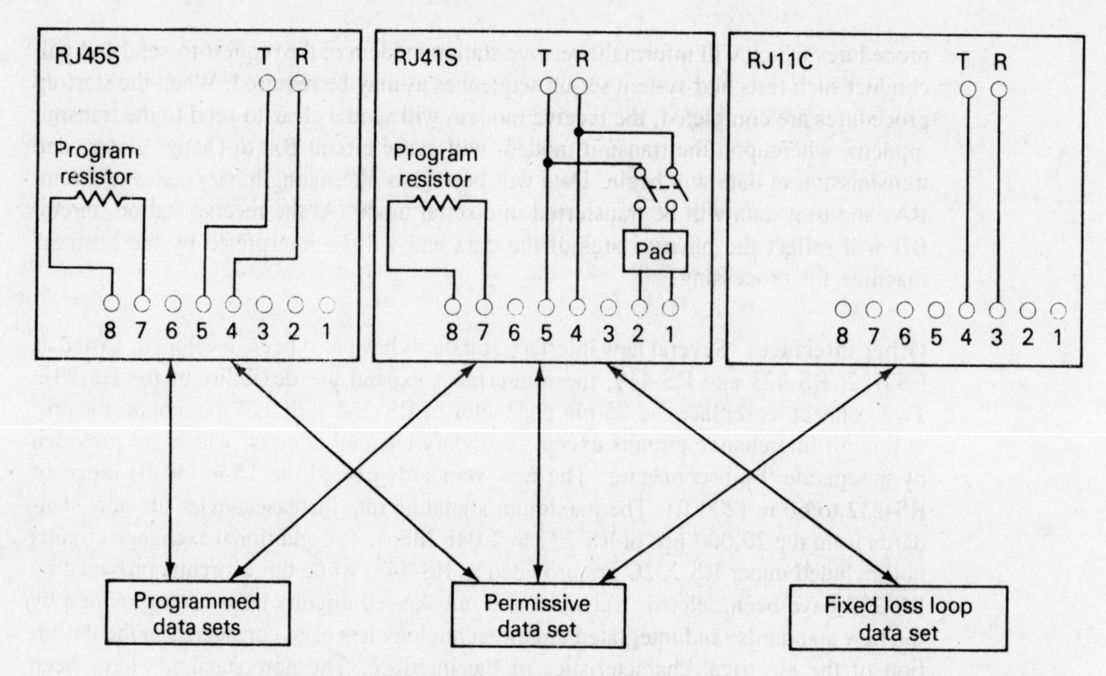

FIGURE 14-34 Standard U.S. Telephone Company jacks showing data set compatibility.

The programmed data set The third type of data set, the programmed data set, can use either the universal jack or the programmed jack, RJ45S. The telephone company installs a resistor in the jack at the time of installation which is used by the programmed data set to determine its signal output level. The value of the programming resistor is selected on the basis of measurements of loop loss made when the data set is installed.

A nonregistered data set can be connected to a telephone circuit in the United States, but it must employ a registered protective device to interface with one of the standard jacks described above (see Fig. 14-34).

14-4 NETWORK AND CONTROL CONSIDERATIONS

Connecting the vast numbers of data facilities which are in existence today requires careful design and organization of transmission networks. Systems now involve many users and remote facilities; large networks interconnect several large computers with networking and essential requirement. The technologies to accomplish these new modes of interconnection have been developed and refined to satisfy the ever increasing demands of a data-hungry society.

14-4.1 Network Organization

As data systems have increased in number and complexity, it has become increasingly important to provide for their proper and orderly interconnection. Small, simple systems could dedicate individual lines for each piece of equipment which was connected in the system. For intraplant connections, this was a practical method; the lines were short and could be installed by the data system user. Leasing was not involved and installation costs were relatively low.

Dedicated lines for each user become less feasible for out-of-plant operations. Such systems normally lease capacity in existing transmission facilities of telephone carriers. Using many full-time dedicated lines for extended periods would result in unacceptable costs, since few remote locations require full-time interconnection with other sites. More typically, connections between sites are established for short periods to obtain and convey data, while the rest of the time is spent interpreting, updating or otherwise processing the data locally. Modern data systems depend on network techniques of interconnection to reduce the expense of data transfer.

The efficiency of networking for data users who do not require full-time interconnection can be illustrated by a simple example. A system consisting of eight data user sites which require interconnection at various times would, as shown in Fig. 14-35, require 28 dedicated lines to connect each user site with every other one. This may also be calculated from a simple formula. Noting that the first user must be connected to seven others, the second one to six (he or she already has a connection to the first one), the third one to five, and so on, we deduce that:

$$N = \sum_{1}^{U-1} A \qquad \text{(14-4)}$$

where N = number of lines
U = number of sites

Here, $U = 8$, so that:

$$N = 7 + 6 + 5 + 4 + 3 + 2 + 1 = 28$$

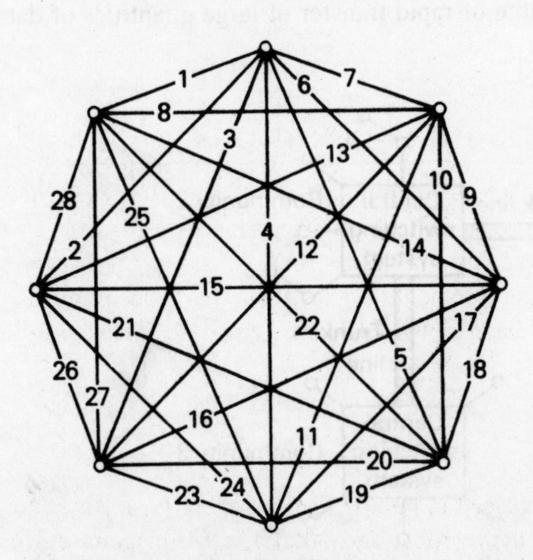

FIGURE 14-35 Interconnection of an eight-user dedicated line network.

It may also be shown that:

$$\sum_{1}^{U-1} A = (U^2 - U)/2$$

Checking, we get:

$$N = (8^2 - 8)/2 = (64 - 8)/2 = 28$$

Centralized switching A better way to provide the required interconnections is to use a central switching system, which will have one line connected to each remote site. Interconnections will be made between remote sites by the central system on a demand basis. If each remote site can handle only one interconnection at a time, this system will provide the same capabilities as the previous system but will require only eight dedicated lines.

Data systems which depend on central switching facilities are referred to as *centralized* networks. Telephone networks in small towns are typically centralized networks. Each customer has a line to the central office, where automated switching equipment interconnects one user with another as required. In turn, central offices are interconnected by means of trunk lines, and in this fashion each centralized network now becomes part of a larger network which can make interconnection between individual users from different centralized networks. Figure 14-36 shows this type of network, further discussed in Chap. 15.

Since the central switch of each centralized network distributes the data between that network and other networks, this type of system is called a *distributed* network. For computer systems, the centralized facilities may consist of large computers which interconnect to permit users access to any of the computers. This type of arrangement can greatly improve the efficiency of the computers by making a computer which is underused by its local subscribers available to subscribers from computer centers which are in heavy demand at that time. The routes which interconnect the centers are normally capable of rapid transfer of large quantities of data, whereas the

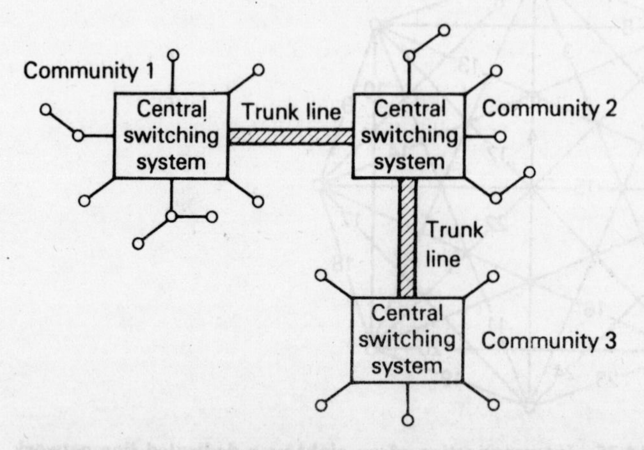

FIGURE 14-36 Telephone network using centralized switching.

lines from users to the central offices do not need to convey these large amounts of data and can therefore be less expensive lines. Data flow within networks is carefully controlled through system protocols to ensure maximum efficiency and minimum interference between users. Network switching systems, line types and network protocols are important considerations for data transmission.

14-4.2 Switching Systems

If only two sites are to be connected, switching is not required. The two facilities are interconnected on a point-to-point basis, as shown in Fig. 14-34. However, switching is likely to be required where three or more sites need to be interconnected. The various types of systems described earlier can all be used for data transmission over networks.

Circuit types A single pair of wires (two-wire circuit) can be used for a unidirectional transfer of data in the simplex mode. In a half-duplex mode, for data to pass in both directions on a two-wire circuit, it is necessary for the two sites to take turns in transmitting over the circuit. A full-duplex system will use a four-wire circuit with one pair of wires for each direction of transmission. The best type of system for a particular application will depend on the nature of the data requirements and the operation of the equipment.

Network interconnection In addition to the type of circuit, the type of connection must be chosen. If the site has continuous or very frequent interconnection requirements, a dedicated line is appropriate. Many users find that their usage is not continuous, and they are able to realize significant savings by using a switched or dial-up system, be it in the public switched network or a private network. This method of operation can be very economical and efficient for users who need access to the computer or other data sites on an infrequent basis or from changing locations.

An extension of the point-to-point system is the *polled multipoint system,* which interconnects a common source such as a computer with a remote location having several users. A simple polled system (Fig. 14-37) is seen to be similar to the two-station system, except that each of the several users now connects to the common circuit through a modem. The computer checks (polls) each user in turn to determine whether one of them is requesting interconnection. When the request is received by the

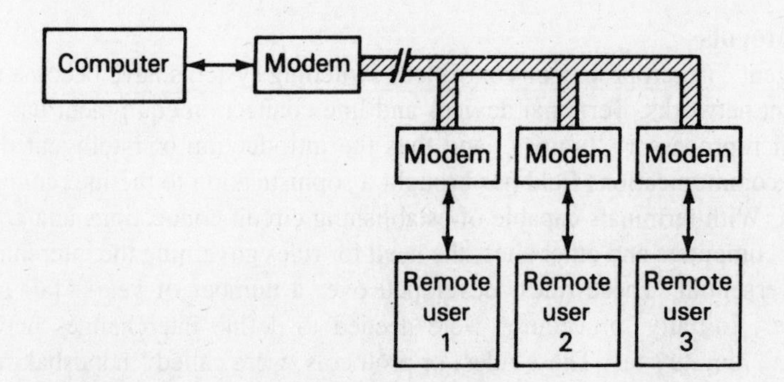

FIGURE 14-37 Polled multipoint communication network.

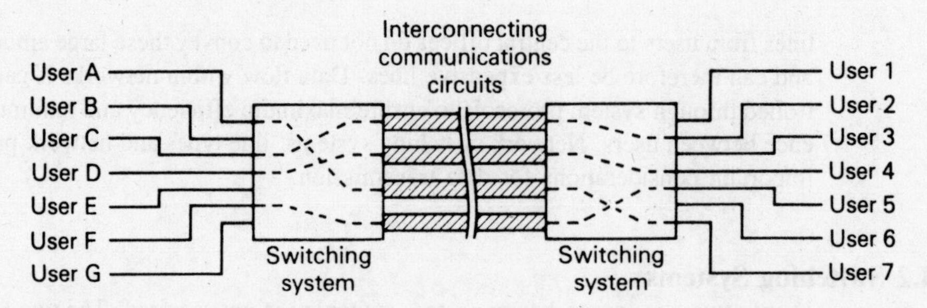

FIGURE 14-38 Network switching showing User A switched through circuit 3 to User 7.

computer, the requesting user's modem is given control of the circuit and data is transferred. If the source has data to transfer to one of the users, it seizes control of the circuit and sends the appropriate command to interconnect the desired user to the line. The polling process and the transfer of data follow specific procedures called *protocols*, which are discussed in Sec. 14-4.3.

Networks can be used to interconnect a large number of users through only a few transmission circuits. While only a few users on either end of the network are connected at any one time, the switching capability contributes to significant economies. A typical switching system of this type is shown in Fig. 14-38.

Modern switching systems benefit from microprocessor technology and can be termed "smart switchers." In large networks, multiple-trunk interconnections and interuser circuits are available. The switching system not only interconnects users but also determines the best and quickest routing to be used for connecting one particular user to a second one, in a manner identical to that described for telephone exchanges in Chap. 15. Indeed, most data interchange utilizes public telephone networks, the switching being accomplished by telephone switching equipment, described in Chap. 15.

It will be seen that processor-controlled switches are beginning to predominate in advanced countries, to the great benefit of switched data systems. Being microprocessor-based and therefore "smart," the electronic switching system can provide services such as redial if busy, automatic dial forward, conferencing, and "camp-on" if busy.

14-4.3 Network Protocols

"Intelligent" (microprocessor-controlled) switching systems have become the hubs of intelligent networks. Terminal devices and line connection equipment have also been given microprocessor "brains," and thus the introduction of intelligent devices into the data communications field has brought a sophistication to the interconnection possibilities. With terminals capable of establishing circuit connections and communicating with computers and other sites, the need for rules governing the interchange of data became essential. These rules, developed over a number of years, fall into several categories. Initially, procedures were needed to define interchanges between computers and remote sites. These rules, or protocols, were called "handshaking." As the systems grew, procedures became necessary to determine standard methods of communicating within data channels, and so protocols for integration of control signals with data in standard formats and sequences were developed. Also, the expansion of network complexity permitted numerous stations access to transmission circuits. To prevent interference between users, protocols were devised which established communications priorities and control sequences to be used to initiate and terminate switched interconnections.

Protocol phases Data communications protocols typically have three phases: establishment, message transfer and termination. The contents of these phases differ for different system arrangements and equipment types. In point-to-point systems which involve a master station and one or more slave stations, the flow of data is determined by the master station. The master station has direct control of each slave station; it establishes the connection, controls the transfer of data and terminates the connection.

Polling protocols Systems which interconnect several stations on a shared basis can use either *polling protocol* or *contention protocol*. In polling systems, one station is designated the master station, and queries, or polls, the other stations to determine which interconnections are to be established. This type of polling is referred to as *roll calling*. The master station remains at the center of the system. It polls each remote station in turn, retains control of the circuit and directs the other stations to send or receive data as required.

Contention protocols Contention systems do not designate a master station; instead, the interconnected stations contend for the role of master station. Whichever station seizes control of the communication channel first directs the flow of data until it terminates the communication. The channel will remain vacant until the next station with data to transmit seizes the line and establishes communication. The protocol must provide for instances of simultaneous line seizure attempts by several stations as well as establishing priority schemes among the users.

Switched or dial-up systems must have protocols which direct the establishment of communication via dial-in requests. These systems are very popular and often involve the use of automatic circuits at both send and receive stations to effect the dial-up interconnection. This requires that protocols be standardized so that equipment from different systems can communicate without intervention.

Some networks interconnect the stations in the form of a loop, with each station connecting to the next station. Data to be transferred to a station around the loop must pass through each intermediate station. The loop arrangement has the benefit of reducing the number and length of data circuits required as compared to a central master station network. Protocols for the loop system must provide for data direction and system control. Polling can be used in loop systems. When used, it is referred to as forward polling, in that each station polls the next station in line.

Character insertion It was indicated earlier that protocols must provide for integration of control characters within the data stream. Control characters are indicated by specific bit patterns, but it is possible that these patterns could accidentally occur in the data stream at places where control characters are not intended. This is particularly true when the data represents digitization of an analog function or some similar situation in which the data is not alphanumeric in nature. To prevent this problem, a data transmission protocol called *character insertion* (also referred to as *character stuffing*) is sometimes used. Under this protocol, the transmitting equipment checks the data stream as it is transmitting, to determine whether character patterns identical to control characters exist in the data. If these patterns are encountered, the control character pattern is inserted into the data stream after the data pattern. The result is to have the control character pattern occur twice. At the receive site, the data is evaluated two characters at a time. If a control character is detected, the receiver checks the following character to see whether it duplicates the control character. If it does, the control character pattern is recognized as false, and the second character is removed from the data stream. If the pattern occurs only once, it is a valid control character, and the appropriate action is taken. This method of control character recognition is called *transparency*.

14-5 SUMMARY

Digital electronics and the binary number system opened a wide door to the field of data communications. This chapter has explored the development of data communications. The problems of data transmission through various media have been considered, and the equipment and encoding schemes which represent the solutions to those problems have been discussed.

Data communications emerged only recently, but the field has expanded rapidly and now presents a large body of specialized techniques and technology. The treatment of the subject in this chapter must necessarily be general and introductory. In-depth treatment of the field is provided by several of the references. Davenport [4] presents a general but comprehensive look at data communications. Techo [2], Doll [3] and Lacky [5] discuss the field from different viewpoints, and all are instructive.

PROBLEMS

For self-testing questions on this chapter, see p. 706.

14-1. Convert the decimal number 146 to its binary equivalent.
[Ans. 10010010]

14-2. Convert the decimal number 146 to its binary-coded decimal equivalent.
[Ans. 0000 0001 0000 0100 0000 0110]

14-3. Convert the binary number 01100010 to its decimal equivalent.
[Ans. 98]

14-4. Calculate the signal-to-noise ratio in dB which would be required for an ideal channel with a bandwidth of 4000 Hz.
[Ans. 4.8 dB]

QUESTIONS

14-1. Explain the difference between analog and digital signals.
14-2. Describe the binary number system and compare it to the decimal system.
14-3. What are the four rules of binary addition?

14-4. Define the *2s complement* and demonstrate how it would be derived for a binary number. Use the 2s complement to effect subtraction of one binary number from another. Check the result by using standard binary subtraction.
14-5. Draw a schematic of a transistor switch and explain the circuit operation which causes it to function as an inverter.
14-6. Draw the symbols of five different digital logic gates and show their truth tables.
14-7. What is the function of a latch?
14-8. What is the function of a shift register?
14-9. Draw the functional diagram of a simple microprocessor system and describe the purposes of the various circuit components.
14-10. When a system is referred to as being "bus-oriented," what does it mean?
14-11. Describe the evolution of the computer and indicate what advances served as the important milestones in this development.
14-12. What events served to spur the advancement of the data communications field?
14-13. Explain baud rate and describe how it may differ from information bit rate.
14-14. What is multilevel encoding, and what are its benefits and limitations?
14-15. What is the Nyquist rate?
14-16. What aspect of the transmission channel is defined by the Shannon-Hartley law?
14-17. How does noise affect channel capacity?
14-18. Describe crosstalk and give some possibilities for reducing its effect.
14-19. Explain how an echo suppressor may interfere with data transmission. What steps are normally taken to prevent this interference?
14-20. What is phase-delay distortion and how does it affect data transmission?
14-21. Describe how equalization can improve the ability of a transmission channel to carry data.
14-22. Describe four different codes used for data transmission and discuss their strengths and weaknesses.
14-23. Describe three kinds of error-detection codes and explain how they detect data errors.
14-24. What penalty is paid when an error-detection code is used? How may circuit efficiency be defined? What is the efficiency of a completely nonredundant code?
14-25. Explain parity and discuss its use for data transmission systems.
14-26. What is a forward error-correcting code? How do such codes function?
14-27. What is a data set? Where is it used in a data transmission system?
14-28. Discuss the differences between various modems, and explain the significance of the differences.
14-29. Describe the RS-232 interface and explain its value for data transmission.
14-30. Discuss the interconnection requirements for data sets when they are connected to telephone company circuits.
14-31. Discuss several methods of network organization, giving benefits and weaknesses of each.
14-32. What is the purpose of a switching system? What impact do switching systems have on data transmission?
14-33. What are network protocols? Why are they necessary?

REFERENCES

1. Shannon, C. E., and W. Weaver, *The Mathematical Theory of Communication*, University of Illinois Press, Urbana, Ill., 1949.

2. Techo, Robert, *Data Communications: An Introduction to Concepts and Design*, Plenum Press, New York, 1980.

3. Doll, D. R., *Data Communications: Facilities, Networks, and Systems Design*, John Wiley & Sons, New York, 1978.

4. Davenport, W. P., *Modern Data Communications: Concepts, Language, and Media*, Hayden Book Company, Inc., New York, 1971.

5. Lacky, R. W., Salz, J., and Weldon, E. J., Jr., *Principles of Data Communications*, McGraw-Hill Book Company, New York, 1968.

6. Vazsonyi, Andrew, *Introduction to Electronic Data Processing*, Richard D. Irwin, Inc., Homewood, Ill., 1973.

15 BROADBAND COMMUNICATIONS SYSTEMS

In our world of direct intercontinental telephone subscriber dialing and instant worldwide telecasts, it is perhaps hard to realize how recent broadband long-distance communications are. True, some form of transoceanic communication has been going on for quite a long time, ever since the first transatlantic telegraph cable in the 1850s. Again, the next milestone was 1901, this being Marconi's first transatlantic radio transmission. However, the bandwidths of these early systems were very low, and hence the rate of transmission of information was painfully slow.

The first real development in broadband communications came in 1915, when vacuum-tube repeaters were first used, together with carrier telephony, to provide a coast-to-coast telephone service in the United States, featuring a few channels. By 1941, a coaxial cable system with 480 channels was in operation over a distance of 320 km from Minneapolis to Stevens Point, Wisconsin.

Transcontinental communications became broadband and "took off" in 1956, the year in which the *TAT-1* cable was laid from Scotland to Newfoundland. This was really two cables, one for each direction of transmission, and had a capacity for 48 simultaneous telephone conversations. By 1984, there were nine major transatlantic cables, with the two biggest each having a capacity of 4000 two-way circuits.

Communications satellites came next on the scene but have taken giant strides and currently provide a large proportion of international circuits, as well as being the only means of transmitting intercontinental television. The first transatlantic transmission involved the *Telstar* satellite, in 1962. This satellite was placed in an elliptical orbit, designed to bring it down relatively low (950 km at its lowest) over the Atlantic. It lasted for 6 months and during that time was used for communications between the United States and Great Britain, France and Italy.

The first geostationary satellite was *Early Bird,* launched in 1965, again over the Atlantic. It had a capacity of 66 telephone channels and one television bearer. It was subsequently replaced by *INTELSAT*[1] *II* and *INTELSAT III*, and

[1] International Telecommunications Satellite Consortium.

expanded to cover the three oceans. Currently *INTELSAT V-A* satellites are in service, with capacities of up to 5000 telephone circuits (depending on the configuration) as well as several simultaneous TV transmissions. Meanwhile, short- and medium-haul broadband systems have become longer, more widespread, more reliable and much more capacious. For example, it will be seen in this chapter that systems with capacities in excess of 100,000 circuits are now in service.

Fiber optics are the most recent development for long-distance communications, and it is the current "growth industry" in the field. The topic was briefly introduced in Chap. 12 and will be elaborated here. Terrestrial fiber-optic cables first began to be used—in North America, Europe and Japan—in the late 1970s. Submarine cables pose an entirely different set of problems, and so the first fiber-optic ones will not go into service until late 1988, when the first transatlantic and Hawaii-California fiber-optic cables are scheduled to be laid.

Growth in trunk and international telephony has been no less spectacular. Indeed, a little reflection shows that all these high-capacity systems would not be in service unless they were needed! Signaling systems, too, have improved. At first, trunk calls were operator-connected, but, as volume grew, trunk telephone exchanges were provided and enabled subscribers to dial their own trunk calls. This, of course, increased the volume of trunk calls, because of increased convenience. Nowadays, trunk and international telephone and telex communications would grind to a halt if exchanges suddenly failed. As an illustration, it is worth pointing out that the volume of trunk telephone calls in the United States reached a milestone in the early 1960s. Indeed, the level was then such that, if the calls had to be connected manually, the number of operators required would have been in excess of the total population of the United States! The same ludicrous situation might soon have been reached with international communications, noting that international telephone calls grew at least 50-fold from 1960 to 1980, except that nowadays international subscriber dialing is in widespread use, and its use is continually expanding. It is worth pointing out that new trunk and international telephone and telex exchanges are computer-controlled, and most of them are digital.

This chapter deals with each of the systems whose historical introduction was given above. It begins with multiplexing, which is a technique of combining channels to ensure that a large number of them can be carried on the one bearer without interference. "Continental" (as opposed to intercontinental) broadband systems are then treated, followed by coaxial cables, fiber-optic cables, microwave links and troposcatter systems. The next major section covers submarine cable (both coaxial and fiber-optic) and satellite communications. Finally, long-distance telephony is treated briefly, in a section which introduces signaling systems, telephone exchanges and traffic engineering.

15-1 MULTIPLEXING

Multiplexing is the sending of a number of separate signals together, over the same cable or bearer, simultaneously and without interference. There are broadly two kinds. *Time-division multiplex,* or *TDM,* is a method of interleaving, in the time domain, pulses belonging to different transmissions. That is to say, use is made of the fact that pulses are generally narrow, and separation between successive pulses is rather wide (see also Chap. 13). That being the case, it is possible, provided the two ends of a link are synchronized, to use the wide spaces for pulses belonging to other transmissions.

On the other hand, *frequency-division multiplex,* or *FDM,* concerns itself with combining continuous (or *analog*) signals. It may be thought of as an outgrowth of independent-sideband transmission,[2] on a much-enlarged scale. As will be seen, 12 or 16 channels are combined into a group, 5 groups into a supergroup, and so on, using frequencies and arrangements that are standard on a worldwide scale. Each group, supergroup or larger aggregate is then sent as a whole unit on one microwave link, cable or other broadband system.

15-1.1 Frequency-Division Multiplex

It is often necessary to send a large number of independent telephone or telegraph channels from one point to another. Between any two major cities in advanced countries, there may be requirements for thousands or even tens of thousands of simultaneous telephone, telex and data transmissions. Clearly, it would be unthinkably expensive to devote a separate cable or radio link to each transmission, and thus some kind of combination of channels (without mutual interference) is indicated. This is done in FDM by taking a bandwidth adequate for the number of channels required and allocating each channel to a frequency ''slot'' adjacent to the previous channel. However, for reasons of flexibility, economy and simplicity, such frequency translations are not performed in one step. Instead, standardized groupings of channels are used, and several steps of frequency translation take place before all the channels have been placed in their locations in the frequency spectrum that is transmitted in a particular link.

Group formation The basic group is the smallest standard agglomeration of channels. It generally consists of 12 adjacent 4-kHz channels, occupying the frequency range from 60 to 108 kHz. A low-level pilot is transmitted at 104.08 kHz, for regulating and monitoring purposes. Narrower channels are used in many submarine cables, and so here a basic group consists of sixteen 3-kHz channels, occupying the same 48-kHz range as the 12-channel basic group. Figure 15-1 shows the channel arrangement for a basic group *B* in each case and also makes it apparent why the pilot in a 16-channel basic group cannot be at 104.08 kHz—84 kHz is used instead. Note that the basic group *A* occupies the frequency range of 12 to 60 kHz but is not normally used. See also [1].

[2]Treated in Sec. 4-4.3.

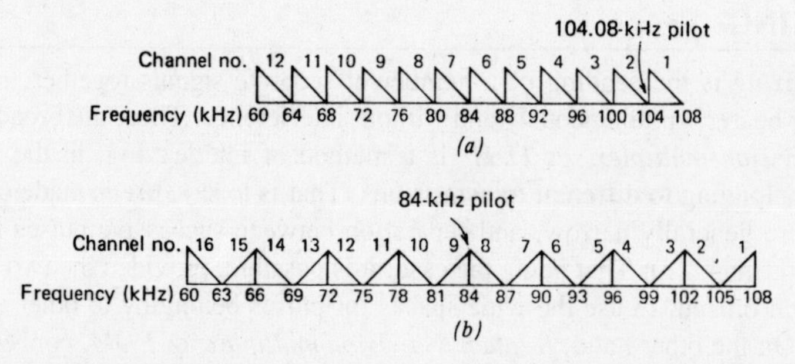

FIGURE 15-1 Channel arrangement in basic group B: (*a*) for 12-channel group; (*b*) for 16-channel group.

It is seen that all the channels in the basic 12-channel group *B* are *inverted* (and the group is therefore also said to be inverted). That is to say, the lowest frequencies in each channel are at the upper end of the allocated frequency ''slot'' for that channel. As shown in Fig. 15-1, the method of producing the basic group is a process of extension from single-sideband, suppressed carrier. It may thus be said that all 12 channels in the basic group are lower sidebands (see also Sec. 4-4.3). On the other hand, the basic 16-channel group is a mixture of inverted and *erect* channels. The reasons for such arrangements are partly practical and partly historical.

Figure 15-2 is a simplified block diagram of *channel translating equipment (CTE)* and shows how a basic group is assembled. It is seen that the process is a repetitive one of producing adjacent lower sidebands, with a frequency separation of 900 Hz between adjoining channels. It should be noted that Fig. 15-2 is a simplification, in that practical CTEs generally have four pregroup modulators, in which subgroups of three channels are produced and then combined into a group. A 16-channel group is produced in a similar fashion, in a 16-channel CTE.

Formation of higher-order groupings The next step up from a group is the basic supergroup, consisting of five groups, and occupying the frequency range of 312 to 552 kHz, i.e., a bandwidth of 240 kHz, as might be expected. Figure 15-3 shows the location of channels and groups in the basic supergroup. Note that the basic supergroup is erect and that, now that they have been translated higher up into the frequency spectrum, the groups are no longer called ''basic.'' The basic supergroup is formed in a *group translating equipment (GTE)*, in a process akin to group formation. The supergroup pilot is injected at 547.94 kHz.

Supergroups may be combined to form mastergroups, supermastergroups, and so on. Students are referred to [1] for details of groupings and the frequencies involved.

It will be noted that all the descriptions so far have been related to only one direction of transmission, at least by default. What happens in a practical system, of course, is that the supergroup, etc., assembly for the reverse direction of transmission is performed in precisely the same fashion. However, supergroups belonging to opposite directions of transmission are allocated differing frequencies in the spectrum,

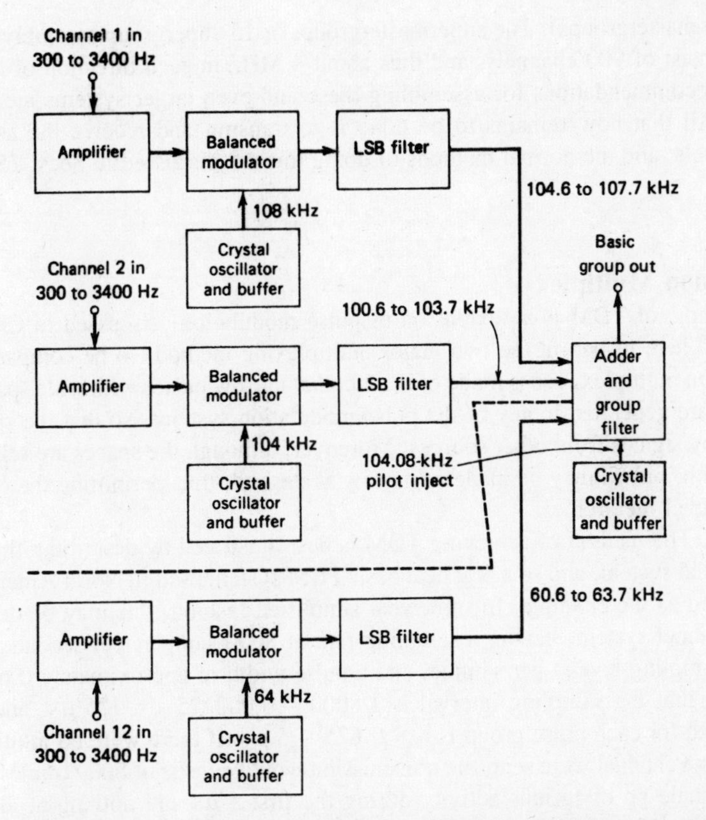

FIGURE 15-2 Channel translating equipment (CTE) showing the formation of a basic 12-channel group B.

different coaxial tubes, or different optic fiber pairs, so that no confusion or interference will take place. For example, in a system where only one supergroup is required, the supergroup in one direction is allocated the frequency range from 12 to 252 kHz, and the supergroup in the other direction occupies 312 to 552 kHz, the latter corresponding to the frequency range of the basic supergroup. The next assemblage up in the hierarchy is the mastergroup (five supergroups) and then the supermastergroup

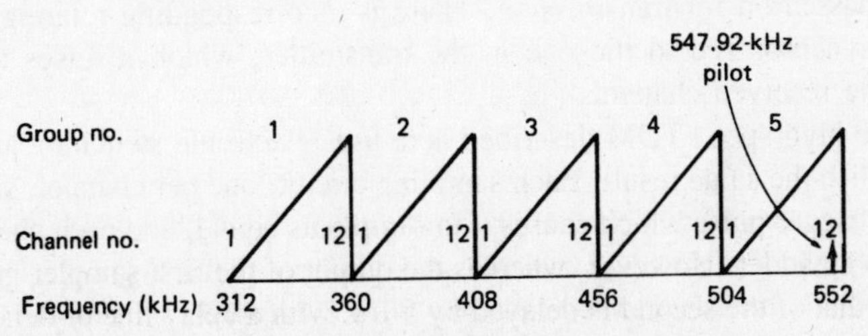

FIGURE 15-3 Group and channel arrangement in basic supergroup. (*Note:* The supergroup pilot lies between channels 11 and 12 in group 5.)

(three mastergroups). The supermastergroup, or 15-supergroup assembly, is thus seen to consist of 900 channels, and thus about 4 MHz in each direction of transmission. The recommendations for assembling these and even larger systems are contained in [1]. All that now remains to be done is to transmit and receive the assemblage of channels, and the normal methods of doing this are discussed in Secs. 15-2 and 15-3.

15-1.2 Time-Division Multiplex

The topic of TDM is an extension of pulse modulation, discussed in Chap. 13; it is treated here to permit the two major multiplexing methods to be compared. In time-division multiplex, use is made of the fact that narrow pulses with wide spaces between them are generated in any of the pulse modulation systems, so that the spaces can be used by signals from other sources. Moreover, although the spaces are relatively fixed in width, pulses may be made as narrow as desired, thus permitting the generation of high-level hierarchies.

The method of achieving TDM is best illustrated by describing the makeup of an actual system, and so a practical basic PCM system used in North America has been selected as the example. In somewhat simplified fashion, this may be described as a 24-channel system, having a sampling rate of 8000 samples per second, 8 bits (i.e., 256 sampling levels) per sample, and a pulse width of approximately 0.625 μs. This means that the sampling interval is $1/8000 = 0.000125$ s $= 125$ μs, and the period required for each pulse group is $8 \times 0.625 = 5$ μs. If there were no multiplexing and only one channel were sent, the transmission would consist of 8000 frames per second, each made up of furious activity during the first 5 μs and nothing at all during the remaining 120 μs. This would clearly be wasteful and would represent an unnecessarily complicated method of encoding a single channel, and so this system exploits the large spaces between the pulse groups. In fact, each 125-μs frame is used to provide 24 adjacent channel time slots, with the twenty-fifth slot assigned for synchronization. Each frame consists of 193 bits—24×8 for each channel, plus 1 for sync, and since there are 8000 frames per second, the bit rate is 1.544 Mbit/s.

Slow-speed TDM, as often used in radiotelemetry, is produced simply with rotating mechanical switches. A number of channels are fed simultaneously to the switch in the transmitter—one channel to each switch contact—while the output is taken from the moving rotor. This rotates slowly and remains in contact with each channel for a predetermined period, during which time the output of that channel is the only one passed on for transmission. There is a corresponding rotating switch in the receiver, synchronized to the one in the transmitter, which reverses the process to separate the received channels.

The high-speed TDM described here uses electronic switching and delay lines to accomplish the same result. Each sampling circuit, one per channel, simultaneously receives a trigger pulse which causes it to sample its signal, and each channel output is then fed to an adder. However, whereas the output of the first sampler goes straight to the adder, that of the second is delayed by 5 μs, with a delay line or delay circuit. The output of the third sampling circuit is similarly delayed but by 10 μs, and so on, until the twenty-fourth channel is delayed by 115 μs. In this way, each successive interval during the 125-μs frame is occupied by the transmission of a different channel, and the process is repeated 8000 times per second.

In the receiver, the output of the main detector is fed simultaneously to 24 AND gates. An *AND gate,* or coincidence circuit, is a simple device having one output and two or more input terminals, so arranged that an output is obtained only if all (in this case both) input signals are present. In this case each gate has two input terminals, and the second input to each gate is provided from a clock-synchronized gating generator, which is a monostable multivibrator providing rectangular pulses of 5 μs duration, 8000 times per second. Delay lines or circuits are used once again, with the gating pulse to the first gate not delayed at all, that to the second gate delayed by 5 μs and so forth. In this fashion each gate is open only during the appropriate time intervals, and the 24 channels are duly separated.

If transmission is by wire, the 1.544-Mbit/s pulse train is the signal sent, but if cable or radio communication is used, the pulse train either modulates the carrier or else is further multiplexed, with similar pulse trains, all combined together into a higher TDM hierarchical level.

The system thus far treated is described and defined in [2] and [3] in considerably more detail; the descriptions include an explanation of the pre-emphasis and de-emphasis used, which in this case happens to be μ-law. It should be mentioned that the 1.544-Mbit/s, μ-law system is by no means in worldwide use. As a matter of fact, it represents yet another of those instances where CCITT has had to produce two sets of parallel recommendations, one for the United States (and Japan, in this case) and another for the rest of the world! This other TDM system (also detailed in [1]) also uses 8 bits per sample and a frame rate of 8000 per second, but has A-law pre-emphasis and 32 channels, of which 30 are used for transmission and the remaining two (channels 0 and 16) for signaling and synchronization—the pulse rate is 2.048 Mbit/s. In Recommendation G.711, CCITT states that:

> Digital paths between countries which have adopted different encoding laws should carry signals encoded in accordance with the A-law. . . . Any necessary conversion should be done by the countries using the μ-law.

Higher-order digital multiplexing The two TDM systems thus far described are generally called "primary PCM" and represent the lowest order of multiplexing, akin to a group in FDM. As in FDM, higher orders of multiplexing have been defined and are in use, corresponding to supergroups, mastergroups and so on. As can be expected, they are in use between places which have sufficient mutual traffic to warrant using such large groupings.

The secondary multiplex level, in both systems, is obtained from combining four primary-level signals. Accordingly, it provides 96 channels in the μ-law system and 120 channels in the A-law system. The bit rates are, respectively, 6.312 Mbit/s and 8.448 Mbit/s. The discerning student will note that each of these rates is somewhat more than four times the corresponding primary bit rate—the additional bits are necessary for synchronization and other "housekeeping" duties. The method of obtaining secondary multiplex levels consists essentially in dividing by 4 the pulse widths in the primary level signal and using the slots thus vacated to combine four primary streams,

TABLE 15-1 Digital Multiplex Hierarchies

MULTIPLEX ORDER	μ-LAW BIT RATE (Mbit/s)	μ-LAW NO. OF TELEPHONE CHANNELS	A-LAW BIT RATE (Mbit/s)	A-LAW NO. OF TELEPHONE CHANNELS
1st	1.544	24	2.048	30
2nd	6.312	96	8.448	120
3nd	44.736*	672	34.368	480
4th	91†	1344	140†	1920
5th	274†	4032	565†‡	7680

*32.064 Mbit/s (= 384 channels) available as an alternative.
†Rounded to the nearest megabit.
‡An intermediate level of 280 Mbit/s (= 3840 channels) is also in use.

using delay lines or circuits in much the same way as was applied when the primary multiplex level was being produced. Still-higher TDM levels are obtained by the extension of this process, and Table 15-1 shows the levels in common use in both systems.

The methods of transmitting and receiving digitally multiplexed signals are discussed in Secs. 15-2 and 15-3.

15-2 SHORT- AND MEDIUM-HAUL SYSTEMS

To provide the required large number of telephone and other channels in national trunk routes, broadband systems are universally employed, consisting of coaxial cables, fiber-optic cables, microwave links, domestic satellites or occasionally tropospheric scatter links.

Coaxial cable is preferred to wire pairs in these circumstances, for its much greater available bandwidth, lower losses and much lower crosstalk. Fiber-optic cable, or "lightguide" is a logical extension of coaxial cable, to higher (infrared) frequencies and even greater bandwidths. Microwave links, in turn, are preferred to lower-frequency links for a variety of reasons, the major ones being the requirement for large bandwidths and highly directional antennas of manageable size. Such antennas reduce interference to and by the system, as well as providing high effective radiated powers in the wanted directions. Taking all factors into consideration, there is not too much to choose between microwave links and coaxial cables (except that generally cables are more expensive), so that the national grids of developed countries generally consist of a mixture of the two transmission media. Fiber optics have come on the scene more recently—in the late 1970s—and so their use is not yet as extensive, but it is rapidly expanding because of lower costs, especially over shorter distances and where very large bandwidths are needed.

Domestic satellite systems are in use in a great number of physically large countries, and regional satellites are employed by groups of closely connected neighboring countries, such as those in Western Europe and the Arab world. They have the advantage of great flexibility, being independent of difficult terrain, and lower costs for greater distances, because costs are essentially independent of distance, whereas they are proportional to distance for terrestrial systems. Finally, tropospheric scatter links are used in sparsely populated, difficult terrain, to interconnect islands or oil rigs, or in situations where territorial or political considerations prevent the use of the other terrestrial systems.

Each of the media described provides good-quality broadband communications, and each will now be treated in turn, except that, for convenience, domestic satellites will be treated with international satellites in Sec. 15-3.

15-2.1 Coaxial Cables

A coaxial cable system consists of a tube carrying a number of coaxial cables of the type treated in Chap. 7, together with repeaters and other ancillary equipment. Separate cables are used for the two directions of transmission, and a pair of spare cables is also provided for protection in case of failure. The number of cables per tube may be as low as four in smaller systems or as high as 22 in major systems, as illustrated in Fig. 15-4. The typical number of channels per cable varies from 600 in a 3-MHz system to 3600 in an 18-MHz system.

Because signals are attenuated as they travel along the cable (see Sec. 7-1.3), amplifying repeaters must be placed at suitable intervals along the route. The distance

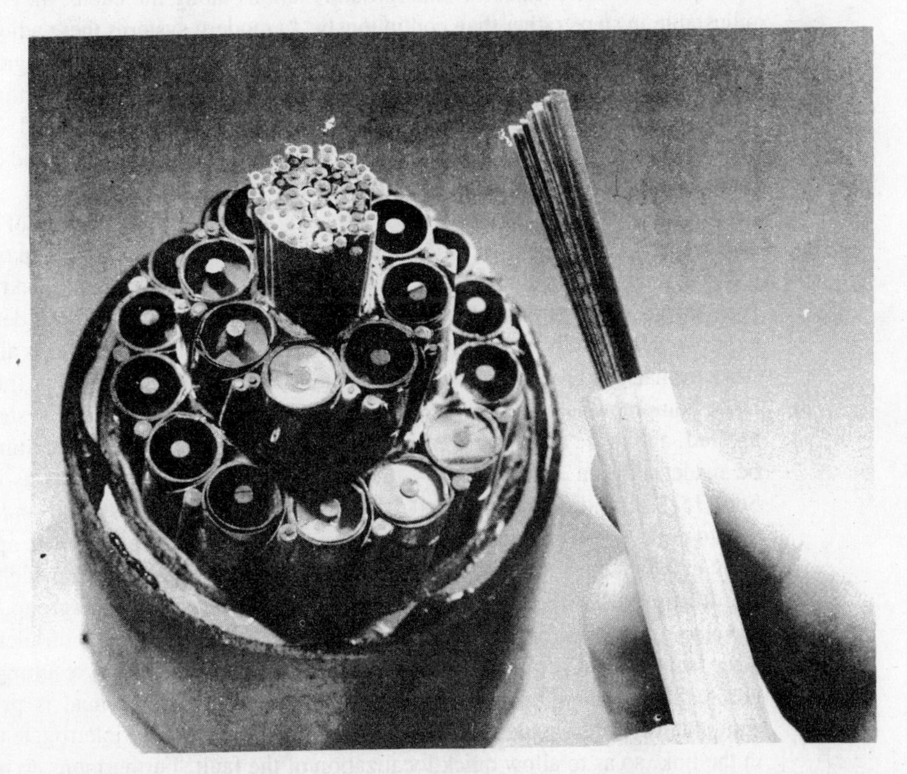

FIGURE 15-4 Coaxial cable used in the L5 system for carrying up to 108,000 simultaneous two-way telephone conversations. *(By permission of AT&T Long Lines.)*

varies, being roughly inversely proportional to the bandwidth of the system. It may be as much as 10 km between repeaters for a small system, but in the L5 system of Fig. 15-4, where bandwidths for all cables are nearly 58 MHz, repeaters are placed at 1.6-km intervals. Since there are repeaters, a dc supply must be fed to the cable to power them. In the L5 system, dc power-feeding stations are located 120 km apart, i.e., 75 repeaters apart. Assuming an 18-V drop across each repeater, and noting that repeaters are in series for direct current since otherwise the required currents would be too high, this means that the dc voltage applied at each station must be 1350 V. To minimize insulation problems, what is done in practice is to apply voltages of half that value, but of opposite polarities, at the two adjoining dc feed stations. A station at one end may thus feed +675 V to the cable, while the next station along feeds −675 V toward the first station and +675 V toward the next station down the cable.

Broadband systems must have excellent frequency and phase-delay responses to be of use. This cannot be achieved by cables and repeaters unaided, so that equalizers are also located along the cable, 60 km apart in the L5 system. It should be noted that there is need for two kinds of equalizers. The fixed type compensates for constant, known deviations in frequency and phase response which are inherent in each particular system. Adjustable equalizers, generally provided at the two ends of the system, are used to compensate for the variables and the unpredictable variations. Where adjustable equalizers are located in underground stations along the cable, they are normally adjustable in steps rather than continuously. In modern systems these adjustments may be made from the control stations at the ends, by sending appropriate signals down the cable. Finally, to ensure constant gain along the system, thus preventing excessive noise and intermodulation distortion, the gain of repeaters is regulated. This may be done by having adjustable-gain repeaters at intervals along the cable and altering their gain as required with suitable control signals.

Multiplexing and demultiplexing bays form the major portion of the terminal equipment. It is in these bays that FDM, as described in Sec. 15-1, takes place. The exact arrangement of supergroups in various systems is shown in [1]. Additionally, [4] is a very good description of the L5 system, and therefore modern high-density coaxial cable systems, and is highly recommended. Dc power feed equipment is also located at the terminals, as are interconnections to other systems, be they local or trunk. Surveillance equipment is also provided at terminal stations. It is here that system pilots are applied, and those that were applied at the other end are extracted. A distinction should be made between a supergroup—or even supermastergroup—pilot, as described in Sec. 15-1.1, and a system pilot. The latter belongs to the system and is used for end-to-end system regulation and monitoring. The supergroup pilot is applied at the point at which the supergroup is formed and extracted at the point at which it is broken up. It is used for regulating and monitoring that particular supergroup, which may traverse many different links. Although each is regulated, small, in-tolerance departures from correct response in the various links may be additive, resulting in a supergroup that is out of tolerance end to end. Finally, each terminal is provided with equipment which, should there be a cable failure, permits it to interrogate the repeaters in the link, so as to allow quick localization of the fault. Furthermore, to minimize the effects of outages, terminal stations may be provided with redundant and/or duplicated systems, allowing their staff to patch rapidly around any breaks.

Some students may wonder why communications systems tend to have more and more capacity. The answer is that long-distance telephony, telex and television transmissions in most countries have been increasing at high rates, for over two decades, while data transmission in developed countries is growing at very high annual rates of close to 50 percent. Coupled to this demand growth is the fact that a (say) 10,800-channel system is decidedly cheaper to install and maintain than three 3600-channel systems. Also, fortunately, such broadband links are manufactured by some of the world's most modern, efficient and reliable companies. In turn, this means that the fourth law of thermodynamics (''everything takes longer and costs more!'') does not usually apply to such systems.

15-2.2 Fiber-Optic Links

It was shown in Sec. 12-9 how coherent waves at light and infrared frequencies may be generated (with lasers or light-emitting diodes) and how they may be detected (with photodiodes). It now remains to discuss the intervening medium, which unfortunately cannot be open space—at least not on the earth's surface. This is because light or infrared is subject to far too much absorption in open space, be it by the moisture content and dust in the air or, worse still, fog or rain. Similarly, plenty of interference can be expected from the many light sources in constant use. Accordingly, optical fibers are used for light and infrared transmissions, in a manner virtually identical to waveguides at microwave frequencies.

An optic fiber is a piece of very thin, highly pure glass, with an outside cladding of glass that is similar but, because of a slightly different chemical composition, has a different refractive index. This is known as a *step-index* fiber, with a core diameter in the range of 2 to 200 μm. The use of fibers with a gradually changing refractive index—reducing from the center to the outside of the fiber cross section—are also employed, but the use of these *graded-index* fibers is greatly diminishing, because lower attenuations are possible with step-index fibers; graded-index fibers were initially easier to manufacture, and so were first on the scene. Step-index fiber may be used for *multimode* or *single-mode* propagation, and it is this latter mode, akin to single-mode propagation in a circular waveguide, that yields the best performance and has just about taken over in new systems. Its introduction was delayed by the fact that it requires extremely thin fibers, with core diameters in the range 2 to 10 μm, whereas diameters of the order of 50 to 200 μm were suitable for multimode propagation. Students are referred to [5] and pp. 55–60 in [6] for additional details of light propagation in these hair-thin fibers.

Optical losses in fibers Figure 15-5 shows attenuation in optic fibers, gives an indication of the improvement in this area over the past few years and introduces the concept of transmission ''windows.'' It is seen that in 1975, the lowest attenuation available was 2.5 dB/km at 1.06 μm (but lasers operating at that wavelength were not readily available, so that the 0.85-μm window was used in practice, giving an attenuation of about 3.2 dB/km). By 1983, the lowest attenuation available from commercial fibers was 0.25 dB/km at the third window of 1.55 μm. It should be recalled that coaxial cables have a lowest attenuation of just under 10 dB/km. What Fig. 15-5 does

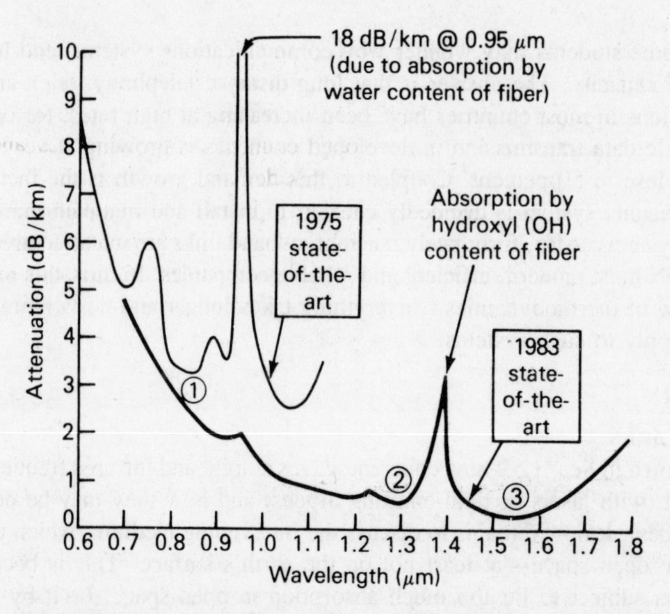

FIGURE 15-5 Attenuation in high-silica, low-loss optical fibers, showing transmission windows: ①, first window at 0.85 μm; ②, second window at 1.3 μm; ③, third window at 1.55 μm.

not show is that, in the late 1960s, the attenuation of optical fibers was measured in thousands of decibels per kilometer. In other words, the total attenuation in 1 km of current optical fiber would have taken only 80 m of fiber to provide in 1975, and only about 20 cm of fiber in the late 1960s!

The startling reduction in fiber losses has been brought about by a clear understanding of what causes them and corresponding refinement in manufacture, as well as selection of the optimum materials. By looking at the attenuation curve in Fig. 15-5, it can be appreciated that many factors must contribute to its shape. One of the many possible causes of losses in fibers is the presence of foreign materials, and another is variation in the density of fiber. Both have been overcome to a very large extent through manufacturing quality control improvements. So-called microbending can result in transmission losses, but this has also been overcome, in that manufacturing refinements have greatly reduced the presence of these minute bends in the fiber, which earlier manufacturing processes had introduced. What remains now is absorption by the glass itself, together with scattering and losses caused by irreducible impurities, such as hydroxyl ions at a wavelength of about 1.38 μm. In [5] there is much additional information on losses in optic fibers, while [8] describes their manufacture in detail.

Advantages of optical fibers An optic fiber cable is quite similar in appearance to the coaxial cable system shown in Fig. 15-4. However, whereas the L5 coaxial system has a diameter of 115 mm (4½ in), the diameter of an equivalent fiber system is only about 38 mm (1½ in)—this immediately shows an important advantage of fiber optics. Telecommunications carriers are able to increase the utilization of their existing cable ducts very significantly, when faced with increased demand for capacity, if they substitute fiber-optic cables for copper ones. Again, optical systems are virtually immune to crosstalk and electromagnetic interference, unlike orthodox cables and microwave links. Finally, the point has now been reached where, for large capacities, fiber-optic systems are cheaper than their coaxial cable or microwave link counterparts, a situation which will spread to lower-capacity links as more fiber-optic systems are manufactured, and unit costs fall because of the increased volume of sales. In [7] and [8] are detailed discussions of the relative economics of fiber-optic and other broadband links, with cost extrapolations to the late 1980s.

Two other points should be made in connection with the relative appeal of fiber-optic links. The first is that laser and LED modulation methods lend themselves ideally to digital operation and, as discussed in Chap. 13, the world trend in broadband communications is decidedly toward digital transmission and switching. The second point is that all of the world's fiber-optic systems so far use single-frequency operation of each fiber, but that is unlikely to last much longer. *Wavelength multiplexing* is currently under close investigation [9]. This is a system in which two or more signals, of differing but nearby wavelengths, are sent along the same fiber. There are no such practical systems in existence, but laboratory results are encouraging, and one may confidently expect fiber-optic links with wavelength multiplexing to be in service before the end of the 1980s. It goes without saying that this would significantly increase the information-carrying capacity of optical fibers, at little additional cost.

Practical fiber-optic systems The majority of the world's commercial fiber-optic links operates in the first window, at 0.85 μm, but the picture is changing rapidly—most if not all links currently being installed operate in the second window, at 1.3 μm. When the first links were being installed in the mid and late 1970s, in North America, Western Europe, Japan and Hong Kong, 0.85 μm was a real window, whereas a glance at Fig. 15-5 reveals that this wavelength is by now a mere "blip" on the attenuation curve; improvements in fiber manufacture have seen to that. More importantly, although it was recognized that the second window would yield lower attenuations (though still much higher than available today), there were no reliable lasers operating at the higher wavelengths, as was explained in Sec. 12-9.4. A quite similar situation currently prevails: it is recognized that the third window would yield even lower attenuation, and thus permit greater repeater separations, but the InGaAsP lasers in use at 1.3 μm have only recently been brought to a level of acceptable reliability, and time must necessarily pass before 1.6-μm devices are similarly developed. However, nothing is surer than that they ultimately will be!

AT&T Long Lines currently has nearly 600 km of fiber-optic links in operation in the United States, nearly all working at 0.85 μm. By 1985 the company expects to complete Northeast Corridor Lightwave System, stretching over 1200 km from Cambridge, Massachusetts, to Mosely, Virginia, via New York City, Philadelphia, Baltimore and Washington, D.C. This will be an all-digital system operating at 1.3 μm, with multiple fiber pairs (one fiber in each pair for each direction of transmission) housed in a 12-mm (½-in) cable, with repeaters about 6.5 km apart. Each fiber pair will use fourth-order (see Sec. 15-1.2) multiplexing, providing 1344 or 2016 channels per fiber pair. The system will have an overall capacity of up to 240,000 channels in parts, i.e., up to 120 fiber pairs. Systems of this type use InGaAsP lasers transmitting −3 dBm (0.5 mW) into the fiber, while the avalanche photodiode at the receiving end operates at a level of −40 dBm. Similar developments are taking place elsewhere in the United States, notably the 1000-km system being built between San Diego and Sacramento in California.

Developments in fiber optics outside the United States have been just as intensive, if not more so. In Western Europe, Great Britain has led the field, already having an extended and well-integrated fiber-optic network. West Germany, the Netherlands, Italy, Canada, Japan and Hong Kong are at a similar stage of development. In [6] and [10] through [14] are details of practical fiber-optic links and their use in the world's broadband routes.

15-2.3 Microwave Links

A microwave link performs the same functions as a copper or optic fiber cable, but in a different manner, by using point-to-point microwave transmission between repeaters. Many links operate in the 4- and 6-GHz region, but some links operate at frequencies as low as 2 GHz and others at frequencies as high as 13 GHz. Propagation is of course by means of the space wave and therefore limited to line of sight. Consequently, typical repeater spacings are close to 50 km, unless a city repeater is located on top of a special tower, or a country one on a hill. Even then, much larger repeater spacings cannot be used because of the very high attenuation with distance (see Sec. 16-1.2) to which radio waves are subject.

A microwave link terminal has a number of similarities to a coaxial cable terminal. For example, the multiplex equipment will be very similar, if not identical, as will be the channel capacity. Where a cable system uses a number of coaxial cable pairs, a microwave link will use a number of carriers at various frequencies within the bandwidth allocated to the system. The effect is much the same, and once again a spare carrier is used as a "protection" bearer in case one of the working bearers fails. Finally, there are interconnections at the terminal to other microwave or cable systems, local or trunk.

As can be seen, the similarities are in what is done, and the differences lie in the specific detail of how it is done. To illustrate the latter point, the simplified block diagram of a typical microwave repeater is shown in Fig. 15-6. Essentially, the repeater receives a modulated microwave signal from one repeater and transmits to the next one, and an identical chain is provided for working in the other direction. The only difference here is that the transmissions in the two directions are somewhat different in frequency to avoid interference; the frequency difference is typically a few hundred megahertz at the 4- or 6-GHz operating frequencies.

The block diagram in Fig. 15-6 shows no amplification of the received signal at the radio frequency. Rather, there is conversion down to an IF which is almost invariably 70 MHz, and this is the frequency at which the bulk of amplification takes place in the link shown. Indeed, low-power links have a modulated output oscillator rather than a power output amplifier, and in those links *all* of the amplification will take place at

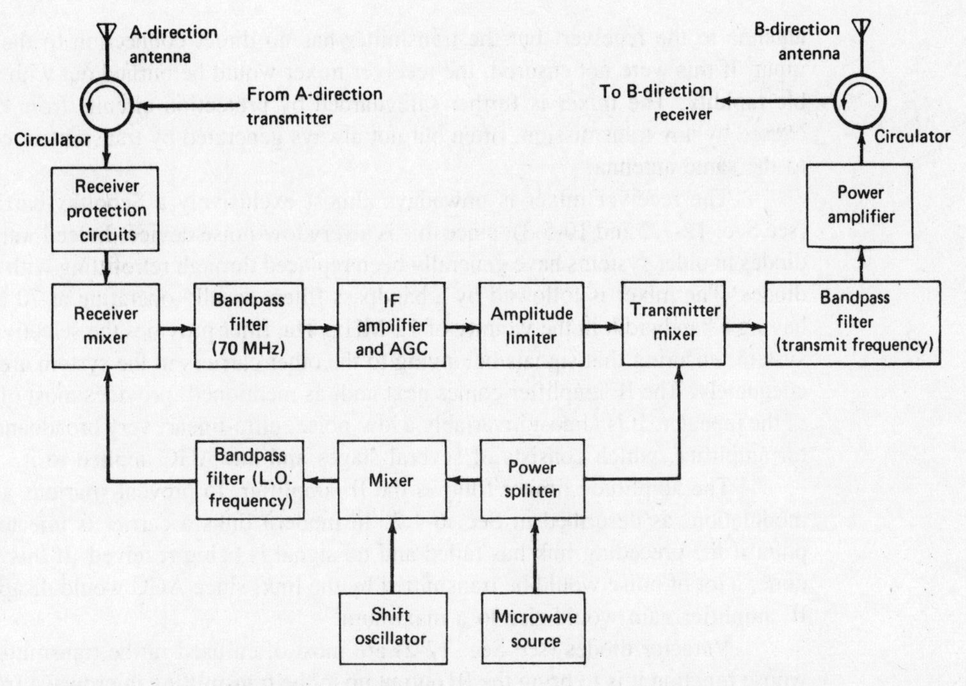

FIGURE 15-6 Simplified block diagram of microwave link carrier chain, shown receiving from A direction and transmitting in B direction.

70 MHz. The reason for this frequency conversion in existing links is noise reduction: until recently, it has been a lot easier to produce a very low-noise amplifier at 70 MHz than 4 GHz or above. It should be noted in this context that a typical microwave link consists of several repeaters between the end points, and of course noise is additive for analog systems. On the other hand, the latest developments in microwave transistors have dramatically reduced their noise figures, and so microwave links (especially digital ones) are beginning to appear with RF preamplifiers.

One must not lose sight of the fact that having a low-noise, sensitive receiver allows the designer to reduce transmit power in proportion: if receiver noise figure can be halved, so can the required link output power. In turn, this allows cost and size reductions in every repeater of what might be a very long chain.

The antennas most frequently used are those with parabolic reflectors. Hoghorn antennas (see Sec. 9.7) are preferred for high-density links, since they are broadband and low-noise. They also lend themselves to so-called frequency reuse, by means of separation of signals through vertical and horizontal polarization. Hoghorns are widely used in the very common United States microwave links in the TD-2C and TD-3C series.

The circulator, as will be recalled from Sec. 10-5.2, ensures a connection between the adjoining ports in the direction of the arrow but not between any other ports. In Fig. 15-6, this means that the transmitter is connected to the antenna and the antenna to the receiver, but the transmitter has no direct connection to the receiver input. If this were not ensured, the receiver mixer would be burned out with remarkable rapidity. The mixer is further safeguarded by protection circuits from overloads caused by any transmission, often but not always generated by transmitters connected to the same antenna.

The receiver mixer is nowadays almost exclusively a Schottky-barrier diode (see Sec. 12-7.2 and 10-5.3), since this is a very low-noise device. Indeed, other mixer diodes in older systems have generally been replaced through retrofitting with Schottky diodes. The mixer is followed by a bandpass filter, usually operating at 70 MHz and having a bandwidth in the vicinity of 12 MHz. The filter provides the selectivity of the system, ensuring that signals belonging to the other carriers in the system are rejected adequately. The IF amplifier comes next and, as mentioned, provides most of the gain of the repeater. It is almost invariably a low-noise, ultra-linear, very broadband transistor amplifier, which consists of several stages and has AGC applied to it.

The amplitude limiter follows the IF amplifier, to prevent spurious amplitude modulation, as described in Sec. 6-4.2. In modern links a carrier is injected at this point if the preceding link has failed and no signal is being received. If this were not done, a lot of noise would be transmitted by the link, since AGC would disappear and IF amplifier gain would rise to a maximum.

Varactor diodes (see Sec. 12-2) are most often used in the transmitter mixer, whose function it is to bring the IF output up to the transmitting microwave frequency. This mixer is followed by a bandpass filter to prevent any straying into unauthorized portions of the frequency spectrum or interference to other carriers in the link.

The output power of a link varies, depending on the bandwidth and therefore the number of circuits per carrier, and on the distance to the next repeater. In most cases powers between 0.25 and 10 W are transmitted, with 2 to 5 W most common. For powers of 0.5 W or less, a power amplifier is not required, and a power oscillator is used instead. This is most likely to be a reflex klystron (see Sec. 11-3) in older equipment, a Gunn diode (see Sec. 12-5) or an IMPATT diode (see Sec. 12-6) in more modern equipment. As previously discussed, the semiconductor devices are preferred for their greater reliability, lower power consumption and simpler power supply requirements. For powers of 1 to 5 W, at frequencies not exceeding 6 GHz, output amplifiers are used, being most commonly push-pull metal-ceramic disk-seal triodes (see Sec. 11-1.2) or single-ended TWT amplifiers (see Sec. 11-5). Equipment installed during the 1980s is most likely to use FET power amplifiers. For powers in excess of about 5 W, and certainly at frequencies above 6 GHz, traveling-wave tubes are almost universal as power amplifiers. They are then preferred to semiconductor devices because of their much higher available output powers.

The microwave source was a klystron up to the 1960s, and a Gunn oscillator with AFC in the 1970s, but it is nowadays most likely to be a VHF transistor crystal oscillator, with a varactor multiplier. Multiplication factors are of the order of 20 to 40, and the power output is in the vicinity of 200 mW. The power splitter sends approximately 75 percent of the power to the transmitter mixer, and the rest to the mixer which is also fed by the shift oscillator. The function of this circuit is to ensure that the receiver mixer is fed with a frequency 70 MHz higher than the incoming signal, so as to provide the 70-MHz frequency difference for the IF amplifier. This assumes that the receive and transmit frequencies are the same and thus implies that the receive and transmit frequencies in the *A* direction in Fig. 15-6 are a few hundred megahertz higher or lower than in the *B* direction for which the figure is drawn. Some links operate slightly differently, and their receive and transmit frequencies in a given direction are somewhat different. In that case, the shift oscillator provides the appropriately different frequency, to ensure still that an IF of 70 MHz is available. The function of the bandpass filter is to remove the unwanted frequencies from the output of the balanced mixer which precedes it.

The typical number of carriers (in each direction) in a microwave link is at least four, and sometimes as many as 12. There are normally 600 to 2700 channels per carrier. In difficult locations, diversity may be used, in which case it is most likely to be space diversity incorporating pairs of antennas for the same direction. Also, it must be reiterated that the repeaters are not directly involved in the modulation process. This is because they are simply *repeaters*; their function is to receive, amplify and retransmit. The fact that frequency changing takes place is extraneous to their function and should certainly not be confused with IF amplification in ordinary receivers (where IF amplifiers are followed by demodulators). Modulation does of course take place, as does demodulation, but only at the terminals, not at repeaters.

The towers used for microwave links range in height up to about 25 m, depending on the terrain, length of that particular link and location of the tower itself. Such link repeaters are unattended, and, unlike coaxial cables where direct current is fed down the cable, repeaters must have their own power supplies. The 200 to 300 W of dc power required by a link is generally provided by a battery. In turn, the power is replenished by a generator, which may be diesel, wind-driven or, in some (especially desert) locations, solar. The antennas themselves are mounted near the top of the tower, a few meters apart in the case of space diversity. They must be accurately aligned to the next repeater in the link, because beamwidths are less than 2°, and any misalignment causes a power loss. Alignment is one of the many items checked at each periodical maintenance visit to a repeater.

It was stated at the beginning of this section that microwave links and coaxial cables perform essentially the same functions. Given that, it may be thought that the two media are in competition. So they are, up to a point, but not to the extent that any one system is likely to oust the other. Basically, microwave links are cheaper and have better properties for TV transmission, although coaxial cable is much less prone to interference.[3] The preference for microwave links in transmitting TV programs to distant stations for rebroadcasting is due to the lesser number of repeaters for a given distance, as compared with a coaxial cable. In turn, this reduces the cumulative phase and amplitude distortion over the large bandwidth occupied by TV. On the other hand, a microwave link is far more subject to impulse noise, or "hits," than the cable, which is protected and a closed-circuit system. The overall result of these considerations is

[3]Coaxial cables are, however, more prone to the kind of industrial interference caused by people using bulldozers and other digging appliances without first checking a map!

that the two media are complementary over the "backbone" routes in most developed countries, although microwave links predominate over the lesser routes.

For those students desiring more information on microwave links, [15–17] are recommended; the first of these describes an analog (FDM/FM) system, and the other two deal with digital systems.

15-2.4 Tropospheric Scatter Links

Troposcatter propagation was described in Sec. 8-2.4. Its application is in troposcatter communications links, which are now treated.

A troposcatter link terminal is rather similar to a microwave link terminal, and indeed a typical block diagram is sufficiently like Fig. 15-6 that a separate block is not shown. The main differences lie in the very much higher output powers and lower receiver noise figures in troposcatter links. Typical output powers are 1 to 10 kW, but powers as high as 100 kW have been used for broadband links, although as little as 5 W may be sufficient for a short link designed to carry only eight voice channels. Powers of 1 to 5 kW are achieved with either high-power TWTs or multicavity klystrons, and klystrons are used to provide the higher powers. At 790 to 960 MHz, perhaps the most common frequency range, receivers have low-noise transistor RF amplifiers. In the 2- and 5-GHz ranges, tunnel-diode or parametric amplifiers are common (see Secs. 12-4 and 12-3, respectively); receiver noise figures under 2 dB are the norm. It will be recalled that the attenuation over a troposcatter path is fearful; hence the high transmitting powers used. Everything else being equal, a 3-dB improvement in receiver noise figure may permit a 3-dB reduction in the transmitted power.

Diversity is *always* used in troposcatter links. It may be space, polarization, or frequency diversity, or quadruple diversity—a combination of any two of those—where fading is particularly severe, i.e., on most longer links. This causes added terminal complexity, but it results in greatly improved reliability. For example, most modern systems are unavailable, because of fading, for an average of less than 0.1 percent of the time during the worst month of the year.

A high proportion of troposcatter links is single-span, although others may have up to 20 spans. This depends on circumstances: a point-to-point link over inaccessible terrain is likely to be single-span, with a length of 300 to 1000 km. On the other hand, a link designed to provide communications for a group of islands, such as in the Caribbean, Indonesia or the Philippines, will have several spans, with baseband access at each point. Antenna diameters vary correspondingly, with typical diameters of 15 m for broadband links. However, longer paths may require parabolic reflectors with diameters as large as 40 m, making them even larger than satellite earth station antennas.

A typical broadband link may carry 192 two-way voice channels, i.e., three supergroups plus one group. Capacities in excess of five supergroups are, however, available, and indeed some shorter links can even carry TV.

Finally, it should be noted that the capital cost of troposcatter links, in dollars per circuit-kilometer, is perhaps four times that of coaxial cable, making it about 12 times that of microwave links. Operating costs are roughly in the same proportion, being high for troposcatter because of the high powers required. Accordingly, troposcatter links are used where special considerations so dictate, rather than interchangeably with the other two broadband transmission media. Hill [18] is an excellent reference on both the theoretical and practical aspects of the subject; see also [7] in Chap. 8.

15-3 LONG-HAUL SYSTEMS

Submarine cables and satellites are the two available means of intercontinental broadband communication. They bear the same competitive and complementary relationship to each other as coaxial cables and microwave links on land. Being historically first, by a dozen or so years, submarine cables are treated first here.

15-3.1 Submarine Cables

Submarine cables use principles very much like those of coaxial cables. Thus they are coaxial, have repeaters and equalizers and have dc power fed to them, with opposite polarities fed from opposite ends to reduce insulation problems. However, submarine cables use a single coaxial tube for both directions of transmission, with frequency techniques akin to those of microwave links to separate the two directions. The extent to which cables have spread out around the world, since *TAT-1* in 1956, is shown in Fig. 15-7.

Cables such as the 48-circuit *TAT-1* and the 80-circuit *CANTAT 1* (1961) are often referred to as ''first-generation'' cables. They feature vacuum-tube repeaters, at intervals of 50 to 60 km. Second-generation cables, such as the *SAT-1* (1968) cable from Portugal to South Africa, have up to 360 circuits, with vacuum-tube repeaters at 18-km intervals. Vacuum tubes were used as late as 1968 because of their proven reliability. Submerged cable or repeater repair is perfectly feasible, but is a complex and costly process. It involves sending cableships to the affected area and dragging the sea bottom for the cable, while the interrupted circuits are restored via another cable or a satellite (at no small cost). It can therefore be appreciated that reliability is the keynote, and vacuum tubes had certainly established a reputation for that in submarine systems.

However, increased bandwidths mean reduced repeater gains and increased cable losses, and so repeaters must be placed closer together. For long cable segments, this results in unduly high dc voltages required at the two ends to accommodate the 70-V drop per vacuum tube repeater. Thus the third- and subsequent-generation cables have used transistor repeaters exclusively, with voltage drops of only 12 V per repeater. The *TASMAN* cable (1974, 480 circuits from Australia to New Zealand) and the *TAT-5* cable (1970, 845 circuits from the United States to Spain), both shown on Fig. 15-7, are typical examples of third-generation cables.

CANTAT 2 is typical of fourth-generation cables. It was laid in 1974 and provides 1840 circuits between Canada and Great Britain. Fig. 15-8 shows the cable, both lightweight and armored, used in *CANTAT 2*, and a repeater from the system is shown in Fig. 15-9. The repeaters are, of course, all solid-state, with separations of about 11 km in practice. This is a very successful design, first used in 1971 for a cable between Spain and the Canary Islands and subsequently employed in the Mediterra-

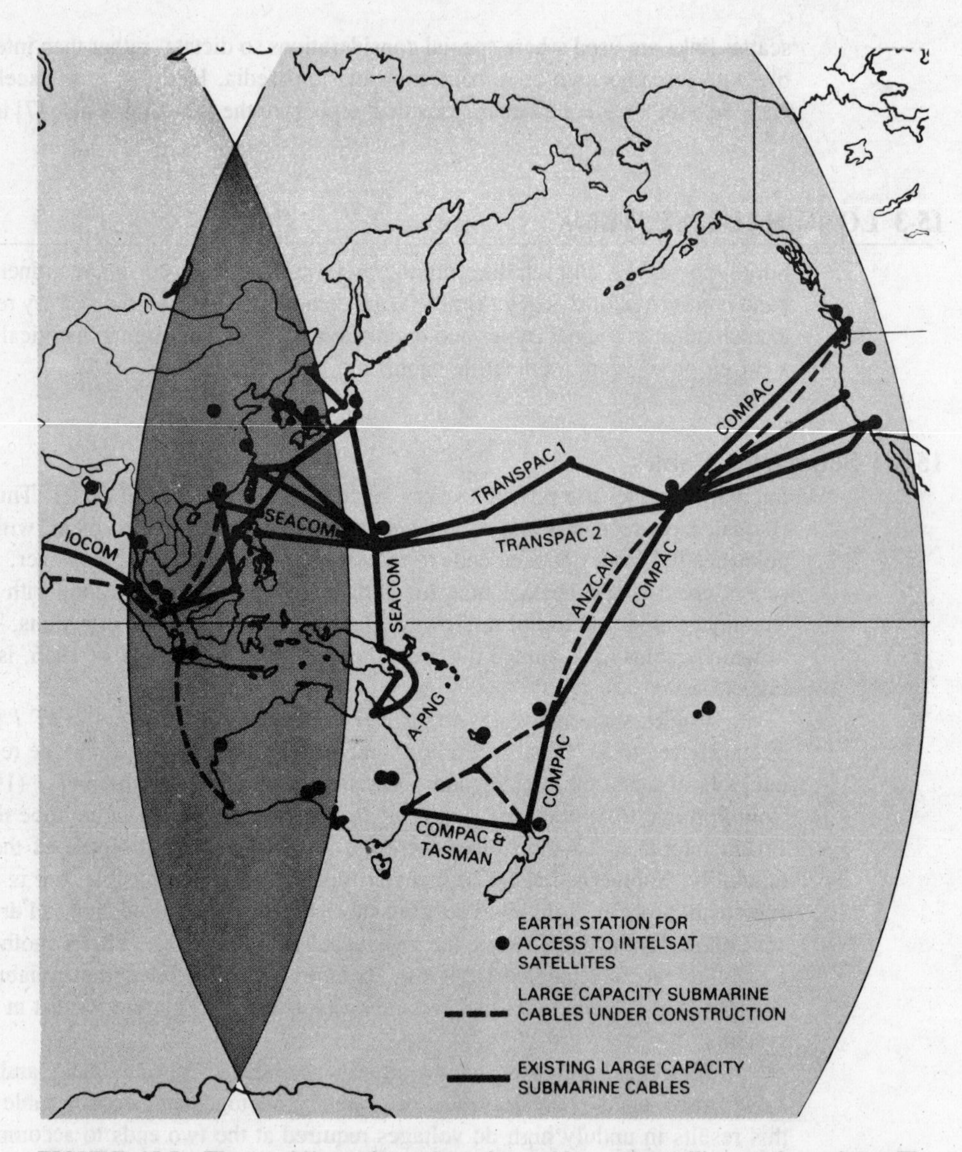

FIGURE 15-7 The world's major submarine cables and satellite earth stations. The curved lines indicate the coverage area limits of the satellites shown along the equator. (Map continues on p. 555.) ***(Courtesy of Overseas Telecommunications Commission, Australia.)***

nean (several cables), the Atlantic (*COLUMBUS*, southern segment of *ATLANTIS*, in 1982) and the Pacific (*ANZCAN*, 1984), as well as several shorter cables in Europe and southeast Asia. All these are shown in Fig. 15-7, except the many Mediterranean cables, which are omitted for lack of space. Detailed maps showing the majority of the world's submarine cables, technical specifications of several cable systems and descriptions of their manufacture and laying are found in [19]—it is thoroughly recommended.

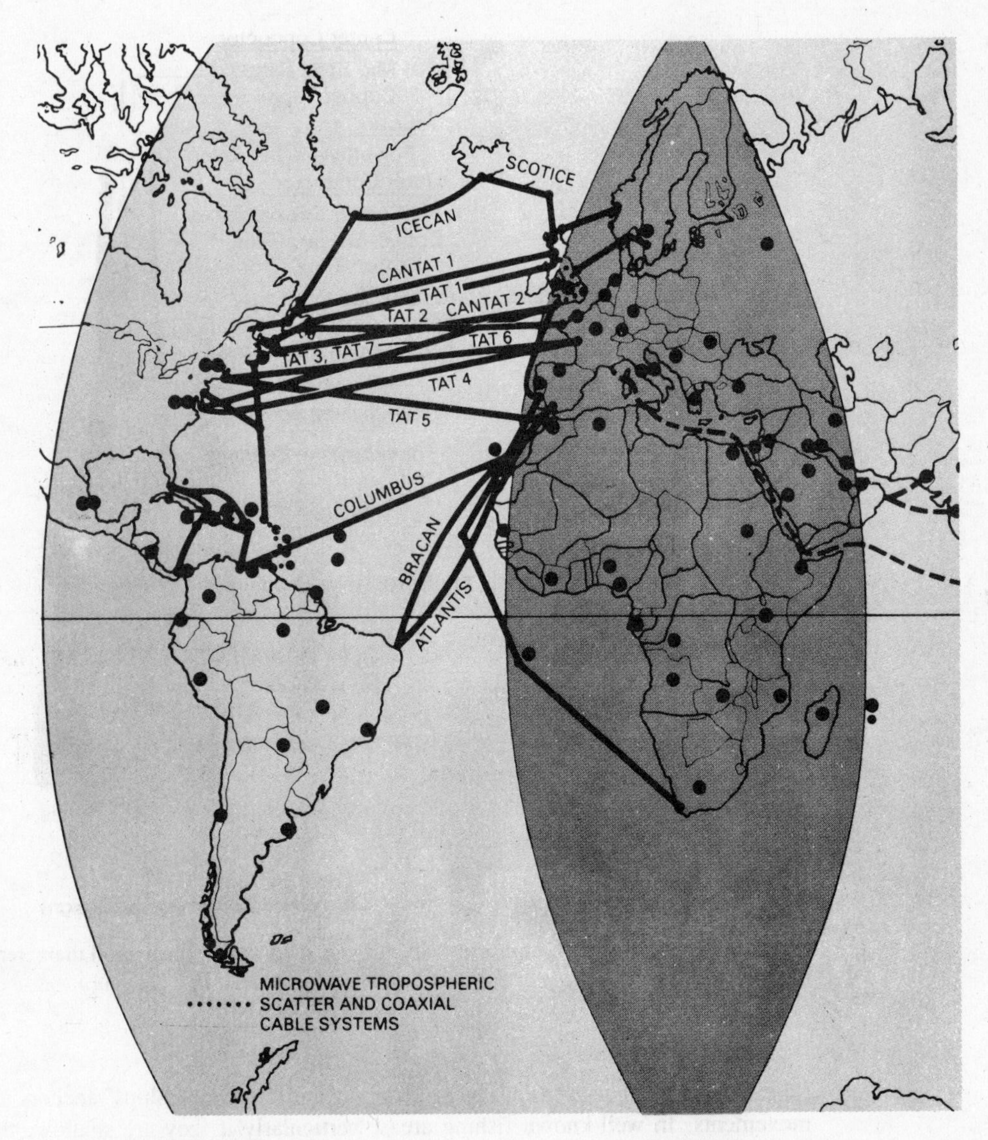

(*Map continued from p. 554.*)

Cable is laid by cableships operating from the two ends separately and sometimes simultaneously, moving at typical speeds of about 8 knots (about 15 km/h)—the final splice is thus the midocean one. Lightweight cable is used for most of the length, including all deep sea portions. Sometimes, where great depths are involved, the cable is laid with sea parachutes, to slow its descent and therefore the rate of temperature change undergone by the cable and electronic components. The repeaters are rigid, and ingenious methods of bypassing shipboard sheaves have been developed. Armored

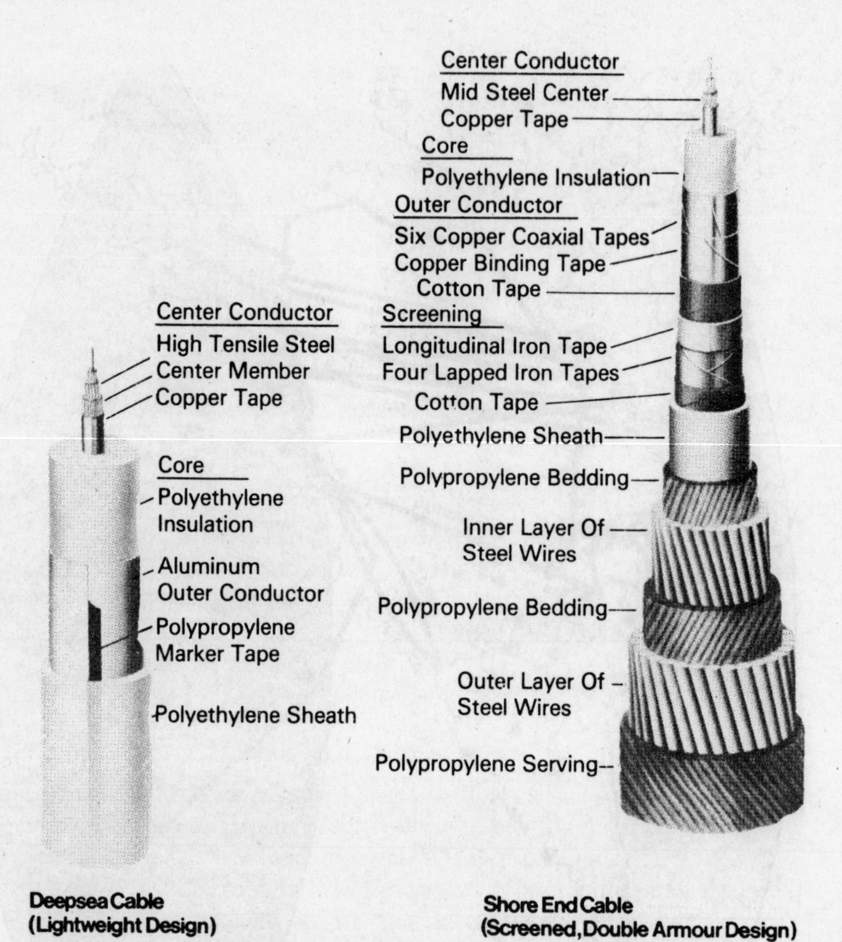

FIGURE 15-8 Display of submarine cable used in *CANTAT 2;* the overall diameter of each cable is 44.5 mm. *(Courtesy of Standard Telephones and Cables, PLC, London.)*

cable is used for the shore ends as protection against trawlers, ships' anchors and tidal movements. In well-known fishing areas, particularly if they are shallow, the technique of ploughing-in is used if the sea bottom permits. As the cable is paid out from the ship, a specially designed submarine, towed by a wire, cuts a 60-cm-deep trench for the cable to fall into; the trench is then covered. This was in fact done for the first 220 km of the *CANTAT 2* cable off the Canadian continental shelf, except for the repeaters, which were too thick to be buried.

The *CANTAT 2* repeaters, typical in this regard, are 25 cm in diameter and nearly 3 m long. Their function, as might be gathered, is simply to amplify. This must be done for both directions. The function of the power-separating and the directional filters in Fig. 15-9 is to help in this regard. In the *CANTAT 2* cable, the 23 supergroups are accommodated in the frequency band 312 to 6012 kHz in one direction, and 8000

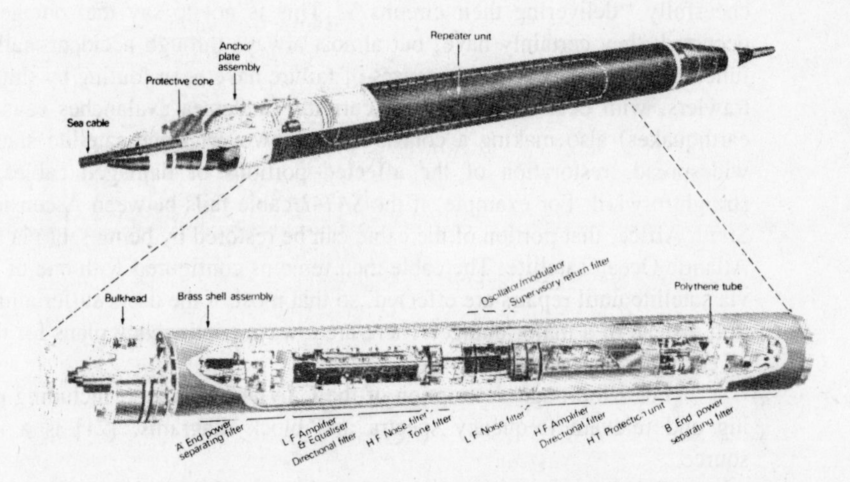

FIGURE 15-9 Construction of *CANTAT 2* submerged repeater. ***(By courtesy of Standard Telephones and Cables PLC, London.)***

to 13,700 kHz in the other direction. Inquisitive students who perform the appropriate calculations will realize that the above figures correspond to 3-kHz circuits and 80-circuit supergroups. It will be recalled that submarine cables are expensive, and 3-kHz voice circuits are often used. Supervisory tones and cable and system pilots are assigned various portions of the nearly 14-MHz spectrum, leaving 940 kHz for separation between the two directions; this is quite adequate in practice.

Reliability is the keynote of a submarine cable project. This point cannot be stressed enough. Whether it is the cable itself, repeaters, equalizers, cable station terminal equipment or power feed equipment, everything is engineered for a long life and slight, predictable aging. All cable and repeater welding is done by specially trained personnel, and all welds are checked by x-ray. The electronic components are assembled and tested under dustfree, laboratory conditions. All the components are used at well below their maximum ratings, and key components are duplicated. The performance of the system is monitored by the cableship during laying, and from the terminals for the rest of the cable life. Power feed arrangements are complex, with mains supplies rectified and regulated at the terminals and then used to float-charge the banks of batteries which feed dc/ac converters whose rectified output is actually fed to the cable at constant current. Duplicate batteries and standby diesel generators are provided, as are complicated interlock arrangements. All this is done to prevent the worst crime that can be perpetrated on a submarine cable: the sudden removal of the dc power feed.

The precautions as outlined are severe, but they have certainly paid off. For instance, all the submarine cables that have been laid since 1956 are still operating, cheerfully "delivering their circuits."[4] This is not to say that outages have never occurred; they certainly have, but almost always through accidents rather than malfunctions. The most common causes of failure have been fouling by ships' anchors or trawlers, with occasional turbidity currents (undersea avalanches caused by nearby earthquakes) also making a contribution. However, since satellite stations are now widespread, restoration of the affected portions of damaged cables is relatively straightforward. For example, if the *SAT-1* cable fails between Ascension Island and South Africa, that portion of the cable can be restored by being sent via an *INTELSAT* Atlantic Ocean satellite. The cable then remains configured with one of its legs going via satellite until repairs are effected, so that most of the users suffer a minor interruption instead of a major outage. There are always contingency plans for the restoration of each leg of every cable.

For a thorough description of the *CANTAT 2* project, including planning, laying and relevant frequency spectra and block diagrams, [21] is a recommended source.

Cables larger than the 14-MHz, 23-supergroup *CANTAT 2* type are also available. They include a 43-supergroup French cable, a 45-supergroup Japanese cable, a 50.8-supergroup American cable and a 69-supergroup British cable (capable of providing 5520 telephone circuits). They are used for a number of high-density applications, but only the American cable is used in intercontinental systems, for example, *TAT-6* and *TAT-7*, described in [20]. It is almost as though users were awaiting the advent of fiber optics.

Fiber-optic submarine cables Apart from a number of experimental, mainly unrepeatered cables in Europe, the United States and Japan (see [22]), there are no submarine fiber-optic cables at the moment, but the situation is due to change shortly. There will be at least two repeatered commercial systems in service in 1985, one across the North Sea from Britain to probably the Netherlands, and the other in the Mediterranean from the south of France to Corsica. The world's first major fiber-optic undersea cables are scheduled for service in 1988. The first will be *TAT-8*, across the Atlantic. It will consist of two working fiber pairs, each with a basic capacity of nearly 4000 circuits. The routing will be from Tuckerton, New Jersey to a submerged join off the coast of France, with one fiber pair continuing to Cornwall (Land's End) in the United Kingdom and the other pair to Brittany (Penmarc'h) in France. There will also be a fiber pair between the British and French landing points.

Although there will be some British and French contributions to the manufacture, the major portion of *TAT-8* will be the (currently experimental) American SL design (see [23]), as will the next likely fiber-optic system. This will be *HAW-4*, scheduled to be ready for service in late 1988, between California (Point Arena) and Hawaii (Sandy Beach, near Honolulu). The companion system, also scheduled for late

[4] Except for the few whose use has been discontinued for economic rather than technical reasons; *TAT-1*, turned off in 1981, falls into this category.

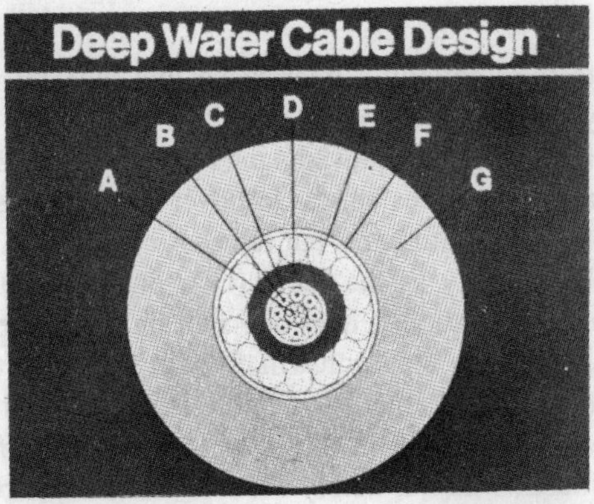

FIGURE 15-10 Cross section of lightweight experimental submarine fiber-optic cable. (A) Fiber strength member, (B) optical fibers, (C) fiber package sheath, (D) aluminium tube, (E) high tensile wires, (F) copper tube, and (G) polyethylene insulant. *(Courtesy of Standard Telephones and Cables PLC, London.)*

1988, will be the *TRANSPAC-3* cable, from Sandy Beach to Guam and Japan, via a submerged join.

The early fiber-optic submarine systems are likely to have a cross-sectional appearance similar to that shown in Fig. 15-10. They will operate at 1.3 μm like their terrestrial counterparts, and like them, they will use separate fibers in each pair for the two directions. The sources are overwhelmingly likely to be lasers, and similarly the detecting devices will almost certainly be avalanche photodiodes. Repeater separations of the order of 40 to 50 km are forecast for the early systems, though this might well be extended for later systems as experience is gained and design techniques improve. These systems will of course be digital, with bit streams of the order of 280 Mbits/s per fiber pair at first, though a doubling of this may be feasible in follow-on systems.

Fiber optics will be applied to submarine systems several years later than their first use in terrestrial systems, because submarine systems have design criteria which are far more difficult to meet, not least of these being reliability. Consequently, it is likely that the lasers in all the submerged repeaters (or *regenerators*, to give them their correct name) will be duplicated or even triplicated, with automatic switchover in case of failure. However, the problems are being overcome, and the financial benefits of fiber optics will make their contributions to the world's submarine communications. It is worth noting that the cost of a fiber-optic cable is likely to be less than that of a copper cable of the same capacity. Additionally, the use of *digital speech interpolation* (see [24] and [25]) is likely to multiply voice circuit capacities by a factor of 5:1, whereas the equivalent multiplication factor in analog systems is only about 2.2:1, so that the initial cost of a *derived* digital submarine circuit is expected to be a good deal less than an equivalent analog cable circuit.

As an overview of current achievements and likely developments in submerged fiber-optic cables, [26] is recommended.

15-3.2 Satellite Communications

A communications satellite is essentially a microwave link repeater. It receives the energy beamed up at it by an earth station and amplifies and returns it to earth at a frequency of about 2 gigahertz away; this prevents interference between the uplink and the downlink. Communications satellites appear to hover over given spots above the equator. This does not make them stationary, but rather *geostationary*. That is to say, they have the same angular velocity as the Earth (i.e., one complete cycle per 24 hours), and so they appear to be stationed over one spot on the globe. Celestial mechanics shows that a satellite orbiting the Earth will do so at a velocity that depends on its distance from the Earth, and on whether the satellite is in a circular or an elliptical orbit. For example, a satellite in a low circular orbit, as was *Sputnik 1*, will orbit the Earth in 90 minutes. The moon, which is nearly 385,000 km away, orbits in 28 days. A satellite in circular orbit 35,800 km away from the Earth will complete a revolution in 24 hours, as does the Earth below it, and this is why it *appears* stationary.[5]

Whether to use a stationary satellite or a succession of satellites in low, elliptical orbits for global communications is a question that exercised the minds of communications engineers in the early 1960s. It was really a case of convenience versus distance, and convenience won. That is to say, satellites in close elliptical orbits require relatively low transmitting powers and receiver sensitivities but must be tracked by the antennas of the ground stations. Stationary satellites present no tracking problems but are so far away that large antennas, high powers and high receiver sensitivities are essential. With the sole exception of the USSR's *Molniya* satellite system, all other communications satellites use the synchronous orbits which all but eliminate satellite tracking.

The major communications satellite systems include those operated by INTELSAT, whose satellites are used for global point-to-point communications; INMARSAT, which serves a similar role for ships at sea; and finally the various regional and domestic satellite systems being operated in a number of regions or by individual countries. Fig. 15-11 shows the geostationary satellites in orbit or planned in late 1982, while [47] gives performance details for all satellites in orbit in mid-1984. Each group will now be discussed in turn.

INTELSAT satellites COMSAT (Communications Satellite Corporation) of the United States, the Overseas Telecommunications Commission (Australia) and nine other world communications agencies met in Washington, D.C., in 1964, to sign a document that made them founder members of the International Telecommunications Satellite Consortium (i.e., INTELSAT). When *INTELSAT I*, better known as *Early Bird*, was launched over the Atlantic in 1965, there were just five earth stations to make use of the 66 telephone circuits it offered. Today, there are over one dozen *INTELSAT IV, IV-A, V* and *VA* satellites in the Atlantic, Indian and Pacific Ocean regions, offering capacities up to 12,500 two-way telephone circuits and two one-way TV channels per satellite. The *INTELSAT VI* satellites, expected to be launched in the late 1980s, will be capable of providing up to 20,000 telephone circuits each. Over 500 earth stations in nearly 150 countries make use of the *INTELSAT* satellites in the three

[5] The actual *orbital* velocity of a geostationary satellite is 11,000 km/per hour, or nearly 2 mi per second.

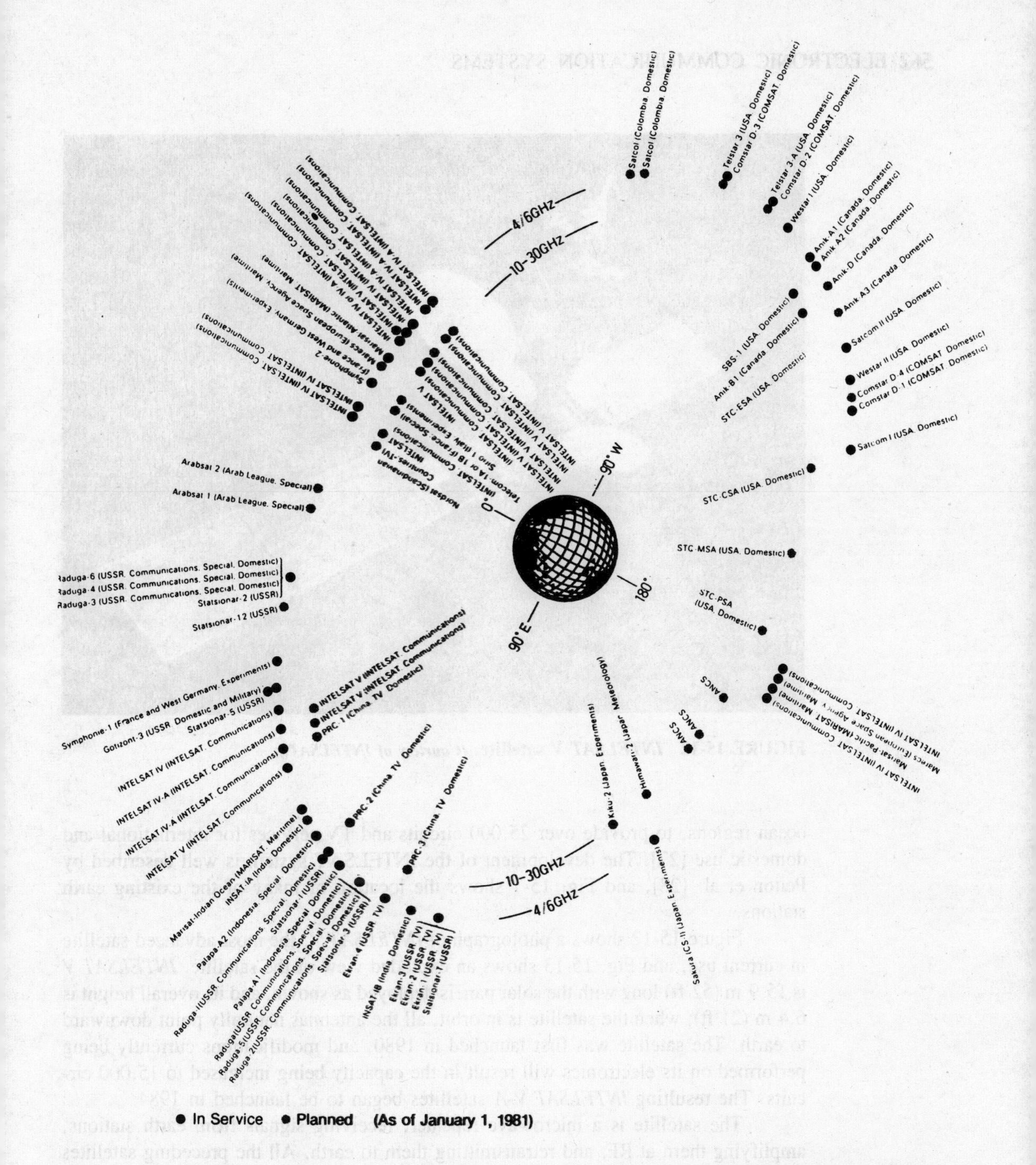

FIGURE 15-11 Satellites in geostationary orbit. Communications: International communications satellite; Experiments: Experimental satellite; Maritime: Maritime communications satellite; Domestic: Domestic communications satellite; Meteorology: Meteorological observation satellite; Special: Satellite for special regions; TV: Direct TV broadcast. ***(Courtesy of Kokusai Denshin Denwa Ltd. (KDD), Tokyo.)***

FIGURE 15-12 ***INTELSAT V* satellite.** *(Courtesy of INTELSAT.)*

ocean regions, to provide over 25,000 circuits and TV services for international and domestic use [27]. The development of the INTELSAT system is well described by Pelton et al. [28], and Fig. 15-7 shows the location of many of the existing earth stations.

Figure 15-12 shows a photograph of *INTELSAT V,* the most advanced satellite in current use, and Fig. 15-13 shows an exploded view of the satellite. *INTELSAT V* is 15.9 m (52 ft) long with the solar panels deployed as shown, and its overall height is 6.4 m (21 ft); when the satellite is in orbit, all the antennas naturally point downward to earth. The satellite was first launched in 1980, and modifications currently being performed on its electronics will result in the capacity being increased to 15,000 circuits. The resulting *INTELSAT V-A* satellites began to be launched in 1984.

The satellite is a microwave repeater, receiving signals from earth stations, amplifying them at RF, and retransmitting them to earth. All the preceding satellites utilized the 5.925- to 6.425-GHz frequency range for the uplink and the 3.7- to 4.2-GHz range for the downlink. *INTELSAT V* does this also, but additionally uses the 4.0- to 14.5-GHz range for a second uplink and the ranges 10.95- to 11.20-GHz and 11.45- to 11.70-GHz for the corresponding downlink. The use of the 14/11-GHz range significantly increases the available system capacity.

An *INTELSAT V* satellite has 11 low-noise 6-GHz receivers, consisting of a four-stage silicon bipolar transistor amplifier and a low-noise mixer. Five of these

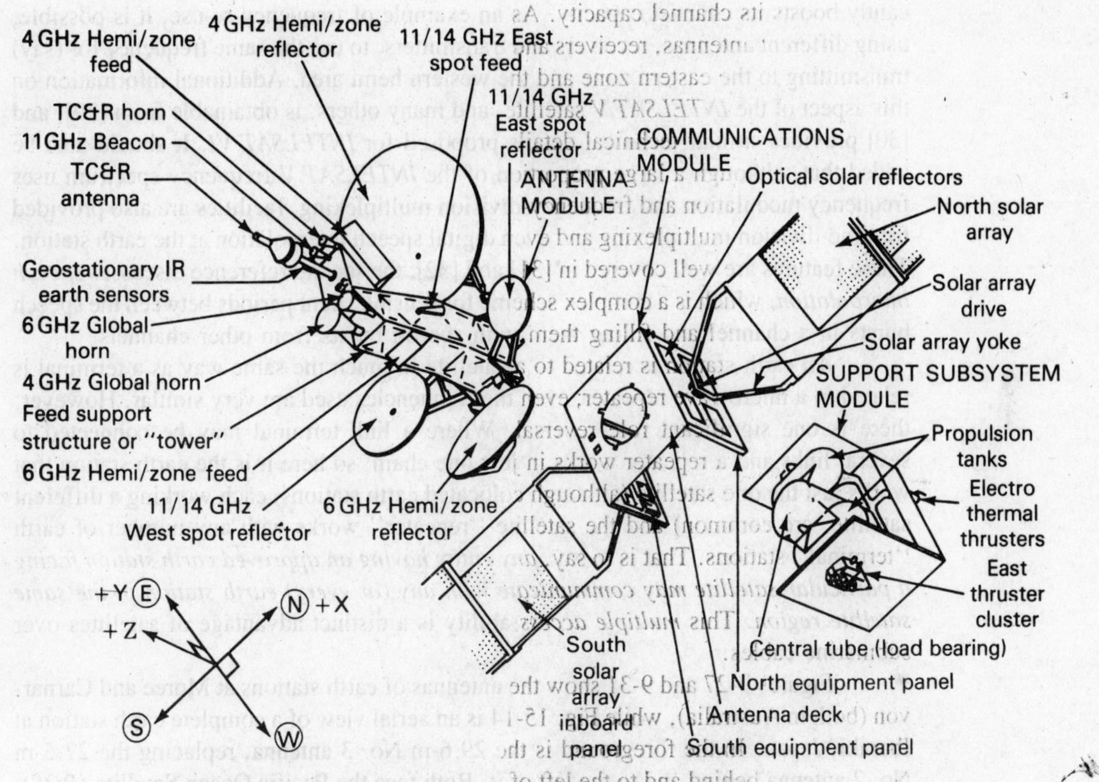

FIGURE 15-13 Exploded view of *INTELSAT V* satellite. ***(Courtesy of INTELSAT.)***

receivers are operational at any given time, with the remainder on standby. The output of each operational receiver, at 4 GHz, is fed to another four-stage bipolar transistor amplifier, and then to traveling-wave tube, whose output of 4.5 to 8.5 W (depending on application—see below) is fed to one of the antennas for retransmission to earth. Much the same arrangement is used at 14/11 GHz, except that this time there are four receivers. The front end in each case consists of a germanium tunnel-diode amplifier, followed by a Schottky-diode mixer, and a five-stage 11-GHz bipolar transistor amplifier feeding a TWT.

With its multiple receivers and antennas, the *INTELSAT V* satellite employs a complex operational pattern of hemispherical, zone and spot beams. For example, in the Indian Ocean Region (IOR), the western hemi beam covers Europe and most of Africa and the Middle East, and its eastern counterpart covers Asia east of Pakistan, and a large portion of Australia—the whole IOR is also covered by a global beam. In the Atlantic Ocean Region (AOR), the western zone is the east coast of Canada, the United States, Mexico and the Caribbean, while the eastern zone consists of Western Europe, North Africa and the Middle East. Finally, the IOR western spot covers a portion of Western Europe, and the eastern spot covers Japan and some surrounding areas. This beam arrangement permits *frequency reuse* with *INTELSAT V* and significantly boosts its channel capacity. As an example of frequency re-use, it is possible, using different antennas, receivers and transmitters, to use the same frequency for (say) transmitting to the eastern zone and the western hemi area. Additional information on this aspect of the *INTELSAT V* satellite, and many others, is obtainable from [29], and [30] provides similar technical details proposed for *INTELSAT VI*. It should also be added that, although a large proportion of the *INTELSAT V* frequency spectrum uses frequency modulation and frequency-division multiplexing, facilities are also provided for time-division multiplexing and even digital speech interpolation at the earth station. These features are well covered in [31] and [32]; the second reference discusses *speech interpolation,* which is a complex scheme for sensing silent periods between the speech bursts in a channel and filling them with speech bursts from other channels.

An earth station is related to a satellite in much the same way as a terminal is related to a microwave repeater; even the frequencies used are very similar. However, there is one significant role reversal. Where a link terminal may be connected to several links and a repeater works in just one chain, so here it is the earth station that works just the one satellite (although colocated earth stations, each working a different satellite, are common) and the satellite "repeater" works with any number of earth "terminal" stations. That is to say, *any entity having an approved earth station facing a particular satellite may communicate with any (or every) earth station in the same satellite region.* This *multiple access* ability is a distinct advantage of satellites over submarine cables.

Figures 9-27 and 9-31 show the antennas of earth stations at Moree and Carnarvon (both in Australia), while Fig. 15-14 is an aerial view of a complete earth station at Ibaraki, Japan. In the foreground is the 29.6-m No. 3 antenna, replacing the 27.5-m No. 2 antenna behind and to the left of it. Both face the Pacific Ocean Satellite *(POS);* the Indian Ocean Satellite *(IOS)* earth station is elsewhere in Japan, at Yamaguchi. The farthest antenna, pointing upward, is the original No. 1 22-m-diameter dish. It is now

FIGURE 15-14 Ibaraki POS satellite earth station. ***(Courtesy of Kokusai Denshin Denwa Ltd. (KDD), Tokyo.)***

used for training, while No. 3 is the main antenna, and No. 2 is the standby when maintenance on No. 3 is required. Behind the main antenna is the central building, housing broadband equipment, power and other plant; behind it is the first repeater of the microwave link to Tokyo. To the left of the main antenna, and just outside the picture, is a 7-m antenna and research building for 15- to 30-GHz investigations. This is a major earth station, and the only other feature such a station might have is an antenna for *TTC&M* (Telemetry, Tracking, Command and Monitoring used in launching, testing and monitoring satellites), such as may be found at Carnarvon (Fig. 9-31), or Paumalu, Hawaii. Indeed, two additional antennas have been provided at Ibaraki since this photograph was taken, one each for maritime communications and maritime TTC&M. Aside from that, some stations for various reasons may be colocated. For example, multiple stations at Hong Kong work with the *POS* and *IOS*, and Goonhilly on the west coast of England works with the *AOS* and *IOS*.

Earth stations must be acceptable to INTELSAT before being allowed to work a given satellite and must undergo exhaustive tests prior to commercial operation. Standard A stations have antenna diameters in the range of 27.5 to 30 m and are nowadays invariably parabolic reflectors with Cassegrain feeds, as shown in Figs. 9-27 and 15-14. They need be steerable only to the extent of being able to follow, automatically, the 20-km figure eight performed daily by the satellite (for complex reasons the satellite is not *quite* geostationary, but a 20-km movement at a distance of 36,000 km is not very significant). However, most antennas are capable of considerably greater motion than that. This applies particularly to antennas in tropical regions, which must be capable of stowage vertically upward when cyclone winds exceed predetermined velocities. Also, they must be made with minimum distortions, both in still air and in high winds. For example, the Goonhilly *AOS* antenna is designed so that its maximum deviation from a true paraboloidal shape does not exceed 5 mm at any point on the dish in a 120-km/h wind. Standard B antennas have diameters of 11 m. The same restrictions apply to them as to standard A stations. In addition, however, they are restricted in other respects, since they place a greater requirement for gain and power from the satellite. They are generally in use at locations where communications requirements are relatively slight, as, for example, in Gibraltar, Mauritius or American Samoa. They can also be portable (actually, *transportable*) and thus useful for emergencies.

Standard C earth stations are designed to operate at the new 14/11-GHz frequency range and have antenna diameters between 14 and 19 m. INTELSAT has also authorized the use of a number of nonstandard earth stations for special purposes such as domestic leases.

The maximum power output of a standard A earth station is up to 8 kW over the total band allocated to satellite communications. However, that would be only if the station transmitted over the complete spectrum of a satellite. In practice, each station is allocated a portion of the total bandwidth for its transmission, in proportion to its requirements and overall availability. It may typically transmit a number of 132-, 252- or 972-channel carriers, together with special TDMA and TV carriers, and so the transmitted power is a good deal less than the 8-kW possible maximum. The station high-power amplifier (HPA), of which a standard A station will have at least two, is generally a water-cooled traveling-wave tube of multicavity klystron, with a saturated maximum output power of about 3 kW. This is often driven by a lower-power TWT, and all the preceding amplifiers are solid-state.

The station receivers are superheterodyne, with low-noise parametric pre-amplifiers known as low-noise amplifiers (LNAs). The LNA is located close to the waveguide in the center of the antenna and is as a rule a multistage traveling-wave amplifier. In older earth stations, the paramp will be cryogenically cooled to a temperature of about 4 K, with a reflex klystron or varactor chain pump. Its output is likely to be fed to a tunnel-diode amplifier, and then perhaps a low-noise TWT amplifier. In newer stations, the paramp will be thermoelectrically cooled to about 230 K (−43°C), and its output will be fed to a multistage FET amplifier; the pump for the paramp is likely to be a transistor oscillator with crystal frequency stabilization (see Chap. 12 for descriptions of the various solid-state devices).

The foregoing amplifiers produce an overall gain of about 60 to 70 dB and are all located close to the antenna receiving point. The signal is fed to the main station below via waveguide. After still further amplification, the signal goes to a power divider and a series of filters. At this point, it must be recalled that, whereas a station must be capable of receiving signals anywhere within the 500-MHz passband of the downlink transmission, it does not have to receive all the signals. Rather, it must be capable of receiving only the transmissions corresponding to the carriers which communicate with this particular station.

Just as a station is allocated carriers which it transmits, so a station allocates receive chains for the carriers which it must receive. Thus the output of the above-mentioned power divider is fed to a series of bandpass filters, each of which is of a bandwidth sufficient to pass the wanted carrier. Each filter is followed by a mixer which downconverts the signal from the wanted carrier to an IF of 70 MHz, where the signal is further amplified and then demodulated.

The output of the receive chain is the baseband of that particular carrier, from which the wanted channels (if a so-called multiuser group was transmitted, with different channels to different countries) are extracted. Sometimes the whole group is destined for this particular station, and often a supergroup or more. Either way, the signals are suitably assembled into supergroups for sending via the terrestrial broadband link to the international terminal in the appropriate gateway city.

Most of the critical gear on a station is duplicated. Also, a number of other transmitting, receiving and monitoring functions are performed at an earth station. These are outside the scope of this book, but students are referred to Miya [33] for such additional information.

A comparison of the properties and advantages of submarine cables and satellite communications reveals that, while each has its own advantages, the two systems are essentially complementary. For example, satellites may be accessed by any earth station within a given region, whereas cables are of primary use only to the areas between which they are connected. This is an oversimplification but holds true in general. Again, all intercontinental television (in practice, thousands of hours per month) goes via satellite, although the advent of fiber-optic cables could change this.

Reliability is similar, in that the high reliability of satellites is marred somewhat by station outages for causes such as cyclones and maintenance or failure of terrestrial links. Conversely, cables are more prone to damage, while cable stations have an excellent record. Finally, the shorter propagation times (typically 20 to 150 ms) on cables, as compared with 300 ms via satellite, form a significant advantage for cables. Some people find it difficult to adjust, in an international telephone call, to the fact that a total of 600 ms will elapse from the time they have finished speaking on a satellite circuit, to the time when the reply begins to be heard.

The reason for the delay is of course the distance involved, a round trip of 72,000 km. Thus tandem satellite hops are avoided, where possible, for interregional calls. For example, New Zealand and Great Britain do not face a common satellite. Thus a double-satellite hop could be involved in their mutual telephone circuits. This is avoided by having these circuits go from Auckland to Sydney via the *Tasman* cable (propagation time 14 ms), and then to London via the Australian and British *IOS* earth stations, Ceduna and Madley.

Current economic forecasts indicate that fiber optic submarine cables are likely to provide cheaper circuits than satellites during the mid-1990s, for all but the longest distances. If this eventuates, we can expect a significant rebalancing of utilization in favor of cables.

INMARSAT satellites Until 1976, all communications with ships at sea went via HF radio. While this is still used a lot for maritime communications, 1976 saw the inauguration of ship-to-shore and shore-to-ship communications via a dedicated geostationary satellite system, providing high-quality telephony, data and telex/telegraphy circuits. This was the MARISAT system, operated by COMSAT and initially intended for use by the U.S. Navy, but with some capacity for commercial use. There were eventually three MARISAT satellites, one in each ocean region, operating at 1.5/1.6 GHz for the uplink and 6/4 GHz for the downlink. A MARISAT satellite is shown in Fig. 15-15.

There were initially three MARISAT earth stations, one for each ocean region: Southbury, Connecticut (Atlantic), Santa Paula, California (Pacific), and Ibaraki, Japan (Indian). A ship wishing to make a call would dial the operator at the appropriate earth station via its shipboard terminal, if the relatively few MARISAT channels in its region were free, and the operator would complete the call to its destination, anywhere

FIGURE 15-15 The *MARISAT* satellite. ***(Courtesy of COMSAT General Corporation.)***

in the world. A call in the reverse direction was completed similarly. By early 1981, over 500 ships of the world's merchant fleet were equipped for MARISAT communications, and congestion was being felt.

Around the time when INTELSAT was formed, the Intergovernmental Maritime Consultative Organization (IMCO), commissioned a group of experts to consider the introduction of satellite communications to the maritime sphere, with the aim of improving communications with ships, particularly for safety and distress purposes. The panel of experts completed its deliberations and made its recommendations just as the MARISAT system was introduced. The recommendation was for the establishment of a maritime satellite organization akin to INTELSAT, and so in July 1979, the International Maritime Satellite Organization (INMARSAT) was born, very much along the INTELSAT lines, with COMSAT (on behalf of the United States) once again the largest shareholder. The formation and operation of INMARSAT are well described in [34] and [35].

Over 20 INMARSAT earth stations are now in service, in a majority of the developed nations. The space segment consists of capacity leased from MARISAT, additional capacity leased from the European Space Agency in two of their *Marecs* satellites, and finally more capacity leased from INTELSAT, in the three *INTELSAT V* satellites equipped with maritime communications subsystems (MCS). The shore stations have antennas with diameters of the order of 13 m, and the shipboard antennas are 1.2 m in diameter and generally contained in radomes. The method of operation is very much along INTELSAT lines and is well described in [35], while [36] gives details of a specific shore station at Goonhilly.

Regional and domestic satellites As the name suggests, a regional satellite system is a kind of mini-INTELSAT designed to serve a region with community interests, especially in communications. The world's first regional satellite system was the Indonesian *Palapa* network, inaugurated in the mid-1970s, initially for domestic services (Indonesia consists of over 3000 islands, with some 1800 of them inhabited), but by the late 1970s it had expanded to neighboring countries such as the Philippines. The Conference of European Post and Telegraph Administrations (CEPT) was next on the scene, with EUTELSAT created in the early 1980s, under the auspices of the European Space Agency (ESA), whose other main function is the development and operation of the *Ariane* satellite launcher (used by a number of organizations, including INTELSAT). EUTELSAT provides and maintains the space segment for the European Communication Satellite (ECS), and individual countries provide their own earth stations, as with INTELSAT.

The ECS system came into service in 1983, operating in the 14/12-GHz band, with ground antennas very much like the INTELSAT standard C antennas, but with lower ground and satellite transmit powers, for reasons which are outlined below. The system is used for intra-European telephone, data and telex/telegraph services, and also by the European Broadcasting Union, for the distribution of its *EUROVISION* programs. European satellite communications are well described in [37].

The next regional satellite network to go into service is likely to be the ARABSAT system in the Middle East, but some problems need to be ironed out before it goes on air.

There is conceptually not a great deal of difference between a regional satellite system used by a group of neighboring countries and a domestic system used by a large or dispersed country. Indeed, they share a common characteristic which makes them quite different from the global INTELSAT system, in requiring a much smaller coverage area. Each INTELSAT satellite must have a beam accessible to roughly one-third of the globe, resulting in a coverage of almost exactly 170 million km^2. On the other hand, a circular beam could cover the whole of India, for example, if it had a radius on the ground of 1450 km. The resulting 6.6-million-km^2-coverage area represents a 26-fold reduction when compared with the global beam. All else being equal, it means that the satellite antenna gain can, in this case, be increased by a factor of 26. The result is a very significant gain increase compared with the global system, and consequently much smaller receiving antennas and simpler receivers on the ground.

Although the conceptual difference between a regional and a domestic satellite system is not great, the *political* difference is enormous! No international conferences are needed; there are no language barriers, no requirements to correlate different national technical standards (making the usual compromises), no necessity to make allowances for the least developed entity in the group, and so on (students will gather from all this that the author speaks from long personal experience!). Moreover, in all the world's countries except one (the United States) there is just one satellite organization, normally government-owned, so that even domestic friction is avoided. It should come as no suprise, therefore, that domestic satellite systems preceded regional ones by several years and, as might be expected, North America led the field.

Telesat Canada was established in 1969 and in January 1973 inaugurated the Canadian domestic satellite system, using *ANIK A1* satellites for the space segment. The United States followed soon afterward, with the launching of the *Westar* system in 1974, and then the competing *Comstar, Satcom, SBS, STC* and *Telstar* networks. [38] outlines the satellite services available in the United States; the orbital locations of the various North American and other domestic satellites are shown in Fig. 15-11, while Fig. 15-16 is a photograph of a Comstar satellite. The Comstar series is jointly owned by AT&T and GTE and operated on their behalf by COMSAT.

Many other countries now have domestic satellite systems using their own satellites, notably the U.S.S.R., China, Indonesia, India, the Scandinavian countries and Colombia; Australia's domestic system's inauguration date is 1985. In addition, nearly 20 countries operate domestic services by means of leasing spacecraft capacity from INTELSAT, among them Algeria, Australia, Brazil, Nigeria and Saudi Arabia. Full details of Australia's "domsat" system are given in [39].

Domestic satellite systems generally use the same frequency ranges as INTELSAT satellites, viz., 6/4 and 14/12 GHz, with similar parameters. In the earth segment, there are usually two sets of earth stations: ones with 5- to 15-m diameters, owned and operated by the provider of the satellite system, and simpler stations with smaller antennas, owned and operated by customers. The resulting network provides point-to-point telephone, data and other services, in a fashion complementary to terrestrial services. Additionally, radio and TV broadcasting are available, by means of a signal originated at a major station and rebroadcast by the satellite to a large number of fairly small and simple, receive-only stations located throughout a country. The rest of the system then works in the same way as community antenna TV, with receivers

FIGURE 15-16 ***Comstar* satellite during manufacture.** ***(By permission of AT&T Long Lines.)***

connected by cable to the receiving station. It is also possible for individual receivers to have their own satellite antennas and downconverters, as is done in the Australian outback and elsewhere.

It can be seen that a parallel exists between domestic and international services, in that each can be achieved by means of competing and yet complementary terrestrial and satellite systems. In each case the terrestrial systems came first, to be followed by mushrooming[6] satellite systems which provided many additional services, as well as access to remote communities. Finally, in each case the terrestrial systems have "hit back" with fiber-optic technology, and the competition remains intense while facilities available to the customer expand and improve—this is clearly a very healthy situation.

15-4 ELEMENTS OF LONG-DISTANCE TELEPHONY

It has been possible since World War I to make a continental telephone call (via an open-wire system) or an intercontinental one (via HF radio). However, long-distance telephony did not take off until after World War II, when it became possible to dial

[6]The second edition of this book included the words ". . . we have not yet progressed to the point where every farm, suburban house, office and apartment block has its own earth station and antenna. However, the author is confident he will be describing such systems in the third edition! (And perhaps personal earth stations with head-mounted antennas in the edition after that!)" Well, a lot of farms, offices and apartment blocks now do have earth stations; perhaps, at the turn of the century, personal ones will be surgically implanted at birth?

such calls without having to go through every operator en route. Some aspects of long-distance telephony will now be treated.

15-4.1 Routing Codes and Signaling Systems

When dialing a subscriber in another part of the world, it is essential to identify the wanted telephone number uniquely, so that the international telephone network selects that number and no other. It simply would not do if a subscriber dialed the number (say) 2345678 in New York from Boston and got the number 2345678 in Antwerp, Belgium, instead. Thus each country (or continent, in the case of North America) has a numbering scheme with unique area codes. For example, the area code for New York is 212, that for Montreal, Canada, is 514, and so on. Again, countries must also have their unique codes, and these have been allocated in the CCITT World Plan [40]. For example, North America has the country code 1, Australia 61 and Israel 972. Thus, a subscriber in Sydney dialing a counterpart in New York would dial:

0011[7]	1	212	921	ABCD
access digits	country code	NPA number	central office code	sub's code

And, needless to say, the number is dialed smoothly and continuously, such as: 00111212921ABCD. The access digits are to tell the outgoing national network that this will be an international call, and the country code states where the call is going. The rest of the number is the same as would be dialed by a subscriber in North America residing outside the New York local zone.

In order for the wanted subscribers in the call described above to be interconnected, signaling systems must exist to send on the appropriate digits, ensuring that correct routing is achieved. A number of signaling systems are in use around the world. The most common ones for national signaling are the decadic and *multifrequency coding (MFC),* while for international signaling CCITT No. 5 and No. 6 are internationally agreed [41]. In the *decadic* system, which is on the way out in most countries, dc pulses are sent on the signaling circuit connected to the telephone, with a number of pulses equal to the digit dialed. In *MFC,* combinations of two tones out of 700, 900, 1100, 1300, 1500 and 1700 Hz are used to define each digit, and such supervisory signals as subscriber busy or no circuits available. The signaling system is *compelled,* in that the receiving office acknowledges each digit sent. Such a system is not practicable for international dialing because of the propagation delays mentioned previously, which would tie up signaling and common equipment of telephone exchanges far too long. Thus the CCITT No. 5 system is used instead. This is also an MFC system, but here only the control signals are compelled, not the actual digits sent.

[7] 0011 is the digit sequence that an Australian subscriber must dial when making an international telephone call; different access digits are required in other countries, often consisting of fewer numbers.

All the systems so far described use the actual telephone circuits for the signaling functions, before and after the telephone call. CCITT No. 6 is the first international signaling system which uses common-channel signaling. Here, signaling circuits are established between the computers controlling each pair of interworking telephone exchanges. These common channels are used exclusively for signaling, and telephone circuits themselves are used only for voice (or data). The international use of CCITT No. 6 was pioneered by the United States, Australia and Japan, during the late 1970s; CCITT is currently evolving a new common-channel signaling system, No. 7. Finally, it should be noted that the foregoing remarks generally apply also to telex, although the signaling systems themselves are somewhat different from those used for telephony.

15-4.2 Telephone Exchanges (Switches) and Routing

The function of a telephone exchange (switch) is to interconnect four-wire lines, so as to permit a call to be established correctly. If both the calling and the called subscriber are connected to the same exchange, it merely has to interconnect them. If the wanted subscriber is connected to some other exchange, the call from calling subscriber must be routed correctly, so that it will reach the wanted number.

There have been basically three generations of exchanges. The first was the step-by-step, or Strowger type, which had an incredible number of relays that made interconnections step by step, i.e., after each digit was received. The second generation was the *crossbar* exchange, which had even more relays but miniaturized and arranged so that up to 20 connections were made simultaneously by the crossbar switch, after all the digits were received. The *processor-controlled* exchange represents the third generation. Here, all the interconnections are made by the exchange processor or computer, and as a result the space occupied is very much smaller. It is worth pointing out that a telephone (or telex) exchange is an incredibly complex piece of equipment, and a 2000-line crossbar exchange may occupy the whole floor of a rather large building. In countries such as the United States and Australia, there are very few Strowger exchanges left, processor-controlled exchange capacities have outstripped those of crossbar exchanges, and most of the latest exchanges are digital.

If the originating and wanted subscribers are not connected to the same exchange, the originating exchange must participate in the correct routing of the call. This is done by analyzing the called number and examining the paths available through and outside the exchange to route the call. The local exchange must establish the group of first-choice trunks to which the call is routed, and which of these is free. If all are occupied, the call is routed to the second-choice trunks, and so on. If no trunks are available, the appropriate signal must be sent to the calling subscriber, in this case perhaps a ''plant engaged'' tone. The same process is performed in each exchange in the hierarchy of exchanges, which is essentially local office—toll center—primary or regional center—international center, and then the same chain in reverse.

As an example, let us examine the routing that may be taken by a call from the small town of Daylesford in Victoria (Australia) to New York. The call will be routed to the toll office in Ballarat, directly or via some intermediate point, and then to the regional center in Melbourne. From there it is routed via any one of a number of paths to its opposite number in Sydney, whence it is sent to one of the two international exchanges in Sydney. A Denver–Sydney satellite or cable circuit is then selected, and in Denver the call is routed from the international exchange to a regional one, then perhaps to the New York No. 6 regional office, then to a toll center, the correct local office and finally to the wanted subscriber. Had all the Denver–Sydney circuits been busy, the Sydney exchange would have selected a Sacramento–Sydney circuit, and the consequent trunk routing to New York would have been different. It is worth noting that the process just described should not take more than a few seconds.

These, then, are some of the functions of telephone exchanges. Others include self-monitoring, the provision of statistical data on traffic and performance, and even customer charging. These topics are discussed in greater detail in [42] and [43].

15-4.3 Miscellaneous Practical Aspects

International gateways An international gateway is the center at which the international exchange, multiplex equipment and ancillary equipment for international telephony and/or telegraphy, telex, data, television and facsimile are located. There are, for example, six such gateways in London, two in Sydney and Tokyo, and only one in lesser centers; in the United States the gateways are geographically separate, with major intercontinental telephone ones being located at Sacramento, Denver, Pittsburgh, New York City and White Plains, New York. It is here that the various International Maintenance, Switching and Coordination centers are located, and from here new circuits, groups and supergroups are lined up, while existing ones are maintained. Such centers are quite often stations for submarine cables, also. International gateways and their functions are described in [1] in great detail, while [43] describes a particular gateway.

Echo and echo suppressors It was shown in Sec. 7-1 that reflections will take place from an imperfect termination on a transmission line. This of course means that, in a telephone system, any imperfect matching between the speaking subscriber and the distant telephone will result in the reflection, to the earpiece of this subscriber, of an attenuated version of what the speaker is saying. This is known as echo. Unless great round-trip delays are involved, this echo is actually beneficial, since it ensures that the earpiece does not sound ''dead''; sidetone is used for the same purpose. However, in long-distance calls, particularly those involving satellite hops, hearing a loud echo several hundred milliseconds after one has spoken is enervating. It may even be a total impediment to the conversation.

To combat this, international circuits (and long cross-continental ones) are fitted with *echo suppressors*. These devices are connected at each end of the circuit, sense the direction of speech and place of the order of 50 dB of attenuation in the listening leg, thus ensuring that echo is thoroughly attenuated. If both parties speak at once, 6 dB of attenuation is placed in each direction. Although wanted speech is thus attenuated by 6 dB, the unwanted echo is attenuated by 12 dB, and so its nuisance value is somewhat reduced.

Echo cancelers are becoming available. These are complex electronic devices which analyze the outgoing speech and the incoming echo and try to cancel the echo by feeding into the circuit a suitably diminished signal from the speaking end, 180° out of phase with the received echo. Their advantage over echo suppressors is that they function as well when both ends are speaking, unlike the suppressors. Such devices are described in [44].

15-4.4 Introduction to Traffic Engineering

Traffic engineering is a most fascinating and complex topic, just as applicable to telephone traffic as to any other kind of traffic. It is related to measuring such traffic and its fluctuations and growth, as well as optimum traffic routing arrangements. It will be briefly introduced here.

Measurement of traffic To find out how many circuits are needed on a given route, it is first necessary to know how much traffic there is. To do that, one must be able to measure traffic. The unit of measurement is the *erlang*, which is a dimensionless quantity (actually, it is minutes per minute). Suppose that four telephone circuits exist between a pair of places, and it is found that, in a particular half-hour period, the circuits carried respectively 25, 15, 5 and 24 minutes of traffic. That is to say, each circuit was busy for the period indicated, and so the total occupied time was 25 + 15 + 5 + 24 = 69 minutes. The average occupancy during the half-hour was thus 69/30 = 2.3 erlangs. Needless to say, the traffic may have fluctuated during this period. At instants when all four circuits were busy, the carried traffic was 4 erlangs, and there may have also been instants of no occupancy at all, i.e., 0 erlangs.

Grade of service The expression "carried traffic" was carefully used above. This is not the same as *offered* traffic. For example, 20 erlangs may be offered to 10 circuits, in which case a lot of the offered traffic will fail to secure a circuit, and *congestion* will result. It is possible to calculate statistically the degree of congestion, or *grade of service,* as it is known, given the amount of traffic in erlangs and the number of circuits and their arrangement. However, it is a lot easier to look up the information in erlang tables [45]. Such tables are used to calculate the grade of service for a particular number of erlangs on a given group of circuits, or to calculate the number of circuits required for a particular traffic level and design grade of service. It should be noted that to provide enough circuits to ensure zero grade of service is virtually impossible, prohibitively expensive and unnecessary. It would be rather like providing an eight-lane highway between two small towns, because of the small but finite probability that all four lanes in one direction might one day have parallel cars in them, and a fifth vehicle will want to pass them. The internationally accepted worst grades of service are 3 percent if a route carries no subscriber-dialed traffic, and 1 percent otherwise. That is to say, on the 10 busiest days of the year (not counting special occasions such as Christmas, or catastrophes) the grade of service may approach, but should not exceed, the design figure. Wilkinson [46] is a highly recommended reference for traffic engineering.

PROBLEMS

For self-testing questions on this chapter, see p. 708.

QUESTIONS

15-1. What is *multiplexing*? Why is it needed? What are the two basic forms of multiplexing?

15-2. Show, diagrammatically and with an explanation, how channels are combined into *groups,* and groups into *supergroups,* and so on, when FDM is generated in a practical system.

15-3. What are the major advantages of the piecemeal method of generating FDM, as in Question 15-2, compared with a method of directly translating each channel, in one step, into its final position in the baseband?

15-4. Explain the principles of time-division multiplex, with a sketch to show how the interleaving of channels takes place.

15-5. Show how first-order TDM signals may be generated and then demultiplexed in the receiver.

15-6. Explain briefly how higher-order TDM multiplexing is achieved. Draw up a table comparing the channel capacities of the first four orders of TDM and FDM.

15-7. Describe a typical terrestrial coaxial cable system. Why are separate cables in the one tube used for the two directions of transmission?

15-8. Sketch the supergroup distribution spectrum of a coaxial cable carrying 900 circuits.

15-9. What are the typical operating frequencies, bandwidths and repeater gains and spacings in a coaxial cable system?

15-10. Sketch an attenuation-versus-wavelength diagram for optical fibers, briefly explaining the factors governing its appearance; label the "windows."

15-11. Briefly describe optical fibers and the factors governing losses in fibers.

15-12. What are the advantages of optical fibers over coaxial cables? Why do most existing systems operate at a wavelength of 0.85 μm, whereas all new systems operate at 1.3 μm? Why is the 1.55-μm wavelength not used?

15-13. Explain in detail why changing down to an intermediate frequency takes place in a microwave link repeater. What part does the link play in the modulation process?

15-14. Draw the block diagram of a microwave link repeater, indicating the function of each block.

15-15. What is the purpose of the circulator found in a microwave link repeater?

15-16. A microwave link repeater has a number of bandpass filters. Describe the function of each one.

15-17. What is the difference between coaxial cable and microwave link repeaters from the point of view of supplying the necessary dc power?

15-18. Compare and contrast the performance and advantages of coaxial cable and microwave links as broadband "continental" transmission media. Explain why microwave links tend to be preferred for long-distance television transmissions. Is it a question of capacity, i.e., bandwidth?

15-19. Where and why are troposcatter links used in preference to the other two medium-distance broadband transmission media?

15-20. Draw a very basic block diagram of a tropospheric scatter link, showing the interconnections required to provide *quadruple diversity*.

15-21. With the aid of outside references as required, draw up a tabular history of submarine cables since 1956, stressing cable capacities, bandwidths, repeater types and spacings.

15-22. Describe the method of laying a submarine cable. What are the respective functions of lightweight and armored cables?

15-23. Compare the salient operating methods of submarine cables with those of land-based coaxial cables. What are the reasons for some of the differences?

15-24. With reliability being so important for submarine cables, describe some of the methods used to achieve it, during both manufacture and laying.

15-25. Discuss the major practical aspects of fiber-optic submarine cables, especially the advantages they might have over conventional copper cables.

15-26. Explain what is meant by saying that a satellite is "stationary." Why are such satellites used for worldwide communications, in preference to any other kind?
15-27. How do the functions of a communications satellite compare with those of a microwave link repeater? What is the most significant difference in their functions.?
15-28. Describe the appearance and layout of a typical INTELSAT earth station. What are colocated stations? Check Fig. 15-7 to note where in the world colocated stations are possible.
15-29. What are the "carriers" allocated to a particular earth station? Correspondingly, what are the functions of receive chains? Sketch the block diagram of a receive chain, from the power divider to the terrestrial multiplex equipment.
15-30. Describe some of the circuits likely to be found aboard an INTELSAT satellite.
15-31. What devices and circuits are likely to be used as the HPAs and LNAs of a satellite earth station?
15-32. How do the three major types of INTELSAT satellite earth stations differ from each other, in general appearance and applications?
15-33. Describe the maritime satellite facilities currently available, stressing the INMARSAT organization.
15-34. Under what circumstances are regional or domestic satellite systems likely to be used? In what ways do they differ from worldwide satellite systems? How do their applications compare with those of domestic terrestrial systems?
15-35. Compare the advantages and disadvantages of submarine cables and communications satellites for intercontinental telephony and television. Show how the two media may be complementary.
15-36. What is done to ensure that international telephone (or telex) calls are not misrouted? Explain in some detail.
15-37. With a line sketch showing the appropriate exchange hierarchy, show how a telephone call may be routed from a city in the United States to one in another country, indicating how alternative routings play a part in determining the overall path of the call.
15-38. What is the difference in basic philosophy between an echo *canceler* and a *suppressor?*
15-39. In a given 1-hour period, the five circuits connecting two small towns carry respectively 55, 45, 35, 20 and 10 minutes of traffic. What can you say about the method used by the exchange to select these circuits, and the erlangs carried?
15-40. Relate *offered* traffic and *carried* traffic, and define *grade of service*.

REFERENCES

1. CCITT, *Yellow Book,* vol. III, Fascicle III.2: *International Analogue Carrier Systems; Transmission Media—Characteristics,* ITU, Geneva, 1981.
2. *Ibid.*, Fascicle II.3: *Digital Networks; Transmission Systems and Multiplexing Equipment.*
3. "PCM Update, pt. 1," *GTE Lenkurt Demodulator,* January 1975.
4. Bangert, J. T., "L5—A Jumbojet Coaxial System," *Bell Lab. Rec.*, November 1973, pp. 291–299.
5. Gardner, W. B., "Fundamental Characteristics of Optical Fibres," *(ITU) Telecommun. J.* **48**(11):638–642 (November 1981).
6. *Transmission Cost Comparison for Satellite, Optical Fiber and Microwave Radio Communications,* FSI Report No. 112, Future Systems Incorporated, Gaithersburg, Md., February 1982.
7. Hull, J. A., *Fiber Optic Communications—An Update,* Special Report NTIA-SP-81-14, National Telecommunications and Information Administration, U.S. Department of Commerce, Washington, October 1981.

8. Nakahara, M., "Optical Fibre Fabrication Techniques," *(ITU) Telecommun. J.* **48**(11):643–648, November 1981.

9. Koonen, A. M. J., and A. Wismeijer, "Optical Devices for Wavelength Division Multiplexing Systems," *Philips Telecommun. Rev.* **40**(2):102–110, (July 1982).

10. Schwartz, M. I., "Design and Performance of the FT3 Lightguide Trunk Transmission Medium," *National Conference on Communications, IEEE,* Conference Record, vol.1, 1981, pp. 6.1.1–6.1.8.

11. Horsley, A. W., and E. S. Usher, "Optical Fiber Communication Systems in PTT Networks," *Electr. Commun.* **55**(4):268–275 (1980).

12. Bhatia, S. A. L., et al., "FO Electronic Equipment and Hierarchy," *Telephony,* July 19, 1982, pp. 60–70 & 134–135.

13. Lilly, C. J., "The Application of Optical Fibres in the Trunk Network," *(ITU) Telecommun. J.* **49**(2):109–117 (February 1982).

14. Buijs, N. A., "Philips Optical Fibre Transmission Systems: II. The Equipment Family," *Philips Telecommun. Rev.* **40**(2):71–88, (July 1982).

15. McNulty, J. C., and A. C. Oestrich, "Transmitting Toll Calls—An International Way," *Bell Lab. Rec.*, February 1980, pp. 46–52.

16. Schwartz, W. J., Jr., R. P. Slade and J. J. Kenny, "Radio Repeater Design for 16 QAM," *National Conference on Communications, IEEE,* Conference Record, vol. 1, 1981, pp. 13.5.1–13.5.7.

17. Lunan, A. R., I. Grier and J. C. Y. Huang, "Technical Consideration in the Design of a 2 GHz DS-3 Digital Radio for Overbuilding Existing Analog Routes," in *ibid.*, pp. 13.7.1–13.7.6.

18. Hill, S. J., "British Telecom Transhorizon Radio Services to Offshore Oil/Gas Production Platforms," pt. 1, "Service Requirements and Propagation Considerations," *Br. Telecommun. Eng.*, April 1982, pp. 42–48; pt. 2, "Radio Techniques and Networking Arrangements," in *ibid.*, July 1982, pp. 70–79.

19. *Submarine Telecommunication Systems,* Submarine Systems Division, Standard Telephone and Cables PLC, London, 1982.

20. Harper, D. N., "TAT-7 Takes Shape," *Br. Telecommun. J.*, Summer 1982, pp. 15–18.

21. *Electr. Commun.*, vol. 49, no. 4, 1974. (*whole issue*)

22. Furukawa, K., et al., "Recent Developments in Optical Fiber Submarine Cable Systems," *Konichiwa,* No. 48, Spring 1983, pp. 20–26.

23. Runge, P. K., and P. R. Trischitta, "Light in the Depths of the Sea with FO Undersea Cable System," *Telephony,* August 23, 1982, pp. 32–37.

24. Easton, R. L., et al., "TASI-E Communications System," *National Conference on Communications, IEEE,* Conference Record, vol. 3, 1981, pp. 49.3.1–49.3.5.

25. Rieser, J. H., H. G. Suyderhoud and Y. Yatsuzuka, "Design Considerations for Digital Speech Interpolation," in *ibid.*, pp. 49.4.1–49.4.7.

26. Yamashita, I., et al., "The Application of Optical Fibres in Submarine Cable Systems," *(ITU) Telecommun. J.*, **49**(2):118–124 (February 1982).

27. For the latest (and many additional) details, see current Annual Report, *International Telecommunications Satellite Organization,* INTELSAT Headquarters, Washington D.C.

28. Pelton, J. N., M. Perras and A. Sinha, "INTELSAT—The Global Telecommunications Network," *Pacific Telecommunications Conference Record,* Honolulu, January 1983.

29. Eaton, R. J., and R. J. Kirkby, "The Evolution of the INTELSAT V System and Satellite," pt. 2, "Spacecraft Design," *Post Off. Electr. Eng. J.*, vol. 70, part 2, July 1977.

30. Ghais, A., et al., "Summary of INTELSAT VI Communications Performance Specifications," *COMSAT Tech. Rev.*, vol. 12, no. 2, Fall 1982.

31. Deal, J., "Increased Transponder Efficiency by Using TDMA," *Commun. Int.*, November 1981, pp. 48–50.

32. Terrell, P. M., ''Application of Digital Speech Interpolation,'' *Commun. Int.*, February 1979, pp. 22–30.

33. Miya, K. (ed.), *Satellite Communications Engineering*, Kyowa Book Company, Tokyo, 1975.

34. Nicolaidis, E., ''INMARSAT: A New Global Maritime Satellite Communications System,'' *Satellite Commun.*, July 1982, pp. 44–48.

35. Lundberg, O., ''INMARSAT: The First Year and the Next Decade,'' *(ITU) Telecommun. J.* **50**(9):469–475, (September 1983).

36. Bridgewater, D. C., ''Goonhilly V Goes Live!,'' *Br. Telecommun. J.*, Summer 1982, pp. 22–24.

37. Wooster, C. B., ''A Bird's-Eye View of Developments in European Satellite Communications,'' *Telephony*, August 23, 1982, pp. 38–42 and 75.

38. Shaw, S. J., ''A Shopper's Guide to Business Satellite Services,'' *Satellite Commun.*, October 1981, pp. 39–44.

39. Nowland, W. L., ''AUSSAT: A milestone in Australia's Communication History,'' *Space Communication and Broadcasting* **1**(1):73–89, (April 1983).

40. CCITT, *Yellow Book*, vol. II, Fascicle II.2: *Recommendations E. 160–161 and E. 163*, ITU, Geneva, 1981.

41. CCITT, *Yellow Book*, vol. VI, Fascicle VI.2: *Recommendations Q.140–164;* Fascicle VI.3; *Recommendations Q.251–300*, ITU, Geneva, 1981.

42. Articles in the *Bell Laboratories Record*, January 1974, as follows:

(*a*) Macurdy, W. B., and A. E. Ritchie, ''The Network: Forging Nationwide Telephone Links,'' pp. 4–15.

(*b*) Joel, A. E., ''Switching,'' pp. 30–39.

43. Puccioni, G., ''The Rome Intercontinental Telephone Centre and Its Transit Facilities,'' *(ITU) Telecommun. J.* **41**(6):374–386, (June 1974).

44. Gould, R. G., and G. K. Helder, ''Transmission Delay and Echo Suppression,'' *IEEE Spectrum*, April 1970, pp. 47–54.

45. Palm, C., *Table of the Erlang Loss Formula*, 3d ed., Kungl. Telestyrensen, Stockholm, 1964.

46. Wilkinson, R. I., ''Theories for Toll Traffic Engineering in the U.S.A.,'' *Bell System Tech. J.*, March 1956, pp. 421–514. (*original and classical reference*)

47. Morgan, W. L., and M. Petronchak, ''1984 Satellite Performance Reference Chart,'' *Satellite Commun.*, March 1984, pp. 31–34.

16 RADAR SYSTEMS

Radar is basically a means of gathering information about distant objects, or *targets,* by sending electromagnetic waves at them and analyzing the echoes. It was evolved during the years just before World War II, independently and more or less simultaneously in Great Britain, the United States, Germany and France. At first, it was used as an all-weather method of detecting approaching aircraft, and later for many other purposes. The word itself is an acronym, coined in 1942 by the U.S. Navy, from the words *radio detection* and *ranging*.

It was radar that gave birth to microwave technology, as early workers quickly found that the highest frequencies gave the most accurate results. Since the majority of components which it uses have been described in preceding chapters, radar will be treated here mainly from the point of view of general methods and systems.

The chapter begins with a basic description and then a historical introduction, followed by a discussion of fundamentals and performance factors. The basic version of the radar range equation is derived at this point. Pulsed systems are then treated, including antenna scanning and the various data display methods. The specific requirements of the several different types of pulsed radars are discussed next, and this is followed by a treatment of more advanced radar concepts, such as *moving-target indication* (MTI) radars and *radar beacons.* The chapter concludes with a description of *CW radars,* which may use the *Doppler effect,* and finally with the relatively recent development of *phased array radar.*

16-1 BASIC PRINCIPLES

In essence, a radar consists of a transmitter and a receiver, each connected to a directional antenna. The transmitter is capable of sending out a large UHF or microwave power through the antenna. The receiver collects as much energy as possible from the echoes reflected in its direction by the target and then treats and displays this information in a suitable way. The receiving antenna is very often the same as the transmitting antenna. This is accomplished through a kind of time-division multiplexing arrangement, since the radio energy is very often sent out in the form of pulses.

16-1.1 Fundamentals

Basic radar system The block diagram of an elementary pulsed radar set is shown in Fig. 16-1. For each transmitted pulse, the cycle of events is as follows.

In response to an internally generated trigger signal, the transmitter generates a short, rectangular pulse. As soon as a small fraction of the pulse power is fed to the duplexer, this device disconnects the receiver from the antenna and connects the transmitter to it, as explained in Sec. 10-5.4. In most radars, though by no means in all, the antenna moves in a predetermined pattern, i.e., it *scans*. Either way, it is normally directional and sends out the generated pulse in the direction in which it is pointing at the time. The scanning speed may be mechanically high, but it is small compared with the time taken by pulses to return from a normal range of targets. Thus, when such echoes are received, the antenna still points in the right direction to collect them.

As soon as the transmitted pulse terminates, the duplexer disconnects the transmitter from the antenna. The duplexer also reconnects the receiver to the antenna, allowing the returning echoes to be correctly processed. The received pulses are amplified and demodulated by the receiver (which is almost invariably superheterodyne, as had been discussed in detail in Chap. 6). The pulses from the returning echoes (and noise, of course) are then fed to the device on which they are to be displayed, as will be described in Sec. 16-2.3. The cycle is complete, and the set is once again ready for the transmission of the next pulse and the succeeding ones, while the antenna scans along its predetermined path.

The radar set is able to show the position of the target, because information about the *azimuth* (horizontal direction) and the *elevation* (vertical direction) of the antenna is available. In addition, the distance to the target may be calculated (automatically) from the total time taken by the pulse on its forward and return journeys. Since the velocity of electromagnetic radiations is very nearly 300,000 km/s, and since the pulse must cover the distance to the target twice, it is easy to calculate that the return distance covered in 1 μs is 300 m. Thus the target must be 150 m away if a pulse is received from it 1 μs after sending. The nautical mile (nmi) (1852 m) is a unit of distance often used in connection with radar distances, and it is seen that the time required for the return of a pulse is close to 12.4 μs/nmi. Finally, the velocity of the target relative to the radar set can be calculated from the positions of successive echoes, or automatically by somewhat different scanning methods in some systems.

The foregoing examples imply that the transmitted pulses are quite short compared to the intervals between successive pulses. Although this subject will be discussed in far more detail in Sec. 16-2.1, it is worth pointing out at this stage that a radar set must be capable of receiving echoes from targets located at the maximum distance

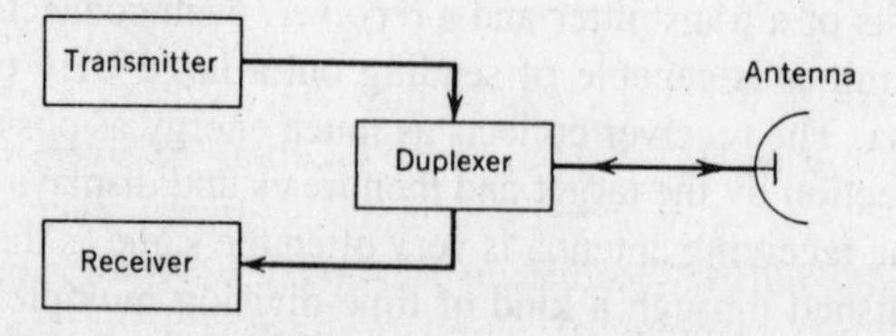

FIGURE 16-1 Block diagram of an elementary pulsed radar.

for which it has been designed. Thus, if the range is to be 25 nmi, at least 310 μs must elapse from the time a pulse is sent out to the time at which the duplexer disconnects the receiver from the antenna because the next pulse is due. On the other hand, the pulse repetition frequency must be quite high compared to the scanning speed. Hence a limit exists on the smallest number of pulses that can be sent per second. The practical result is that typical systems send out several hundred pulses per second, having durations in the vicinity of 1 μs.

Commenting briefly on the other aspects of the radar set, we find that pulse-modulated magnetrons, klystrons, TWTs or CFAs are normally used as transmitter output tubes, and the first stage of the receiver is often a diode mixer. The antenna generally uses a parabolic reflector of some form, as will be mentioned in Sec. 16-2.2.

Development of radar From its inception, radar has used a system of sending short, powerful pulses of radio energy and then analyzing the returned echoes to determine the position, distance and possibly velocity of the target. However, the methods of doing so have evolved and become far more refined and sophisticated as time has gone by. The primary incentive, as in so many other things, was the imminence of war. Radar was made possible by a technology which, at the time war broke out, was just beginning to show promise. This technology itself took great strides forward to meet the new challenges imposed by war.

The first radars worked at much lower frequencies than present systems (as low as 60 MHz for the original British coastal air-warning radar) because of a lack of sufficiently powerful transmitting tubes at higher frequencies. This was changed in 1940 with the appearance of the cavity magnetron (Sec. 11-4.1), and the stage was then set for the development of modern radar. As can be appreciated, one of the prime requirements of a radar system is that it should have a fair degree of accuracy in its indication of target direction. This is possible only if the antennas used are narrow-beam ones, i.e., have dimensions of several wavelengths. That requirement cannot be fulfilled satisfactorily unless the wavelengths themselves are fairly short, corresponding to the upper UHF or microwave frequencies.

The advent of the magnetron also made possible the next steps in the evolution of radar, namely, airborne radar for the detection of surface vessels and then airborne aircraft interception radar. In each of these, particularly the former, tight beams are necessary to prevent the receiver from being swamped by ground reflections, which would happen if insufficient discrimination between adjacent targets existed. Microwave radar for antiaircraft fire control was quickly developed, of which the most successful ground-based version was the U.S. Army's SCR-584. It was capable of measuring the position of enemy aircraft to within 0.1°, and the distance, or *range,* to within 25 m. Such radars were eventually capable of tracking targets by *locking* onto them, with the aid of servomechanisms controlling the orientation of the antennas. Anti-surface vessel (ASV) radars became very common and quite accurate toward the end of the war. So did airborne radar for navigation, bombing or bomber protection; electronic navigation systems were also developed. Radar countermeasures were instituted, consisting mainly of *jamming* (transmission of confusing signals at enemy radar) or the somewhat more effective dropping of aluminum foil, in strips of about a half-wavelength, to cover approaching aircraft by producing false echoes. This "chaff"

(American) or ''window'' (British) proved very effective, but its use in the war was considerably delayed. Each side thought that the other did not know about it and so it was kept secret; however, it eventually came to be used on a very large scale. One of the indications of the enormous growth in the importance of radar in World War II is the increase in the staff of the U.S. Army's Radiation Laboratory. It started with about 40 people in 1941, and numbers multiplied tenfold by 1945.

The subsequent developments of radar have also been numerous. They have included the use of wavelengths well into the millimeter range, at which atmospheric interference becomes noticeable—but see also Fig. 8-3 for the presence of radar ''windows.'' We have witnessed the use of greater powers at all wavelengths and the use of computers for a number of applications (especially fire control) to improve accuracy and reduce the time lag of manual operation. Long-range, fixed early-warning radars have been built, including the MEWS and BMEWS systems. These radars use huge antennas and enormous transmitting powers and are supplemented by radar-carrying high-flying aircraft, which have an extended radar horizon because of their height. Satellites carrying radar have been employed for military purposes, such as early detection of ballistic missiles, and civilian uses, notably in meteorology and mapping. Other important civilian uses of radar have included coastal navigation for shipping, position finding for shipping and aircraft, and air-traffic control at airports. This has extended the use of the landing facilities to weather conditions which would have made them unusable without radar and its allied systems. Last, but not least, is the use of radar by various police forces, for the control of traffic speed and the prosecution of offenders.

Numerous scientific advances have been made with the aid of radar; for instance, as early as in 1945 an error of 900 m was found (by accident) in the map position of the island of Corsica. More recent scientific uses of radar on an interplanetary scale have yielded much useful information about the sun and the rest of the solar system, and especially about the distances and rotations of the various planetary bodies. For example, it is now known that the planet Mercury rotates with a speed not equal to its angular orbital velocity, so that it does not always present the same face to the sun.

Frequencies and powers used in radar As has already been discussed, the frequencies employed by radar lie in the upper UHF and microwave ranges. As a result of wartime security, names grew up for the various frequency ranges, or bands, and these are still being used. One such term has already been met (the *X band*), and the others will now be tabulated. Since there is not a worldwide agreement on radar band nomenclature, the names used in Table 16-1 are the common American designations.

16-1.2 Radar Performance Factors

Quite apart from being limited by the curvature of the earth, the maximum range of a radar set depends on a number of other factors. These can now be discussed, beginning with the classical radar range equation.

TABLE 16-1 Radar Bands*

BAND NAME	FREQUENCY RANGE, GHz	MAXIMUM AVAILABLE PEAK POWER† MW
UHF	0.3–1.0	5.0
L	1.0–1.5	30.0
S	1.5–3.9	25.0
C	3.9–8.0	15.0
X	8.0–12.5	10.0
Ku	12.5–18.0	2.0
K	18.0–26.5	0.6
Ka	26.5–40.0	0.25
V	40.0–80.0	0.12
N	80.0–170.0	0.01
A	Above 170	—

*Note that the frequency ranges corresponding to the band names are not quite as widely accepted as the frequency spectrum band names of Fig. 8-11.
†This column shows the maximum available power *per tube*. Nothing prevents the use of several tubes in a transmitter to obtain a higher output power.

Radar range equation To determine the maximum range of a radar set, it is necessary to determine the power of the received echoes, and to compare it with the minimum power that the receiver can handle and display satisfactorily. If the transmitted pulsed power is P_t (peak value) and the antenna is isotropic, then the power density at a distance r from the antenna will be as given by Eq. (8-1), namely,

$$\mathcal{P} = \frac{P_t}{4\pi r^2} \tag{16-1}$$

However, antennas used in radar are directional, rather than isotropic. If A_p is the maximum power gain (see Sec. 9-3.1) of the antenna used for transmission, so the power density at the target will be

$$\mathcal{P} = \frac{A_p P_t}{4\pi r^2} \tag{16-2}$$

The power intercepted by the target depends on its *radar cross section*, or effective area (discussed later). If this area is S, the power impinging on the target will be

$$P = \mathcal{P}S = \frac{A_p P_t S}{4\pi r^2} \tag{16-3}$$

The target is not, of course, an antenna; thus, its radiation may be thought of as being omnidirectional. Accordingly, the power density of its radiation at the receiving antenna will be

$$\mathcal{P}' = \frac{P}{4\pi r^2} = \frac{A_p P_t S}{(4\pi r^2)^2} \tag{16-4}$$

Like the target, the receiving antenna intercepts a portion of the reradiated power, which is proportional to the cross-sectional area of the receiving antenna. However, it is the *capture area* of the receiving antenna that is used here, as defined and explained in Eq. (9-8). The received power is

$$P' = \mathcal{P}'A_0 = \frac{A_p P_t S A_0}{(4\pi r^2)^2} \tag{16-5}$$

where A_0 = capture area of the receiving antenna.

If (as is usually the case) the same antenna is used for both reception and transmission, we have from Eq. (9-7') that the maximum power gain is given by

$$A_p = \frac{4\pi A_0}{\lambda^2} \tag{16-6}$$

Substituting Eq. (16-6) into (16-5) gives

$$P' = \frac{4\pi A_0}{\lambda^2}\,\frac{P_t S A_0}{16\pi^2 r^4} = \frac{P_t A_0^2 S}{4\pi r^4 \lambda^2} \tag{16-7}$$

The maximum range r_{max} will be obtained when the received power is equal to the minimum receivable power of the receiver, P_{min}. Substituting this into Eq. (16-7), and making r the subject of the equation, we have

$$r_{max} = \left(\frac{P_t A_0^2 S}{4\pi \lambda^2 P_{min}}\right)^{1/4} \tag{16-8}$$

Alternatively, if Eq. (16-6) is turned around so that $A_0 = A_p\lambda^2/4\pi$ is substituted into Eq. (16-8), we have

$$r_{max} = \left[\frac{P_t A_p^2 \lambda^2 S}{(4\pi)^3 P_{min}}\right]^{1/4} \tag{16-8a}$$

Equations (16-8) and (16-8*a*) represent two convenient forms of the *radar range equation*, simplified to the extent that the minimum receivable power P_{min} has not yet been defined. It should also be pointed out that *idealized conditions have been employed*. Since neither the effects of the ground nor other absorption and interference have been taken into account, the maximum range in practice is often less than that indicated by the radar range equation.

Factors influencing maximum range A number of very significant and interesting conclusions may be made if the radar range equation is examined carefully. The first and most obvious is that *the maximum range is proportional to the fourth root of the peak transmitted pulse power*. In other words, the peak power must be increased sixteenfold, all else being constant, if a given maximum range is to be doubled. Eventually, such a power increase obviously becomes uneconomical in any particular radar system.

Equally obviously, a decrease in the minimum receivable power will have the same effect as raising the transmitting power and is thus a very attractive alternative to it. However, a number of other factors are involved here. Since P_{min} is governed by the sensitivity of the receiver (which in turn depends on the noise figure), the minimum receivable power may be reduced by a gain increase of the receiver, accompanied by a reduction in the noise at its input. Unfortunately, this may make the receiver more susceptible to jamming and interference, because it now relies more on its ability to amplify weak signals (which could include the interference), and less on the sheer power of the transmitted and received pulses. In practice, some optimum between transmitted power and minimum received power must always be reached.

The reason that the range is inversely proportional to the *fourth power* of the transmitted peak power is simply that the signals are subjected twice to the operation of the inverse square law, once on the outward journey and once on the return trip. By the same token, any property of the radar system that is used twice, i.e., for both reception and transmission, will show a double benefit if it is improved. Equation (16-8) shows that the maximum range is proportional to the square root of the capture area of the antenna, and is therefore directly proportional to its diameter if the wavelength remains constant. It is thus apparent that possibly the most effective means of doubling a given maximum radar system range is to double the effective diameter of the antenna. This is equivalent to doubling its real diameter if a parabolic reflector is used. Alternatively, a reduction in the wavelength used, i.e., an increase in the frequency, is almost as effective. There is, unfortunately, a limit here also. As will be remembered from Eq. (9-6), the beamwidth of an antenna is proportional to the ratio of the wavelength to the diameter of the antenna. Consequently, any increase in the diameter-to-wavelength ratio will reduce the beamwidth. This is very useful in some radar applications, in which good discrimination between adjoining targets is required, but it is a disadvantage in some *search radars*. It is their function to sweep a certain portion of the sky, which will naturally take longer as the beamwidth of the antenna is reduced.

Finally, Eq. (16-8) shows that the maximum radar range depends on the target area, as might be expected. Also, ground interference will limit this range, as shown in Fig. 8-8. The presence of a conducting ground, it will be recalled, has the effect of creating an interference pattern such that the lowest lobe of the antenna is some degrees above the horizontal. A distant target may thus be situated in one of the interference zones, and will therefore not be sighted until it is quite close to the radar set. This explains the development and emphasis of ''ground-hopping'' military aircraft, which are able to fly fast and close to the ground and thus remain undetectable for most of their journey.

Effects of noise The previous section showed that noise affects the maximum radar range insofar as it determines the minimum power that the receiver can handle; the extent of this can now be calculated exactly. From the definition of noise figure in Eq. (2-11), it is possible to calculate the equivalent noise power generated at the input of the receiver, N_r. This is the power required at the input of an ideal receiver having the same noise figure as the practical receiver. We then have

$$F = \frac{(S/N)_i}{(S/N)_o} = \frac{S_i N_o}{S_o N_i} = \frac{S_i}{GS_i}\frac{G(N_i + N_r)}{N_i}$$

$$= 1 + \frac{N_r}{N_i} \qquad \textbf{(16-9)}$$

where S_i = input signal power
N_i = input noise power
S_O = output signal power
N_o = output noise power
G = power gain of the receiver

We thus have

$$\frac{N_r}{N_i} = F - 1$$

$$N_r = (F - 1)N_i = kT_0 \, \delta f(F - 1) \tag{16-10}$$

where $kT_0 \, \delta f$ = noise input power of receiver
k = Boltzmann's constant = 1.38×10^{-23} J/K
T_0 = standard ambient temperature = 17°C = 290 K
δf = bandwidth of receiver

It has been assumed that the antenna temperature is equal to the standard ambient temperature, which may or may not be true, but the actual antenna temperature is of importance only if a very low-noise amplifier is used. Reference may be made to Sec. 2-4 for the reasoning behind the substitution for N_i.

The minimum receivable signal for the receiver, under so-called *threshold detection* conditions, is equal to the equivalent noise power at the input of the receiver, as just obtained in Eq. (16-10). This may seem a little harsh, especially since much higher ratios of signal to noise are used in continuous modulation systems. However, it must be realized that *the echoes from the target are repetitive, whereas noise impulses are random*. An integrating procedure thus takes place in the receiver, and meaningful echo pulses may be obtained although their amplitude is no greater than that of the noise impulses. This may be understood by considering briefly the display of the received pulses on the cathode-ray tube screen. The signal pulses keep recurring at the same spot if the target is stationary, so that the brightness at this point of the screen is maintained (whereas the impulses due to noise are quite random and therefore not additive). If the target itself is in rapid motion, i.e., moves significantly between successive scans, a system of *moving-target indication* (see Sec. 16-3) may be used. Substituting these findings into Eq. (16-8), we have

$$r_{\max} = \left[\frac{P_t A_0^2 S}{4\pi\lambda^2 kT_0 \, \delta f(F - 1)} \right]^{1/4} \tag{16-11}$$

Equation (16-11) is reasonably accurate in predicting maximum range, provided that a number of factors are taken into account when it is used. Among these are system losses, antenna imperfections, receiver nonlinearities, anomalous propagation, proximity of other noise sources (including deliberate jamming) and operator errors and/or fatigue (if there is an operator). It would be safe to call the result obtained with the aid of this equation *the maximum theoretical range*, and to realize that the maximum practical range varies between 10 and 100 percent of this value. However, range is sometimes capable of exceeding the theoretical maximum under unusual propagating conditions, such as superrefraction (see Sec. 8-2.3).

It is possible to simplify Eq. (16-11), which is rather cumbersome as it stands. Substituting for the capture area in terms of the antenna diameter ($A_0 = 0.65\pi D^2/4$) and for the various constants, and expressing the maximum range in kilometers, allows simplification to

$$r_{max} = 48\left[\frac{P_t D^4 S}{\delta f\, \lambda^2 (F-1)}\right]^{1/4} \tag{16-12}$$

where r_{max} = maximum radar range, km
P_t = peak pulse power, W
D = antenna diameter, m
S = effective cross-sectional area of target, m^2
δf = receiver bandwidth, Hz
λ = wavelength, m
F = noise figure (expressed as a ratio)

Example 16-1 Calculate the minimum receivable signal in a radar receiver which has an IF bandwidth of 1.5 MHz and a 9-dB noise figure.

$$F = \text{antilog}\,\frac{9}{10} = 7.943$$

$$\begin{aligned} P_{min} &= kT_0\, \delta f(F-1) = 1.38 \times 10^{-23} \times 290 \times 1.5 \times 10^6 (7.943 - 1) \\ &= 1.38 \times 2.9 \times 1.5 \times 6.943 \times 10^{-15} = 4.17 \times 10^{-14}\ \text{W} \end{aligned}$$

Example 16-2 Calculate the maximum range of a radar system which operates at 3 cm with a peak pulse power of 500 kW, if its minimum receivable power is 10^{-13} W, the capture area of its antenna is 5 m^2, and the radar cross-sectional area of the target is 20 m^2.

$$\begin{aligned} r_{max} &= \left(\frac{P_t A_0^2 S}{4\pi\lambda^2 P_{min}}\right)^{1/4} = \left[\frac{5 \times 10^5 \times 5^2 \times 20}{4\pi \times (0.03)^2 \times 10^{-13}}\right]^{1/4} = \left(\frac{2.5}{3.6\pi} \times 10^{24}\right)^{1/4} \\ &= 10^5 \times \sqrt[4]{2{,}210} = 6.86 \times 10^5\ \text{m} \\ &= 686\ \text{km}\ (= 370\ \text{nmi}) \end{aligned}$$

Example 16-3 A low-power, short-range radar is solid-state throughout, including a low-noise RF amplifier which gives it an overall noise figure of 4.77 dB. If the antenna diameter is 1 m, the IF bandwidth is 500 kHz, the operating frequency is 8 GHz and the radar set is supposed to be capable of detecting targets of 5-m^2 cross-sectional area at a maximum distance of 12 km, what must be the peak transmitted pulse power?

From Eq. (16-12) we have

$$\left(\frac{r_{max}}{48}\right)^4 = \frac{P_t D^4 S}{\delta f \lambda^2 (F-1)} = \left(\frac{12}{48}\right)^4 = \frac{1}{256}$$

Thus the power required here is

$$P_t = \frac{\delta f \lambda^2 (F-1)}{256 D^4 S}$$

$$\text{where } \lambda = \frac{30}{8} = 3.75 \text{ cm} = 3.75 \times 10^{-2} \text{ m}$$

$$F = \text{antilog} \frac{4.77}{10} = 3.0$$

Substituting these gives

$$P_t = \frac{5 \times 10^5 (3.75 \times 10^{-2})^2 \times (3 - 1)}{2.56 \times 10^2 \times 1^4 \times 5} = 1.1 \text{ W}$$

It will be noted that this power is well within the ability of Gunn effect or IMPATT oscillators (see Secs. 12-6 and 12-7). Even if the vagaries of the system reduce this range to half of its value, as may well happen, the resulting sixteenfold increase of the peak pulse power to 17.5 W (required to restore the maximum range to its original value) is still quite feasible with those devices.

Target properties In connection with the derivation of the radar range equation, a quantity was used but not defined. This was the *radar cross section*, or effective area, of the target; some explanation is now in order. For targets whose dimensions are large compared to the wavelength, as for aircraft if microwave radar is used, the radar cross section may be defined as the projected area of a perfectly conducting sphere which would reflect the same power as the actual target reflects, if it were located at the same spot as the target. However, the practical situation is far from simple.

First of all, the radar cross section depends on the frequency used. If this is such that the target is small compared to a wavelength, its cross-sectional area for radar appears much smaller than its real cross section. Such a situation is referred to as the *Rayleigh region*. When the circumference of a spherical target is between 1 and 10 wavelengths, the radar cross section oscillates about the real one (see p. 27-21 in [1]). This is the so-called *resonance region*. Finally, for shorter wavelengths (in the *optical region*) the radar and true cross sections are equal.

Quite apart from variations with frequency, the radar cross section of a target will depend on the polarization of the incident wave, the degree of surface roughness (if it is severe), the use of special coatings on the target and, most importantly of all, the *aspect of the target*. For instance, a large jet aircraft, measured at 425 MHz, has been found to have a radar cross section varying between 0.2 and 300 m^2 for the fuselage, depending on the angle at which the radar pulses arrived on it. The situation is seen to be complex because of the large number of factors involved, so that a lot of the work is, in fact, empirical. Further information, much of it in graphical form, may be obtained from chap. 27 in [1].

16-2 PULSED SYSTEMS

Pulsed systems can now be described in some detail, starting with a block diagram of a typical pulsed radar set and its description, followed by a discussion of scanning and display methods. Pulsed radars can then be divided broadly into *search radars* on the one hand and *tracking radars* on the other. Finally, some mention can be made of auxiliary systems, such as *beacons* and *transponders*.

16-2.1 Basic Pulsed Radar System

A very elementary block diagram of a pulsed radar set was shown in Fig. 16-1. A more detailed block diagram will now be given, and it will then be possible to compare some of the circuits used with those treated in other contexts and to discuss in detail those circuits peculiar to radar.

Block diagram and description The block diagram of Fig. 16-2 shows the arrangement of a typical high-power pulsed radar set. The trigger source provides pulses for the modulator, in a manner that will be described in more detail. In turn, the modulator provides rectangular voltage pulses used as the supply voltage for the output tube, thus switching it ON and OFF as required. This tube may be a magnetron oscillator or an amplifier such as the klystron, traveling-wave tube or crossed-field amplifier, depending on specific requirements. All those devices were treated in detail in Chap. 11. If an amplifier is used, a source of microwaves is also required. While an amplifier may be modulated at a special grid, the magnetron cannot. If the radar is a low-powered one, it may use IMPATT or Gunn oscillators, or TRAPATT amplifiers. Below *C* band, power transistor amplifiers or oscillators may also be used. All these devices have already been discussed at length in Chap. 12. Finally, in this sequence, the transmitter portion of the radar is terminated with the duplexer, which passes the output pulse to the antenna for transmission.

The receiver is connected to the antenna at suitable times (i.e., when no transmission is instantaneously taking place). As previously explained, this is also done by the duplexer. As shown here, a (semiconductor diode) mixer is the most likely first stage in the receiver, since it has a fairly low noise figure, but of course it shows a conversion loss. An RF amplifier can also be used, and this would most likely be a transistor or IC, or perhaps a tunnel diode or paramp (see Chap. 12). A better noise

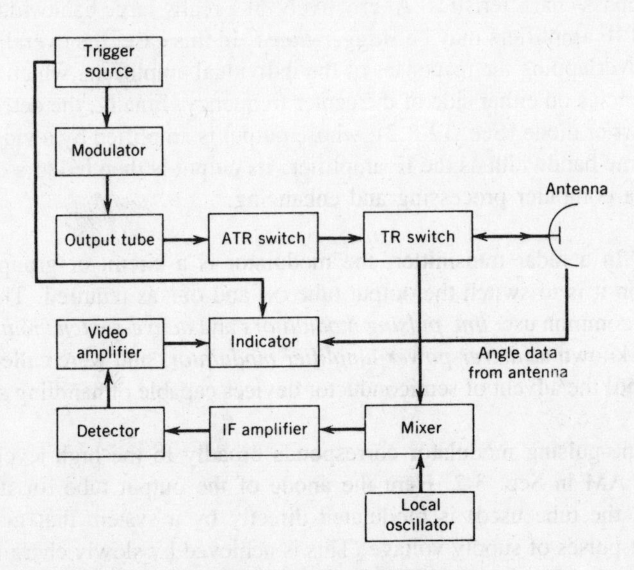

FIGURE 16-2 Pulsed radar block diagram.

figure is thus obtained, and the RF amplifier may have the further advantage of saturating for large signals, thus acting as a limiter that prevents mixer diode burnout from strong echoes produced by nearby targets. The main receiver gain is provided at an intermediate frequency that is typically 30 or 60 MHz. However, it may take two or more downconversions to reach that IF from the initial microwave RF, to ensure adequate image frequency suppression.

If a diode mixer is the first stage, the (first) IF amplifier must be designed as a low-noise stage to ensure that the overall noise figure of the receiver does not deteriorate unduly. A noisy IF amplifier would play havoc with the overall receiver performance, especially when it is noted that the "gain" of a diode mixer is in fact a conversion loss, typically 4 to 7 dB. A *cascode* connection is quite common for the transistor amplifiers used in the IF stage, because it removes the need for *neutralization* to avoid the *Miller effect*.

Another source of noise in the receiver of Fig. 16-2 may be the local oscillator, especially for microwave radar receivers. One of the methods of reducing such noise is to use a varactor or step-recovery diode multiplier (Sec. 12-3). Another method involves the connection of a narrowband filter between the local oscillator and the mixer to reduce the noise bandwidth of the mixer. However, in receivers employing automatic frequency correction this may be unsatisfactory. The solution of the oscillator noise problem may then lie in using a balanced mixer and/or a cavity-stabilized oscillator, as described on pp. 5-9 to 5-16 of [1]. If used, AFC may simply consist of a phase discriminator which takes part of the output from the IF amplifier and produces a dc correcting voltage if the intermediate frequency drifts. The voltage may then be applied directly to a varactor in a diode oscillator cavity.

The IF amplifier is broadband, to permit the use of fairly narrow pulses, as will be shown. This, in turn, means that cascaded rather than single-stage amplifiers are used. These can be *synchronous,* that is, all tuned to the same frequency and having identical bandpass characteristics. Alternatively, if a really large bandwidth is needed, the individual IF amplifiers may be *stagger-tuned*. In this case, the overall response is achieved by overlapping the responses of the individual amplifiers, which are tuned to nearby frequencies on either side of the center frequency. Finally, the detector is often a Schottky-barrier diode (Sec. 12-8.2), whose output is amplified by a video amplifier having the same bandwidth as the IF amplifier. Its output is then fed to a display unit, directly or via computer processing and enhancing.

Modulators In a radar transmitter, the modulator is a circuit or group of circuits whose function it is to switch the output tube ON and OFF as required. There are two main types in common use: *line-pulsing modulators* and *active-switch modulators*. The latter are also known as *driver-power-amplifier modulators* and were called *hard-tube modulators* until the advent of semiconductor devices capable of handling some modulator duties.

The line-pulsing modulator corresponds broadly to the high-level modulator described for AM in Sec. 3-2. Here the anode of the output tube (or its collector, depending on the tube used) is modulated directly by a system that generates and provides large pulses of supply voltage. This is achieved by slowly charging and then rapidly discharging a transmission line. The charging is made slow to reduce the current requirements and is generally done through an inductance. The transmission line is able to store energy in its distributed inductance and capacitance. If the line is charged to a voltage V from a high-impedance source, this voltage will drop to $\frac{1}{2}V$ when a load is connected (the output tube) whose impedance is equal to the characteristic impedance of the line. However, at the instant of load connection the voltage across the line is $\frac{1}{2}V$ only at the input; it is still V everywhere else. The voltage drop now propagates along the line to the far end, from which it is reflected to the input terminals. It is thus seen that a voltage $\frac{1}{2}V$ will be maintained across the load for a time $2t$, where t is the time taken by an electromagnetic wave to travel from one end of the line to the other.

If the pulse duration ($2t$) is to be 1 μs, the line length must be 150 m. This is far too long for convenience, and consequently a *pulse-forming network* (PFN) is almost always substituted for the transmission line. As shown in Fig. 16-3, which illustrates a very basic line modulator, the PFN looks just like the equivalent circuit of a transmission line. It also behaves identically to the transmission line for frequencies below $f = 1/\pi\sqrt{LC}$, where L and C are the inductance and capacitance, respectively, per section. In high-power radars, the device most likely for use as a switch is a hydrogen *thyratron,* because it is capable of switching very high powers and of rapid deionization. *Silicon-controlled rectifiers* (SCRs) may also be used to good advantage.

The advantages of the line modulator are that it is simple, compact, reliable and efficient. However, it has the disadvantage that the PFN must be changed if a different pulse length is required. Consequently, line modulators are not used at all in radars from which variable pulse widths are required, but they are often used otherwise. The pulses that are produced have adequately steep sides and flat tops.

The active-switch modulator is one that can also provide high-level modulation of the output tube, but this time the pulses are generated at a low power level and then amplified. The driver is often a *blocking oscillator,* triggered by a timing source and driving an amplifier. Depending on the power level, this may be a transistor amplifier or a powerful tube such as a shielded-grid triode. The amplifier then controls the dc power supply for the output RF tube. This type of modulator is less efficient, more complex and bulkier than the line modulator, but it does have the advantage of easily variable pulse length, repetition rate or even shape. It is thus often used in practice. Various modulators are described in detail, and their merits are compared, in [1], pp. 7-68 to 7-87.

Finally, low-level modulation is also sometimes possible. This may be done in UHF radar, which uses orthodox vacuum tubes, or at higher frequencies if a velocity-

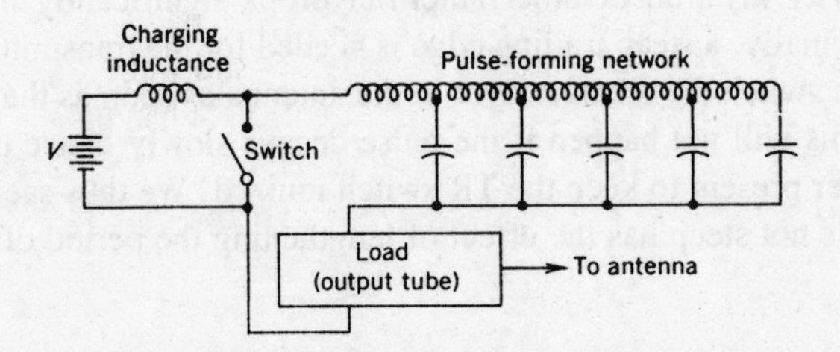

FIGURE 16-3 Simple line-pulser using a pulse-forming network.

modulated amplifier is used. Such possibilities were explored when those tubes were described in Chap. 11, and indeed Fig. 11-7 shows the basic circuit of such a modulator. In these cases, and also in some low-power radars, it becomes possible to apply the output of the blocking oscillator directly to the output tube, thus simplifying the modulator circuitry.

Receiver bandwidth requirements As will be appreciated on the basis of the work done in Chap. 1, the receiver frequency response is governed by the width of the pulses which it is desired to receive. The narrower the pulses, the greater is the IF (and video) bandwidth required, whereas the RF bandwidth is normally greater than these, as in other receivers. With a given pulse duration T, the receiver bandwidth may still vary, depending on how many harmonics of the pulse repetition frequency are needed to provide a received pulse having a suitable shape. Thus, if vertical sides are required for the pulses in order to give a good resolution (as will be seen), a large bandwidth is required. It is seen that the bandwidth must be increased if more *information* about the target is required, but too large a bandwidth will reduce the maximum range by admitting more noise, as shown by Eq. (16-11).[1]

The IF bandwidth of a radar receiver is made n/T, where T is the pulse duration and n is a number whose value ranges from under 1 to over 10, depending on the circumstances. Values of n from 1 to about 1.4 are the most common. Because pulse widths normally range from 0.1 to 10 μs, it is seen that the radar receiver bandwidth may lie in the range from about 200 kHz to over 10 MHz. Bandwidths from 1 to 2 MHz are the most common.

Factors governing pulse characteristics We may now consider why flat-topped rectangular pulses are preferred in radar and what it is that governs their amplitude, duration and repetition rate. These factors are of the greatest importance in specifying and determining the performance of a radar system.

As it happens, there are several reasons (not just one) why radar pulses ideally ought to have vertical sides and flat tops. The leading edge of the transmitted pulse must be vertical to ensure that the leading edge of the received pulse is also close to vertical. Otherwise, ambiguity will exist as to the precise instant at which the pulse has been returned, and therefore inaccuracies will creep into the exact measurement of the target range. This requirement is of special importance in fire-control radars. A flat top is required for the voltage pulse applied to the magnetron anode, otherwise its frequency will be pushed (see Sec. 11-4.3). It also is needed because the efficiency of the magnetron, multicavity klystron or other amplifier drops significantly if the supply voltage is reduced. Finally, a steep trailing edge is needed for the transmitted pulse, so that the duplexer can switch the receiver over to the antenna as soon as the body of the pulse has passed. This will not happen if the pulse decays slowly, since there will be sufficient pulse power present to keep the TR switch ionized. We thus see that a pulse trailing edge which is not steep has the effect of lengthening the period of time which

[1] Note how similar all these considerations are to those that were discussed in Chap. 13 in connection with information theory and pulse modulation. See also Sec. 17-1.2, for a basic calculation of bandwidth required for television.

the receiver is disconnected from the antenna. Therefore it limits the *minimum* range of the radar; this will be discussed in connection with pulse width.

The pulse repetition frequency, or PRF, is governed mainly by two conflicting factors. The first is the maximum range required, since it is necessary not only to be able to detect pulses returning from distant targets but also to allow them time to return before the next pulse is transmitted. For example, if a given radar is to have a range of 50 nmi (92.6 km), at least 620 μs must be allowed between successive pulses; this period is called the *pulse interval*. Ambiguities will result if this is not done. For example, if only 500 μs is used as the pulse interval, an echo received 120 μs after the transmission of a pulse could mean either that the target is 120/12.4 = 9.7 nmi (18 km) away or else that the pulse received is a reflection of the previously sent pulse, so that the target is (120 + 500)/12.4 = 50 nmi away. From this point of view, it is seen that the pulse interval should be as large as possible. On the other hand, the greater the number of pulses reflected from a target, the greater the probability of distinguishing this target from noise. This follows from a statement made in the previous section, which showed how an integrating effect takes place if echoes repeatedly come from the same target, whereas noise is random. Since the antenna moves at a significant speed in many radars, and yet it is necessary to receive several pulses from a given target, a lower limit on the pulse repetition frequency clearly exists. Values of PRF from 200 to 10,000/s are commonly used in practice, corresponding to pulse intervals of 5000 to 100 μs and therefore to maximum ranges from 400 to 8 nmi (740 to 15 km). However, when the targets are very distant (satellites and space probes, for example), lower PRFs may have to be used (as low as 30 pps). (See also chap. 8, pt. VI, in [2]).

If a short minimum range is required, then short pulses must be transmitted. This is really a continuation of the argument in favor of a vertical trailing edge for the transmitted pulse. Since the receiver is disconnected from the antenna for the duration of the pulse being transmitted (in all radars using duplexers), it follows that echoes returned during this period cannot be received. For instance, if the total pulse duration is 2 μs, then no pulses can be received during this period. That is, no echoes can be received from targets closer than 300 m away, and this is thus the minimum range of the radar. Another argument in favor of short pulses is that they improve the *range resolution*, which is the ability to separate targets whose distance from the transmitter differs only slightly.[2] If a pulse duration of 1 μs is used, this means that echoes returning from separate targets that are 1 μs apart in time (i.e., about 300 m in distance) will merge into one returned pulse and will not be separated. It is thus seen that the range resolution in this case is no better than 300 m.

It is now necessary to consider some arguments in favor of long pulse durations. The main one is simply that the receiver bandwidth must be increased as pulses are made narrower, and Eq. (16-11) shows that this tends to reduce the maximum range by admitting more noise into the system. This may, of course, be counteracted by increasing the peak pulse power, but only at the expense of cost, size and power

[2] *Angular resolution*, as the name implies, is dictated by the beamwidth of the antenna. If the beamwidth is (say) 2°, then two separate targets that are less than 2° apart will appear as one target and will therefore not be *resolved*.

consumption. In fact, a careful look at the situation reveals that *the maximum range depends on the pulse energy rather than on its peak power*. Since one of the terms of Eq. (16-11) is $P_t/\delta f$, and the bandwidth δf is inversely proportional to the pulse duration, we are entitled to say that range depends on the product of P_t and T, and this product is equal to the pulse *energy*. On the other hand, it must be borne in mind that increasing the pulse width while keeping a constant PRF has the effect of increasing the *duty cycle* of the output tube, and therefore its average power. As the name implies, the duty cycle is the fraction of time during which the output tube is ON. For example, if the PRF is 1200 and the pulse width is 1.5 μs, the period of time actually occupied by the transmission of pulses is $1200 \times 1.5 = 1800$ μs/s, or 0.0018 (0.18 percent). Increasing the duty cycle thus increases the dissipation of the output tube. It may also have the effect of forcing a reduction in the peak power, because the peak and average powers are closely related for any type of tube. If large duty cycles are required, it is worth considering a traveling-wave tube or a crossed-field amplifier as the output tube, since both are capable of duty cycles in excess of 0.02 (see also Chap. 11).

16-2.2 Antennas and Scanning

As has so far been assumed, the majority of radar antennas use dipole or horn-fed paraboloid reflectors, or at least reflectors of a basically paraboloid shape, such as those of Fig. 9-28 (namely, the cut paraboloid, parabolic cylinder or pillbox). In each of the latter, the beamwidth in the vertical direction (and therefore the angular resolution) will be much worse than in the horizontal direction, but this is immaterial in ground-to-ground or even air-to-ground radars. It has the advantages of allowing a significantly reduced antenna size and weight, reduced wind loading and smaller drive motors.

Antenna scanning As already mentioned, radar antennas are often made to scan a given area of the surrounding space, but the actual scanning pattern depends on the application. Figure 16-4 shows some typical scanning patterns.

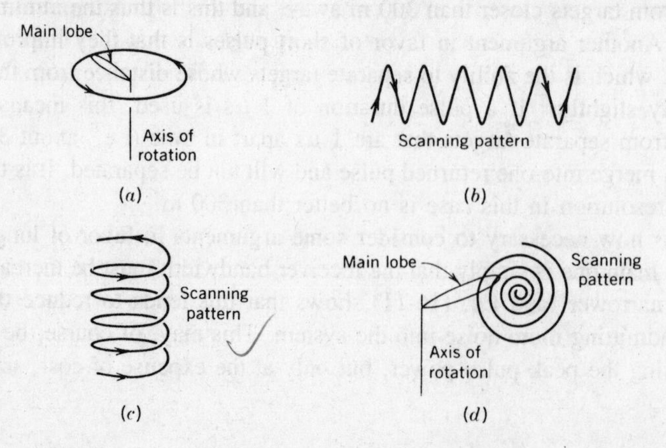

FIGURE 16-4 Representative antenna scanning patterns. (*a*) Horizontal; (*b*) nodding; (*c*) helical; (*d*) spiral.

The first of these is the simplest but has the disadvantage of scanning in the horizontal plane only. However, there are many applications for this type of scan in searching the horizon, e.g., in ship-to-ship radar. The nodding scan of Fig. 16-4*b* is an extension of this; the antenna is now rocked rapidly in elevation while it rotates more slowly in azimuth, and thus scanning in both planes is obtained. The system can be used to scan a limited sector, as shown, or else it can be extended to cover the complete hemisphere. Another system capable of search over the complete hemisphere is the helical scanning system of Fig. 16-4*c*, in which the elevation of the antenna is raised slowly while it rotates more rapidly in azimuth. The antenna is returned to its starting point at the completion of the scanning cycle, as shown, and typical speeds are a rotation of 6 rpm accompanied by a rise rate of 20°/minute (World War II SCR-584 radar). Finally, if a limited area of more or less circular shape is to be covered, spiral scan may be used, as shown in Fig. 16-4*d*; it is very popular for this purpose.

Antenna tracking Having *acquired* a target through a scanning method as just described, it may then be necessary to locate it very accurately, perhaps in order to bring weapons to bear upon it. Having an antenna with a narrow, pencil-shaped beam helps in this regard, but the accuracy of even this type of antenna is generally insufficient in itself.[3] Thus auxiliary methods of tracking or precise location must be employed. The simplest of these is the *lobe-switching* technique illustrated in Fig. 16-5*a*, which is also called *sequential lobing*. The direction of the antenna beam is rapidly switched between two positions in this system, as shown, so that the strength of the echo from the target will fluctuate at the switching rate, unless the target is exactly midway between the two directions. In the latter case, the echo strength will be the same for both antenna positions, and the target will have been tracked with much greater accuracy than would be achieved by merely pointing the antenna at it.

Conical scanning is a logical extension of lobe switching and is shown in Fig. 16-5*b*. It is achieved by mounting the parabolic antenna slightly off center and then rotating it about the axis of the parabola, as shown; the rotation is slow compared to the PRF. The name *conical scan* is derived from the surface described in space by the pencil radiation pattern of the antenna, as the tip of the pattern moves in a circle. The same argument applies with regard to target positioning as for sequential lobing, except

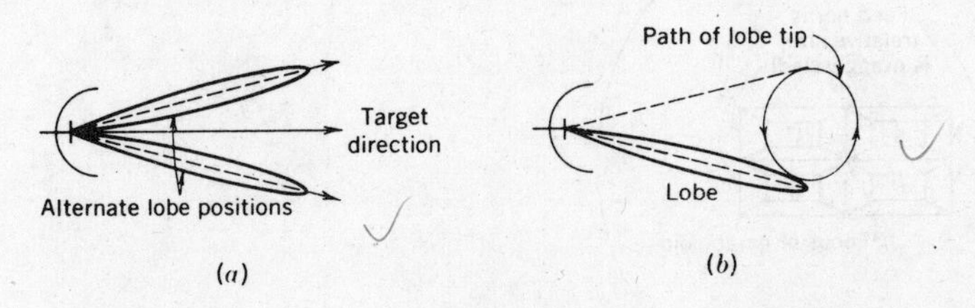

FIGURE 16-5 Antenna tracking. (*a*) Lobe switching; (*b*) conical scanning.

[3] An error of only 1° seems slight, until one realizes that a weapon so aimed would miss a nearby target, only 10 km away, by 175 m (i.e., completely!).

that the conical scanning system is just as accurate in elevation as in azimuth, whereas sequential lobing is accurate in one plane only.

There are two disadvantages of the use of either sequential lobing or conical scanning. The first and most obvious is that the motion of the antenna is now more complex, and additional servomechanisms are required. The second drawback is due to the fact that more than one returned pulse is required to locate a target accurately (a minimum of four are required with conical scan, one for each extreme displacement of the antenna). The difficulty here is that if the target cross section is changing, because of its change in attitude or for other reasons, the echo power will be changing also. Hence the effect of conical scanning (or sequential lobing, for that matter) will be largely nullified. From this point of view, the ideal system would be one in which all the information obtained by conical scanning could be achieved with just one pulse. Such a system fortunately exists and is called *monopulse*.

In an amplitude-comparison monopulse system, four feeds are used with the one paraboloid reflector. A system using four horn antennas displaced about the central focus of the reflector is shown in Fig. 16-6. The transmitter feeds the horns simultaneously, so that a sum signal is transmitted which is little different from the usual pulse transmitted by a single horn. In reception, a duplexer using a rat race, as discussed in Sec. 10-3.4, is employed to provide the following three signals: the sum $A + B + C + D$, the vertical difference $(A + C) - (B + D)$ and the horizontal difference $(A + B) - (C + D)$.

Each of the four feeds produces a slightly different beam from the one reflector, so that in transmission four individual beams ''stab out'' into space, being centered on the direction a beam would have had from a single feed placed at the focus of the reflector. As in conical scanning and sequential lobing, no differences will be recorded if the target is precisely in the axial direction of the antenna. However, once the target has been acquired, any deviation from the central position will be shown by the presence of a vertical difference signal, a horizontal difference signal, or both. The receiver has three separate input channels (one for each of the three signals) consisting of three mixers with a common local oscillator, three IF amplifiers and three detectors (see also

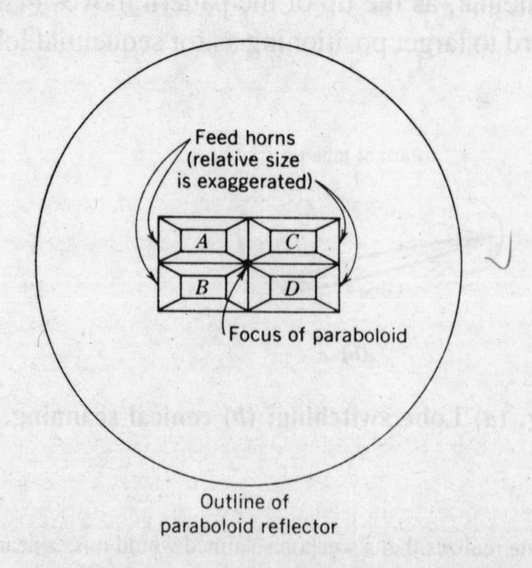

FIGURE 16-6 Feed arrangements for monopulse tracking.

the block diagram on p. 21-13 in [1]). The output of the sum channel is used to provide the data generally obtained from a radar receiver, while each of the difference or error signals feeds a servoamplifier and motor, driving the antenna so as to keep it pointed exactly at the target. Once this has been done, the output of the sum channel can be used for the automatic control of gunnery if that is the function of the radar.

The advantage of monopulse, as previously mentioned, is that it obtains with one pulse the information which required several pulses in conical scanning. Thus monopulse is not subject to errors due to the variation in target cross section. On the other hand, it requires two extra receiving channels and a more complex duplexer and feeding arrangement and will thus be bulkier and more expensive. Further information about antenna scanning and tracking patterns may be found in chap. 2 of [3] and chap. 21 of [1].

Finally, some antennas are required to provide a certain amount of tracking but are themselves too bulky to move, e.g., the 120-by-50-m BMEWS antennas at Thule, Greenland. In such cases, the feed is scanned on either side of the focus of the reflector. In simple systems, the feed horn may actually move, but in others a multiple-feed arrangement is used. This is rather similar to the monopulse feed but contains far more horns, and the signal is then applied to each horn in turn. For details of such an "organ-pipe" scanner, see [1] (pp. 22-6 and 22-7). An alternative to this system, which is rather similar to an interferometer, consists of using a number of fairly closely spaced *fixed* antennas and varying the direction of the scanning beam by changing the relative phase of the signals fed to the various antennas. The name given to this is *phased array*. Note that no antenna movement is required for scanning with either the phased array or the organ-pipe scanner. A description of various aspects of phased array radars is given in Sec. 16-3.3

16-2.3 Display Methods

The output of a radar receiver may be displayed in any of a number of ways, the following three being the most common: *deflection modulation* of a cathode-ray-tube screen as in the *A scope, intensity modulation* of a CRT as in the *plan-position indicator* (PPI) or direct feeding to a computer. Additional information, such as height, speed or velocity, may be shown on separate displays.

A scope As can be seen from Fig. 16-7, the operation of this display system is rather similar to that of an ordinary oscilloscope. A sweep waveform is applied to the horizontal deflection plates of the CRT and moves the beam slowly from left to right across the face of the tube, and then back to the starting point. The *flyback* period is rapid and occurs with the beam blanked out.[4] In the absence of any received signal, the display is simply a horizontal straight line, as with oscilloscopes. The demodulated receiver output is applied to the vertical deflection plates and causes the departures from the horizontal line, as seen in Fig. 16-7. The horizontal deflection sawtooth waveform is synchronized with the transmitted pulses, so that the width of the CRT screen corre-

[4] Just as in TV, as will be seen in the next chapter.

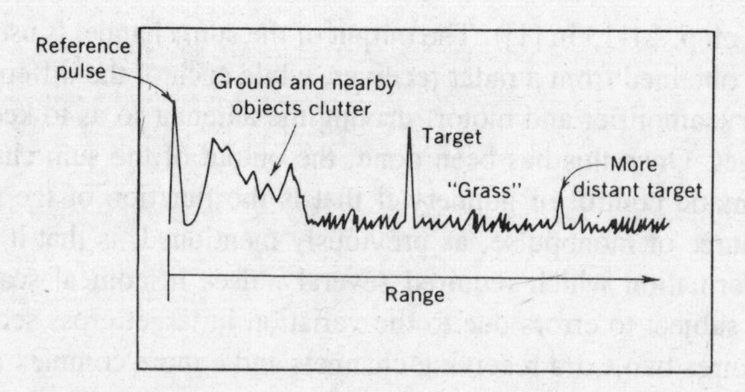

FIGURE 16-7 A scope display.

sponds to the time interval between successive pulses. Displacement from the left-hand side of the CRT thus corresponds to the range of the target. As indicated, the first "blip" is due to the transmitted pulse, part of which is deliberately applied to the CRT for reference. Then come various strong blips due to reflections from the ground and nearby objects, followed by noise, which is here called *grass* (the name is very descriptive, although the pips due to noise are not, of course, constant in amplitude or position). The various targets then show up as (ideally) large blips, again interspersed with grass. The height of each blip corresponds to the strength of the returned echo, while its distance from the reference blip is a measure of its range. This is why the blips on the right of the screen have been shown smaller than those nearer to the left. It would take a very large target indeed at a range of 40 km to produce the same size of echo as a normal target only 5 km away!

Of the various indications and controls for the A scope, perhaps the most important is the range calibration, shown horizontally across the tube. In some radars only one may be shown, corresponding to a fixed value of (say) 1 km per cm of screen deflection, although in others several scales may be available, with suitable switching for more accurate range determination of closer targets. In others still, it is possible to expand any section of the scan to allow more accurate indication of that particular area (this is rather similar to bandspread in communications receivers). It is also often possible to introduce pips derived from the transmitted pulse, which have been passed through a time-delay network. The delay is adjustable, so that the *marker* blip can be made to coincide with the target. The distance reading provided by the marker control is more accurate than could have been estimated from a direct reading of the CRT. Finally, a gain control for vertical deflection is provided, which allows the sensitivity to be increased for weak echoes or reduced for strong ones. In the case of strong signals, reducing the sensitivity will reduce the amplitude of the grass.

By its very nature, the A scope presentation is more suitable for use with tracking than with search antennas, since the echoes returned from one direction only are displayed; the antenna direction is generally indicated elsewhere.

Plan-position indicator As shown in Fig. 16-8, the PPI display shows a map of the target area. The CRT is now intensity-modulated, so that the signal from the receiver after demodulation is applied to the grid of the cathode-ray tube. The CRT is biased slightly beyond cutoff, and thus only blips corresponding to targets permit beam cur-

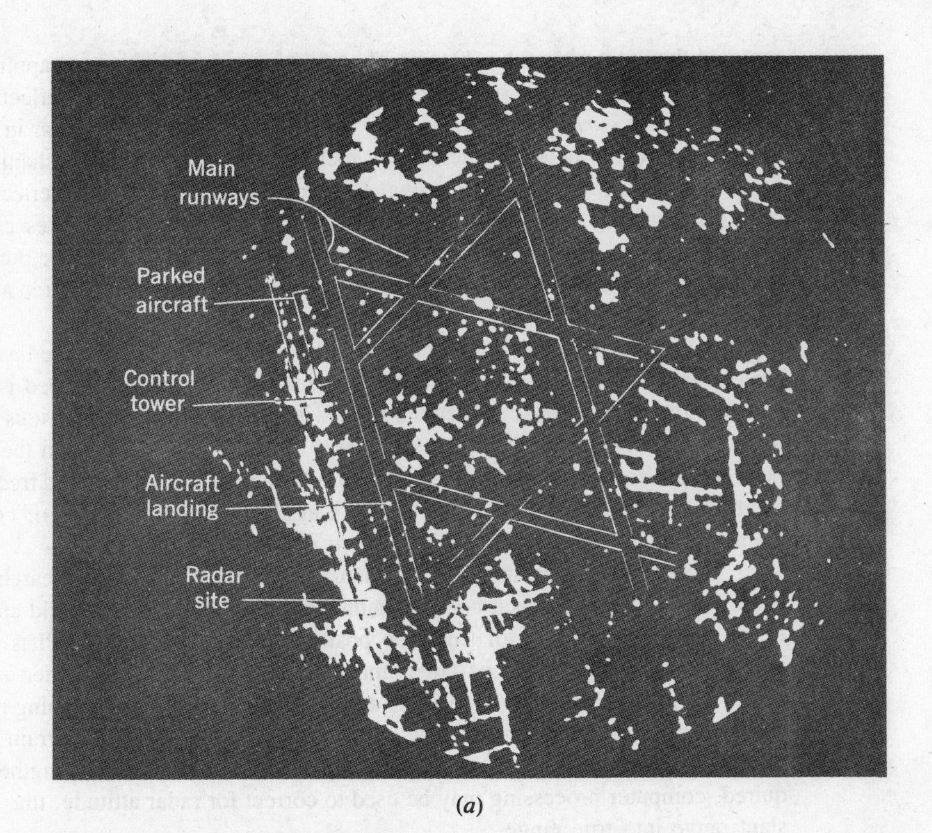

(*a*)

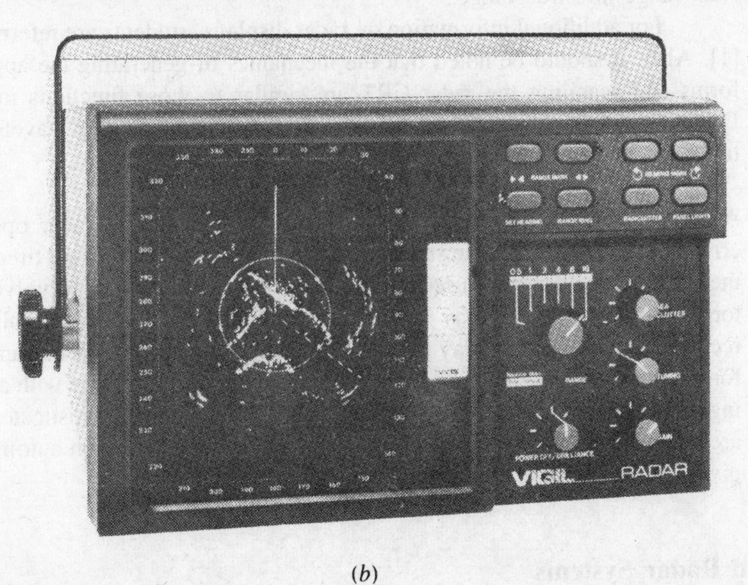

(*b*)

FIGURE 16-8 PPI display. (*a*) Radar map of London's Heathrow Airport ***(British Information Services (BIS) Pictures)*****; (*b*) Portable modern marine radar set.** ***(Courtesy of AWA Australia.)***

rent and therefore screen brightness. The scanning waveform is now applied to a pair of coils on opposite sides of the neck of the tube, so that magnetic deflection is used, and a sawtooth *current* is required. The coils, situated in a *yoke* similar in appearance to that around the neck of a television picture tube, are rotated mechanically at the same angular velocity as the antenna. Hence the beam is not only deflected radially outward from the center and then back again rapidly but also rotates continuously around the tube. The brightness at any point on the screen indicates the presence of an object there, with its position corresponding to its actual physical position and its range being measured radially out from the center.

Long-persistence phosphors are normally used to ensure that the face of the PPI screen does not flicker. It must be remembered that the scanning speed is rather low compared to the 60 fields per second used with television, so that various portions of the screen could go dim between successive scans. The resolution on the screen depends on the beamwidth of the antenna, the pulse length, the transmitted frequency, and even on the diameter of the CRT beam. Circular screens are used, of course, with diameters ranging up to 40 cm, but 30 cm is more common.

As can be appreciated, the PPI display lends itself to use with search radars and is particularly suitable when conical scanning is employed. Note should also be taken of the fact that distortion of true map positions will take place if PPI is used on an aircraft, and its antenna does not point straight down. The range then seen on the screen is called the *slant range*. For example, if the antenna of a mapping radar points straight down from the aircraft body, but the aircraft is climbing, the terrain behind will appear foreshortened, while the area ahead is distorted by being lengthened. If required, computer processing may be used to correct for radar attitude, thus converting slant range into true range.

For additional information on radar displays, students are referred to chap. 6 of [1]. Also, it should be noted that the mechanics of generating the appropriate waveforms and scanning the radar CRT are similar to those functions in TV receivers. Discussion of these, including the need for sawtooth scanning waveforms, is undertaken in Chap. 17, in conjunction with television receivers.

Automatic target detection Because the performance of radar operators may be erratic or inaccurate (people staring at screens for long hours *do* get tired), the output of the radar receiver may be used in a number of ways that do not involve human operators. One such system may involve computer processing and simplification of the received data prior to display on the radar screen. Other systems use analog computers for the reception and interpretation of the received data, together with automatic tracking and gun laying (or missile pointing). Some of the more sophisticated radar systems are discussed later in this chapter, while further information on automatic tracking is given in Chap. 21 of [1].

16-2.4 Pulsed Radar Systems

As has probably been realized by now, a radar system is generally required to perform one of two tasks: It must either search for targets or else track them once they have been acquired. Sometimes the same radar performs both functions, whereas in other installations separate radars are used. Within each broad group, further subdivisions are possible, depending on the specific application. The most common of these will now be described.

Search radar systems The general treatment of radar so far in this chapter has revealed the basic features of search radars, including block diagrams, antenna scanning methods and display systems. Consequently, it has been seen that such a radar system must acquire a target in a large volume of space, regardless of whether its presence is known. To do this, the radar must be capable of scanning its region rapidly; therefore, the narrow beam is not the best antenna pattern for this purpose, because scanning a given region would take too long. Once the approximate position of a target has been obtained with a broad beam, the information can be passed on to a tracking radar, which quickly acquires and then follows the target. Another solution to the problem consists in using two fan-shaped beams (from a pair of connected cut paraboloids), oriented so that one is directional in azimuth and the other in elevation. The two rotate together, using helical scan, so that while one searches in azimuth, the other antenna acts as a height finder, and a large area is thus covered rapidly. Perhaps the most common application of this type is the air-traffic-control radar used at both military and civilian airports.

If the area to be scanned is relatively small, a pencil beam and spiral scanning can be used to advantage, together with a PPI display unit. Weather avoidance and airborne navigation radars are two examples of this type. Marine navigation and ship-to-ship radars are of a similar type, except that here the scan is simply horizontal, with a fan-shaped beam. A typical marine radar installation is shown in Fig. 16-9.

Early-warning and aircraft surveillance radars are also acquisition radars with a limited search region, but they differ from the other types in that they use UHF wavelengths to reduce atmospheric and rain interference. They thus are characterized not only by huge powers, but also by equally large antennas. As mentioned previously, the antennas are stationary, so that scanning is achieved by moving-feed or similar methods.

Tracking radar systems Once a target has been *acquired*, it may then be *tracked*, as discussed in the section dealing with antennas and scanning. The most common tracking methods used purely for tracking are the conical scan and monopulse systems described previously. A system that gives the angular position of a target accurately is said to be *tracking in angle*. If range information is also continuously obtained, *tracking in range* (as well as in angle) is said to be taking place, while a tracker that continuously monitors the relative target velocity by *Doppler shift* (see next section) is said to be *tracking in Doppler* as well. If a radar is used purely for tracking, then a search radar must be present also. Because the two together are obviously rather bulky, they are often limited to ground or shipborne use and are employed for tracking hostile aircraft and missiles. They may also be used for fire control, as previously mentioned, in which case information is fed to a computer as well as being displayed. The computer directs the antiaircraft batteries or missiles, keeping them pointed not at the target, but at the position in space where the target will be intercepted by the dispatched salvo (if all goes well) some seconds later.

(*a*) (*b*)

FIGURE 16-9 Marine radar installation. (*a*) Shipboard antenna arrangement; (*b*) receiver and display console. ***(Courtesy of Soc. Selenia, S.P.A., Rome.)***

Airborne tracking radars differ from those just described in that there is usually not enough space for two radars, so that the one system must perform both functions. One of the ways of doing this is to have a radar system, such as the World War II SCR-584 radar, capable of being used in the search mode and then switched over to the tracking mode, once a target has been acquired. The difficulty, however, is that the antenna beam must be a compromise, to ensure rapid search on the one hand and accurate tracking on the other. In addition, after the switchover to the tracking mode, no further targets can be acquired, and the radar is ''blind'' in all directions except one.

Track-while-scan (TWS) radar is a partial solution to the problem, especially if the area to be searched is not too large, as often happens with airborne interception. Here a small region is searched by using spiral scanning and PPI display. A pencil beam can be used, since the targets arrive from a general direction that can be predicted. Blips can be marked on the face of the CRT by the operator, and thus the path of the target can be reconstructed and even extrapolated, for use in fire control. The advantage of this method, apart from its use of only the one radar, is that it can acquire some targets while tracking others, thus providing a good deal of information simulta-

1.a) Derive the radar range eqn of pulse radar.

b) discuss the factors governing it

2) Name the radar freq band.

3) Explain the Block diagram principle of opern of radar, and state its appns.

4) Define. (i) Second time around freq.

(ii) Max unambiguous range

(iii) Gain of antenna

(5) If the PRF of radar transmitter is 1500 ppf what is max unambiguous range.

II

1) Define foll & explain

Parasympathetic Nervous System
Diaspora.
Javaholism Major
dramatic
Disposition

(i) Min detectable signals ✓.

(ii) false alarm. ✓

(ii) Receiver noise ✓

Noise figure. ✓

false alarm probability ✓

Integration ~~Indication~~ of radar pulses.

Pre + Post ~~Pre~~ - detection Integration

~~Post~~ " "

Integration Improvement factor.

false alarm no.

2) Derive Expn for Radar range in terms of Noise fig.

3) Define Radar crossection of targets

4) Give radar range Expn in terms of Transmitter power.

5) What is the antenna pattern. What are the types of beam shapes. Explain with applications.

once I get into the scheme of things

then I'm going to zoom away.

neously. If this becomes too much for an operator, automatic computer processing can be employed, as in the semiautomatic ground environment (SAGE) system used for air defense. The disadvantage of the system, as compared with the pure tracking radar, is that although search is continuous, tracking is not, so that the accuracy is less than that obtained with monopulse or conical scan.

Tracking of extraterrestrial objects, such as satellites or spacecraft, is another specialized form of tracking. Because the position of the target is usually predictable, only the tracker is required. However, the difficulty lies in the small size and great distance of the targets. This does not necessarily apply to satellites in low orbits up to 600 km, but it certainly is true of satellites in synchronous orbits (see also Sec. 8-2.5) 36,000 km up, and also of space vehicles. Huge transmitting powers, extremely sensitive receivers and enormous fully steerable antennas are required, as may be illustrated with the following example.

Example 16-4 Calculate the maximum range of a deep-space radar operating at 2.5 GHz and using a peak pulse power of 25 MW. The antenna diameter is 64 m, the target cross section 1 m^2 and, because a maser amplifier is used, the receiver noise figure is only 1.1. Furthermore, because of the low PRF to allow the pulses to return from long distances (and thus the wide pulses used), the receiver bandwidth is only 5 kHz.

We have $\lambda = 30/2.5 = 12$ cm $= 0.12$ m, which gives

$$r_{\max} = 48\left[\frac{P_t D^4 S}{\delta f \lambda^2 (F-1)}\right]^{1/4} = 48\left[\frac{2.5 \times 10^7 \times 64^4 \times 1}{5 \times 10^3 \times 0.12^2 \times (1.1 - 1)}\right]^{1/4}$$

$$= 48\left[\frac{2.5 \times 10^7 \times 1.68 \times 10^7}{5 \times 10^3 \times 1.44 \times 10^{-2} \times 10^{-1}}\right]^{1/4} = 48\sqrt[4]{58.3 \times 10^{12}}$$

$$= 48 \times 2.76 \times 10^3 = 132{,}700 \text{ km}$$

The 64-m-diameter "dishes" at Goldstone, Calif., and near Parkes, N.S.W. (the Australian radiotelescope "borrowed" for the occasion) were able to track the Apollo missions all the way to the moon, which is at a mean distance of 385,000 km away. Yet the transmitted peak pulse power was well under 25 MW; a question thus arises as to how they were able to do it. The answer is rather simple and is connected with the use of radar *transponders*, as described in Sec. 16-2.6.

Finally, also in connection with deep-space tracking, it should be mentioned that not all radars are *monostatic* (transmitting and receiving antennas located at the same point), although the vast majority of them are. Some radars may for convenience be *bistatic*, with the transmitter and receiver separated by quite large distances. The example described may perhaps be the principal use of bistatic radar. Further information on bistatic and extraterrestrial radar applications may be obtained from chap. VI-8 in [2] and chap. 36 in [1].

16-2.5 Moving-Target Indication (MTI)

It is possible to remove from the radar display the majority of *clutter*, that is, echoes corresponding to stationary targets, thus showing only the moving targets. This is often required, although of course not in such applications as radar used in mapping or navigational applications. One of the methods of eliminating clutter is the use of MTI, which employs the *Doppler effect* in its operation.

Doppler effect The apparent frequency of electromagnetic or sound waves depends on the relative radial motion of the source and the observer. If source and observer are moving away from each other, the apparent frequency will decrease, while if they are moving toward each other, the apparent frequency will increase. This was postulated in 1842 by Christian Doppler and put on a firm mathematical basis by Armand Fizeau in 1848. The Doppler effect is observable for light and is responsible for the so-called *red shift* of the spectral lines from stellar objects moving away from the solar system. It is equally noticeable for sound, being the cause of the change in the pitch of a whistle from a passing train. It can also be used to advantage in several forms of radar.

Consider an observer situated on a platform approaching a fixed source of radiation, with a relative velocity $+v_r$. A stationary observer would note f_t wave crests (or troughs) per second if the transmitting frequency were f_t. However, because the observer is moving toward the source, that person of course encounters more than f_t crests per second. In fact, the number observed under these conditions is given by

$$f_t + f'_d = f_t\left(1 + \frac{v_r}{v_c}\right) \tag{16-13}$$

Consequently,

$$f'_d = \frac{f_t v_r}{v_c} \tag{16-14}$$

where $f_t + f'_d$ = new observed frequency

f'_d = Doppler frequency difference

Note that the foregoing holds if the relative velocity, v_r, is less than about 10 percent of the velocity of light, v_c. If the relative velocity is higher than that (most unlikely in practical cases), relativistic effects must be taken into account, and a somewhat more complex formula must be applied. However, the principle still holds under those conditions, and it holds equally well if the observer is stationary and the source is in motion. Equation (16-14) was calculated for a positive radial velocity, but if v_r is negative, f'_d in Eq. (16-14) merely acquires a negative sign. In radar involving a moving target, the signal undergoes the Doppler shift when impinging upon the target. This target becomes the "source" of the reflected waves, so that we now have a moving source and a stationary observer (the radar receiver). The two are still approaching each other, and so the Doppler effect is encountered a second time, and the overall effect is thus double. Hence the Doppler frequency for radar is

$$f_d = 2f'_d = \frac{2f_t v_r}{v_c}$$

$$= \frac{2v_r}{\lambda} \tag{16-15}$$

since $f_t/v_c = 1/\lambda$, where λ is the transmitted wavelength.

The same magnitude of Doppler shift is observed regardless of whether a target is moving toward the radar or away from it, with a given velocity. However, it will represent an increase in frequency in the former case and a reduction in the latter. Note also that the Doppler effect is observed only for radial motion, not for *tangential* motion. Thus no Doppler effect will be noticed if a target moves across the field of view of a radar. However, a Doppler shift will be apparent if the target is rotating, and the resolution of the radar is sufficient to distinguish its leading edge from its trailing edge. One example where this has been employed is the measurement of the rotation of the planet Venus (whose rotation cannot be observed by optical telescope because of the very dense cloud cover).

On the basis of this frequency change, it is possible to determine the relative velocity of the target, with either pulsed or CW radar, as will be shown. One can also distinguish between stationary and moving targets and eliminate the blips due to stationary targets. This may be done with pulsed radar by using moving-target indication.

Fundamentals of MTI Basically, the moving-target indicator system compares a set of received echoes with those received during the previous sweep. Those echoes whose phase has remained constant are then canceled out. This applies to echoes due to stationary objects, but those due to moving targets do show a phase change; they are thus not canceled—nor is noise, for obvious reasons. The fact that clutter due to stationary targets is removed makes it much easier to determine which targets are moving and reduces the time taken by an operator to "take in" the display. It also allows the detection of moving targets whose echoes are hundreds of times smaller than those of nearby stationary targets and which would otherwise have been completely masked. MTI can be used with a radar using a power oscillator (magnetron) output, but it is easier with one whose output tube is a power amplifier; only the latter will be considered here.

The transmitted frequency in the MTI system of Fig. 16-10 is the sum of the outputs of two oscillators, produced in mixer 2. The first is the *stalo,* or stable local oscillator (note that a good case can be made for using a varactor chain here). The second is the *coho*, or coherent oscillator, operating at the same frequency as the intermediate frequency and providing the *coherent* signal, which is used as will be explained. Mixers 1 and 2 are identical, and both use the same local oscillator (the stalo); thus phase relations existing in their inputs are preserved in their outputs. This makes it possible to use the Doppler shift at the IF, instead of the less convenient radio frequency $f_0 + f_c$. The output of the IF amplifier and a reference signal from the coho are fed to the phase-sensitive detector, a circuit very similar to the phase discriminator of Sec. 16-4.3.

The coho is used for the generation of the RF signal, as well as for reference in the phase detector, and the mixers do not introduce differing phase shifts. Because of this, the transmitted and reference signals are locked in phase and are said to be *coherent;* hence the name of the coho. Since the output of this detector is phase-sensitive, an output will be obtained for all fixed or moving targets. However, the phase difference between the transmitted and received signals will be constant for fixed targets, whereas it will vary for moving targets. This variation for moving targets is due to the Doppler frequency shift, which is naturally accompanied by a phase shift,

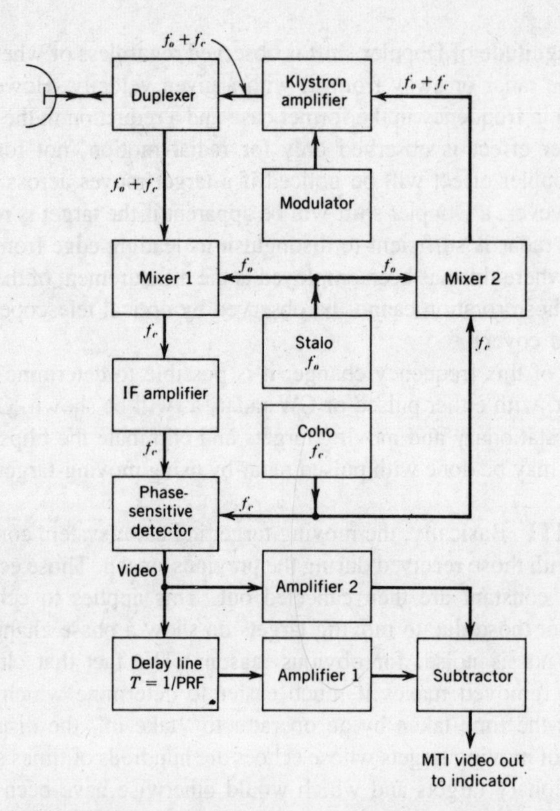

FIGURE 16-10 Block diagram of MTI radar using power amplifier output.

but this shift is not constant if the target has a radial component of velocity. If the Doppler frequency is, for example, 2000 Hz and the return time for a pulse is 124 μs (10 nmi), the phase difference between the transmitted and received signals will be some value ϕ (the same as for a stationary target at that point) plus $2000/124 = 16.12$ complete cycles, or $16.12 \times 2\pi = 101.4$ rad. When the next pulse is returned from the moving target, the latter will now be closer, perhaps only 123 μs away, giving a phase shift of $101.4 \times {}^{123}\!/_{124} = 100.7$ rad. Thus the phase shift is definitely not constant for moving targets. The situation is illustrated graphically, for a number of successive pulses, in Fig. 16-11.

It is seen from Fig. 16-11 that those returns of each pulse that correspond to stationary targets are identical with each pulse, but those portions corresponding to moving targets keep changing in phase. It is thus possible to subtract the output for each pulse from the preceding one, by delaying the earlier output by a time equal to the pulse interval, or 1/PRF. Since the delay line also attenuates heavily and since signals must, of course, be of the same amplitude if permanent echoes are to cancel, an amplifier follows the delay line. To ensure that this does not introduce a spurious phase shift, an amplifier is placed in the undelayed line, which has exactly the same response characteristics (but a much lower gain) than amplifier 1. The delayed and undelayed signals are compared in the subtractor (adder with one input polarity re-

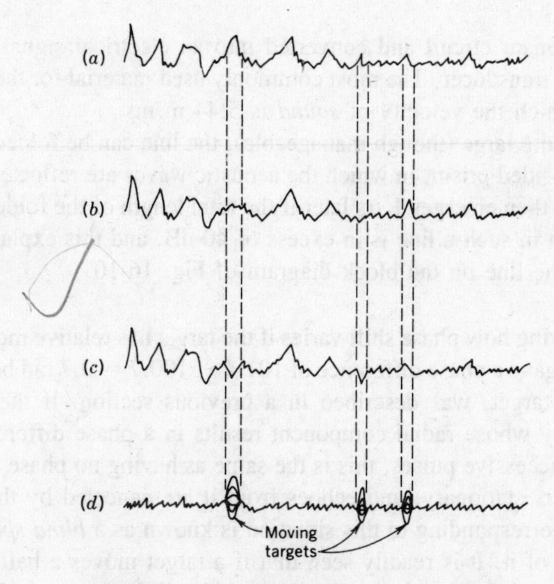

FIGURE 16-11 Operation of MTI radar. (*a*), (*b*), (*c*) Phase detector output for three successive pulses; (*d*) subtractor output.

versed), whose output is shown in Fig. 16-11*d*. This can now be rectified and displayed in the usual manner.

Other analog MTI systems These include *area MTI*, which involves subtracting a complete scan from the previous one and displaying only the difference; it is done with storage CRTs. Another system is almost identical to the one described but uses a pulsed magnetron oscillator instead of an amplifier. A different technique must be employed here to achieve coherence, because each cycle of the magnetron oscillations begins with a phase quite unrelated to the previous pulsed cycle. *Noncoherent MTI* is also sometimes used, deriving the required phase variations by comparing the returns from stationary and radially moving targets. However, this method suffers from the disadvantage of requiring stationary targets in each scan in addition to the moving target. Finally, note that all coherent systems require a fairly high pulse repetition frequency to ensure the return of several pulses from each target. The various MTI systems are excellently described in chaps. 17 and 18 of [1]. This also describes *airborne moving-target indication* (AMTI), in which compensation for the motion of the radar set is an added requirement. The performance of MTI and that of allied radar systems is described and compared in [4].

Delay lines Because of the delay times required, it would be unthinkable to use electromagnetic delay means in MTI. If the PRF is 1000, then the delay required is 1 ms, in which time an electromagnetic wave in an air-dielectric line travels 300 km! The method adopted to provide the requisite delay in practice is rather similar to that used with mechanical filters. The signal is converted into acoustic vibrations, passed through a mechanical resonant circuit and converted into an electrical signal at the output end, with a suitable transducer. The most commonly used material for the delay line is fused quartz, in which the velocity of *sound* is 5.44 m/ms.

Since this is still quite large (though manageable), the line can be folded. This consists in having a many-sided prism, in which the acoustic waves are reflected from the plane sides. The signal then emerges 1 μs later if the total length of the folded path is 5.44 m. The attenuation in such a line is in excess of 40 dB, and this explains the amplifier accompanying the line on the block diagram of Fig. 16-10.

Blind speeds When showing how phase shift varies if the target has relative motion, a fictitious situation, which gave a phase difference of 101.4 − 100.7 = 0.7 rad between successive pulses on the target, was described in a previous section. If the target happens to have a velocity whose radial component results in a phase difference of exactly 2π rad between successive pulses, this is the same as having no phase shift at all. The target thus appears stationary, and echoes from it are canceled by the MTI action. A radial velocity corresponding to this situation is known as a *blind speed,* as are any integral multiples of it. It is readily seen that if a target moves a half-wavelength between successive pulses, the change in phase shift will be precisely 2π rad. We may thus state that

$$v_b = \text{PRF}\frac{n\lambda}{2} \tag{16-16}$$

where v_b = blind speed
λ = wavelength of transmitted signal
n = any integer (including 0!)

Example 16-5 An MTI radar operates at 5 GHz, with a pulse repetition frequency of 800 pps. Calculate the lowest three blind speeds of this radar.

$$\lambda = \frac{v_c}{f} = \frac{3 \times 10^8}{5 \times 10^9} = 0.06 \text{ m}$$

The lowest blind speed corresponds to $n = 1$. Thus

$$v_b = 800 \times 0.06 = 48 \text{ m/s}$$
$$= 48 \times 60 \times 60 \times 10^{-3} = 172.8 \text{ km/h}$$

Consequently the lowest three blind speeds will be 172.8, 345.6 and 518.4 km/h (for n = 1, 2, and 3).

The fact that blind speeds exist need not be a serious problem and does not normally persist beyond a small number of successive pulses. This could be caused by a target flying directly toward the radar set at a constant velocity, but it would be sheer coincidence, and a far-fetched one at that, for a target to do this accidentally. However, we do live in a world that produces sophisticated electronic countermeasures, and it is not beyond the realm of possibility that a target may be flying at a blind speed on purpose. A wideband receiver and microprocessor on board the target aircraft or missile could analyze the transmitted frequency and PRF and adjust radial velocity accordingly. The solution to that problem is to have a variable PRF. That presents no difficulty, but varying the delay in the MTI radar does. It can, however, be done by having two delay lines and compensating amplifiers. One of these can be a small delay line, having a delay that is (say) 10 percent of the main delay. This second line will then be switched in and out on alternate pulses, changing the blind speed by 10 percent each time.

Digital MTI It is possible to replace the delay line and amplifier arrangement of an analog MTI system with digital-to-analog conversion of the received signal. After the signal has been digitally coded, it can be stored in a computer memory. The echoes received from each pulse are now subtracted in the memory from those received from the previous pulse, whereupon the difference is converted to analog form and displayed as before. With digital MTI (or DMTI), no difficulties arise in varying the PRF. Indeed, it may be varied almost randomly from one pulse to the next. In [5] a digital MTI system is described in detail, and [4] compares it with analog MTI systems. Interestingly enough, the resolution limit in DMTI is governed, in part, by quantizing noise. Just like in pcm, as discussed in Chap. 13, analog signals must be quantized before their conversion to digital form.

16-2.6 Radar Beacons

A radar beacon is a small radar set consisting of a receiver, a separate transmitter and an antenna which is often omnidirectional. When another radar transmits a coded set of pulses at the beacon, i.e., *interrogates* it, the beacon *responds* by sending back its specific pulse code. The pulses from the beacon, or *transponder* as it is often called, may be at the same frequency as those from the interrogating radar, in which case they are received by the main station together with its echo pulses. They may alternatively be at a special beacon frequency, in which case a separate receiver is required by the interrogating radar. Note that the beacon does not transmit pulses continuously in the same way as a search or tracking radar but only responds to the correct interrogation.

Applications One of the functions of a beacon may be to identify itself. For example, the beacon may be installed on a target, such as an aircraft, and will transmit a specific pulse code when interrogated. These pulses then appear on the PPI of the interrogating radar and inform it of the identity of the target. The system is in use in airport traffic control and also for military purposes, where it is called identification, friend or foe (IFF). However, whereas the IFF operates satisfactorily with naval targets, it tends to fail if large numbers of fast planes are involved (some friend and some foe), even with modern radars.

Another use of radar beacons is rather similar to that of lighthouses, except that radar beacons can operate over much larger distances. An aircraft or ship, having interrogated a number of beacons of whose exact locations it may be unaware (on account of being slightly lost), can calculate its position from the coded replies accurately and automatically.

The presence of a beacon on a target increases enormously the distance over which a target may be tracked. Such *active* tracking gives much greater range than the *passive* tracking so far described, because the power transmitted by the beacon (modest though it normally is) is far in excess of the power that this target would have reflected had it not carried a beacon. This is best demonstrated quantitatively, as in the next section.

Beacon range equation Following the reasoning used to derive the general radar range equation, we may change Eq. (16-3) slightly to show that the power intercepted by the beacon antenna is given by

$$P_B = \frac{A_{pT}P_{tT}A_{0B}}{4\pi r^2} \tag{16-17}$$

where all symbols have their previously defined meanings, except that the subscript T is now used for quantities pertaining to the transmitter of the main radar, and B is used for the beacon functions. A_{0B} is thus the capture area of the beacon's antenna.

If $P_{\min,B}$ is the minimum power receivable by the beacon, the maximum range for the *interrogation link* will be

$$r_{\max,I} = \sqrt{\frac{A_{pT}P_{tT}A_{0B}}{4\pi P_{\min,B}}} \tag{16-18}$$

Substituting into Eq. (16-17) for the power gain of the transmitter antenna from Eq. (16-6), and for the minimum power receivable by the beacon from Eq. (16-10), and then canceling, we obtain the final form of the maximum range for the interrogation link. This is

$$r_{\max,I} = \sqrt{\frac{A_{0T}P_{tT}A_{0B}}{\lambda^2 kT_0\delta f(F_B - 1)}} \tag{16-19}$$

It has been assumed in Eq. (16-19) that the bandwidth and antenna temperature of the beacon are the same as those of the main radar. By an almost identical process of reasoning, the maximum range for the *reply link* is

$$r_{\max,R} = \sqrt{\frac{A_{0B}P_{tB}A_{0T}}{\lambda^2 kT_0\delta f(F_T - 1)}} \tag{16-20}$$

To calculate the maximum (theoretical) range for active tracking, both Eqs. (16-19) and (16-20) are solved, and *the lower of the two values obtained is used.*

It is now possible to solve the puzzle of the radar tracking of the Apollo moon missions, with the aid of an example based on Example 16-4.

Example 16-6 Calculate the maximum active tracking range of a deep space radar operating at 2.5 GHz and using a peak pulse power of 0.5 MW, with an antenna diameter of 64 m, a noise figure of 1.1 and a 5-kHz bandwidth, if the beacon antenna diameter is 1 m, its noise figure is 13 dB and it transmits a peak pulse power of 50 W. (Note the reduced transmitting power as compared with Example 16-4, as well as the very low beacon power.)

Preliminary calculations reveal that the 13-dB noise figure of the beacon receiver is equal to a ratio of 20, and applying $A_0 = 0.65\pi D^2/4$ gives capture areas of 2090 m^2 for the ground radar and 0.51 m^2 for the beacon. Substituting the relevant data into

Eq. (16-19) gives

$$r_{\max,I} = \sqrt{\frac{2.09 \times 10^3 \times 5 \times 10^5 \times 5.1 \times 10^{-1}}{1.2^2 \times 10^{-2} \times 1.38 \times 10^{-23} \times 2.9 \times 10^2 \times 5 \times 10^3(20 - 1)}}$$
$$= 9.87 \times 10^{12} \text{ m}$$
$$= 9870 \text{ million km } (= 5330 \text{ million nmi})$$

Since this is almost one and a half times the diameter of the solar system (out to Pluto), there should be no difficulty in tracking the beacon over the relatively short distance to the moon. For the reply link, the maximum range is

$$r_{\max,R} = \sqrt{\frac{5.1 \times 10^{-3} \times 5 \times 10 \times 2.09 \times 10^3}{1.2^2 \times 10^{-2} \times 1.38 \times 10^{-23} \times 2.9 \times 10^2 \times 5 \times 10^3(1.1 - 1)}}$$
$$= 1.36 \times 10^{11} \text{ m}$$
$$= 136 \text{ million km } (= 73.4 \text{ million nmi})$$

Being the shorter of the two, 136 million km is the maximum tracking range here.

The results of Example 16-6 should be taken with a grain of salt, because system losses, clutter and other vagaries of nature can reduce this range by as much as tenfold. To compensate for this, however, the range could be tripled if the diameter of the beacon antenna is also tripled. A fold-out, metallized umbrella spacecraft antenna with a 3-m (10-ft) diameter is certainly feasible. Again, the 13-dB noise figure for the beacon receiver is conservative, and reducing it to 10 dB (still fairly conservative) would further increase the range. A slower PRF and less insistence on pulses with steep sides would permit a tenfold bandwidth reduction and a similar pulse power increase from the beacon. A total range for the reply link could thus comfortably exceed 1000 million km, even allowing for the degradations mentioned above. That distance puts within range all the planets up to and including Saturn.

16-3 OTHER RADAR SYSTEMS

A number of radar systems are sufficiently unlike those treated so far to be dealt with separately. They include first of all *CW radar* which makes extensive use of the Doppler effect for target speed measurements. Another type of CW radar is frequency-modulated to provide range as well as velocity. Finally, *phased array* and *planar array* radars will be treated in this "separate" category. Here, the transmitted (and receiving) beam is steered not by moving an antenna but by changing the phase relationship in the feeds for a vast array of small individual antennas. These systems will now be described in turn.

16-3.1 CW Doppler Radar

A simple Doppler radar, such as the one shown in Fig. 16-12, sends out continuous sine waves rather than pulses. It uses the Doppler effect to detect the frequency change caused by a moving target and displays this as a relative velocity.

Example 16-7 With a (CW) transmit frequency of 5 GHz, calculate the Doppler frequency seen by a stationary radar when the target radial velocity is 100 km/h (62.5 mph).

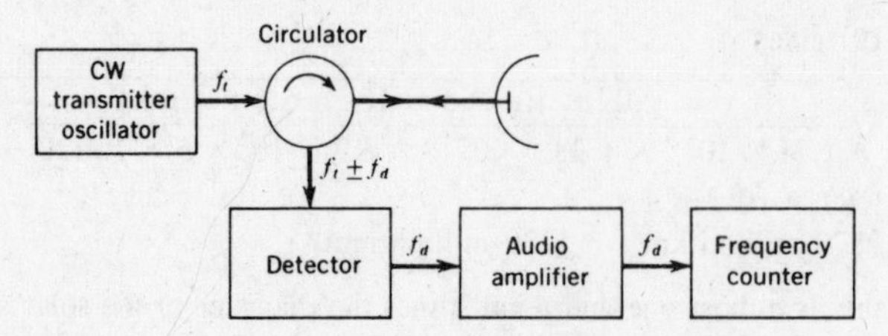

FIGURE 16-12 Simple Doppler CW radar.

Before using Eq. (16-15), it is necessary to calculate the wavelength, and also the target speed in meters per second.

$$\lambda = \frac{3 \times 10^8}{5 \times 10^9} = 0.06 \text{ m}$$

$$v_r = \frac{100 \times 10^3}{60 \times 60} = 27.8 \text{ m/s}$$

$$f_d = \frac{2v_r}{\lambda} = \frac{2 \times 27.8}{0.06} = 927 \text{ Hz}$$

It is seen that, with C-band radar frequencies, the speeds which motorists may be ticketed for exceeding give Doppler frequencies in the audio range.

Since transmission here is continuous, the circulator of Fig. 16-12 is used to provide isolation between the transmitter and the receiver. Since transmission *is* continuous, it would be pointless to use a duplexer. The isolation of a typical circulator is of the order of 30 dB, so that some of the transmitted signal leaks into the receiver. This is quite convenient: the signal can be mixed in the detector with returns from the target, and the difference is the Doppler frequency. Being generally in the audio range in most Doppler applications, the detector output can be amplified with an audio amplifier before being applied to a frequency counter. The counter is a normal one, except that its output is shown as kilometers or miles per hour, rather than the actual frequency in hertz. The main disadvantage of a system as simple as this is its lack of sensitivity. The type of diode detector that is used to accommodate the high incoming frequency is not a very good device at the audio output frequency, because of the *modulation* noise which it exhibits at low frequencies (see Sec. 2-2.4). The receiver whose block diagram is shown in Fig. 16-13 is an improvement in that regard.

As shown, a small portion of the transmitter output is mixed with the output of a local oscillator, and the sum is fed to the receiver mixer. This also receives the Doppler-shifted signal from its antenna and produces an output difference frequency that is typically 30 MHz, plus or minus the Doppler frequency. The output of this mixer is amplified and demodulated again, and the signal from the second detector is just the Doppler frequency. However, its sign is lost, so that it is not possible to tell whether the target is approaching or receding. The overall receiver system is rather similar to the superheterodyne. Extra sensitivity is provided by the lowered noise, because the output of the diode mixer is now in the vicinity of 30 MHz, at which FM noise has disappeared.

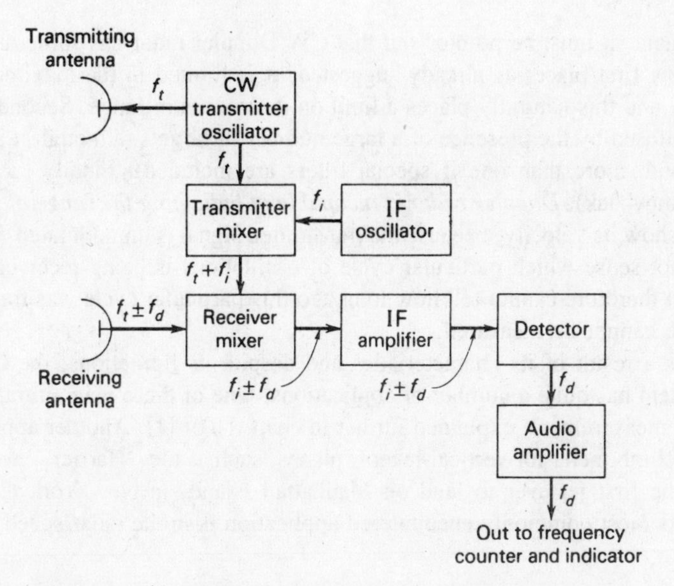

FIGURE 16-13 CW Doppler radar with IF amplification.

Separate receiving and transmitting antennas have been shown, although this arrangement is not compulsory; a circulator could be used, as in the simpler set of Fig. 16-12. Separate antennas are used to increase the isolation between the transmitter and receiver sections of the radar, especially since there is no longer any need for a small portion of the transmitter output to leak into the receiver mixer, as there was in the simpler set. To the contrary, such leakage is highly undesirable, because it brings with it the hum and noise from the transmitter and thus degrades the receiver performance. In fact, the problem of isolation is the main determining factor, rather than any other single consideration in the limiting of the transmitter output power. As a consequence, the CW power from such a radar seldom exceeds 100 W and is often very much less. Gunn or IMPATT diodes or, for the highest powers, CW magnetrons are used as power oscillators in the transmitter. They operate at much the same frequencies as in pulsed radar.

Advantages, applications and limitations CW Doppler radar is capable of giving accurate measurements of relative velocities, using low transmitting powers, simple circuitry, low power consumption and equipment whose size is much smaller than that of comparable pulsed equipment. It is unaffected by the presence of stationary targets, which it disregards in much the same manner as MTI pulsed radar (it also has blind speeds, for the same reason as MTI). In addition, it can operate (theoretically) down to zero range because, unlike in the pulsed system, the receiver is ON at all times. It is also capable of measuring a large range of target speeds quickly and accurately. With some additional circuitry, CW radar can even measure the direction of the target, in addition to its speed.

Before the reader begins to wonder why pulsed radar is still used in the majority of equipment, it must be pointed out that CW Doppler radar has some disadvantages also. In the first place, as already suggested, it is limited in the maximum power it transmits, and this naturally places a limit on its maximum range. Second, it is rather easily confused by the presence of a large number of targets (although it is capable of dealing with more than one if special filters are included). Finally (and this is its greatest drawback), *Doppler radar is incapable of indicating the range of the target;* it can only show its velocity, because the transmitted signal is unmodulated. The receiver thus cannot sense which particular cycle of oscillations is being received at the moment, and therefore cannot tell how long ago this particular cycle was transmitted, so that range cannot be measured.

As a result of its characteristics and despite its limitations, the CW Doppler radar system has quite a number of applications. One of these is in aircraft navigation for speed measurement, explained further in chap. 16 of [1]. Another application is in a rate-of-climb meter for vertical-takeoff planes, such as the "Harrier," which in 1969 became the first jet ever to land on Manhattan Island, in New York City. Finally, perhaps its most commonly encountered application is in the radar speed meters used by police.

16-3.2 Frequency-Modulated CW Radar

The greatest limitation of Doppler radar, i.e., its inability to measure range, may be overcome if the transmitted carrier is frequency-modulated. If this is done, it should be possible to eliminate the main difficulty with CW radar in this respect, namely, its inability to distinguish one cycle from another. Using FM will require an increase in the bandwidth of the system, and thus once again it is seen that a bandwidth increase in a system is required if more information is to be conveyed (in this case, information with regard to range).

Figure 16-14 shows the block diagram of a common application of the FM CW radar system, the airborne altimeter. Sawtooth frequency modulation is used for simplicity, although in theory any modulating waveform might be adequate. If the target (in this case, the Earth) is stationary with respect to the plane, a frequency difference

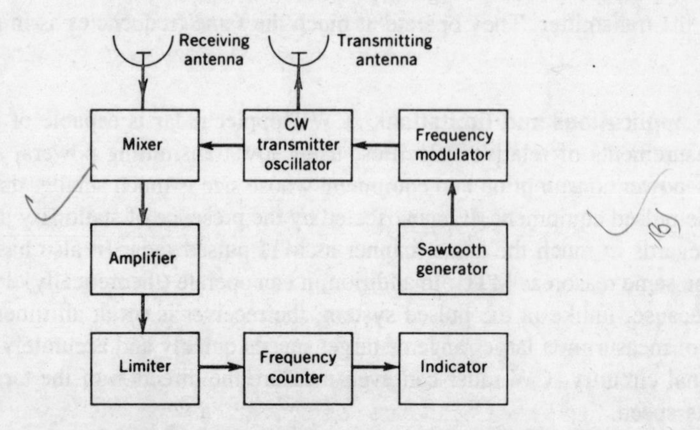

FIGURE 16-14 Block diagram of simple FM CW radar altimeter.

proportional to the height of the plane will exist between the received and the transmitted signals. It is due to the fact that the signal now being received was sent at a time when the instantaneous frequency was different. If the rate of change of frequency with time due to the FM process is known, the time difference between the sent and received signals may be readily calculated, as can the height of the aircraft. Thus the output of the mixer in Fig. 16-14, which produces the frequency difference, can be amplified, fed to a frequency counter and then to an indicator whose output is calibrated in meters or feet.

If the relative velocity of the radar and the target is not zero, another frequency difference, or beat, will superimpose itself on top of the frequency difference just discussed, because of the Doppler frequency shift. However, the *average* frequency difference will be constant and due to the time difference between the sending and return of a particular cycle of the signal. Thus correct height measurements can still be made on the basis of the average frequency difference. The beat superimposed on this difference can now be used, as with ordinary Doppler radar, to measure the velocity of (in this case) the aircraft, when due allowance has been made for the slant range.

The altimeter is a major application of FM CW radar. It is used in preference to pulsed radar because of the short ranges (i.e., heights) involved, since CW radar has no limit on the minimum range, whereas pulsed radar does have such a limit, as previously discussed. Furthermore, fairly simple low-power equipment can be used, as with CW Doppler radar. Because of the size and proximity of the Earth, small antennas can also be used, reducing the bulk of the equipment even further. A typical altimeter operates in the C band, uses a transmitter power typically from 1 to 2 W, easily obtained from an IMPATT or a Gunn diode, and has a range of up to 10,000 m or more, with a corresponding accuracy of about 5 percent. Further information on FM CW radar may be obtained in chap. 16 of [1].

16-3.3 Phased Array Radars

Introduction With some notable exceptions, the vast majority of radars have to cover an area in searching and/or tracking, rather than always being pointed in the same direction. This implies that the antenna will have to move, although it was seen in Sec. 16-2.2 that some limited beam movement can be produced by multiple feeds or by a moving feed antenna. As long as antenna motion is involved in moving the beam, limitations caused by inertia will always exist. For example, a limit on the maximum scanning speed will be imposed by antenna mechanics.

Additionally, the problem encountered with a single antenna of fixed shape is that the shape of the beam it produces is also constant, unless some rather complex modifications are introduced. Finally, there is the difficulty (too obvious to mention until now) caused by the fact that a single antenna can point in only one direction at a time, therefore sending out only one beam at a time. This makes it rather difficult to track a large number of targets simultaneously and accurately. A similar difficulty is encountered when trying to track some targets while acquiring others. Such problems could be overcome, and a very significant improvement in versatility would result, if a moving beam could be produced by a stationary antenna. Although this cannot be done readily with a single antenna, it can be done with an array consisting of a large number of individual radiators. Beam steering can be achieved by the introduction of variable phase differences in the individual antenna feeders, and electronic variation of the phase shifts.

Possibilities It was shown in Sec. 9-6.1 that a collinear dipole array can have either broadside or end-fire action. It will be recalled that the direction of the beam will be at right angles to the plane of the array if all the dipoles are fed in phase, whereas feeding them with a progressive phase difference results in a beam that is in the plane of the array, along the line joining the dipole centers. It will thus be appreciated that if the phase differences between the dipole feeds are varied between these two extremes, the direction of the beam will also change accordingly. Extending this principle one step further, it can be appreciated that a plane dipole array, with variable phase shift to the feeders, will permit moving the direction of the radiated beam in a plane rather than a line. Nor do the individual radiators have to be dipoles: slots in waveguides and other arrangements of small omnidirectional antennas will do as well. Finally, it is possible to arrange four such antenna arrays, thus obtaining a full hemispherical coverage.

Each plane array would, for hemispherical coverage, point 45° upward. In this way, the beam issuing from each face would have to move ±45° in elevation and ±45° in azimuth in order to cover its quadrant. In practical systems, vast numbers of individual radiators are involved. One tactical radar has, in fact, 4096 (2^{12}) radiating slots per face.

Types There are broadly two different types of phased arrays possible. In the first, one high-power tube feeds the whole array; as a variation on this theme, the array is split into a small number of subarrays, and a separate tube feeds each of these. The feeding is done through high-level power dividers (hybrids) and high-power phase shifters. The phase shifters are often ferrite. Indeed, most of the advances in ferrite technology in the 1960s were spin-offs from phased array military contracts. It will be recalled that the phase shift introduced by a suitable piece of ferrite depends on the magnetic field to which the ferrite is subjected; this was shown in Sec. 10-5.2. By adjusting this magnetic field, a full 360° phase change is possible.

Digital phase shifters are also available, using PIN diodes in distributed circuits. A particular section will give a phase shift that has either of two values, depending on whether the diode is ON or OFF. A typical "4-bit" digital phase shifter may consist of four PIN phase shifters in series. The first will produce a shift of either 0 or 22½°, depending on the diode bias. The second offers the alternatives of 0 or 45°, the third 0 or 90° and the fourth 0 or 180°. By using various combinations, a phase shift anywhere between 0 and 360° (in 22½° steps) may be provided. The ferrite phase shifters have the advantages of continuous phase shift variation and the ability to handle higher powers. PIN diode phase shifters, on the other hand, although they cannot handle quite such high powers, are able to provide much faster variations in phase shift and therefore beam movement. As a good guide, the phase variations that take a few milliseconds with ferrite shifters (Fig. 16-15) can be accomplished in the same number of microseconds with digital shifters.

FIGURE 16-15 Two 32-element array modules, including radiation, phase shifters, RF power divider, and bias harnesses. The unit on the left is an MIC phase shifter; on the right is a ferrite phase shifter. ***(Courtesy of RCA.)***

A second broad type of phased array radar uses many RF generators, each of which drives a single radiating element or bank of radiating elements. Semiconductor diode generators are normally used, with phase relationships closely controlled by means of phase shifters. The use of YIG and microwave integrated circuit (MIC) phase shifters has enhanced several aspects of the phased array radar. The YIG phase shifter, when coupled with irises for matching purposes, results in a radiating element which is compact, easy to assemble and relatively inexpensive. The MIC phase shifter greatly reduces the size of arrays, since it is itself small and integrated into the radiating element.

These multigenerator arrays provide wide-angle scanning over an appreciable frequency range. Scanning may be accomplished through a combination of mechanical and electronic means, or through electronic means alone. The array shown in Fig. 16-16 employs RF generators to drive each horizontal bank of radiators. Elevation scanning can therefore be accomplished electronically, although horizontal scanning uses traditional mechanical techniques. The array shown in Fig. 16-17 provides one generator for each radiating element, and this makes electronic scanning for both

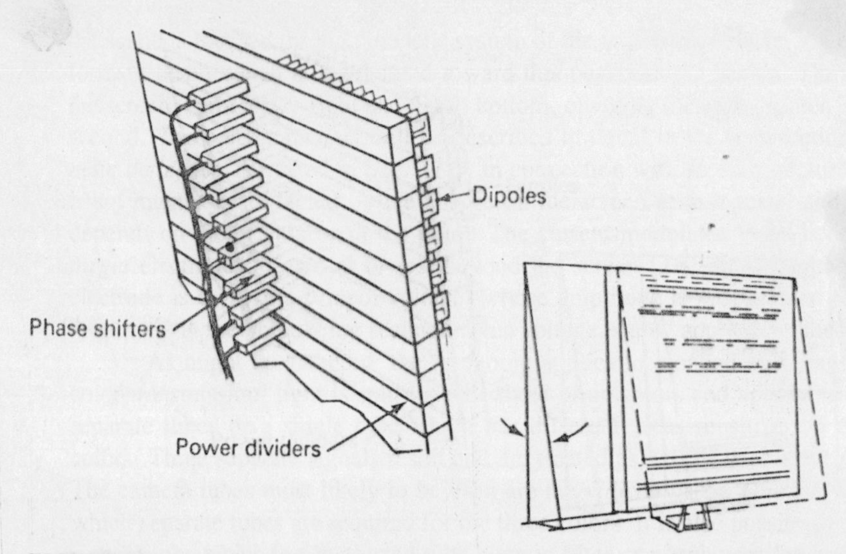

FIGURE 16-16 A phased array antenna that provides for elevation scanning by feeding each horizontal row of elements with a separate phase shifter. *(RCA Engineer, courtesy of RCA.)*

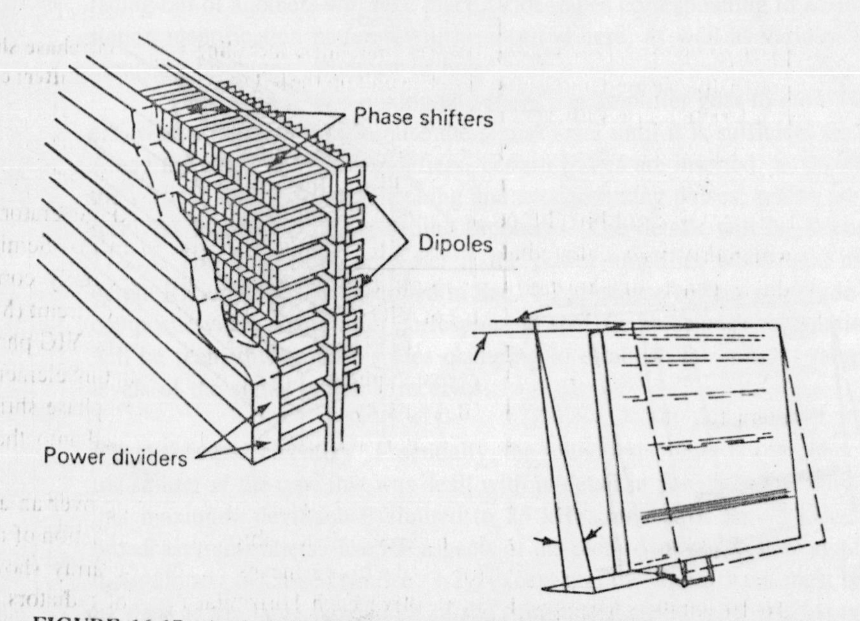

FIGURE 16-17 A phased array antenna that provides for both azimuth and elevation scanning. A separate phase shifter feed each radiating element. *(RCA Engineer, courtesy of RCA.)*

horizontal and vertical planes possible, although the cost for this type of array is of course significantly higher. The number of phaser/generator elements increases from 70 for a typical array of the first type to 4900 for an array of the second type.

Arrays using multiple semiconductor diode generators have several advantages. The generators operate at much lower power levels and are therefore cheaper and more reliable. With so many independent RF generators, any failures that occur will be individual rather than total, and their effect will thus be merely a gradual deterioration, not a catastrophic failure. The disadvantages of the second system include the high cost of so many Gunn or IMPATT or even TRAPATT oscillators. The lower available powers at higher frequencies are yet another problem: even 4096 oscillators producing 100-W pulses each give out only a little over 400 kW, much less than a medium-large tube. The power dissipation is more of a problem than with tubes, since efficiencies of diode RF generators are noticeably lower.

Practicalities In a sense, phased array radars have been the "glamour" systems, in terms of development money spent and space devoted in learned journals. Certainly, there is no doubt that they can work and currently do so in quite a number of establishments (see [6]). They can be astonishingly versatile. For example, the one array can rapidly locate targets by sending out two fan-shaped beams simultaneously. One is vertical and moves horizontally, while the other is horizontal and moves vertically. Once a target has been located, it can then be tracked with a narrow beam, while other wide beams meanwhile acquire more targets. The phased array radar utilizing electronic techniques benefits from inertialess scanning. Since the beam can be redirected and reconfigured in microseconds, one array can be programmed to direct pulses to various locations in rapid succession. The result is that the array can simultaneously undertake acquisition and tracking operations for multiple targets. The possibilities are almost endless and are well covered in the literature ([1], chap. 11; [6], [7], [8], [9]).

Because phased array radars have to perform complex tasks, they must themselves also be complex. This, in turn, makes them atrociously expensive. One authority [7] quotes a typical cost of $1 million for just the phase shifters and their drivers in one system. That still leaves the computer control, RF sources, power dividers and the arrays themselves, as well as the costs of testing and installation. A significant cost reduction could be achieved by mass production, if demand for phased array radars increases. It is to be hoped, however, that this situation does not develop into a vicious circle.

Related technology Signal processing is one aspect of radar technology which has resulted in a significant improvement in radar capabilities. Signal processing systems currently in use with radar systems depend heavily on computer and microchip technology. These systems perform the functions of analyzing, evaluating and displaying radar data, as well as controlling the subsequent pulse emissions.

Signal processing used with radar systems includes filtering operations of the full bandwidth signal to separate signal waveforms from noise and interfering background signals. This accommodation to the electromagnetic environment in which the radar system operates is further enhanced by the ability to utilize computer algorithms to alter pulse frequency and other characteristics, in response to the transmissions of other systems. By varying the transmitted signals, it is possible for the system to attain significant immunity from interference (from other signals). Computer evaluation and control prevent interference to the system since the interfering signal cannot track the frequency changes and the subpulses generated by the system at the direction of the signal-processing computer. Usable images can be obtained even in adverse or very active electromagnetic environments. This enhancement of the radar system capability is of particular value to military and other systems which must operate in close proximity to other radars. The improvement of displays resulting from the use of computer recognition of moving targets within ground clutter was discussed in broad terms in Sec. 16-2.5. With sophisticated computer systems available to the radar, additional display manipulations and improvements can be achieved.

Radar systems benefit from large scale integration in the same way as other electronic fields. As a "signal processor on a chip" becomes a reality, the cost, complexity and size of even a complex radar system will decrease. Digital simulation of analog filters and other devices will also contribute to reduction of system costs. Because real-time radar signal processing needs to execute instructions rates exceeding 2×10^7 operations per second, the current digital switching speed has become a limiting factor. As digital technology improves in speed, signal processing will become even more important for radar systems.

16-3.4 Planar Array Radars

The planar array radar uses a high-gain planar array antenna. A fixed delay is established between horizontal arrays in the elevation plane. As the frequency is changed, the phase front across the aperture tends to tilt, with the result that the beam is moved in elevation.

Figure 16-18 shows a planar antenna array to which a burst of five subpulses,

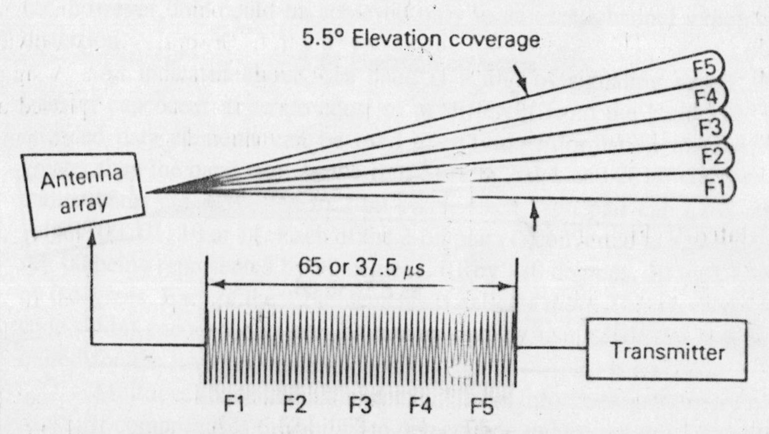

FIGURE 16-18 Frequency scanning as used by planar array radar causes radar beams to be elevated slightly above one another.

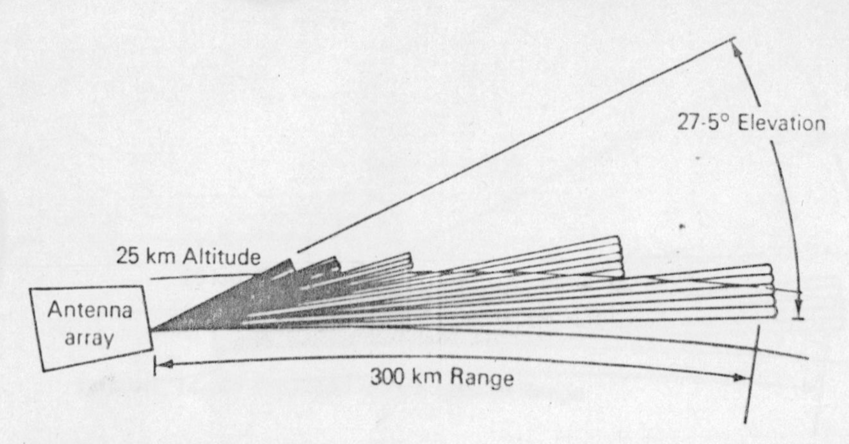

FIGURE 16-19 Planar array radar showing five separate groups of fine beams which permit scanning of 27.5° of elevation.

each at a different frequency, is applied. The differing frequencies cause each successive beam to be elevated slightly more than the previous beams. A 27.5° elevation is scanned by the radar illustrated in Fig. 16-19 with five of the five beam groups used. The planar array system has several advantages in that each beam group has full transmitter peak power, full antenna gain and full antenna sidelobe performance. The use of frequency changes provides economical, simple and reliable inertialess elevation scanning.

OBLEMS

For self-testing questions on this chapter, see p. 711.

16-1. A radar is to have a maximum range of 60 km. What is the maximum allowable pulse repetition frequency for unambiguous reception?
[Ans.: 2500 pps]

16-2. An L-band radar operating at 1.25 GHz uses a peak pulse power of 3 MW and must have a range of 100 nmi (185.2 km) for objects whose radar cross section is 1 m^2. If the minimum receivable power of the receiver is 2×10^{-13} W, what is the smallest diameter the antenna reflector could have, assuming it to be a full paraboloid with $k = 0.65$?
[Ans.: 3.84 m]

16-3. The noise figure of a radar receiver is 12 dB, and its bandwidth is 2.5 MHz. What is the value of P_{min} for this radar?
[Ans.: 1.59×10^{-13} W]

16-4. The AN/FPS-16 guided-missile tracking radar operates at 5 GHz, with a 1-MW peak power output. If the antenna diameter is 3.66 m (12 ft), and the receiver has a bandwidth of 1.6 MHz and an 11-dB noise figure, what is its maximum detection range for 1-m^2 targets?
[Ans.: 345.6 km (this is 186.6 nmi, whereas the officially quoted figure for this radar is 180 nmi)]

16-5. A radar transmitter has a peak pulse power of 400 kW, a PRF of 1500 pps and a pulse width of 0.8 μs. Calculate (*a*) the maximum unambiguous range, (*b*) the duty cycle, (*c*) the average transmitted power (*d*) a suitable bandwidth.
[Ans.: (a) 100 km, (b) 0.12%, (c) 480 W, (d) 1.25 to 1.75 MHz]

16-6. An 8-GHz police radar measures a Doppler frequency of 1788 Hz, from a car approaching the stationary police vehicle, in an 80-km/h (50-mph) speed limit zone. What should the police officer do?
[Ans.: Give chase; that car is traveling at 120.7 km/h (75 mph)!]

16-7. An MTI radar operates at 10 GHz with a PRF of 3000 pps. Calculate its lowest blind speed.
[Ans.: 162 km/h]

16-8. Repeat Prob. 16-7 for a frequency of 3 GHz and a PRF of 500 pps.
[Ans.: 90 km/h]

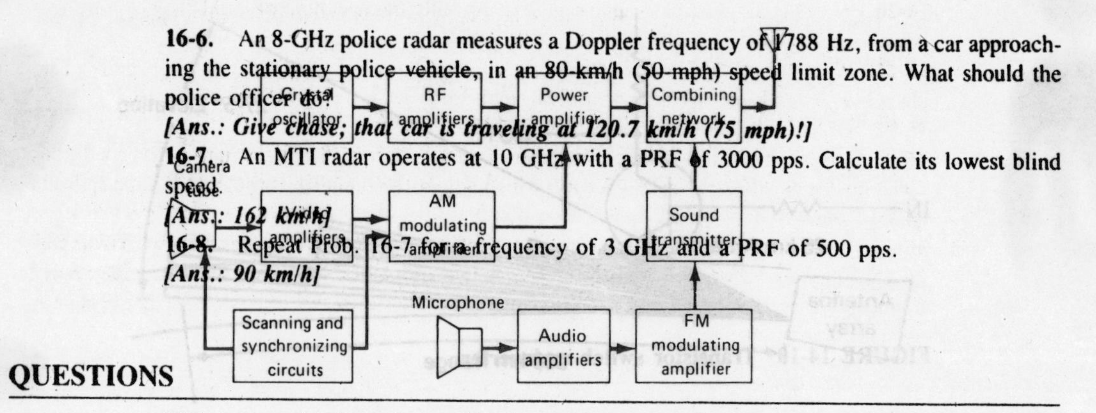

QUESTIONS

16-1. Draw the block diagram of a basic radar set, and explain the essentials of its operation.

16-2. What are the basic functions of radar? In indicating the position of a target, what is the difference between *azimuth* and *elevation?*

16-3. What is the difference between the *pulse interval* and the PRF? What are the factors that govern the selection of the PRF for a particular radar?

16-4. What are some of the ways of *jamming,* or confusing, enemy radar?

16-5. Derive the basic radar range equation, as governed by the minimum receivable echo power.

16-6. Describe briefly some of the factors governing the relation between the radar cross section of a target and its true cross section.

16-7. Draw a functional block diagram of a pulsed radar set, and describe the function of each block.

16-8. Describe the operation of a line-pulsing radar modulator. Why is a *thyratron* never used? What is used instead? What are the advantages of this modulator? What is its most significant drawback?

16-9. What are the factors influencing the bandwidth of a radar receiver? What are the advantages and disadvantages of a very large bandwidth?

16-10. By what factors is the pulse repetition frequency governed? What is meant by *ambiguous reception?* Give a numerical example of this.

16-11. With diagrams, describe the motion of the antenna beam in some of the more common antenna scanning patterns.

16-12. Describe the method of *lobe switching,* as used to track a target after it has been acquired. In what way is lobe switching an improvement over merely pointing an antenna accurately at the target?

16-13. Describe, with the aid of a sketch, the *conical scanning* method of tracking an acquired target. How is this an improvement over lobe switching?

16-14. With the aid of a sketch, describe the equipment and technique used in the *monopulse* method of target tracking.

16-15. Describe the functions of the more important controls that may be provided with an A-scope radar display.

16-16. With the aid of a sketch showing a typical display, explain fully the PPI radar indicator. Why is this method called *intensity modulation?*

16-17. Describe the essential characteristics, functions and major applications of search radar systems.

16-18. How does *track-while-scan* radar operate? In what ways is it a compromise?

16-19. What is the *Doppler effect?* What are some of the ways in which it manifests itself? What are its radar applications?

Picture tube
Tuner
IF amplifiers
Video detector
Video amplifiers
Sync
Sound IF amplifiers
Sound demodulator
Audio amplifiers
Scanning and synchronizing circuits
Loudspeaker

FIGURE 17-1 Basic monochrome television system. (a) Transmitter; (b) receiver.

17-1.2 Television Systems and Standards

It is clear that a large amount of information must be broadcast by a television transmitter and that there are a variety of ways in which this can be done. Accordingly, a need exists for uniform standards for TV transmission and reception. Regrettably, no agreement has been reached for the adoption of worldwide standards, and it seems unlikely in the extreme that such a standard will ever be reached. Thus several different systems exist, necessitating standards conversion for many international television transmissions.

TV systems Although some variety of standards and systems is in some evidence, there are five essentially different television systems in use. The main ones are the American[1] and the European[2] systems.

The American system is used in the whole of North and South America (except for Argentina and Venezuela) and in the Philippines and Japan. With some exceptions, the European system is used by the rest of the world. One of these exceptions is

[1] Federal Communications Commission (FCC) system for monochrome and National Television Standards Committee (NTSC) system for color.

[2] Comité Consultatif International de Radio (CCIR) system for monochrome and Phase Alternation by Line (PAL) system for color.

16-20. With the aid of a block diagram, explain fully the operation of an MTI system using a power amplifier in the transmitter.

16-21. What does an MTI radar actually do? Give instances of situations where it is indispensable. Give at least one instance of a radar application for which MTI cannot be used.

16-22. Describe briefly the various analog MTI systems.

16-23. Explain what is meant by the term *blind speed* in MTI radar. Under what conditions could this be an embarrassment? What is a method of overcoming the problems of blind speed in a radar?

16-24. What is the major problem with analog MTI? How can digital MTI overcome it?

16-25. Why are very much greater ranges possible with active, than with passive, tracking? Derive the equation for the maximum range for the *reply* link when a radar beacon is present on a target.

16-26. Draw the block diagram and explain the operation of a CW Doppler radar using an intermediate frequency in the receiver. How have the drawbacks of the basic CW radar been overcome?

16-27. With the aid of a block diagram explain the operation of an FM CW radar altimeter.

16-28. List the major difficulties occasioned by the use of moving radar antennas. How can phased arrays overcome these difficulties?

16-29. Describe briefly the two different types of phased array radars, and compare their relative merit.

16-30. List some of the functions that phased array radars could perform with ease, but which moving-antenna radars could perform with difficulty, or not at all. On the other hand, what are the main problems with phased array radars?

16-31. Explain the operation of the planar array radar. What are some of the advantages of this scanning method?

REFERENCES

1. Skolnik, Merrill I. (editor-in-chief), *Radar Handbook*, McGraw-Hill Book Company, New York, 1970.
2. Berkowitz, R. S. (ed.), *Modern Radar*, John Wiley & Sons, Inc., New York, 1965.
3. Wheeler, G. J., *Radar Fundamentals*, Prentice-Hall, Inc., Englewood Cliffs, N.J., 1967.
4. Dax, Peter R., "Eliminating Clutter in Radar Systems," *Microwaves*, April 1975, pp. 35–43, 67.
5. Ellis, J. G., "Digital MTI: A New Tool for the Radar User," *Marconi Rev.*, 4th quarter, 1973, pp. 237–248.
6. Delaney, William P., "Phased Array Radars: How Did We Get Here? Where Should We Be Going?" *Microwave J.*, October 1975, pp. 10 and 15.
7. White, (Dr.) Joseph F., "Phased Arrays Can Be Practical from a Cost Standpoint," *Microwave J.*, October 1974, pp. 49–52.
8. Sebring, J. R., and J. K. Ruth, "MLS Scanning-Beam Antenna Implementation," *Microwave J.*, January 1974, pp. 41–46.
9. Fu Peng, Boris Sheleg, and Bing Chiang, "Millimeter-Wave Phased Array Transmitters for ECM," *Microwave J.*, June 1983, pp. 81–89.

function is the same, but in the horizontal plane. It is here, also, that the very high voltage for the anode of the picture tube is generated.

Having dealt with monochrome television, the chapter now takes a look at its color counterpart. For this purpose, it will be assumed that students are already familiar with color and realize that it is not necessary to transmit every color of the rainbow to obtain a satisfactory reproduction in the receiver. Instead, three fundamental colors are transmitted, and in the receiver all others are reconstructed from them. We shall thus be looking at what a TV system must transmit and receive, in addition to monochrome information, in order to reproduce correct colors in the receiver.

17-1 REQUIREMENTS AND STANDARDS

The main body of this chapter deals with the transmission and reception of television signals. However, before concentrating on that, it is necessary to look at what information must be transmitted in a TV system and how it can be transmitted. The work involves an examination of the most important television standards and their reasons for existence.

17-1.1 Introduction to Television

Television means seeing at a distance. To be successful, a television system may be required to reproduce faithfully:

1. The shape of each object, or structural content
2. Shade of each object, or tonal content
3. Motion, or kinematic content
4. Sound
5. Color, or chromatic content
6. Perspective, or stereoscopic content

If only the first of these, i.e., the shape of each object in a scene, were shown, we would have merely black-and-white TV (without any shades of gray). If tonal content were added, we would have black-and-white still pictures. With items 3 and 4 added, we would have, respectively, "movies" and "talkies." The last two items are not essential, although most people consider color TV desirable, and the next generation will probably get item 6 and talk blithely about "three-dee."

The human eye contains many millions of photosensitive elements, in the shape of rods and cones, which are connected to the brain by some 800,000 nerve fibers (i.e., channels). A similar process by the camera tube is used at the transmitting station and the picture tube in the TV receiver. Indeed, some 150,000 effective elements are displayed in each scene. However, the use of that number of channels is out of the question; a single channel is used instead, each element being scanned in succession, to convey the total information in the scene. This is done at such a high rate that the eye

17 TELEVISION FUNDAMENTALS

Everyone has seen the front of a television receiver—for students of communications, it is important also to look at the inside, and in fact at the television system as a whole. This chapter treats television fundamentals. However, because this is a wide-ranging and extensive topic, the treatment is abbreviated to give an outline of the subject. As is well known, specialized books are devoted to television exclusively. For example, Grob [1] is a highly recommended reference. It is a book much the same size as this one, on TV only, and even then it concentrates its treatment on reception and on the American system rather than on all the others that exist.

The treatment of television in this chapter is similar. Other systems will be mentioned, but only in passing. The chapter begins with a description of what TV must reproduce, and the standards that must be defined in order to produce this. Students will learn about *line, frames* and *fields,* the need to define their speeds and the means of transmitting the picture and the sound information in the one channel.

The elements of monochrome transmission are discussed next, beginning with the fundamentals, which include a block diagram of a monochrome transmitter. Scanning is then considered, and finally we look at all the various pulses that must be transmitted, as well as the reasons for their existence, widths and repetition rates.

The next section deals with black-and-white TV reception in some detail, again beginning with a typical block diagram. Students will find that this is a rather large and complicated block diagram, and yet there are a large number of blocks and functions with which they are already familiar through having studied earlier portions of this book (especially Chap. 7). For example, it will be seen very early that TV receivers are invariably superheterodyne.

After familiar circuits (but in a new context) have been discussed, we begin the treatment of circuits specific to television receivers. The first of these are *sync separation* circuits, in which the *synchronizing information* transmitted along with *video voltages* is extracted and correctly applied to other portions of receiver circuitry. The *vertical deflection* circuits come next. They generate and apply to the picture tube the waveforms which are needed to make the electron beam move vertically up and down the tube as required. The *horizontal circuits* follow—their

function is the same, but in the horizontal plane. It is here, also, that the very high voltage for the anode of the picture tube is generated.

Having dealt with monochrome television, the chapter now takes a look at its color counterpart. For this purpose, it will be assumed that students are already familiar with color and realize that it is not necessary to transmit every color of the rainbow to obtain a satisfactory reproduction in the receiver. Instead, three fundamental colors are transmitted, and in the receiver all others are reconstructed from them. We shall thus be looking at what a TV system must transmit and receive, in addition to monochrome information, in order to reproduce correct colors in the receiver.

17-1 REQUIREMENTS AND STANDARDS

The main body of this chapter deals with the transmission and reception of television signals. However, before concentrating on that, it is necessary to look at what information must be transmitted in a TV system and how it can be transmitted. The work involves an examination of the most important television standards and their reasons for existence.

17-1.1 Introduction to Television

Television means seeing at a distance. To be successful, a television system may be required to reproduce faithfully:

1. The shape of each object, or structural content
2. The relative brightness of each object, or tonal content
3. Motion, or kinematic content
4. Sound
5. Color, or chromatic content
6. Perspective, or stereoscopic content

If only the structural content of each object in a scene were shown, we would have truly black-and-white TV (without any shades of gray). If tonal content were added, we would have black-and-white still pictures. With items 3 and 4 we would have, respectively, "movies" and "talkies." The last two items are not essential, although most people consider color TV desirable, and the next generation will probably get item 6 and talk blithely about "three-dee."

The human eye contains many millions of photosensitive elements, in the shape of rods and cones, which are connected to the brain by some 800,000 nerve fibers (i.e., channels). A similar process by the camera tube is used at the transmitting station and the picture tube in the TV receiver. Indeed, some 150,000 effective elements are displayed in each scene. However, the use of that number of channels is out of the question; a single channel is used instead, each element being scanned in succession, to convey the total information in the scene. This is done at such a high rate that the eye sees the whole scene, without being aware of the scanning motion. A single static picture thus results.

The problem of indicating motion is solved as in the cinema. A succession of pictures is shown, each with the scene slightly altered from the previous one—the eye is thus fooled into seeing continuous motion, through the property known as the persistence of vision. There are 30 pictures (or "frames," as they are called) per second in the U.S. television system. The number of frames is related to the 60-Hz frequency of the ac voltage system and is above the minimum required (about 18 frames per second) to make the eye believe that it sees continuous motion. Commercial films are run at 24 frames per second; while the perception of smooth motion still results, the flicker due to the light cutoff between frames would be obvious and distracting. In motion pictures, this is circumvented by passing the shutter across the lens a second time, while the frame is still being screened, so that a light cutoff occurs 48 times per second. This is too fast for the eye to notice the flicker. The same effect could be obtained by running film at 48 frames per second, but this would result in all films being twice as long as they need be (to indicate smooth motion).

To explain how flicker is avoided in TV, it is first necessary to look at the scanning process in a little detail. The moving electron beam is subjected to two motions simultaneously. One is fast and horizontal, and the other is vertical and slow, being 262½ times slower than the horizontal motion. The beam gradually moves across the screen, from left to right, while it simultaneously descends almost imperceptibly. A complete frame is covered by 525 horizontal lines, which are traced out 30 times per second. However, if each scene were shown traced thus from top to bottom (and left to right), any given area of the picture tube would be scanned once every one-thirtieth of a second, too slowly to avoid flicker. Doubling the vertical speed, to show 60 frames per second, would do the trick but would double the bandwidth.

The solution, as will be explained, consists in subdividing each frame into two fields. One field covers half the picture, from top to bottom, and the second field fills in the "missing lines." This is known as *interlaced scanning,* and all the world's TV systems use it. We still have 30 frames per second, but any given area of the display tube is now illuminated 60 times per second, and so flicker is too fast to be registered by the eye.

The scene elements at the transmitting station are produced by a mosaic of photosensitive particles within the camera tube, onto which the scene is focused by optical means. They are scanned by an electronic beam, whose intensity is modulated by the brightness of the scene. A varying voltage output is thus obtained, proportional to the instantaneous brightness of each element in turn. The varying voltage is amplified, impressed as modulation upon a VHF or UHF carrier, and radiated. At the receiver, after amplification and demodulation, the received voltage is used to modulate the intensity of the beam of a CRT. If this beam is made to cover each element of the display screen area exactly in step with the scan of the transmitter, the original scene will then be synthesized at the receiver.

The need for the receiver picture tube to be exactly in step with that of the transmitter requires that appropriate information be sent. This is *synchronizing,* or *sync* information, which is transmitted in addition to the picture information. The two sets of signals are interleaved in a kind of time-division multiplex, as will be seen, and the picture carrier is amplitude-modulated by this total information. At the receiver, signals derived from the transmitted sync control the vertical and horizontal scanning circuits, thus ensuring that the receiver picture tube is in step with the transmitter camera tube.

Black-and-white television can be transmitted in this manner, but color TV requires more information still. As well as indicating brightness or *luminance,* as is done in black-and-white TV, color (or actually *hue*) must also be shown. That is, for each picture element we must show not only how bright it is, but also what hue this element should have, be it white, yellow, red, black or any other. The hue is indicated by a *chrominance,* or *chroma,* signal.

The colors actually indicated are red, green and blue, but all other colors can be synthesized from these three, as is well known. Separate signals for each of the three colors are produced by the transmitter camera tube. In the receiver, these signals are applied to the three guns of the picture tube, or *kinescope*. The screen consists of adjacent green, blue and red dots, which luminesee in that color when the scanning beam falls on them. Needless to say, the beams themselves are not colored! They merely indicate to each colored dot on the screen how bright it should be at any instant of time, and the combination of brightnesses of these three colors reproduces the actual hues we see. Because of the smallness of the color dots and our distance from the screen, we see color combinations instead of the individual dots.

Color TV will be discussed in more detail later in this chapter, but it is worth mentioning at this stage that FDM is used to interleave the chrominance signal with luminance. The process is quite complex. The chroma signal is assigned portions of the total frequency spectrum which luminance does not use. The situation is complicated by the fact that color and black-and-white TV must be *compatible*. That is to say, the chroma signals must be coded in such a way that a satisfactory picture will be produced (in black and white) by a *monochrome* receiver tuned to that channel. Conversely, color TV receivers must be designed so that they are able to reproduce satisfactorily (in black and white) a transmitted monochrome signal.

The simplest item has been left until last; this is the sound transmission. A separate transmitter is used for sound, connected to the same antenna as the picture transmitter. However, it is a simple matter to have a receiver with common amplification for all signals up to a point, at which the various signals go to their respective sections for special treatment. This separating point is almost invariably the video detector, whose output consists of picture, sync and sound information. The sound signal is amplified, applied to its own detector, amplified again and fed to a loudspeaker. The modulating system used for sound in the U.S. system, and most other major systems around the world, is wideband FM. It is not quite as wideband as in FM radio transmissions, but it is quite adequate for good sound reproduction. The transmitting frequency for the sound transmitter is quite close to the picture transmitting frequency. Thus, as just outlined, the one tuning mechanism and amplifiers can handle both. A block diagram of a rudimentary television system is now shown in Fig. 17-1, indicating basically how the requirements of monochrome TV transmission and reception may be met.

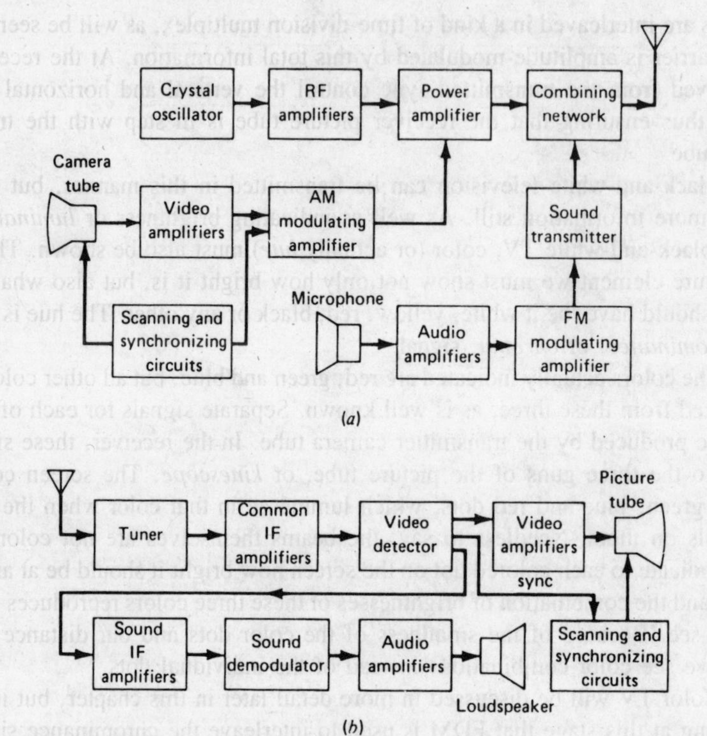

FIGURE 17-1 Basic monochrome television system. (*a*) Transmitter; (*b*) receiver.

17-1.2 Television Systems and Standards

It is clear that a large amount of information must be broadcast by a television transmitter and that there are a variety of ways in which this can be done. Accordingly, a need exists for uniform standards for TV transmission and reception. Regrettably, no agreement has been reached for the adoption of worldwide standards, and it seems unlikely in the extreme that such a standard will ever be reached. Thus several different systems exist, necessitating standards conversion for many international television transmissions.

TV systems Although agreement in certain respects is in some evidence, there are five essentially different television systems in use around the world. The two main ones are the American[1] and the European[2] systems.

The American system is used in the whole of North and South America (except for Argentina and Venezuela) and in the Philippines and Japan. With some exceptions, the European system is used by the rest of the world. One of these exceptions is

[1] Federal Communications Commission (FCC) system for monochrome and National Television Standards Committee (NTSC) system for color.

[2] Comité Consultatif International de Radio (CCIR) system for monochrome and Phase Alternation by Line (PAL) system for color.

TABLE 17-1 Selected Standards of Major Television Systems

STANDARD	AMERICAN SYSTEM	EUROPEAN SYSTEM
Number of lines per frame	525	625
Number of frames per second	30	25
Field frequency, Hz	60	50
Line frequency, Hz	15,750	15,625
Channel width, MHz	6	7
Video bandwidth, MHz	4.2	5
Color subcarrier, MHz	3.58*	4.43*
Sound system	FM	FM
Maximum sound deviation, kHz	25	50
Intercarrier frequency, MHz	4.5	5.5

* As a good approximation. The precise frequency in the American system is 3.579545 MHz, for reasons that will be explained in Sec. 17-4.1.

France, which, together with a part of Belgium, uses its own system, SECAM (sequential technique and memory storage), for color. The USSR and Eastern Europe use a system for monochrome that is almost identical to CCIR, but they use SECAM for color. With its greater line frequency, the French system has superior definition, but it requires a bandwidth twice as great as for the major systems. Table 17-1 shows the most important standards in the American and European systems. This is done for comparison; all subsequent detailed work will refer to the American system exclusively.

Apart from the differences, the two major TV systems have the following standards in common.

1. Vestigial sideband amplitude modulation for video, with most of the lower sideband removed. This is done to save bandwidth; the system was described in detail in Sec. 4-4.4 and pictured in Fig. 4-9.
2. Negative video modulation polarity. In both systems black corresponds to a higher modulation percentage than white.
3. 2:1 interlace ratio. This can be seen from the table, which shows that the field frequency is twice the frame frequency. Interlacing will be described fully in Sec. 17-2.2.
4. 4:3 aspect ratio. This is the ratio of the horizontal to the vertical dimension of the receiver picture (or transmitter camera) tube. The absolute size is not limited, but the aspect ratio must be; otherwise the receiving screen would not reproduce all the transmitted picture (or else a portion of the receiving screen would have nothing to show).

Notes on the major American standards The field frequency is purposely made equal to the 60-Hz frequency of the ac supply system, so that any supply interference will produce stationary patterns, and will thus not be too distracting. This automatically makes the frame frequency equal to 30 per second. The number of lines per frame, 525, was chosen to give adequate definition without taking up too large a portion of the frequency spectrum for each channel. The line frequency is, of course, the product of 30 frames per second and 525 lines per frame, i.e., 15,750 Hz.

As shown in Fig. 6-9*a*, the channel width of 6 MHz is required to accommodate the wanted upper sideband, the necessary portion of the unwanted lower sideband, the FM sound frequency spectrum and, as will be shown in Sec. 17-4.1, the color subcarrier and its sidebands. The difference in frequency between the picture carrier and the sound carrier is precisely 4.5 MHz. This was shown in Fig. 6-9*a* and is given in Table 17-1 as the *intercarrier frequency*. The fact that this frequency difference is 4.5 MHz is used in extracting the sound information from the video detector. This will be explained in Sec. 17-3.2.

In each TV channel, the picture carrier frequency is 1.25 MHz above the bottom edge of the channel, and the color subcarrier frequency is 3.58 MHz higher still. The sound carrier frequency is 4.5 MHz above the picture carrier frequency. Channels 2 to 13 are in the VHF band, with channels 2 to 6 occupying the frequency range 54 to 88 MHz, while channels 7 to 13 occupy the 174- to 216-MHz range. Note that the frequencies between 88 and 174 MHz are allocated to other services, including FM broadcasting. Channels 14 to 83 occupy the continuous frequency range from 470 to 890 MHz, in the UHF band. The frequency details of each channel may be obtained in [2], appendixes C and D.

Video bandwidth requirement The frequency band needed for the video frequencies may be estimated (actually, *overestimated*) as follows. Consider at first that the lowest frequency required corresponds to a line across the screen which is of uniform brightness. This represents a period of 1/15,750 = 0.0000635 = 63.5 μs during which the brightness of the beam does not change. If a large number of lines of that brightness followed in succession, the frequency during that time would be zero. This is too awkward to arrange, since it requires dc coupling. Thus the lowest frequency transmitted in practice is higher than zero, approximately 60 Hz in fact. As regards the highest required frequency, this will of course correspond to the highest possible variation in the brightness of the beam along a line.

Consider now that the picture has been divided into 525 lines from top to bottom, so that the maximum resolution in the vertical direction corresponds to 525 changes (e.g., from black to white) down the picture. It is desirable that horizontal and vertical resolution be the same. However, because of the 4:3 aspect ratio, the picture is 4/3 times as wide as it is high, so that $525 \times 4/3 = 700$ transitions from black to white during the length of a horizontal line is the maximum required. This, of course, corresponds to 700/2 = 350 complete (black-white-black) transitions along the line, occurring in 63.5 μs. The period of this maximum transition is thus 63.5/350 = 0.1814 μs. If each transition is made gradual (i.e., sine wave) rather than abrupt (square wave), 0.01814 μs is the period of this sine wave, whose frequency therefore is $1/0.1814 \times 10^{-6} = 5.51$ MHz.

As was mentioned, this figure is an overestimate, and the video bandwidth of 4.2 MHz quoted in Table 17-1 is quite enough. The reason for the difference is mainly that not all the 525 lines are visible. Several of them occur during the vertical *retraces* and are *blanked* out; this will be explained in Sec. 17-2.2. Consequently, neither the vertical nor the horizontal resolution needs to be as good as assumed above, and so the maximum video frequency may be lower than the rough 5.51-MHz calculation. However, this calculation yields a reasonable approximation, and it does show that the bandwidth required is very large. This explains why vestigial sideband modulation is used.

17-2 BLACK-AND-WHITE TRANSMISSION

It is now proposed to look at the significant aspects of monochrome television transmission in some detail. During this examination, the reasons for, and the effects and implications of, the most important TV standards will emerge.

17-2.1 Fundamentals

As shown in the block diagram of Fig. 17-2, a monochrome TV transmission system is quite unlike any of the transmission systems studied previously. This section will deal with the fundamental, "straightforward" blocks, while the functions specific to television transmitters are described in more detail in the succeeding sections.

Camera tubes The video chain at the transmitting station begins with a transducer which converts light into (video) electric signals, i.e., a camera tube. Detailed descriptions of the various camera tubes are outside the scope of this chapter, but they may be found in chap. 3 of [1]. Very basically, a camera tube has a mosaic screen, onto which

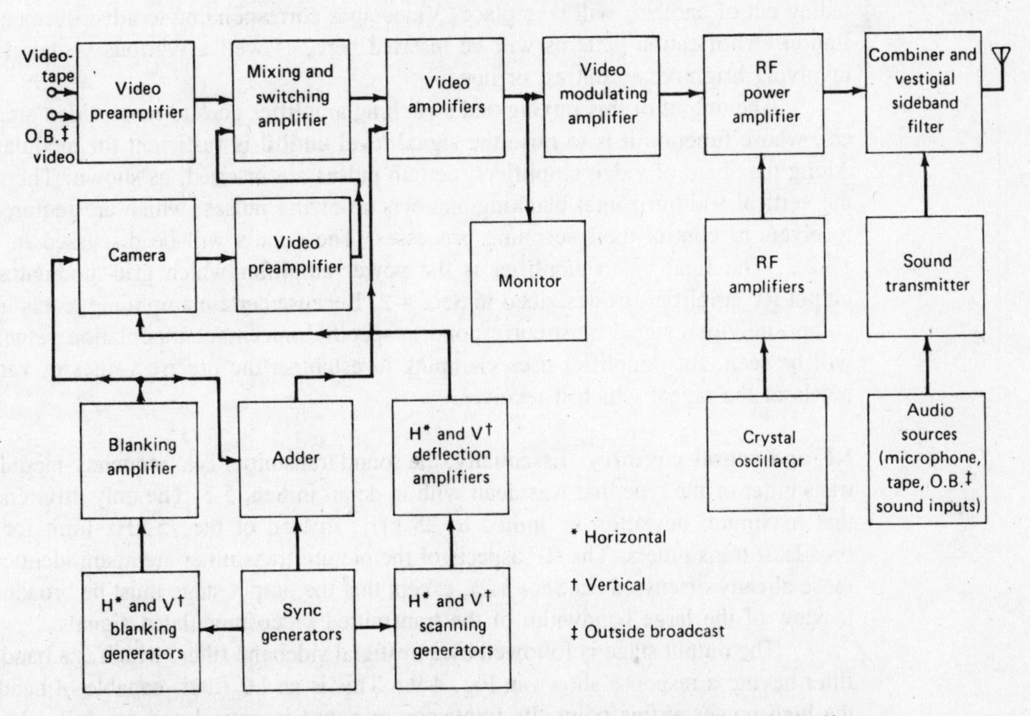

FIGURE 17-2 Simplified monochrome television transmitter block diagram.

the scene is focused through the lens system of the television camera. An electron gun forms a beam which is accelerated toward this photoelectric screen. The beam scans the screen, from left to right and top to bottom, covering the entire screen 30 times per second. The precise manner will be described in detail in the next section, and *magnetic deflection* is treated in Sec. 17-3, in connection with receiver picture tubes. The beam intensity is affected by the charge on the screen at that point, and this in turn depends on the brightness of the point. The current-modulated beam is collected at a *target* electrode, located at or just beyond the screen. The output voltage from this electrode is a varying (video) voltage, whose amplitude is proportional to the screen brightness at the point being scanned. This voltage is now applied to video amplifiers.

As might be gathered, the foregoing applies to monochrome transmission. In color transmission, light is split into the three basic colors and applied to either three separate tubes or a single tube which has different areas sensitized to the different colors. Three separate signals result and are treated as will be described in Sec. 17-4. The camera tubes most likely to be used are the *vidicon* or the *plumbicon,* in both of which separate tubes are required for the three colors. It is also possible to use a single camera tube which is constructed with a stripe filter or which uses three electron guns to produce all three colors at once.

Video stages The output of the camera is fed to a video switcher which may also receive videotape or outside broadcast video signals at other inputs. The function of this switching system is to provide the many video controls required. It is at this point that mixing or switching of the various inputs, such as fading in of one signal and fading out of another, will take place. Videotapes corresponding to advertisements or station identification patterns will be inserted here, as well as various visual effects involving brightness, contrast or hue.

The output of this mixing and switching amplifier goes to more video amplifiers, whose function it is to raise the signal level until it is sufficient for modulation. Along the chain of video amplifiers, certain pulses are inserted, as shown. These are the vertical and horizontal blanking and synchronizing pulses, which are required by receivers to control their scanning processes. The details will be discussed in Sec. 17-2.3. The final video amplifier is the power amplifier which grid-modulates the output RF amplifier, as described in Sec. 4-2. Because certain amplitude levels in the composite video signal must correspond to specific precentage modulation values, as will be seen, this amplifier uses clamping to establish the precise values of various levels of the signal which it receives.

RF and sound circuitry Essentially, the sound transmitter is a frequency-modulated transmitter of the type that was dealt with in detail in Sec. 5-3. The only difference is that maximum deviation is limited to 25 kHz, instead of the 75-kHz limit for FM broadcast transmitters. The RF aspects of the picture transmitter are again identical to those already discussed (in Sec. 3-2), except that the output stage must be broadband, in view of the large bandwidth of the transmitted video modulated signals.

The output stage is followed by a vestigial sideband filter, which is a bandpass filter having a response shown in Fig. 4-9*a*. This is an *LC* filter, capable of handling the high power at this point. Its frequency response is critical and carefully shaped.

Finally, the output of the sound and picture transmitters is fed to the antenna via a combining network. This is in fact an *LC* equivalent of the circulator discussed in Sec. 10-5.2. Its function is to ensure that, although both the picture and sound transmitters are connected to the antenna with a minimum of loss, neither is connected to the other.

17-2.2 Scanning

As already discussed, the complete frame of a television picture is scanned 30 times per second, in a manner that is very similar to reading this page. The beam in the camera or picture tube moves at a constant velocity across the screen and, when it reaches the end of the screen on the right-hand side, it ''whips back'' to the left-hand edge of the screen and starts again. Meanwhile, it has descended down the screen, so that the next line traced out is somewhat below the first one. This process continues until the bottom of the screen is reached. When this happens to the eyes of a person reading this page, the person will then glance up quickly to the top of the next page (if still interested!). In the case of TV, the information at the top of the existing ''page'' (screen) has changed; so the beam ''whips up'' to the top of the screen, and the whole process begins again. The whole procedure is, however, a little more complex than just described, and a more detailed examination is now required.

Horizontal scanning As previously explained, the total time taken from the beginning of one line to the instant when the next line begins to be scanned is 63.5 μs. This time obviously includes not only the scan of the picture from left to right but also the rapid return, or *retrace*, from right to left. Clearly, the retrace cannot take an infinitesimally short time, and in fact a period of 10.2 μs is allocated to it, i.e., 16 percent of the time allocated to scanning one line. That is to say, the retrace time is $0.16H$, and the *active* time is $0.84H$, where $H = 63.5$ μs.

It is clearly undesirable for the retrace to be visible. We cannot have a pattern of bright horizontal lines across the previously scanned picture pattern. The method of preventing this disturbance is a very simple one. It consists in reducing the scanning beam current to zero, from just before the beginning of the retrace until just after its end. The process is known as (horizontal, in this case) *blanking*. This consists in adding a pulse to the video waveform, at the right time and for the correct period, to ensure that the signal level has been raised to that corresponding to black. It will be recalled that negative modulation is used, in which the voltage corresponding to black is much higher than the voltage that indicates white. The sequence of events is as follows:

1. As the active beam is about to reach the right-hand side of the picture, the blanking voltage is applied.
2. Immediately afterward, the horizontal scanning generator receives a (horizontal) sync pulse, which initiates the retrace.
3. The retrace continues for a period of time that is governed by the time constant of the oscillator generating the scanning waveform but must be less than $0.16H$.
4. The retrace ends, and scanning of the next line begins.
5. Immediately afterward, after a total time of $0.16H$, the horizontal blanking pulse ends, and the picture becomes visible once again.

Note that the size of the displayed image is arranged so that the brief blanked-out periods, just before the retrace begins and after it ends, are just beyond the edges of the screen. This also applies to vertical blanking, and may be observed by making the picture in a receiver roll down slowly with the vertical hold control. Black horizontal bars will be seen, which normally belong above and below the picture seen on the screen. Blanking is discussed further in Sec. 17-2.3, where the relevant waveforms are shown.

Vertical scanning Basically, vertical scanning is similar to horizontal scanning, except for the obvious difference in the direction of movement and the fact that everything happens much more slowly (i.e., 60 rather than 15,750 times per second). However, interlacing introduces a complication which must now be explored further.

The sequence of events in vertical scanning is as follows:

1. Line 1 starts at the top left-hand corner of the picture, at point F. As this line and the succeeding lines are scanned horizontally, the beam gradually moves downward. This continues until, midway through line 242, vertical blanking is applied. The situation is illustrated in Fig. 17-3a. Note that active horizontal lines are solid, the horizontal retraces are dashed, and the point at which vertical blanking is applied is labeled A.

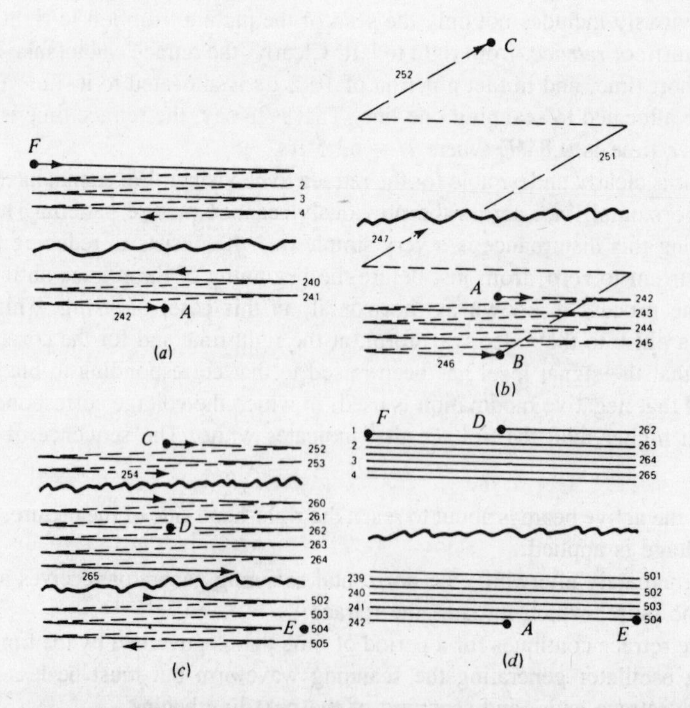

FIGURE 17-3 Interlaced scanning. (*a*) First field; (*b*) blanking and retrace after first field; (*c*) second field, showing removal and application of vertical blanking; (*d*) first and second fields interlaced.

2. Soon, but not immediately, after the application of vertical blanking, the vertical scanning generator receives a (vertical) sync pulse. This causes vertical retrace to commence, at point *B* in Fig. 17-3*b*.
3. Vertical retrace continues, for a time corresponding to several *H*, until the beam reaches the top of the picture, point *C* in Fig. 17-3*b*. Note that horizontal scanning continued during the vertical retrace—it would be harmful to stop the horizontal oscillator just because vertical retrace is taking place.
4. The beam, still blanked out, begins its descent. The precise point is determined by the time constants in the vertical scanning oscillator, but it is usually 5 or 6*H* between points *B* and *C*. The situation is shown in Fig. 17-3*c*.
5. Precisely 21*H* after it was applied, i.e., midway through line 263, vertical blanking is removed. The first (odd) field is now completed, and the second (even) field begins. This is also illustrated in Fig. 17-3*c*; note that *D* is the point at which vertical blanking is removed.
6. The visible portion of line 263 begins at the same height as did line 1, i.e., at the top of the screen. However, line 263, when it becomes visible, is already halfway across the screen, whereas line 1 began at the left-hand edge of the screen. Thus line 263 lies *above* line 1, line 264 is *between* lines 1 and 2, and so on. This is illustrated in Fig. 17-3*d*.
7. The second field continues, until vertical blanking is applied at the beginning of the retrace after line 504. This is point *E* in Fig. 17-3*c*.
8. The sequence of events which now takes place is identical to that already described, for the end of the first field. The only difference is that, after the 21 lines of vertical blanking, the beam is located at the top left-hand corner of the picture tube, at point *F*. Thus, when vertical blanking is now removed, the next odd field is traced out, as in Fig. 17-3*a*.

Regrettably, the vertical scanning procedure is complicated by the use of interlacing. However, it is basically simple, in that blanking is applied some time before retrace begins and removed some time after it has ended. Both margins are used for safety and to give individual designers of receivers some flexibility. As explained, horizontal scanning continues during vertical retrace, complicating the drawings and the explanation, but actually simplifying the procedure. To stop the horizontal oscillator for precisely 21 lines, and then to restart it exactly in sync, would simply not be practical. Finally, beginning one field at the start of a line and the next field at the midpoint of a line is a stratagem that ensures that interlacing will take place. If this were not done, the lines of the second field would coincide with those of the first, and vertical resolution would immediately be halved!

Please note that the scanning waveforms themselves are sawtooth. The means of generating them and applying them to the picture tube are discussed in Sec. 17-3.

17-2.3 Blanking and Synchronizing Pulses

Blanking Video voltage is limited to certain amplitude limits. Thus, for example, the white level corresponds to 12.5 percent (±2.5 percent) modulation of the carrier, and the black level corresponds to approximately 67.5 percent modulation. Thus, at some point along the video amplifier chain, the voltage may vary between 1.25 and 6.75 V, depending on the relative brightness of the picture at that instant. The darker the picture, the higher will be the voltage, within those limits. At the receiver, the picture tube is biased to ensure that a received video voltage corresponding to 12.5 percent modulation yields whiteness at that particular point on the screen, and an equivalent arrangement is made for the black level. Besides, set owners are supplied with *brightness* and *contrast* controls, to make final adjustments as they think fit. Note that the lowest 12.5 percent of the modulation range (the whiter-than-white region) is not used, to minimize the effects of noise.

When the picture is blanked out, before the vertical or horizontal retrace, a pulse of suitable amplitude and duration is added to the video voltage, at the correct instant of time. Video superimposed on top of this pulse is clipped, the pulses are clamped, and the result is video with blanking, shown in Fig. 17-4. As indicated, the blanking level corresponds to 75 percent (±2.5 percent) of maximum modulation. The black level is actually defined relatively rather than absolutely. It is 5 to 10 percent below the blanking level, as shown in Fig. 17-4. If in a given transmission the blanking level is exactly 75 percent, then the black level will be about 7.5 percent below this, i.e., approximately 67.5 percent as previously stated. At the video point mentioned previously, we thus have white at 1.25 V, black at about 6.75 V and the blanking level at 7.5 V.

The difference between the blanking level and the black level is known as the *setup* interval. This is made of sufficient amplitude to ensure that the black level cannot possibly "poke up" above the blanking level. If it did so, it would intrude into the region devoted exclusively to sync pulses, and it might conceivably interfere with the synchronization of the scanning generators.

Synchronizing pulses As shown in Fig. 17-5, the procedure for inserting synchronizing pulses is fundamentally the same as used in blanking pulse insertion. Horizontal and vertical pulses are added appropriately on top of the blanking pulses, and the resulting waveform is again clipped and clamped. It is seen that the tips of horizontal and vertical synchronizing pulses reach a level that corresponds to 100 percent modulation of the picture carrier. At the hypothetical video point mentioned previously, we may thus have video between 1.25 and 6.75 V, the blanking level at 7.5 V and the sync pulse tips at 10 V. The overall arrangement may be thought of as a kind of voltage-division multiplex.

Although this will be explored in further detail in Sec. 17-3.3, it should be

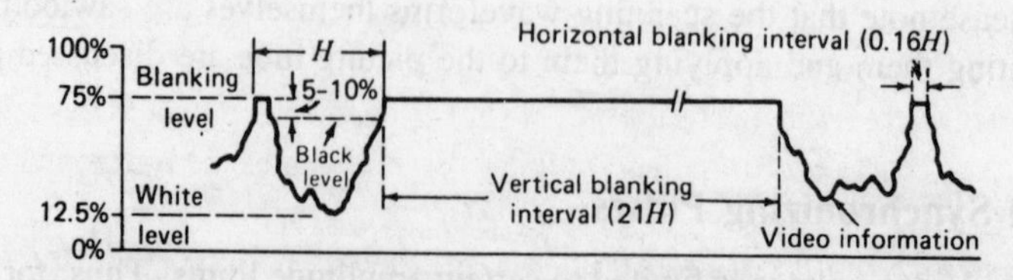

FIGURE 17-4 TV video waveform, showing video information and horizontal and vertical blanking pulses (at the end of an even field).

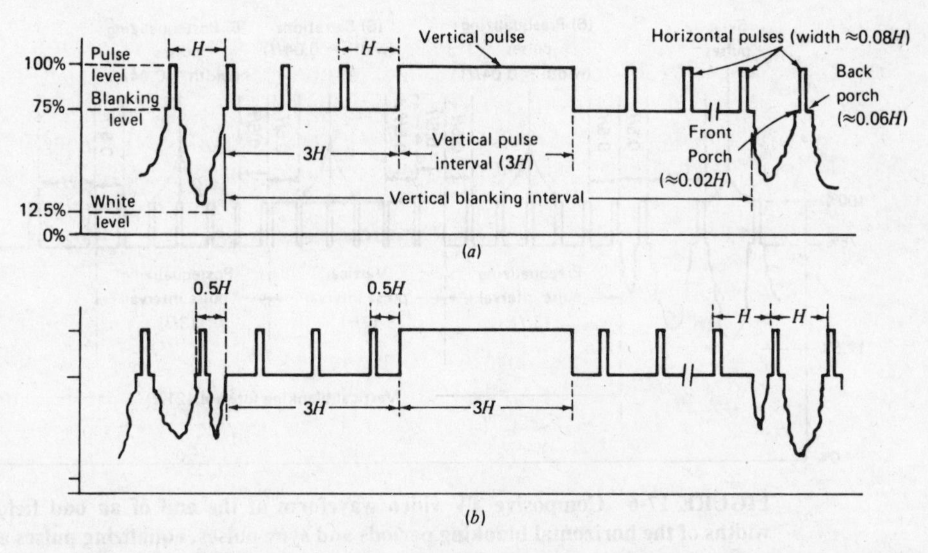

FIGURE 17-5 TV video waveform, showing horizontal and basic vertical sync pulses, at the end of an (*a*) even field; (*b*) odd field. (Note: The width of the horizontal blanking intervals and sync pulses is exaggerated.)

noted that the horizontal sync information is extracted from the overall waveform by differentiation. Indeed, pulses corresponding to the differentiated leading edges of sync pulses are actually used to synchronize the horizontal scanning oscillator. This is the reason why, in Figs. 17-4 to 17-6, all time intervals are shown between pulse leading edges. Furthermore, receivers often use monostable-type circuits to generate horizontal scan, so that a pulse is required to initiate each and every cycle of horizontal oscillation in the receiver. With these points in mind, it should be noted that there are two things terribly wrong with the sync pulses shown in Fig. 17-5.

The first and more obvious shortcoming of the waveforms shown may be examined with the aid of Fig. 17-5*a*. After the start of the vertical blanking period, the leading edges of the three horizontal sync pulses and the vertical sync pulse shown will trigger the horizontal oscillator in the receiver. However, there are no leading edges for a time of 3*H* after that, as shown, so that the receiver horizontal oscillator will either lose sync or stop oscillating, depending on the design.

It is obvious that three leading edges are required during this 3*H*-period. By far the easiest way of providing these leading edges is to cut slots in the vertical sync pulse. The beginning of each slot has no effect, but the end of each provides the desired leading edge. These slots are known as *serrations*. They have widths of 0.04*H* each and are shown exaggerated in Fig. 17-6 (to ensure that they are visible). Note that, at the end of an even field, serrations 2, 4 and 6 or, to be precise, the leading edges following these three serrations, are actually used to trigger the horizontal oscillator in the receiver.

The situation after an odd field is even worse. As expected, and as shown in Fig. 17-5*b*, the vertical blanking period at the end of an odd field begins midway through a horizontal line. Consequently, looking further along this waveform, we see

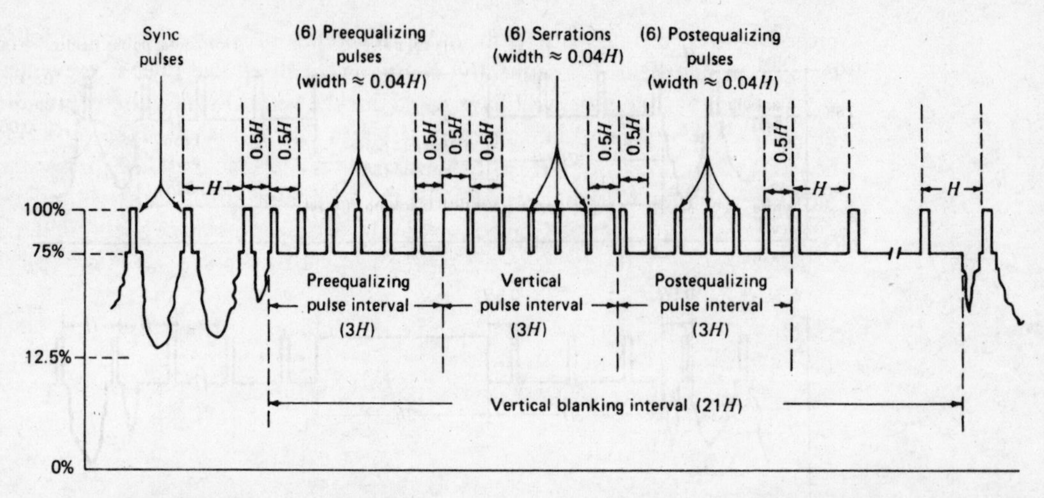

FIGURE 17-6 Composive TV video waveform at the end of an odd field. (Note: The widths of the horizontal blanking periods and sync pulses, equalizing pulses and serrations are exaggerated.)

that the leading edge of the vertical sync pulse comes at the wrong time to provide synchronization for the horizontal oscillator. The obvious answer is to have a serration such that the leading edge following it occurs at time H after the leading edge of the last horizontal sync pulse. Two more serrations will be required, at H intervals after the first one. In fact, this is the reason for the existence of the first, third and fifth serrations in Fig. 17-6. The overall effect, as shown, is that there are six serrations altogether, at $0.5H$ intervals from one another.

Note that the leading edges which now occur midway through horizontal lines do no harm. All leading edges are used sometime, either at the end of an even field or at the end of an odd one. Those that are not used in a particular instance come at a time when they cannot trigger the horizontal waveform, and they are thus ignored. This behavior will be further discussed in Sec. 17-3.5.

We must now turn to the second shortcoming of the waveforms of Fig. 17-5. First, it must be mentioned that synchronization is obtained in the receiver from vertical sync pulses by integration. The integrator produces a small output when it receives horizontal sync pulses, and a much larger output from vertical sync pulses, because their energy content is much higher. What happens is that as a result of receiving a vertical pulse, the output level from the integrator eventually rises enough to cause triggering of the vertical oscillator in the receiver. This will be discussed further in Sec. 17-3.3.

However, we must note at this stage that the residual charge on this integrating circuit will be different at the start of the vertical sync pulses in Fig. 17-5*a* and 17-5*b*. In the former, the vertical sync pulse begins a time H after the last horizontal pulse. In the latter, this difference is only $0.5H$, so that a higher charge will exist across the capacitor in the integrating circuit. The equalizing pulses shown in the composite video waveform of Fig. 17-6 take care of this situation. It is seen that the period immediately preceding each vertical pulse is the same, regardless of whether this pulse follows an even or an odd field. Consequently, charge is equalized and jitter is prevented.

Observant students will have noted that the vertical sync pulse begins $3H$ after the start of the vertical blanking period, although Fig. 17-3 showed the vertical retrace beginning four lines (i.e., $4H$) after the start of vertical blanking. The discrepancy can now be explained. It is simply caused by the integrating circuit taking a time approximately equal to H, from the moment when the vertical sync pulse begins to the instant when its output is sufficient to trigger the vertical retrace.

Summary It is seen that the provision of blanking and synchronizing pulses, to ensure that TV receivers scan correctly, is a very involved process. It is also seen how important it is to have adequate television transmission standards. In retrospect, Table 17-1 is seen as decidedly incomplete, and this is why it was entitled "selected standards." Needless to say, the composite video waveforms in other TV systems are different from those shown, but they are, of course, as carefully defined and observed.

In essence, however, all systems have the same general principles in common. In each, blanking is applied before, and removed after, synchronizing pulses. A front porch precedes a horizontal sync pulse, and a back porch follows such a pulse, in all the systems. All systems have equalizing pulses, though not necessarily the same number as in the FCC system. In all cases serrations are used to provide horizontal sync during vertical pulses, with some minor differences as applicable. For example, the width of a vertical pulse in the CCIR system is $2.5H$, and the H itself is different from H in the American system.

Three final points should now be mentioned. The first is that many people refer to a set of six vertical sync pulses, which this section has been consistent in referring to a single pulse with six serrations. The difference in terminology is not very significant, as long as the user explains what is meant. Second, it is a moot point whether the vertical pulse has five or six serrations. This section has referred to the no-pulse region between the trailing edge of the vertical pulse and the first postequalizing pulse as a serration. This is done because, if there were no serrations, this period would be occupied by the final portion of the vertical sync pulse, whose trailing edge has now been cut into. Other sources do not consider this as a serration, but again the point is not significant, as long as the terms are adequately defined.

The third item is related to the fact that the one crystal-controlled source is used for all the various pulses transmitted. It operates at 31,500 Hz; this is twice the horizontal frequency and is also the repetition rate of the equalizing pulses and serrations. As expected, the horizontal frequency is obtained by dividing 31,500 Hz by 2. Similarly, the 60-Hz field frequency is achieved by dividing 31,000 Hz by 525 (i.e., $7 \times 5 \times 5 \times 3$). This point acquires added significance in color television.

The precise details of the complete FCC sync waveforms may be found on p. 421 of [3]. This shows the vertical blanking interval at the end of each field, enlarged pulses to indicate their precise shapes (they are not *quite* rectangular) and all amplitude and width tolerances.

Finally, the methods of producing and applying the scanning waveforms are discussed in Sec. 17-3.

17-3 BLACK-AND-WHITE RECEPTION

In this section are described the receiver counterparts of the transmission processes just studied. Circuits common to transmitters and receivers are also treated.

17-3.1 Fundamentals

As shown in Fig. 17-7 and previously implied in Fig. 17-1*b*, TV receivers use the superheterodyne principle. In addition, however, there is extensive pulse circuitry, to ensure that the demodulated video is displayed correctly. To that extent the TV receiver is quite similar to a radar receiver, but radar scan is generally simpler, nor are sound and color normally required for radar. It is also worth making the general comment that TV receivers of current manufacture are likely to be either solid-state or hybrid. In the latter case, all stages are transistor or integrated-circuit, except for the high-power scanning (and possibly video) output stages. It is now proposed to discuss briefly those stages which television receivers have in common with those types of receivers already discussed in previous chapters, and then to concentrate on the stages that are peculiar to TV receivers.

17-3.2 Common, Video and Sound Circuits

Tuners A modern television receiver has two tuners. This arrangement was left out of Fig. 17-7 for simplification but is shown in detail in Fig. 17-8.

The VHF tuner must cover the frequency range from 54 to 216 MHz. The

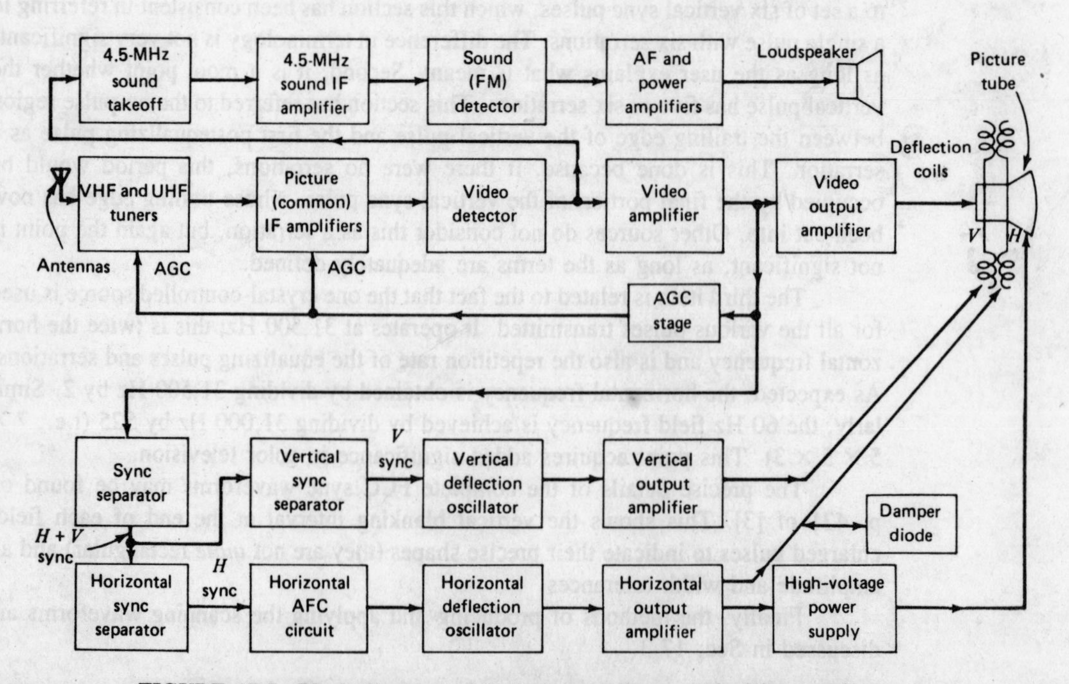

FIGURE 17-7 Block diagram of typical monochrome television receiver.

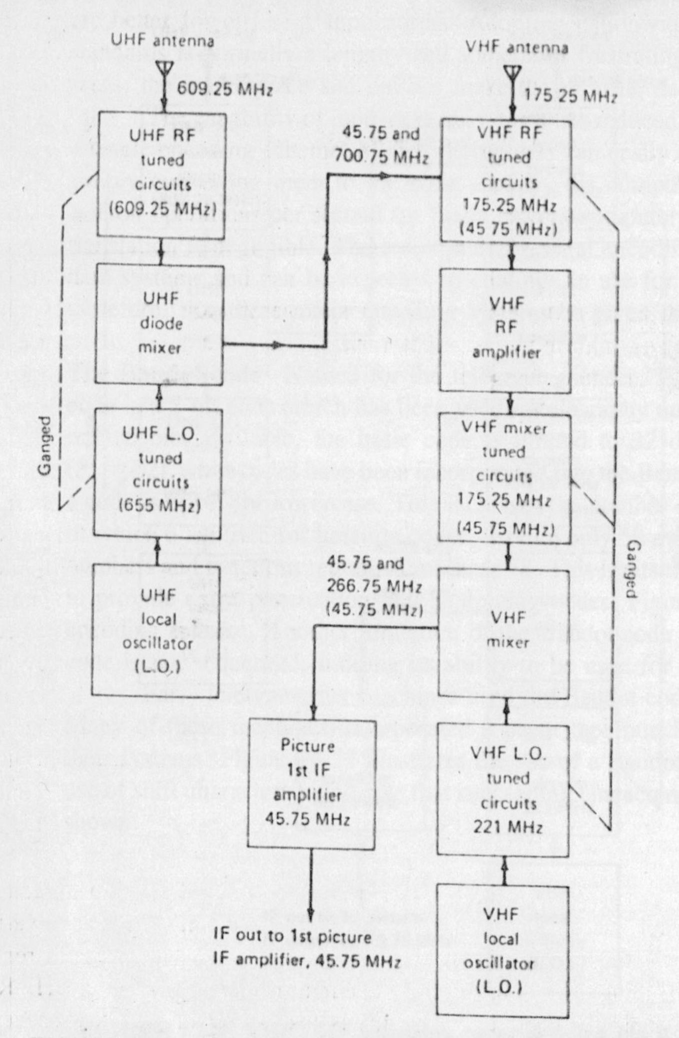

FIGURE 17-8 VHF/UHF television tuner detailed block diagram. VHF section shown receiving channel 7 (UHF local oscillator is then disabled). UHF section shown receiving channel 37 (VHF local oscillator is then disabled, and frequencies in parentheses apply). Note that, where applicable, only picture (not sound) carrier frequencies are shown (see text).

antenna most frequently used for reception the Yagi-Uda, consisting at its simplest of a reflector, a folded dipole for the five low channels and a shorter dipole for the upper seven channels. More elaborate Yagis ma ave a reflector, four dipoles and up to six directors. (See also Sec. 9-6.2.)

The frequency range covered by th UHF tuner is the 470- to 890-MHz band, and here the antenna used is quite likely to be a log-periodic, with the one antenna covering the whole band (see Sec. 9-8.4). I s also possible to cover the VHF and UHF bands with the one antenna. This is then likely to be similar to the discone antenna of

Sec. 9-8.3 but with the disk bent out to form a second cone. This *biconical* antenna is then used for UHF, with wire extensions for the two cones increasing the antenna dimensions for VHF. Other antennas used or specially adapted for use as TV receiving antennas are described in [1], pp. 59[illegible]599.

VHF tuners often use a *turret* [illegible]inciple, in which 12 sets of (RF, mixer and local oscillator) coils are mounted in sp[illegible]g-loaded brackets around a central shaft. The tuning knob is connected to this shaft, a[illegible] channels are changed by means of switching in the appropriate set of coils for the fi[illegible]d tuning capacitor. This automatically means that the tuned circuits for these three stages are ganged together, as shown in Fig. 17-8. Fine tuning is achieved by a slight variation of the tuning capacitance in the local oscillator. Many newer-model television receivers use PLL (phase-locked loop) circuitry to replace switch-type tuners with electronic tuners. Reliability is much better with these tuners, which have no mechanical parts.

The UHF tuner's active stages are a diode (point-contact or Schottky-barrier) mixer and a bipolar or FET local oscillator. This, like its VHF counterpart, is likely to be a Colpitts oscillator, as previously discussed in Sec. 11-1.2. That section also explained why VHF or UHF RF amplifiers are likely to be grounded-gate (or base). The diode mixer is used here as the first stage to lower the UHF noise figure—adequate gain is available from the remaining RF circuits, as will be seen. Coaxial transmission lines are used instead of coils in the UHF tuner, and they are tuned by means of variable capacitors. These are continuously variable (and of course ganged) over the whole range, but click stops are sometimes provided for the individual channels. Since the IF is quite small compared to the frequency at which the UHF local oscillator operates, AFC is provided. This takes the form of a dc control voltage applied to a varactor diode in the oscillator circuit. Such an arrangement was shown, in a slightly different context, in Sec. 5-3.3 and also Fig. 6-21.

An alternative means of UHF tuning consists of having varactor diodes to which fixed dc increments are applied to change capacitance, instead of variable capacitors. One of the advantages of this arrangement is that it facilitates remote-control channel changing. The remainder of the circuit is unchanged, but a UHF RF amplifier is normally added. The reason for this is the low Q of varactors, necessitating an additional tuned circuit to sharpen up the RF frequency response.

Figure 17-8 shows the VHF channel 7 being received. When any VHF channel is received, the UHF local oscillator is disabled, so that the output of the UHF mixer is a rectified UHF signal (channel 37 in this case), applied to the VHF tuner. This signal is a long way from the VHF radio frequency and thus has no effect. The significant carriers appearing at the input to the VHF RF amplifier are the picture (P), chroma (C) and sound (S) carriers of channel 7, of which only P is shown in Fig. 17-8.

We thus have P = 175.25 MHz, C = 178.83 MHz and S = 179.75 MHz applied to the RF amplifier, and hence to the mixer. These three are then mixed with the output of the local oscillator operating at the standardized frequency of 45.75 MHz above the picture carrier frequency. The resulting carrier signals fed to the first IF amplifier are thus P = 45.75 MHz, C = 42.17 MHz and S = 41.25 MHz. The IF bandpass is large enough to accommodate these signals and their accompanying modulating frequencies.

When the VHF tuner is set to the UHF position, the following three things happen:

1. The UHF local oscillator is enabled (dc supply voltage connected).
2. The VHF local oscillator is disabled (dc removed).
3. The VHF tuner RF and mixer tuned circuits are switched to (a picture carrier frequency of) 45.75 MHz.

The UHF tuner is now able to process the channel 37 signal from its antenna. The relevant frequencies, P = 609.25 MHz, C = 612.83 MHz and S = 613.75 MHz, are mixed with the local oscillator frequency of 655 MHz. The resulting outputs from the mixer diode are thus P = 45.75 MHz, C = 42.17 MHz and S = 41.25 MHz, being of course identical to the IF signals that the VHF tuner produces when receiving channel 7 (or any other channel). These are now fed to the VHF amplifier, which, together with the VHF mixer, acts as an IF amplifier for UHF. It is to be noted that the VHF mixer uses a transistor and not a diode and therefore becomes an amplifier when its local oscillator signal is removed. Since the UHF tuner has a (conversion) loss instead of a gain, this extra IF amplification is convenient.

The block diagram of Fig. 17-8 was drawn in a somewhat unorthodox fashion, tuned circuits being shown separately from the active stages to whose inputs they belong. This is not due to any particular quirk of TV receivers. Rather, it was done to show precisely what circuits are ganged together and to enable all relevant (picture carrier and local oscillator) frequencies to be shown precisely where they occur with either VHF or UHF reception. In turn, this means that it was possible to show the sum and difference frequencies at the outputs of the two mixers, with only the difference signals surviving past the next tuned circuit.

As shown in Fig. 17-9, the frequency response of a tuner is quite wide, being similar to, but broader than, the picture IF response. Note that the frequencies in Fig. 17-9*a* apply for channel 7, although those of Fig. 17-9*b* are of course fixed. Finally, it should be mentioned that chap. 24 in [1] is devoted entirely to TV tuners and is highly recommended to all those who wish to pursue the subject further.

Picture IF amplifiers The picture (or common) IF amplifiers are almost invariably double-tuned, because of the high percentage bandwidth required. As in other receivers, the IF amplifiers provide the majority of the sensitivity and gain before demodulation. Consequently, three or four stages of amplification are normally used. The IF stages provide amplification for the luminance, chrominance and sound information. However, as shown in Fig. 17-9*b*, the IF bandwidth is somewhat lower than might be expected—three factors govern this. At the upper end, relative response is down to 50 percent at the picture carrier frequency, to counteract the higher powers available at the lowest video frequencies because of the vestigial sideband modulation used. This was explained in Fig. 4-9*b* and the discussion in Sec. 4-4.4. At the lower end, relative amplitude is also down to 50 percent at the chroma subcarrier frequency, to minimize interference from this signal. At the sound carrier frequency of 41.25 MHz, response is down to about 10 percent, also to reduce interference. If a TV receiver is misaligned or purposely mistuned (with the fine-tuning control), the sound carrier may correspond

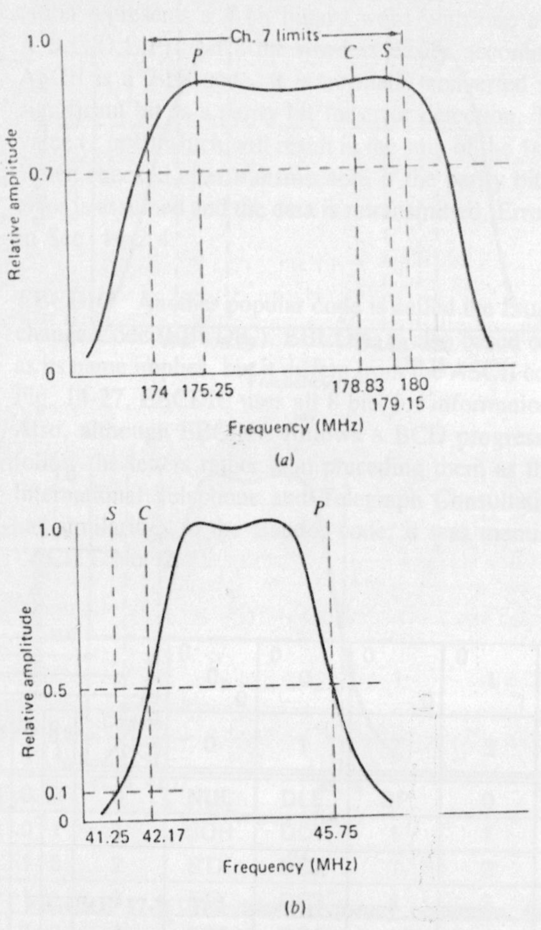

FIGURE 17-9 Television frequency responses. *(a)* **RF (shown for channel 7);** *(b)* **IF.**

to a point higher on the IF response curve. If this happens, the extra gain at this frequency will counteract the subsequent 4.5 MHz filtering, and some of the sound signal will appear in the output of the video amplifiers. This will result in the appearance of distracting horizontal sound bars across the picture, moving in tune with sound frequency changes.

The result of the foregoing considerations is that the picture IF bandwidth is approximately 3 MHz, as compared with the transmitted video bandwidth of 4.2 MHz. There is a consequent slight reduction in definition because of this compromise, but interference from the other two carriers in the channel is reduced, as is interference from adjacent channels. As anyone who has watched a good TV receiver will know, the resulting picture is perfectly acceptable. Reference should be made to chap. 23 of [1], if additional information on this topic is required.

Video stages It will be seen that the last picture IF amplifier is followed by the video detector and (customarily) two video amplifiers, whose output drives the (cathode of the) picture tube. At various points in this sequence, signals are taken off for sound IF, AGC and sync separation. The circuit of Fig. 17-10 shows these arrangements in detail.

As indicated, the circuit has a lot in common with detector-AGC circuits described previously, notably those in Figs. 6-12 and 6-19. Only the differences will be mentioned here. The first of these is the presence of coils L_1 and L_2. They are, respectively, series and shunt *peaking coils*, needed to ensure an adequate frequency response for the video amplifier shown. The second video amplifier also uses such an arrangement. Note that all C_T capacitors in Fig. 17-10 are fixed tuning capacitors, with values of a few picofarads. As shown, the coils are adjustable for alignment. Similarly, all components with F subscripts are used for (in this case, low-pass) filtering.

The transformer in the emitter of the first video amplifier, tuned to 4.5 MHz, has two functions. The more obvious of these is to provide the sound IF takeoff point. Since the video detector is a nonlinear resistance, the FM sound signal beats with the picture carrier, to produce the wanted 4.5-MHz frequency difference. This is extracted across the 4.5-MHz tuned transformer and applied to the first sound IF amplifier. At 4.5 MHz, this tuned circuit represents a very high unbypassed emitter impedance, much higher than the load resistance R_D. Consequently, the first video amplifier has a very low gain at the sound intermediate frequency. In fact, this is the second function of this arrangement. The sound IF transformer thus acts as a trap, to attenuate 4.5-MHz signals in the video output, preventing the appearance of the previously mentioned sound bars. Note finally that a portion of the video output voltage is also taken from here and fed to the sync separator, and another portion is rectified for AGC use. Since the AGC is delayed, a separate diode must be used. Other AGC systems are also in use, including *keyed AGC*. These and also video amplifiers are described further in chap. 14 of [3] and chaps. 13 to 16 of [1].

The video amplifiers of the TV receiver have an overall frequency response as

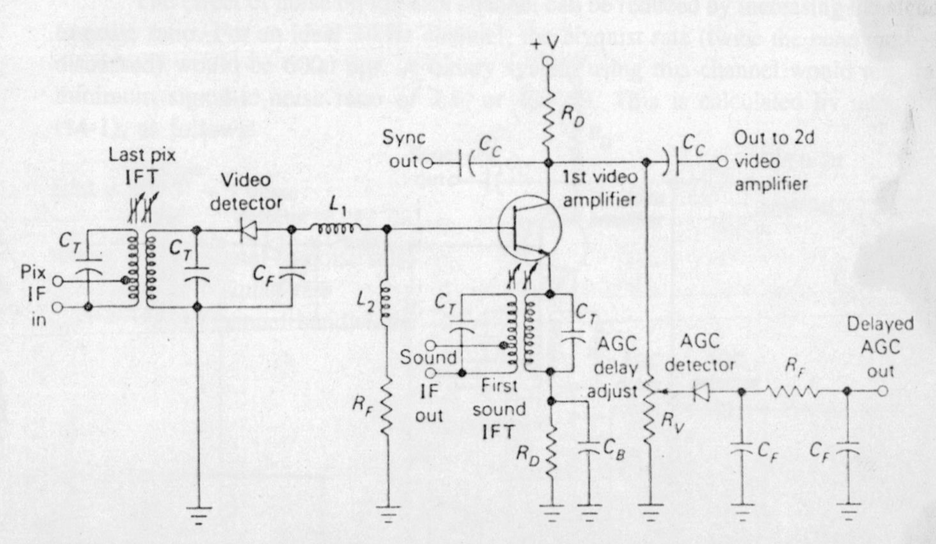

FIGURE 17-10 TV receiver video detector, first video amplifier and AGC detector.

shown and discussed in Sec. 4-4.4. The second stage drives the picture tube, adjusting the instantaneous voltage between its cathode and grid in proportion to the video voltage. This modulates the beam current and results in the correct degree of whiteness appearing at the correct point of the screen, which in turn is determined by the deflection circuits. The blanking pulses of the composite video signal drive the picture tube beyond cutoff, correctly blanking out the retraces. Although the sync pulses are still present, their only effect is to drive the picture tube even further beyond cutoff. This is quite harmless, so that the removal of the sync pulses from the composite video signal is not warranted.

The contrast and brightness controls, as should be expected, are located in the circuitry of the output video amplifier. The contrast control is in fact the direct video equivalent of the volume control in a radio receiver. When contrast is varied, the size of the video output voltage is adjusted, either directly or through a variation in the gain of the video output stage. Note that a typical picture tube requires about 100 V peak to peak of video voltage for good contrast. When an elderly picture tube begins to fade away (like an old soldier!), it is because it has lost sensitivity, and even maximum contrast is no longer sufficient to drive it fully. The brightness control varies the grid-cathode dc bias on the picture tube, compensating for the average room brightness.

Some receivers perform this function automatically, using a photodiode which is sensitive to ambient brightness, in addition to an adjustable potentiometer. Receivers with a single "picture" control normally have twin potentiometers for brightness and contrast, mounted on the one shaft and therefore adjustable together. This arrangement should not be decried too much: it has the advantage of giving the customer fewer knobs to adjust (i.e., *misadjust*).

The sound section As shown in the block diagram of Fig. 17-8, the sound section of a television receiver is identical to the corresponding section of an FM receiver. This was described in great detail in Sec. 6-4, to which students are referred. Note that the ratio detector is used for demodulation far more often than not. Note further that the intercarrier system for obtaining the FM sound information is always used, although it is slightly modified in color receivers, as will be seen.

17-3.3 Synchronizing Circuits

The task of the synchronizing circuits in a television receiver is to process received information, in such a way as to ensure that the vertical and horizontal oscillators in the receiver work at the correct frequencies. As shown in Fig. 17-7, this task is broken down into three specific functions, namely:

1. Extraction of sync information from the composite waveform
2. Provision of vertical sync pulses (fr[illegible]n the transmitted vertical sync pulses)
3. Provision of horizontal sync pulses [illegible]om the transmitted horizontal, vertical and equalizing pulses)

These individual functions are n[illegible] described, in that order.

Sync separation (from composite wavef[illegible]rm) The "clipper" portion of the circuit in Fig. 17-11*a* shows the normal method of removing the sync information from the

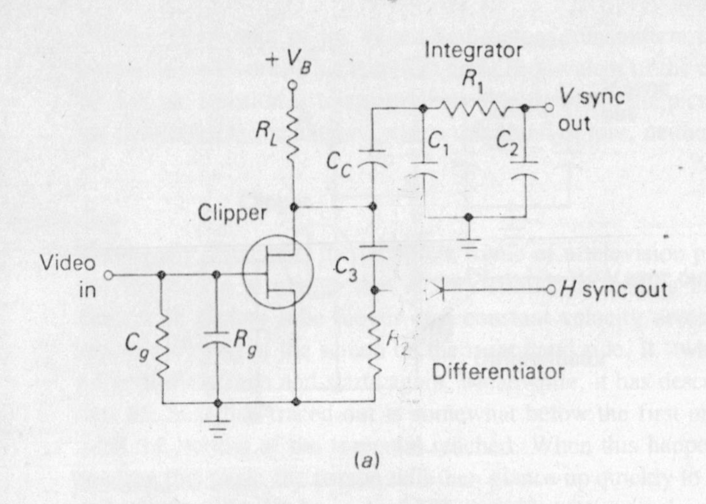

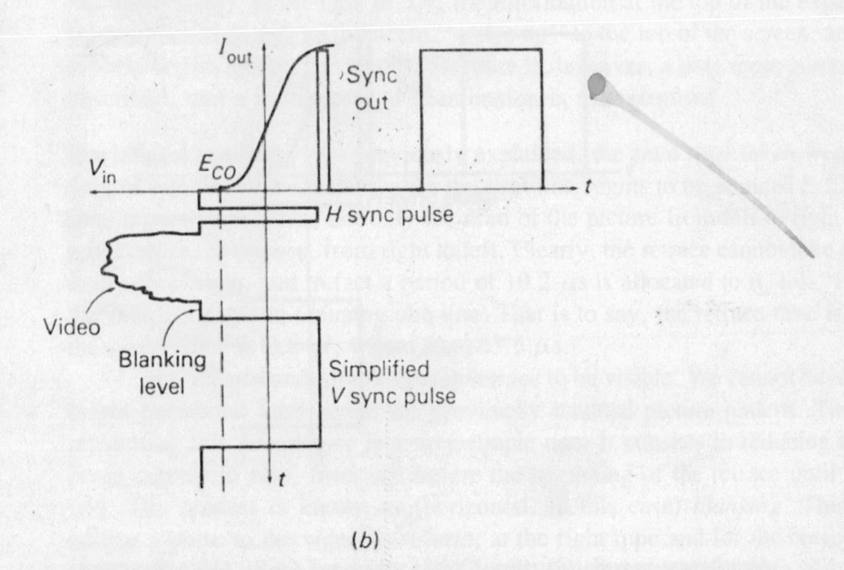

FIGURE 17-11 Sync separator. (*a*) Circuit; (*b*) clipper waveforms.

composite waveform received. The clipper uses leak-type bias and a low drain supply voltage to perform a function that is rather similar to amplitude limiting, as described in Sec. 6-4.2. It is seen from the waveforms of Fig. 17-11*b* that video voltage has been applied to an amplifier biased beyond cutoff, so that only the tips of the sync pulses cause output current to flow. It would not be practicable to use fixed bias for the sync clipper, because of possible signal voltage variation at the clipper input. If this happened, the fixed bias could alternate between being too high to pass any sync, or so low that blanking and even video voltages would be present in the output for strong signals. A combination of fixed and leak-type bias is sometimes used.

Horizontal sync separation The output of the sync clipper is split, as shown in Fig. 17-11*a*, a portion of it going to the combination of C_3 and R_2. This is a differentiating

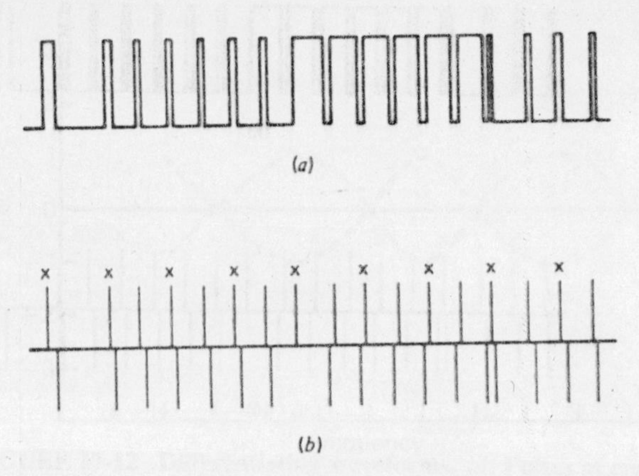

FIGURE 17-12 Differentiating waveforms. (*a*) **Pulses at end of even field;** (*b*) **(simplified) differentiator output. [Note: The pulses marked (x) are the only ones needed at the end of this field.]**

circuit, whose input and output waveforms are indicated in Fig. 17-12. As expected, a positive pulse is obtained for each sync pulse leading edge, and a negative pulse for each trailing edge. When the input sync waveform has constant amplitude, no output results from the differentiating circuit. The time constant of the differentiating circuit is chosen to ensure that, by the time a trailing edge arrives, the pulse due to the leading edge has just about decayed. However, the output does not consist of pulses that are quite as sharp as the simplified ones shown.

The output of the differentiator, at the junction of C_3 and R_2 in Fig. 17-12, is seen to contain negative pulses as well as the wanted positive ones. These negative-going triggers may be removed with a diode such as the one shown. In practice, the problem is taken care of, more often than not, by the diodes in the horizontal AFC circuit, as will be seen. Note finally that not all the positive triggers at the end of a vertical field are actually needed each time. However, if Fig. 17-12 is redrawn to show the situation at the end of an odd field, it will be seen that the pulses not used at the end of the even field will be needed then.

Vertical sync separation The coupling capacitor C_c in Fig. 17-11*a* is taken to a circuit consisting of C_1, R_1 and C_2, which should be recognized as a standard integrating circuit. Its time constant is made such as to yield the waveforms of Fig. 17-13. That is, its time constant is made long compared with the duration of horizontal pulses but not with respect to the width of the vertical sync pulse. When one considers that the former have widths of about 8 μs, and the width of the latter is just over 190 μs, the task is not seen as a very difficult one. Conversely, this situation just goes to show how much thought went into the design of the standards themselves. The integrating circuit, incidentally, may equally well be looked upon as a low-pass filter, with a cutoff frequency such that the horizontal sync pulses produce very little output, and the vertical pulses have a frequency that falls into the passband of the filter.

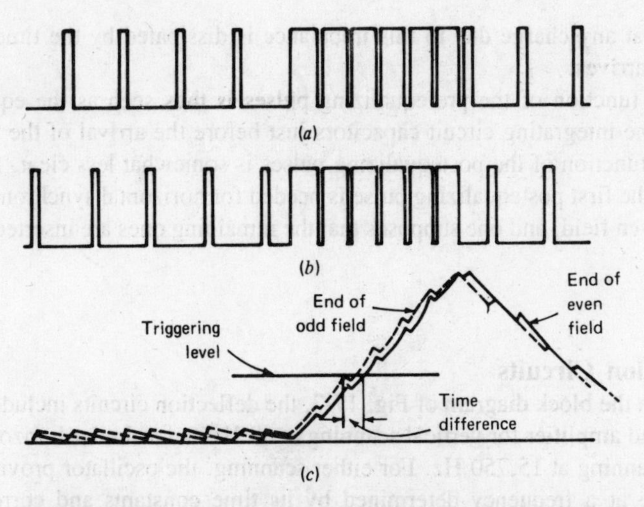

FIGURE 17-13 Integrating waveforms. (*a*) Pulses at end of even field; (*b*) pulses at end of odd field; (*c*) integrator outputs. (Note: *These waveforms have purposely been drawn as though there were no equalization pulses.*)

The waveforms of Fig. 17-13 explain the operation of the vertical integrator, but they do not represent a real-life situation. Instead, they have purposely been drawn to show what would happen if there were no equalizing pulses. As shown by means of the dotted line in Fig. 17-13*c*, without pre-equalizing pulses the charge remaining in the integrating circuit would be greater at the end of the odd field, because the preceding horizontal pulse would have been significantly closer than at the end of an even field.

As is well known, an oscillator is triggered not because an infinitely thin sync pulse arrives, but when a sync pulse of finite width reaches a particular amplitude. This is shown in Fig. 17-13*c*. It is also seen that the integrated pulse at the end of an odd field would reach this level sooner than the pulse produced at the end of an even field. If this were allowed, the odd field would become somewhat shorter (and thus the even field somewhat longer) than the required 262½ lines. A glance at Fig. 17-3 reveals that this would have a harmful effect on the interlace mechanism; the lines of one field would no longer be midway between the lines of the other field. The problem could possibly be solved by using an integrating circuit with a much longer time constant, to ensure that it was virtually uncharged by the horizontal pulses. However, this would have the effect of significantly reducing the integrator output for vertical pulses, so that a vertical sync amplifier would have to be used.

In a broadcasting situation, there are always thousands of receivers for every transmitter. Consequently, it is much more efficient to cure a potential problem in one transmitter than in thousands of receivers. This is achieved here by transmitting pre-equalizing pulses. Because they are transmitted, the appearance of the pulse train immediately preceding the vertical pulse is now the same at the end of either field, and the resulting output is the same in both cases. Admittedly, prior to the pre-equalizing pulses there is still an imbalance at the end of the two fields. However, this is so far upstream that any charge due to this imbalance is dissipated by the time the vertical sync pulse arrives.

The function of the pre-equalizing pulses is thus seen as the equalization of charge on the integrating circuit capacitors just before the arrival of the vertical sync pulse. The function of the postequalizing pulses is somewhat less clear. Figure 17-12 shows that the first postequalizing pulse is needed for horizontal synchronization at the end of an even field, and one supposes that the remaining ones are inserted for symmetry.

17-3.4 Vertical Deflection Circuits

As shown in the block diagram of Fig. 17-7, the deflection circuits include the vertical oscillator and amplifier for vertical scanning at 60 Hz and a similar horizontal arrangement for scanning at 15,750 Hz. For either scanning, the oscillator provides a deflection voltage at a frequency determined by its time constants and corrected by the appropriate sync pulses. This voltage is used to drive the corresponding output amplifier, which provides a current of the correct waveform, and at the right frequency, for the deflection coils. Magnetic deflection is always used for TV picture tubes and requires a few watts of power for the complete 90° or 110° (measured diagonally) deflection across the tube. Two pairs of deflection coils are used, one pair for each direction, mounted in a *yoke* around the neck of the picture tube, just past the electron gun.

This section is devoted to the vertical deflection circuits in a TV receiver but, before these can be discussed, it is necessary to look at the waveforms required and the means of producing them.

Sawtooth deflection waveform The scanning coils require a linear current change for gradually sweeping the beam from one edge of the screen to the other. This must be followed by a rapid (not necessarily linear) return to the original value for rapid retrace. The process must repeat at the correct frequency, and the average value must be zero to ensure that the picture is correctly centered. The waveform just described is in fact a *sawtooth* current, obtainable from a sawtooth voltage generator. It is shown in Fig. 17-14*a*.

If a capacitor is allowed to charge through a resistance to some high voltage (solid line in Fig. 17-14*c*), the voltage rise across it will at first be linear. However, as the voltage rises across the capacitor, so the remaining voltage to which it can charge is thus diminished, and the charging process slows down (dashed line in Fig. 17-14*c*). The process is useful nonetheless, because it shows that a linear voltage rise can be achieved if the charging process can be interrupted before its exponential portion. If, at this point, the capacitor is discharged through a resistor smaller than the charging one, a linear voltage drop will result (solid line in Fig. 17-14*c*). Although linearity is not quite so important for the discharge, speed is important, so that the discharge process is not allowed to continue beyond its linear region, as shown in Fig. 17-14*c*. If the ratio of charge time to discharge time is made about 8:1, we have the correct relationship for sweep and flyback of the vertical scanning waveform.

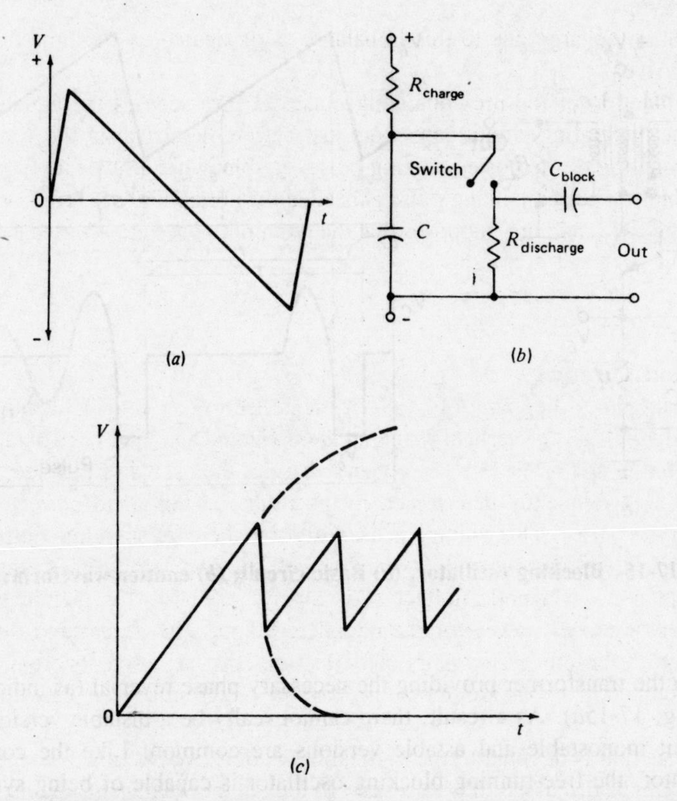

FIGURE 17-14 The sawtooth wave. (*a*) Waveform; (*b*) simple generator; (*c*) capacitor charge-discharge waveforms.

Figure 17-14*b* shows the simplest method of obtaining the charge/discharge sequence just described. Note, incidentally, that the charge process is not actually interrupted. The capacitor thus continues to charge (slowly) while it is being discharged, but this presents no problem. All that happens is that the discharge resistor is made slightly smaller to speed up discharge than it would have been if charge had been interrupted. To stop the slow charging during discharge would require a second switch synchronized with the first one, a needless complication. Note finally that C_{block} in Fig. 17-14*b* ensures that an ac sawtooth voltage is obtained from this circuit, being identical to Fig. 17-14*a*.

Blocking oscillator Having determined what waveform is required for scanning, and the basic process for obtaining it, we must now find a suitable switch. A multivibrator will fill the bill, but not really at a frequency as low as 60 Hz. The blocking oscillator, which, as shown in Fig. 17-15*a*, uses an iron-cored transformer, is perfectly capable of operating at frequencies even lower than 60 Hz. Consequently, it is almost invariably used as the vertical oscillator in TV receivers and is also sometimes used as the horizontal oscillator.

The blocking oscillator, unlike a multivibrator, uses only one amplifying de-

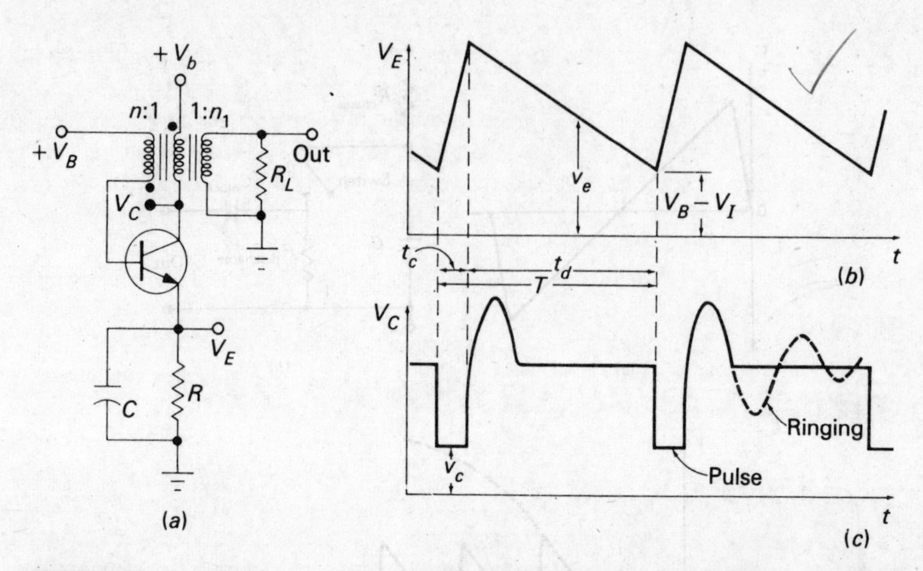

FIGURE 17-15 Blocking oscillator. (a) Basic circuit; (b) emitter waveform; (c) collector waveform.

vice, with the transformer providing the necessary phase reversal (as indicated by the dots in Fig. 17-15*a*). As a result, there cannot really be a bistable version of such a circuit, but monostable and astable versions are common. Like the corresponding multivibrator, the free-running blocking oscillator is capable of being synchronized. The circuit shown is an astable blocking oscillator; a careful look reveals its similarity to the Armstrong oscillator. Although the operation could be explained from that point of view, it is more common, and probably easier, to understand the operation from a step-by-step, pulse-type treatment.

The blocking oscillator uses an iron-cored pulse transformer, with a turns ratio having an n:1 voltage stepdown to the base, and a 1:n_1 voltage stepup to R_L. R_L is the load resistor with the subsidiary function of damping out undesired oscillations. As will be seen, such oscillations are likely to break out at the end of each collector pulse.

The circuit diagram shows the base winding returned to a positive voltage V_B. It is thus evident that this oscillator must be free-running, since there is no potential present which could cut the base off permanently. Note that the circuit can be converted to a triggered or monostable blocking oscillator by the simple expedient of turning V_B into a negative voltage. Trigger pulses are then required to make the circuit oscillate.

Assume initially that there is a voltage on C, v_e, larger than $V_B - V_I$, where V_I is the cut-in base-to-emitter voltage. Such a situation is in fact shown at the beginning of the waveform in Fig. 17-15*b*. Since this is the emitter-ground voltage of the transistor at that instant, the transistor is quite clearly OFF, and C is therefore discharging exponentially toward ground, with a time constant RC. When v_e is reduced to equal $V_B - V_I$, the base starts to draw current, as does the collector, and regenerative action begins

The increase (from an initial value of zero) in collector current lowers collector voltage, which in turn raises the base voltage. Hence, still more collector current flows, resulting in a further drop in collector current. In practical circuits loop gain exceeds unity, so that regeneration takes place and the transistor is very quickly driven into saturation. (The base waveform, which is not shown here, has exactly the same appearance as the collector waveform of Fig. 17-15*c*. However, it is inverted and scaled down by the factor n:1.)

The very short period of time just described marks the beginning of the collector output pulse. The base voltage is positive and saturated, while the collector voltage is at its minimum and also saturated. However, this cannot be a permanent state of affairs. After the transition to ON, the transistor collector impedance is low, and it forms an integrating circuit with the magnetizing inductance of the transformer ($v = L\ di/dt$, so that $i = 1/L \int v\ dt$). Therefore the collector current begins to rise and continues to do so linearly, while the collector voltage remains low and constant. After a time t_c, nonlinearities prevent collector current from increasing any further, and therefore the voltage across the transformer starts to fall (since $v = L\ di/dt$, and di/dt is dropping). This makes the collector more positive and the base less positive; the transistor is quickly switched off by regenerative action. Although the pulse duration is determined basically by the magnetizing inductance of the transformer and the total resistance across it, the calculation is decidedly complex. This is because the resistance itself is complex; it includes the transistor output resistance, its input resistance reflected from the secondary and the load resistance reflected from the tertiary winding.

The voltage across C cannot change instantaneously, and so it was unaffected by the rapid switching ON of the transistor. However, although v_c remains saturated, charging current flows through C, which becomes more positive gradually. It reaches its maximum as the switching OFF transient begins.[3] C charges toward V_b, but this charging is abruptly terminated by the disappearance of collector current when the transistor switches off. The maximum value of v_e is the top of the sawtooth shown in Fig. 17-15*b*. After the switching OFF transient, C discharges through R, eventually reaching once again the value $v_e = V_B - V_I$, whereupon the base cuts in and the process repeats. It is seen that the OFF period, t_d, and therefore the pulse repetition rate, is governed by the time constant RC to a large extent.

The period of the sawtooth free-running oscillation is $T = t_c + t_d$. As with other relaxation oscillators, the period may be shortened, thus making the oscillator a synchronized one, by the application of positive pulses to the base just before the transistor would have switched ON of its own accord. Like multivibrators, blocking oscillators have periods that can be shortened, but not lengthened, by trigger pulses. A switching-on pulse arriving at the base just *after* the transistor has switched itself ON is of no use whatever.

The rapid current change through the transformer at the end of the switching OFF transient induces a large overshoot in the collector waveform. Because of trans-

[3]Note that in a normal blocking oscillator it is not the rise in emitter voltage v_e which cuts off the transistor. This is because, even when v_e reaches its maximum during the transistor ON period, the base voltage is higher still, being the inverse of the low collector voltage, as previously mentioned. What initiates the switching OFF transient is quite definitely the drop di/dt, as described above.

former action, a large negative-going overshoot is also induced in the base waveform. Unless properly damped, this can cause *ringing* (decaying oscillations at the resonant frequency of the transformer and stray capacitances), as shown by the dashed line in Fig. 17-15*c*. It is the function of R_L to damp this oscillation, so that it does not persist after the first half-cycle. If this were not done, the transistor could switch itself ON too early. Also, care must be taken to ensure that the half-cycle overshoot that does occur is not so large as to exceed the base or collector breakdown voltage. A diode across the primary winding of the blocking oscillator transformer is sometimes used to provide limiting.

Vertical oscillator A television receiver vertical oscillator, together with a typical output stage, is shown in Fig. 17-16. It is seen to be a blocking oscillator quite similar to the one just discussed, but with some components added to make it a practical proposition. The first thing to notice is the resistor R which, together with the capacitor C, has been shifted to the collector circuit. Furthermore, this resistor has been made variable in part, and this part is labeled V_{height}. This is in fact the *vertical height control* in the TV receiver and is virtually a vertical size gain control. It will be recalled that the charging period of C is governed by the blocking oscillator transformer $Tr1$ and its associated resistances. By adjusting R, we thus vary the charging rate of the capacitor C, during the conduction time of the transistor $T1$. If R is adjusted to its maximum, a long RC time constant will result, and consequently C will not charge very much during this time. The output of the blocking oscillator will therefore be low. Since this is the voltage driving the vertical output stage, the yoke deflection current will also be low, yielding a small height. If, on the other hand, the value of R is reduced, C will charge to a higher voltage during conduction time, and a greater height will result. The height control is generally located around the back of the TV receiver, to lessen knob twiddling by its owner.

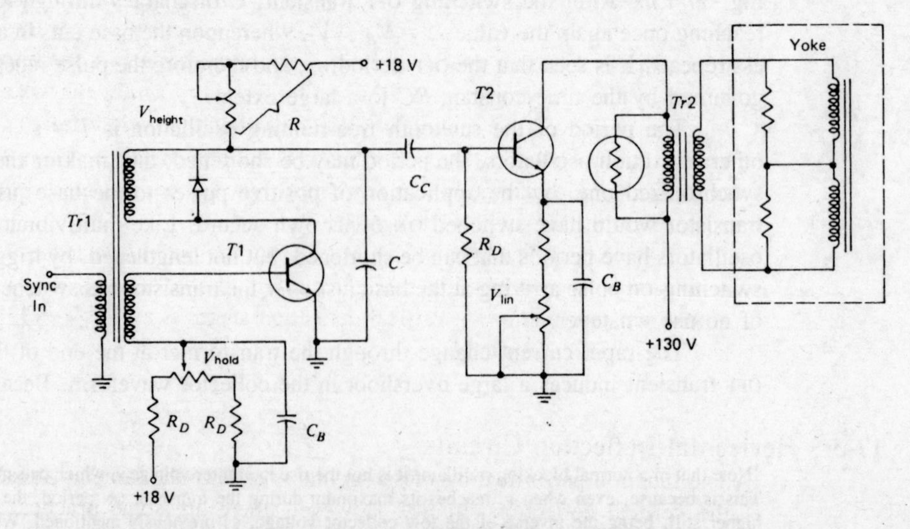

FIGURE 17-16 TV receiver basic vertical oscillator and output stage.

V_{hold} is the *vertical hold control*, with which positive bias on the base of $T1$ is adjusted. A glance at Fig. 17-15 shows that this has the effect of adjusting $V_B - V_I$. In this fashion the voltage through which RC must discharge is varied, and so is the discharge period (indirectly). Therefore, the vertical frequency, i.e., vertical hold, is varied.

As envisaged in the preceding section, the blocking oscillator transformer tertiary winding is used for the application of sync pulses. Here, they are positive-going and used to initiate prematurely the conduction period of $T1$. As discussed, this has the effect of controlling the period of the sawtooth, so that this is made equal to the time difference between adjoining vertical sync pulses. Note finally that, as foreshadowed, a protective diode is used across the primary winding of $Tr1$, in lieu of the load resistor across the tertiary in Fig. 17-15.

Vertical output stage The vertical output stage is a power output stage with a transformer-coupled output, as shown by $T2$ and its associated circuitry in Fig. 17-16. An additional amplifier is often used between the vertical oscillator and output stage. This driver generally takes the form of an emitter-follower, whose function is to isolate the oscillator and provide additional drive power for the output stage.

The deflection voltage from the vertical oscillator provides a linear rise in base voltage for the output stage, to produce a linear rise in collector current during trace time. The drive voltage cuts off the amplifier during retrace, causing the output current to drop to zero rapidly. The result is a sawtooth output current in the primary and secondary windings of the vertical output transformer $Tr2$, and this induces the sawtooth deflection current in the vertical coils in the yoke. In actual practice, the situation is a little more complicated. The inductance of the coils and transformers must be taken into account, so that a certain amount of wave shaping must take place, with R–C components which have not been shown. Their function is to predistort the driving waveform, to produce the correct sawtooth deflection current in the yoke coils.

The V_{lin} potentiometer is the *vertical linearity control* of the receiver, again located at the back of the receiver. Its adjustment varies the bias on the output transistor to obtain the optimum operating point. The thermistor across the primary winding of $Tr2$ stabilizes the collector of $T2$, and the resistors across the yoke coils have the function of preventing ringing immediately after the rapid retrace. Their values are typically a few hundred ohms. If ringing is not prevented, the beam will trace up and down in the (approximately) top one-third of the screen, producing broad, bright horizontal bars in that area of the screen.

Note lastly the high supply voltage for the output transistor. This is needed to provide the large deflection swing required, of the order of 100 V peak to peak. Additional information on vertical deflection stages is available in [3], pp. 453–456, and [1], chaps. 20 and 21.

17-3.5 Horizontal Deflection Circuits

The function performed by these circuits is exactly the same as already described for the vertical deflection circuits. However, there are some practical differences; the major one is the much higher horizontal frequency. This makes a lot of difference to the circuitry used by the horizontal oscillator and amplifier. Another important difference, as shown in the block diagram of Fig. 17-7, is that the horizontal output stage is used to provide the anode voltage for the picture tube. The current requirement is quite low, of the order of 800 μA. However, the voltage required is 10 to 18 kV. It must be produced somewhere in the receiver, and the horizontal output stage happens to be the most convenient point. The final difference between this and the vertical output section is quite minor but worth mentioning here. This is the fact that, since the aspect ratio of the picture tube favors the horizontal side by 4:3, the horizontal deflection current must be greater by the same amount.

Horizontal oscillator and AFC Being much narrower than vertical sync pulses, and occurring at a much higher rate, horizontal pulses are a lot more susceptible to noise interference than vertical sync pulses. The latter contain a fair amount of power (25 percent modulation for just over 190 μs), and it is thus unlikely that random or impulse noise could duplicate this. Accordingly, the output of the vertical sync separator may be used directly to synchronize the vertical oscillator, as was shown in the preceding section. Here, however, the situation is different. A noise pulse arriving at the horizontal oscillator could quite easily upset its synchronization, through being mistaken for a horizontal sync pulse. The horizontal oscillator would be put out of synchronism, and the picture would break up horizontally; this is clearly undesirable. It is avoided in a practical TV receiver by the use of an AFC system which isolates the horizontal oscillator so that neither sync nor noise pulses actually reach it.

The AFC loop uses a Foster-Seeley discriminator in a manner similar to that described in Sec. 5-3.3. The output of the horizontal sync separator is compared with a small portion of the signal from the horizontal output stage. If the two frequencies differ, a dc correcting voltage is present at the output of the discriminator. When the two frequencies are the same, the output is zero. Note that the system depends on average frequencies instead of individual pulses.

Since the output of the horizontal AFC system is a dc voltage, the horizontal oscillator must be capable of being dc-controlled. This is certainly true of the blocking oscillator, which is one of the forms of the horizontal oscillator. If, in this so-called *synchro-phase system,* a dc voltage is applied instead of +18 V at the top of the V_{hold} control in Fig. 17-16, frequency control with a dc voltage will be obtained. The reasoning is identical to that used in explaining the operation of the vertical hold control, in the preceding section.

Multivibrators are also quite often used as horizontal oscillators, and their manner of synchronization by a dc voltage is very similar to the blocking oscillator's. The system is called *synchro-guide*. Finally, recognizing that sinusoidal oscillators are somewhat more stable in frequency than pulse oscillators, some receivers use them. The system is then called *synchro-lock,* and the control voltage is applied to a varactor diode in the oscillator's tank circuit.

Horizontal output stage As in the vertical system, there is generally a driver between the horizontal oscillator and the horizontal output stage. Again, its function is to isolate the oscillator and to provide drive power for the horizontal amplifier. It also matches the relatively high output impedance of the oscillator to the very low input impedance

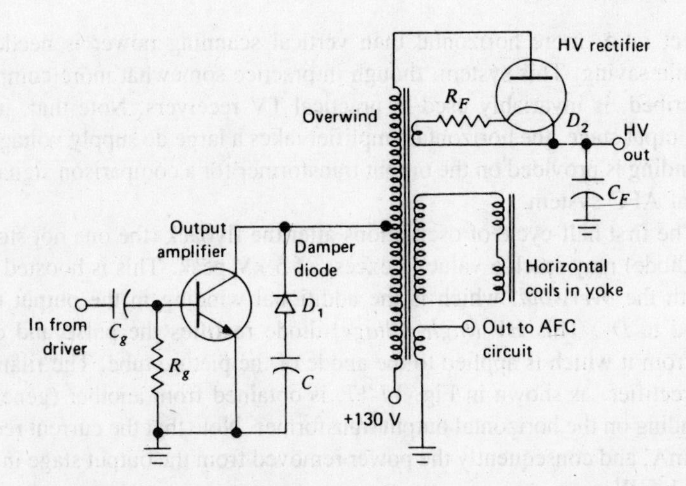

FIGURE 17-17 Simplified TV receiver horizontal output stage.

of the horizontal output stage, which is a high-power (about 25 W output) amplifier. The circuit diagram of a very simplified horizontal output amplifier is shown in Fig. 17-17.

This is a highly complex stage, whose operation is now briefly indicated. The output transistor is biased in class *C*, so as to conduct only during the latter two-thirds of each line. It is driven with a sawtooth voltage, which is large enough to drive the output transistor into conduction from roughly one-third along the horizontal line to just beyond the start of the flyback. While the output stage is conducting, a sawtooth current flows through the output transformer and the horizontal yoke coils, so that the beam is linearly deflected. Meanwhile the *damper diode*, D_1, is nonconducting, since its cathode is positive with respect to its anode.

The onset of the flyback promptly and vigorously switches off the output amplifier. If it were not for the damper diode, ringing would now begin, as previously explained in connection with the blocking oscillator. The typical frequency in the horizontal output transformer would be of the order of 50 kHz. However, what happens instead is that, as soon as flyback begins, the damper diode begins to conduct. This does not prevent the initial, negative-going half-cycle of oscillations. But, since D_1 is conducting, the capacitor *C* is charged, and in this manner energy is stored in it, instead of being available for the ringing oscillations. It is thus seen that the damper diode prevents all but the first half-cycle of oscillations and charges the capacitor *C*. The fact that the initial oscillatory swing took place is all to the good, because it helps to speed up the retrace.

At the end of the flyback *C* begins to discharge, through D_1 and the primary of the horizontal output transformer. If conditions are suitably arranged, the current due to the discharge of this capacitor provides the scanning current to the horizontal yoke coils, for the ''missing'' first one-third of each line. The advantage of doing this, instead of letting the output stage handle the whole scan (as was done in the vertical output stage), is that the maximum voltage rating and power handled by the horizontal output transistor are reduced by about one-third. Bearing in mind that, because of the

4:3 aspect ratio, more horizontal than vertical scanning power is needed, this is a worthwhile saving. This system, though in practice somewhat more complicated than just described, is invariably used in practical TV receivers. Note that, just as in the vertical output stage, the horizontal amplifier takes a large dc supply voltage, and that a small winding is provided on the output transformer for a comparison signal used in the horizontal AFC system.

The first half-cycle of oscillations after the flyback (the one not stopped by the damper diode) may reach a value in excess of 5 kV peak. This is boosted to 15 kV or more with the *overwind,* which is the additional winding in the output transformer, connected to D_2. This *HV (high-voltage)* diode rectifies the pulse and derives a dc voltage from it which is applied to the anode of the picture tube. The filament voltage for this rectifier, as shown in Fig. 17-17, is obtained from another (generally single-turn) winding on the horizontal output transformer. Note that the current requirement is under 1 mA, and consequently the power removed from the output stage in this manner is under 1.5 W.

The filtering of the HV rectifier output is obtained in a rather cunning manner. The filter resistor R_F is generally very small, of the order of a few ohms; the filter capacitance C_F is typically about 800 pF. Although these are quite small values, it must be remembered that the frequency is 15,750 Hz, and so these small values are sufficient. The cunning part of the proceedings is that C_F is not a capacitor. It is in fact the stray capacitance between the inner and outer (earthed) aluminized coatings of the picture tube. Note that if any of the horizontal stages fails, so will this scheme, and the picture will disappear, since the picture tube anode voltage will have disappeared also.

The above treatment of the horizontal output stage was, it must be reiterated, a very simplified one. For further information, pp. 456–460 of [3], and Chap. 22 of [1], are recommended.

17-4 COLOR TRANSMISSION AND RECEPTION

The subject of color transmission and reception was introduced in Secs. 17-1.1 and 17-2.1. It was seen that the color TV system requires the transmission and reception of the monochrome signals that have already been discussed, and in addition specific color information must be sent and decoded. It now remains to specify the requirements in more detail and to show how they are met.

17-4.1 Introduction

It is a great pity for all concerned that color TV came after monochrome TV. If it had not, the system would be far simpler than it actually has to be now. Since only the three *additive primary colors* (red, blue and green) need be indicated for all colors to be reproduced, one visualizes three channels, similar to the video channel in monochrome, transmitted and received. One further visualizes FDM rather than three separate transmissions, with signals corresponding to the three hues side by side in the one channel. Regrettably, color TV does not work that way: if it did, it would not be compatible.[4]

Compatibility Color television must have two-way compatibility with monochrome television; either system must be able to handle the other. That is to say, color transmissions must be reproducible in black and white on a monochrome receiver, just as a color receiver must be capable of displaying monochrome TV in black and white. The pity of it all is that this is a never-ending problem. The day all monochrome transmissions are superseded, which has already arrived in the industrialized countries, it will still not be possible to simplify transmission systems, because too many sets are already using the existing ones.

It is clear that a color television system must, in order to be compatible:

1. Transmit, and be capable of receiving, a luminance signal which is either identical to a monochrome transmission, or easily converted to it
2. Use the same 6-MHz bandwidth as monochrome TV
3. Transmit the chroma information in such a way that it is sufficient for adequate color reproduction, but easy to ignore by a monochrome receiver in such a way that no interference is caused to it

Color combinations White may be synthesized by the addition of blue (B), green (G) and red (R). It may equally well be synthesized by the addition of voltages that correspond to these colors in the receiver picture tube. However, it is not just a simple matter of saying white (Y) equals 33⅓ percent each of B, G and R. This is because, optically, our eyes have a color frequency response curve which is very similar to the response curve of a single-tuned circuit. Red and blue are at the two edges, and green is right in the middle of the response curve. Our eyes are thus most sensitive to green. Indeed, they are about twice as sensitive to green as to red, and three times as sensitive to red as to blue. The result is that "100 percent white" is given by

$$Y = 0.30R + 0.59G + 0.11B \tag{17-1}$$

Equation (17-1) in fact gives the proportions of the three primary colors in the luminance transmission of an NTSC color TV transmitter. Note, however, that it refers to the *proportions, not absolute values*. That is to say, if Y, as given by Eq. (17-1), has an amplitude that corresponds to 12.5 percent modulation of the carrier, the receiver will reproduce white. If the amplitude of the Y video voltage yields 67.5 percent modulation, a black image results. Any value in between gives varying shades of gray.

Since three primary colors must be capable of being indicated, two more signals must be sent. These clearly cannot be pure colors, since Y is already a mixture. In the NTSC system, the remaining two signals are

$$I = 0.60R - 0.28G - 0.32B \tag{17-2}$$

$$Q = 0.21R - 0.52G + 0.31B \tag{17-3}$$

[4] However, there is nothing to prevent a nonbroadcast color TV system, such as closed-circuit TV, from working this way.

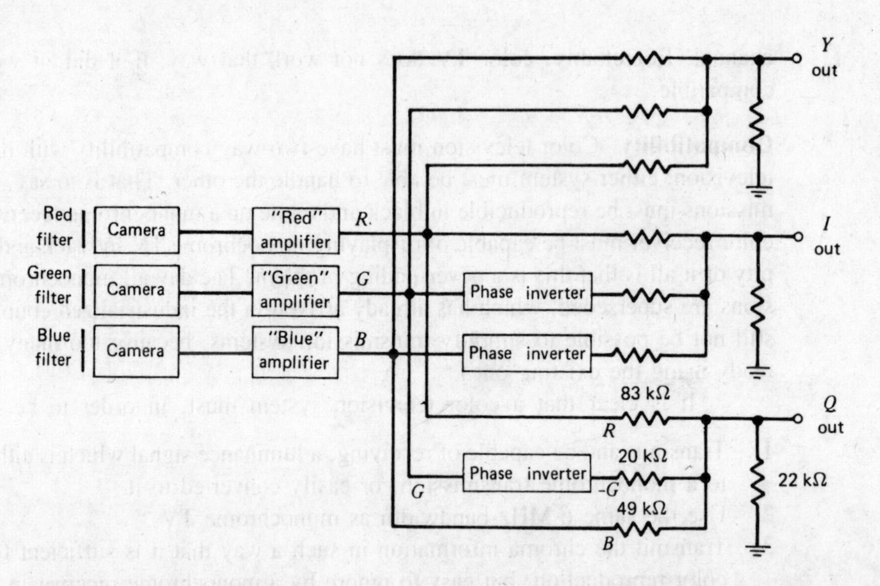

FIGURE 17-18 Color camera tube and matrix arrangements, showing typical resistor values for the Q matrix. (It is assumed that the phase inverters shown have unity gain.)

I stands for "in phase," and Q for "quadrature phase"; both terms are related to the manner of transmission, as will be seen shortly. Figure 17-18 shows how the Y, I and Q signals are generated, and Fig. 17-19 is a color disk (in monochrome!) showing how the various signals and colors are interrelated. For example, the color disk shows that if the received Q signal is instantaneously zero and I is maximum, a saturated

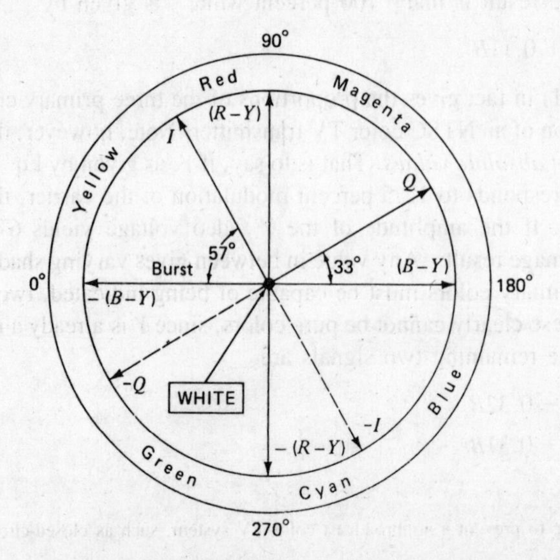

FIGURE 17-19 Color phase relationships and NTSC chroma vectors.

reddish-orange will be reproduced at that instant. Had I been less than maximum, a paler (i.e., less saturated) color of the same reddish-orange would have been reproduced. To take another example, consider $I = 0$ and $Q =$ negative maximum; the resulting color is a saturated yellowish-green. Most colors are in fact obtainable from vector addition. For instance, it may be checked by vector addition on the color disk that $0.8Q - 0.6I$ yields saturated, almost pure blue. Thus various combinations of the transmitted I and Q signals may be sent to represent whatever color is desired.

In addition to showing the phase relations of the I and Q signals of either polarity, the color disk also indicates three other vectors. The first of these is the *color burst,* which, as the name suggests, is a short burst of color subcarrier. It is sent once each horizontal line, as will be shown, and is used in the receiver as a phase reference. This is required to ensure that the absolute phase of the I and Q vectors is correct. For example, if it were not sent and a spurious $+90°$ phase shift of the color subcarrier in the receiver occurred, I would be mistaken for Q, and Q for $-I$. The resulting reproduced colors would have the correct relationship to each other, but they would be absolutely wrong. The $(R - Y)$ and $(B - Y)$ vectors are not transmitted but are often used in the receiver, as will be seen.

17-4.2 Color Transmission

Having discussed the manner of indicating luminance and the two components of chrominance in color TV, it is now necessary to investigate how they may be modulated and sent in the 6-MHz channel, without interference to monochrome TV.

Color subcarrier and chroma modulation The actual transmission methods used for the chroma components of the color TV system were determined by the following requirements and observations:

1. The sound carrier frequency must remain 4.5 MHz above the picture carrier frequency, because all TV receivers used the intercarrier system of sound detection, as explained in Sec. 17-3.2.
2. The energy dispersal of monochrome TV was found to be concentrated, clustered in fact, at harmonics of the line frequency. Thus, for example, significant video energy would be found at frequencies such as 15,750, 31,500, 47,250, 63,000 Hz, . . . 1.575000, 1.590750 MHz, and so on to the 4.2-MHz upper frequency limit for video.
3. There was very little video energy at frequencies midway between adjoining line frequency sidebands, such as 39,375 Hz (midway between the second and third sidebands) or at 1.582875 MHz (midway between the 100th and 101st sidebands. Note that these are odd harmonics of one-half the horizontal scanning frequency.
4. To arrange for the video voltages due to the chroma signals to fall within these "vacant slots," it would be necessary to have a color subcarrier frequency which was also an odd multiple of one-half the horizontal scanning frequency.
5. To minimize further any possible interference between the chroma and luminance video voltages, it would be a good idea to have the color subcarrier frequency as high as possible.

6. The color subcarrier frequency must not be too high, or else:
   (a) it would tend to interfere with the sound subcarrier (at 4.5 MHz).
   (b) the video voltages due to chroma would fall outside the 0- to 4.2-MHz video passband of the TV system.
7. To reduce further the possibility of interference between the sound subcarrier and video voltages due to color, it would be a good idea to make the sound subcarrier frequency a multiple of the horizontal scanning frequency.
8. Since the 4.5-MHz frequency was ''untouchable,'' it would be necessary to work the other way. The 286th submultiple of 4.5 MHz is 4,500,000/286 = 15,734.26 Hz. This is in fact the horizontal scanning frequency of color TV transmitters and receivers. It is within 0.1 percent of 15,750 Hz as used in monochrome TV and quite acceptable to that system.
9. Since the vertical field frequency is derived from the same oscillator as the horizontal line frequency (see Sec. 17-2.3), this would have to be altered correspondingly. Thus the vertical frequency used in practice by color systems is 59.94 Hz. Again, this is so close to the monochrome frequency as to be perfectly acceptable.
10. The eye has much poorer resolution for color than for brightness. That is, it is able to distinguish brightness variation between two adjacent points which are too close for it to be able to note a hue variation between them (as long as their brightness is the same). Thus the chroma video bandwidth need not be as large as the luminance bandwidth.
11. The eye's resolution for colors along the Q axis (reddish-blue--yellowish-green) is only about one-eighth of its luminance resolution, so that a 0.5-MHz bandwidth for the Q signal would suffice. However, it is able to resolve the colors along the I axis (yellowish-red–greenish-blue) about three times better than that. Thus a 1.5-MHz bandwidth for the I signal would be needed.
12. Bandwidth could be saved, and interference minimized, if the I signal were sent by using vestigial-sideband modulation, with the top 1 MHz of its upper sideband suppressed.
13. Interference would be further reduced if the color subcarrier frequency were suppressed.
14. The best method of combining the I and Q signals seemed to be the modulation of the same subcarrier by them, with a 90° phase difference between the I and Q signals.
15. The (suppressed) color subcarrier should be located so high that the upper sidebands of the signals modulating it (both extending 0.5 MHz from this subcarrier) should come just below the 4.2-MHz upper frequency limit of the video channel.
16. Since the color subcarrier is suppressed, some other form of color synchronization will have to be employed, to ensure correct absolute phases of the I and Q signals in the receiver (as explained in Sec. 17-4.1).

The foregoing considerations have resulted in the use of a color subcarrier frequency that is the 455th harmonic of half the horizontal scanning frequency. Another way of putting it is to say that the color subcarrier frequency is the 277th harmonic of the horizontal frequency plus one-half of the horizontal frequency. Either way, we have

$$f_c = \frac{15{,}734.26}{2} \times 455 = 3{,}579{,}545 = 3.579545 \text{ MHz}$$

This is the actual frequency generated. However, for simplicity, it is normally quoted as 3.58 MHz.

The 3.58-MHz reference signal is sent in the form of a brief pulse, or burst. It is superimposed on top of the back porch of each horizontal sync pulse. It will be recalled that the duration of this period of horizontal blanking is approximately 6 μs. The burst of 3.58 MHz consists of 8 to 11 complete cycles. These occupy a period not longer than 3.1 μs, so that adequate time is available for its sending. The peak-to-peak amplitude of the burst signal is approximately 15 percent of the percentage modulation range of video. Since it is superimposed on the 75 percent modulation blanking level, its peak-to-peak amplitude range stretches from 67.5 percent at the lowest point (top of the black level) to 82.5 percent at the highest point (one-third of the way from blanking to sync tops). It does not interfere with monochrome TV and is usable by a color receiver, as will be seen. Note that the color burst is not sent during the vertical blanking period, during which it is not needed.

Color transmitters The block diagram of a color TV transmitter is shown in Fig. 17-20. This is a simplified block diagram, in which the sections not directly related to color TV (and hence previously discussed in Sec. 17-2) have been "attenuated." Note that each block represents a function, not just a single circuit.

The *Y*, *I* and *Q* outputs from the color matrix are fed to their respective low-pass filters, as shown. These filters attenuate the unwanted frequencies, but they also introduce unwanted phase shifts. Phase-compensating networks (not shown) are inserted after the filters, to produce the correct phase relationships at the balanced modulators.

The output of the color subcarrier generator is sent in three directions. One of the three outputs is used to synchronize the blanking and sync pulse generators. Their output, in turn, is transmitted as in monochrome TV, and a portion of it is used to synchronize the transmitter cameras, as well as introducing blanking into the transmitted video. The second path for the 3.58-MHz oscillator output is to the color burst generator, which is a fairly complex piece of equipment that ensures the correct transmission (and phase preservation) of the color burst. Finally, the last output from this oscillator is fed to a 57° phase shifter, to provide the necessary shift for the *I* signal. A further 90° phase shift is produced, giving a total of 147° (180° − 33° in Fig. 17-19) for the *Q* signal. Note the 90° phase difference between the *I* and *Q* signals.

The *I* balanced modulator produces a double-sideband (suppressed-carrier) signal stretching 1.5 MHz on either side of the 3.58-MHz subcarrier. The vestigial-sideband filter then removes the top 1 MHz from that. The output of the *Q* balanced modulator is a signal occupying the range of 0.5 MHz below and above the suppressed 3.58-MHz subcarrier. The added 90° phase shift puts this signal in quadrature with the *I* component; hence the name "*Q* signal," as previously mentioned.

All these signals are fed to the adder, whose output therefore contains:

1. The *Y* luminance signal, occupying the band from 0 to 4.2 MHz, and virtually indistinguishable from the video signal in monochrome TV

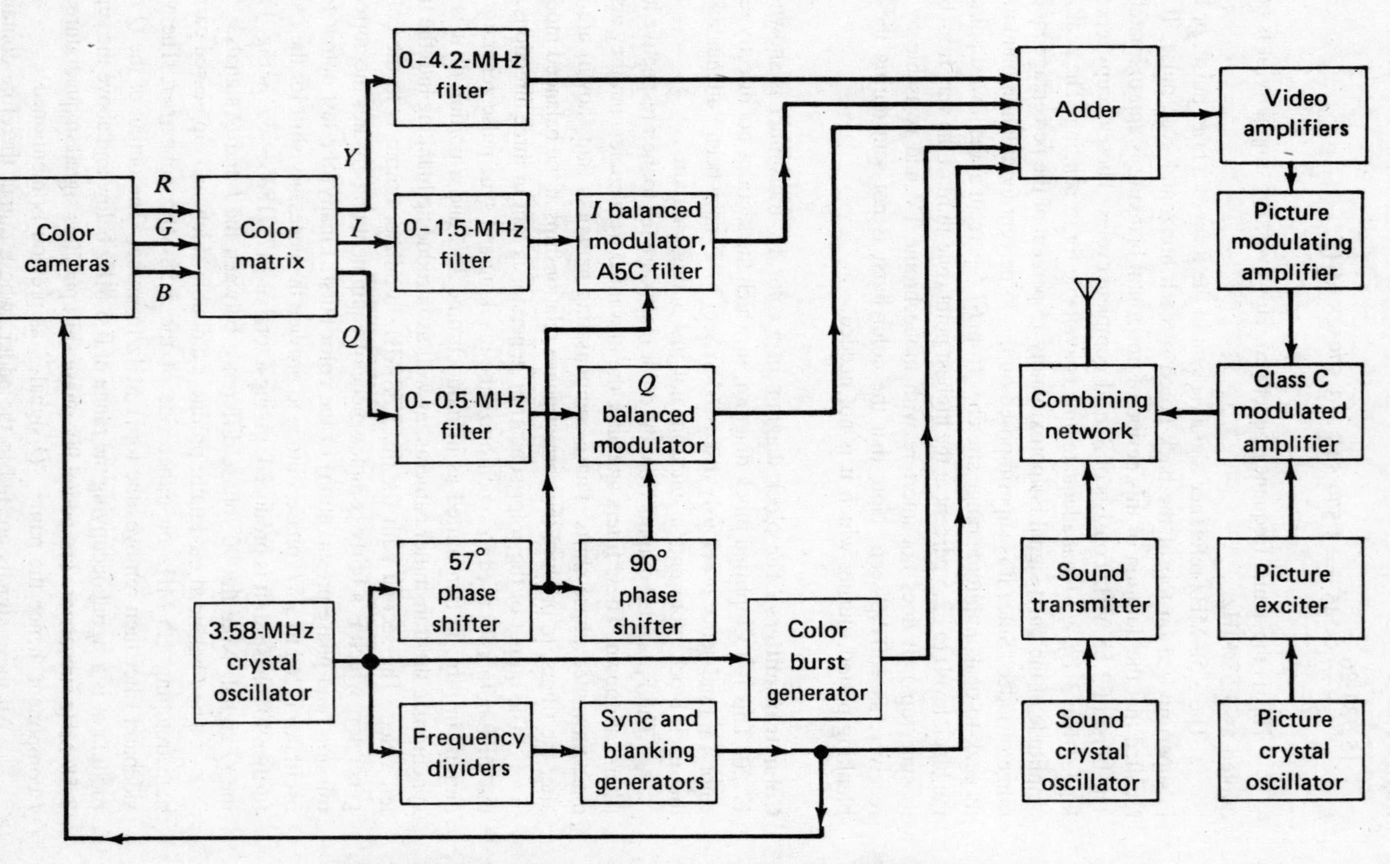

FIGURE 17-20 Basic block diagram of color television transmitter.

2. Synchronizing and blanking pulses, identical to those in monochrome TV, except that the scanning frequencies have been slightly shifted as discussed, to 15,734.26 Hz for the horizontal frequency and 59.94 Hz for the vertical frequency
3. (Approximately) 8 cycles of the 3.579545-MHz color subcarrier reference burst superimposed on the front porch of each horizontal sync pulse, with an amplitude of ± 7.5 percent of peak modulation
4. An *I* chroma signal, occupying the frequency range from 1.5 MHz below to 0.5 MHz above the color subcarrier frequency, and an energy dispersal occupying the frequency clusters not used by the luminance signal
5. A *Q* chroma signal, occupying the frequency range from 0.5 MHz below to 0.5 MHz above the color subcarrier frequency, and an energy dispersal occupying the same frequency clusters as the *I* signal, but with a 90° phase shift with respect to the *I* signal

The output of the adder then undergoes the same amplifying and modulating processes as did the video signal at this point in a black-and-white transmitter. The signal is finally combined with the output of an FM sound transmitter, whose carrier frequency is 4.5 MHz above the picture carrier frequency, as in monochrome TV.

It is worth pointing out at this stage that one of the main differences between the PAL system and the NTSC system thus far described is that in the former the phase of the *I* and *Q* signals is switched after every line. This tends to average out any errors in the phase of hue that may be caused by distortion or noise and tends to make this system somewhat more noise-immune. Needless to say, this phase alternation by line is what gives this system its name.

Students seeking information additional to the necessarily abbreviated treatment in Secs. 17-4.1 and 17-4.2 are referred to chap. 15 in [3] and chap. 8 in [1].

17-4.3 Color Reception

There are a large number of circuits and functions which monochrome and color television receivers have in common. Thus, for example, a color TV receiver (like its monochrome counterpart) requires a tuner, picture and sound IF stages, a sound demodulator section, horizontal and vertical deflection currents through a yoke, a picture tube anode high dc voltage, and finally video amplifiers (luminance amplifiers in this case). Where the construction and operation of these circuits are virtually the same as in monochrome receivers, they will not be mentioned further. If they differ somewhat from their black-and-white counterparts, the differences will be explained. Those circuits that are specific to color TV receivers will be described in some detail.

The sections of the color TV receiver that are most likely to be quite new are the picture tube and the circuits associated with it. Although the picture tube is the final point in the color receiver, it actually makes an ideal starting point in the discussion of color receivers.

Color picture tube and its requirements A color picture tube requires correct sweep currents, input voltages and drive voltages. Having said this very quickly, it is now a good idea to examine the circuit block of Fig. 17-21, to gauge the complexity of those

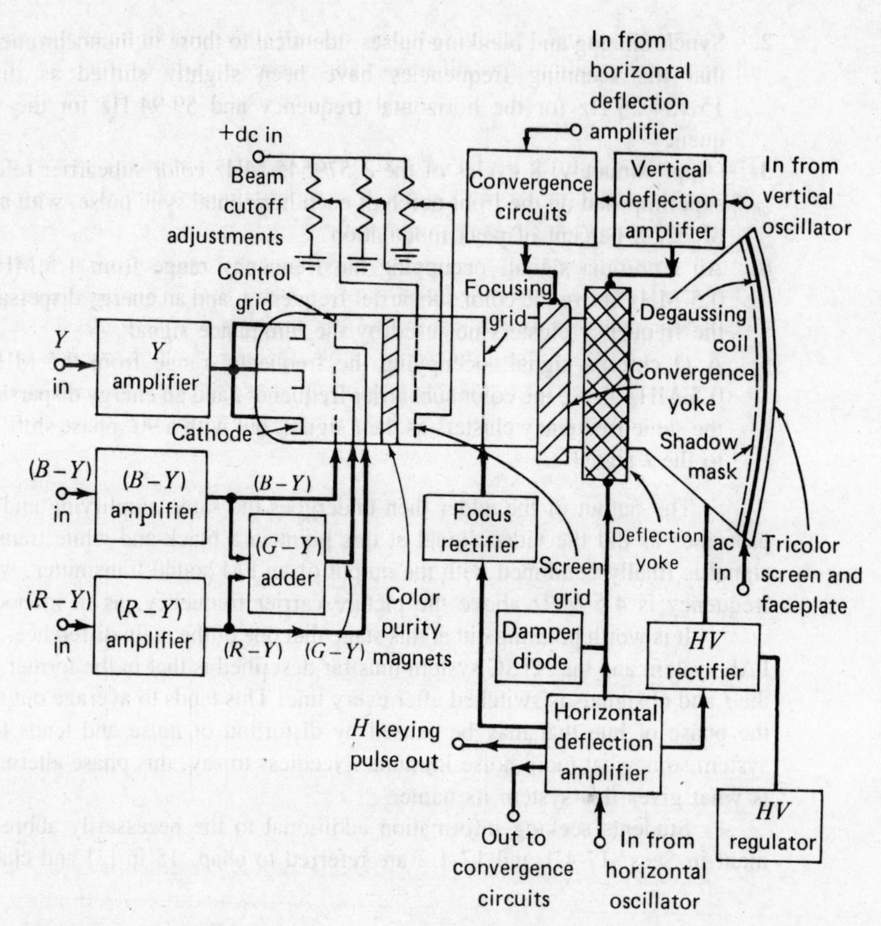

FIGURE 17-21 Television color picture tube and associated circuitry.

requirements. As already mentioned, the tube has three cathodes, or electron guns; they may be in-line or in a delta formation. It is the function of each cathode to produce an electron beam which, having been affected by various voltages and magnetic fields along its path, eventually reaches the correct part of the screen at precisely the right time. So far, the main difference between this tube and a monochrome tube is that three beams are formed, instead of just one.

Some color TV receivers, such as the one whose partial block is shown in Fig. 17-21, use the picture tube as a matrix. In others, voltages fed to each of the three hue amplifiers correspond to the pure primary colors, blue, green and red. In the receiver type shown, the output of the color demodulators has two channels, with voltages corresponding to $(B - Y)$ provided in one of the channels, while the other channel provides $(R - Y)$. The next section will show how and why these two signals are obtained. Each of the signals is amplified separately, and they are then added in the correct proportions to produce the $(G - Y)$ video voltages. Reference to the color vector disk of Fig. 17-19 will show that, as a good approximation, the vector addition of $-0.5R$ and $-0.2B$ produces the G vector. If the same voltage, Y, is subtracted from all three, the relationship still holds, and we have

$$(G - Y) = -0.5(R - Y) - 0.2(B - Y) \qquad \textbf{(17-4)}$$

The $(G - Y)$ adder of Fig. 17-21 performs the function of Eq. (17-4) with the aid of circuits akin to those of Fig. 17-18. The three primary color voltages (with the luminance voltage, Y, subtracted from each) are now applied to their respective grids, as shown in Fig. 17-21. There is a potentiometer in each path (not shown), to provide adjustment ensuring that the three drive voltages have the correct amplitudes. If this were not done, one of the colors on the screen could predominate over the others.

In a monochrome transmission, all three grid voltages would be zero, and the only voltage then modulating the beam currents would be the $-Y$ luminance signal applied to all three cathodes in parallel. In a color transmission, the four drive voltages will all be produced. However, the luminance signal applied to the cathodes will add to each of the grid voltages, canceling the Y component of each and ensuring that only the R, G or B video voltages modulate the respective beams from this point onward. Note that usual 180° phase reversal between grid and cathode takes place here also. Thus the $-Y$ voltage applied to the cathodes is equivalent to $+Y$ at a grid, and addition does take place.

After undergoing the episode just described, the three beams now pass the color purity magnets. These are small, adjustable permanent magnets, which have the task of ensuring that each resultant color is as pure as possible. That is to say, adjustment is made to produce minimum interference between the beams.

The next port of call for the beam is a series of three screen grids. Aside from accelerating the beam, as in any other vacuum tube, these screen grids have a very important function. As shown, each is connected to a positive dc voltage via a potentiometer, which is adjusted to give the same cutoff characteristic for each beam. That is to say, there will be the same input-voltage–beam-current relationship for the small-drive nonlinear portion of each electron gun's operating region. This is necessary to ensure that one beam does not predominate over the others in this low-drive portion of the curve; otherwise white could not be obtained at low light levels. Control of the cutoff characteristics at the screen grid is convenient and common.

It is then necessary to focus each beam, so that it has the correct small diameter. This ensures that fineness of detail is obtainable, like painting a canvas with a fine brush. Focusing is performed with an electrostatic lens, in the form of a grid to which a dc potential of about 5 kV is applied. The current requirement is very low, so that it is possible to obtain the focusing voltage by rectifying the flyback pulse in the horizontal output stage. The operation of the focus rectifier is identical to that of the HV rectifier in monochrome receivers, as described in Sec. 17-3.5.

We must now switch our attention to the color screen end of the picture tube. As previously described, this is a large glass surface with a very large number of phosphor dots on it. Three types of medium-persistence phosphor are used, one for each of the three colors. Thus dots (or sometimes small stripes) of one of the phosphors will glow red when struck by the beam from the "red gun," with an intensity depending on the instantaneous beam current. Dots of the second phosphor will similarly glow green, and those of the third will glow blue. The dots are distributed uniformly all over the screen, in triplets, so that under a powerful magnifying glass one would see three adjacent dots, then a small space, three more adjacent dots, and so on. A correct picture is obtained if the beam for each gun is able to strike only the dots that belong to it. Students will appreciate what an unreal picture would be obtained if, for example, the beam from the "blue gun"[5] were able to strike phosphor dots which could glow green or red.

The *shadow mask* is used to ensure that a beam strikes only the appropriate phosphor dots, anywhere on the screen. It is a thin metal plate with about 200,000 small holes, corresponding to the (approximately) 200,000 screen dots of each color. The holes, or slots, in the shadow mask are thus arranged so that there is one of them for each adjacent trio of phosphor dots (or stripes). Since each beam strikes from a slightly different angle, it is possible to position the shadow mask so that each beam can strike only the correct dots.

The shadow mask is bonded into place during the manufacture of the tube, so there is no question of adjusting it to ensure correct physical alignment. Any adjustments that are performed during the lineup of the receiver must thus be on the beams themselves. The process is known as adjustment of the *convergence*. It is performed with the *convergence yoke,* situated just before the deflection yoke as shown in Fig. 17-21. The convergence yoke has a set of three coils, each with its own permanent magnet, which is adjustable. Convergence for the undeflected beam, or *static convergence,* is obtained by adjusting the permanent magnets. *Dynamic convergence,* when the beam is being deflected, is provided by varying the currents through the convergence coils. As shown, these currents are derived from the horizontal and vertical deflection amplifiers. The currents required and the methods of adjusting the receiver for correct convergence are described in [1], pp. 212–218.

The beams, now more than halfway to the screen, then encounter the vertical and horizontal coils in the deflection yoke. What happens then is exactly what happened at the corresponding point in a monochrome receiver, except that here three beams are simultaneously deflected, whereas previously there had been only one beam. The methods of providing the requisite deflection currents are also as already described.

It is worth mentioning at this point that most color picture tubes now, like their monochrome counterparts for some time, have deflections of the order of 110°, whereas these previously had been 90°. This deflection, it will be recalled, is given as the total corner-to-corner figure, and it thus corresponds to 55° beam deflection away from center, when the beam is in one of the four corners of the picture tube. The greater the deflection, the shorter need the tube be.

Since the length of the picture tube determines the depth of the cabinet, large deflection is thus advantageous. It does have the disadvantage of requiring greater deflection currents, since more work must be done on the beams to deflect them 55°

[5] Needless to say, the gun in question is not blue, and neither is the beam which issues forth from it. Come to think of it, neither is the video voltage applied to it! Quotation marks are used to emphasize that the only "blue" involved is the color of the screen dots upon which this beam must impinge, having had its brightness (or intensity) originally modulated by video voltages which correspond to the instantaneous saturation of the blue hue in the particular picture being scanned.

from center, instead of 45°. The problem is somewhat alleviated by making the 110° deflection tube with a narrower neck, so that the deflection coils are closer to the beams themselves. The magnetic field can thus be made more intense over the smaller area.

The shadow mask, through which the beams now pass, ensures that the correct dots are activated by the right beams, but it also produced three side effects. The first is a reduction in the number of electrons that hit the screen. This results in reduced brightness but is compensated by the use of a higher anode voltage. Thus color tubes require typically 25 kV for the anode, where monochrome tubes needed about 18 kV. In hybrid receivers the higher voltage is obtained by having a larger overwind in the horizontal output transformer, and a rectifier with an appropriately higher rating. In solid-state receivers an additional winding is often used for this purpose, with silicon diodes in a doubling or tripling rectifier configuration. Because color tubes are rather sensitive to anode voltage variations, this voltage is regulated, as shown.

Those electrons that do not hit the screen quite obviously hit the shadow mask. With the high anode accelerating voltage, such electrons are traveling at *relativistic velocities* (i.e., at velocities sufficiently appreciable when compared with the velocity of light that relativity cannot be entirely ignored). When striking the shadow mask, these electrons are liable to produce x-ray emissions from the steel in it. This is problem number two. However, it is not a very serious one, because the *soft* (low-energy) x-rays emitted are stopped by most solid materials. A metal hood around the picture tube is sometimes used to contain the x-rays, but the aquadag coating is generally sufficient. With a properly constructed faceplate, the radiation is negligible unless the anode voltage exceeds the design value. Receivers generally have a circuit designed to disable the horizontal output stage (where this voltage is generated) if anode voltage becomes excessive. Health authorities set limits on the maximum permissible radiation for color TV receivers.

The third problem results from the presence of large metallic areas, especially the shadow mask, near the screen of the picture tube. These can become permanently magnetized by the earth's magnetic field, producing a local magnetic field which can deflect the beam. Such a spurious deflection may not be very large, but even so it is likely to affect the convergence. The standard method used for demagnetization, or *degaussing,* is the application of a gradually reducing ac magnetic field. This explains the presence of the *degaussing coil* around the rim of the picture tube near the screen. A spiral coil is used, and has the mains ac voltage applied to it when the set is switched on. This takes place automatically, and a thermistor is used in such a way that the current soon decays and eventually drops to zero. Meanwhile the tube has been degaussed, in more or less the time it takes to warm up. Needless to say, the coil is shielded for safety.

Common color TV receiver circuits Figure 17-22 shows the block diagram of a color television receiver, but for simplicity the circuits shown in Fig. 17-21 are omitted. However, interconnection points are shown on both diagrams, so that there should be no difficulty in reconciling the two figures. It is now proposed to look first at the (remaining) common circuits in the color receiver, i.e., those circuits which have direct counterparts in monochrome receivers, commenting on those differences that exist.

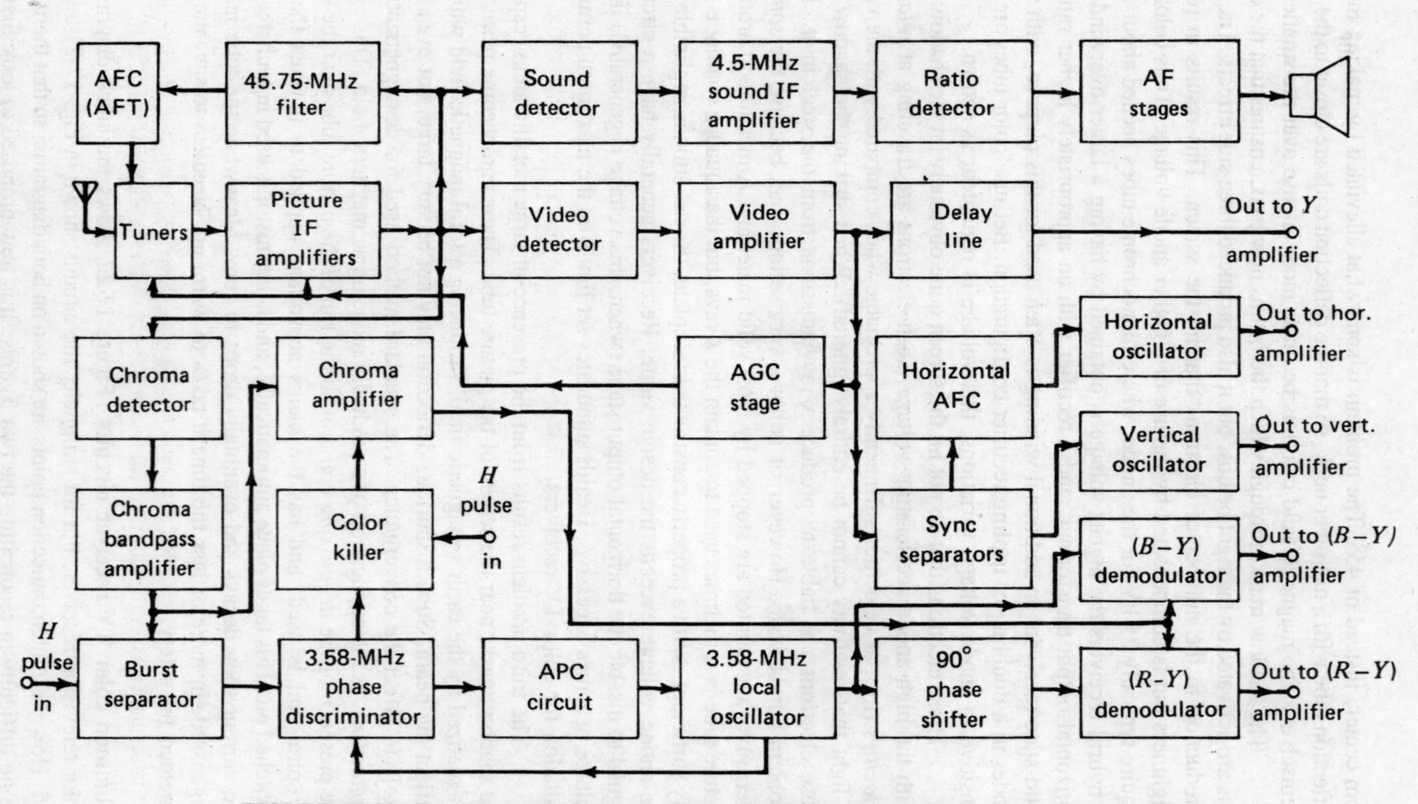

FIGURE 17-22 Partial block diagram of color television receiver, showing connections to corresponding points on Fig. 17-21.

A color TV receiver almost invariably has an AFC circuit, as indicated in Fig. 17-22. It is often called automatic fine tuning (*AFT*) and is used automatically to minimize mistuning, particularly to too high a frequency. This would produce added amplification of the sound carrier, and hence 920-kHz interference between the chroma and sound carriers. Alternatively, if the receiver is misadjusted to too low a frequency, insufficient gain will be available in the IF amplifiers at the chroma subcarrier frequency, and the output will be lacking in color. The AFT circuit consists basically of a 45.75-MHz filter, whose output is fed to a phase discriminator. This produces a dc correcting voltage whenever its input frequency differs from 45.75 MHz, and this voltage is applied to a varactor diode in the circuit of the appropriate local oscillator in the tuner. Such a system was described in detail in Sec. 6-3.2. It is normally possible to switch out the AFT circuit, so as to permit manual fine tuning.

The next point of difference from monochrome receivers arises in connection with sound demodulation. The intercarrier system is still used, but this time sound is extracted at an earlier point, again to reduce interference between it and chroma. The output of the last IF amplifier is fed to three separate, but more or less identical, diode detectors. Each of these acts as a nonlinear resistance, with the usual difference frequencies appearing in its output. The frequency selected from the output of the sound detector is 4.5 MHz, and this is then followed by exactly the same circuitry as in a monochrome receiver.

The output of the video detector undergoes the same treatment as in black-and-white receivers, with two differences. The first of these is that additional sound traps are provided, and the bandwidth of the video amplifiers is somewhat narrower than in a monochrome receiver. The latter is to reduce interference between the *Y* signal, which these amplifiers handle, and the lowest *I* sidebands of the chroma signal. The second difference is denoted by the presence of the delay line in Fig. 17-22. It will be recalled that the *Y* signal is subtracted from *R*, *G* and *B* in or just before the picture, so that a correct phase relation there is essential. As will be seen in the next section, the chroma signal undergoes more phase delay than the luminance signal before reaching the picture tube, and so a correction is required. The simplest method of equalizing the phase differences is by introducing a delay into the *Y* channel. This delay is normally just under 1 μs.

Color circuits We have reached the stage where we know how the luminance signal is delivered to the cathode of the picture tube, and the sound signal to the loudspeaker. We also know what deflection currents are required, and how they are obtained. We further know what other inputs the picture tube requires and at what point the chroma subcarrier is divorced from the luminance voltages. What we must now do is to determine what happens in the circuitry between the chroma takeoff point and the $(B - Y)$ and $(R - Y)$ inputs to the appropriate amplifiers preceding the picture tube grids in Fig. 17-21.

The output of the chroma detector is fed to a bandpass amplifier, having a frequency response designed to reject the lower video frequencies representing *Y* signals, as well as the 4.5-MHz sound carrier. In more elaborate receivers the bandpass stretches from 1.5 MHz below to 0.5 MHz above the 3.58-MHz chroma subcarrier. In most receivers, however, this bandpass is only 3.58 $\pm$ 0.5 MHz, so that some of the transmitted I information is lost. The resulting loss in color definition is not too serious, and the advantage is a reduction in interference from the distant Y sidebands. Consequently the use of this arrangement is widespread. The chroma signal is now amplified again and fed to the color demodulators. Because the chroma amplifiers have a much narrower bandwidth than the Y video amplifiers, a greater phase delay is introduced here; hence the delay line used in the Y channel.

It was shown in the preceding section that two color signals, such as $(B - Y)$ and $(R - Y)$, are sufficient, because the third one can be obtained from them by vector addition. It is thus necessary to decide which two color signals should be obtained, by the appropriate demodulation of the chroma output. At first sight, it would seem obvious that the two signals should be I and Q, for which R, G and B would be obtained by a matrixing process that is likely to be the reverse of the one shown in Fig. 17-18. However, this is rather awkward to do and requires sufficient bandwidth to make all of the I signal available in the first place—an unlikely situation, as just described. The next logical thought is to try to obtain the R, G and B signals directly, but this is also awkward, because the required phase differences between these three vectors and the reference burst (77°, 299° and 193°) are also difficult to produce. These values are, incidentally, obtainable from the color disk of Fig. 17-19.

The result of the foregoing considerations is that most receivers produce the $(R - Y)$ and $(B - Y)$ voltages from their color demodulators. This results in the loss of a little color information, but this loss is outweighed by two important considerations. The first is the easy production of the requisite phase differences with respect to the color burst, being 90° for $(R - Y)$ and 180° for $(B - Y)$. The second reason for using this arrangement is that the resulting signals can be matrixed by the picture tube without any further ado, as was seen earlier.

Synchronous demodulators are used for detecting the $(R - Y)$ and $(B - Y)$ signals. As shown in the block diagram of Fig. 17-22, each such detector has two input signals: the chroma which it is required to demodulate and the output of the local 3.58-MHz crystal oscillator. The second signal is used to gate the detector, producing the correct output when the chroma signal is in phase with the local oscillator. If the phase of the local oscillator corresponds to the $(B - Y)$ vector, the demodulated voltages will also be $(B - Y)$. When, as in the other color demodulator of Fig. 17-22, a 90° phase change is introduced into the 3.58-MHz oscillator signal, its phase will now correspond to that of the $(R - Y)$ vector, and $(R - Y)$ chroma voltages will be the only ones produced. In this fashion, the 90° phase difference between the two sets of voltages is used to separate them in the outputs of their respective demodulators. Circuits of synchronous detectors are shown and explained in detail on pp. 376–382 of [1] and pp. 526–531 of [3].

The *burst separator* has the function of extracting the 8 to 11 cycles of reference color burst which, it will be recalled, are transmitted on the back porch of every horizontal sync pulse. This is done by having an amplifier biased so that only signals having amplitudes corresponding to the burst level (or higher) are passed. Moreover, this amplifier is capable of amplifying only during the back porch, so that only the burst information is amplified. This is achieved by keying it with pulses derived from the horizontal output stage, as shown in the block diagram. The situation then is that the burst separator will amplify only when such a keying pulse is present, and then it will amplify only signals whose level is as high as the 67.5 percent modulation point, so that ordinary video voltages are rejected.

The output of the burst separator is applied to the 3.58-MHz phase discriminator, as is a portion of the signal from the local 3.58-MHz crystal oscillator. With the aid of the APC circuits, the phase discriminator output controls the phase and frequency of this local oscillator. This is done to provide the correct signals for the color demodulators. Note that the phase of the chroma carrier oscillator must be controlled, because the color TV system depends on absolute phase relationships to ensure that correct colors are reproduced at all times.

The final circuit that must be considered is the *color killer*. This circuit is used by the color television receiver to prevent video voltages received in a black-and-white program from entering the chroma amplifier. If they were amplified, the result would be the appearance of random color voltages, or confetti, which would clearly be unwanted.

The function of the color killer is thus to disable the chroma amplifier by cutting it off during monochrome reception. It is done by noting the presence or absence of the color burst and acting accordingly. As shown in Fig. 17-22, the color killer receives the same keying pulses from the horizontal output stage as did the burst separator. Here, however, the pulses are used as the dc supply for the transistor in the color killer stage. Accordingly, it can conduct only when these pulses are present. During color reception, color bursts are present at the same time as the gating pulses. This results in a dc output from the 3.58-MHz phase discriminator, which is used to bias off the color killer. Thus this circuit does not conduct at all during color reception. However, during monochrome reception, the color burst is absent, no dc issues forth from the phase discriminator, and the color killer is able to conduct. Its output is used to bias off the second chroma amplifier, or sometimes the color demodulators, so that no spurious signals in the chroma channel are amplified during monochrome program reception. The recommended references for further study of color TV receivers are chap. 16 of [3], pp. 240–249 of [2], and chaps. 8, 9, 10, 11, 17 and 18 of [1].

PROBLEMS

For self-testing questions on this chapter, see p. 713.

QUESTIONS

17-1. Explain how television is capable of displaying complete moving pictures, despite the fact that at any instant of time only a tiny portion of the picture tube screen is active.

17-2. Briefly describe camera and picture tubes, and explain what actually happens in them when a picture is being scanned. Why is *sync* transmitted?

17-3. Explain briefly the difference between *chrominance* and *luminance*. How is a color picture tube able to display white?

17-4. Explain (*a*) how television sound is transmitted; (*b*) what is meant by saying that color television must be *compatible*.

17-5. Why are television standards required? What are the major U.S. TV standards? What other TV systems are there in other parts of the world?

17-6. Draw the block diagram of a monochrome TV transmitter, and describe the camera tube, video amplifiers and sound circuits shown.

17-7. Fully explain what happens in horizontal scanning, giving a step-by-step account of all events from the time when the beam starts at the left-hand edge of the screen to the instant when it is ready to repeat the journey.

17-8. With appropriate sketches showing lines scanned and the vertical retrace, explain fully what happens from the beginning of the first field to the start of scanning for the second field.

17-9. Draw a waveform at the end of one of the vertical fields, showing a horizontal and a vertical blanking pulse. Indicate the durations and relative amplitudes of the two pulses, and explain their functions. Does it matter that there are no horizontal blanking pulses during vertical blanking period?

17-10. With the aid of a sketch, explain the function of the *serrations* in the vertical sync pulse.

17-11. Draw the composite video waveform at the end of either field, labeling all the pulses shown.

17-12. Draw a block diagram of the tuner arrangement in a VHF/UHF television receiver, and fully explain how the arrangement works. Indicate the various frequencies present at all points in both tuners when the receiver is tuned to (*a*) channel 3, and (*b*) channel 15.

17-13. Draw the block diagram of a monochrome television receiver, and explain the function and operation of all the blocks other than those corresponding to the tuners and the pulse circuits.

17-14. Using a circuit diagram, explain how sync pulses are obtained from the composite video waveform, and how, in turn, horizontal sync pulses are extracted.

17-15. Use waveforms in an explanation of how vertical sync pulses are obtained and then used to trigger the vertical oscillator in a TV receiver.

17-16. With the aid of a circuit diagram and the appropriate waveforms, explain how a sawtooth voltage may be obtained in a simple manner.

17-17. Sketch the circuit of a simple blocking oscillator, and explain how it may be synchronized with either sync pulses or a dc voltage.

17-18. Draw the circuit diagram of a TV receiver vertical deflection oscillator and amplifier. Use it to explain how the vertical hold, height and linearity controls operate.

17-19. Draw the circuit diagram, and explain the operation of the horizontal output stage of a television receiver.

17-20. How is the high-voltage supply for the anode of the picture tube generated in a television receiver?

17-21. Explain what is meant by the Y, I and Q signals in color TV, and why they are generated.

17-22. With the aid of the circuit diagram of a simple matrix, show how the I, Q and Y signals are generated in a color TV transmitter. Show typical values for the Y and I components on your matrix.

17-23. Draw a simplified color disk, showing only the colors around the periphery. Using the appropriate vectors, indicate on your disk the location of fully saturated magenta, 50 percent saturated cyan, 25 percent saturated orange, and pure white.

17-24. Explain why 3.58 MHz was selected as the color subcarrier frequency.

17-25. Why and how is the *color burst* transmitted? When is it *not* sent? Why not?

17-26. Draw the basic block diagram of a color television transmitter, and briefly explain the function of each block.

17-27. Sketch a color picture tube, and indicate its signal voltage inputs. Explain how the tube may be used as a matrix for the R, G and B voltages.

17-28. Explain fully what is done to ensure that the beams in a color picture tube all fall on only the correct phosphor dots or strips on the screen. Include in your explanation the function of the shadow mask. What precautions should be taken to ensure that the beams do not interfere with one another as they simultaneously scan different portions of the screen? In other words, what prevents beam criss-crossing?

17-29. Draw the block diagram of a color TV receiver, showing all the important functions from the tuners to the picture tube.

17-30. Describe the functions of the *chroma* stages in a television receiver, from the chroma detector to the picture tube inputs.

17-31. Discuss the operation of the color demodulators in a TV receiver. Explain why $(R - Y)$ and $(B - Y)$ are normally selected as their outputs. How is the $(G - Y)$ signal derived?

17-32. Describe the function and operation of the *color killer* stage in a color TV receiver. What would happen if, in the block diagram of Fig. 17-22, the output of the color killer were applied to the *first* chroma amplifier?

REFERENCES

1. Grob, Bernard, *Basic Television: Principles and Servicing,* 4th ed., McGraw-Hill Book Company, New York, 1975.

2. Mandl, Matthew, *Principles of Electronic Communications,* Prentice-Hall, Inc., Englewood Cliffs, N.J., 1973.

3. Herrick, Clyde N., *Introduction to Electronic Communication,* Charles E. Merrill Publishing Company, Columbus, Ohio, 1969.

SELF-TESTING QUESTIONS

Each of the following multiple-choice questions consists of an incomplete statement followed by four possible completions, or of a question followed by four possible answers. In each case the wanted response (be it the true or false one) must be selected; this may be done by circling the letter belonging to it, *a, b, c,* or *d,* as the case may be. *There is only one "correct" answer for each question.* The customary marking scale gives one mark for each correct answer and no marks for a question which is not answered or which has more than one answer; it subtracts one-third of a mark for each incorrect answer (this is done to discourage guessing).

CHAPTER 1

1. In a communications system, noise is most likely to affect the signal
a. at the transmitter
b. in the channel
c. in the information source
d. at the destination

2. Indicate the *false* statement. Fourier analysis shows that a sawtooth wave consists of
a. fundamental and subharmonic sine waves
b. a fundamental sine wave and an infinite number of harmonics
c. fundamental and harmonic sine waves whose amplitude decreases with the harmonic number
d. sinusoidal voltages, some of which are small enough to ignore in practice

3. Indicate the *false* statement. Modulation is used to
a. reduce the bandwidth used
b. separate differing transmissions
c. ensure that intelligence may be transmitted over long distances
d. allow the use of practicable antennas

CHAPTER 2

4. One of the following types of noise becomes of great importance at high frequencies. It is the
a. shot noise

b. random noise
c. impulse noise
d. transit-time noise

5. Indicate the *false* statement.
a. HF mixers are generally noisier than HF amplifiers.
b. Impulse noise voltage is independent of bandwidth.
c. Thermal noise is independent of the frequency at which it is measured.
d. Industrial noise is usually of the impulse type.

6. The value of a resistor creating thermal noise is doubled. The noise power generated is therefore
a. halved
b. quadrupled
c. doubled
d. unchanged

7. One of the following is *not* a useful quantity for comparing the noise performance of receivers:
a. Input noise voltage
b. Equivalent noise resistance
c. Noise temperature
d. Noise figure

8. Indicate the noise whose source is in a category different from that of the other three.
a. Solar noise
b. Cosmic noise
c. Atmospheric noise
d. Galactic noise

9. Indicate the *false* statement. The square of the thermal noise voltage generated by a resistor is proportional to
a. its resistance
b. its temperature
c. Boltzmann's constant
d. the bandwidth over which it is measured

CHAPTER 3

10. If the plate supply voltage for a plate-modulated class C amplifier is V, the maximum plate-cathode voltage could be almost as high as
a. $4V$
b. $3V$
c. $2V$
d. V

11. In a low-level AM system, amplifiers following the modulated stage must be
a. linear devices
b. harmonic devices
c. class C amplifiers
d. nonlinear devices

12. If the carrier of a 100 percent modulated AM wave is suppressed, the percentage power saving will be
a. 50
b. 150
c. 100
d. 66.66

13. Leak-type bias is used in a plate-modulated class C amplifier to
a. prevent tuned circuit damping
b. prevent excessive grid current
c. prevent overmodulation
d. increase the bandwidth

14. The output stage of a television transmitter is most likely to be a
a. plate-modulated class C amplifier
b. grid-modulated class C amplifier
c. screen-modulated class C amplifier
d. grid-modulated class A amplifier

15. The modulation index of an AM wave is changed from 0 to 1. The transmitted power is
a. unchanged
b. halved
c. doubled
d. increased by 50 percent

16. One of the advantages of base modulation over collector modulation of a transistor class C amplifier is
a. the lower modulating power required
b. higher power output per transistor
c. better efficiency
d. better linearity

17. A carrier is simultaneously modulated by two sine waves with modulation indices of 0.3 and 0.4; the total modulation index
a. is 1
b. cannot be calculated unless the phase relations are known
c. is 0.5
d. is 0.7

18. Amplitude modulation is used for broadcasting because
a. it is more noise immune than other modulation systems.

b. compared with other systems it requires less transmitting power.
c. its use avoids receiver complexity.
d. no other modulation system can provide the necessary bandwidth for high fidelity.

CHAPTER 4

19. Indicate the *false* statement regarding the advantages of SSB over double-sideband, full-carrier AM.
a. More channel space is available.
b. Transmitter circuits must be more stable, giving better reception.
c. The signal is more noise-resistant.
d. Much less power is required for the same signal strength.

20. When the modulation index of an AM wave is doubled, the antenna current is also doubled. The AM system being used is
a. Single-sideband, full carrier (H3E)
b. Vestigial sideband (C3F)
c. Single-sideband, suppressed carrier (J3E)
d. Double sideband, full carrier (A3E)

21. Indicate which one of the following advantages of the phase cancellation method of obtaining SSB over the filter method is *false:*
a. Switching from one sideband to the other is simpler.
b. It is possible to generate SSB at any frequency.
c. SSB with lower audio frequencies present can be generated.
d. There are more balanced modulators; therefore the carrier is suppressed better.

22. The most commonly used filters in SSB generation are
a. mechanical
b. *RC*
c. *LC*
d. low-pass

23. In an SSB transmitter, one is most likely to find a
a. class C audio amplifier
b. tuned modulator
c. class B RF amplifier
d. class A RF output amplifier

24. Indicate in which one of the following only one sideband is transmitted:
a. H3E
b. A3E
c. B8E
d. C3F

25. One of the following *cannot* be used to remove the unwanted sideband in SSB. This is the
a. filter system
b. phase-shift method
c. third method
d. balanced modulator

26. R3E modulation is sometimes used to
a. allow the receiver to have a frequency synthesizer
b. simplify the frequency stability problem in reception
c. reduce the power that must be transmitted
d. reduce the bandwidth required for transmission

27. To provide two or more voice circuits with the same carrier, it is necessary to use
a. ISB
b. carrier reinsertion
c. SSB with pilot carrier
d. Lincompex

28. Vestigial sideband modulation (C3F) is normally used for
a. HF point-to-point communications
b. monaural broadcasting
c. TV broadcasting
d. stereo broadcasting

CHAPTER 5

29. In the stabilized reactance modulator AFC system,
a. the discriminator must have a fast time constant to prevent demodulation
b. the higher the discriminator frequency, the better the oscillator frequency stability
c. the discriminator frequency must not be too low, or the system will fail
d. phase modulation is converted into FM by the equalizer circuit

30. In the spectrum of a frequency-modulated wave
a. the carrier frequency disappears when the modulation index is large
b. the amplitude of any sideband depends on the modulation index
c. the total number of sidebands depends on the modulation index
d. the carrier frequency cannot disappear

31. The difference between phase and frequency modulation
a. is purely theoretical because they are the same in practice
b. is too great to make the two systems compatible
c. lies in the poorer audio response of phase modulation
d. lies in the different definitions of the modulation index

32. Indicate the *false* statement regarding the Armstrong modulation system.
 a. The system is basically phase, not frequency, modulation.
 b. AFC is not needed, as a crystal oscillator is used.
 c. Frequency multiplication must be used.
 d. Equalization is unnecessary.

33. An FM signal with a modulation index m_f is passed through a frequency tripler. The wave in the output of the tripler will have a modulation index of
 a. $m_f/3$
 b. m_f
 c. $3m_f$
 d. $9m_f$

34. An FM signal with a deviation δ is passed through a mixer, and has its frequency reduced fivefold. The deviation in the output of the mixer is
 a. 5δ
 b. indeterminate
 c. $\delta/5$
 d. δ

35. A pre-emphasis circuit provides extra noise immunity by
 a. boosting the bass frequencies
 b. amplifying the higher audio frequencies
 c. preamplifying the whole audio band
 d. converting the phase modulation to FM

36. Since noise phase-modulates the FM wave, as the noise sideband frequency approaches the carrier frequency, the noise amplitude
 a. remains constant
 b. is decreased
 c. is increased
 d. is equalized

37. When the modulating frequency is doubled, the modulation index is halved, and the modulating voltage remains constant. The modulation system is
 a. amplitude modulation
 b. phase modulation
 c. frequency modulation
 d. any one of the three

38. Indicate which one of the following is *not* an advantage of FM over AM:
 a. Better noise immunity is provided.
 b. Lower bandwidth is required.
 c. The transmitted power is more useful.
 d. Less modulating power is required.

39. One of the following is an indirect way of generating FM. This is the
a. reactance FET modulator
b. varactor diode modulator
c. Armstrong modulator
d. reactance bipolar transistor modulator

40. In an FM stereo multiplex transmission, the
a. sum signal modulates the 19 kHz subcarrier
b. difference signal modulates the 19 kHz subcarrier
c. difference signal modulates the 38 kHz subcarrier
d. difference signal modulates the 67 kHz subcarrier

CHAPTER 6

41. Indicate which of the following statements about the advantages of the phase discriminator over the slope detector is *false:*
a. Much easier alignment
b. Better linearity
c. Greater limiting
d. Fewer tuned circuits

42. Show which of the following statements about the amplitude limiter is *untrue:*
a. The circuit is always biased in class C, by virtue of the leak-type bias.
b. When the input increases past the threshold of limiting, the gain decreases to keep the output constant.
c. The output must be tuned.
d. Leak-type bias must be used.

43. In a radio receiver with simple AGC
a. an increase in signal strength produces more AGC
b. the audio stage gain is normally controlled by the AGC
c. the faster the AGC time constant, the more accurate the output
d. the highest AGC voltage is produced between stations

44. In a broadcast superheterodyne receiver, the
a. local oscillator operates below the signal frequency
b. mixer input must be tuned to the signal frequency
c. local oscillator frequency is normally double the IF
d. RF amplifier normally works at 455 kHz above the carrier frequency

45. To prevent overloading of the last IF amplifier in a receiver, one should use
a. squelch
b. variable sensitivity

c. variable selectivity
d. double conversion

46. A superheterodyne receiver with an IF of 450 kHz is tuned to a signal at 1200 kHz. The image frequency is
a. 750 kHz
b. 900 kHz
c. 1650 kHz
d. 2100 kHz

47. In a ratio detector
a. the linearity is worse than in a phase discriminator
b. stabilization against signal strength variations is provided
c. the output is twice that obtainable from a similar phase discriminator
d. the circuit is the same as in a discriminator, except that the diodes are reversed

48. The typical squelch circuit
a. cuts off an audio amplifier when the carrier is absent
b. eliminates the RF interference when the signal is weak
c. cuts off an IF amplifier when the AGC is maximum
d. cuts off an IF amplifier when the AGC is minimum

49. Indicate the *false* statement in connection with communications receivers.
a. The noise limiter cuts off the receiver's output during a noise pulse.
b. A product demodulator could be used for the reception of Morse code.
c. Double conversion is used to improve image rejection.
d. Variable sensitivity is used to eliminate selective fading.

50. The controlled oscillator synthesizer is sometimes preferred to the direct one because
a. it is a simpler piece of equipment
b. its frequency stability is better
c. it does not require crystal oscillators
d. it is relatively free of spurious frequencies

51. The frequency generated by each decade in a direct frequency synthesizer is much higher than the frequency shown; this is done to
a. reduce the spurious frequency problem
b. increase the frequency stability of the synthesizer
c. reduce the number of decades
d. reduce the number of crystals required

52. Indicate which of the following circuits could *not* demodulate SSB:
a. Balanced modulator
b. Product detector

c. BFO
d. Phase discriminator

53. If an FET is used as the first AF amplifier in a transistor receiver, this will have the effect of
a. improving the effectiveness of the AGC
b. reducing the effect of negative-peak clipping
c. reducing the effect of noise at low modulation depths
d. improving the selectivity of the receiver

54. Indicate the *false* statement. The superheterodyne receiver replaced the TRF receiver because the latter suffered from
a. gain variation over the frequency coverage range
b. insufficient gain and sensitivity
c. inadequate selectivity at high frequencies
d. instability

55. The image frequency of a superheterodyne receiver
a. is created within the receiver itself
b. is due to insufficient adjacent channel rejection
c. is not rejected by the IF tuned circuits
d. is independent of the frequency to which the receiver is tuned

56. One of the main functions of the RF amplifier in a superheterodyne receiver is to
a. provide improved tracking
b. permit better adjacent-channel rejection
c. increase the tuning range of the receiver
d. improve the rejection of the image frequency

57. A receiver has poor IF selectivity. It will therefore also have poor
a. blocking
b. double-spotting
c. diversity reception
d. sensitivity

58. Three-point tracking is achieved with
a. variable selectivity
b. the padder capacitor
c. double spotting
d. double conversion

59. The local oscillator of a broadcast receiver is tuned to a frequency higher than the incoming frequency
a. to help the image frequency rejection
b. to permit easier tracking

c. because otherwise an intermediate frequency could not be produced
d. to allow adequate frequency coverage without switching

60. If the intermediate frequency is very high (indicate *false* statement)
a. image frequency rejection is very good
b. the local oscillator need not be extremely stable
c. the selectivity will be poor
d. tracking will be improved

61. A low ratio of the ac to the dc load impedance of a diode detector results in
a. diagonal clipping
b. poor AGC operation
c. negative-peak clipping
d. poor AF response

62. One of the following *cannot* be used to demodulate SSB:
a. Product detector
b. Diode balanced modulator
c. Bipolar transistor balanced modulator
d. Complete phase-shift generator

63. Indicate the *false* statement. Noting that no carrier is transmitted with J3E, we see that
a. the receiver cannot use a phase comparator for AFC
b. adjacent-channel rejection is more difficult
c. production of AGC is a rather complicated process
d. the transmission is not compatible with A3E

64. When a receiver has a good blocking performance, this means that
a. it does not suffer from double-spotting
b. its image frequency rejection is poor
c. it is unaffected by AGC derived from nearby transmissions
d. its detector suffers from burnout

65. An AM receiver uses a diode detector for demodulation. This enables it satisfactorily to receive
a. single-sideband, suppressed-carrier
b. single-sideband, reduced carrier
c. ISB
d. single-sideband, full carrier

CHAPTER 7

66. Indicate the *false* statement. The SWR on a transmission line is infinity; the line is terminated in
a. a short circuit

b. a complex impedance
c. an open circuit
d. a pure reactance

67. A (75 − *j*50)-Ω load is connected to a coaxial transmission line of $Z_0 = 75\ \Omega$, at 10 GHz. The *best* method of matching consists in connecting
a. a short-circuited stub at the load
b. an inductance at the load
c. a capacitance at some specific distance from the load
d. a short-circuited stub at some specific distance from the load

68. The velocity factor of a transmission line
a. depends on the dielectric constant of the material used
b. increases the velocity along the transmission line
c. is governed by the skin effect
d. is higher for a solid dielectric than for air

69. Impedance inversion may be obtained with
a. a short-circuited stub
b. an open-circuited stub
c. a quarter-wave line
d. a half-wave line

70. Short-circuited stubs are preferred to open-circuited stubs because the latter are
a. more difficult to make and connect
b. made of a transmission line with a different characteristic impedance
c. liable to radiate
d. incapable of giving a full range of reactances

71. For transmission-line load matching over a range of frequencies, it is best to use a
a. balun
b. broadband directional coupler
c. double stub
d. single stub of adjustable position

72. The main disadvantage of the two-hole directional coupler is
a. low directional coupling
b. poor directivity
c. high SWR
d. narrow bandwidth

73. To couple a coaxial line to a parallel-wire line, it is best to use a
a. slotted line
b. balun
c. directional coupler
d. quarter-wave transformer

CHAPTER 8

74. Indicate which one of the following terms applies to troposcatter propagation:
a. SIDs
b. Fading
c. Atmospheric storms
d. Faraday rotation

75. VLF waves are used for some types of services because
a. of the low powers required
b. the transmitting antennas are of convenient size
c. they are very reliable
d. they penetrate the ionosphere easily

76. Indicate which of the following frequencies *cannot* be used for reliable beyond-the-horizon terrestrial communications without repeaters:
a. 20 kHz
b. 15 MHz
c. 900 MHz
d. 12 GHz

77. High-frequency waves are
a. absorbed by the F_2 layer
b. reflected by the D layer
c. capable of use for long-distance communications on the moon
d. affected by the solar cycle

78. Distances near the skip distance should be used for sky-wave propagation
a. to avoid tilting
b. to prevent sky-wave and upper ray interference
c. to avoid the Faraday effect
d. so as not to exceed the critical frequency

79. A ship-to-ship communications system is plagued by fading. The best solution seems to be the use of
a. a more directional antenna
b. a broadband antenna
c. frequency diversity
d. space diversity

80. A range of microwave frequencies more easily passed by the atmosphere than are the others is called a
a. window
b. critical frequency
c. gyro frequency range
d. resonance in the atmosphere

81. Frequencies in the UHF range normally propagate by means of
a. ground waves
b. sky waves
c. surface waves
d. space waves

82. Tropospheric scatter is used with frequencies in the following range:
a. HF
b. VHF
c. UHF
d. VLF

83. The ground wave eventually disappears, as one moves away from the transmitter, because of
a. interference from the sky wave
b. loss of line-of-sight conditions
c. maximum single-hop distance limitation
d. tilting

84. In electromagnetic waves, polarization
a. is caused by reflection
b. is due to the tranverse nature of the waves
c. results from the longitudinal nature of the waves
d. is always vertical in an isotropic medium

85. As electromagnetic waves travel in free space, only one of the following can happen to them:
a. absorption
b. attenuation
c. refraction
d. reflection

86. The absorption of radio waves by the atmosphere depends on
a. their frequency
b. their distance from the transmitter
c. the polarization of the waves
d. the polarization of the atmosphere

87. Electromagnetic waves are refracted when they
a. pass into a medium of different dielectric constant
b. are polarized at right angles to the direction of propagation
c. encounter a perfectly conducting surface
d. pass through a small slot in a conducting plane

88. Diffraction of electromagnetic waves
a. is caused by reflections from the ground

b. arises only with spherical wavefronts
c. will occur when the waves pass through a large slot
d. may occur around the edge of a sharp obstacle

89. When microwave signals follow the curvature of the earth, this is known as
a. the Faraday effect
b. ducting
c. tropospheric scatter
d. ionospheric reflection

90. Helical antennas are often used for satellite tracking at VHF because of
a. troposcatter
b. superrefraction
c. ionospheric refraction
d. the Faraday effect

CHAPTER 9

91. An ungrounded antenna near the ground
a. acts as a single antenna of twice the height
b. is unlikely to need an earth mat
c. acts as an antenna array
d. must be horizontally polarized

92. One of the following consists of nonresonant antennas:
a. The rhombic antenna
b. The folded dipole
c. The end-fire array
d. The broadside array

93. One of the following is very useful as a multiband HF receiving antenna. This is the:
a. conical horn
b. folded dipole
c. log-periodic
d. square loop

94. Which of the following antennas is best excited from a waveguide?
a. Biconical
b. Horn
c. Helical
d. Discone

95. Indicate which of the following reasons for using a counterpoise with antennas is *false*:
a. Impossibility of a good ground connection

b. Protection of personnel working underneath
c. Provision of an earth for the antenna
d. Rockiness of the ground itself

96. One of the following is *not* a reason for the use of an antenna coupler:
a. To make the antenna look resistive
b. To provide the output amplifier with the correct load impedance
c. To discriminate against harmonics
d. To prevent reradiation of the local oscillator

97. Indicate the antenna that is *not* wideband:
a. Discone
b. Folded dipole
c. Helical
d. Marconi

98. Indicate which one of the following reasons for the use of an earth mat with antennas is *false*:
a. Impossibility of a good ground connection
b. Provision of an earth for the antenna
c. Protection of personnel working underneath
d. Improvement of the radiation pattern of the antenna

99. Which one of the following terms does *not* apply to the Yagi-Uda array?
a. Good bandwidth
b. Parasitic elements
c. Folded dipole
d. High gain

100. An antenna that is circularly polarized is the
a. helical
b. small circular loop
c. parabolic reflector
d. Yagi-Uda

101. The standard reference antenna for the directive gain is the
a. infinitesimal dipole
b. isotropic antenna
c. elementary doublet
d. half-wave dipole

102. Top loading is sometimes used with an antenna in order to increase its
a. effective height
b. bandwidth
c. beamwidth
d. input capacitance

103. Cassegrain feed is used with a parabolic reflector to
a. increase the gain of the system
b. increase the beamwidth of the system
c. reduce the size of the main reflector
d. allow the feed to be placed at a convenient point

104. Zoning is used with a dielectric antenna in order to
a. reduce the bulk of the lens
b. increase the bandwidth of the lens
c. permit pin-point focusing
d. correct the curvature of the wavefront from a horn that is too short

105. A helical antenna is used for satellite tracking because of its
a. circular polarization
b. maneuverability
c. broad bandwidth
d. good front-to-back ratio

106. The discone antenna is
a. a useful direction-finding antenna
b. used as a radar receiving antenna
c. circularly polarized like other circular antennas
d. useful as a UHF receiving antenna

107. One of the following is *not* an omnidirectional antenna:
a. Half-wave dipole
b. Log-periodic
c. Discone
d. Marconi

CHAPTER 10

108. When electromagnetic waves are propagated in a waveguide
a. they travel along the broader walls of the guide
b. they are reflected from the walls but do not travel along them
c. they travel through the dielectric without touching the walls
d. they travel along all four walls of the waveguide

109. Waveguides are used mainly for microwave signals because
a. they depend on straight-line propagation which applies to microwaves only
b. losses would be too heavy at lower frequencies
c. there are no generators powerful enough to excite them at lower frequencies
d. they would be too bulky at lower frequencies

110. The wavelength of a wave in a waveguide
a. is greater than in free space

b. depends only on the waveguide dimensions and the free-space wavelength
c. is inversely proportional to the phase velocity
d. is directly proportional to the group velocity

111. The main difference between the operation of transmission lines and waveguides is that
a. the latter are not distributed, like transmission lines
b. the former can use stubs and quarter-wave transformers, unlike the latter
c. transmission lines use the principal mode of propagation, and therefore do not suffer from low-frequency cutoff
d. terms such as *impedance matching* and *standing-wave ratio* cannot be applied to waveguides

112. Compared with equivalent transmission lines, 3-GHz waveguides (indicate *false* statement)
a. are less lossy
b. can carry higher powers
c. are less bulky
d. have lower attenuation

113. When a particular mode is excited in a waveguide, there appears an extra electric component, in the direction of propagation. The resulting mode is
a. transverse-electric
b. transverse-magnetic
c. longitudinal
d. transverse-electromagnetic

114. When electromagnetic waves are reflected at an angle from a wall, their wavelength along the wall is
a. the same as in free space
b. the same as the wavelength perpendicular to the wall
c. shortened because of the Doppler effect
d. greater than in the actual direction of propagation

115. As a result of reflections from a plane conducting wall, electromagnetic waves acquire an apparent velocity greater than the velocity of light in space. This is called the
a. velocity of propagation
b. normal velocity
c. group velocity
d. phase velocity

116. Indicate the *false* statement. When the free-space wavelength of a signal equals the cutoff wavelength of the guide
a. the group velocity of the signal becomes zero
b. the phase velocity of the signal becomes infinite

c. the characteristic impedance of the guide becomes infinite
d. the wavelength within the waveguide becomes infinite

117. A signal propagated in a waveguide has a full wave of electric intensity change between the two further walls, and no component of the electric field in the direction of propagation. The mode is
a. $TE_{1,1}$
b. $TE_{1,0}$
c. $TM_{2,2}$
d. $TE_{2,0}$

118. The dominant mode of propagation is preferred with rectangular waveguides because (indicate *false* statement)
a. it leads to the smallest waveguide dimensions
b. the resulting impedance can be matched directly to coaxial lines
c. it is easier to excite than the other modes
d. propagation of it without any spurious generation can be ensured

119. A choke flange may be used to couple two waveguides
a. to help in the alignment of the waveguides
b. because it is simpler than any other join
c. to compensate for discontinuities at the join
d. to increase the bandwidth of the system

120. In order to couple two generators to a waveguide system without coupling them to each other, one could *not* use a
a. rat-race
b. *E*-plane T
c. hybrid ring
d. magic T

121. Which one of the following waveguide tuning components is *not* easily adjustable?
a. Screw
b. Stub
c. Iris
d. Plunger

122. A piston attenuator is a
a. vane attenuator
b. waveguide below cutoff
c. mode filter
d. flap attenuator

123. Cylindrical cavity resonators are not used with klystrons because they have
a. a *Q* that is too low
b. a shape whose resonant frequency is too difficult to calculate

c. harmonically related resonant frequencies
d. too heavy losses

124. A directional coupler with three or more holes is sometimes used in preference to the two-hole coupler
a. because it is more efficient
b. to increase coupling of the signal
c. to reduce spurious mode generation
d. to increase the bandwidth of the system

125. A ferrite is
a. a nonconductor with magnetic properties
b. an intermetallic compound with particularly good conductivity
c. an insulator which heavily attenuates magnetic fields
d. a microwave semiconductor invented by Faraday

126. Manganese ferrite may be used as a (indicate *false* answer)
a. circulator
b. isolator
c. garnet
d. phase shifter

127. The maximum power that may be handled by a ferrite component is limited by the
a. Curie temperature
b. saturation magnetization
c. line width
d. gyromagnetic resonance

128. A PIN diode is
a. a metal semiconductor point-contact diode
b. a microwave mixer diode
c. often used as a microwave detector
d. suitable for use as a microwave switch

129. A duplexer is used
a. to couple two different antennas to a transmitter without mutual interference
b. to allow the one antenna to be used for reception or transmission without mutual interference
c. to prevent interference between two antennas when they are connected to a receiver
d. to increase the speed of the pulses in pulsed radar

130. For some applications, circular waveguides may be preferred to rectangular ones because of
a. the smaller cross section needed at any frequency

b. lower attenuation
c. freedom from spurious modes
d. rotation of polarization

131. Indicate which of the following cannot be followed by the word ''waveguide''
a. Elliptical
b. Flexible
c. Coaxial
d. Ridged

132. In order to reduce cross-sectional dimensions, the waveguide to use is
a. circular
b. ridged
c rectangular
d. flexible

133. For low attenuation, the best transmission medium is
a. flexible waveguide
b. ridged waveguide
c. rectangular waveguide
d. coaxial line

CHAPTER 11

134. A microwave tube amplifier uses an axial magnetic field and a radial electric field. This is the
a. reflex klystron
b. coaxial magnetron
c. traveling-wave magnetron
d. CFA

135. One of the following is unlikely to be used as a pulsed device. It is the
a. multicavity klystron
b. BWO
c. CFA
d. TWT

136. One of the reasons why vacuum tubes eventually fail at microwave frequencies is that their
a. noise figure increases
b. transit time becomes too short
c. shunt capacitive reactances become too large
d. series inductive reactances become too small

137. Indicate the *false* statement. Transit time in microwave tubes will be reduced if
a. the electrodes are brought closer together
b. a higher anode current is used

c. multiple or coaxial leads are used
d. the anode voltage is made larger

138. The multicavity klystron
a. is not a good low-level amplifier because of noise
b. has a high repeller voltage to ensure a rapid transit time
c. is not suitable for pulsed operation
d. needs a long transit time through the buncher cavity to ensure current modulation

139. Indicate the *false* statement. Klystron amplifiers may use intermediate cavities to
a. prevent the oscillations that occur in two-cavity klystrons
b. increase the bandwidth of the device
c. improve the power gain
d. increase the efficiency of the klystron

140. The TWT is sometimes preferred to the multicavity klystron amplifier, because the former
a. is more efficient
b. has a greater bandwidth
c. has a higher number of modes
d. produces a higher output power

141. The transit time in the repeller space of a reflex klystron must be $n + 3/4$ cycles to ensure that
a. electrons are accelerated by the gap voltage on their return
b. returning electrons give energy to the gap oscillations
c. it is equal to the period of the cavity oscillations
d. the repeller is not damaged by striking electrons

142. The cavity magnetron uses strapping to
a. prevent mode jumping
b. prevent cathode back-heating
c. ensure bunching
d. improve the phase-focusing effect

143. A magnetic field is used in the cavity magnetron to
a. prevent anode current in the absence of oscillations
b. ensure that the oscillations are pulsed
c. help in focusing the electron beam, thus preventing spreading
d. ensure that the electrons will orbit around the cathode

144. To avoid difficulties with strapping at high frequencies, the type of cavity structure used in the magnetron is the
a. hole-and-slot
b. slot

c. vane
d. rising-sun

145. The primary purpose of the helix in a traveling-wave tube is to
a. prevent the electron beam from spreading in the long tube
b. reduce the axial velocity of the RF field
c. ensure broadband operation
d. reduce the noise figure

146. The attenuator is used in the traveling-wave tube to
a. help bunching
b. prevent oscillations
c. prevent saturation
d. increase gain

147. Periodic permanent-magnet focusing is used with TWTs to
a. allow pulsed operation
b. improve electron bunching
c. avoid the bulk of an electromagnet
d. allow coupled-cavity operation at the highest frequencies

148. The TWT is sometimes preferred to the magnetron as a radar transmitter output tube because it is
a. capable of a longer duty cycle
b. a more efficient amplifier
c. more broadband
d. less noisy

149. A magnetron whose oscillating frequency is electronically adjustable over a wide range is called a
a. coaxial magnetron
b. dither-tuned magnetron
c. frequency-agile magnetron
d. VTM

150. Indicate which of the following is *not* a TWT slow-wave structure:
a. Periodic-permanent magnet
b. Coupled cavity
c. Helix
d. Ring-bar

151. The glass tube of a TWT may be coated with aquadag to
a. help focusing
b. provide attenuation
c. improve bunching
d. increase gain

152. A backward-wave oscillator is based on the
a. rising-sun magnetron
b. crossed-field amplifier
c. coaxial magnetron
d. traveling-wave tube

CHAPTER 12

153. A parametric amplifier must be cooled
a. because parametric amplification generates a lot of heat
b. to increase bandwidth
c. because it cannot operate at room temperature
d. to improve the noise performance

154. A ruby maser amplifier must be cooled
a. because maser amplification generates a lot of heat
b. to increase bandwidth
c. because it cannot operate at room temperature
d. to improve the noise performance

155. A disadvantage of microstrip compared with stripline is that microstrip
a. does not readily lend itself to printed circuit techniques
b. is more likely to radiate
c. is bulkier
d. is more expensive and complex to manufacture

156. The transmission system using two ground planes is
a. microstrip
b. elliptical waveguide
c. parallel-wire line
d. stripline

157. Indicate the *false* statement. An advantage of stripline over waveguides is its
a. smaller bulk
b. greater bandwidth
c. higher power-handling capability
d. greater compatibility with solid-state devices

158. Indicate the *false* statement. An advantage of stripline over microstrip is its
a. easier integration with semiconductor devices
b. lower tendency to radiate
c. higher isolation between adjacent circuits
d. higher Q

159. Surface acoustic waves propagate in
a. gallium arsenide

b. indium phosphide
c. stripline
d. quartz crystal

160. SAW devices may be used as
a. transmission media like stripline
b. filters
c. UHF amplifiers
d. oscillators at millimeter frequencies

161. Indicate the *false* statement. FETs are preferred to bipolar transistors at the highest frequencies because they
a. are less noisy
b. lend themselves more readily to integration
c. are capable of higher efficiencies
d. can provide higher gains

162. For best low-level noise performance in the X-band, an amplifier should use
a. a bipolar transistor
b. a Gunn diode
c. a step-recovery diode
d. an IMPATT diode

163. The biggest advantage of the TRAPATT diode over the IMPATT diode is its
a. lower noise
b. higher efficiency
c. ability to operate at higher frequencies
d. lesser sensitivity to harmonics

164. Indicate which of the following diodes will produce the highest pulsed power output:
a. Varactor
b. Gunn
c. Schottky barrier
d. RIMPATT

165. Indicate which of the following diodes does not use negative resistance in its operation:
a. Backward
b. Gunn
c. IMPATT
d. Tunnel

166. One of the following is *not* used as a microwave mixer or detector:
a. Crystal diode
b. Schottky-barrier diode

c. Backward diode
d. PIN diode

167. One of the following microwave diodes is suitable for very low-power oscillators only:
a. Tunnel
b. avalanche
c. Gunn
d. IMPATT

168. The transferred-electron bulk effect occurs in
a. germanium
b. gallium arsenide
c. silicon
d. metal semiconductor junctions

169. The gain-bandwidth frequency of a microwave transistor, f_T, is the frequency at which the
a. alpha of the transistor falls by 3 dB
b. beta of the transistor falls by 3 dB
c. power gain of the transistor falls to unity
d. beta of the transistor falls to unity

170. For a microwave transistor to operate at the highest frequencies, the (indicate the *false* answer)
a. collector voltage must be large
b. collector current must be high
c. base should be thin
d. emitter area must be large

171. A varactor diode may be useful at microwave frequencies (indicate the *false* answer)
a. for electronic tuning
b. for frequency multiplication
c. as an oscillator
d. as a parametric amplifier

172. If high-order frequency multiplication is required from a diode multiplier,
a. the resistive cutoff frequency must be high
b. a small value of base resistance is required
c. a step-recovery diode must be used
d. a large range of capacitance variation is needed

173. A parametric amplifier has an input and output frequency of 2.25 GHz, and is pumped at 4.5 GHz. It is a
a. traveling-wave amplifier
b. degenerate amplifier

c. lower-sideband up-converter
d. upper-sideband up-converter

174. A nondegenerate parametric amplifier has an input frequency f_i and a pump frequency f_p. The idler frequency is
a. f_i
b. $2f_i$
c. $f_i - f_p$
d. $f_p - f_i$

175. Traveling-wave parametric amplifiers are used to
a. provide a greater gain
b. reduce the number of varactor diodes required
c. avoid the need for cooling
d. provide a greater bandwidth

176. A parametric amplifier sometimes uses a circulator to
a. prevent noise feedback
b. allow the antenna to be used simultaneously for transmission and reception
c. separate the signal and idler frequencies
d. permit more efficient pumping

177. The nondegenerate one-port parametric amplifier should have a high ratio of pump to signal frequency because this
a. permits satisfactory high-frequency operation
b. yields a low noise figure
c. reduces the pump power required
d. permits satisfactory low-frequency operation

178. The tunnel diode
a. has a tiny hole through its center to facilitate tunneling
b. is a point-contact diode with a very high reverse resistance
c. uses a high doping level to provide a narrow junction
d. works by quantum tunneling exhibited by gallium arsenide only

179. A tunnel diode is loosely coupled to its cavity in order to
a. increase the frequency stability
b. increase the available negative resistance
c. facilitate tuning
d. allow operation at the highest frequencies

180. The negative resistance in a tunnel diode
a. is maximum at the peak point of the characteristic
b. is available between the peak and valley points
c. is maximum at the valley point
d. may be improved by the use of reverse bias

181. The biggest advantage of gallium antimonide over germanium for tunnel-diode use is that the former has a
a. lower noise
b. higher ion mobility
c. larger voltage swing
d. simpler fabrication process

182. Negative resistance is obtained with a Gunn diode because of
a. electron transfer to a less mobile energy level
b. avalanche breakdown with the high-voltage gradient
c. tunneling across the junction
d. electron domains forming at the junction

183. For Gunn diodes, gallium arsenide is preferred to silicon because the former
a. has a suitable empty energy band, which silicon does not have
b. has a higher ion mobility
c. has a lower noise at the highest frequencies
d. is capable of handling higher power densities

184. The biggest disadvantage of the IMPATT diode is its
a. lower efficiency than that of the other microwave diodes
b. high noise
c. inability to provide pulsed operation
d. low power-handling ability

185. The magnetic field is used with a ruby maser to
a. provide sharp focusing for the electron beam
b. increase the population inversion
c. allow room-temperature operation
d. provide frequency adjustment

186. The ruby maser has been preferred to the ammonia maser for microwave amplification, because the former has
a. a much greater bandwidth
b. a better frequency stability
c. a lower noise figure
d. no need for a circulator

187. Parametric amplifiers and masers are similar to each other in that both (indicate *false* statement)
a. must have pumping
b. are extremely low-noise amplifiers
c. must be cooled down to a few kelvins
d. generally require circulators, since they are one-port devices

188. A maser RF amplifier is not really suitable for
a. radioastronomy
b. satellite communications
c. radar
d. troposcatter receivers

189. The ruby laser differs from the ruby maser in that the former
a. does not require pumping
b. needs no resonator
c. is an oscillator
d. produces much lower powers

190. The output from a laser is monochromatic; this means that it is
a. infrared
b. polarized
c. narrow-beam
d. single-frequency

191. For a given average power, the *peak* output power of a ruby laser may be increased by
a. using cooling
b. using Q spoiling
c. increasing the magnetic field
d. dispensing with the Fabry-Perot resonator

192. Communications lasers are used with optical fibers, rather that in open links, to
a. ensure that the beam does not spread
b. prevent atmospheric interference
c. prevent interference by other lasers
d. ensure that people are not blinded by them

193. Indicate the *false* statement. The advantages of semiconductor lasers over LEDs include
a. monochromatic output
b. higher power output
c. lower cost
d. ability to be pulsed at higher rates

CHAPTER 13

194. Indicate which of the following is *not* a binary code.
a. Morse
b. Baudot
c. CCITT-2
d. ARQ

195. To permit the selection of 1 out of 16 equiprobable events, the number of bits required is
a. 2
b. $\log_{10} 16$
c. 8
d. 4

196. A signaling system in which each letter of the alphabet is represented by a different symbol is not used because
a. it would be too difficult for an operator to memorize
b. it is redundant
c. noise would introduce too many errors
d. too many pulses per letter are required

197. The Hartley law states that
a. the maximum rate of information transmission depends on the channel bandwidth
b. the maximum rate of information transmission depends on the depth of modulation
c. redundancy is essential
d. only binary codes may be used

198. Indicate the *false* statement. In order to combat noise,
a. the channel bandwidth may be increased
b. redundancy may be used
c. the transmitted power may be increased
d. the signaling rate may be reduced

199. The most common modulation system used for telegraphy is
a. frequency-shift keying
b. two-tone modulation
c. pulse-code modulation
d. single-tone modulation

200. Pulse-width modulation may be generated
a. by differentiating pulse-position modulation
b. with a monostable multivibrator
c. by integrating the signal
d. with a free-running multivibrator

201. Indicate which of the following systems is digital.
a. Pulse-position modulation
b. Pulse-code modulation
c. Pulse-width modulation
d. Pulse-frequency modulation

202. Quantizing noise occurs in
a. time-division multiplex
b. frequency-division multiplex
c. pulse-code modulation
d. pulse-width modulation

203. The modulation system inherently most noise-resistant is
a. SSB, suppressed-carrier
b. frequency modulation
c. pulse-position modulation
d. pulse-code modulation

204. In order to reduce quantizing noise, one must
a. increase the number of standard amplitudes
b. send pulses whose sides are more nearly vertical
c. use an RF amplifier in the receiver
d. increase the number of samples per second

205. The Hartley-Shannon theorem sets a limit on the
a. highest frequency that may be sent over a given channel
b. maximum capacity of a channel with a given noise level
c. maximum number of coding levels in a channel with a given noise level
d. maximum number of quantizing levels in a channel of a given bandwidth

206. Indicate which of the following pulse modulation systems is analog.
a. PCM
b. Differential PCM
c. PWM
d. Delta

207. Companding is used
a. to overcome quantizing noise in PCM
b. in PCM transmitters, to allow amplitude limiting in the receivers
c. to protect small signals in PCM from quantizing distortion
d. in PCM receivers, to overcome impulse noise

208. The biggest disadvantage of PCM is
a. its inability to handle analog signals
b. the high error rate which its quantizing noise introduces
c. its incompatibility with TDM
d. the large bandwidths that are required for it

CHAPTER 14

209. Digital signals
a. do not provide a continuous set of values

b. represent values as discrete steps
c. can utilize decimal or binary systems
d. all of the above

210. The event which marked the start of the modern computer age was
a. design of the ENIAC computer
b. development of the Hollerith code
c. development of the transistor
d. development of disk drives for data storage

211. The baud rate
a. is always equal to the bit transfer rate
b. is equal to twice the bandwidth of an ideal channel
c. is not equal to the signaling rate
d. is equal to one-half the bandwidth of an ideal channel

212. The Shannon-Hartley law
a. refers to distortion
b. defines bandwidth
c. describes signaling rates
d. refers to noise

213. The code which provides for parity checks is
a. Baudot
b. ASCII
c. EBCDIC
d. CCITT-2

214. A forward error-correcting code corrects errors by
a. requiring partial retransmission of the signal
b. requiring retransmission of the entire signal
c. requiring no part of the signal to be retransmitted
d. using parity to correct the errors in all cases

215. Full duplex operation
a. requires two pairs of cables
b. can transfer data in both directions at once
c. requires modems at both ends of the circuit
d. all of the above

216. The RS-232 interface
a. interconnects data sets and transmission circuits
b. uses several different connectors
c. permits custom wiring of signal lines to the connector pins as desired
d. all of the above

217. Switching systems
a. improve the efficiency of data transfer
b. are not used in data systems
c. require additional lines
d. are limited to small data networks.

CHAPTER 15

218. Broadband long-distance communications were originally made possible by the advent of
a. telegraph cables
b. repeater amplifiers
c. HF radio
d. geostationary satellites

219. A scheme in which several channels are interleaved and then transmitted together is known as
a. frequency-division multiplex
b. time-division multiplex
c. a group
d. a supergroup

220. A basic group *B*
a. occupies the frequency range from 60 to 108 kHz
b. consists of erect channels only
c. is formed at the group translating equipment
d. consists of five supergroups

221. Time-division multiplex
a. can be used with PCM only
b. combines five groups into a supergroup
c. stacks 24 channels in adjacent frequency slots
d. interleaves pulses belonging to different transmissions

222. The number of repeaters along a coaxial cable link depends on
a. whether separate tubes are used for the two directions of transmission
b. the bandwidth of the system
c. the number of coaxial cables in the tube
d. the separation of the equalizers

223. A supergroup pilot is
a. applied at each multiplexing bay
b. used to regulate the gain of individual repeaters
c. applied at each adjustable equalizer
d. fed in at a GTE

224. Microwave link repeaters are typically 50 km apart
a. because of atmospheric attenuation
b. because of output tube power limitations
c. because of the Earth's curvature
d. to ensure that the applied dc voltage is not excessive

225. Microwave links are generally preferred to coaxial cable for television transmission because
a. they have less overall phase distortion
b. they are cheaper
c. of their greater bandwidths
d. of their relative immunity to impulse noise

226. Armored submarine cable is used
a. to protect the cable at great depths
b. to prevent inadvertent ploughing-in of the cable
c. for the shallow shore ends of the cable
d. to prevent insulation breakdown from the high feed voltages

227. A submarine cable repeater contains, among other equipment,
a. a dc power supply and regulator
b. filters for the two directions of transmission
c. multiplexing and demultiplexing equipment
d. pilot inject and pilot extract equipment

228. A geostationary satellite
a. is motionless in space (except for its spin)
b. is not really stationary at all, but orbits the Earth within a 24-hr period
c. appears stationary over the Earth's magnetic pole
d. is located at a height of 35,800 km to ensure global coverage

229. Indicate the correct statement regarding satellite communications.
a. If two earth stations do not face a common satellite, they should communicate via a double-satellite hop.
b. Satellites are allocated so that it is impossible for two earth stations not to face the same satellite.
c. Colocated earth stations are used for frequency diversity.
d. A satellite earth station must have as many receive chains as there are carriers transmitted to it.

230. Satellites used for intercontinental communications are known as
a. Comsat
b. Domsat
c. Marisat
d. Intelsat

231. Identical telephone numbers in different parts of a country are distinguished by their
a. language digits
b. access digits
c. area codes
d. central office codes

232. Telephone traffic is measured
a. with echo cancelers
b. by the relative congestion
c. in terms of the grade of service
d. in erlangs

233. In order to separate channels in a *TDM* receiver, it is necessary to use
a. AND gates
b. bandpass filters
c. differentiation
d. integration

234. To separate channels in an FDM receiver, it is necessary to use
a. AND gates
b. bandpass filters
c. differentiation
d. integration

235. Higher order TDM levels are obtained by
a. dividing pulse widths
b. using the a-law
c. using the μ-law
d. forming supermastergroups

236. Losses in optical fibers can be caused by (indicate the *false* statement)
a. impurities
b. microbending
c. attenuation in the glass
d. stepped index operation

237. The 1.55 μm ''window'' is not yet in use with fibre optic systems because
a. the attenuation is higher than at 0.85 μm
b. the attenuation is higher than at 1.3 μm
c. suitable laser devices have not yet been developed
d. it does not lend itself to wavelength multiplexing

238. Indicate which of the following is *not* a submarine cable
a. TAT-7
b. INTELSAT V

c. ATLANTIS
d. CANTAT 2

239. Indicate which of the following is an American domsat system
a. INTELSAT
b. COMSAT
c. TELSTAR
d. INMARSAT

CHAPTER 16

240. If the peak transmitted power in a radar system is increased by a factor of 16, the maximum range will be increased by a factor of
a. 2
b. 4
c. 8
d. 16

241. If the antenna diameter in a radar system is increased by a factor of 4, the maximum range will be increased by a factor of
a. $\sqrt{2}$
b. 2
c. 4
d. 8

242. If the ratio of the antenna diameter to the wavelength in a radar system is high, this will result in (indicate the *false* statement)
a. large maximum range
b. good target discrimination
c. difficult target acquisition
d. increased capture area

243. The radar cross section of a target (indicate the *false* statement)
a. depends on the frequency used
b. may be reduced by special coating of the target
c. depends on the aspect of a target, if this is nonspherical
d. is equal to the actual cross-sectional area for small targets

244. Flat-topped rectangular pulses must be transmitted in radar to (indicate the *false* statement)
a. allow a good minimum range
b. make the returned echoes easier to distinguish from noise
c. prevent frequency changes in the magnetron
d. allow accurate range measurements

245. A high PRF will (indicate the *false* statement)
a. make the returned echoes easier to distinguish from noise
b. make target tracking easier with conical scanning
c. increase the maximum range
d. have no effect on the range resolution

246. The IF bandwidth of a radar receiver is inversely proportional to the
a. pulse width
b. pulse repetition frequency
c. pulse interval
d. square root of the peak transmitted power

247. If a return echo arrives after the allocated pulse interval,
a. it will interfere with the operation of the transmitter
b. the receiver might be overloaded
c. it will not be received
d. the target will appear closer than it really is

248. After a target has been acquired, the best scanning system for tracking is
a. nodding
b. spiral
c. conical
d. helical

249. If the target cross section is changing, the best system for accurate tracking is
a. lobe switching
b. sequential lobing
c. conical scanning
d. monopulse

250. The biggest disadvantage of CW Doppler radar is that
a. it does not give the target velocity
b. it does not give the target range
c. a transponder is required at the target
d. it does not give the target position

251. The A scope displays
a. the target position and range
b. the target range, but not position
c. the target position, but not range
d. neither range nor position, but only velocity

252. The Doppler effect is used in (indicate the *false* statement)
a. moving-target plotting on the PPI
b. the MTI system
c. FM radar
d. CW radar

253. The *coho* in MTI radar operates at the
a. intermediate frequency
b. transmitted frequency
c. received frequency
d. pulse repetition frequency

254. The function of the quartz delay line in an MTI radar is to
a. help in subtracting a complete scan from the previous scan
b. match the phase of the coho and the stalo
c. match the phase of the coho and the output oscillator
d. delay a sweep so that the next sweep can be subtracted from it

255. A solution to the ''blind speed'' problem is to
a. change the Doppler frequency
b. vary the PRF
c. use monopulse
d. use MTI

256. Indicate which one of the following applications or advantages of radar beacons is *false*:
a. Target identification
b. Navigation
c. Very significant extension of the maximum range
d. More accurate tracking of enemy targets

257. Compared with other types of radar, phased array radar has the following advantages (indicate the *false* statement)
a. very fast scanning
b. ability to track and scan simultaneously
c. circuit simplicity
d. ability to track many targets simultaneously

CHAPTER 17

258. The number of lines per field in the United States TV system is
a. 262½
b. 525
c. 30
d. 60

259. The number of frames per second in the United States TV system is
a. 60
b. 262½
c. 4.5
d. 30

260. The number of lines per second in the United States TV system is
a. 31,500
b. 15,750
c. 262½
d. 525

261. The channel width in the United States TV systém, in MHz, is
a. 41.25
b. 6
c. 4.5
d. 3.58

262. Interlacing is used in television to
a. produce the illusion of motion
b. ensure that all the lines on the screen are scanned, not merely the alternate ones
c. simplify the vertical sync pulse train
d. avoid flicker

263. The signals sent by the TV transmitter to ensure correct scanning in the receiver are called
a. sync
b. chroma
c. luminance
d. video

264. In the United States color television system, the *intercarrier* frequency, in MHz, is
a. 3.58
b. 3.579545
c. 4.5
d. 45.75

265. Indicate which voltages are *not* found in the output of a normal monochrome receiver video detector.
a. Sync
b. Video
c. Sweep
d. Sound

266. The carrier transmitted 1.25 MHz above the bottom frequency in a United States TV channel is the

a. sound carrier
b. chroma carrier
c. intercarrier
d. picture carrier

267. In television, 4:3 represents the
a. interlace ratio
b. maximum horizontal deflection
c. aspect ratio
d. ratio of the two diagonals

268. Equalizing pulses in TV are sent during
a. horizontal blanking
b. vertical blanking
c. the serrations
d. the horizontal retrace

269. An odd number of lines per frame forms part of every one of the world's TV systems. This is
a. done to assist interlace
b. purely an accident
c. to ensure that line and frame frequencies can be obtained from the same original source
d. done to minimize interference with the chroma subcarrier

270. The function of the *serrations* in the composite video waveform is to
a. equalize the charge in the integrator before the start of vertical retrace
b. help vertical synchronization
c. help horizontal synchronization
d. simplify the generation of the vertical sync pulse

271. The width of the vertical sync pulse in the United States TV system is
a. $21H$
b. $3H$
c. H
d. $0.5H$

272. Indicate which of the following frequencies will *not* be found in the output of a normal TV receiver tuner:
a. 4.5 MHz
b. 41.25 MHz
c. 45.75 MHz
d. 42.17 MHz

273. The video voltage applied to the picture tube of a television receiver is fed in
a. between grid and ground
b. to the yoke
c. to the anode
d. between grid and cathode

274. The circuit that separates sync pulses from the composite video waveform is
a. the keyed AGC amplifier
b. a clipper
c. an integrator
d. a differentiator

275. The output of the vertical amplifier, applied to the yoke in a TV receiver, consists of
a. direct current
b. amplified vertical sync pulses
c. a sawtooth voltage
d. a sawtooth current

276. The HV anode supply for the picture tube of a TV receiver is generated in the
a. mains transformer
b. vertical output stage
c. horizontal output stage
d. horizontal deflection oscillator

277. Another name for the horizontal retrace in a TV receiver is the
a. ringing
b. burst
c. damper
d. flyback

278. Indicate which of the following signals is *not* transmitted in color TV:
a. Y
b. Q
c. R
d. I

279. The *shadow mask* in a color picture tube is used to
a. reduce x-ray emission
b. ensure that each beam hits only its own dots
c. increase screen brightness
d. provide degaussing for the screen

280. In a TV receiver, the *color killer*

a. cuts off the chroma stages during monochrome reception

b. ensures that no color is transmitted to monochrome receivers

c. prevents color overloading

d. makes sure that the color burst is not mistaken for sync pulses, by cutting off reception during the back porch

ANSWERS TO SELF-TESTING QUESTIONS

1. b	41. c	81. d	121. c	161. b	201. b	241. c
2. a	42. a	82. c	122. b	162. c	202. c	242. d
3. a	43. a	83. d	123. c	163. b	203. d	243. d
4. d	44. b	84. b	124. d	164. d	204. a	244. b
5. b	45. b	85. b	125. a	165. a	205. b	245. c
6. d	46. d	86. a	126. c	166. d	206. c	246. a
7. a	47. a	87. a	127. a	167. a	207. c	247. d
8. c	48. a	88. d	128. d	168. b	208. d	248. c
9. b	49. d	89. b	129. b	169. d	209. d	249. d
10. a	50. d	90. d	130. b	170. d	210. c	250. b
11. a	51. a	91. c	131. c	171. c	211. b	251. b
12. d	52. d	92. a	132. b	172. c	212. d	252. a
13. b	53. b	93. c	133. c	173. b	213. b	253. a
14. b	54. b	94. b	134. d	174. d	214. c	254. a
15. d	55. c	95. b	135. b	175. d	215. d	255. b
16. a	56. d	96. d	136. a	176. a	216. a	256. d
17. c	57. a	97. d	137. c	177. b	217. a	257. c
18. c	58. b	98. c	138. a	178. c	218. b	258. a
19. b	59. d	99. d	139. a	179. a	219. a	259. d
20. c	60. d	100. a	140. b	180. b	220. a	260. b
21. d	61. c	101. b	141. b	181. c	221. d	261. b
22. a	62. c	102. a	142. a	182. a	222. b	262. d
23. c	63. b	103. d	143. d	183. a	223. d	263. a
24. a	64. c	104. a	144. d	184. b	224. c	264. c
25. d	65. d	105. a	145. b	185. d	225. a	265. c
26. b	66. b	106. d	146. b	186. a	226. c	266. d
27. a	67. d	107. b	147. c	187. c	227. b	267. c
28. a	68. a	108. b	148. a	188. c	228. b	268. b
29. c	69. c	109. d	149. d	189. c	229. d	269. a
30. b	70. c	110. a	150. a	190. d	230. d	270. c
31. d	71. c	111. c	151. b	191. b	231. c	271. b
32. d	72. d	112. c	152. d	192. b	232. d	272. a
33. c	73. b	113. b	153. d	193. c	233. a	273. d
34. d	74. b	114. d	154. c	194. a	234. b	274. b
35. b	75. c	115. d	155. b	195. d	235. a	275. d
36. b	76. d	116. c	156. d	196. c	236. d	276. c
37. c	77. d	117. d	157. c	197. a	237. c	277. d
38. b	78. b	118. b	158. a	198. a	238. b	278. c
39. c	79. c	119. c	159. d	199. a	239. c	279. b
40. c	80. a	120. b	160. b	200. b	240. a	280. a

INDEX

A

B

C

D

E

G

H

I

J

K

N

O

P

Q

R

S

T

V

W

Y

Z

TATA McGRAW-HILL